Setsuo Arikawa   Ayumi Shinohara (Eds.)

Progress in Discovery Science

Final Report of the Japanese Dicsovery Science Project

Series Editors

Jaime G. Carbonell,Carnegie Mellon University, Pittsburgh, PA, USA
Jörg Siekmann, University of Saarland, Saarbrücken, Germany

Volume Editors

Setsuo Arikawa
Ayumi Shinohara
Kyushu University
Department of Informatics
6-10-1 Hakozaki, Higashi-ku, Fukuoka 812-8581, Japan
E-mail: {arikawa/ayumi}i.kyushu-u.ac.jp

Cataloging-in-Publication Data applied for

Die Deutsche Bibliothek - CIP-Einheitsaufnahme

Progress in discovery science : final report of the Japanese discovery science project ; Setsuo Arikawa ; Ayumi Shinohara (ed.). - Berlin ; Heidelberg ; New York ; Barcelona ; Hong Kong ; London ; Milan ; Paris ; Tokyo : Springer, 2002
(Lecture notes in computer science ; 2281 : Lecture notes in artificial intelligence)
ISBN 3-540-43338-4

CR Subject Classification (1998): I.2, H.2.8, H.3, G.3, F.2.2, I.5, J.1, J.2

ISSN 0302-9743
ISBN 3-540-43338-4 Springer-Verlag Berlin Heidelberg New York

Springer-Verlag Berlin Heidelberg New York
a member of BertelsmannSpringer Science+Business Media GmbH

http://www.springer.de

Printed in Germany

Typesetting: Camera-ready by author, data conversion by Steingräber Satztechnik GmbH
Printed on acid-free paper SPIN 10846270 06/3142 5 4 3 2 1 0

Preface

This volume contains the research reports of the Discovery Science project in Japan (No. 10143106), in which more than 60 scientists participated. It was a three-year project sponsored by Grant-in-Aid for Scientific Research on Priority Areas from the Ministry of Education, Culture, Sports, Science, and Technology (MEXT) of Japan. This project mainly aimed to (1) develop new methods for knowledge discovery, (2) install network environments for knowledge discovery, and (3) establish Discovery Science as a new area of study in Computer Science / Artificial Intelligence.

In order to attain these aims we set up five groups for studying the following research areas:

(A) Logic for/of Knowledge Discovery
(B) Knowledge Discovery by Inference/Reasoning
(C) Knowledge Discovery Based on Computational Learning Theory
(D) Knowledge Discovery in Huge Databases and Data Mining
(E) Knowledge Discovery in Network Environments

These research areas and related topics can be regarded as a preliminary definition of Discovery Science by enumeration. Thus Discovery Science ranges over philosophy, logic, reasoning, computational learning, and system developments.

In addition to these five research groups we organized a steering group for planning, adjustment, and evaluation of the project. The steering group, chaired by the principal investigator of the project, consists of leaders of the five research groups and their subgroups as well as advisors from outside of the project. We invited three scientists to consider Discovery Science and the five above mentioned research areas from viewpoints of knowledge science, natural language processing, and image processing, respectively.

Group A studied discovery from a very broad perspective, taking account of historical and social aspects, and computational and logical aspects of discovery. Group B focused on the role of inference/reasoning in knowledge discovery, and obtained many results on both theory and practice in statistical abduction, inductive logic programming, and inductive inference. Group C aimed to propose and develop computational models and methodologies for knowledge discovery mainly based on computational learning theory. This group obtained some deep theoretical results on the boosting of learning algorithms and the minimax strategy for Gaussian density estimation, and also methodologies specialized to concrete problems such as algorithms for finding best subsequence patterns, biological sequence compression algorithms, text categorization, and MDL-based compression. Group D aimed to create computational strategies for speeding up the discovery process in total. For this purpose, group D was made up of researchers from other scientific domains and researchers from computer science so that real issues in the discovery process could be exposed and practical computational techniques could be devised and tested for solving these real issues. This group handled many kinds of data: data from national projects such as genomic data and satellite observations, data generated from laboratory experiments, data collected from personal interests such as literature and medical records, data collected in business and marketing areas, and data for proving the efficiency of algorithms such as the UCI repository. So many theoretical and practical results were obtained on such a variety of data. Group E aimed to develop a unified media system for knowledge discovery and network agents for knowledge discovery. This group obtained practical results on a new virtual materialization of DB records and scientific computations to help scientists make a scientific discovery, a convenient visualization interface that treats web data, and an efficient algorithm that extracts important information from semi-structured data in the web space.

We would like to express our immense gratitude to the members of the Discovery Science project, listed on the subsequent pages. The papers submitted to this volume were reviewed both by peers and external referees. We would like to express our sincere gratitude to the following external referees:

Jean-Francois Boulicaut
Koji Eguchi
Toshirou Ejima
Tatsuaki Hashimoto
Yutaka Hata
Masafumi Hirahara
Koichi Hirata
Koichi Hishida
Takashi Iida
Daisuke Ikeda
Katsumi Inoue
Makio Ishiguro
Osamu Katai
Hajime Kato
Iwane Kimura
Yoshiki Kinoshita
Kazuo Kitahara
Satoshi Kobayashi
Shozo Makino
Kazunori Matsumoto
Aart Middeldorp
Tsuyoshi Murata
Tadas Nakamura
Kazuki Nishi
Osamu Nishizawa
Takashi Okada
Seishi Okamoto
Ryuji Omori
Chiaki Sakama
Hiroshi Sakamoto
Takafumi Sakurai
Hiroyuki Sato
Tetsuo Shibuya
Tetsuro Shimizu
Wataru Sunayama
Jun-ichi Takeuchi
Kiyotaka Uchida
Shinji Yoshioka

We would also like to thank to the external advisors, Raymond Greenlaw, Carl Smith, and Thomas Zeugmann, for their valuable comments.

September 2001

Setsuo Arikawa
Ayumi Shinohara

Organization

Steering Group

In addition to the five groups listed below we had a steering group (Soukatsu-han in Japanese) for planning, adjustment, and evaluation of the project, consisting of the following members:

Setsuo Arikawa (Chair, Kyushu University)
Masahiko Sato (Leader of Group A, Kyoto University)
Taisuke Sato (Leader of Group B, Tokyo Institute of Technology)
Akira Maruoka (Leader of Group C, Tohoku University)
Satoru Miyano (Leader of Group D, University of Tokyo)
Yasumasa Kanada (Leader of Group E, University of Tokyo)
Yuichiro Anzai (Keio University)
Setsuo Ohsuga (Waseda University)
Kinji Ono (NII)
Hiroakira Ono (JAIST)
Takuya Katayama (JAIST)
Yahiko Kambayashi (Kyoto University)
Tadao Saito (Chuo University)
Hidehiko Tanaka (University of Tokyo)
Yuzuru Tanaka (Hokkaido University)
Jun'ichi Tsujii (University of Tokyo)
Jun-ichiro Toriwaki (Nagoya University)
Makoto Nagao (Kyoto University)
Teruyuki Nakajima (University of Tokyo)
Shozo Makino (Tohoku University)
Keiichi Noe (Tohoku University)
Koichi Furukawa (Keio University)
Hiroshi Motoda (Osaka University)
Ayumi Shinohara (Kyushu University)

Group A: Logic of Knowledge Discovery

Masahiko Sato (Leader, Kyoto University)
Tetsuo Ida (University of Tsukuba)
Hiroakira Ono (JAIST)
Mitsuhiro Okada (Keio University)
Keiichi Noe (Tohoku University)
Masami Hagiya (University of Tokyo)
Yukiyoshi Kameyama (University of Tsukuba)
Shun Tsuchiya (Chiba University)

Group B: Knowledge Discovery by Inference

Taisuke Sato (Leader, Tokyo Institute of Technology)
Hiroki Arimura (Kyushu University)
Mutsumi Imai (Keio University)
Masako Sato (Osaka Prefecture University)
Takeshi Shinohara (Kyushu Institute of Technology)
Makoto Haraguchi (Hokkaido University)
Hiroshi Tsukimoto (Toshiba)
Chiaki Sakama (Wakayama University)
Ken Sato (Hokkaido University)
Yukio Ohsawa (University of Tsukuba)
Akihiro Yamamoto (Hokkaido University)
Koichi Furukawa (Keio University)
Hiroshi Tanaka (Tokyo Medical and Dental University)
Masahiko Yachida (Osaka University)
Katsumi Inoue (Kobe University)

Group C: Knowledge Discovery by Learning Algorithms

Akira Maruoka (Leader, Tohoku University)
Naoki Abe (C&C, NEC)
Hiroshi Imai (University of Tokyo)
Ayumi Shinohara (Kyushu University)
Atsuhiro Takasu (NII)
Osamu Watanabe (Tokyo Institute of Technology)
Eiji Takimoto (Tohoku University)
Sadao Kurohashi (Kyoto University)
Kuniaki Uehara (Kobe University)

Group D: Knowledge Discovery in Databases

Satoru Miyano (Leader, University of Tokyo)
Tohru Araki (Kyoto University)
Genshiro Kitagawa (Institute for Statistical Mathematics)
Shinichi Morishita (University of Tokyo)
Hiroshi Motoda (Osaka University)
Ryohei Nakano (Nagoya Institute of Technology)
Koichi Niijima (Kyushu University)
Katsutoshi Yada (Kansai University)
Yuji Ikeda (Kobe University)
Yoshiya Kasahara (Kyoto University)
Einoshin Suzuki (Yokohama National University)
Takehiko Tanaka (Kyushu University)

Shusaku Tsumoto (Shimane Medical University)
Hiroshi Hasegawa (Ibaragi University)
Rie Honda (Kochi University)
Takahira, Yamaguchi (Shizuoka University)
Toyofumi Saito (Nagoya University)
Masayuki Takeda (Kyushu University)
Osamu Maruyama (Kyushu University)
Kenichi Yoshida (Hitachi)
Shozo Makino (Tohoku University)
Shogo Nishida (Osaka University)
Tomoyuki Higuchi (Institute for Statistical Mathematics)
Shinobu Machida (Kyoto University)

Group E: Knowledge Discovery in Network Environments

Yasumasa Kanada (Leader, University of Tokyo)
Setsuo Arikawa (Kyushu University)
Shigeki Goto (Waseda University)
Etsuya Shibayama (Tokyo Institute of Technology)
Yuji Takada (Fujitsu Lab.)
Yuzuru Tanaka (Hokkaido University)
Hiroyuki Kawano (Kyoto University)
Sachio Hirokawa (Kyushu University)
Fumio Mizoguchi (Science University of Tokyo)
Hiroki Ishizaka (Kyushu Institute of Technology)

Table of Contents

Searching for Mutual Exclusion Algorithms Using BDDs 1
Koichi Takahashi, Masami Hagiya

Reducing Search Space in Solving Higher-Order Equations 19
Tetsuo Ida, Mircea Marin, Taro Suzuki

The Structure of Scientific Discovery: From a Philosophical Point of View . 31
Keiichi Noé

Ideal Concepts, Intuitions, and Mathematical Knowledge Acquisitions
in Husserl and Hilbert 40
Mitsuhiro Okada

Theory of Judgments and Derivations 78
Masahiko Sato

Efficient Data Mining from Large Text Databases 123
Hiroki Arimura, Hiroshi Sakamoto, Setsuo Arikawa

A Computational Model for Children's Language Acquisition
Using Inductive Logic Programming 140
Ikuo Kobayashi, Koichi Furukawa, Tomonobu Ozaki, Mutsumi Imai

Some Criterions for Selecting the Best Data Abstractions 156
Makoto Haraguchi, Yoshimitsu Kudoh

Discovery of Chances Underlying Real Data 168
Yukio Ohsawa

Towards the Integration
of Inductive and Nonmonotonic Logic Programming 178
Chiaki Sakama

EM Learning for Symbolic-Statistical Models in Statistical Abduction 189
Taisuke Sato

Refutable/Inductive Learning from Neighbor Examples
and Its Application to Decision Trees over Patterns 201
Masako Sato, Yasuhito Mukouchi, Mikiharu Terada

Constructing a Critical Casebase to Represent a Lattice-Based Relation... 214
Ken Satoh

On Dimension Reduction Mappings
for Approximate Retrieval of Multi-dimensional Data 224
Takeshi Shinohara, Hiroki Ishizaka

Rule Discovery from fMRI Brain Images by Logical Regression Analysis .. 232
Hiroshi Tsukimoto, Mitsuru Kakimoto, Chie Morita, Yoshiaki Kikuchi

A Theory of Hypothesis Finding in Clausal Logic 246
Akihiro Yamamoto, Bertram Fronhöfer

Efficient Data Mining by Active Learning 258
Hiroshi Mamitsuka, Naoki Abe

Data Compression Method Combining Properties of PPM and CTW 268
Takumi Okazaki, Kunihiko Sadakane, Hiroshi Imai

Discovery of Definition Patterns by Compressing Dictionary Sentences 284
Masatoshi Tsuchiya, Sadao Kurohashi, Satoshi Sato

On-Line Algorithm to Predict Nearly as Well as the Best Pruning
of a Decision Tree 296
Akira Maruoka, Eiji Takimoto

Finding Best Patterns Practically 307
Ayumi Shinohara, Masayuki Takeda, Setsuo Arikawa, Masahiro Hirao,
Hiromasa Hoshino, Shunsuke Inenaga

Classification of Object Sequences Using Syntactical Structure 318
Atsuhiro Takasu

Top-Down Decision Tree Boosting and Its Applications 327
Eiji Takimoto, Akira Maruoka

Extraction of Primitive Motion and Discovery of Association Rules
from Human Motion Data 338
Kuniaki Uehara, Mitsuomi Shimada

Algorithmic Aspects of Boosting 349
Osamu Watanabe

Automatic Detection of Geomagnetic Jerks by Applying a Statistical Time
Series Model to Geomagnetic Monthly Means 360
Hiromichi Nagao, Tomoyuki Higuchi, Toshihiko Iyemori, Tohru Araki

Application of Multivariate Maxwellian Mixture Model
to Plasma Velocity Distribution 372
Genta Ueno, Nagatomo Nakamura, Tomoyuki Higuchi, Takashi Tsuchiya,
Shinobu Machida, Tohru Araki

Inductive Thermodynamics from Time Series Data Analysis 384
Hiroshi H. Hasegawa, Takashi Washio, Yukari Ishimiya

Mining of Topographic Feature from Heterogeneous Imagery
and Its Application to Lunar Craters 395
Rie Honda, Yuichi Iijima, Osamu Konishi

Application of Neural Network Technique
to Combustion Spray Dynamics Analysis 408
Yuji Ikeda, Dariusz Mazurkiewicz

Computational Analysis of Plasma Waves and Particles
in the Auroral Region Observed by Scientific Satellite 426
Yoshiya Kasahara, Ryotaro Niitsu, Toru Sato

A Flexible Modeling of Global Plasma Profile Deduced from Wave Data .. 438
Yoshitaka Goto, Yoshiya Kasahara, Toru Sato

Extraction of Signal from High Dimensional Time Series:
Analysis of Ocean Bottom Seismograph Data 449
Genshiro Kitagawa, Tetsuo Takanami, Asako Kuwano, Yoshio Murai, Hideki Shimamura

Foundations of Designing Computational Knowledge Discovery Processes . 459
Yoshinori Tamada, Hideo Bannai, Osamu Maruyama, Satoru Miyano

Computing Optimal Hypotheses Efficiently for Boosting 471
Shinichi Morishita

Discovering Polynomials to Fit Multivariate Data
Having Numeric and Nominal Variables 482
Ryohei Nakano, Kazumi Saito

Finding of Signal and Image
by Integer-Type Haar Lifting Wavelet Transform 494
Koichi Niijima, Shigeru Takano

In Pursuit of Interesting Patterns
with Undirected Discovery of Exception Rules 504
Einoshin Suzuki

Mining from Literary Texts:
Pattern Discovery and Similarity Computation 518
Masayuki Takeda, Tomoko Fukuda, Ichirō Nanri

Second Difference Method Reinforced by Grouping:
A New Tool for Assistance in Assignment of Complex Molecular Spectra .. 532
Takehiko Tanaka

Discovery of Positive and Negative Knowledge in Medical Databases Using Rough Sets 543
Shusaku Tsumoto

Toward the Discovery of First Principle Based Scientific Law Equations ... 553
Takashi Washio, Hiroshi Motoda

A Machine Learning Algorithm for Analyzing String Patterns Helps to Discover Simple and Interpretable Business Rules from Purchase History .. 565
Yukinobu Hamuro, Hideki Kawata, Naoki Katoh, Katsutoshi Yada

Constructing Inductive Applications by Meta-Learning with Method Repositories 576
Hidenao Abe, Takahira Yamaguchi

Knowledge Discovery from Semistructured Texts 586
Hiroshi Sakamoto, Hiroki Arimura, Setsuo Arikawa

Packet Analysis in Congested Networks 600
Masaki Fukushima, Shigeki Goto

Visualization and Analysis of Web Graphs 616
Sachio Hirokawa, Daisuke Ikeda

Knowledge Discovery in Auto-tuning Parallel Numerical Library 628
Hisayasu Kuroda, Takahiro Katagiri, Yasumasa Kanada

Extended Association Algorithm Based on ROC Analysis for Visual Information Navigator 640
Hiroyuki Kawano, Minoru Kawahara

WWW Visualization Tools for Discovering Interesting Web Pages 650
Hironori Hiraishi, Fumio Mizoguchi

Scalable and Comprehensible Visualization for Discovery of Knowledge from the Internet 661
Etsuya Shibayama, Masashi Toyoda, Jun Yabe, Shin Takahashi

Meme Media for Re-editing and Redistributing Intellectual Assets and Their Application to Interactive Virtual Information Materialization .. 672
Yuzuru Tanaka

Author Index 683

Searching for Mutual Exclusion Algorithms Using BDDs

Koichi Takahashi and Masami Hagiya

National Institute of Advanced Industrial Science and Technology
and
University of Tokyo
k.takahashi@aist.go.jp
http://staff.aist.go.jp/k.takahashi/

Abstract. The impact of verification technologies would be much greater if they could not only verify existing information systems, but also synthesize or discover new ones. In our previous study, we tried to discover new algorithms that satisfy a given specification, by first defining a space of algorithms, and then checking each algorithm in the space against the specification, using an automatic verifier, i.e., model checker. Needless to say, the most serious problem of this approach is in search space explosion. In this paper, we describe case studies in which we employed symbolic model checking using BDD and searched for synchronization algorithms. By employing symbolic model checking, we could speed up enumeration and verification of algorithms. We also discuss the use of approximation for reducing the search space.

1 Introduction

Verification technologies have been successfully applied to guarantee the correctness of various kinds of information systems, ranging from abstract algorithms to hardware circuits. Among them is that of model checking, which checks the correctness of a state transition system by traversing its state space [2]. The success of model checking is mainly due to its ability to automatically verify a state transition system without human intervention.

However, the impact of verification technologies would be much greater if they could not only verify existing systems, but also synthesize or discover new ones.

In our previous study [6, 7], we tried to discover new algorithms that satisfy a given specification, by first defining a space of algorithms, and then checking each algorithm in the space against the specification, using a verifier, i.e., model checker. Note that this approach is possible only if the employed verifier is fully automatic. By the approach, we discovered new variants of the existing algorithms for concurrent garbage collection, and a new algorithm for mutual exclusion under some restrictions on parallel execution.

Perrig and Song have also taken a similar approach in the field of protocol verification [10]. They enumerated protocols for asynchronous and synchronous

S. Arikawa and A. Shinohara (Eds.): Progress in Discovery Science 2001, LNAI 2281, pp. 1–18.

mutual authentication, and successfully discovered new ones by checking enumerated protocols using their protocol verifier, Athena [11].

Superoptimization in the field of compiler technologies is very close to the above mentioned approach, though it does not employ a complete verifier [9, 5]. It automatically synthesizes the code generation table of a compiler by enumerating and checking sequences of machine instructions.

Needless to say, the most serious problem of this approach is in search space explosion. The space of algorithms explodes if the size of enumerated algorithms is not carefully bounded. It is therefore vital for the approach to efficiently traverse the space of algorithms, employing various kinds of search heuristics.

In this paper, we investigate the possibility of simultaneously checking all algorithms in the algorithm space by a single execution of the verifier. Algorithms in the space are symbolically represented by a template containing parameters. Each instantiation of the parameters in the template corresponds to a specific algorithm. By symbolically verifying the template, we obtain constraints on the parameters for the template to be correct.

By symbolic verification, it is possible to share computations for verification among different algorithms, because computations with some parameters uninstantiated are shared by algorithms that correspond to instantiations of those parameters.

By symbolically representing a template of algorithms, it is also possible to apply approximation or abstraction [4] on the template. Before or during verification, algorithms can be approximated by another having similar but smaller representation. If abstraction could be applied on symbolic representation, it would greatly reduce the search space.

In this paper, we employ BDDs (binary decision diagrams) for symbolic representation and verification [1, 2]. As a case study, we take the problem of searching for synchronization algorithms for mutual exclusion without using semaphores, which was also taken in our previous study [6, 7].

In our case studies, a template of algorithms is represented by a sequence of pseudo-instructions including boolean parameters. We define two templates. The first template has the original version of Peterson's algorithm as an instance. The second template has both Peterson and Dekker's algorithms as instances. The predicate defining the initial state, that of error states, and the state transition relation are all represented by BDDs.

We checked the safety and liveness of the template using BDDs, and successfully obtained the constraints on the parameters. We compared the time required for verifying a single concrete algorithm — Peterson's algorithm (or Dekker's algorithm) — with that for checking the template, and gained speed-up of more than two hundred times.

Finally, we made an experiment in which we collapsed BDDs with a small Hamming distance. This is a first step towards using approximation and abstraction in the approach.

In this paper, we only assume that the reader has basic knowledge about state transition systems and their properties, because the essence of our work can be understood without deep knowledge about model checking.

The rest of the paper is organized as follows. In the next section, we explain the original versions of Peterson and Dekker's algorithms for mutual exclusion, and their safety and liveness properties. In Sections 3 and 4, we describe the results of searching for variants of those algorithms by the above mentioned approach. In Section 5, we report the result of a small experiment in which we tried approximation during search. Section 5 is for conclusion.

2 Mutual Exclusion

The target of these case studies is to synthesize synchronization algorithms for mutual exclusion without using semaphores.

Peterson and Dekker's algorithms realize mutual exclusion among processes without semaphores. Figure 1 shows an instance of Peterson's algorithm for two processes. An instance of Dekker's algorithm is shown in Figure 2.

```
for (;;) {
    // beginning of the entry part
    flags[me] = true;
    turn = you;
    while (flags[you] == true) {
        if (turn != you) break;
    } // end of the entry part

    // the critical section

    // beginning of the finishing part
    flags[me] = false;
    // end of the finishing part

    // the idle part
}
```

Fig. 1. Peterson's algorithm.

In the figures, `me` denotes the number of the process that is executing the code (`1` or `2`), and `you` denotes the number of the other process (`2` or `1`). The entry part realizes mutual execution before entering the critical section, and the finishing part is executed after the critical section. The idle part represents a process-dependent task.

The safety property of mutual exclusion algorithm is:

Two processes do not simultaneously enter the critical section.

```
for (;;) {
    // beginning of the entry part
    flags[me] = true;
    while (flags[you] == true) {
       if (turn != me) {
           flags[me] = false;
           while (turn != me) {}
           flags[me] = true;
       }
    } // end of the entry part

    // the critical section

    // beginning of the finishing part
    turn = you;
    flags[me] = false;
    // end of the finishing part

    // the idle part
}
```

Fig. 2. Dekker's algorithm.

Liveness is:

> There does not exist an execution path (loop) that begins and ends with the same state, at which one process is in its entry part, and satisfies the following conditions.
> – The process stays in the entry part on the execution path, i.e., it does not enter the critical section.
> – Both processes execute at least one instruction on the execution path.

The results of searching for variants of these algorithms are described in the next two sections.

3 Search for Variants of Peterson's Algorithm

In this section, we describe the first case study. We make a template which has Peterson's algorithm as an instance, and check the safety and liveness of the template using BDDs.

3.1 Pseudo-code

We represent Peterson's algorithm using pseudo-code consisting of pseudo-instructions. The pseudo-code may refer to three variables, each of which holds a boolean value.

- FLAG1: This variable corresponds to flags[1] in Figure 1.
- FLAG2: This variable corresponds to flags[2] in Figure 1.
- FLAG0: This variable corresponds to turn in Figure 1. FLAG0=true means turn=2, and FLAG0=false means turn=1,

Each instruction in the pseudo-code is in one of the following three forms.

SET_CLEAR(p),L
IF_WHILE(p),L_1,L_2
NOP,L_1,L_2

The operands L, L_1 and L_2 in the instructions denote addresses in the pseudo-code. In SET_CLEAR(p),L, the operand L should point to the next address in the pseudo-code.

The operators SET_CLEAR and IF_WHILE have a three-bit parameter, denoted by p. Each value of the parameter results in a pseudo-instruction as defined in Figures 3 and 4 for each process. In the figures, b denotes 0 or 1.

The instruction

NOP,L_1,L_2

jumps to either L_1 or L_2 nondeterministically.

	Process 1	Process 2
SET_CLEAR(b00),L	goto L	goto L
SET_CLEAR(b01),L	FLAG0 := b; goto L	FLAG0 := not(b); goto L
SET_CLEAR(b10),L	FLAG1 := b; goto L	FLAG2 := b; goto L
SET_CLEAR(b11),L	FLAG2 := b; goto L	FLAG1 := b; goto L

Fig. 3. SET_CLEAR

The original version of Peterson's algorithm for mutual exclusion can be represented by the following pseudo-code.

```
0: SET_CLEAR(110),1
1: SET_CLEAR(101),2
2: IF_WHILE(111),3,4
3: IF_WHILE(001),4,2
4: NOP,5,5
5: SET_CLEAR(010),6
6: NOP,6,0
```

The first column of the pseudo-code denotes the address of each instruction. The fourth instruction, 4: NOP,5,5, represents the critical section, and the sixth

	Process 1	Process 2
IF_WHILE(b00),L_1,L_2	goto L_1	goto L_1
IF_WHILE(b01),L_1,L_2	IF FLAG0=b then goto L_1 else goto L_2	IF FLAG0=not(b) then goto L_1 else goto L_2
IF_WHILE(b10),L_1,L_2	IF FLAG1=b then goto L_1 else goto L_2	IF FLAG2=b then goto L_1 else goto L_2
IF_WHILE(b11),L_1,L_2	IF FLAG2=b then goto L_1 else goto L_2	IF FLAG1=b then goto L_1 else goto L_2

Fig. 4. IF_WHILE

instruction, 6: NOP,6,0, the part that is specific to each process. Each process is allowed to loop around the sixth instruction.

We then parameterize five instructions in Peterson's algorithm as in Figure 5. The safety and liveness of this parameterized code were verified as described in the next section.

```
0: SET_CLEAR(p0),1
1: SET_CLEAR(p1),2
2: IF_WHILE(p2),3,4
3: IF_WHILE(p3),4,2
4: NOP,5,5
5: SET_CLEAR(p4),6
6: NOP,6,0
```

Fig. 5. Template 1

3.2 Verification

The initial state of the state transition system is defined as a state that satisfies the following condition.

$$\mathtt{PC1} = \mathtt{PC2} = 0$$
$$\mathtt{FLAG0} = \mathtt{FLAG1} = \mathtt{FLAG2} = 0$$

In the above condition, PC1 and PC2 denote the program counter of Process 1 and Process 2, respectively. Let $I(x)$ denote the predicate expressing the condition, where x ranges over states. $I(x)$ holds if and only if x is the initial state.

The safety of the state transition system is defined as unreachability of an error state that satisfies the following condition.

$\texttt{PC1} = \texttt{PC2} = 4$

In an error state, both processes enter their critical section simultaneously. The system is safe unless it reaches an error state from the initial state. Let $E(x)$ denote the predicate expressing the condition.

Let $T(x, y)$ mean that there is a one-step transition from state x to state y, and $T^*(x, y)$ denote the reflexive and transitive closure of $T(x, y)$. The safety of the system is then expressed by the following formula.

$$\neg\exists xy.\ I(x) \wedge T^*(x, y) \wedge E(y)$$

We can describe the liveness for Process 1 of the system as non-existence of an infinite path on which

$$0 \leq \texttt{PC1} \leq 3$$

is always satisfied though both processes are infinitely executed. The condition, $0 \leq \texttt{PC1} \leq 3$, means that Process 1 is trying to enter its critical section. $S(x)$ denote the predicate expressing the condition.

We verify the liveness as follows. Let $T_1(x, y)$ denote the one-step transition relation for Process 1. $T_1(x, y)$ holds if and only if state y is obtained by executing Process 1 for one step from x. Similarly, let $T_2(x, y)$ denote the one-step transition relation for Process 2. Note that $T(x, y)$ is equivalent to $T_1(x, y) \vee T_2(x, y)$. We then define the following three predicates.

$$T_1'(x, y) = T_1(x, y) \wedge S(x)$$
$$T_2'(x, y) = T_2(x, y) \wedge S(x)$$
$$T'(x, y) = T(x, y) \wedge S(x)$$

For any predicate $Z(x)$ on state x, and any binary relation $R(x, y)$ on states x and y, we define the relation, denoted by $Z \circ R$, as follows.

$$(Z \circ R)(x) = \exists y.\ Z(y) \wedge R(x, y)$$

$Z \circ R$ is also a predicate on states.

For verifying the liveness of the system, we compute the following limit of predicates.

$$S \circ T_1' \circ T'^* \circ T_2' \circ T'^* \circ$$
$$T_1' \circ T'^* \circ T_2' \circ T'^* \circ$$
$$T_1' \circ T'^* \circ T_2' \circ T'^* \circ \cdots$$

This limit always exists, because the sequence of predicates

$$S \circ T_1' \circ T'^*$$
$$S \circ T_1' \circ T'^* \circ T_2'$$
$$S \circ T_1' \circ T'^* \circ T_2' \circ T'^*$$
$$S \circ T_1' \circ T'^* \circ T_2' \circ T'^* \circ T_1'$$
$$S \circ T_1' \circ T'^* \circ T_2' \circ T'^* \circ T_1' \circ T'^*$$
$$\cdots$$

is monotonically decreasing. For example, we can prove that the second predicate is smaller than the first one as follows: $S \circ T_1' \circ T'^* \circ T_2' \subseteq S \circ T_1' \circ T'^* \circ T' \subseteq S \circ T_1' \circ T'^*$.

Let us denote this limit by S'. It expresses the beginning of an infinite path on which S always holds and both processes are infinitely executed, i.e., a state satisfies the limit if and only if there exists such an infinite path from the state. The liveness is then equivalent to unreachability from the initial state to a state satisfying the limit.

$$\neg\exists xy.\ I(x) \wedge T^*(x,y) \wedge S'(y)$$

The liveness for Process 2 can be symmetrically described as that for Process 1. The whole system satisfies the liveness if it holds for both processes.

Since there are 7 pseudo-instructions in the pseudo-code, the program counter of each process can be represented by three bits. Therefore, a state in the state transition system can be represented by nine boolean variables; three are used to represent the program counter of each process, and three for the three shared variables.

There are five parameters in the pseudo-code, so fifteen boolean variables are required to represent the parameters. Let p denote the vector of the fifteen boolean variables. All the predicates and relations introduced so far are considered indexed by p. For example, we should have written $T_p(x,y)$.

Predicates such as $I_p(x)$ and $S_p(x)$ contain 24 boolean variables, and relations such as $T_p(x,y)$ contain 33 boolean variables. All these predicates and relations are represented by OBDD (ordered binary decision diagrams)

We employed the BDD package developed by David E. Long and distributed from CMU [8]. The verification program was written in C by hand. The essential part of the program is shown in the appendix.

Variables in predicates and relations are ordered as follows.

- variables in x first,
- variables in y (if any) next, and
- variables in p last.

We have tried the opposite order but never succeeded in verification.

By the above order, predicates such as $S_p(x)$ are decomposed into a collection of predicates on p as follows.

if ...x... **then** ...p...
else if ...x... **then** ...p...
else ...

The value of x is successively checked by a series of conditions, and for each condition on x, a predicate on p is applied.

Similarly, relations such as $T_p(x,y)$ are decomposed as follows.

if ...x...y... **then** ...p...
else if ...x...y... **then** ...p...
else ...

3.3 Result

We checked the safety and liveness of the system as described in the previous section. The experiments are made by Vectra VE/450 Series8, from Hewlett-Packard.

- It took 0.853 seconds to check the original version of Peterson's algorithm. The result was, of course, true.
- It took 117.393 seconds to check the template of algorithms containing a fifteen-bit parameter.

Since the template contains fifteen boolean variables, checking the template amounts to checking 2^{15} instances of the template simultaneously. If we simply multiply 0.853 by 2^{15} and compare the result with 117.393, we found speed up of about 238 times.

The size of the BDD that represents the constraints on the parameters is 93. Only 16 instances of parameters satisfy the constraints. So, we found 15 variants of Peterson's algorithm. But they are essentially equivalent to Peterson's algorithm. In the variants, the interpretation of some shared variables is simply reversed.

4 Search for Variants of Peterson's or Dekker's Algorithm

In this section, we show the second case study. We apply the method described in the previous section to another template. Since Dekker's algorithm is very similar to Peterson's, we make a template which cover both algorithms.

4.1 Pseudo-code

We define another template which has both Dekker and Peterson's algorithms as instances. In order to define such a template with a small number of parameters, we modify instructions as follows.

SET_CLEAR(p),L
JUMP_IF_TURN(p),L_{00},L_{01},L_{10},L_{11}
JUMP_IF_FLAG_YOU(p),L_{00},L_{01},L_{10},L_{11}
NOP,L_1,L_2

The instruction NOP is the same as in the previous section.

The operator SET_CLEAR is slightly changed to reduce parameters. The parameter p of SET_CLEAR(p) has two-bit value. Each value of the parameter results in a pseudo-instruction as defined in Figure 6.

The operators JUMP_IF_TURN and JUMP_IF_FLAG_YOU have a three-bit parameter and four operands. The operands denote addresses in the pseudo-code. The intuitive meaning of the operands is that L_{00} points to the next address, L_{01} points to the address of the critical section, L_{10} points to the address of the beginning of the loop of the entry part, and L_{10} points to the current address. Each value of the parameter results in a pseudo-instruction as defined in Figures 7 and Figures 8.

	Process 1	Process 2
SET_CLEAR(b0),L	FLAG0 := b; goto L	FLAG0 := not(b); goto L
SET_CLEAR(b1),L	FLAG1 := b; goto L	FLAG2 := b; goto L

Fig. 6. Modified SET_CLEAR

	Process 1	Process 2
JUMP_IF_TURN(b00),L_{00},L_{01},L_{10},L_{11}	IF FLAG0 = b THEN goto L_{00} ELSE goto L_{00}	IF FLAG0 = not(b) THEN goto L_{00} ELSE goto L_{00}
JUMP_IF_TURN(b01),L_{00},L_{01},L_{10},L_{11}	IF FLAG0 = b THEN goto L_{01} ELSE goto L_{00}	IF FLAG0 = not(b) THEN goto L_{01} ELSE goto L_{00}
JUMP_IF_TURN(b10),L_{00},L_{01},L_{10},L_{11}	IF FLAG0 = b THEN goto L_{10} ELSE goto L_{00}	IF FLAG0 = not(b) THEN goto L_{10} ELSE goto L_{00}
JUMP_IF_TURN(b11),L_{00},L_{01},L_{10},L_{11}	IF FLAG0 = b THEN goto L_{11} ELSE goto L_{00}	IF FLAG0 = not(b) THEN goto L_{11} ELSE goto L_{00}

Fig. 7. JUMP_IF_TURN

	Process 1	Process 2
JUMP_IF_FLAG_YOU(b00),L_{00},L_{01},L_{10},L_{11}	IF FLAG2 = b THEN goto L_{00} ELSE goto L_{00}	IF FLAG1 = b THEN goto L_{00} ELSE goto L_{00}
JUMP_IF_FLAG_YOU(b01),L_{00},L_{01},L_{10},L_{11}	IF FLAG2 = b THEN goto L_{01} ELSE goto L_{00}	IF FLAG1 = b THEN goto L_{01} ELSE goto L_{00}
JUMP_IF_FLAG_YOU(b10),L_{00},L_{01},L_{10},L_{11}	IF FLAG2 = b THEN goto L_{10} ELSE goto L_{00}	IF FLAG1 = b THEN goto L_{10} ELSE goto L_{00}
JUMP_IF_FLAG_YOU(b11),L_{00},L_{01},L_{10},L_{11}	IF FLAG2 = b THEN goto L_{11} ELSE goto L_{00}	IF FLAG1 = b THEN goto L_{11} ELSE goto L_{00}

Fig. 8. JUMP_IF_FLAG_YOU

We define a template as in Figure 9. Note that we fixed the values of some parameters to reduce the search space. In the template, each parameter has two-bit value. So the template contains sixteen boolean variables.

```
 0: SET_CLEAR(11),1
 1: SET_CLEAR(p1),2
 2: JUMP_IF_FLAG_YOU(0p2),3,8,2,2
 3: JUMP_IF_TURN(1p3),4,8,2,3
 4: JUMP_IF_TURN(0p4),5,8,2,4
 5: SET_CLEAR(p5),6
 6: JUMP_IF_TURN(1p6),7,8,2,6
 7: SET_CLEAR(p7),2
 8: NOP,9,9
 9: SET_CLEAR(p8),10
10: SET_CLEAR(01),11
11: NOP,0,11
```

Fig. 9. Template 2

The template contains both algorithms as instances. Dekker's algorithm can be represented by the following pseudo-code.

```
0: SET_CLEAR(11),1
1: SET_CLEAR(11),2
2: JUMP_IF_FLAG_YOU(001),3,8,2,2
3: JUMP_IF_TURN(100),4,8,2,3
4: JUMP_IF_TURN(010),5,8,2,4
5: SET_CLEAR(01),6
6: JUMP_IF_TURN(111),7,8,2,6
7: SET_CLEAR(11),2
8: NOP,9,9
9: SET_CLEAR(10),10
10: SET_CLEAR(01),11
11: NOP,0,11
```

The Peterson's algorithm can be represented by the following pseudo-code.

```
0: SET_CLEAR(11),1
1: SET_CLEAR(10),2
2: JUMP_IF_FLAG_YOU(001),3,8,2,2
3: JUMP_IF_TURN(110),4,8,2,3
4: JUMP_IF_TURN(001),5,8,2,4
5: SET_CLEAR(00),6
6: JUMP_IF_TURN(100),7,8,2,6
7: SET_CLEAR(00),2
8: NOP,9,9
9: SET_CLEAR(01),10
10: SET_CLEAR(01),11
11: NOP,0,11
```

4.2 Verification

Verification of the template goes almost in the same manner as in the previous section.

In this verification, the condition of an error state is

$$\mathtt{PC1} = \mathtt{PC2} = 8.$$

For the liveness, we modify the condition of the starvation loop of Process 1 as follows.

$$0 \leq \mathtt{PC1} \leq 7$$

Since there are twelve pseudo-instructions in the pseudo-code, the representation of the program counter of each process requires four bits. Therefore, a state in the state transition system can be represented by eleven boolean variables. Sixteen boolean variables are required to represented the parameters.

4.3 Result

We checked the safety and liveness of the system on the same machine as in the previous section.

- It took 10.195 seconds to check the Peterson's algorithm. The result was, of course, true.
- It took 19.506 seconds to check the Dekker's algorithm. The result was, of course, true.
- It took 741.636 seconds to check the template of algorithms containing a sixteen-bit parameter.

We found speed up of about 900 times at least.

The size of the BDD that represents the constraints on the parameters is 62. There are about 400 solutions of the constraints. We found that just one solution represents the original Dekker's algorithm, and the remaining solutions essentially represent Peterson's algorithm.

5 Approximation

Remember that according to the variable ordering we adopted in BDDs, a predicate on states indexed by p is represented as a collection of sub-predicates on p as follows.

if $C_0(x)$ **then** $P_0(p)$
else if $C_1(x)$ **then** $P_1(p)$
else if $C_2(x)$ **then** $P_2(p)$
else ...

The sub-predicates, $P_0(p), P_1(p), P_1(p), \cdots$, occupy the branches of the entire predicate. We tried to reduce the size of such a predicate by collapsing some of the sub-predicates on p with a small Hamming distance. This is considered a first step towards using approximation in our approach.

The Hamming distance between two predicates $P_i(p)$ and $P_j(p)$ is the fraction of assignments to p that make $P_i(p)$ and $P_j(p)$ different. For example, the Hamming distance between $p_0 \wedge p_1$ and $p_0 \vee p_1$ is 1/2, since they agree on the assignments ($p_0 = 0,\ p_1 = 0$) and ($p_0 = 1,\ p_1 = 1$), while they do not agree on the assignments ($p_0 = 0,\ p_1 = 1$) and ($p_0 = 1,\ p_1 = 0$).

Let θ be some threshold between 0 and 1. By collapsing sub-predicates $P_i(p)$ and $P_j(p)$, we mean to replace both $P_i(p)$ and $P_j(p)$ with $P_i(p) \vee P_j(p)$, provided that their Hamming distance is less than θ. After $P_i(p)$ and $P_j(p)$ are replaced with $P_i(p) \vee P_j(p)$, the size of the entire predicate is reduced because the BDD node representing $P_i(p) \vee P_j(p)$ is shared. This replacement has the following property: if the predicate $R'(x)$ is obtained by collapsing the original predicate $R(x)$ by the replacement, then $R'(x)$ is implied by $R(x)$, i.e., $\forall x.\ R(x) \Rightarrow R'(x)$.

We made the following experiment. We first computed the reachability predicate $R_p(y)$ defined as follows.

$$\exists x.\ I_p(x) \wedge T_p^*(x, y)$$

We then collapsed $R_p(y)$ according to some threshold θ, and continued verification using the collapsed $R'_p(y)$.

Algorithms discovered by using the collapsed predicate $R'_p(x)$ always satisfy the safety and the liveness. The safety is expressed as $\neg\exists x.\ R_p(x) \wedge E(x)$. Since $R'_p(x)$ is implied by the original predicate $R_p(x)$, we have $\forall p.\ (\neg\exists x.\ R'_p(x) \wedge E(x)) \Rightarrow (\neg\exists x.\ R_p(x) \wedge E(x))$. Therefore, if an algorithm with parameter p satisfies the safety under the collapsed $R'_p(x)$, it satisfies the safety under the original $R'_p(x)$. A similar argument holds for the liveness, because the liveness is expressed as $\neg\exists x.\ R_p(x) \wedge S'(x)$.

We successfully obtained constraints on the parameters. We summarize the results in Figure 10. In the figure, if the size of the final result is marked as '–', it means that verification failed, i.e., no instantiation of parameters could satisfy the safety and liveness.

Template	Threshold θ	Size of $R_p(y)$		Size of final result	
		w/ collapse	w/o collapse	w/ collapse	w/o collapse
Template 1	0.03	175	278	38	93
Template 1	0.05	145	278	–	93
Template 2	0.15	50	429	30	62
Template 2	0.20	39	429	–	62

Fig. 10. Results of collapsing

Using abstract BDDs [3] is another approach to reduce the size of BDDs. It is similar to ours in that abstract BDDs are obtained by merging BDD nodes whose abstract values coincide. Abstract values of BDD nodes are given in advance. In our case, since it is difficult to define such abstraction before the search, we dynamically collapse the algorithm space according to the Hamming distance of BDDs.

6 Conclusion

We searched for mutual exclusion algorithms using BDDs. Since the algorithm space was given by a template, we could simultaneously check all algorithms in the algorithm space by a single execution of the verifier. We gained speed-up of several hundred times compared with verifying all algorithms one by one.

The result of the second case study might suggest that Peterson's algorithm could be discovered as a variant of Dekker's algorithm. In the previous study [7], we searched for variants of Dekker's algorithm, but we could not find Peterson's. This was because we fixed the finishing part as that of the original version of Dekker's. Designing an algorithm space that contains interesting algorithms is not easy. In these case studies, we found only known algorithms.

Compared with the case studies, the success of superoptimization and protocol discovery is due to the fact that the algorithm space is generated by a simple measure, i.e., the length of programs or protocols. In our case studies, the algorithm spaces could only be designed by selecting some instructions so that the spaces included the original algorithms. In order to find new algorithms, we need to define some simple measures that are independent from the original one.

Analysis of the constraints obtained by verification was also a difficult task. We examined all solutions of the constraints by hand, and we found out that the discovered algorithms are equivalent to the original ones. Automatic check of equivalence of algorithms would be a great help. We might be able to use bisimilarity checkers for the purpose.

Acknowledgments

The work reported in this paper was supported, in part, by the Ministry of Education, Science, Sports and Culture, under the grants 11480062 and 10143105.

References

1. R. E. Bryant. Graph Based Algorithms for Boolean Function Manipulation. *IEEE Transactions on Computers,* Vol.C-35, No.8, pp.677–691, 1986.
2. Edmund M. Clarke, Jr., Orna Grumberg, and Doron A. Peled. *Model Checking,* The MIT Press, 1999.
3. Edmund M. Clarke, Somesh Jha, Yuan Lu, and Dong Wang. Abstact BDDs: A Technique for Using Abstraction in Model Checking. *Correct Hardware Design and Verification Methods*, Lecture Notes in Computer Science, Vol.1703, pp.172–186, 1999.

4. Patrick Cousot and Radhia Cousot. Abstract interpretation: a unified lattice model for static analysis of programs by construction or approximation of fixpoints. *Conference Record of the 4th ACM Symposium on Principles of Programming Languages,* pp.238–252, 1977.
5. Torbjörn Granlund and Richard Kenner. Eliminating Branches using a Superoptimizer and the GNU C Compiler. *PLDI'92, Proceedings of the conference on Programming language design and implementation,* pp.341–352, 1992.
6. Masami Hagiya. Discovering Algorithms by Verifiers. Programming Symposium, Information Processing Society of Japan, pp.9–19, 2000, in Japanese.
7. Masami Hagiya and Koichi Takahashi. Discovery and Deduction, *Discovery Science, Third International Conference, DS 2000* (Setsuo Arikawa and Shinichi Morishita Eds.), Lecture Notes in Artificial Intelligence, Vol.1967, pp.17–37, 2000.
8. David E. Long. bdd - a binary decision diagram (BDD) package, 1993. `http://www.cs.cmu.edu/~modelcheck/code.html`
9. Henry Massalin. Superoptimizer: A Look at the Smallest Program. *Proceedings of the Second International Conference on Architectural Support for Programming Languages and Operating Systems, ASPLOS II,* pp.122–126, 1987.
10. Adrian Perrig and Dawn Song. A First Step on Automatic Protocol Generation of Security Protocols. *Proceedings of Network and Distributed System Security,* 2000 Feb.
11. Dawn Xiaodong Song. Athena: a New Efficient Automatic Checker for Security Protocol Analysis, *Proceedings of the 12th IEEE Computer Security Foundations Workshop,* pp.192–202, 1999.

Appendix

In this appendix, we show the essential part of the verification program of Section 3.2. It employs the BDD package developed by David E. Long and distributed from CMU [8]. Functions beginning with "`bdd_`" are from the bdd package. They include, for example, the following functions.

- `bdd_not`: returns the negation of a BDD.
- `bdd_and`: returns the conjunction of BDDs.
- `bdd_or`: returns the disjunction of BDDs.

Their first argument is the bdd manager as defined in the bdd package [8] and is of the type `bdd_manager`. The second and third arguments are BDDs of the type `bdd`.

The function `bdd_exists` computes the existential quantification of a given BDD. Before calling `bdd_exists`, the array of variables to be quantified should be specified as follows.

```
bdd_temp_assoc(bddm, y, 0);
bdd_assoc(bddm, -1);
```

The function `bdd_rel_prod` computes the relational product of given BDDs. Semantically, it first computes the conjunction of the second and third arguments and then computes the existential quantification as `bdd_exists` does.

We also defined some auxiliary functions described below.

The C function `bdd_closure_relation` takes the following seven arguments and computes the closure of a given relation.

- `bdd_manager bddm`: the bdd manager.
- `int AND`: the flag specifying whether the closure is computed by conjunction (if it is 1) or disjunction (if it is 0).
- `bdd g`: the initial predicate.
- `bdd f`: the relation whose closure is computed. It contains variables in `x` and `y`.
- `bdd *x`: the array of the variables in `f` and `g`.
- `bdd *y`: the array of the variables in `f`.
- `bdd *z`: the array of temporary variables.

Let x denote the vector of variables in `x`, y the vector of variables in `y`. It is assumed that `g` is a predicate on x and `f` is a relation between x and y. If the value of the flag `AND` is 0, then the above function computes and returns the following predicate on y.

$$\exists x.\ \mathtt{g}(x) \wedge \mathtt{f}^*(x, y)$$

The return value of the function is of the type `bdd`.

Following are some more auxiliary functions.

- `bdd_rename`: renames variables in a BDD and returns the resulting BDD.
- `bdd_equal_int`: constrains an array of variables as a binary number. For example,

  ```
  bdd_equal_int(bddm, 3, pc11, 4)
  ```

  returns the bdd expressing the following condition.
  $\mathtt{pc11[0]} = \mathtt{pc11[1]} = 0 \wedge$
  $\mathtt{pc11[2]} = 1$
- `bdd_and3`: returns the conjunction of three BDDs.
- `bdd_or4`: returns the disjunction of four BDDs.

The function `smash` makes approximation as described in Section 5. It takes the bdd manager, the BDD to be collapsed, and the threshold value. The identifier `smash_threshold` must have been defined as a C macro.

Following is the main function of the verification program. In the initialization part, omitted from the following code, the transition relations are initialized as follows.

- `t`: the one-step transition relation.
- `t1`: the one-step transition relation for Process 1.
- `t2`: the one-step transition relation for Process 2.

States are represented by nine variables. In the main function, the arrays `x`, `y` and `z` store nine variables representing states. Since they are null terminated, their size is 10. `x` denotes the state before a transition and `y` the state after a transition. `z` consists of temporary variables. The pointers `pc01`, `pc02`, `pc11`, and `pc12` point to the program counters in `x` and `y`.

```
int main()
{
  bdd_manager bddm;
  bdd x[10];
  bdd *pc01 = &x[0];
  bdd *pc02 = &x[3];
  bdd *vs0 = &x[6];
  bdd y[10];
  bdd *pc11 = &y[0];
  bdd *pc12 = &y[3];
  bdd *vs1 = &y[6];
  bdd z[10], *p, *q;
  bdd t, t1, t2,
      initial, reachable, conflict;
  bdd starving, starving_t,
      starving_t1, starving_t2;
  bdd starvation, s, s0;
  bdd e;

  bddm = bdd_init();

  ...
  /* initialization */
  ...

  initial = bdd_equal_int(bddm, 9, y, 0);
  reachable = bdd_closure_relation(bddm, 0,
                                   initial,
                                   t, x, y, z);

#ifdef smash_reachable
  reachable = smash(bddm, reachable,
                    smash_reachable);
#endif

  conflict = bdd_and3(bddm, reachable,
                      bdd_equal_int(bddm, 3,
                                    pc11, 4),
                      bdd_equal_int(bddm, 3,
                                    pc12, 4));
  bdd_temp_assoc(bddm, y, 0);
  bdd_assoc(bddm, -1);
  conflict = bdd_exists(bddm, conflict);

  starving = bdd_or4(bddm,
                     bdd_equal_int(bddm, 3,
                                   pc01, 0),
                     bdd_equal_int(bddm, 3,
                                   pc01, 1),
                     bdd_equal_int(bddm, 3,
```

```
                                         pc01, 2),
                          bdd_equal_int(bddm, 3,
                                         pc01, 3));

  starving_t = bdd_and(bddm, starving, t);
  starving_t1 = bdd_and(bddm, starving, t1);
  starving_t2 = bdd_and(bddm, starving, t2);

  s = starving;
  do {
    printf("iteration for fairness\n");
    s0 = s;

    s = bdd_rename(bddm, s, x, y);
    bdd_temp_assoc(bddm, y, 0);
    bdd_assoc(bddm, -1);
    s = bdd_rel_prod(bddm, starving_t1, s);
    s = bdd_closure_relation(bddm, 0, s,
                             starving_t,
                             y, x, z);

    s = bdd_rename(bddm, s, x, y);
    bdd_temp_assoc(bddm, y, 0);
    bdd_assoc(bddm, -1);
    s = bdd_rel_prod(bddm, starving_t2, s);
    s = bdd_closure_relation(bddm, 0, s,
                             starving_t,
                             y, x, z);

  } while (s0 != s);

  s = bdd_rename(bddm, s, x, y);
  bdd_temp_assoc(bddm, y, 0);
  bdd_assoc(bddm, -1);
  starvation = bdd_rel_prod(bddm, reachable, s);

  e = bdd_and(bddm,
              bdd_not(bddm, conflict),
              bdd_not(bddm, starvation));
}
```

Reducing Search Space in Solving Higher-Order Equations*

Tetsuo Ida[1], Mircea Marin[1], and Taro Suzuki[2]

[1] Institute of Information Sciences and Electronics, University of Tsukuba, Tsukuba 305-8573, Japan
{ida,mmarin}@score.is.tsukuba.ac.jp
[2] Department of Computer Software, University of Aizu, Aizu Wakamatsu 965-8580, Japan
taro@u-aizu.ac.jp

Abstract. We describe the results of our investigation of equational problem solving in higher-order setting. The main problem is identified to be that of reducing the search space of higher-order lazy narrowing calculi, namely how to reduce the search space without losing the completeness of the calculi. We present a higher-order calculus HOLN_0 as a system of inference rules and discuss various refinements that enable the reduction of the search space by eliminating some sources of non-determinism inherent in the calculus.

1 Introduction

With the advancement of high speed computers, it becomes possible to make significant discoveries in some area where in the past the search space involved is so large that exhaustive search was infeasible. Those successful discoveries are realized by the combination of development of methods for reducing the search space by careful analysis of properties of the subject domain and by appealing to brute force of powerful computers. Many techniques have been developed to reduce the search space in the discipline of computational intelligence, such as simulated annealing and genetic algorithms. Most of techniques to date rely in part on statistical methods.

Contrary to the above approach, in our study we only use a logical framework and investigate the reduction of search space. We further focus on equational problem solving. To treat the problem abstractly we use simply-typed lambda calculus. Our problem is then to find solutions of higher-order equations over the domain of simply-typed λ-terms. The problem is not only of theoretical interest but also of practical importance since the formulation is used to model computation of functional logic programming and constraint solving.

The paper is organized as follows. We devote Section 2 to the background of our study, focusing on narrowing. We try to explain it in an intuitive way,

* This work has been also supported in part by JSPS Grant-in-Aid for Scientific Research (B) 12480066, 2000-2002, and (C) 13680388, 2001-2002. Mircea Marin has been supported by JSPS postdoc fellowship 00096, 2000-2001.

S. Arikawa and A. Shinohara (Eds.): Progress in Discovery Science 2001, LNAI 2281, pp. 19–30.

without giving formal definitions of notions. The "undefined" notions in this section are defined either in Section 3 or readers are referred to a standard textbook [BN98]. In particular, we assume readers' knowledge of basic theories of rewriting. In Section 3 we give main results of our research. We first present a higher-order lazy narrowing calculus HOLN_0 as a system of inference rules that operate on goals. Then we give various refinements of HOLN_0 together with the completeness and soundness results. In Section 4 we summarize our results and point out directions of further research.

2 Background

Our problem is as follows. Given an equational theory represented as a set $\mathcal{R}$ of rewrite rules,[1] prove a formula

$$\exists X_1, \ldots, X_m.(s_1 \approx t_1, \ldots, s_n \approx t_n). \tag{1}$$

The sequence of equations in (1) is called a *goal*. We are interested not only in the validity of the formula, but the values, say $v_1, \ldots, v_m$, that instantiate the variables $X_1, \ldots, X_m$. Those values are represented as a substitution $\tau = \{X_1 \mapsto v_1, \ldots, X_m \mapsto v_m\}$. Obtaining such a substitution τ is called *solving* the problem (1), and τ is called the *solution*.

There arise naturally questions about the intended domain over which a solution is sought, and a method for finding it. The notion of solution could be formally defined. For the present purpose, it suffices to note that the intended domain can be fixed to be that of terms generated from a given signature, i.e. the domain of term algebra. It coincides with Herbrand universe in the framework of the first-order predicate logic.

A solution is then defined as follows.

Definition 1. An equation $s \approx t$ is called *immediately solvable* by a substitution τ if $s\tau = t\tau$.

Definition 2. Let G be a goal and $\mathcal{R}$ be a rewriting system. A substitution τ is called a solution of G for $\mathcal{R}$ if

$$G\tau \rightarrow^*_{\mathcal{R}} \boldsymbol{F}\tau,$$

where $\boldsymbol{F}\tau$ is a sequence of immediately solvable equations by an empty substitution.

The next question is how to find the solution; the question to which we address in this paper. An exhaustive search in this domain is clearly infeasible. A systematic method called *narrowing* for searching the solution in the first-order case was devised in the late 70s [Fay79,Hul80].

The narrowing method is the following. We define narrowing for an equation. Extension to a goal is straightforward.

[1] Reader can assume that our discussion in this section is in the framework of first-order term rewrite systems, although it can be carried over to simply-typed (higher-order) pattern rewrite systems.

Definition 3. (Narrowing)

(i) If the given equation e is immediately solvable, we employ the most general substitution σ. We denote this as $e \rightsquigarrow_\sigma \texttt{true}$.

(ii) We rewrite e to e' in the following way:

$$e \rightsquigarrow_\theta e' \tag{2}$$

if there exist

(a) a fresh variant $l \to r$ of a rewrite rule in $\mathcal{R}$, and

(b) a non-variable position p in e

such that $e|_p\theta = l\theta$ by the most general unifier θ and $e' = (e[r]_p)\theta$.

The relation defined above is called *one step narrowing*. We say that e is narrowed to e' in one step.

In case (i), the substitution σ is clearly a solution since it satisfies Definition 2. In case (ii), θ approximates a solution. We narrow the equation e repeatedly by (ii) until it is an immediately solvable one e_n by a substitution σ, and then rewrite e_n to `true` by (i). Namely,

$$e(= e_0) \rightsquigarrow_{\theta_1} e_1 \rightsquigarrow_{\theta_2} \cdots \rightsquigarrow_{\theta_n} e_n \rightsquigarrow_\sigma \texttt{true}. \tag{3}$$

Here we explicitly write the substitution that we employ in each step of narrowing. We will further denote the derivation (3) as:

$$e(= e_0) \rightsquigarrow_\theta^* \texttt{true} \tag{4}$$

where θ is the composition of substitutions employed in each one step narrowing, i.e. $\theta_1 \cdots \theta_n\sigma$. The sequence (4) of one step narrowing is called a *narrowing derivation.*

It is easy to prove that θ is a solution of e. More generally, we have the following soundness theorem.

Theorem 1. *Let G be a goal, and $\mathcal{R}$ be a rewrite system. A substitution θ in a narrowing derivation*

$$G \rightsquigarrow_\theta^* \top, \tag{5}$$

is a solution of G, where $\top$ is a goal consisting only of `true`*s.*

Soundness is not much of a problem in the quest for reduction of the search space. What is intriguing is the property of completeness which is defined as follows. Let $\mathcal{V}ars(t)$ denote the set of free variables in term t.

Definition 4. Narrowing is *complete* if for any solution τ of a goal G there exists a substitution θ such that

$$G \rightsquigarrow_\theta^* \top \text{ and } \theta \preceq \tau[\mathcal{V}ars(G)].$$

Intuitively, completeness of narrowing is the ability to find all solutions. The completeness depends on the kind of solutions that we would like to have and on the kind of rewrite systems. Understanding to what extent generalization of these factors is possible requires deep investigation. Our problem then becomes how to reduce the search space while ensuring the completeness.

There are two sources of non-determinism in narrowing, i. e. choices of rewrite rules at Definition 3(ii)(a) and of positions in an equation at Definition 3(ii)(b). Non-determinism at (a) is unavoidable since the choice determines the solutions, and we have to explore all the possibilities. Therefore we focused on eliminating unnecessary visits of positions in an equation subjected to narrowing. Note that the restriction to non-variable positions at (b) produces already big reduction of the search space. The method without the condition of non-variable-ness is *paramodulation.* It is easy to see that in the case of paramodulation the search space expands greatly since $e|_p\theta = l\theta$ means simply an instantiation of a variable $e|_p$ to the left hand side of the variant of a rewrite rule. Narrowing in the first-order case eliminates this possibility. However, the problem of variable instantiation pops up in the higher-order case.

There are certain assumptions and restrictions that we have to make in order to make our problem tractable. We assume that

- $\mathcal{R}$ is confluent pattern rewrite system[2], and
- solutions are $\mathcal{R}$-normalized.

Confluence is fundamental for our system to be regarded as a sound computing system. A pattern rewrite system is a suitable subset of higher-order systems to which several important notions and properties are carried over from the first-order systems; e.g., pattern unification is unitary.

A crucial observation in our study of the reduction of the search space is the lifting lemma introduced originally in order to prove the completeness [Hul80], and the standardization theorem [HL92].

Lemma 1 (Lifting lemma). *Let G be a goal and τ be a solution of G. For*

$$G\tau \to^*_{\mathcal{R}} \boldsymbol{F}\tau,$$

there exists a narrowing derivation

$$G \leadsto^*_{\theta} \boldsymbol{F} \text{ such that } \theta \preceq \tau[\mathcal{V}ars(G)].$$

$\boldsymbol{F}\tau$ is reached from $G\tau$ by repeated rewriting of $G\tau$ without backtracking if $\mathcal{R}$ is confluent. The lifting lemma tells that for any such rewrite sequence, we can find a corresponding narrowing derivation. Here "corresponding" means that rewrite sequence and the narrowing derivation employ the same rewrite rules at the same positions. There are many paths to reach $\boldsymbol{F}\tau$, but we do not have to explore all these paths if there is a "standard" path to $\boldsymbol{F}\tau$. This means that we only have to consider positions that make the rewrite sequence standard. Then,

[2] The definition of pattern rewrite system is given in Section 3

by the lifting lemma we can focus on the corresponding narrowing derivation. Indeed, for a class of left-linear confluent systems such a standard derivation exists [Bou83,Suz96]. In the standard rewrite sequence, terms are rewritten in an outside-in manner, and hence equations are narrowed in an outside-in manner [IN97].

The next task is how to define a calculus that generates a narrowing derivation that corresponds to the standard derivation. Such a calculus can easily be shown to hold the completeness property. In our previous research we designed lazy narrowing calculus called LNC with completeness proofs [MO98,MOI96]. LNC is a rule-based specification of lazy narrowing in the first-order framework. In particular LNC with the eager variable elimination strategy serves as our starting point. LNC is designed to be more general than the one that generates the narrowing derivation corresponding to the standard rewrite sequence, however.

3 Higher-Order Lazy Narrowing Calculi

We now present higher-order lazy narrowing calculi to be called HOLN. HOLN are higher-order extensions of LNC as well as a refinement of Prehofer's higher-order narrowing calculus LN [Pre98] using the results obtained in the study of LNC.

We developed a family of HOLN from the initial calculus HOLN_0. The rest of the calculi are derived by refining HOLN_0. Each calculus is defined as a pair consisting of a domain of simply-typed λ-terms and a set of inference rules. We always work with $\beta\eta^{-1}$ normal forms, and assume that transformation of a simply-typed λ-term into its $\beta\eta^{-1}$ normal form is an implicit meta operation.

Since the domain is the same in all the calculi, we only specify the set of inference rules. The inference rules operate on pairs consisting of a sequence G of equations and a set W of variables to which normalized values of the solution is bound. This pair is denoted by $E\lfloor_W$, and is called *goal of HOLN*. The reason for this new definition of a goal is that we usually specify our problem as a pair of equations together with the set of variables for which we intend to compute normalized bindings. This pair can be carried along throughout the process of equation solving by the calculi.

We present four calculi of HOLN together with the soundness and completeness results. The results are given without proofs. Interested readers should refer to [MSI01].

Before describing the calculi, we briefly explain our notions and notations.

- X, Y and H, possibly subscripted, denote free variables. Furthermore, whenever H (possibly subscripted) is introduced, it is meant to be fresh.
- x, y, z (possibly subscripted) denote bound variables.
- v denotes a bound variable or a function symbol.
- $\overline{t_n}$ denotes a sequence of terms $t_1, \ldots, t_n$.
- A term of the form $\lambda\overline{x_k}.X(\overline{t_n})$ is called *flex*, and *rigid* otherwise.
- An equation is *flex* if both sides are flex terms.

- A *pattern* t is a term such that every occurrence of the form $X(\overline{t_n})$ in t satisfies that $\overline{t_n}$ is a sequence of distinct bound variables. Furthermore, the pattern t is called *fully-extended* if $\overline{t_n}$ is a sequence of all the bound variables in the scope of X.
- A pattern is *linear* if no free variable occurs more than once in it.

Definition 5. (Pattern Rewrite System)

- A *pattern rewrite system* (PRS for short) is a finite set of rewrite rules $f(\overline{l_n}) \to r$, where $f(\overline{l_n})$ and r are terms of the same base type, and $f(\overline{l_n})$ is a pattern.
- A *left-linear* (resp. *fully-extended*) PRS is a PRS consisting of rewrite rules which have a linear (resp. fully-extended) pattern on their left hand side. A fully-extended PRS is abbreviated to EPRS, and a left-linear EPRS is abbreviated to LEPRS.

3.1 HOLN$_0$

The inference rules are relations of the form

$$(E_1, e, E_2)\downharpoonleft_W \Rightarrow_{\alpha,\theta} (E_1\theta, E, E_2\theta)\downharpoonleft_{W'}$$

where e is the selected equation, α is the label of the inference rule, θ is the substitution computed in this inference step, $W' = \bigcup_{X \in W} \mathcal{V}ars(X\theta)$, and E is a sequence of equations whose elements are the *descendants* of e. If an equation e' is an equation in E_1 or E_2, then e' has only one descendant in the inference step, namely the corresponding equation $e'\theta$ in $E_1\theta$ or $E_2\theta$.

HOLN distinguish between *oriented* equations written as $s \rhd t$, and *unoriented* equations written as $s \approx t$. In oriented equations, only the left hand sides are allowed to be rewritten; for a solution θ, $s\theta \rhd t\theta \to_{\mathcal{R}}^{*} t\theta \rhd t\theta$. For brevity, we introduce the following notation: $s \trianglerighteq t$ denotes $s \approx t$, $t \approx s$, or $s \rhd t$; $s \approxeq t$ denotes $s \approx t$, $t \approx s$, $s \rhd t$, or $t \rhd s$.

3.2 The Calculus HOLN$_0$

HOLN$_0$ consists of three groups of inference rules: *pre-unification rules, lazy narrowing rules,* and *removal rules of flex equations.*

Pre-unification Rules

[i] Imitation.

$$(E_1, \lambda\overline{x_k}.X(\overline{s_m}) \approxeq \lambda\overline{x_k}.g(\overline{t_n}), E_2)\downharpoonleft_W \Rightarrow_{[\mathsf{i}],\theta} (E_1, \overline{\lambda\overline{x_k}.H_n(\overline{s_m}) \approxeq \lambda\overline{x_k}.t_n}, E_2)\theta\downharpoonleft_{W'}$$

where $\theta = \{X \mapsto \lambda\overline{y_m}.g(\overline{H_n(\overline{y_m})})\}$.

[p] Projection.
If $\lambda\overline{x_k}.t$ is rigid then

$$(E_1, \lambda\overline{x_k}.X(\overline{s_m}) \approxeq \lambda\overline{x_k}.t, E_2)\lfloor_W \Rightarrow_{[\mathrm{p}],\theta} (E_1, \lambda\overline{x_k}.X(\overline{s_m}) \approxeq \lambda\overline{x_k}.t, E_2)\theta\lfloor_{W'}$$

where $\theta = \{X \mapsto \lambda\overline{y_m}.y_i(\overline{H_n(\overline{y_m})})\}$

[d] Decomposition.

$$(E_1, \lambda\overline{x_k}.v(\overline{s_n}) \trianglerighteq \lambda\overline{x_k}.v(\overline{t_n}), E_2)\lfloor_W \Rightarrow_{[\mathrm{d}],\epsilon} (E_1, \overline{\lambda\overline{x_k}.s_n \trianglerighteq \lambda\overline{x_k}.t_n}, E_2)\lfloor_{W'}$$

Lazy Narrowing Rules

[on] Outermost narrowing at non-variable position.
If $f(\overline{l_n}) \to r$ is an $\overline{x_k}$-lifted rewrite rule of $\mathcal{R}$ then

$$(E_1, \lambda\overline{x_k}.f(\overline{s_n}) \trianglerighteq \lambda\overline{x_k}.t, E_2)\lfloor_W \Rightarrow_{[\mathrm{on}],\epsilon} (E_1, \overline{\lambda\overline{x_k}.s_n \rhd \lambda\overline{x_k}.l_n}, \lambda\overline{x_k}.r \trianglerighteq \lambda\overline{x_k}.t, E_2)\lfloor_{W'}$$

[ov] Outermost narrowing at variable position.

If $f(\overline{l_n}) \to r$ is an $\overline{x_k}$-lifted rewrite rule of $\mathcal{R}$ and $\begin{cases} \lambda\overline{x_k}.X(\overline{s_m}) \text{ is not a pattern} \\ \text{or} \\ X \notin W \end{cases}$

then

$$(E_1, \lambda\overline{x_k}.X(\overline{s_m}) \trianglerighteq \lambda\overline{x_k}.t, E_2)\lfloor_W \Rightarrow_{[\mathrm{ov}],\theta} (E_1\theta, \overline{\lambda\overline{x_k}.H_n(\overline{s_m\theta}) \rhd \lambda\overline{x_k}.l_n}, \lambda\overline{x_k}.r \trianglerighteq \lambda\overline{x_k}.t\theta, E_2\theta)\lfloor_{W'}$$

where $\theta = \{X \mapsto \lambda\overline{y_m}.f(\overline{H_n(\overline{y_m})})\}$.

Removal Rules

[t] Trivial equation.

$$(E_1, \lambda\overline{x_k}.X(\overline{s_n}) \approxeq \lambda\overline{x_k}.X(\overline{s_n}), E_2)\lfloor_W \Rightarrow_{[\mathrm{t}],\epsilon} (E_1, E_2)\lfloor_W$$

[fs] Flex-same.
If $X \in W$ then

$$(E_1, \lambda\overline{x_k}.X(\overline{y_n}) \trianglerighteq \lambda\overline{x_k}.X(\overline{y'_n}), E_2)\lfloor_W \Rightarrow_{[\mathrm{fs}],\theta} (E_1, E_2)\theta\lfloor_{W'}$$

where $\theta = \{X \mapsto \lambda\overline{y_n}.H(\overline{z_p})\}$ with $\{\overline{z_p}\} = \{y_i \mid y_i = y'_i, 1 \leq i \leq n\}$.

[fd] Flex-different.
If $\begin{cases} X \in W \text{ and } Y \in W & \text{if } \trianglerighteq \text{ is } \approx \\ X \in W & \text{if } \trianglerighteq \text{ is } \rhd \end{cases}$ then

$$(E_1, \lambda\overline{x_k}.X(\overline{y_m}) \trianglerighteq \lambda\overline{x_k}.Y(\overline{y'_n}), E_2)\lfloor_W \Rightarrow_{[\mathrm{fd}],\theta} (E_1, E_2)\theta\lfloor_{W'}$$

where $\theta = \{X \mapsto \lambda\overline{y_m}.H(\overline{z_p}), Y \mapsto \lambda\overline{y'_n}.H(\overline{z_p})\}$ with $\{\overline{z_p}\} = \{\overline{y_m}\} \cap \{\overline{y'_n}\}$.

The pre-unification rules and the removal rules constitute a system equivalent to the transformation system for higher-order pre-unification of Snyder and Gallier [SG89]. The lazy narrowing rules play the role of one step narrowing in more elementary way. Namely narrowing is applied at the outermost position (excluding the λ-binder) of either side of the terms of the selected equation, leaving the computation of the most general unifier in later steps of the narrowing derivation. Note also that the system performs narrowing at variable positions as well by the rule [ov], in contract to the first-order narrowing, where narrowing at variable position is unnecessary.

In HOLN we do not have to search for positions at which subterms are narrowed, since terms are processed at the outermost position. Narrowing defined in Definition 3(ii) is decomposed into more elementary operations of the eight inference rules and the non-determinism associated with search for positions have disappeared. This facilitates the implementation and analysis of the behavior of narrowing. However, we now introduced a new source of non-determinism. Namely, the non-determinism associated with the choices of the inference rules. More than one rule can be applied to the same goal.

Let $\mathcal{C}$ denote a calculus in the family of HOLN. We call an inference step of calculus $\mathcal{C}$ $\mathcal{C}$*-step*. A $\mathcal{C}$*-derivation* is a sequence of $\mathcal{C}$-steps

$$E\lfloor_W (= E_0\lfloor_{W_0}) \Rightarrow_{\alpha_1,\theta_1} E_1\lfloor_{W_1} \Rightarrow_{\alpha_2,\theta_2} \cdots \Rightarrow_{\alpha_n,\theta_n} E_n\lfloor_{W_n},$$

where E_n consists of flex equations and there is no $\mathcal{C}$-step starting with $E_n\lfloor_{W_n}$. We write this derivation as $E_0\lfloor_{W_0} \Rightarrow^*_\theta E_n\lfloor_{W_n}$, where $\theta = \theta_1 \cdots \theta_n$, as before.

We redefine completeness for calculus $\mathcal{C}$.

Definition 6. A substitution τ is a $\mathcal{C}$-solution of a goal $E\lfloor_W$ for $\mathcal{R}$ if

- $\tau\lceil_W$ is $\mathcal{R}$-normalized, and
- τ is a solution of E.

Definition 7. Calculus $\mathcal{C}$ is $\mathcal{C}$*-complete* if for any $\mathcal{C}$-solution τ of a goal $E\lfloor_W$ there exists a substitution θ such that

$$E\lfloor_W \Rightarrow^*_\theta \boldsymbol{F}\lfloor_{W'} \text{ and } \theta\sigma \preceq \tau[\mathcal{V}ars(E)], \text{ where } \sigma \text{ is a } \mathcal{C}\text{-solution of } \boldsymbol{F}\lfloor_{W'}.$$

The prefix "$\mathcal{C}$-" is omitted whenever it is clear from the context.

We have for HOLN_0 the following result.

Theorem 2. *Let $\mathcal{R}$ be a confluent PRS. The calculus* HOLN_0 *is sound and complete.*

3.3 HOLN_1

When $\mathcal{R}$ is a confluent LEPRS, we can design a calculus that has more deterministic behavior with respect to the application of rule [on]. HOLN_1 is a refinement of HOLN_0 in that HOLN_1 does not apply rule [on] to parameter-passing descendants of the form $\lambda\overline{x_k}.f(\overline{s_n}) \rhd \lambda\overline{x_k}.X(\overline{y_k})$ where f is a defined function sysmbol. Here a parameter-passing descendant is defined as follows.

Definition 8. A *parameter-passing equation* is any descendant, except the last one, of a selected equation in [on]-step or [ov]-step. A *parameter-passing descendant* is either a parameter-passing equation or a descendant of a parameter-passing descendant.

Note that parameter-passing descendants are always oriented equations. To distinguish them from the other oriented equations, we mark parameter-passing descendants as $s \blacktriangleright t$.

We define $[\mathrm{on}]_r$ as follows.

$[\mathrm{on}]_r$ Restricted outermost narrowing at non-variable position.
If $f(\overline{l_n}) \to r$ is an $\overline{x_k}$-lifted rewrite rule of $\mathcal{R}$ and the selected equation $\lambda\overline{x_k}.f(\overline{s_n}) \trianglerighteq \lambda\overline{x_k}.t$ is not of the form $\lambda\overline{x_k}.f(\overline{s_n}) \blacktriangleright \lambda\overline{x_k}.X(\overline{y_k})$ then

$$(E_1, \lambda\overline{x_k}.f(\overline{s_n}) \trianglerighteq \lambda\overline{x_k}.t, E_2)\downharpoonleft_W \Rightarrow_{[\mathrm{on}],\epsilon} (E_1, \overline{\lambda\overline{x_k}.s_n \rhd \lambda\overline{x_k}.l_n}, \lambda\overline{x_k}.r \trianglerighteq \lambda\overline{x_k}.t, E_2)\downharpoonleft_{W'}$$

We replace [on] of HOLN_0 by $[\mathrm{on}]_r$, and let HOLN_1 be the resulting calculus.

We note that HOLN_1 simulates the rewriting in an outside-in manner by the combination of the lazy narrowing rules and the imitation rule.

We have for HOLN_1 the following result.

Theorem 3. *Let $\mathcal{R}$ be a confluent LEPRS. Calculus* HOLN_1 *is sound and complete.*

The proof of this theorem is based on the standardization theorem for confluent LEPRSs, which has been proved by van Oostrom [vO00]. This result is the higher-order version of the eager variable elimination for LNC.

3.4 HOLN_2

Calculus HOLN_2 is a specialization of HOLN_0 for confluent constructor LEPRSs, i.e., confluent LEPRSs consisting of rewrite rules $f(\overline{l_n}) \to r$ with no defined symbols in $l_1, \ldots, l_n$. We introduce

- a new inference rule [c] for parameter-passing descendants, and
- an equation selection strategy $\mathcal{S}_{\mathrm{left}}$.

We define inference rule [c] as follows:

[c] Constructor propagation.
If $\exists s \blacktriangleright \lambda\overline{x_k}.X(\overline{y_k}) \in E_1$ and $s' = \lambda\overline{y_k}.s(\overline{x_k})$ then

$$(E_1, \lambda\overline{z_n}.X(\overline{t_k}) \trianglerighteq \lambda\overline{z_n}.u, E_2)\downharpoonleft_W \Rightarrow_{[\mathrm{c}],\epsilon} (E_1, \lambda\overline{z_n}.s'(\overline{t_k}) \trianglerighteq \lambda\overline{z_n}.u, E_2)\downharpoonleft_{W'}.$$

We then define the calculus HOLN_2 as consisting of the inference rules of HOLN_0 and [c], where [c] has the highest priority. We can show that HOLN_2 does not produce parameter-passing descendants if it respects the equation selection strategy $\mathcal{S}_{\mathrm{left}}$.

We say that calculus $\mathcal{C}$ respects equation selection strategy $\mathcal{S}_{\text{left}}$ if every $\mathcal{C}$-step is applied to the leftmost selectable equation. As a consequence, there is no non-determinism between inference rules [on] and [d] for parameter-passing descendants. Thus, solving parameter-passing descendants becomes more deterministic.

We have for HOLN_2 the following result.

Theorem 4. *Let $\mathcal{R}$ be a confluent constructor LEPRS. Calculus* HOLN_2 *that respects strategy $\mathcal{S}_{left}$ is complete. Moreover,* HOLN_2 *with strategy $\mathcal{S}_{left}$ is sound for goals consisting only of unoriented equations.*

3.5 HOLN_3

Suppose we have a HOLN_0-derivation

$$E\lfloor_W \Rightarrow_\theta^* (E_1, e, E_2)\lfloor_{W'}$$

and a solution γ' of $(E_1, E_2)\lfloor_{W'}$. If for any such γ' there exists a solution γ of $(E_1, e, E_2)\lfloor_{W'}$ such that $\theta\gamma\lceil_{\mathcal{V}ars(E)} = \theta\gamma'\lceil_{\mathcal{V}ars(E)}$, then e does not contribute to the solution of $E\lfloor_W$. We call such an equation *redundant.* One of redundant equations is $\lambda\overline{x_k}.s \blacktriangleright \lambda\overline{x_k}.X(\overline{y_k})$ in $(E_1, \lambda\overline{x_k}.s \blacktriangleright \lambda\overline{x_k}.X(\overline{y_k}), E_2)\lfloor_W$, where $\overline{y_k}$ is a permutation of the sequence $\overline{x_k}$, and $X \notin \mathcal{V}ars(E_1, \lambda\overline{x_k}.s, E_2)$.

Calculus HOLN_3 is a refinement of HOLN_0 for LEPRSs which avoids solving the above redundant equation. We introduce the following inference rule [rm] and define HOLN_3 as consisting of the inference rules of HOLN_0 and [rm] that has to be applied with the highest priority.

[rm] Removal of redundant equations.
If $\overline{y_k}$ is a permutation of the sequence $\overline{x_k}$, and $X \notin \mathcal{V}ars(E_1, \lambda\overline{x_k}.s, E_2)$ then

$$(E_1, \lambda\overline{x_k}.s \blacktriangleright \lambda\overline{x_k}.X(\overline{y_k}), E_2)\lfloor_W \Rightarrow_{[\text{rm}],\epsilon} (E_1, E_2)\lfloor_W .$$

We have for HOLN_3 the following result.

Theorem 5. *Let $\mathcal{R}$ be a confluent LEPRS. Calculus* HOLN_3 *is sound and complete.*

3.6 Rationale for Refinements

After presenting the family of HOLN, we can now see more clearly the rationale of our refinements. First, we noted that [ov] is the main source of non-determinism, since it allows us to compute approximations of any solution, be it $\mathcal{R}$-normalized or not. Our first investigation was how to restrict the application of [ov] without losing completeness. We achieved this by redefining the notion of goal. A goal of HOLN is a pair $E\lfloor_W$ where E is a sequence of equations, and W is a finite set of variables for which we want to compute $\mathcal{R}$-normalized bindings. By keeping track of W, we can restrict the application of [ov] without losing completeness. The resulted calculus is HOLN_0.

Next, we studied how to reduce the non-determinism between the inference rules applicable to the same selected equation. We found two possible refinements of HOLN_0 which reduce this source of non-determinism. HOLN_1 is a refinement inspired by the eager variable elimination method of LNC. This refinement reduces the non-determinism between the inference rules applicable to parameter-passing descendants of the form $\lambda\overline{x_k}.f(\ldots) = \lambda\overline{x_k}.X(\ldots)$, where f is a defined function symbol. More precisely, [on] is not applied to such an equation, when it is selected. This refinement is valid for confluent LEPRSs.

HOLN_2 is a refinement inspired by the refinement of LNC for left-linear constructor rewrite systems. The idea is to avoid producing parameter-passing descendants with defined function symbols in the right-hand side. As a consequence, there is no non-determinism between [d] and [on] when the selected equation is a parameter-passing descendant. Thus, the calculus is more deterministic. We achieved this property by introducing a new inference rule [c] which has the highest priority, and a suitable equation selection strategy (which is the higher-order version of the leftmost equation selection strategy). This refinement is valid for confluent constructor LEPRSs.

HOLN_3 is a refinement aiming at detecting redundant equations. It is useful to detect redundant equations, because they can be simply eliminated from the goal, thus saving time and resources to solve them. We identified a criterion to detect redundant equations and introduced [rm] which simply eliminates a redundant equation from the goal.

4 Conclusions and Future Work

We have shown the results of our investigation of reducing the search space in solving higher-order equations. We presented a family of sound and complete lazy narrowing calculi HOLN designed to compute solutions which are normalized with respect to a given set of variables. All the calculi have been implemented as part of our distributed constraint functional logic system CFLP [MIS00].

An interesting direction of research is to extend HOLN to conditional PRSs. A program specification using conditions is much more expressive because it allows the user to impose equational conditions under which rewrite steps are allowed. Such an extension is quite straightforward to design, but it introduces many complexities for proving completeness.

References

[BN98] F. Baader and T. Nipkow. *Term Rewriting and All That.* Cambridge University Press, 1998.

[Bou83] G. Boudol. Computational Semantics of Term Rewriting Systems. Technical report, INRIA, February 1983.

[Fay79] M. Fay. First-Order Unification in Equational Theories. In *Proceedings of the 4th Conference on Automated Deduction*, pages 161–167, 1979.

[HL92] G. Huet and J.-J. Lévy. Computations in Orthogonal Rewriting Systems, I and II. In *Computational Logic, Essays in Honor of Alan Robinson*, pages 396–443. The MIT Press, 1992.

[Hul80] J.-M. Hullot. Canonical Forms and Unification. In *Proceedings of the 5th Conference on Automated Deduction*, volume 87 of *LNCS*, pages 318–334. Springer, 1980.

[IN97] T. Ida and K. Nakahara. Leftmost Outside-in Narrowing Calculi. *Journal of Functional Programming*, 7(2):129–161, 1997.

[MIS00] M. Marin, T. Ida, and T. Suzuki. Cooperative constraint functional logic programming. In T. Katayama, T. Tamai, and N. Yonezaki, editors, *International Symposium on Principles of Software Evolution (ISPSE 2000)*, Kanazawa, Japan, November 1-2, 2000. IEEE Computer Society.

[MO98] A. Middeldorp and S. Okui. A Deterministic Lazy Narrowing Calculus. *Journal of Symbolic Computation*, 25(6):733–757, 1998.

[MOI96] A. Middeldorp, S. Okui, and T. Ida. Lazy Narrowing: Strong Completeness and Eager Variable Elimination. *Theoretical Computer Science*, 167(1,2):95–130, 1996.

[MSI01] M. Marin, T. Suzuki, and T. Ida. Refinements of lazy narrowing for left-linear fully-extended pattern rewrite systems. Technical Report ISE-TR-01-180, Institute of Information Sciences and Electronics, University of Tsukuba, Tsukuba, Japan, to appear, 2001.

[Pre98] C. Prehofer. *Solving Higher-Order Equations. From Logic to Programming.* Foundations of Computing. Birkhäuser Boston, 1998.

[SG89] W. Snyder and J. Gallier. Higher-order unification revisited: Complete sets of transformations. *Journal of Symbolic Computation*, 8:101–140, 1989.

[Suz96] T. Suzuki. Standardization Theorem Revisited. In *Proceedings of 5th International Conference, ALP'96*, volume 1139 of *LNCS*, pages 122–134, 1996.

[vO00] V. van Oostrom. Personal communication, August 2000.

The Structure of Scientific Discovery: From a Philosophical Point of View

Keiichi Noé

Tohoku University, Graduate School of Arts and Letters,
Kawauchi, Aoba-ku, Sendai 980-8576, Japan
Noe@sal.tohoku.ac.jp

Abstract. The term "discovery" can be defined as "the act of becoming aware of something previously existing but unknown." There are many kinds of discoveries from geographical to mathematical ones. Roughly speaking, we can classify discoveries into two types: the factual discovery and the conceptual one. In the historical development of scientific inquiry, the former typically occurs during the period of "normal science" in Kuhnian term. The latter emerges in the opportunity of "scientific revolution," i.e., the time of a paradigm-shift. It is necessary for scientific discoveries to use imagination as well as reasoning. Based on the study of aphasic disturbances from a viewpoint of using figurative expressions by Roman Jakobson, we would like to discriminate between the metonymical imagination from the metaphorical one. While the former is related to discoveries of facts and laws which complete an unfinished theory, the latter is concerned with conceptual discoveries which change a viewpoint from explicit facts to implicit unknown relations. Considering these points, it is useful to examine Poincaré's description of his own experience concerning mathematical discovery. He emphasized the working of "the subliminal ego" and "the aesthetic sensibility" in his discovery of Fuchsian function. These activities can be compared to the function of metaphorical imagination. However, it is not easy to formulate the unconscious process of "aspect-dawning" in the form of explicit rules or algorithm. This suggests the difficulty in realizing scientific discovery by computers.

1 Introduction

What is scientific discovery? The very term of "discovery" or "to discover" etymologically stemmed from "to get rid of a cover." A discovery arises when something wearing a veil becomes explicit for us through removing obstacles. Generally speaking, it can be defined as "the act of becoming aware of something previously existing but unknown." Here, the qualification of "previously existing" is the characteristic of discovery in distinction from invention. Roentgen's discovery of X-rays and Edison's invention of electric light provide typical examples of both. However, this demarcation might become ambiguous in some fields of science. For example, is synthesizing a new substance in chemistry a discovery or an invention? How about mathematical theorems? Thus, there are

S. Arikawa and A. Shinohara (Eds.): Progress in Discovery Science 2001, LNAI 2281, pp. 31–39.

many kind of discoveries from geographical to mathematical ones, but here I limit the discussion to discoveries in theoretical sciences.

Then, is there a difference between the discovery of a new continent by Columbus and the discovery of the law of gravitation by Newton? The former discovery was nothing but a mere accident, because Columbus intended to reach India in the beginning of his sailing. The latter included a change of viewpoint "from the closed world to the infinite universe" in Koyré's term.[4] Although the Aristotelian world-view presupposes the distinction between the heaven and the earth, Newton is convinced that the universe is one and the same, and is combined by the same kind of natural laws. In this sense, it is quite likely that great scientific discoveries depend more or less upon a change of viewpoints in the given data and the existing theoretical context.

Our concern is to elucidate the mechanism of such conceptual transformation caused by changing a viewpoint in scientific discoveries. For this purpose, we begin our study with distinguishing two kinds of discovery. Then we will examine Henri Poincare's analysis of mathematical discovery as a case study. Finally, we would like to consider the structure of scientific imagination which is closely related to our linguistic ability, especially the use of metaphor.

2 Two Types of Scientific Discoveries

The simplest example of changing viewpoint is a trompe l'oeil like Rubin's goblet-profile figure. It is an optical experience that a figure seen as a goblet suddenly changes into a figure of profile. Wittgenstein calls such perceptual experience as "aspect-dawning" or "a change of aspect" and analyses the essential feature of our perception as "seeing-as structure."[13] According to Gestalt psychology, such experiences are characterized as the reversal between "figure" and "ground." In the context of scientific inquiry, the figure corresponds to regularities in nature and the ground to experimental data. The figure- ground relation is influenced by scientists' interests, their background knowledge and many kinds of theoretical viewpoints. Though these conditions are intricately interrelated each other, it is nothing other than a change of viewpoint that brings about the figure-ground reversal. Many discoveries in the history of science give good examples which emphasize the importance of changing viewpoint, not to mention the formation process of the quantum theory by Planck as well as the relativity theory by Einstein. Most of these discoveries are based on an "aspect-dawning" which suggests a new kind of order or structure in established facts and preconceived ideas.

On that basis, as is well known, Norwood Hanson draws out the concept of "theory-ladenness of observation" in the philosophy of science. It means that we cannot observe any scientific fact without making any theoretical assumptions. Observations must always reflect these assumptions. Thus he concludes that "There is a sense, then, in which seeing is a 'theory-laden' undertaking. Observation of x is shaped by prior knowledge of x."[1] His philosophical analysis casts doubt on the observation-theory dichotomy supported by logical positivists, and opens a door to the New Philosophy of Science in the 1960's.[9]

Although logical positivists lay emphasis on the role of sense-data in the verification of a scientific theory, observation is not a simple experience of accepting sense-data. It presupposes a complicated theoretical background to which most of scientists commit themselves. Excellent scientists can observe in familiar objects what no one else has observed before in light of a new theoretical hypothesis. In this sense, making observations and forming a hypothesis are one and inseparable. Relevant to this point is Hanson's following remark:

> Physical theories provide patterns within which data appear intelligible. They constitute a 'conceptual Gestalt'. A theory is not pieced together from observed phenomena; it is rather what makes it possible to observe phenomena as being of a certain sort, and as related to other phenomena. Theories put phenomena into systems. [2]

The words "patterns" and "conceptual Gestalt" are important in this context, because they make it possible to observe a meaningful fact. Scientists do not observe phenomena as a brute fact but as a patterned fact organized in a conceptual Gestalt. There is no observation without any theoretical framework. We should rather suggest that a scientific observation is usually preceded by a theoretical hypothesis. The discovery of the elementary particle called "meson" predicted by Hideki Yukawa's physical theory is a notable example. He intended to explain the phenomena of nuclear force but finally introduces a new particle in terms of analogy with the photon's role in mediating electromagnetic interactions. It was two years after the publication of Yukawa's meson theory that Anderson experimentally observed the concerned meson in cosmic rays. No one can identify the existence of meson in cosmic rays without theoretical hypothesis. Astronomical discovery of Neptune based on theoretical prediction of a new planet by Adams and Le Verrier is another illustration of the same point.

For reasons mentioned above, we can distinguish two kinds of discoveries, i.e., the factual discovery and the conceptual discovery. Here, the word "factual" is sometimes extended to cover the notion of mathematical facts. The former is the discovery of a new fact guided by an established theory. Identification of unknown elements based on Mendeleev's periodic table is a good example to illustrate this kind of discovery. Examples of the latter are Newton's gravitational theory and Einstein's relativity theory, which propose systematic explanations of the anomalous phenomena by reinterpreting pre-existing facts and laws from a new point of view. Of course, both discoveries are inseparable from each other, but historically speaking, each type of discovery corresponds to the different phase of scientific development.

As is well known, Thomas Kuhn introduces a famous distinction between the period of *normal science* and that of *scientific revolution* into the historiography of science.[5] Normal science means scientific researches conducted by a particular *paradigm* and is compared to a puzzle-solving activity following preestablished norms. On the contrary, scientific revolution is no other than an event of paradigm-shift, which radically changes presuppositions of normal scientific researches. Generally speaking, it seems reasonable to suppose that factual discoveries are generally made during the period of normal science, and the greater part of conceptual discoveries emerge in the midst of scientific revolutions. In the period of normal science, most of discoveries have to do with novelties of fact and partial revision of theories. Therefore normal scientific research is a rather cumulative enterprise to extend the scope and precision of scientific knowledge. On the other hand, in the period of scientific revolution, discoveries are usually accompanied by a radical change of concept as well as theory. Both fact and assimilation to theory, observation and conceptualization, are closely interrelated in this kind of discovery. Scientific revolutions consist of a complicated series of events accompanied by conflicts between the old and the new paradigm.

These points will lead us further into a consideration of scientific imagination which gives a clue to scientific discovery. Roughly speaking, two types of discoveries above mentioned correspond to different functions of imagination. I would like to call them *metonymical imagination* and *metaphorical imagination* respectively. While the former is related to factual discoveries in the period of normal science, the latter is concerned with conceptual discoveries in the opportunity of scientific revolutions. These two kinds of imagination are based on our ability of using figurative expressions, i.e., metonymy and metaphor. From this point, we might go on to an even more detailed examination of the relation between scientific discovery and imagination by appealing to linguistic achievements. Before turning to this problem, I would like to examine Poincaré's analysis of a methematical discovery in the following section.

3 Poincaré's Self-examination

In order to inquire into the process of scientific discovery, it is important to listen to the testimonies of working scientists. However, they are usually too busy with producing a research paper to describe and evaluate their Eureka experience. Fortunately, Henri Poincaré, a famous French mathematician, left a valuable document in his book entitled *Science and Method* published in 1908. He pays attention to the mechanism of mathematical discovery and explicates its structure on the basis of his own experience. Looking back on his discovery of a new theorem concerning Fuchsian function, he trys to formulate a general principle of mathematical discovery. This is what he has to say on this matter:

> What, in fact, is mathematical discovery? It does not consist in making new combinations with mathematical entities that are already known. That can be done by any one, and the combinations that could be so formed would be infinite in number, and the greater part of them would be absolutely devoid of interest. Discovery consists precisely in not constructing useless combinations, but in constructing those that are useful, which are an infinitely small minority. Discovery is discernment, selection. [10]

First, Poincaré suggests that the essence of mathematical discovery does not belong to making new combinations of mathematical entities, because the number of combination is infinite and moreover most of them are useless. Only excellent mathematicians can choose a few useful combination. Therefore, he tentatively concludes that "discovery is discernment, selection." However, such an operation is not carried out by enumeration of all the possible examples. It would take a mathematician's lifetime to achieve such a work. They rather intuitively find a mathematical fact which deserves sincere research through specific sensibility about the mathematical order. Of course, nowadays, such a simple enumeration might be easily done by a computer in a moment. Still, the difficult work is to tell whether it is a significant combination or not.

As experimental facts form a basis of discovering physical laws, mathematical facts can also become a clue to discovering mathematical laws. Furthermore, to echo Poincaré, such heuristic facts must be "those which reveal unsuspected relations between other facts, long since known, but wrongly believed to be unrelated to each other." [11] Here we can compare the words "reveal unsuspected relations" to the change of aspect between figure and ground mentioned above. That is to say, as a figure seen as a goblet changes into a figure of profile in an instant, the new mathematical relation suddenly emerges from well-known facts which have been regarded as unrelated to each other. Mathematician's intuition can make "aspect-dawning" or "a change of viewpoint" possible in the conceptual Gestalt of mathematical entities. It is difficult to describe this kind of intuition in precise language, because a change of aspect occurs in the unconscious process rather than in the conscious one. After emphasizing a great contribution of unconscious activities to mathematical discoveries, Poincaré continues:

> One is at once struck by these appearances of sudden illumination, obvious indications of a long course of previous unconscious work. The part played by this unconscious work in mathematical discovery seems to me indisputable, and we shall find traces of it in other cases where it is less evident. (.....) Such are the facts of the case, and they suggest the following reflections. The result of all that precedes is to show that the unconscious ego, or, as it is called, the subliminal ego, plays a most important part in mathematical discovery. But the subliminal ego is generally thought of as purely automatic. Now we have seen that mathematical work is not a simple mechanical work, and that it could not be entrusted to any machine, whatever the degree of perfection we suppose it to have been brought to. [12]

Needless to say, he did not know a powerful function of computers in our times. Nowadays computers can easily accomplish not only a mechanical work like calculation but also a complicated operation like translation. It would be a favorite work for computers to enumerate as many combinations as possible by applying certain rules, but the combinations so obtained are extremely numerous and a mixture of chaff and grain. The essential part of a methematical discovery consists in sorting out the useful combinations from the useless. This point is similar to the so-called "frame problem" in Artificial Intelligence, which has to do with the ability of noticing the relevant information and ignoring the irrelevant. Such discernment requires the understanding of the concerned context, namely the conceptual Gestalt. In this sense, it is no exaggeration to say that the possibility of mechanical discovery by computers depends upon the solution of frame problem. That is the reason why automatizing scientific discovery is still difficult. In the age before the emergence of a computer, Poincaré has to entrust the task of mathematical discovery to the unconscious activity.

While Poincaré points out the important role of "the unconscious ego" or "the subliminal ego", he does not explain the function and mechanism of this ego in detail. He only suggests the indispensability of "the special aesthetic sensibility" of a real discoverer. However, this ability should not be regarded as a mysterious intuition. We would like to elucidate such unconscious activities by considering the function of scientific imagination, which is inseparable from the use of figurative expression such as metonymy and metaphor.

4 Two Types of Scientific Imagination

In order to explicate the function of imagination concerning scientific discoveries, turning now to the well-known distinction between "syntagmatic relation" and "paradigmatic relation" in F. de Saussure's linguistics. The former means a linear combination of linguistic components (i.e., words) which build up a sentence. This relation forms a perceptible explicit context which is expressed by sounds or letters. On the other hand, the latter means an opposition between words which are chosen and ones which are not chosen in the process of making up a sentence. We unconsciously choose a suitable word from a cluster of mutual exclusive words in speaking or writing a sentence. This relation forms an implicit context in the dimension of simultaneous consciousness.

Based on the above contrast, our linguistic competence can be divided into two coordinates, namely the perceptible explicit context and the inperceptible implicit context. One may notice that the latter is combined with our ability of memory and association. What is important is that this distinction is not only basic for linguistics, but also fundamental for epistemology. For it shows two types of human mental activities. The syntagmatic relation corresponds to a rational thinking and has characteristics of "real, explicit and united." On the contrary, the paradigmatic relation corresponds to an activity of imagination, and has characteristics of "imaginative, implicit and dispersive." We may say that these two types of mental activity are compared to the contrast between the conscious activity governed by obvious rules and the function of unconscious ego in mathematical discovery as pointed out by Poincaré. This analogy will become more explicit through considering Roman Jakobson's famous study of aphasic disturbances, because aphasia is not losing a stock of words but a defect of ability to select and combine words in order.

Jakobson interpreted Saussure's distinction as the difference between ability of "combination" and that of "selection" concerning our use of language. These abilities are affected in aphasia. One is a disorder of the ability of combination which destroys the syntagmatic relation. In this type of aphasia, the syntactic rules organizing words into propositions are lost. A patient speaks a mere "word heap" and cannot make an ordered sentence. The other is a disorder of the ability of selection which destroys the paradigmatic relation. Patients of this kind can grasp the literal meaning of words, but cannot understand the metaphorical meaning. According to Jakobson's analysis, these two types of aphasia are closely related to two kinds of use of figurative expression, i.e., metonymy and metaphor. Relevant to this point is his following remark:

> The varieties of aphasia are numerous and diverse, but all of them lie between the two polar types just described. Every form of aphasic disturbance consists in some impairment, more or less severe, of the faculty either for selection and substitution or for combination and contexture. The former affliction involves a deterioration of metalinguistic operations, while the latter damages the capacity for maintaining the hierarchy of linguistic units. The relation of similarity is suppressed in the former, the relation of contiguity in the latter type of aphasia. Metaphor is alien to the similarity disorder, and metonymy to the contiguity disorder. [3]

These points will lead us further into a consideration of the scientific imagination which gives a clue to scientific discovery. Metaphors set up a relation of similarity between two kinds of objects, for example, "life is a journey." On the other hand, metonymies find a relation of contiguity among objects like "from the cradle to the grave," which means "from the birth to the death." Both of them are the workings to discover an usually unnoticed relation on the basis of the function of imagination. Roughly speaking, two types of discoveries (factual and conceptual ones) above mentioned correspond to different functions of imagination. I would like to call them *metonymical imagination* and *metaphorical imagination* respectively. While the former is related to discoveries which complete an unfinished theory, the latter is concerned with discoveries which change a viewpoint from explicit facts to implicit unknown relations. What Poincaré called "the aesthetic sensibility" or "the subliminal ego" can be compared to this metaphorical imagination.

Therefore, I would like to focus attention on a metaphorical usage of scientific concepts and terms in discovering a new theory. Scientists, as Kuhn says, have to devote themselves to normal scientific research in order to be aware of an anomaly which is a valuable key to a discovery.[7] Even when they deal with anomalies from a revolutionary viewpoint, they cannot help using the existing concepts and terms in the established theory. In such a case, it is very useful for them to employ metaphor and analogy to express unfamiliar notions. Scientists involved in scientific revolution often try to develop their new ideas by using old terms metaphoricaly. This process has a connection with the function of metaphorical imagination mentioned above. One may say from what has been said that an important symptom of theoretical discovery is an appearance of metaphors in the context of scientific explanation.

As we have seen, the metaphorical use of existing terms and concepts to deal with the anomalies takes an important role in the initial stage of scientific discoveries. One can find a good example to illustrate this in Max Planck's metaphorical use of the acoustic concept "resonators" to solve the black-body problem.[8] Especially at times of scientific revolution, as Kuhn pointed out, there is no explicit logical rules to invent a new concept or theory, because scientific revolutions include drastic changes of our basic premisses of scientific knowledge..[6] Logical procedures are certainly useful for normal scientific research. By contrast, in the period of paradigm-shift, scientists cast doubt on the existing conceptual scheme and grope for a new categorization of phenomena. At this poit, metaphor shows its ability fully to alter the criteria by which terms attach to nature. It opens up a new sight of research through juxtapositions of different kind of ideas. From this viewpoint, we may say that the metaphorical imagination is indispensable to conceptual discoveries.

5 Conclusion

In the history of science, the event called discovery have been often ascribed to scientific geniuses' inspiration or intuition. In fact, scientific discoveries require not only the ability of rational reasoning but also that of creative imagination. As I mentioned in the previous section, one should notice that the workings of imagination include two features. We have characterized them as the metaphorical imagination and metonymical one by considering the use of figurative expressions. To borrow Kuhn's well-known terms, while the metonymical imagination functions in the period of "normal science", the metaphorical one displays its power during "scientific revolutions." The former is a kind of "puzzle-solving" activity and contributes to the discovery of unknown facts and laws on the basis of the fixed paradigm. The latter plays a role of switches from the old paradigm to the new one. Such a paradigm-shift stimulates the discovery of new concepts and theories.

Therefore, in the case of discoveries based on the metonymical imagination, it would be possible to formulate some explicit rules and to write a program on computers. The logical analysis of induction can be seen as a preliminary study of a computational discovery. The philosophical consideration concerning abduction and analogy will make a contribution to knowledge discovery by computers. However, it is rather difficult to specify the process where the metaphorical imagination works. For this procedure consists in finding a new relation of similarity among objects, which cannot be reduced to a ready-made analogy or a typical regularity. It stands in need of the conceptual "jump" which makes a change of viewpoint possible. If "necessity is the mather of invention", one may say that "a conceptual jump is the father of discovery." As a matter of fact, we can find many examples of such conceptual jumps in the history of science. Further research on the concrete cases of scientific discovery would clarify this interesting but difficult problem.

References

1. Hanson, N., *Patterns of Discovery*, Cambridge, 1969[1958], p.19.
2. Hanson, N., *Ibid.*, p. 90.
3. Jakobson, R., *On Language*, Cambridge Mass., 1990, p.129.
4. Koyre, A., *From the Closed World to the Infinite Universe*, Baltimore, 1970[1957]
5. Kuhn, T., *The Structure of Scientific Revolutions*, Chicago, 1970.
6. Kuhn, T., *Ibid.*, p.94f.
7. Kuhn, T., *The Essential Tension*, Chicago, 1977., p. 227ff.
8. Kuhn, T., "What Are Scientific Revolutions?", in Lorenz Krueger et al. (eds.), *The Probabilistic Revolution*, Vol. 1, Cambridge, Mass., 1987., p. 18.
9. Noé, K., "Philosophical Aspects of Scientific Discovery: A Historical Survey", in Setsuo Arikawa and Hiroshi Motoda (eds.), *Discovery Science*, Berlin, 1998, p.6ff.
10. Poincaré, H., *Science and Method*, London, 1996[1914], pp. 50-51.
11. Poincaré, H., *Ibid.*, p.51.
12. Poincaré, H., *Ibid.*, pp.55-57.
13. Wittgenstein, L., *Philosophical Investigation*, Oxford, 1997[1953], p. 193ff.

Ideal Concepts, Intuitions, and Mathematical Knowledge Acquisitions in Husserl and Hilbert

(A Preliminary Report) *

Mitsuhiro Okada

Department of Philosophy Keio University, Tokyo
Department of Philosophy, Université Paris-I (Sorbonne) **

Abstract. We analyze Husserl's and Hilbert's studies on the role of "ideal concepts" and that of the notion of "mathematical intuition" in the context of mathematical knowledge acquisitions. We note a striking similarity between Husserl's analysis in 1901 (presented at Hilbert's seminar) and Hilbert's analysis in the 1920's on the problem justifying the ideal concepts. We also analyze the similarity and difference on Husserl's standpoint and Hilbert's standpoint on mathematical objectivities and on mathematical intuitions. In the course of this analysis we compare these with Gödel's and some Hilbert scool members' standpoints. We also propose a view on mathematical knowledge acquisitions along a "dynamic" interpretation of the Husserlian philosophy of mathematics, which provides a view different from the traditional views such as realist, nominalist, constructivist, conventionalist, empiricist views.

1 Introduction: The Scope of This Paper

In this paper we shall consider some logical aspects on the nature of knowledge-acquisitions in mathematics with the framework of formal mathematical theories, from the Husserlian and Hilbertian points of view. We shall, in particular, investigate in the role of introduction of "ideal concepts" and that of the notion of "(mathematical) intuition" in the context of mathematical knowledge acquisitions.

* The original version of the paper was presented at the "Epistemology and Sciences" Colloquium at the Poincaré Institute in Paris on Dec. 5th 2000. Parts of this paper were also presented in some meetings, including colloques "Categorial Themes of Husserl's Logic" organized by the Husserl Archives of Paris in April 2001, "Husserl et Wittgenstein" organized by Université de Picardie and l'Institut Universitaire de France in April 2001, "Logic and Philosophy of Science" organized by Academie de Paris and Université Paris-I Sorbonne in March 2000.

** This work was partly done when the author stayed at Université Paris-I as an invited professor in 2000/2001. The author would like to express his thanks for their support on this work.

S. Arikawa and A. Shinohara (Eds.): Progress in Discovery Science 2001, LNAI 2281, pp. 40–77.

To clarify the logical role of ideal concepts in mathematics would provide an aspect of the nature of our problem-solving activities in mathematics from a logical point of view. In Section 2 we shall show that both Husserl and Hilbert took this clarification problem as the basis of their philosophy of mathematics. We shall also note a striking similarity between Husserl's analysis in 1901 (presented at Hilbert's seminar) and Hilbert's analysis in the 1920's on this problem.

In the course of the clarification of the role of ideal concepts, both Husserl and Hilbert gave an evidence theory for mathematical knowledge by using the notion of intuition. Although their uses of the notion of intuition are different (the categorial intuition in Husserl's case and the (Kantian) space-intuition in Hilbert's case), there are some common features in their uses. For example, both take the linguistically decidable or verifiable (i.e., and syntactically complete) fragment of arithmetic (which in this paper we shall often refer to the "syntactically lowest level" of propositions) as the objectivity domain for intuition. In Section 3 we shall examine the Husserlian notion of intuition and the Hilbertian notion of intuition in the relation with their theories on justifying the ideal concepts.

The underlying idea of mathematical intuition which is based on the linguistically (or nominalistically) introduced mathematical domain would, in our opinion, leads us to a new standpoint of philosophy of mathematics, which is different from the traditional naive standpoints, such as mathematical realism, nominalism, conventionalism, mental-constructivism or empiricism. This standpoint is very much oriented by the side of our linguistic activities of mathematics while the mathematical knowledge acquisitions can be explained as the activities of "discovery" at the same time; here, a mathematical objectivity domain of a theory is determined linguistically by conventionally introduced rules or axioms (as a minimal term-model), then such a domain plays the role of the basis of our mathematical intuition for the theory for our linguistic proving activities in the theory. In Section 4 we shall give such a standpoint following Husserl's view.

In our opinion, the bridge between the linguistic structure and the corresponding objectivity or denotational interpretation in the context of (formalized) mathematics can be well-explained by means of a hidden logical operator (linear-logical modality, which we also call "stablizor" operator) in formal mathematics. We shall not enter into this logical analysis in this paper since we discuss it in details elsewhere [1]. Further discussion on Husserl's meaning theory and its relationship with Hilbert (and With wittgenstein) may be found elsewhere [2].

[1] M.Okada, "Intuitionism and Linear Logic", revue internationale de Philosophie, an invited article, to appear.

[2] M.Okada, "Some correspondence theory and contextual theory of meaning in Husserl and Wittgenstein, ed.S. Laugier, (Olms, Paris, to appear), and M. Okada ".Husserl's 'Concluding Theme of the Old Philosophico-Mathematical Studies' and the Role of his Notion of Multiplicity (A revised English and French versions of the original appeared paper in Japanese in Tetsugaku vol.37, pp.210-221, Journal of the Philosophical Society of Japan 1987).

2 The Role of Ideal Concepts in the Mathematical Activities

2.1 Leibniz' Analysis on the Role of Ideal Concepts in Mathematics

Although we cannot go into the details to look at the historical background of the problem of the ideal concepts here, we mention a typical proposal on the problem by Leibniz.

For Leibniz, the problem was serious since he introduced the notion of "infinitesimal" (for example, dx in a differentiation and integral) and its dual "infinity" in his theory of differential calculus. In order to justify the use of infinitsimals in the course of differential calculus to obtain some result of real numbers without any occurrence of infinitsimals, he emphasized the following points;
1) these ideal notions are introduced only linguistically and operationally (by rules) in the course of algebraic manipulations in the calculations, and should not be considered as any kind of existing entities;
2) one could eliminate these notions contextually in principle. Such ideal concepts shorten the reasoning in order to obtain a result containing no ideal concept, and make the proofs significantly shorter. [3]
Although Leibniz did not give the (contextual) elimination procedure of the ideal concepts systematically (in fact, he repeatedly expressed that the time was not matured enough for the study of the foundations of mathematics), his insight had a close relationship with the solution given by Cauchy-Weierstrass where the infinitesimals are eliminated using the $\epsilon - \delta$ argument systematically [4].

Here, it is important to notice that Leibniz gives the typical view on the problem of justifying the ideal concepts; namely, (1) every ideal concepts in mathematics should be eliminable in principle and (2) but necessary to use them in our mathematical reasoning in practice (to shorten our reasoning). We shall now look at Husserl's and Hilbert's systematic approaches on their claims of (1) above, and Gödel's denial on (1) but at the same time his agreement on (2), in the next subsections.

---

[3] "[The infinitesimals and infinites of the calculus are] imaginary, yet fitted for determining reals, as are imaginary roots. They have their place in the ideal reasons by which things are ruled as by laws, even though they have no existence in the parts of matter"[Leibniz, Mathematische Schriften, Gerhart, III 499-500 (June 7, 1698).] " [T]hese incomparable magnitudes themselves, as commonly understood, are not at all fixed or determined but can be taken to be as small as we wish in our geometrical reasoning and so have the effect of the infinitely small in the rigorous sense. ...[E]ven if someone refuses to admit infinite and infinitesimal lines in a rigorous metaphysical sense and real things, he can still use them with confidence as ideal concepts which shorten his reasoning, similar to what we call imaginary roots in the ordinary algebra.[Letter to Varignon, Feb.2, 1702.] (The English Translations above are in L.E.Loemker, Leibniz, Philosophical Papers and Letters, 1969, Reidel.)"

[4] See H.Ishiguro, La notion dite confuse de l'infinitesimal chez Leibniz, Studia Leibnitiana Sonderheft 15 (1987)

2.2 Husserl's Solution on the Problem of Justification of the Use of Imaginary Concepts in Mathematics

Husserl asks himself how one can justify the use of imaginary mathematical concepts in the context of mathematical proofs, namely how one can justify a proof which uses some imaginary concepts in the middle but whose conclusion does not refer to any imaginary concepts. (Note that Husserl uses the word "imaginary" rather than "ideal", and we use the word "imaginary" in this subsection.) A typical example of this situation is the use of imaginary numbers in the course of calculation to conclude a statement only about real numbers as Leibniz considered, but Husserl considers the question in a more general setting. For example, the negative integers are imaginaries for arithmetic of the natural numbers; the national numbers are the imaginaries for arithmetic of the integers, etc. [5]

---

[5] "Under what conditions can one operate freely, in a formally defined deductive system ..., with the concepts that ... are imaginary? and when can one be sure that deductions that involve such an operating, but yield propositions free from the imaginary, are indeed 'correct'-that is to say, correct consequences of the defining forms of axioms?" (Sec.31, 85, *Formale und transzendentale Logik* (FTL), 1929).

Husserl identified this question, namely question to "clarify the logical sense of the computational transition through the imaginaries", as "the concluding theme of my old philosophico-mathematical studies"[Sec.31, 85, FTL]. In fact, he seems to have been obsessed by this problem throughout his early philosophical research as he confessed that he had been motivated by this theme when his writing Philosophie der Arithmetik. Before publication of the Volume I he already had a project to include this theme in the volume II of Philosophie der Arithmetik. cf. Letter to C. Stumpf, Husserliana XXI, pp.244-251. (We shall abbreviate Husserliana by H.) . He concluded his last chapter of the volume I of Philosophie der Arithmetik (1891) with a closely related topic, and mentioned the problem of the imaginaries in the very beginning chapter of Vol.I (Prolegomena) of Logische Untersuchungen (1900) p.10, in Ch.1, Sect.4 (1928 edition). According to him (FTL Sec.31; also cf. Ideen I. Sec.72), the clue for his solution to this problem was missing in the frst volume (Prolegomena) of the Logische Untersuchungen. However, Husserl already discussed the use of the notion of multiplicity [Mannigfaltigkeit] and state the problem of the imaginaries in Chapter 11 (in Section 70, cf. also Section 66) in a specific setting. In fact, he confessed that his study on multiplicities had been a long work starting 1890 "clearly with the purpose of finding a solution .....of the problem of imaginary quantities" (Sec.72 Ideen I (p.135)). His attitude to solve this problem seemed to have changed in the 1980's. He used to use the notions of inverse functions and arithmetical equations when studying some closely related topics to the problem of the imaginaries in the volume I of Philosophie der Arithmetik (1891), but when he connected his problem to the notion of multiplicity he presupposed the notion of formal axiomatic mathematical theory in general, and his scope of the problem had been widened much beyond pure arithmetic, to cover geometry and set theory, for example. In fact, his solution appeared in the series of unpublished manuscripts written around in the winter 1901. The period was overlapped with when he had contacts with D. Hilbert, and gave lectures at Hilbert's seminar, the Göttingen Mathematical Society, where D.Hilbert set the initiatives for a new movement of the studies in the foundations of mathematics.

Husserl's solution (to justify the use of the imaginaries in mathematics) is based on the assumption that an axiomatic mathematical theory determines its intended mathematical domain (multiplicity [Mannigfaltigkeit]) "definitely", without leaving any ambiguity in the structure of the domain. Such an axiomatic theory is "complete" in the sense that for any proposition of the theory either it is deducible or the negation of it is deducible from the axioms.

We shall look through Husserl's argument on his solution to the problem of the imaginaries.

Husserl's argument on this solution can be formalized as the following two conditions;

(a) if a mathematical theory has a definite multiplicity, the use of imaginary concepts is always justified,

(b) one can actually find an example of complete axiomatic theories in mathematics, which is a theory of arithmetic.

The above (formal-ontological) condition (a) of a definite multiplicity is often replaced by an equivalent [6] (syntactic-deductive) condition;

(a)' if a mathematical theory has a (syntactically) complete [7] axiomatization, the use of imaginary concepts is always justified.

Actually Husserl's presentation of the solution was more precise than (a) above; he even gave the break-downs of the condition into two parts in some manuscripts; For example, Husserl essentially gave the following two conditions in the manuscripts for his lectures at Hilbert's seminar (the Göttingen Mathematical Society [8]);

[6] The meaning of this equivalence, as well as other equivalent expressions, will be analyzed later in this section.

[7] Here, we often use the word "syntactic completeness" in this paper when we would like to distinguish clearly the notion of semantic completeness in the sense of Gödel's completeness theorem (1930), although Husserl always use the word "complete" for our "syntactic complete" in his texts. It is known that most of mathematical axiomatic systems are not syntactic complete (by the Gödel's incompleteness theorem (1931)) but all first order mathematical axiomatic systems (including first order pure logic) are semantic complete (in the sense of Gödel's completeness theorem (1930)).

[8] p.441, Appendix VI, H. XII

I) the extended system with the imaginaries is consistent;
II) the domain of the axiomatic theory (namely, the multiplicity of the original theory) is definite. (Namely, the original system is (syntactically) complete) [9].

These are the most precise conditions Husserl gave regarding his solution on the problem of the imaginaries. However, he repeatedly gave the solution with a more emphasis on the condition (II) above for both the original and the enlarged systems in his series of unpublished manuscripts during the winter semester 1901/1902 as well as in *Formale und transzendentale Logik* (FTL) (1929) [10]. Namely, Husserl often presents less precise version (a) (or (a)' or (a)" in the footnote below), by replacing the two explicit conditions (I) and (II) above; The condition (a) tells the both the original and enlarged domains are definite, or equivalently, both the original system and the extended system are (syntactically) complete. [11]

The condition (a) means that the considering axiomatic system determines the intended domain of the theory without ambiguity. Here, a domain of a theory is called a multiplicity[12], and the domain determined by an axiomatic system is called a definite multiplicity. One could, in our opinion, interpret a multiplicity as a model in the sense of Tarski school of model theory since the notion of model is defined set-theoretically in the framework of ZF set theory where the domain of a model does not refer to any concrete objects but to sets in the standard interpretation in the universe of ZF set theory, which does not assume

[9] In fact, his expression for the condition (2) here is a mixture of (a) and (a)' above. Husserl writes:
"From this the following general law seems to result: The passage through the imaginaries is allowed (1) if the imaginary can be defined in a consistent wide[r] deduction system and if (2) [t]he original deduction-domain, [once] formalized, has the property that any proposition falling into this domain [jader in dieses Gebiet fallende Satz] is either true based on the axioms of the domain or false based on them, that is, contradictory with the axioms." (p.441, Appendix VI, H. XII)
Then he gives further clarification of the meaning of this condition (especially, further clarification of the meaning of "any proposition falling into the domain") to reach essentially the notions of (a) and (a)' above. The clarified form of the notions are also explained in other manuscripts, (eg. Appendix VII, H XII), which also includes another equivalent expression (a)"
(a)" If the deducible propositions of the axiomatic theory constitute a maximal consistent set of propositions, the use of imaginary concepts is always justified.

[10] Appendices VII, VIII, X, H. XII, and Sec.31, FTL. For example, in FTL (p.85) Husserl states the solution as "the [both original and enlarged] systems are "definite" [namely, syntactic complete]."

[11] For example, in FTL Section 32, when he resumes his result of 1901 he writes "How far does the possibility extend of "enlarging" a "multiplicity", a well-defined deductive system, to make a new one [the new deductive system] that contains the old one [the old deductive system] as a part? The answer is as follows: If the systems are "definite", then calculating with imaginary concepts can never lead to contradictions."

[12] Husserl identifies those two; eg. H.XII, Appendix VI p.440).

any concrete urelements on the universe, hence which fits Husserlian notion of formal ontology (whose urelements are "something whatever") [13].

Husserl's solution can be resumed as follows;
An axiomatic formal system, say S, is called (syntactically) complete if for any proposition, say A, of the system S, either A is provable or the negation of A is provable. Here, "or" needs to be understood as the "exclusive or" [14].

Consider axiomatic systems $S_1 \subset S_2$ where

S_1: original system [of real elements]

S_2: enlarged system with imaginaries

Husserl showed that the above conditions I and II guarantee for any real proposition A,

$$S_2 \vdash A \Longrightarrow S_1 \vdash A$$

This relation is often called the "conservation" relation (and S_2 is called a conservative extension of S_1, and S_1 is called a conservative subsystem of S_2.) We shall give an important difference of the above specific relation from the usual conservative relation later.

Husserl also gave a proof that there exists such a complete system S_1 One can see that Husserl's presented system S_1 is the quantifier-free numerical arithmetic with decidable predicates (and a fragment of the decidable Σ_1^0-sentences). He

[13] There is a slight difference between a model of the Tarskian model theory and a multiplicity of Husserlian formal ontology; when an axiomatic theory is not syntactically complete, in model theory one could consider the (possible infinitely many) different models of the theory while Husserl seems to consider a model which is still unique but whose structure is partly unspecified. In other words, for incomplete axiomatic theory, there are many (nonstandard) models satisfying the axioms while there is still a unique but "unclear" multiplicity of the axioms from Husserl's view point. However, since we shall mainly consider the definite multiplicity of a syntactically complete axiomatic theory in the Husserl's sense in this paper, in our opinion we could identify the intended standard model in the sense of model theory and the definite multiplicity. We shall return to this topic in the last section.

[14] The reason why Husserl often replace the precise conditions (I), (II) by the less precise single condition (a) is that (syntactic) completeness in Husserl's sense implies consistency. This is because the "or" in the definition (a)' above needs to be understood as the "exclusive or" which excludes the possibility of contradiction.

Husserl also seemed to often omit (I) and state only (II) as the condition for the justification of the imaginaries. This is mainly because the consistency, manely non-contradiction, was a primary requirement for his notion of deductive system.

actually sketched the decision procedure of the system. [15]; The detailed textual analysis on Husserl's proof of the completeness is given elsewhere [16].

[15] Husserliana XII,443. Then Husserl asks himself whether there is a definite multiplicity (or equivalently, a syntactic complete axiomatization by a (first order) axiomatic system or not, and claims that it is the case of arithmetic; he then argues that all arithmetical propositions are reducible into the basic arithmetical numerical equations, of either quantifier-free form (of s=t) or of (solvable) existential equations with indeterminate variables; the latter case occurs especially when he apply the reduction procedure to ("apparently non-quational") arithmetical primitive predicates. In particular, he claims such a non-equational primitive predicate in arithmetic should be reduced to an equation with indeterminate (but solvable) variables (of existential proposition), by giving an example of the reduction of $a < b$ into $\exists u(a+u = b)$ (in the modern logical terminology of existential quantifier). Here, Husserl lacks the explanation why this case can be solved. But, in fact, this case has an obvious upperbound for this quantifier, namely, $\exists u < b(a + u = b)$. In fact, We could rather interpret Husserl's intention as follows; every primitive predicate of arithmetic should be decidable in the sense that it can be reduced to the solvable equations if the formal axioms determines the primitive predicate "definitely". Then, he claims that every numerical equations are decidable by checking whether the equation becomes the identity of the form of $a = a$ or not This checking procedure is well defined here; for any arithmetical terms s and t without any variable, s and t can always be reduced to concrete numerals, say a and b respectively. Here, the condition of definiteness is used; since all operators occurring in s and t are calculable according to the axioms (since the axioms do not leave any ambiguity for calculating the operators by the assumption of his notion of completeness), the terms, s and t, should be eventually reduced to two concrete numerals, a and b. Then, if a exactly coincides with b, the original numerical equation s = t is provable, otherwise it implies a contradiction. (The last claim is obvious since s=a and t=b are provable according to Husserl's reduction argument, while if b is of the form of a, then s=a and t=a are provable and by the transitivity of the equality, s=t are provable; if b is different from a, then not (a=b) is provable in any standard formulation of arithmetic theory, hence the provability of s=a and t=b and a few standard equality rules imply "not (s=t)", hence s=t implies a contradiction.) So, under the standard setting of axioms of equality and the assumption that the arithmetic axioms define arithmetical operators completely, (namely every operators used are defined on the domain and calculable according to the rules given by the axioms), Husserl's proof of syntactic completeness of the fragment of arithmetic for the numerical equations (without quantifiers) works well and is correct. And, although Husserl included only one example of the non-equational arithmetical primitive predicate/primitive relation (namely that for the "greater-than" relation $<$) in his reduction procedure, his argument works with the appearance of any further additional arithmetical primitive predicates/relations in the usual sense since the usual primitive predicates are in fact primitive recursive predicates in the modern logical sense and those can be axiomatized so that the decidability of such a primitive proposition can be shown numerical-wise; in fact, such a decidable primitive arithmetical predicate can be defined in the manner that the decidability can be reduced to the solvable numerical equations In the modern logical sense, this solvable existential proposition can be expressed using bounded existential quantifiers. as essentially in the same manner that Husserl suggested with using a reduction procedure for the "greater-than" relation $<$. Hence, one could

One can understand that Husserl characterized the complete system and the corresponding definite multiplicity (the " canonical model" in the contemporary sense of logic, which is determined by the complete system as the term-model composed of the canonical terms (or normal terms) calculated by the given (axiomatic) rules). we shall analyze this in section 3 in details. The only wrong part in his "completeness proof" (from the modern logical point of view) is the very last one line added at the end of his "proof" of (syntactic) completeness: Namely, after giving a correct transformation procedure for each numerical equation to be transformed to an identity [of the form a=a] or its negation, he concluded;

"Any numerical equation is true if it is transformed in an identity [of the form a=a], otherwise it is false. Any algebraic formula is therefore also decidable, because it is decided for any numerical case"[17].

Here, Husserl uses "algebraic" to express indeterminate free-variable occurrences in the arithmetical proposition, which is logically equivalent to existentially quantified arithmetical propositions. [18]. The decidability of the existentially quantified propositions (Σ^0_1 arithmetical sentences, in the Kleene's arithmetical hierarchy) is logically equivalent to the decidability of universally quantified propositions (Π^0_1 arithmetical sentences). Hence, this Husserl's claim is equivalent to the assumption of ω-rule [19]. It is only on this last line where he made a mistake from the modern logical point of view. In fact, Gödel exactly showed a counterexample of this last line in a specific form with his first incompleteness theorem [20] . In fact, in the last section we shall give an alternative interpretation of this last line of Husserl; we shall give such an interpretation that this last line refers to the evidence for syntactically higher-level of propositions

say that Husserl's suggested decision procedure works for any reasonable setting of arithmetical primitive relations/predicates.

[16] See the footnote on p.2.

[17] Husserliana XII, 443

[18] Husserl often uses the word "algebraic" for indeterminate variables. Husserl's very restricted use of quantified propositions is examined in two papers of ours mentioned in the footnote 2.

[19] ω-rule is the rule with infinitely many premises; from $A(0)$, $A(1)$, $A(2)$, ... one concludes $\forall x A(x)$, which cannot be expressed in any linguistic deductive system in the usual sense, and is essentially the same as the semantic definition of $\forall$ of Tarski.

[20] Namely, Gödel, in his proof of the incompleteness theorem, constructed a proposition of the form $\forall x R(x)$ where R(x) is a decidable (technically speaking, primitive-recursive) arithmetical predicate or proposition, and showed that neither $\forall x R(x)$ nor its negation is provable in arithmetic. Here, the negation can be expressed by an existential numerical formula of the form $\exists x P(x)$ (for another decidable arithmetical proposition to express the negation of R(x)) or of the form $\exists x f(x) = 0$ where f is composed calculable (in fact, primitive recursive) operators, hence f can be completely determined by axioms in Husserl's sense. Therefore, with the appearance of only one indeterminate variables (equivalently, existential quantifiers), the last line of the argument of Husserl above breaks down.

(namely, quantifiered propositions) which is conceived by Husserlian "categorial intuition".

It is important to notice that the above Husserlian conservation relation should not be understood as the conservation relation which we usually understand in the context of contemporary logic; this Husserlian conservation relation is meant only for the very restricted class of propositions A (the class of quantifier-free arithmetical sentences or at most of the bounded Σ_1^0-arithmetical sentences), not for the arbitrary arithmetical sentences A. In other words, Husserl restricted his attention only on the decidable fragment of language, which is syntactically the lowest level.

As long as this syntactically lowest class of propositions (for which Husserl's reduction procedure holds) is concerned, the class of propositions is out of the scope of Gödel's incompleteness theorem. This is because a true and unprovable formula (often referred to "Gödel sentence") never appears in this range of the propositions, but appears in the level of the universal formulas, i.e., one-step higher level.

There is no conflict between Husserl's solution and Gödel's incompleteness theorem. Gödel's incompleteness theorem shows that there is always a universal arithmetical sentence, namely a universally quantifiered arithmetical sentence, (or equivalently, its dual, namely an existential arithmetical sentence) which is neither provable nor refutable in any axiomatic theory (including some arithmetic). On the other hand, whatever the concrete axiomatization of arithmetic (of the natural numbers) may be, for which Husserl seldom explains us what it is explicitly or concretely except that there is a hint that he admits Hilbert's axiomatization of arithmetic without his "completeness axiom" [21] as an example of a part of some axiomatization, Husserl's "completeness theorem" for the syntactically lower class of the propositions (namely, the quantifier-free numerical propositions, or Π_0^0 arithmetical sentences) is correct and compatible with Gödel's incompleteness theorem [22].

[21] This exclusion of the completeness axiom of Hilbert was very essential for Husserl because the inclusion will destroy Husserl's setting of the problem of the imaginaries in general !, as he correctly repeatedly remarked.

[22] To summarize, we can formulate as follows;

1. Husserl's problem of imaginaries seemed to be expressed as the conservation relation in the modern terminology.

(1) For any proposition, say A, of the original system S_1,
if A is provable in S_2 then A is already provable in S_1.

2. However, if we understand Husserl's problem as formulated in (1) above with the modern notion of logical language, this cannot give the real setting of the problem by Husserl. This is because with the typical examples of the imaginaries, the situation is;

The notion of completeness for which Husserl's argument works is very restricted from the point of view of modern logic. Nevertheless, this meant for Husserl's eye that there were some meaningful concrete examples of complete axiomatic systems and definite multiplicities.

---

(2) $\exists x P(x)$ is provable in S_2, but the same logical formula $\exists x P(x)$ is not probable in S_1.

3. Hence, it is clear that Husserl's intended problem of the imaginaries cannot be formulated by (1) above using the modern notion of logical language. In fact, our analysis of Husserl's "completeness proof" shows that the correct part of his proof (except for the wrong one line) gives the conservation relation for a more restricted class of propositions.

(3) For any quantifier-free numerical proposition A of the original system S_1, if A is provable in S_2 then A is already provable in S_1, (In this sense, the completeness which Husserl claims can be seen as a completeness "in the relative sense" from the modern view of the logical formal language. Here the relativeness was introduced by Husserl himself, but it seems that he rather thought that his completeness is in the absolute sense, there. cf. H. XII, Appendix VI.)

4. Recall that if we take the syntactic completeness property for arbitrary logical proposition A in the modern sense of formal logical language, Husserl's completeness condition (a)' was expressed as follows;

(4) For any proposition A, either A or its negation "not A" is provable.

5. This looks contradictory with Gödel's incompleteness theorem since the incompleteness theorem implies that;

(5) There is a universally quantifiered arithmetical (closed) proposition A of the form $\forall x P(x)$ (where P(x) is quantifier-free) such that neither A nor "not A" is provable.

6. In order to guarantee the restricted form of conservation result (3), the condition (II) above needs only for this restricted class of propositions. In fact, as we have seen, Husserl's notion of syntactic completeness property of an axiomatic system can be understood on the restricted class of propositions. Hence, the completeness condition should be expressed by the following, rather than (4) above.

(6) For any quantifier-free numerical proposition A, either A or its negation "not A" is provable.

Because of the restriction of the class of the propositions in (6), now there is no contradiction between (5) and (6).

As we emphasized before, it is important to notice that this fact holds for any formal axiomatization, how strong or weak is the axiomatic system of the number theory.

Husserl seems to have been satisfied with actually proving the completeness or decidability only for such a very restricted class of syntactically lowest level propositions when he discussed the notion of complete axiomatic systems and the notion of definite multiplicities. There seems two main reasons here why Husserl was concerned with only the numerical fragment of arithmetic propositions for his completeness/definiteness properties. One reason is, as we already pointed above, his scope of the logical language was very much restricted. The other, which is closely related to this first reason, is that he was mainly interested in the elementary-basic level (or, the syntactically lowest level) of a multiplicity, namely the ultimate substrates, which he called the "cores" or "ultimate substrates" of the multiplicity (and the elementary judgments, what Husserl called "ultimate judgments", for the cores, in contrast to the higher objectivities constructed categorially (or equivalently syntactically in Husserl's sense). Once the cores (the syntactically lowest level of a model, namely, the domain and the elementary relations and (first order) functions on the domain) is determined linguistically, the syntactically higher structure of the model is uniquely constructed according to the Tarskian truth definition (or the definition of model). Hence, Husserl's emphasis on the linguistic characterization of "cores" naturally leads us to the determination of a unique model when we understand Husserlian multiplicity as the notion of (set-theoretic) model[23].

2.3 Hilbert's Epistemic Re-formulation of His Consistency Problem as the Problem of the Ideal Concepts

Hilbert introduced his idea of axiomatic formalism in mathematics with his publication of "Foundations of Geometry" (1899) and of "On the number-concept" (1900) and a few others at the very beginning of the century and emphasized the need of proving consistency of mathematical theories at the same time. In particular, when Husserl presented his problem and solution on the imaginaries at the mathematical meeting of Hilbert in 1901, Hilbert is in the middle of this first active period of the foundational research. After the series of the work on the foundations of mathematics with the emphasis on his program on "consistency problem" of mathematical theories (which lasted until around 1904), Hilbert became silent for more than a decade about his program on consistency proof. Hilbert returned to the research of his consistency problem in the late 1910's with various new theoretical arms which had lacked in his first period of the foundational research. This second period of his foundational research lasted until the end of his life. The new theoretical arms which he carried upon his return to this field were much more sophisticated and powerful than those in

[23] Husserl's emphasis on the ultimate substrates-cores level is also related to his explanation on the genetic source of the formal ontological framework, which, according to Husserl, connects the formal ontological framework of logic with our experience and the individual objects. We shall not go into this subject in this paper. (Cf. Section 3 of M.Okada, "Some correspondence theory and contextual theory of meaning in Husserl and Wittgenstein", ed.S. Laugier, (Olms, Paris, to appear) for our view point on this

his first period[24]. For example, he had the full knowledge of formal logic calculus of Frege-Russell-Peano-Schröder, he had the distinction between the formal (object) language (of the targeted theory for consistency proof) and the meta-language (to be used by consistency proof) to counter-attack the criticism of Poincaré, the adaptation to Brouwer's intuitionist argument to a certain extent (in particular, the notion of verifiability and his rejection of the law of excluded middle from his finitist standpoint)[25] However, one of the most striking changes between the first period and the second period is the way he explains his consistency problem itself. Instead of just saying the need of consistency as he actually did at the very beginning of the century in his first period, he connects, or in fact identifies, the consistency problem to the problem of the ideal concepts.

Hilbert in his "second foundational study period" explains the consistency problem as follows.

"[W]e must nevertheless not forget the essential prerequisite of our procedure. For there is a condition, a single but absolutely necessary one, to which the use of the method of ideal elements is subject, and that is the *proof of consistency*; for extension by the addition of ideals is legitimate only if no contradiction is thereby brought about in the old, narrow domain, that is, if the relations that result for the old objects whenever the ideal objects are eliminated are valid in the old domain."[Über das Unendliche, 1925]

This is exactly the same form of the problem of the imaginaries of Husserl, which we showed in the previous subsection [26], although Husserl uses the term "imaginary" instead of "ideal". A difference of the setting is that for Husserl the "real" system S_1 (of the original domain) and its enlarged system S_2 with imaginaries (of the enlarged domain) are relative notions, while Hilbert fixes a unique system S_1, for the question.[27] Hilbert calls this fixed system finitary arithmetic or "contentual number theory", which is the system admissible in his finitist standpoint.

Despite of this difference of the global setting of the problem, the actual content of the two arguments, one by Husserl in 1901 and the other by Hilbert

---

[24] For Hilbert's development of the foundational-logical studies between his first and second periods, and at his beginning of the second period, see for example;
Sieg Hilbert's programs: 1917-1922, (Bulletin of Math. Logic, Vol 5, No.1), Mancosu Between Russell and Hilbert: Behmann on the foundations of math., (Bulletin of Math. Logic, Vol.5. No.3), Zach, Completeness before Post: Bernays, Hilbert and the development of propositonal logic, (Bulletin of Math. Logic, Vol. 5, No. 3)

[25] Brouwer later complained that Hilbert criticized his intuitionism while Hilbert's standpoint in fact took many ideas from his intuitionism.

[26] Cf. the footnote 5.

[27] Hilbert considers any mathematical theory for S_2 with an intermediate step of arithmetization (namely, for example, when Hilbert considers the problem of the imaginaries (the ideals) for a theory of geometry, a theory of geometry is translated to the theory of arithmetic for the real numbers first, then the latter is considered as the enlarged system S_2 above.

in the 1920's, are very similar. This is because Husserl's actual argument dealt mainly with one particular type of system S_1. In fact, as we saw in the previous subsection, his completeness argument works only a particular type of S_1. And it was also the case when Husserl presented his theory on the problem of the imaginaries and his completeness proof of S_1 at the meeting of Hilbert in November 1901. In fact, we shall see how close each other between Hilbert's presumed S_1 and Husserl's presumed S_1. This would imply the following;

It is important to notice that the similarity between Husserl and Hilbert is not only on the problem itself (namely on the fact that the both problems can be expressed as the problem of the ideal or imaginary concepts), but also on the actual choice of S_1.

Just after the quotation above, Hilbert explains why this problem of the ideals, namely the problem of the imaginaries in Husserl's terminology, is actually the same as the usual sense of the consistency problem (which he used to use to express his consistency program in his "first period"). He continues that this consistency problem expressed in terms of the problem of the ideals can be reduced to the problem of checking whether or not a specific typical contradiction statement "$1 \neq 1$" can be deduced in the axiomatic system in consideration.

Hilbert emphasizes the consistency condition (I) out of the two conditions (I) and (II) in the previous subsection. This is partly because Hilbert does not consider the "original" system S_1 as a formal axiomatic system. He rather considers it on the more epistemological terms; The "real" propositions are, according to Hilbert in the 1920's, contentual [inhaltich] and its content is conceived by Kantian intuition, and meaningful from the finitist standpoint. We shall discuss on this later in Subsection 3.2 [28] .

[28] Here, a question rises; Is this difference between Husserl's understanding of S_1 as a formal theory and Hilbert's understanding of S_1 as a contentual informal theory very crucial to consider the similarity of the two approaches? In fact, although Hilbert seems to consider the finitistically meaningful (and contentual) propositions at the informal level in his explanation of the finitist standpoint in the strict sense, Hilbert tried to generalize the framework of the real-ideal systems in the development of his proof theory from time to time, in particular in the 30's. (This extension of the finitist standpoint was necessary for Hilbert and his school especially by facing the appearance of Gödel's incompleteness theorem in 1931. We shall discuss the issue relating to Hilbert school later.) For example, in the actual development of his proof theory in the 30th on the line of his finitist consistency program, the distinction between the real propositions and the ideal propositions becomes more or less a relative distinction, rather than the absolute distinction between the finitist propositions in his strict sense and the non-finitist propositions. When he talks about the first ϵ theorem, the underlying relative distinction between the real propositions and the ideal propositions are defined both in the formal deductive systems level. There, the system S_1 for the real propositions is a purely logical or arithmetical deductive system for quantifier-free (namely, ϵ-symbol-free, in his formulation) formulas and the system S_2 for the ideal propositions is a corresponding deductive system with

2.4 Some Consequences on the Role of Ideal Concepts from Gödel's Incompleteness Theorem

The first incompleteness theorem of Gödel (1931) directly implies that there always exists a proposition which is true (in the standard model of the theory in consideration) but not provable in any given mathematical theory (For any ω-consistent theory containing arithmetic), hence requires to introduce some additional ideal (or imaginary in the sense of Husserl) concepts which are not present in the original given theory in order to prove the proposition. This means that the introduction of new ideal concepts is sometimes essential to find a proof or solving a mathematical problem.

Here, it is important in our context that such a proposition has a higher syntactical complexity, namely at least the universally quantifiered sentence. This was the point which, as we remarked in Subsection 2.2, makes compatible with Husserl's claim on the existence of complete theories in the syntactically lower level. (With the Hilbertian distinction between the real and the ideal, one could say that a new ideal concept may be needed only for an ideal proposition.)

The second incompleteness theorem of Gödel (1931), on the other hand, shows that the consistency statement itself is an example of such a syntactically higher level and unprovable proposition. Since, as Hilbert showed (cf. the previous subsection), the consistency statement is logically equilavent to the relative conservation relation expressing the justification of the ideal concepts (as we formulated in Subsection 2.2), both consistency problem and the relative conservation problem are not provable within the techniques allowed in the original narrower system, which seems to block the program originally intended by Hilbert in the 1920's. [29]

It is worth noting that Gödel himself did not deny the Hilbertian program of finitist consistency proof when he proved his incompleteness theorem (1931). In his paper on the incompleteness theorem, as well-known, Gödel even explicitly remarked that his theorem did not directly mean the impossibility of Hilbert's finitist program on consistency proof. He expressed his negative view on the

the full-use of quantifiers (with epsilon symbols). The first ϵ theorem of Hilbert-Bernays (Hilbert-Bernays, Grundlagen der Mathematik, I, (1934)) claims that the logical system S_2 is a conservative extension of the system S_1 of real propositions. be a contentual and informal proposition any more. Hence, under this distinction approach between the real system and the ideal systems, both S_1 and S_2 are formal deductive systems as Husserl took it.

[29] We also remark that the notion of "consistency" and "conservation" themselves should be understood as highly ideal concepts compared with the finitist arithmetic. Hence, Hilbertian finitist consistency program first faces a failure not because of the methodological limitation stated by the second incompleteness theorem, but because of the linguistic limitation of stating the consistency problem (or stating the problem of ideal concepts) in a finitist way; the consistency statement and the conservation-relation statement themselves are universally quantified propositions, hence meaningless finitistically from the Hilbertian finitist standpoint in the 20's.

Hilbertian finitist program only after his detailed examination of Genzen's "finitist" consistency proof of arithmetic (1936). [30] There, Gödel first claimed that any consistency proof required some abstract notion beyond the notions justifiable by Hilbertian finitist intuition (see the next section on this point). Hence, the consistency problem provides us with a good example of the effect of Gödel's incompleteness theorem into the problem-solving or proof-discovery process; in order to prove the consistency statement (which is of a syntactically higher level, namely at least Π^0_1), one needs to take some ideal concepts beyond the original system in consideration. The typical way to choose an enlarged system in this case is a higher order extension of an original system since the enlarged higher order system can easily construct the truth-definition of the original (lower-order) system, which implies consistency of the original system. On the other hand, Gödel himself proposed one criteria of finding a necessary ideal concept from the mental-constructivist point of view; He claimed that consistency proof requires "abstract" notions beyond finitist standpoint, which are "mental constructs"; his proposal was to use an economic mental construct, and he claimed that Genzen's notion of "accessibility" of ordinals introduced in his consistency proof (of arithmetic) is a mental construct too, but that his introduced notion, the primitive recursion of finite types, is a more economical mental construct than the one introduced by Gentzen, for the same purpose of proving consistency of arithmetic. [31]. We shall discuss the Hilbertian finitist intuition and Gödel's attack against it in the next section.

Another important remarked by Gödel on the incompleteness theorem is that the theorem also gives a direct implication to the provable theorems (namely, the theorems already provable in the narrower system without introduction of any ideal concept); He proved the following fact; the use of enlarged mathematical system (with ideal propositions) makes many already-existing proofs (in an original narrower system) much shorter; in other words, an introduction of ideal propositions may speed-up many existing proofs [32]. This is the exactly same point which Leibniz emphasized for the role of the introduction of ideal concepts in mathematics. Both Gödel and Leibniz seem to consider such a speed-up is essential in our mathematical activities of proving theorems.

---

[30] Gödel's consistency proof by means of primitive recursive functionals of finite types (Dialectica 1958) was the result of his detailed studies in Gentzen's consistency proof by means of the proof-normalization method.

[31] M.Okada (J. Symbolic Logic 1988) gave a clarification of this claim of Gödel's; Genzen's inference rule on accessibility implies the well-ordering of ordinals up to $\phi_\omega 0$ of the Veblin hierarchy of transfinite ordinals while as Gödel claimed the logical strength of primitive recursive functionals is ϵ_0 which is equivalent to $\phi_1 0$. Hence, Gödel's ideal concept is much more economical than Gentzen's from the point of view of the logical strength.

[32] K. Gödel, Über die Länge von Beweisen (1936), also in Collected Works od k. Gödel, Vol.I, 396-399.

3 The Role of Intuition in the Knowledge Acquisitions in Mathematics

We shall now consider the role of intuition in the mathematical activities. Here we especially consider the two typical notions of mathematical intuitions, one of Hilbertian finitist intuition, the other Husserlian categorial intuition.

3.1 Husserl's Notion of Categorial Intuition in Mathematics[33]

We start our discussion by recalling the debate between Frege and Hilbert[34] on the role of intuition in geometry in the period (1899-1900) when Hilbert just published his book on axiomatic geometry and gave his view of contextual meaning theory (or linguistic-monist meaning-network theory) where he claimed that meaning of any term in an axiomatic theory should be understood in relation with the whole system of axioms, which is independent of any intuition, while Frege took a traditional correspondence theory of meaning based on the notion of space-intuition of the geometrical figures. [35]

From Husserl's note on his occasion to study the correspondences (1899-1900) between Frege and Hilbert that it is clear that Husserl in this period adapted himself to and accepted Hilbert's view of axiomatism and against Frege's view. Against Frege who insists that the geometric axioms should have the evidence by our spatial intuition, Hilbert claims that consistency of the whole system is an only requirement for evidence and the whole system of axioms contributes to the meaning of any concept appearing in the theory. [36].

Husserl stood for Hilbert and against Frege and said;

[33] In this paper we restrict our attention on the notion of intuition in mathematics, and we do not intend to consider Husserlian notion of categorial intuition in a general framework; we consider his notion of categorial intuition only in the context of formal mathematical theories. Husserl himself consider the role of categorial intuitions in the general framework of the fulfillment model of signification intentions of judgments. cf. The Sixth Investigation (1901)

[34] Frege, Wissenschaftlicher Briefwechsel, 1976, p. 66, cf. also, Husserliana XII p. 449

[35] Frege claims; "Axioms are the positions which are true but whose truth are not proved, because their knowledge follows from a source of knowledge which is entirely different from the logical source and which one can call intuition of the space. It results from the truth of the axioms that they don't contradict each other. Therefore, another proof [of consistency] is not needed." "Only one way which I know [for proving consistency] is to indicate an object which has its all properties, to show a case in which all requirements [of the axioms] are fulfilled. It would be impossible to prove the consistency in any other way."

On the other hand, Hilbert claims; "If arbitrarily given axioms do not contradict ..., they [axioms] are true, the things defined by the axioms exist."

[36] Husserliana XII p. 449

Frege does not understand the meaning of the foundation of the Hilbertian "axiomatic" geometry, namely that it is concerned with a purely formal system of conventions which coincides, regarding the form of theories, with the Euclidian system. [H. XII, p. 448.]

Husserl echoed Hilbert in many manuscripts just after he received Hilbert's view of axiomatic contextual theory of meaning [37] .

Husserl agrees that (1) axioms of a formal mathematical theory need no justification by intuition or any other evidences except for "consistency" of the system as a whole in the deductive-level, and that (2) the meaning of a concept appearing in such a system should be understood in the relation with other concepts used in the axioms.

For Husserl, insight into consistency is not the matter of spacial-geometrical intuition, but the matter of "extra-(spatial) geometrical" relationships, which Husserl identifies with the "categorial" relationships [38] .

Husserl often uses "categorial" and "syntactical" interchangeably, from this period throughout his life [39]. What he calls "categorial" is a syntactic categorial structures in the linguistic side. Since he uses both linguistic-judgment form side and the objectivity-correlatives side soon after this period, the reader might be mislead to some confusion. But, in the setting of the study of formal axiomatic mathematical theories with Hilbert's linguistic monist standpoint, it is clear that

---

[37] For the Frege-Hilbert debate and Husserl's influence from Hilbert, see 4.3 of Keiichi Noe "Mukonkyo kara no Shuppatsu" (in Japanese), Keisou-Shyobou, 1993, and the two papers by Mitsuhiro Okada at footnote 2.

[38] [H.XXII, p.429] "The consistency of axioms is also not a matter of "axiomatic" insight, but is certified only from freedom from contradiction in the consequences derived. This of course does not preclude insight into the consistency of particular propositions or sub-set of axioms. But such insight will, then, not rest upon the intuition of geometric relationships, since we lack that with reference to Ideal concepts, but rather upon insight into certain extra-geometrical relationships: namely, universal and categorial ones." Also Cf. the following text; "The geometric shape, on the contrary, has merely represented existence, existence in virtue of definition and valid deduction from the axiomatic principles. The possibility of *geometric* structures - thus, the consistency of the determinations unified in representations of them - is not guaranteed by means of the intuition of those structures (for we do in fact lack such intuition). Rather it is guaranteed through the consistency of the elementary determinations brought together in the axiomatic principles, and through the pure deduction that supplies the 'existence proof'." [H. XXII, p.327 (Here we use the English translation by D. Willard, Early writingsthe philosophy of logic and mathematics, Kluwer)]

[39] Cf. Section 2.1 of M.Okada, "Some correspondence theory and contextual theory of meaning in Husserl and Wittgenstein", ed.S. Laugier, (Olms, Paris, to appear). For example, in Section 42. d [FTL(1929), p.102], he says "some categorial (or, as we also say, syntactical) forms".

Husserl's use of "categorial" refers to the syntactic structure of the linguistic-apophantic side.

As is suggested from Husserl from H. XII p.448 (quoted in the footnote below), Husserl's contextual meaning theory was based on a conventionalist view of logical and mathematical categories. In fact, according to Husserl, consistency is the only requirement of his conventionalist view of categories, as he says "Which logical categories are to be introduced by definitions is a matter of choice, though the choice is restricted by the requirement of non-contradiction"[FTL Section 34].

The important move of Husserl after he received the Hilbertian linguistic monist doctorine is that Husserl depatred Hilbert's linguistic monist doctrine and placed his philosophy of mathematics into the phenomenologist doctrine, by considering the ontological correlatives of every formal deductive theory. Such an objectivity domain is called multiplicity. Here, a multiplicity is not just a set of formal elements "something whatever", but is structured (in a modern terminology of logic, a (set-theoretic) model of the corresponding axiomatic system). This structure is determined or specified by the linguistic axioms. This means that the structure of a multiplicity for an axiomatic system is induced from the linguistic categorial structure of the corresponding axiomatic system. Namely, The structure of the multiplicity is mapped from the linguistic structure, which Husserl called "categorial" forms. Hence, with the transition from the monist Hilbertian standpoint into the correlative-theory between the deductive-side and the (formal) objectivity side, Husserl obtained a clear idea about the ontologically-categories, which was just the correlative ontological syntactic structure of the linguistic syntactic structure, namely, the grammatical categories. the

For the basic paradigm (of the *Logical Investigations*) of Husserl's meaning theory based on the signification-intention and its fulfillment, the fulfillment is given with a certain intuition in general. This paradigm obviously comes from the perception model of sensible objects, where the signification-intention of a judgment can be fulfilled (or cannot be fulfilled) with an intuition by a concrete material content in our perceptional experience. It is more delicate when Husserl faces to adjust this paradigm of "the signification-intention" and "the fulfillment by intuition" into the framework of purely formal logic and formal mathematics. Now in the purely formal logic and mathematics there is no concrete material domain for which we experience and by which the signification-intention is fulfilled into true knowledge. This is why Husserl reaches his notion of formal ontology, which provides a formal (non-material) domain of objectivity. However, a question remains here. Husserl, on the one hand, refuses the material content-ness of the formal obectivity (multiplicity), while he seems to keep the basic paradigm of his evidence theory/meaning theory, (namely the theory that a signification-intention of a judgment is fulfilled by an intuition-content to be an adequate judgment with evidence or true judgment). Are these two underlying ideas in harmony with each other? Some authors seem to consider it impossible since they take the priority on the deductive side or linguistic side to understand Husserl's notion of multiplicity and seems to admit that Husserl abandons the "signification-intention and fulfillment-by-intuition" paradigm of his basic theory of evidence (of *Logical Investigations*) in the context of his theory of multiplicities (especially in FTL), or at least reduce the role of intuitions when understanding the notion of multiplicities. [40] The other extreme interpretation is the realist interpretation of Husserl's notion of mathematical theory, where the notion of intuition gets the primary role and the role of conventionalist contextual theory of meaning in the second level is ignored or significantly reduced [41]. However, the abandonment of the the "signification-intention and fulfillment-by-intuition" paradigm destroys/damages Husserl's basic framework of meaning theory/evidence theory too much. In fact, Husserl seems to keep the paradigm by introducing the notion of "definite" multiplicity. Here, a judgment on a formal mathematical theory can be conceived as true or as false with the presence of the definite multiplicity of the theory in his "signification-intention and fulfillment-by-intuition" paradigm framework. Although Husserl does not usually attach the word "intuition" explicitly when he talks about multiplicities, the framework of his explanation is, in our view, nothing but in the the "signification-intention and fulfillment-by-intuition" paradigm. Then, the next question is; what kind of intuition is involved in this presence of the definite multiplicity? The natural candidate is the categorial intuition since it is an intuition of the non-material, non-sensible and formal nature and its role had been very important in the fulfillment for logical/formal significations in his basic meaning theory of *Logical Investigations.*

In fact, Husserl endorses in FTL that any categorial formations of mathematical objectivities of a mathematical multiplicity must be verified by the categorial intuition;

> "The categorial formations which were simply existing objectivities for the judger ... must be verified by going over to the evidence, the "categorial intuition" (FTL, Section 44 b-β)

Now we shall discuss what is the actual role of the categorial intuition in the mathematical knowledge acquisition process.

In the 6th Investigation [42] Husserl gives two different interpretations of fulfillment in his "intuition" theory of meaning; the static and the dynamic. The static fulfillment refers to the state of unity between the signification-intention

[40] For example, Cavaillès' characterization on Husserl's notion of mathematical theory as a nomology and S.Bachelard's identification of definiteness (of multiplicity) as syntactical completeness are such examples. Cf. also Jean-Michel Salanskis on p.95 in "Husserl", Les Belles Lettres, 1998

[41] For example, R. Tragesser, Husserl and Realism in Logic and Mathematics, Cambridge University Press

[42] Husserl, Section 8 of the 6th Investigation (1901)

and the fulfillment-by-intuition, while the dynamic fulfillment refers to the "process" of fulfilling the signification-intention by intuition. The static interpretation of the fulfillment-model of meaning theory gives just a correspondence-theory of meaning, namely a meaning theory where the linguistic representations and the intuitive objectivities correspond each other. On the other hand, the dynamic interpretation clearly gives the direction of this correspondence, the direction from the linguistic representation-side to the intuitive objectivity side; the linguistic representations (and their mental-acts of significations) are primary, and the fulfillments by intuitions achieve the completion of this meaning-acquisition and knowledge acquisition processes. However, Husserl's intuition theory for formal mathematics seems to go beyond this distinction and to give another higher level of dynamic view. This is because the only direction from the linguistic side (or axiomatic deductive theory side) into the mathematical domain side (the multiplicity or model side) does not give any real significance of the role of Husserlian "fulfillment-by-intuition" in a formal mathematics theory. (Hence, Husserlian theory of mathematics is often considered from the purely linguistic side as a theory of formal deductions. According to Husserl, an axiomatic formal mathematical theory should be syntactically complete for the lowest syntactical structure. Hence, as long as the syntactically lowest class of propositions, the deduction activities can be self-contained, and no need of any intuition nor objectivity for the proving activities of those. The situation becomes different when considering the syntactically higher level; for example, when we consider the Gödel's unprovable (but true) proposition $\forall x G(x)$ where $G(x)$ is a recursive, hence decidable, syntactically lowest-level proposition. Then, we can see from the purely linguistical deduction each $G(n)$ for natural number n is deducible, hence true in the standard (minimal) model of natural number domain (the definite multiplicity in Husserl's sense). But, as Gödel showed, $\forall x G(x)$ is not deducible from the linguistic side. According to the natural (extensional or Tarskian) interpretation of the universal quantifier, one could conceive the truth of $\forall x G(x)$ in the standard model, which can be considered an example of fulfillment by categorial intuition. Then, one could add this unprovable but true proposition $\forall x G(x)$ in the axiomatic deductive theory side. The extended theory is in this way still a conservative extension of the original arithmetic theory and determines exactly the same standard (minimal) model of arithmetic domain, while this extension makes possible to prove new theorems as well as to significantly shorten the formerly existing proofs of other theorems [43]. This example shows the direction of interaction from the intuition or objectivity side into the linguistic (or deductive-theory) side. The determination of the objectivity domain (as the unique minimal or standard model of a theory) provides an evidence theory

[43] One could find examples of new theorems and new shortened proofs by means of introduction of the Gödel's unprovable proposition above in the field of proof-theoretic finite combinatorics. Cf. M.Hamano-M.Okada, Relationships among Gentzen's Proof-Reduction, Kirby-Paris' Hydra Game and Buchholz's Hydra Game, Mathematical Logic Quarterly, 1997.

(the criterion of truth based on the determined basic structure of the domain), then that evidence theory could provide new tools for the proving activities.

In Husserl's view the back-and-forth zig-zag and try-and-error processes of developing a theory contribute to identifying objectivities in the formal-ontological domain, namely, a multiplicity. In other words, the dynamic process of inferring new judgments from already known judgments and modifying and changing the steps of proof by finding some difficulties in one way or another—-namely, a real discovery process of creating activities of a mathematical proof, (rather than a clean rearrangement of formal proof after the first proving attempt is completed,) reveals the presence of more structure of multiplicity before the judger/mathematician.

This view of Husserl shows a good contrast with Descartes' view of the evidence theory based on intuitions and deductions. In the case of Descartes, the only sources of the true scientific knowledge are intuitions and deductions [44]; the evidence (truth) of the axioms (first principles) of the deductions are obtained directly from intuitions while a deduction has the role of preserving the truth from the first axioms. Descartes needed the proof of existence of God to guarantee this preservation of truth through a long deduction process since if the process is long the evidence of such a long chain of inference steps of the deduction is not given by a direct intuition but via memories [45]. On the other hand, for Husserl, finding process of a deduction itself gives more and more intuition, which is the categorial intuition of a categorial ontological formation of multiplicity.

With this dynamic view, the deductive logic, the second level of Husserl's three-layer structure of logic in FTL is concerned with a process of judging activities with a formal theory, eg. inferring a new proposition from the formerly obtained propositions, developing an argument to prove a targeted theorem, changing a conjecture because of the presence of a contradiction with the current argument, etc. Along with this dynamic judging activities, one can recognize more and more structures of the objectivity counterpart, namely the multiplicity of the theory, in other words, the judger is given the presence of more and more structures of the multiplicity. But, not only the direction from the judging activities towards the presence of more structures of the objectivity before the judger is the part of this dynamic judging activities process but also the reverse direction is another part. Namely, the presence of more structures of the multiplicity (which the judger is intending as the objectivity domain in his judging activities with a formal deductive theory) means that the judger gets more and more intuitions of (the logical-formal structure of) her/his objectivity, namely categorial intuitions of the multiplicity. Hence, the reverse direction means that

[44] Oeuvres de Descartes X, eds. Adam-Tanery, Regulae 3, 369-370, Vrin.

[45] Cf. Mitsuhiro Okada, The Cartesian notion of proof and the criteria of metaphysical proofs, in "Philosophy", the Mita Philosophical Society, Tokyo, Vol. 100, the special issue, 1995

the direction of the categorial intuitions towards the judging activities on the deductive theory. This means that the more clear presence of the logical structures of the multiplicity. namely, the more categorial intuitions on the multiplicity the judger conceives, the more clear direction towards scientific knowledge on the theory the judger obtains in her/his judging-proving activity side. And these two directions go back and forth. This go-back-and-forth relation between the judging activities in the apophantic side and the (categorial) intuitions in the formal ontology side seems very essential in the dynamic view of the second level of logic, logic of non-contradiction/logic of consequence[46] .

This back-and-forth interactions and interference between the linguistic side and the objectivity (the intended domain or standard model) side can be well contrasted with the two extreme standpoints of Frege and of Hilbert in their debate in 1899-1900, where Hilbert only considered the direction from the linguistic consistency to the existence of the standard model while Frege only considered the direction from the standard model (of geometry) to linguistic consistency. The debate was based on the traditional (Cartesian and Kantian) distinction between the intuition side and the linguistic (or logical deduction) side. In the Cartesian and Kantian traditions intuition is considered completely independent of the linguistic or logical side. Within this tradition, Frege stood on the intuition side and Hilbert stood on the linguistic side at the debate. On the other hand, Husserl's "categorial intuitions" and "logical deductions" are interacted each other, which provides Husserlian (phenomenological) correspondence theory of meaning for formal mathematics.

3.2 Mathematical Intuition in the Hilbert School and Gödel

Although he refused any role of intuition in his consistency problem during his first period of the foundational research, as we have seen from the Frege-Hilbert debate at the beginning of the previous subsection, Hilbert started to explain his standpoint with using the notion of intuition during the second period (the period of his finitist standpoint) [47]. The evidence of the finitary propositions (real propositions) is, according to him, conceived by space intuition.

After raising the problem of justifying ideal concepts in mathematics, he continues;

"[W]e find ourselves in agreement with the philosophers, especially with Kant ...., as a condition for the use of logical inferences and the performance of logical operations, something must already be given to our faculty of representation

[46] Heffernan considers that each of three levels of Husserl's logic corresponds to the different level fulfillment. (cf. George Heffernan, Am Anfang war die Logik, 1988, Grüner, Amsterdam).

[47] Note that Hilbert uses the notion of spatial-intuition mainly for the evidence theory of finitist arithmetic, while Frege at the debate used it for the evidence theory of geometry, and Frege remained in his logicist standpoint for arithmetic.

[Vorstellung], certain extralogical concrete objects that are intuitively present as immediate experience prior to all thought [48]."

Hilbert explains his finitist standpoint based on the evidence theory using this notion of intuition.

For an equation with concrete numerals, namely finitary arithmetic proposition, (in the Husserl's terminology, a numerical equation), Hilbert says; "we recognize that we can obtain and prove its truths through contentual intuitive considerations." On the other hand, for a number-theoretic equation with "a genuine number-theoretic variable" "already goes considerably beyond contentual number theory" [49]. Here, Hilbert gives

$$1 + 3 = 3 + 1$$

and

$$1 + 7 = 7 + 1$$

as examples of the former, namely finitary arithmetical propositions of contentual number theory, and

$$1 + a = a + 1$$

as examples of the latter, namely number-theoretic equations with "a genuine number-theoretic variable".

Moreover, Hilbert says that the former numerical equations "can be verified by contentual considerations" [50].

This is one of the sources why Hilbert's finitist standpoint is widely interpreted as a version of the verification theory of meaning, where a proposition is meaningful only if it is verifiable in finite steps (of calculation). [51]

From the finitist point of view, the latter example and other examples such as "$a + b = b + a$ are meaningless in themselves. Here, since the above propositions with number theoretic (open) variable are logically equivalent to the universally quantifiered form $\forall x(1 + x = x + 1)$ and $\forall x \forall y(x + y = y + x)$, respectively, one could understand that the universal quantifier is an ideal concept (an imaginary concept) in Hilbert's finitist standpoint in the 20's. Hilbert, in the 20's, explains that an existential proposition is meaningful in the finitist sense only

[48] Hilbert Über das Unedliche, English translation from van Heijennort "Source Book of Logic", Harvard Univ. Press.

[49] Die Grundlagen der Mathematik (1927). An almost same paragraph can be found in Über das Unendliche (1925).

[50] Die Grundlagen der Mathematik (1927)

[51] This way to interpret Hilbert's finitism is an obvious close similarity with the logical positivist view of the verification theory of meaning, although the logical positivist applied their verification theory of meaning to the empirical propositions (such as perceptional propositions) while the finitist did it to the mathematical propositions.

when there is a bound, namely only a bounded existential quantifier is finitistically acceptable [Über das Unendliche, van Heijenoot's "Source Book of Logic" p.377]. Hence, the existential quantifier in general is also an ideal concept. [Über das Unendliche, van Heijenoot's "Source Book of Logic", p.380]. Hilbert replace the word "finitary propositions" with " numerical equations and inequalities" and he gives further examples in this range; "$3 > 2$, $2+3 = 3+2$, $2+3$, and $1 \neq 1$".
As one can see, this range of the finitary propositions is exactly the same as the range which Husserl presumed in his "completeness proof". Namely the propositions which are finitary in Hilbert's sense coincides with the propositions for which Husserl claimed to have the completeness property (namely the propositions which are shown to be decidable by Husserl's completeness proof). This can be seen in the following three points;

Hilbert often refers to the evidence of the real propositions to Kantian (space-) intuition. The important difference between Hilbert and Husserl here is that in Hilbert the concrete computation (in his "real" or finitist mathematical theory) in the finitistically contentual level is independent of any linguistic rule, hence not considered a formal deduction in an axiomatic system, while Husserl put such a concrete computation process as a typical case of formal deductions when he argues the notions of completeness and definiteness. In fact, Husserl often identifies "axioms" with "algorithms"

However, it is important to notice that Hilbert as another sort of the notion of mathematical intuition, which is different from the notion of intuition for the source of evidence for the finitist (real) propositions. At the almost same period of the philosophically peak period of his finitist meaning theory based on the notion of (finitist) intuition in the 1920's, Hilbert also claims that the actual practice of classical mathematics activities (namely, the activities beyond the finitist standpoint) are based on (another type of) mathematical intuition. In fact he stands against the set-theoretical and logicist (higher order logical) foundations of the continuum, such as the set-theoretical construction of the real numbers by means of the Dedekind cuts, and claims that we could conceive the real numbers by intuition.

> This [axiomatic] grounding of the theory of the continuum is not at all opposed to intuition. The concept of extensive magnitude, as we derive it from intuition, is independent of the concept of number [such as "Dedekind cut"] ..., and the only thing that remains to be decided is, whether a system of the requisite sort is thinkable, that is, whether the axioms do not, say, lead to a contradiction. [52]

This quotation suggests that Hilbert in his 1920's, the peak period of his original finitism, holds two different notions of intuitions, the space (and time) intuition in the sense of Kant (for providing the evidence theory of real (or finitist) propositions) on the one hand, and the mathematical intuition (for conceiving

[52] "The new grounding of mathematics" Hilbert, 1922. translated by William Ewald.

the continuum, for example). Then Hilbert's aim on his consistency program is to protect the free use of the working mathematicians' ideal concepts by means of mathematical intuition; In order to protect/guarantee this free use of ideal concepts, the only requirement is, according to Hilbert, to show a consistency proof on a concrete and contentual ground of real (or finitist) mathematics; for this concrete and contentual ground he uses the other notion of intuition, namely the Kantian space intuition.

The similar double meanings of the notion of intuition may be found in Gödel. Although he claims in 1958 that any consistency proof of mathematics requires more than the space-time intuition in the sense of Hilbert and the finitist standpoint needs to admit some additional abstract concepts (namely, ideal concepts in the original Hilbertian terminology), Gödel at the same time claims that one should have another sort of intuition to justify the new axioms (eg. those for existence of large cardinals) in his Platonistic or realist view of absolute set theoretic universe [53].

Hilbert's explanation on the concrete verification "by contentual considerations" with the space-intuition seems to work well with the above-mentioned examples such as $1 + 3 = 3 + 1$, under a certain interpretation of natural numbers with the stroke figures as Hilbert often uses. It is, however, not so clear whether the same explanation works when we take a numerical equation with more complexed calculable (e.g. primitive recursive) functions. It seems very essential to use the linguistically represented rules of those calculable functions in order to verify the numerical equation, hence the verification process should be like the one described by Husserl (see 2.2 and the footnote 17). Moreover, even though we consider only + as Hilbert took in the above example, but with huge numbers, instead of 3 and 1, then how one can verify the equality of the long stroke figures of those huge numbers by the spatial intuition? In his proof-theory, Hilbert emphasizes that the objects of proof-theory are proof-figures and that one can manipulate those proof-figures in the finitist standpoint with the evidence of the finitist intuition. However, the formation of a proof-figure is based on a highly complicated rules. Those considerations lead us to the view that although Hilbert expresses the apriority of the spatial-intuition of his finitist evidence theory there should be some interaction between the linguistic-rule side and spatial intuition in order to keep the Hilbertian theory of finitist evidence.

In fact, Hilbert and his School extend their scope of finitist mathematics after facing Gödel's incompleteness theorem (1931). The Hilbert School in the 1930's in fact presumes the primitive recursive arithmetic (PRA) of Skolem as the finitist arithmetic, where the primitive recursion rules (and inductive-definitions) are essential. Now, with PRA, the open arithmetical equations (arithmetical formulas with free variables) such as $1 + a = a + 1$ enter in the meaningful finitist arithmetic. Those open formulas (the formulas with variables) are inevitable to express the recursive rules and inductive definitions. The real system (the

[53] K. Gödel, "What is Cantor's continuum hypothesis?" (1947), in in Collected Works of Gödel, Vol.I.

finitist arithmetical system PRA) is now expressed only by a formal deductive system.

Here, we return back to Gödel's attack against the finitist notion of intuition [54]. Gödel's philosophical claim in his 1958 *Dialectica* paper [55] includes the following remarks[56].

Gödel claimed that any consistency proof required some abstract notion beyond the finitist intuition. Gödel himself called the non-intuitive "abstract" notions necessary for consistency proof, "mental constructs" (cf. Subsection 2.4). He claimed that Genzen's notion of "accessibility" of ordinals introduced in his consistency proof (of arithmetic) was a mental construct. Note that Genzen pro-

[54] We look through the claims about finitist intuition of the Hilbert school members in the 1930's.

Gentzen gave a consistency proof of first order arithmetic by means of (higher-order) inferences on accessibility of (transifinite) ordinal numbers, and claimed that the "inferences on accessibility of ordinals" are indisputable from the finitist standpoint (1936). He also claimed that the well-foundedness ("no-infinite decreasing sequence") of ordinal numbers up to ε_0 can be conceived concretely. One can survey them.

Ackermann (1951) also endorsed this claim saying that the inferences on accessibility of ordinals are still intuitive. On the other hand, Bernays considers that one needs to be beyond the original notion of Hilbertian intuition in order to conceive the evidence of the extended framework of finitist standpoint based on the primitive recursive arithmetic in the 1930's.

[55] Note that the title of his paper "On a hitherto unutilized extension of the finitary standpoint" already expresses his strong standpoint that any consistency proof requires some extension from the finitist standpoint.

[56] Gödel remarks as follows. "P. Bernays has pointed out ... it is necessary to go beyond the framework of what is, in Hilbert's sense, finitary mathematics if one wants to prove the consistency of classical mathematics, or even that of classical number theory. Consequently , since finitary mathematics is defined as the mathematics in which evidence rests on what is " intuitive ", certain "abstract" notions are required for the proof of the consistency of number theory .... Here, by abstract (or nonintuitive) notions we must understand those that are essentially of second or higher order, that is ... notions that relate to mental constructs ...; and in the [consistency] proofs we must take use of insights into these mental constructs. ... [Gentzen's results show that] the validity of inference by recursion up to ε_0 surely cannot be made immediately intuitive , as it can up to, say ω^2 .... To be sure, W. Ackermann tells us ... that "accessibility" will be intuitively meaningful ... But to this one must reply that ... the validity ... can be demonstrated only by means of abstract notions .... The notion "accessibility" can, however, be replaced, at least for induction up to ε_0, by weaker abstract notions [which means Gödel's primitive recursive functionals of higher types]".

Note that the original Gentzen's "accessibility"-inference can reach up to $\varphi_\omega 0$ [of the Veblen hierarchy of transfinite ordinals] (M. Okada, "Theory of weak implications", *Journal of Symbolic Logic, 1988*)

posed the notion of accessibility of transfinite ordinal numbers of the Contor's second class. Gentzen's criteria to introduce a new concept for his consistency proof was, according to Gentzen, based on Hilbertian "finitist intuition" since Genzen believed that his consistency proof was in harmony with the Hilbertian finitist standpoint, while Gödel considered that the notion of accessibility was beyond the scope of finitist intuition, and, moreover, from the point of view of mental construction the notion of accessibility is less economical than his notion of primitive recursive functionals (cf. Subsection 2.4) .

Gödel introduced system T obtained from PRA by introducing the primitive recursive functionals of higher types. Here, admitting the standpoint of the Hilbert school in the 1930's, Gödel took PRA as the finitist system and extended it by what he considered the "economical" abstract "mental constructs" priminitve recursive functionals [57].

Gödel's system **T** of primitive recursive functionals has influenced contemporary intuitionist movements very much. [58].

Gödel's standpoint and the Hilbert School's standpoint (in the 1930's) can be contrasted clearly by their claims on ordinal numbers; For the Hilbertian finitist spatial intuition, surveyability was one of the important factor. Gentzen claimed that the ordinal numbers up to ϵ_0 was surveyable, while Gödel claimed only the ordinal numbers up to ω^ω was surveyable. (Note that the critical ordinal number of PRA is ω^ω, hence Gödel admitted PRA as an upper-bound border of finitist mathematics based on the finitist intuition. On the other hand, ϵ_0 is the ordinal number whose accessibility implies consistency of first order arithmetic. Hence, Gentzen's claim implies the consistency proof can be performed in the finitist standpoint.

4 Towards the Explanation of Mathematical Objectivity and Mathematical Knowledge Acquisitions Based on the "Definite" Model of a Theory

As we remarked, when considering the determination of a multiplicity (model [59]) by linguistic means, Husserl focuses on the basic substrates level or the cores-level (namely, the syntactically lowest level of the structure) of a multiplicity.

[57] Gödel also admitted that the Π_1^0 arithmetic propositions, the propositions of PRA, were finitary propositions, following the claim of the Hilbert school in the 1930's. in footnote 42 of in "What is Cantor's Continuum Problem?", "Philosophy of Mathematics (eds. Benacerraf and Putnam, 1964)

[58] (1) The "third-generation" Intuitionist School (Troelstra, et al.) (2) Computation models for typed functional programming languages (e.g. Inductive-types in Coq., Girard's introduction of "polymorphic type"), although we shall not go into these topics in this paper. See eg. M. Okada, "Intuitionism and Linear Logic", Revue Internationale de Philosophie, to appear

[59] As we remarked before, we shall use set-theoretic "model" and Husserlian notion of "multiplicity" interchangeably in this paper.

Note that Husserl distinguishes the ultimate substrates (as the basis level of formal ontology) and the other syntactic constitutes (as higher levels of formal ontology) clearly. For example, in section 11 of Chapter 1 of Ideen I he classifies the objectivities into the two kinds; the syntactical objectivities and the ultimate substrates. A higher syntactical objectivity is derived from relatively lower syntactical objectivities which serve as the substrates for this syntactical derivation, while according to Husserl, one can reach the ultmite substrates which are not the result of any other syntactical derivation. Such ultimate substrates are also called the core [Kern] [60]

---

[60] See our paper "Husserl's 'Concluding Theme of Old Philosophico-Mathematical Studies' and the Role of his Notion of Multiplicity" referred in the footnote 2 for more details. We repeat below a few points. When he explains the "transition link (FTL Sec.82, p.179) from the formal deduction level (the second-level) to the level of truth (the third level) with a multiplicity, Husserl needs the most basic stuffs of the formal objectivities. Here, Husserl's main concern was the "definiteness" of the basic entities level, the cores, of a multiplicity (or of model in the modern logic sense).

"[I]t can be seen a priori that any actual or possible judgment leads back to ultimate cores when we follow up its syntaxes ,,,, And always it is clear that, by reduction, we reach a corresponding ultimate, that is: ultimate substrates,......,ultimate universalities, ultimate relations." (FTL Sec. 82).

For the case of arithmetic, the judgments regarding this level is, logically speaking, essentially the quantifier-free closed propositions in the modern logical terminology, namely what Husserl calls numerical equations and numerical propositions, for which Husserl successfully proved the definiteness/syntactical completeness of arithmetic.

And Husserl emphasizes that the attention on the ultimate cores of the objectivity is the source of the notion of truth.

"For mathesis universalis, as formal mathematics [as the formal deduction level], these utlimiates have no particular interest. Quite the contrary for truth-logic: because ultimate substrate-objects are individuals, about which very much can be said in formal truth, and back to which all truth ultimately relates. "[FTL Sec.82] Husserl continues that in order for a proposition to go beyond a merely analytic proposition toward a proposition about truth, "one must make ultimate cores intuited, one must draw fullness of adequation, not from evidence of the judgment-senses" [FTL Sec.82].

Husserl's emphasis here is on this ultimate cores level of the formal objectivity. This view is actually shared with his characterization of the definiteness of a multiplicity when he first reached his solution on the problem of the imaginaries in the winter semester 1901-1902, where he says: "Proof of this assertion [i.e., of the syntactic completeness, equivalently of the definiteness of arithmetic for the natural numbers] lies in the fact that every defined operation-formation [is] a natural number and that every natural number is related to the others in terms of the scale-relation determined by the axioms." [p.443, Appendix VI, H.XII].

However, it is important to notice that this very restricted notion of definiteness, although it keeps open a huge indeterminacy in terms of deducibility and decidability, surprisingly already completely determines a unique model (unique multiplicity), namely this determines the standard model of the natural numbers (up to isomorphisms) with only one additional assumption

In fact, if we take a minimal domain satisfying a typical axiomatization of natural number domains, e.g. first order Peano arithmetic, Primitive Recursive Arithmetic, Robinson's Q (which is one of the smallest axiomatizations), etc., the above condition determines a complete structure of the model (multiplicity). Here, the minimal model is actually "the" intended model for these axiomatization, in the sense that every intended individual element has the unique description (in the sense of Russell), or one could say each element (of this intended domain) has a canonical form of the term. (For example, 3 might be expressed as 1+1+1, in many cases of axiomatizations.) So, as long as the intended minimal model (multiplicity) is concerned, the structure is completely determined. However, in any of the usual axiomatizations, even if the intended minimal model is determined by the axiomatization of the natural number theory, there are infinitely many models (multiplicities) which still satisfy the same axiomatic system. (Such models are usually called "non-standard models" of the axiomatic theory.) In other words, the syntactic completeness restricted to the core-level of language does not implies the uniqueness of the model (multiplicity) of the axiomatic system, but determines uniquely the minimal model. Namely, the following fact (∗) holds.

(∗) The minimal domain of the primitive individual elements (in this case the set of natural numbers) is determined by the rules on the function and constants (which are often called "constructors") syntactically.

This notion of minimal model is also often called the term-model or the language-model of a theory where the notion of constructor-terms is defined. Any closed first order term has a unique representation, which is usually called a normal form or canonical form. This situation is well-described in Section 18, Chapter 3 of the Sixth Investigation of Husserl (1901), where the canonical form of each number is represented as 1+1+...+1, and any expression of number can be normalized to a unique canonical form by the underlying (definite axiomatic) rules; Husserl here in the Sixth Investigation explains that the normalization steps are in parallel with the fulfillment steps.

He emphasizes that "the primitive relations among the elements" are needed to be determined for a definite multiplicity. (H.XII Appendix VII, p.466 and p.468, cf. also p.462.)

This explanation gives only the lowest-level of the structure of a natural number domain. Hence, this does not tell any decidability/definiteness of the propositions of syntactically higher-level in terms of deducibility, namely, this characterization of definiteness by Husserl does not care directly any syntactic objectivities except for the ultimate substrates-level (the cores-level).

This situation is similar to Hilbert's definition of the finitary mathematics domain, where the real (or finitist) propositions are exactly the quantifier-free (or only bounded-quantifiered) closed propositions on such a constructor-term domain, and Hilbert especially considers the natural number domain of arithmetic as the typical example of the real/finitist domain.

The notions of intuition of Husserl and of Hilbert ("categorial intuition" of Husserl in the context of definite multiplicity/complete theory and finitist "intuition" of Hilbert) shares the following points;
1. The intuition gives the content of the meaning, hence the propositions of the real systems are contentual by mean of intuition.
2. Those contentual propositions are also decidable from the linguistic (deductive) theory-side alone. Hence, the intuition is in parallel with a certain algorithmic procedure, or at least backed up by the linguistic contextual meaning.
3. The content of a proposition by intuition gives us the evidence and the truth value. This evidence and truth value of (of the syntactically lowest propositions) are obtained in paralell with the decision procedure or verification procedure of the linguistic-judgment side.
4. In this sense, the class of linguistically decidable propositions and the basis class of intuitive and contentual propositions are the same for both Husserl's and Hilbert's frameworks.

It may be seen at the first look that there is an important difference between Husserl's categorial intuition and Hilbert's (finitist) intuition. For Hilbert, this finitist intuition is the Kantian space intuition; [61], hence Hilbert seems to claim that the intuition is principally independent of the linguistic rules and axioms, and that the evidence and the verification for the real propositions of the decidable (or syntactically complete) finitist system are given solely based on the spatial intuition without any linguistic (or axiomatic) rules, e.g. arithmetic is justified from the intuition-side. Therefore, from this first look Hilbert's evidence theory seems quite different from Husserl's since Husserl claims that the conventionally introduced decidable (or syntactically complete) rules and axioms (which Husserl calls "categories" or "syntax" in the context of his formal mathematics studies) give the basis of the intuitive objectivity domain of a formal mathematical theory.

However, in our opinion Hilbert's notion of finitist intuition is actually based on the linguistic rules and the two notions of intuition seem very close each other. This is because, as we discussed at the end of Subsection 3.2, we needs (certain arithmetic or combinatorial) rules to verify even numerical equations (the typical "real" propositions in Hilbert's strict sense of finitism). We also remark that

[61] As C. Persons remarks Hilbertian finitist intuition for arithmetic is better explained by Kantian theory of geometry, rather than Kantian theory of arithmetic. Cf. C. Persons,"Mathematical Intuition", Proceedings of Aristotelian Society 80 (1980).

the mathematical objectivities of the spatial intuition of Hilbert (in the 20's) are "signs" as he admits;

"[W]hat we consider is the concrete signs themselves, whose shape, according to the conception we have adopted, is immediately clear and recognizable [62] ."

The mathematical domain composed of signs for the canonical terms (or normal terms) is exactly the notion of term-model or canonical model, and the domain can be characterized only through certain rules governing the signs. This is nothing but the intended "definite" multiplicity, which Husserl considers. Here, the canonical representations (or canonical signs) are characterized by means of the primitive relations and functions determined by rules (or axioms). Traditionally, eg. either for Cartesian intuition or for Kantian intuition, "intuition" for evidence is immediately given completely independently of any linguistic rules. Husserl took a radical "linguistic turn" with his notion of categorial intuition (especially for his meaning theory of formal mathematical theories) by making a bridge between the linguistic rules and the categorial intuitions (forming the objectivities of intuition by linguistic rules or what he call "grammatical categories" [63]. According to the above analysis on Hilbert's notion of spatial intuition of signs, Hilbert's notion of intuition also seems to depend on the linguistic rule side very heavily, and the two notions of intuitions share the common feature on this point. Of course, Husserl's definite multiplicity (identified with the canonical or minimal term model) is not composed of the "concrete signs" as Hilbert took, since Husserl's formal ontology (of mathematics) takes "somethings whatever" (which we identify abstract sets of the universe of ZF set theory. Cf. Subsection 2.2.). However, it is natural to understand that the Husserlian definite multiplicity is an (abstract set-theoretic) model isomorphic to the model based on the Hilbertian canonical terms (or canonical signs) domain. Since Husserl considers the isomorphic multiplicities identical, as the usual mathematicians presume, the two objectivity-domains (the cores) for the two notions of intuition can be identified.

However, there is still an important difference between Hilbertian finitist intuition and Husserlian categorial intuition. The scope of the Hilbertian intuition of objectivities is strictly the finitary mathematics [64], while Husserl's categorial intuition for the mathematical evidence has a wider scope than just the "cores";

---

[62] Über das Unendliche (1925), p. 376, in van Heijenoort.

[63] Note again that Husserl uses the words "category" and "syntax" interchangeably and that he took the grammatical (or linguistic) categories conventionally in the context of formal mathematics. Cf. at the beginning of Subsection 3.1

[64] As we remarked before, the Hilbert School's official standpoint was changed in the 1930's after the appearance of Gödel's incompleteness theorem; As Bernays claimed in Hilbert-Bernays (1934), they took the primitive recursive arithmetic (of Skolem) so that the purely universal propositions of arithmetic (or open arithmetical formulas) were included into the finitist propositions, which are beyond the scope of the original finitist intuition, according to Bernays and Gödel (but, as we remarked, Gentzen and Ackermann still consider those inside the finitist intuition).

Husserl considers the syntactic constitutes, or syntactically higher level of objectivities, based on the definite "cores" level (which could be understood naturally by means of Tarskian construction of a model for the syntactically higher level).

In order to see this difference clearer, we first take a further close look at the Husserlian definite multiplicity, and then return back to point out the difference from the objectivity of the Hilbertian finitist intuition. We recall that Husserl shows that a unique mathematical domain (minimal term-model domain), the cores part, is determined by the linguistic rules, i.e., by the (first order) decidable relations and calculable functions. As we remarked before, Husserl added one line (in his sytactic completeness proof of arithmetic) claiming the propositions with algebraic variables (namely, the quantifiered propositions) are also decidable since the quentifier-free arithmetic propositions are decidable. This last line could be interpreted as the determination of syntactically higher structure of the model in the categorial intuition side. This line, in the modern terminology of logic, is equivalent to claim the ω-rule, an infinite inference inferring $\forall x A(x)$ from infinitely many premises $A(0), A(1), A(2), ...$, which cannot be represented as a linguistic rule due to the infinitely many premises; this rule is essentially the same as Tarskian model-theoretic definition of the meaning of universal quantifier (which is essential for defining any syntactically higher structure of the model) [65]. Hence, Husserlian categorial intuition-based meaning theory can be understood as Tarskian model-theoretic (extensional) semantics for the higher syntactic levels [66]. It is important to notice that the lowest syntactic level (the cores-level) Husserl's truth-definition and semantics are determined linguistically (categorially in Husserl's sense) definitely; this is a big difference from the Tarskian model-theoretic semantics because the Tarskian semantics takes arbitrary set-theoretical domains for the semantical definition of the syntactically lowest level of propositions, which leads us to arbitrary models (including many

[65] Tarski originally gives the truth definition of an object-language using a meta-language, and defined the meaning of, eg., logical connectives and of quantifiers of an object language in a meta-language. Tarski and his school later emphasizes the model-theoretic truth definition where a meta-language is replaced by the set-theoretical or ontological structure or a domain, the later of which is now usually referred as the logical semantics or extensional semantics.

[66] It is important to notice that the core level determination (the condition (2)) gives the determination of the intended minimal model, by shifting the truth definition of the core level up to all syntactically higher objectivities, using a standard construction of the truth definition along with the construction of logical formulas (in the terminology of Husserl, the derivation of categorial objectivities along with the logical categories). This situation (of shifting up the truth definition from the cores level) can be well-explained by using the standard construction of the truth definition in the Tarskian semantics/modern model theory; in order to define the full-semantics for one model, namely in order to define truth for all syntactically higher propositions, it is known to be sufficient to provide the two things, namely, (i) a set (which serves as the individual domain of the model), and (ii) the interpretation for primitive operations (functions) and primitive predicates/relations on the domain.

non-standard models) of a theory. On the other hand, given an axiomatic theory Husserl's semantics targets an intended unique standard model from the beginning, and determines the domain and the syntactically lowest structure uniquely linguistically (or categorially) by rules and axioms. Then the syntactically higher level structures of the domain are not determined by linguistically (nor axiomatically) anymore, but determined in the usual sense of model theory (essentially by the ω-rule). So, as soon as the syntactically lowest level of the structure is linguistically determined, one could consider the whole structure of the intended model uniquely (in the modern sense of set-theoretic model theory).

Now, we point out the difference between this Husserlian determination of a unique model and the Hilbertian standpoint on the objectivity of the finitist intuition. For Hilbert's finitist standpoint, the interpretation of the quantifiers is more delicate than Husserl's (phenomenological) corresponding theory of meaning. The interpretation of quantifiers as the Tarskian way requires the assumption of closed infinity of the domain. Hence Hilbert kept the scope of the objectivity of his finitist intuition in the syntactically lowest level in the 1920's. This is how the essential difference on the scope of objectivity between Husserl and Hilbert occurs. Even when the Hilbert School widened the scope of finitist objectibity in the 1930's (to PRA), the syntactical complexity of meaningful propositions were still at the second to the lowest level, namely the class of open arithmetic formulas (or, equivalently, Π^0_1-formulas) [67].

The examples of constructor terms and of constructor-term domain which are different from the arithmetical domains have often appeared in the recent information theories and computer science theories, in particular, for denotational models of programming and formal specification languages, since those theories often require rich recursive data structures as their semantic domains, which are often called "initial algebra models" [68].

In our opinion, this view would give alternative explanation to C.Persons' comment on the choice of standard model and on mathematical intuition of a standard model. C.Persons comments that since the essence of mathematical objects is the structure of the relations presented in a theory, one has no reason to choose one model from another as a unique intended model, beyond the structure itself (which is neutral on the choice of models). [69] For this point, we

[67] There is another syntactic restriction in Hilbert's finitism, namely on the use of classical disjunction, since Hilbert considers the law of excluded middle as an ideal proposition.

[68] Cf. See the standard reference books on this subject such as 3.2, Dershowitz-Jouannaud, Rewrite Systems in Chapter 6 of Handbook of Theoretical Computer Science (ed. van Leeuwen, Elsevier, 1990), and Huet-Oppen "Equations and Rewrite Rules: A Survey" in Formal Language Theory (ed. Book, Academic Press, 1980). Jouannaud-Okada's "Abstract Data Type Systems" (Theoretical Computer Science 173 (1997)) contains some examples of non-recursive higher-type rules which still determine a unique initial algebra model for abstract data structures.

[69] C.Persons, "Mathematical Intuition", Section 2, Proceedings of the Aristotelian Society 80 (1980)

could answer that in a certain circumstances, such as the arithmetic, one could characterize by means of an axiomatic theory a unique intended model as the minimal term model. C.Persons there also pointed out the possibility of nominalist interpretation of a mathematical theory without assuming any objects of intuition. On the other hand, as discussed above, one natural nominalist interpretation such that a term is interpreted as a term itself on the sign-level, actually leads to the unique minimal model under isomorphism. In fact, as we showed, Husserl's categorial intuition for arithmetic is derived from such a nominalist interpretation; in other words, the nominalist interpretation following the linguistic rules gives the stability of the objectivity domain of Husserlian intuition in formal mathematics.

This view also implies that we could talk about discovery of mathematical knowledges since the truth-value of any proposition can be presumed by the definite standard model, independent of our knowledge [70] [71].

[70] This situation is exactly coincides with Husserl's characterization of the third-level, the level of (logic of) truth in his three-levels structure of logic. Namely, if we understand that Husserl's definition of the definiteness of a multiplicity as the axiomatic determination of the core (ultimate substrates) domain, as Husserl claims, the notions of truth and falsehood for a mathematical theory appear if and only if the axiomatic theory has a definite multiplicity as the intended objectivity domain. The latter is, again as Husserl correctly claimed, equivalent to (syntactical) completeness of the axiomatic system in Husserl's sense

[71] This situation is in particular visible, and in fact becomes important, when we consider the meaning of Gödel's incompleteness theorem. See the second paper of ours in footnote 2 for a detailed discussion. We shall recall some points here. Gödel's statement of the incompleteness theorem says that if a mathematical axiomatic theory is consistent, there is a true but unprovable proposition. (Originally, he used a slightly stronger condition, Ω-consistent. but the difference does not affect our argument here.) A weaker version just says that there is a proposition which is neither provable nor refutable (i.e., its negation is provable) under the same condition. How can we get this former epistemic version of Gödel's incompleteness theorem, rather than the latter mere syntactic statement? This first statement is meaningful only when we consider the standard model (the definite multiplicity) of natural numbers, namely, the intended natural numbers domain uniquely determined in the syntactically lowest level. Thanks to this standard model, one can see the Gödel's proposition is true; Gödel's statement can be expressed by a universally-quantified form, $\forall x G(x)$, where G is a unary-decidable proposition and by Gödel's construction of G, G(n) is true on the standard model for any numeral n individually. hence, our understanding of the standard model of natural numbers tells us, "hence, $\forall x G(x)$ is true".

Gödel, moreover, utilized in an essential way in his incompleteness proof the fact that our standard model is definite in Husserl's sense above, namely an axiomatic theory of natural numbers (which is contained in a formal system) determines the standard model. Gödel expressed this fact as his "Representation Theorem" (which is Theorem V in his incompleteness paper in 1931) where for any (essentially quantifier-free except for bounded quantifiers) arithmetical proposition, say A(x), and for any numeral n, A(n) is true in the standard model (the definite multiplicity in Husserl"s

Of course, there are some cases in which there exists no definite standard model associated to an axiomatic theory. A typical example can be found in axiomatic theories of abstract algebra. For example, the axioms of semigroup in the previous section does not have any standard model. Here in abstract algebra, one of the main role of axioms is not to determine a unique standard model, but to give the correlation among many different models which have the same algebraic structure, in other words, which satisfy the same axioms. Hence, in abstract algebra, one does not expect the set of axioms specifies a unique stan-

---

sense) of the natural numbers if and only if A(n) is deducible in a formal system considered; and A(n) is false if and only if not A(n) is provable. This is exactly the same form of Husserl's condition of multiplicity.

Gödel wrote in his "incompleteness" paper about this very essential theorem (Theorem V) to derive his incompleteness result for the syntactically higher level, that he of course use the fact that the axiomatic number theory has the completeness property for the syntactically lowest level, as follows;

"Theorem V, of course, is a consequence of the fact that in the case of a [primitive] recursive relation R it can, for every n-tuple of numbers, be decided on the basis of the axioms of the system P whether the relation R obtains or not."

Gödel's proof of incompleteness depends heavily on that fact that the natural number domain is definite in our sense. His proof depends on the two conditions; (1) The incomplete statement H ($\forall x G(x)$ expressed above) is equivalent to the negation of "there is a proof of H". (2) The proposition "x is a proof of A" can be expressed as a primitive recursive arithmetical formula on n and A, hence by definiteness, If "n is a proof of A" is true (in the definite natural numbers model), "n is a proof of A" is deducible. By putting an existential quantifier on proof n, one gets; if "there is a proof of A" is true then "there is a proof of A" is deducible. From these two, Gödel proves the unprovability of H as follows;
Assume that H is provable. Namely "there is a proof of H" is true . Then by (2) above, "there is a proof of H" is provable. Hence, by (1) above, the negation of H is provable, which is contradictory with our assumption.

In fact, when we understand any version of proof of Gödel's incompleteness theorem, we really need to stay ourself inside the world of the standard model (the definite multiplicity) of the natural numbers. This is because the incompleteness proof requires the reader to identify the number-theoretically described definition of predicates, such as "a is a proof of B", and the definition of the same thing in the ordinary mathematical sense. The switching between the truth of the number theoretic statement "a is a proof of B" and the truth of the "fact" that a is actually a proof of B, is very frequently performed in the standard reading of incompleteness proof. Here, this switching is acceptable because we have the standard model (the definite multiplicity) of the natural numbers which can be described easily by most of natural axiomatizations of the natural numbers, and because we can stay in this intended definite multiplicity when we read an incompleteness proof.

Hence, Gödel's incompleteness theorem tells us the real significance and the meaning of Husserl's theory of definite multiplicities in a very clear way from another point of view, too; When we take a purely formalist standpoint, the undecidable proposition cannot tell true or false. Our cognition of truth of this undecidable proposition is depends on the definite standard model of natural numbers.

dard model. In fact, using Husserl's terminology, one might be able to classify the mathematical axiomatizations into two classes; one by "formalization", the other "abstraction". Husserl considers "formalization" as the principal role of forming a definite multiplicity and a complete axiomatic system in mathematics and he distinguish "formalization" from "abstraction", where the latter leads to the genus-species relation [72] while the former cannot be explained by the genus-species relation. In this context, we could consider that a typical mathematical axiomatic theory which has an intended unique standard model (definite multiplicity) is classified as a theory of "formalization", while the axiomatization of abstract algebras , the exceptional case from Husserl's point of view, could be considered to belong to the class of "abstraction" where abstract relations are shared in many different models (indefinite multiplicities in Husserl's sense) and relations among such models are more central roles, while determining the structure of unique standard model (definite multiplicity) is the central in the case of "formalized" axiomatic theories.

There is another case of mathematics where Husserl's idea of definite multiplicity does not work as it is; when the domain, the set of ultimate substrates in Husserl's sense, is uncountably infinite, there is no hope to determine even the cores level of the intended standard model by any axiomatic system. This is because our language could generate only countably many infinite sentences and only countably many infinite words, hence most of the ultimate elements of the intended model has no definite description nor even no name in the case the cardinality of the domain is uncountable. (So-called "Skolem's paradox" is just a natural effect of this.) This is the cases of, for example, the real number theory (and any theories which contain the real numbers) and set theory. However, we have our intended standard model for such cases. Moreover, all ultimate elements in which we are interested have a definite description in the axiomatic system side. For example, the real number π is not even algebraic real and is transcendental real, but still we have the definite description and a concrete algorithm to calculate it to accuracy in the decimal expression.
For the theories with uncountable standard models, Husserl's standpoint is not applicable directly. Hence. one may need different kinds of intuition, different from the categorial intuition in Husserl's sense, to give some evidence theory. This is the case where Hilbert needed his second meaning of "intuition" (to conceive the real number domain) and Gödel needed his second meaning of "intuition" (to conceive the absolute set-theoretical universe), which we discussed before.

However, although any axiomatic theory whose standard model is of an uncountable cardinal does not determine such a model "definitely" in the sense of Husserl, we consider that Husserl's overall/global theory based on the dynamic view of the mathematical reasoning activities interacted with categorial

[72] It seems natural to interpret the axiomatic abstract algebras as examples of "abstraction" rather than "formalization", hence, for example, the relation between the axiomatic theory of group and a concrete example of domain of group as a genus-species relation.

intuitions of the unique intended multiplicity can work well. If our (dynamic) view of Husserl's theory gives a coherent explanation for the mathematical activities and the mathematical objectivities completely in the framework of logic of Husserl for the case of the typical formal mathematical theories, such as formal arithmetic for a countable many infinite domain, e.g. for the integers, for the rationals, one could consider other cases to a fairly significant extent. We shall discuss this elsewhere. In any case, the failure of definite specification of an intended multiplicity for the case of an uncountable domain does not mean any failure of a logical theory, but it only shows a natural limitation of our language.

It is also interesting to consider as to what is the difference in the logical constitutes between the case in which we have a standard model or stable objectivity model and the case in which the mathematical objects are more dynamically changing. In a subsequent paper [73] we claim that the difference comes from a hidden logical operator which distinguishes the traditional logics whose natural semantics are denotational and the linear logics whose natural semantics are resource-sensitive and concurrency and time-sensitive, and that the usually hidden (but explicitly represented by linear logic) logical operator provides a certain logical structure for some (traditional) mathematical proposition with an interpretation by means of denotational models. This suggests that the ground of understanding our mathematical proving activities as the discovery process is based on a certain logical or categorial framework of the linguistic side.

To summerize, we analyzed Husserl's and Hilbert's researches on the problem of justifying the ideal concepts in mathematics as the source of their notions of intuition. We discussed a striking similarity between the two author's theories. We also pointed out some important differences. The both views lead us to the notion of mathematical intuition backed up with the language-oriented activities of mathematicians. Their idea of combining the role of our language-oriented activity and the role of intuition provides us with a new framework to explain mathematical discoveries, which is different from the traditional ones, such as a realist or nominalist or conventionalist framework. We, in particular, emphasized the Husserlian view on the interactions between the language-oriented meaning theory and the categorial intuitions. This view, on the one hand, gives a linguistic characterization of the mathematical structure of an intended (denumerably infinite) objectivity domain (such as the natural number domain), which forms the basis of (categorial) intuitions for such a denumerably infinite domain, and, on the other hand, explains, in our opinion, the nature of mathematical knowledge acquisition activities as "discoveries" in the course of the above-mentioned interaction between the linguistic activities and categorial intuitions.

[73] M. Okada, "Intuitionism and Linear Logic", Revue Internationale de Philosophie, to appear

Theory of Judgments and Derivations

Dedicated to the Memory of Professor Katuzi Ono

Masahiko Sato

Graduate School of Informatics, Kyoto University
masahiko@kuis.kyoto-u.ac.jp

Abstract. We propose a computational and logical framework NF (Natural Framework) which is suitable for presenting mathematics formally. Our framework is an extendable framework since it is *open-ended* both computationally and logically in the sense of Martin-Löf's theory or types. NF is developed in three steps. Firstly, we introduce a theory of expressions and schemata which is used to provide a universe for representing mathematical objects, in particular, judgments and derivations as well as other usual mathematical objects. Secondly, we develop a theory of judgments within the syntactic universe of expressions. Finally, we introduce the notions of derivation and derivation game and will show that we can develop mathematics as derivation games by regarding mathematics as an open-ended process of defining new concepts and deriving new judgments. Our theory is inspired by Martin-Löf's theory of expressions and Edinburgh LF, but conceptually much simpler. Our theory is also influenced by Gentzen's natural deduction systems.

1 Introduction

The continuous development of mathematics has been supported by the introduction of new *concepts*. For example, the concept *number* has been extended starting from *natural number*, through *integer*, *rational number*, *real number* up to the concept *complex number*. For each object s and a concept P we can form a *judgment* $s : P$ which asserts that the object s falls under the concept P.[1] As is customary in modern mathematics, we can assert a judgment when and only when we could derive the judgment, namely, when we could construct a *derivation* whose conclusion is the judgment. Here the derivation provides an *evidence* for the judgment. Firstly, the derivation makes the judgment evident to the very person who constructed the derivation because, supposedly, he has just constructed it by applying *derivation rules* correctly. Secondly, the derivation makes the judgment evident to any reader of the derivation, because[2] the

[1] See, e.g., Bell [7, page 9] and Frege [22].

[2] Here, we assume that the reader also accepts the rules as evident. So, if the reader is an intuitionist and the derivation contained an application of the law of the excluded middle for an undecidable proposition, then the derivation is not evident to the reader. But, even in such a case, the reader can check whether or not the derivation is a *correct* derivation obeying the rules of classical mathematics.

S. Arikawa and A. Shinohara (Eds.): Progress in Discovery Science 2001, LNAI 2281, pp. 78–122.

reader can read and check the *correctness* of the derivation with respect to a given set of rules and thereby he can *experience* the process of constructing the derivation for himself.

In order to make the above analysis valid, it is necessary that (1) the derivation is written in a common language shared by the writer and reader of the derivation and (2) a derivation rule in the derivation sends evident judgments to an evident judgment. In this paper we are mostly interested in the first condition of the above sentence, and we will provide a general and uniform framework NF (Natural Framework) for defining various formal systems conveniently. We will fulfill the first condition by realizing Kreisel's dictum (see [6, 31, 43–45, 80]) which asks the (primitive) recursive decidability of whether or not a given derivation is a correct derivation of a given judgment. The point is that the *correctness* of a derivation can be checked mechanically against a given set of formal rules while the *evidence* of the judgment proven by the derivation is determined semantically. We will achieve the decidability by defining the possible forms of *judgments* and *derivations* syntactically without referring to the meanings of judgments. Thus our method is along the line of Hilbert's program (see, e.g. [42, 44, 56, 77]), according to which formalistic notions of formulas and theorems must be defined and treated in a *finitistic* manner.

There are already numerous attempts to provide general frameworks for defining formal systems. Among these works, our work is greatly influenced by Martin-Löf's theory of expressions [55] and the Edinburgh Logical Framework [31].[3] Martin-Löf developed a theory of expressions which is used as an underlying language for representing his intuitionistic theory of types. Edinburgh LF was inspired by Martin-Löf's theory of expressions. It is a dependently typed λ-calculus and it can encode various logical systems by using the judgments-as-types principle. We will compare our work with these works in detail later. Here, we only mention that we have designed our framework NF so that it will become much simpler than these frameworks. We think the simplicity of the framework is crucial since we wish to provide a foundational framework for defining various formal systems on top of it.

Martin-Löf [46, 49] classified judgments into *categorical judgments* which are judgments made without hypotheses and *hypothetical judgments* which are judgments made under several hypotheses. We will write $H \Rightarrow J$ for the hypothetical judgment which asserts the judgment J under the hypothesis H. In Martin-Löf's notation, this judgment is written as $J\ (H)$. Thus, the hypothetical judgment $J\ (H_1, \ldots, H_n)$ in Martin-Löf's notation can be written as $H_1 \Rightarrow \cdots \Rightarrow H_n \Rightarrow J$ in our notation. A typical hypothetical judgment takes the form $J\ (x : A)$ where x is a variable and it means that for any x, if $x : A$ holds, then J also holds. Therefore, the variable x is universally quantified in the hypothetical judgment. To make this quantification explicit, we introduce judgments of the form $(x)\ [J]$ and call such judgments *universal judgments*. Using universal judgments, Martin-

[3] Throughout this paper, the Edinburgh Logical Framework will be called Edinburgh LF or, simply, LF.

Löf's hypothetical judgment $J\ (x : A)$ can be written as $(x)\,[x : A \Rightarrow J]$[4] which we will also abbreviate as $(x : A)\,[J]$.

In order to represent judgments and derivations formally, we first introduce a set E of symbolic expressions. The set E contains two infinite denumerable sets of variables and constants, and it is closed under the operation of forming a *pair* $<e\,|\,f>$ of two expressions e and f, and the operation of forming an *abstract* $(x)\,[e]$ which is obtained by binding a specified variable x in a given expression e. The set E is rich enough to represent judgments naturally. For instance, the universal judgment $(x)\,[x : A \Rightarrow J]$ can be written as the expression

$$(x)\,[<\Rightarrow, <:, x, A>, J>]$$

where '$\Rightarrow$' and ':' are constants, and $<:, x, A>$, for example, is a list which we can construct by the pairing operation in a usual manner. We will see that derivations can be represented naturally by expressions in E.

Formal systems are commonly presented by specifying derivation rule-schemata whose instances may be used to construct a new derivation D from already constructed derivations $D_1, \ldots, D_n$. Such a rule-schema can be displayed as a figure of the form:

$$\frac{H_1 \quad \cdots \quad H_n}{J}\ N(\boldsymbol{x}_1, \ldots, \boldsymbol{x}_m)$$

where N is the name of the rule schema (which is a constant), $J, H_1, \ldots, H_n$ are judgment-schemata whose free variables stand for unspecified expressions, and $\boldsymbol{x}_1, \ldots, \boldsymbol{x}_m$ is a sequence of distinct variables containing all the free variables in $J, H_1, \ldots, H_n$. For example, we can define a formal system which defines the set of natural numbers by the two rule-schemata below.

$$\frac{}{\mathtt{0 : Nat}}\ \mathtt{zero()} \qquad \frac{\boldsymbol{n} : \mathtt{Nat}}{\mathtt{s}(\boldsymbol{n}) : \mathtt{Nat}}\ \mathtt{succ}(\boldsymbol{n})$$

We can represent this formal system by the following expression Nat.

$$\begin{aligned}
\mathsf{Nat} &:\equiv <\mathsf{zero}, \mathsf{succ}>,\\
\mathsf{zero} &:\equiv <\mathtt{0 : Nat}, <\mathtt{zero}>, <>>,\\
\mathsf{succ} &:\equiv <\mathtt{s}(\boldsymbol{n}) : \mathtt{Nat}, <\mathtt{succ}, \boldsymbol{n}>, <\boldsymbol{n} : \mathtt{Nat}>>.
\end{aligned}$$

In the formal system Nat we can construct the following derivation which shows that 2, which is represented by $\mathtt{s(s(0))}$, is a natural number.

$$\dfrac{\dfrac{\dfrac{}{\mathtt{0 : Nat}}\ \mathtt{zero()}}{\mathtt{s(0) : Nat}}\ \mathtt{succ(0)}}{\mathtt{s(s(0)) : Nat}}\ \mathtt{succ(s(0))}$$

[4] In [49], Martin-Löf introduces universal judgments, which he calls general judgments. Edinburgh LF can represent general judgments as dependent types.

The rule-schemata for Nat are *elementary* in the sense that all the premises and conclusions are categorical judgments. Such elementary rule-schemata are often sufficient for presenting syntax and operational semantics of *computational* systems. We will need non-elementary rule-schemata to present *logical* systems in natural deduction style. Consider, for example, the following rule of mathematical induction:

$$\frac{P[0] \quad \begin{matrix}(P[x]) \\ P[\mathtt{s}(x)]\end{matrix}}{\forall x.P[x]}\ \mathtt{ind}$$

When this rule is applied, the specified occurrences of the assumption $P[x]$ is discharged, and the rule may be applied only if the variable x does not occur free in any assumptions, other than $P[x]$, on which $P[\mathtt{s}(x)]$ depends. In Edinburgh LF, this rule can be defined by declaring the constant IND as follows:

$$\begin{aligned}&\text{IND} : \Pi\phi : \iota \to o.\ \mathtt{True}(\phi 0) \to \\ &\qquad (\Pi x : \iota.\mathtt{True}(\phi x) \to \mathtt{True}(\phi(\mathtt{s}(x))) \to \mathtt{True}(\forall(\lambda x : \iota.\phi x))).\end{aligned}$$

Thus, in LF, the induction rule is represented by a higher-order function that accepts proofs of the two premises $\mathtt{True}(\phi 0)$ and $\Pi x : \iota.\mathtt{True}(\phi x) \to \mathtt{True}(\phi(\mathtt{s}(x)))$, and returns a proof of the conclusion $\mathtt{True}(\forall(\lambda x : \iota.\phi x))$. The second premise is an LF encoding of a general judgment whose body is a hypothetical judgment.

In NF, the induction rule is given by the following rule-schema:

$$\frac{\boldsymbol{P}[\![\mathtt{0}]\!] : \mathtt{True} \quad (x : \mathtt{Nat})\,[\boldsymbol{P}[\![x]\!] : \mathtt{True} \Rightarrow \boldsymbol{P}[\![\mathtt{s}(x)]\!] : \mathtt{True}]}{\forall(x : \mathtt{Nat})\,[\boldsymbol{P}[\![x]\!]] : \mathtt{True}}\ \mathtt{ind}(\boldsymbol{P})$$

In this rule-schema, $\boldsymbol{P}$ is a *schematic parameter* of the rule, and when we use the rule-schema, we must *instantiate* the rule-schema by assigning an abstract expression $e \in \mathsf{E}$ to $\boldsymbol{P}$. So, if we wish to prove, for instance, the judgment

$$\forall(x : \mathtt{Nat})\,[x + x = 2 \cdot x] : \mathtt{True}$$

by applying the ind-rule, then we must assign the abstract $(x)\,[x + x = 2 \cdot x]$ to $\boldsymbol{P}$. In NF, in order to express such rule-schemata formally, we extend the set E of expressions to the set S of *schemata* by adding *schematic variables* which will be used as parameters of rule-schemata and *application schemata* which are of the form $\boldsymbol{X}[\![S]\!]$ where $\boldsymbol{X}$ is a schematic variable and S is a schema. Although elementary rule-schemata can be represented by expressions in E we use schematic variables so that rule-schemata can be presented uniformly as schemata.

In addition to the mechanism of instantiating rule-schemata, we need a general mechanism to generate valid *derivations* of judgments under a given set of rule-schemata. We will introduce the notion of *derivation game*[5] which provides

[5] We use the term 'game' to reflect the formalistic view of mathematics according to which mathematical theories are presented by spcifying the *rules* of the games (i.e., theories) which determine the derivable judgments of the theories.

such a mechanism. A derivation game G is simply a finite list of rule-schemata and hence it is an element in S. Relative to a derivation game G, the set of derivations valid in G is determined as a recursively enumerable subset of E and it is always decidable whether a given expression is a valid derivation in G or not. We can use derivation games to define various formal systems in a completely formal way.

The paper is organized as follows. In §2, we introduce a theory of expressions and schemata. In §3, we introduce three forms of judgments, namely, categorical, hypothetical and universal judgments syntactically. The set of judgments becomes a primitive recursive subset of E. We then define derivation games, which can be used to assign meanings to these judgments. In this way, one and the same judgment can have different meanings in different derivation games. NF is neutral semantically in this sense, and can define both intuitionistic and classical mathematics equally well. All the definitions and proofs in §§2–3 will be given in a finitistic fashion. Then in §§4–5, we give examples of defining computational and logical systems in NF. These examples will show the expressiveness as well as the naturalness of NF. In §6, we compare our work with related works, in particular, we compare NF with Edinburgh LF and also with the theory of expressions by Martin-Löf. Our theory of schemata is inspired not only by Martin-Löf's theory of expressions but also by Takeuti's definition of formulae in his generalized logic calculus GLC. We will describe this as well in §6. In §7, we conclude the paper by explaining metamathematical motivations for this work and suggesting possible future work.

2 Theory of Expressions and Schemata

We define the set E of *expressions* as a structure which is equipped with the following two operations. The first operation is *pair formation.* If e and f are expressions then we can form an expression $\texttt{<}e\,|\,f\texttt{>}$ which is the pair of e and f. The second operation is *abstraction.* This operation abstracts away a specified variable x from a given expression e yielding an abstract of the form $\texttt{(}x\texttt{)}\,\texttt{[}e\texttt{]}$.

We assume that we have a countably infinite set of atomic symbols which consists of two mutually disjoint infinite sets of *variables* and *constants.* Variables will also be called *object variables* to distinguish them from other kinds of variables we introduce later in this section. Furthermore, we assume that the set of object variables is a union of two infinite and mutually disjoint sets of *general variables* and *derivation variables.* Object variables will be designated by letters x, y, z and derivation variables will be designated by X, Y, Z, and constants will be designated by the letter c. We will also assume that strings of characters in `typewriter` font are included among constants and they are distinct if they are spelled differently.

Definition 1 (Context). *A sequence*

$$x_1, \ldots, x_n$$

of distinct object variables is called a context.

We will say that a variable x is *declared* in a context Γ if x occurs in Γ.

Definition 2 (Expression). *We define* expressions *relative to a context Γ as follows.*

1. Object variable. *If Γ is a context and x is declared in Γ, then x is an expression under Γ.*
2. Constant. *If Γ is a context and* c *is a constant, then* c *is an expression under Γ.*
3. Pair. *If Γ is a context and e, f are expressions under Γ, then $\texttt{<}e\texttt{|}f\texttt{>}$ is an expression under Γ.*
4. Abstract. *If Γ, x is a context and e is an expression under Γ, x, then $\texttt{(}x\texttt{)}\texttt{[}e\texttt{]}$ is an expression under Γ*

In the 4th clause of the above definition, the variable x which was (globally) declared in Γ, x becomes locally declared in the expression $\texttt{(}x\texttt{)}\texttt{[}e\texttt{]}$. The syntax $\texttt{(}x\texttt{)}\texttt{[}e\texttt{]}$ may look heavy compared to more conventional notations like $x.\ e$ or $(x)e$. However, we prefer this notation, since it clearly shows that the variable x in $\texttt{(}x\texttt{)}$ is a binding occurrence and the scope of the binder x is e.

We will identify two expressions e and f if they are defined in the same way except for the choices of variables, and we will write $e \equiv f$ in this case. Then, we can define the capture avoiding substitution operation on expressions as usual, and we will write $[x := d](e)$ for the result of substituting d for x in e. For an expression e, we can define the set $\mathrm{FV}(e)$ of *free variables* in e as usual and can verify that if e is an expression under Γ, then $\mathrm{FV}(e) \subseteq \Gamma$. An expression is *closed* if $\mathrm{FV}(e) = \emptyset$ or, equivalently, if it is an expression under the empty context.

As a notational convention, we will write

$$\texttt{<}e_1, e_2, \ldots, e_n\texttt{>} \quad \text{for} \quad \texttt{<}e_1\texttt{|}\texttt{<}e_2\texttt{|} \cdots \texttt{<}e_n\texttt{|}\texttt{nil}\texttt{>}\cdots\texttt{>}\texttt{>}.$$

Such an expression is called a *list* and we will define concatenation of two lists by:

$$\texttt{<}e_1, \ldots, e_m\texttt{>} \oplus \texttt{<}f_1, \ldots, f_n\texttt{>} :\equiv \texttt{<}e_1, \ldots, e_m, f_1, \ldots, f_n\texttt{>}.$$

We will extend this notational convention to schemata we define below. We will also identify a context $x_1, \ldots, x_n$ with the list $\texttt{<}x_1, \ldots, x_n\texttt{>}$.

We now generalize the notion of an expression to that of a *schema*. We will write S for the set of schemata. We first extend the atomic symbols by adding a countably infinite set V_s of *schematic variables*. Schematic variables are classified into two mutually disjoint subsets of *expression variables*[6] and *abstract variables*[7]. Schematic variables will be written by bold italic letters such as $\boldsymbol{z}$, $\boldsymbol{X}$ etc.

Definition 3 (Schema). *We define a* schema *relative to a context where, as before, a context is a sequence of distinct object variables.*

[6] So called since they can be instantiated to arbitrary expressions.

[7] So called since they can be instantiated to arbitrary abstracts.

1. Expression variable. *If Γ is a context and $\boldsymbol{z}$ is an expression variable, then $\boldsymbol{z}$ is a schema under Γ.*
2. Object variable. *If Γ is a context and x is declared in Γ, then x is a schema under Γ.*
3. Constant. *If Γ is a context and c is a constant, then c is a schema under Γ.*
4. Pair. *If S and T are schemata under Γ, then $\langle S \mid T \rangle$ is a schema under Γ.*
5. Abstract. *If S is a schema under $\Gamma \oplus \langle x \rangle$, then $(x)\,[S]$ is a schema under Γ*
6. Application. *If $\boldsymbol{X}$ is an abstract variable and S is a schema under Γ, then $\boldsymbol{X}\,[\![S]\!]$ is a schema under Γ.*

Any expression is a schema because of clauses 2 – 5. Here again we will identify two schemata S and T if they are defined in the same way except for the choices of binding variables in clause 5. In this case, we will write $S \equiv T$ and say that S and T are *definitionally equal.* It is clear that if $S \equiv T$ and S is an expression, then T is also an expression. A schema is *closed* if it is a schema under the empty environment. A closed schema may contain schematic variables but may not contain free object variables. Schematic variables will be used as *parameters* of rule-schemata and there is no need to abstract over schematic variables. This is the reason why we do not have a clause to bind schematic variables in the above definition. We now define a very important operation of *instantiation.*

Definition 4 (Environment and Instance). *A function*

$$\rho : \mathsf{V}_s \to \mathsf{E}$$

is an environment *if $\rho(\boldsymbol{X})$ is an abstract for any abstract variable $\boldsymbol{X}$ and $\rho(\boldsymbol{z}) \equiv (x)\,[x]$ for all but finitely many schematic variables $\boldsymbol{z}$. An environment ρ can be extended to a function $[\![\,\cdot\,]\!]_\rho : \mathsf{S} \to \mathsf{E}$ as follows.*

1. Expression variable. $[\![\boldsymbol{z}]\!]_\rho :\equiv \rho(\boldsymbol{z})$.
2. Object variable. $[\![x]\!]_\rho :\equiv x$.
3. Constant. $[\![c]\!]_\rho :\equiv c$.
4. Pair. $[\![\langle S \mid T \rangle]\!]_\rho :\equiv \langle [\![S]\!]_\rho \mid [\![T]\!]_\rho \rangle$.
5. Abstract. $[\![(x)\,[S]]\!]_\rho :\equiv (x)\,[[\![S]\!]_\rho]$. *(We assume, without loss of generality, that for any $\boldsymbol{z}$ occurring in S, x is not free in $\rho(\boldsymbol{z})$.)*
6. Application. $[\![\boldsymbol{X}\,[\![S]\!]\,]\!]_\rho \equiv [x := [\![S]\!]_\rho](e)$, *where $\rho(\boldsymbol{X}) \equiv (x)\,[e]$.*

The expression $[\![S]\!]_\rho$ is called the instantiation of S by ρ *or an* instance of S.

We see from clause 6 that an *application* $\boldsymbol{X}\,[\![S]\!]$ is instantiated by a call-by-value manner.

3 Judgments and Derivation Games

Based on the theory of expressions and schemata, we can describe the syntax of *judgment* as follows. In this way, we can characterize the possible *form* a judgment can take. A *general variable context* is a context such that all the variables declared in it are general variables. In the following definition, Γ stands for an arbitrary general variable context.

Definition 5 (Judgment). Judgments *are defined relative to general variable contexts and thereby classified into the following three types.*

1. Categorical Judgment. *If s and P are expressions under Γ, then*

$$s : P$$

   is a judgment under Γ.[8] *In this case, s (P, resp.) is called the* object part (concept part, *resp.) of the categorical judgment.*
2. Hypothetical Judgment. *If H and J are judgments under Γ, then*

$$H \Rightarrow J$$

   is a judgment under Γ.[9]
3. Universal Judgment. *If x is a general variable not declared in Γ, and J is an expression under $\Gamma \oplus \texttt{<}x\texttt{>}$, then*

$$(x)\,[J]$$

   is a judgment under Γ.

One can verify that if J is a judgment under a general variable context Γ, then J is an expression under Γ. We can explain the meanings of judgments informally as follows. A categorical judgment $s : P$ means that the object s falls under the *concept* P. When P is a constant, it is possible to regard P as a *name* of the concept which is characterized by those s for which the judgment $s : P$ holds. For example, if `Nat` is a name given to the concept of the natural number, then `0` : `Nat` means that `0` is a natural number. A hypothetical judgment $H \Rightarrow J$ means that we can derive the judgment J whenever H is derivable. A universal judgment of the form $(x)\,[J]$ means that we can derive the judgment $[x := e](J)$ for any expression e. Since we will often consider judgments of the form $(x)\,[x : P \Rightarrow J]$, we will abbreviate this form by $(x : P)\,[J]$.

Given a context Γ, we put J_Γ to be the set of judgments under Γ. It is easy to see that J_Γ is a primitive recursive subset of E.

Based on the above informal explanation of the intended meanings of judgments, we introduce the notion of derivation game.

Definition 6 (Rule-Schema). *A closed schema of the form:*

$$\texttt{<RS}, J, \texttt{<}c, z_1, \ldots, z_m\texttt{>}, \texttt{<}H_1, \ldots, H_n\texttt{>>}$$

is a rule-schema *if c is a constant (called the* name *of the rule-schema) and $z_1, \ldots, z_m$ is a list (called the* parameter list *of the rule-schema) of distinct schematic variables exactly covering all the schematic variables in the schemata $J, H_1, \ldots, H_n$.*

[8] This is an abbreviation of the expression: $\texttt{<}:, s, P\texttt{>}$.

[9] This is an abbreviation of the expression: $\texttt{<}\Rightarrow, H, J\texttt{>}$.

A rule-schema can be displayed as a figure of the following form:

$$\frac{H_1 \quad \cdots \quad H_n}{J} \; R(z_1, \ldots, z_m)$$

We will also use the following Prolog-like notation for the same rule-schema:

$$J \text{ :- }_{R(z_1,\ldots,z_n)} H_1, \ldots, H_n.$$

We will often omit the rule name and its parameter list in these notations; or we keep the rule name only and omit the parameter list; or we keep the rule name and the list of abstract variables and omit the expression variables from the parameter list. If $J, H_1, \ldots, H_n$ are all of the form $s : c$ where $s \in \mathsf{S}$ and c is a constant, then the rule-schema is said to be an *elementary rule schema.*

Definition 7 (Derivation Game). *A* derivation game *is a schema of the form* $<R_1, \ldots, R_n>$ *where each* R_i *is a rule-schema.*

A derivation game is a closed schema since a rule-schema is closed. We write G for the set of derivation games. The set G is a primitive recursive subset of S. A derivation game is *elementary* if each rule-schema of the game is elementary.

A derivation game G determines a set $\mathsf{D}(G)$ of expressions called the set of *G-derivations*, which is the set of derivations derivable in G. To define $\mathsf{D}(G)$, we must first define the notion of a derivation context.

Definition 8 (Derivation Context). *We define a* derivation context Γ *together with its* general variable part $\mathrm{GV}(\Gamma)$ *which is a context in the sense of Definition 1.*

1. Empty context. *The empty list* $<>$ *is a derivation context and its general variable part is* $<>$.
2. General variable declaration. *If* Γ *is a derivation context, and* x *is a general variable not declared in* Γ, *then* $\Gamma \oplus <x>$ *is a derivation context and its general variable part is* $\mathrm{GV}(\Gamma) \oplus <x>$.
3. Derivation variable declaration. *If* Γ *is a derivation context,* J *is a judgment under* $\mathrm{GV}(\Gamma)$, *and* X *is a derivation variable not declared in* Γ, *then* $\Gamma \oplus <X :: J$[10]$>$ *is a derivation context and its general variable part is* $\mathrm{GV}(\Gamma)$.

Let Γ be a derivation context. An environment ρ is a *Γ-environment* if $\rho(z)$ is an expression under $\mathrm{GV}(\Gamma)$ for any schematic variable z.

Definition 9 (G-derivation). *We define a G-*derivation *(also called a* derivation in G*) relative to a derivation context* Γ *as follows. We define its* conclusion *at the same time. In the following definition,* Γ *stands for an arbitrary derivation context. We can see from the definition below, that if* D *is a G-derivation under* Γ, *then its conclusion is an expression under* $\mathrm{GV}(\Gamma)$.

[10] $X :: J$ is an abbreviation of $<::, X, J>$.

1. Derivation variable. *If X is a derivation variable and $X :: H$ is in Γ, then*

$$X$$

is a G-derivation under Γ and its conclusion is H.

2. Composition. *If $D_1, \ldots, D_n$ are G-derivations under Γ such that their conclusions are $H_1, \ldots, H_n$, respectively, and*

$$\frac{H_1 \quad \cdots \quad H_n}{J} \; R(e_1, \ldots, e_m)$$

is an instantiation of a rule-schema in G by a Γ-environment and J is a categorical judgment under $\mathrm{GV}(\Gamma)$, *then*

$$\frac{D_1 \quad \cdots \quad D_n}{J} \; R(e_1, \ldots, e_m)^{11}$$

is a G-derivation and its conclusion is J.

3. Hypothetical derivation. *If D is a G-derivation under $\Gamma \oplus \texttt{<}X :: H\texttt{>}$ and its conclusion is J, then*

$$\texttt{(}X :: H\texttt{)}\,\texttt{[}D\texttt{]}^{12}$$

is a G-derivation under Γ and its conclusion is $H \Rightarrow J^{13}$.

4. Universal derivation. *If D is a G-derivation under $\Gamma \oplus \texttt{<}x\texttt{>}$ and its conclusion is J, then*

$$\texttt{(}x\texttt{)}\,\texttt{[}D\texttt{]}$$

is a G-derivation under Γ and its conclusion is $\texttt{(}x\texttt{)}\,\texttt{[}J\texttt{]}$.

We will write

$$\Gamma \vdash_G D :: J$$

if D is a derivation in G under Γ whose conclusion is J. For example, for the game Nat we gave in §1 and for the derivation D given there, we have

$$\vdash_{\mathsf{Nat}} D :: \texttt{s(s(0))} : \texttt{Nat}.$$

NF provides another notation which is conveniently used to input and display derivations on a computer terminal. In this notation, instead of writing $\Gamma \vdash_G D :: J$ we write:

$$\Gamma \vdash J \;\texttt{in}\; G \;\texttt{since}\; D.$$

Also, when writing derivations in this notation, a derivation of the form

$$\frac{D_1 \quad \cdots \quad D_n}{J} \; R(e_1, \ldots, e_m)$$

---

[12] This is an abbreviation of the expression: $\texttt{<HD}, H, \texttt{(}X\texttt{)}\,\texttt{[}D\texttt{]>}$.

[13] We sometimes write the hypothetical judgment $H \Rightarrow J$ also by $\texttt{(}X :: H\texttt{)}\,\texttt{[}J\texttt{]}$ to emphasize the intended meaning of the judgment; namely, we can derive J if we have a derivation X of H. This notation also matches well with the notation for the hypothetical derivation we introduced here.

will be written as:

$$J \text{ by } R(e_1, \ldots, e_m) \; \{D_1; \ldots; D_n\}$$

Here is a complete derivation in Nat in this notation.

```
⊢ (x)[x:Nat ⇒ s(s(x)):Nat] in Nat since
(x)[(X::x:Nat)[
    s(s(x)):Nat by succ(s(x)) {
        s(x):Nat by succ(x) {X}
    }
]]
```

The conclusion of the above derivation asserts that for any expression x, if x is a natural number, then so is s(s(x)), and the derivation shows us how to actually construct a derivation of s(s(x)):Nat given a derivation X of x:Nat.

There are certain syntactic structures that are common among G-derivations for all derivation games G. To capture such structures, we introduce the notion of derivation.

Definition 10 (Derivation). *We define a* derivation *and its* conclusion *relative to a derivation context as follows. In the following definition, Γ stands for an arbitrary derivation context.*

1. Derivation variable. *If X is a derivation variable and $X :: H$ is in Γ, then*
$$X$$
is a derivation under Γ and its conclusion is H.
2. Composition. *If $D_1, \ldots, D_n$ are derivations under Γ, R is a constant, $e_1, \ldots, e_m$ are expressions under Γ, and J is a categorical judgment under $\mathrm{GV}(\Gamma)$, then*
$$\frac{D_1 \quad \cdots \quad D_n}{J} \; R(e_1, \ldots, e_m)$$
is a derivation under Γ and its conclusion is J.
3. Hypothetical derivation. *If D is a derivation under $\Gamma \oplus \langle X :: H \rangle$ and its conclusion is J, then*
$$(X :: H)\,[D]$$
is a derivation under Γ and its conclusion is $H \Rightarrow J$.
4. Universal derivation. *If D is a G-derivation under $\Gamma \oplus \langle x \rangle$ and its conclusion is J, then*
$$(x)\,[D]$$
is a derivation under Γ and its conclusion is $(x)\,[J]$.

We will write

$$\Gamma \vdash D :: J$$

if Γ is a derivation context and D is a derivation under Γ whose conclusion is J. We have the following theorem which characterizes derivations in terms of G-derivations.

Theorem 1 (Characterization of Derivations). *For any expressions Γ, D and J, we have $\Gamma \vdash D :: J$ if and only if $\Gamma \vdash_G D :: J$ for some derivation game G.*

Proof. Since *if-part* is trivial, we only show the *only-if-part* by induction on the construction of D. The only non-trivial case is the case where D is a derivation under Γ of the form:

$$\frac{D_1 \quad \cdots \quad D_n}{J} \; R(e_1, \ldots, e_m)$$

We let J_i be the conclusion of D_i for all i such that $1 \leq i \leq n$. In this case, we may assume by induction hypothesis, that we have derivation games G_i $(1 \leq i \leq n)$ such that $\Gamma \vdash_{G_i} D_i :: J_i$ $(1 \leq i \leq n)$. Then, we can obtain the derivation game with the required property by putting:

$$G :\equiv G_1 \oplus \cdots \oplus G_n \oplus \mathtt{<<RS}, \mathtt{<}R, e_1, \ldots, e_m\mathtt{>}, \mathtt{<}J_1, \ldots, J_n\mathtt{>>>}.$$

Note that the last rule-schema of G is actually an expression and not a proper schema. □

Given a context Γ, we let D_Γ be the set of derivations under Γ. For any $e \in \mathsf{E}$, we can decide whether $\Gamma \vdash e :: J$ for some J by induction on the construction of $e \in \mathsf{E}$. So, we see that D_Γ is a primitive recursive subset of E, and there is a primitive recursive function:

$$\mathrm{conc}_\Gamma : \mathsf{D}_\Gamma \to \mathsf{J}_\Gamma$$

such that $\Gamma \vdash_G D :: \mathrm{conc}_\Gamma(D)$ holds for all $D \in \mathsf{D}_\Gamma$. Then, by the characterization theorem of derivations, we see that if $\Gamma \vdash_G D :: J$, then $D \in \mathsf{D}_\Gamma$ and $J \equiv \mathrm{conc}_\Gamma(D) \in \mathsf{J}_\Gamma$.

Theorem 2 (Decidability). *If $G \in \mathsf{S}$ and $\Gamma, D, J \in \mathsf{E}$, then it is primitive recursively decidable whether $\Gamma \vdash_G D :: J$ or not.*

Proof. By induction on the construction of D as an element of E. □

For a derivation game G, we let $\mathsf{D}(G)$ be the set of G-derivations under the empty context. Then, we have the following corollary which fulfills Kreisel's dictum.

Corollary 1. *For any derivation game G, $\mathsf{D}(G)$ is a primitive recursive subset of E.*

Proof. Suppose that an element $e \in \mathsf{E}$ is given. If $e \notin \mathsf{D}_{\mathtt{<>}}$ then, by the characterization theorem, $e \notin \mathsf{D}(G)$. If $e \in \mathsf{D}_{\mathtt{<>}}$, then by the Decidability Theorem 2, we can decide $e \in \mathsf{D}(G)$ or not by checking $\mathtt{<>} \vdash_G e :: \mathrm{concl}(e)$ or not. □

We can also check the following properties of derivations.

Proposition 1. *If $\Gamma \vdash D :: J$, then any free variable in J is declared in* $\mathrm{GV}(\Gamma)$, *and any free variable in D is declared in Γ.*

Derivation games enjoy the following fundamental properties.

Theorem 3. *The following properties hold for any derivation game G.*

1. *Weakening. If $\Gamma \vdash_G D :: J$ and $\Gamma \oplus \Gamma'$ is a context, then $\Gamma \oplus \Gamma' \vdash_G D :: J$.*
2. *Strengthening for general variable. If $\Gamma \oplus \texttt{<}x\texttt{>} \oplus \Gamma' \vdash_G D :: J$, and $x \notin \mathrm{FV}(\Gamma') \cup \mathrm{FV}(D) \cup \mathrm{FV}(J)$, then $\Gamma \oplus \Gamma' \vdash_G D :: J$.*
3. *Strengthening for derivation variable. If $\Gamma \oplus \texttt{<}X :: H\texttt{>} \oplus \Gamma' \vdash_G D :: J$, and $X \notin \mathrm{FV}(D)$, then $\Gamma \oplus \Gamma' \vdash_G D :: J$.*
4. *Substitution for derivation variable. If $\Gamma \oplus \texttt{<}X :: H\texttt{>} \oplus \Gamma' \vdash_G D :: J$ and $\Gamma \vdash_G D' :: H$, then $\Gamma \oplus \Gamma' \vdash_G [X := D'](D) :: J$.*
5. *Substitution for general variable. If $\Gamma \oplus \texttt{<}x\texttt{>} \oplus \Gamma' \vdash_G D :: J$, and $e \in \mathsf{E}$, then $\Gamma \oplus [x := e](\Gamma') \vdash_G [x := e](D) :: [x := e](J)$.*
6. *Exchange. If $\Gamma \oplus \texttt{<}e, f\texttt{>} \oplus \Gamma' \vdash_G D :: J$, and $\Gamma \oplus \texttt{<}f, e\texttt{>} \oplus \Gamma'$ is a derivation context, then $\Gamma \oplus \texttt{<}f, e\texttt{>} \oplus \Gamma' \vdash_G D :: J$.*

These basic properties of derivations and G-derivations imply that it is possible to implement a system on a computer that can manipulate these symbolic expressions and decide the correctness of derivations. At Kyoto University we have been developing a computer environment called CAL (for Computation And Logic) [73] which realizes this idea.

4 Computational Aspects of NF

We show the expressiveness of elementary derivation games by giving a complete description of pure Lisp as described in McCarthy [51] including its syntax and operational semantics. By the following examples, we can see that elementary derivation games contain Horn clause logic and Smullyan's elementary formal systems [79] as special cases.

4.1 Syntax of Pure Lisp Expressions

The syntax of pure Lisp expressions is extremely simple owing to its simple and elegant structures. We will define pure Lisp expressions formally by the derivation game PureLispExp which is a list consisting of the following 42 elementary rule-schemes.[14] The derivation game PureLispExp is defined as the concatenation of five games:

$$\mathsf{PureLispExp} :\equiv \mathsf{Alpha} \oplus \mathsf{Num} \oplus \mathsf{Char} \oplus \mathsf{Atom} \oplus \mathsf{Sexp}.$$

Lisp's symbolic expressions (`Sexp`) are constructed from atomic symbols (`Atom`) by means of the *cons* operation which constructs a cons pair of `Sexp`s out of two already constructed `Sexp`s. It should be clear that our expressions E are inspired from Lisp's symbolic expressions.

Lisp's atomic symbols are constructed by upper case alphabetic characters (`Alpha`) and numeric characters (`Num`). Therefore, we define `Alpha` and `Num` as follows.

[14] Among the 42 rules, 36 are rules for 36 characters used in pure Lisp.

Alpha.

$$\frac{}{\texttt{A} : \texttt{Alpha}} \qquad \frac{}{\texttt{B} : \texttt{Alpha}} \qquad \cdots \qquad \frac{}{\texttt{Z} : \texttt{Alpha}}$$

Num.

$$\frac{}{\texttt{0} : \texttt{Num}} \qquad \frac{}{\texttt{1} : \texttt{Num}} \qquad \cdots \qquad \frac{}{\texttt{9} : \texttt{Num}}$$

Char. We define a character as either an $\texttt{Alpha}$ or a $\texttt{Num}$:

$$\frac{c : \texttt{Alpha}}{c : \texttt{Char}} \qquad \frac{c : \texttt{Num}}{c : \texttt{Char}}$$

These definitions are clearly equivalent to the following definitions by BNF notation.

$$\begin{aligned}
\langle\texttt{Alpha}\rangle &::= \texttt{A} \mid \texttt{B} \mid \texttt{C} \mid \texttt{D} \mid \texttt{E} \mid \texttt{F} \mid \texttt{G} \mid \texttt{H} \mid \texttt{I} \mid \texttt{J} \mid \texttt{K} \mid \texttt{L} \mid \texttt{M} \mid \\
&\quad\;\; \texttt{N} \mid \texttt{O} \mid \texttt{P} \mid \texttt{Q} \mid \texttt{R} \mid \texttt{S} \mid \texttt{T} \mid \texttt{U} \mid \texttt{V} \mid \texttt{W} \mid \texttt{X} \mid \texttt{Y} \mid \texttt{Z} \\
\langle\texttt{Num}\rangle &::= \texttt{0} \mid \texttt{1} \mid \texttt{2} \mid \texttt{3} \mid \texttt{4} \mid \texttt{5} \mid \texttt{6} \mid \texttt{7} \mid \texttt{8} \mid \texttt{9} \\
\langle\texttt{Char}\rangle &::= \langle\texttt{Alpha}\rangle \mid \langle\texttt{Num}\rangle
\end{aligned}$$

Atom. An atomic symbol is a sequence of $\texttt{Char}$s whose first character is an $\texttt{Alpha}$:

$$\frac{c : \texttt{Alpha}}{c : \texttt{Atom}} \qquad \frac{a : \texttt{Atom} \quad c : \texttt{Char}}{\langle a \,|\, c\rangle : \texttt{Atom}}$$

For example, we have

$$\vdash_{\mathsf{PureLispExp}} \langle\texttt{E}\,|\,\texttt{Q}\rangle : \texttt{Atom}$$

and we can convince ourself of this fact by looking at the following derivation:

$$\frac{\dfrac{\dfrac{}{\texttt{E} : \texttt{Alpha}}}{\texttt{E} : \texttt{Atom}} \quad \dfrac{\dfrac{}{\texttt{Q} : \texttt{Alpha}}}{\texttt{Q} : \texttt{Char}}}{\langle\texttt{E}\,|\,\texttt{Q}\rangle : \texttt{Atom}}$$

For the sake of readability, we will write EQ for $\langle\texttt{E}\,|\,\texttt{Q}\rangle$, and will use the same convention for other $\texttt{Atom}$s. For example, CAR stands for $\langle\langle\texttt{C}\,|\,\texttt{A}\rangle\,|\,\texttt{R}\rangle$.

Sexp. Now, we can define Lisp's symbolic expressions. In the following definition, we follow Lisp's convention and write $(s\,.\,t)$ for $\langle\texttt{cons}, s, t\rangle$. We will also use the *list notation* which abbreviates, e.g., $(s\;.\;(t\;.\;(u\;.\;\mathsf{NIL})))$ to $(s\ t\ u)$[15].

$$\frac{a : \texttt{Atom}}{a : \texttt{Sexp}} \qquad \frac{s : \texttt{Sexp} \quad t : \texttt{Sexp}}{\langle\texttt{cons}, s, t\rangle : \texttt{Sexp}}$$

[15] In [51] it is written as (s, t, u), but, instead, we will use the now common notation.

4.2 Equality on Lisp Atoms

Two `Atoms` a and b are *equal* if they are constructed in the same way by means of the rule-schemes given in the game PureLispExp. We can decide the equality as follows. Since we know that a and b are `Atoms`, we have derivations D_a and D_b such that:

$$\vdash_{\mathsf{PureLispExp}} D_a :: a : \mathtt{Atom} \quad \text{and} \quad \vdash_{\mathsf{PureLispExp}} D_b :: b : \mathtt{Atom}.$$

So, we can decide the equality by deciding the definitional equality of D_a and D_b. We can internalize the process of deciding $D_a \equiv D_b$ as follows. We first define games to decide equality of two characters and use them to decide equality of two atoms.

In our framework, an n-ary relation R among expressions $e_1, \ldots, e_n$ can be represented by the categorical judgment $\mathtt{<}e_1, \ldots, e_n\mathtt{>} : R$. We will also write this judgment as $R(e_1, \ldots, e_n)$ which is a common notation for asserting that the relation R holds for $e_1, \ldots, e_n$.

EqAlpha. Equality of two `Alphas` is defined by the following 26 rules:

$$\frac{}{\mathtt{EqAlpha(A,A)}} \qquad \frac{}{\mathtt{EqAlpha(B,B)}} \qquad \cdots \qquad \frac{}{\mathtt{EqAlpha(Z,Z)}}$$

NeqAlpha. Inequality of two `Alphas` is defined by the following 650 rules:

$$\frac{}{\mathtt{NeqAlpha(A,B)}} \qquad \frac{}{\mathtt{NeqAlpha(A,C)}} \qquad \cdots \qquad \frac{}{\mathtt{NeqAlpha(Z,Y)}}$$

EqNum. Equality of two `Nums` is defined by the following 10 rules:

$$\frac{}{\mathtt{EqNum(0,0)}} \qquad \frac{}{\mathtt{EqNum(1,1)}} \qquad \cdots \qquad \frac{}{\mathtt{EqNum(9,9)}}$$

NeqNum. Inequality of two `Nums` is defined by the following 90 rules:

$$\frac{}{\mathtt{NeqNum(0,1)}} \qquad \frac{}{\mathtt{NeqNum(0,2)}} \qquad \cdots \qquad \frac{}{\mathtt{NeqNum(9,8)}}$$

EqChar. Equality of two characters is defined as follows.

$$\frac{\mathtt{EqAlpha}(\boldsymbol{c}, \boldsymbol{d})}{\mathtt{EqChar}(\boldsymbol{c}, \boldsymbol{d})} \qquad \frac{\mathtt{EqNum}(\boldsymbol{c}, \boldsymbol{d})}{\mathtt{EqChar}(\boldsymbol{c}, \boldsymbol{d})}$$

NeqChar. Inequality of two characters is defined as follows.

$$\frac{\mathtt{NeqAlpha}(\boldsymbol{c}, \boldsymbol{d})}{\mathtt{NeqChar}(\boldsymbol{c}, \boldsymbol{d})} \qquad \frac{\mathtt{NeqNum}(\boldsymbol{c}, \boldsymbol{d})}{\mathtt{NeqChar}(\boldsymbol{c}, \boldsymbol{d})} \qquad \frac{\boldsymbol{c} : \mathtt{Alpha} \quad \boldsymbol{d} : \mathtt{Num}}{\mathtt{NeqChar}(\boldsymbol{c}, \boldsymbol{d})} \qquad \frac{\boldsymbol{c} : \mathtt{Num} \quad \boldsymbol{d} : \mathtt{Alpha}}{\mathtt{NeqChar}(\boldsymbol{c}, \boldsymbol{d})}$$

EqAtom. We can define two equal atoms as follows:

$$\frac{\texttt{EqAlpha}(c,d)}{\texttt{EqAtom}(c,d)} \qquad \frac{\texttt{EqAtom}(a,b) \quad \texttt{EqChar}(c,d)}{\texttt{EqAtom}(\langle a \mid c \rangle, \langle b \mid d \rangle)}$$

NeqAtom. Inequality of two atoms is defined as follows.

$$\frac{\texttt{NeqAlpha}(c,d)}{\texttt{NeqAtom}(c,d)}$$

$$\frac{a : \texttt{Atom} \quad c : \texttt{Char} \quad d : \texttt{Alpha}}{\texttt{NeqAtom}(\langle a \mid c \rangle, d)} \qquad \frac{c : \texttt{Alpha} \quad b : \texttt{Atom} \quad d : \texttt{Char}}{\texttt{NeqAtom}(c, \langle b \mid d \rangle)}$$

$$\frac{\texttt{NeqAtom}(a,b) \quad c : \texttt{Char} \quad d : \texttt{Char}}{\texttt{NeqAtom}(\langle a \mid c \rangle, \langle b \mid d \rangle)} \qquad \frac{\texttt{EqAtom}(a,b) \quad \texttt{NeqChar}(c,d)}{\texttt{NeqAtom}(\langle a \mid c \rangle, \langle b \mid d \rangle)}$$

4.3 Lisp Forms

In pure Lisp, forms are expressions that may be evaluated under a given environment, and they are syntactically represented by the so-called *meta-expressions*. Here, we define forms directly as Sexps without using meta-expressions.

Form. Lisp forms consist of constants, variables, five forms for applications of the five Lisp primitive functions, function applications and conditional forms.

```
e : Form :- e : Const.
e : Form :- e : Var.
e : Form :- e : ConsF.
e : Form :- e : CarF.
e : Form :- e : CdrF.
e : Form :- e : AtomF.
e : Form :- e : EqF.
e : Form :- e : AppF.
e : Form :- e : Cond.
```

Const.

```
(QUOTE s) : Const :- s : Sexp.
```

Var.

```
s : Var :- s : Atom.
```

ConsF, CarF, CdrF, AtomF and EqF. These forms are evaluated by the Lisp primitive functions *cons*, *car*, *cdr*, *atom* and *eq*, respectively.

$$(\mathsf{CONS}\ \boldsymbol{e1}\ \boldsymbol{e2}) : \mathtt{ConsF}\ \text{:-}\ \boldsymbol{e1} : \mathtt{Form}, \boldsymbol{e2} : \mathtt{Form}.$$

$$(\mathsf{CAR}\ \boldsymbol{e}) : \mathtt{CarF}\ \text{:-}\ \boldsymbol{e} : \mathtt{Form}.$$

$$(\mathsf{CDR}\ \boldsymbol{e}) : \mathtt{CdrF}\ \text{:-}\ \boldsymbol{e} : \mathtt{Form}.$$

$$(\mathsf{ATOM}\ \boldsymbol{e}) : \mathtt{AtomF}\ \text{:-}\ \boldsymbol{e} : \mathtt{Form}.$$

$$(\mathsf{EQ}\ \boldsymbol{e1}\ \boldsymbol{e2}) : \mathtt{EqF}\ \text{:-}\ \boldsymbol{e1} : \mathtt{Form}, \boldsymbol{e2} : \mathtt{Form}.$$

AppF, Fun, FormList and VarList. The predicate AppF checks if a given Sexp is of the form of a function applied to a list of forms.

$$(\boldsymbol{f}\ .\ \boldsymbol{E}) : \mathtt{AppF}\ \text{:-}\ \boldsymbol{f} : \mathtt{Fun}, \boldsymbol{E} : \mathtt{FormList}.$$

$$\boldsymbol{f} : \mathtt{Fun}\ \text{:-}\ \boldsymbol{f} : \mathtt{Var}, \mathtt{NeqAtom}(\boldsymbol{f}, \mathsf{QUOTE}), \mathtt{NeqAtom}(\boldsymbol{f}, \mathsf{CONS}),$$
$$\mathtt{NeqAtom}(\boldsymbol{f}, \mathsf{CAR}), \mathtt{NeqAtom}(\boldsymbol{f}, \mathsf{CDR}), \mathtt{NeqAtom}(\boldsymbol{f}, \mathsf{ATOM}), \mathtt{NeqAtom}(\boldsymbol{f}, \mathsf{EQ}).$$
$$(\mathsf{LAMBDA}\ \boldsymbol{X}\ \boldsymbol{e}) : \mathtt{Fun}\ \text{:-}\ \boldsymbol{X} : \mathtt{VarList}, \boldsymbol{e} : \mathtt{Form}.$$
$$(\mathsf{LABEL}\ \boldsymbol{x}\ \boldsymbol{f}) : \mathtt{Fun}\ \text{:-}\ \boldsymbol{x} : \mathtt{Var}, \boldsymbol{f} : \mathtt{Fun}.$$

$$() : \mathtt{FormList}\ \text{:-}\ .$$
$$(\boldsymbol{e}\ .\ \boldsymbol{E}) : \mathtt{FormList}\ \text{:-}\ \boldsymbol{e} : \mathtt{Var}, \boldsymbol{E} : \mathtt{FormList}.$$

$$() : \mathtt{VarList}\ \text{:-}\ .$$
$$(\boldsymbol{x}\ .\ \boldsymbol{X}) : \mathtt{VarList}\ \text{:-}\ \boldsymbol{x} : \mathtt{Var}, \boldsymbol{X} : \mathtt{VarList}.$$

Cond and CList.

$$(\mathsf{COND}\ .\ \boldsymbol{L}) : \mathtt{Cond}\ \text{:-}\ \boldsymbol{L} : \mathtt{CList}.$$

$$() : \mathtt{CList}\ \text{:-}\ .$$
$$((\boldsymbol{c}\ \boldsymbol{e})\ .\ \boldsymbol{L}) : \mathtt{CList}\ \text{:-}\ \boldsymbol{c} : \mathtt{Form}, \boldsymbol{e} : \mathtt{Form}, \boldsymbol{L} : \mathtt{CList}.$$

4.4 Evaluation of Lisp Forms

We are ready to define operational semantics of pure Lisp. We define the semantics as a partial function

$$\mathsf{Eval} : \mathsf{Form} \times \mathsf{Env} \to \mathsf{Sexp}$$

where Env realizes an environment as a list of variable-value pair. The overall evaluation strategy is call-by-value.

Eval **and** Ev. The categorical judgment Eval(e, a, v) means that the form e evaluated under the environment a has value v.

```
Eval(e, a, v) :- e:Form, a:Env, Ev(e, a, v).

Ev(e, a, v) :- e:Const, EvConst(e, a).
Ev(e, a, v) :- e:Var, EvVar(e, a, v).
Ev(e, a, v) :- e:ConsF, EvCons(e, a, v).
Ev(e, a, v) :- e:CarF, EvCar(e, a, v).
Ev(e, a, v) :- e:CdrF, EvCdr(e, a, v).
Ev(e, a, v) :- e:AtomF, EvAtom(e, a, v).
Ev(e, a, v) :- e:EqF, EvEq(e, a, v).
Ev(e, a, v) :- e:AppF, EvAppF(e, a, v).
Ev(e, a, v) :- e:Cond, EvCondF(e, a, v).
```

Env. An environment is a list of variable-value pairs known as an *association list.*

```
() : Env :- .
((x . v) . a) : Env :- x : Var, v : Sexp, a : Env.
```

EvConst.

```
EvConst((QUOTE v), v) :- .
```

EvVar **and** Assoc. The value of a variable is computed by looking up the environment using the function Assoc.

```
EvVar(x, a, v) :- Assoc(x, a, v).

Assoc(x, ((x . v) . a), v).
Assoc(x, ((y . w) . a), v) :- NeqAtom(x, y), Assoc(x, a, v).
```

EvCons. *Cons* of s and t is (s . t), so we have the following rule.

```
EvCons((CONS e1 e2), a, (v1 . v2)) :- Ev(e1, a, v1), Ev(e2, a, v2).
```

EvCar **and** EvCdr. *Car* of (s . t) is s and *cdr* of (s . t) is t:

```
EvCar((CAR e), a, v1) :- Ev(e, a, (v1 . v2)).

EvCdr((CDR e), a, v2) :- Ev(e, a, (v1 . v2)).
```

EvAtom. The Lisp primitive function *atom* decides whether a given Sexp is an Atom or not, and returns T (true) if it is an Atom and F[16] (false) if it is not an Atom. Since each of the two rule-schemes in the game Sexp corresponds to one of these cases, we have the following rules.

```
EvAtom((ATOM e), a, T) :- Ev(e, a, v), v : Atom.
EvAtom((ATOM e), a, F) :- Ev(e, a, (v1 . v2)).
```

[16] Modern Lisp systems represent false by NIL.

EvEq. The Lisp primitive function *eq* decides whether given two Sexps are equal Atoms. We can use the games EqAtom and NeqAtom to implement the *eq* function.

```
EvEq((EQ e1 e2),a,T) :- Ev(e1,a,v1),Ev(e2,a,v2),EqAtom(v1,v2).
EvEq((EQ e1 e2),a,F) :- Ev(e1,a,v1),Ev(e2,a,v2),NeqAtom(v1,v2).
```

EvAppF, EvArgs, Apply and PairUp.

```
EvAppF((f . E),a,v) :- EvArgs(E,a,V),Apply(f,V,a,v).

EvArgs((),a,()) :- .
EvArgs((e . E),a,(v . V)) :- Ev(e,a,v),EvArgs(E,a,V).

Apply(x,V,a,v) :- x:Var,Assoc(x,a,f),f:Fun,Apply(f,V,a,v).
Apply((LAMBDA X e),V,a,v) :- PairUp(X,V,a,b),Ev(e,b,v).
Apply((LABEL x f),V,a,v) :- Apply(f,V,((x . f) . a),v).

PairUp((),(),a,a) :- .
PairUp((x . X),(v . V),a,((x . v) . b)) :- PairUp(X,V,a,b).
```

EvCondF and EvCond.

```
EvCondF((COND . L),a,v) :- EvCond(L,a,v).

EvCond((),a,()).
EvCond(((c e) . L),a,v) :- Ev(c,a,T),Ev(e,a,v).
EvCond(((c e) . L),a,v) :- Ev(c,a,F),EvCond(L,a,v).
```

4.5 Comparison with Other Approaches

The definition of the syntax and the operational semanitics of pure Lisp we just gave is basically along the line of specifying programming languages by structural operational semanitcs. For example, we introduced a language called Hyperlisp in [65, 66] and defined its semantics in a similar way as above. We also pointed out in [67] that the definitions given by such an approach can be regarded as PROLOG-like programs. Furthermore, we pointed out in [67] that Smullyan's elementary formal system [79] provides a similar framework for defining formal systems.

Elementary derivation games, by themselves, are as useful as other approaches such as PROLOG or elementary formal systems. The true advantage of our aporoach is that an elementary derivation game can be combined in a seemless way with other derviation games that embody logical and mathematical systems so that, in the combined game, we can formally reason about and even develop programs defined by the elementary derivation game. In other approaches, such a task can only be carried out by developing a meta-system which is quite different from the original systems.

5 Logical Aspects of NF

In this section, we show that we can naturally define various logical systems by means of derivation games. The naturalness of the definition owes very much to the facts that the discharging of an assumption is simply expressed by using a hypothetical judgment as a premise of rule-schemata and that the so-called eigen-variable condition on *local* variables in a derivation is automatically guaranteed by using an application schema in a premise of rule-schemata.

triv. Before we proceed, consider the following trivial rule:

$$\frac{J}{J}\ \texttt{triv}$$

It is clear that for any game G, the derivable judgments in G and $G \oplus \mathsf{triv}$ are the same. Therefore, we will assume that any game contains the `triv` rule-schema. It may seem that the rule is not only harmless but also useless. However, this rule is sometimes useful in practice, since, as we see below later we can use this rule to remind us of the judgment which a derivation variable or a general derivation proves.

5.1 Intuitionistic and Classical Propositional Logics

We first define intuitionistic and classical propositional logics as derivation games.

Prop. We define formulas of propositional logic as follows. The judgment $p : \texttt{Prop}$ means that p is a formula. The derivation game Prop is elementary.

$$\begin{aligned}
\bot : \texttt{Prop} &\ :\!-_{\bot\mathrm{F}}\ .\\
p \supset q : \texttt{Prop} &\ :\!-_{\supset\mathrm{F}}\ p : \texttt{Prop}, q : \texttt{Prop}.\\
p \wedge q : \texttt{Prop} &\ :\!-_{\wedge\mathrm{F}}\ p : \texttt{Prop}, q : \texttt{Prop}.\\
p \vee q : \texttt{Prop} &\ :\!-_{\vee\mathrm{F}}\ p : \texttt{Prop}, q : \texttt{Prop}.
\end{aligned}$$

True. We define provable formulas of propositional logic as follows. The game True is non-elementary because $\supset\mathtt{I}$ rule-schema and $\vee\mathtt{E}$ rule-schema below contain hypothetical judgment schemata in their premises.

$$\begin{aligned}
p \supset q : \texttt{True} &\ :\!-_{\supset\mathrm{I}}\ p : \texttt{Prop},\ p : \texttt{True} \Rightarrow q : \texttt{True}.\\
q : \texttt{True} &\ :\!-_{\supset\mathrm{E}}\ p \supset q : \texttt{True},\ p : \texttt{True}.\\
p \wedge q : \texttt{True} &\ :\!-_{\wedge\mathrm{I}}\ p : \texttt{True},\ q : \texttt{True}.\\
p : \texttt{True} &\ :\!-_{\wedge\mathrm{EL}}\ p \wedge q : \texttt{True}.\\
q : \texttt{True} &\ :\!-_{\vee\mathrm{ER}}\ p \wedge q : \texttt{True}.
\end{aligned}$$

$$p \vee q : \texttt{True} \ \texttt{:-}_{\vee \mathrm{IL}} \ p : \texttt{True},\ q : \texttt{Prop}.$$
$$p \vee q : \texttt{True} \ \texttt{:-}_{\vee \mathrm{IR}} \ p : \texttt{Prop},\ q : \texttt{True}.$$
$$r : \texttt{True} \ \texttt{:-}_{\vee \mathrm{E}} \ p \vee q : \texttt{True},\ p : \texttt{True} \Rightarrow r : \texttt{True},\ q : \texttt{True} \Rightarrow r : \texttt{True}.$$
$$r : \texttt{True} \ \texttt{:-}_{\bot \mathrm{E}} \ \bot : \texttt{True},\ r : \texttt{Prop}.$$

We define the intuitionistic propositional logic NJ by putting

$$\mathsf{NJ} :\equiv \mathsf{Prop} \oplus \mathsf{True}.$$

Here is an example of a derivation in NJ.

$$\cfrac{U \quad (X :: A : \texttt{True}) \left[\cfrac{\cfrac{U \quad V}{A \supset B : \texttt{Prop}} \supset_{\mathrm{F}} \ (Y :: A \supset B : \texttt{True}) \left[\cfrac{X \quad Y}{B : \texttt{True}} \supset_{\mathrm{E}} \right]}{(A \supset B) \supset B : \texttt{True}} \supset_{\mathrm{I}} \right]}{A \supset (A \supset B) \supset B : \texttt{True}} \supset_{\mathrm{I}}$$

This derivation is a derivation under the derivation context:

$$\langle A, U :: A : \texttt{Prop}, B, V :: B : \texttt{Prop} \rangle.$$

Note that in ordinary natural deduction style formulation of the propositional logic, the above context is shifted to meta-level by declaring, at meta-level, that the variables A, B range over propositions. In derivation games such implicit assumptions are all made into explicit assumptions at object-level. So, if we allow ourselves to suppress such assumptions and suppress derivations of categorical judgments asserting A and $A \supset B$ are propositions, then[17] we get the following derivation.

$$\cfrac{(X :: A) \left[\cfrac{(Y :: A \supset B) \left[\cfrac{Y \quad X}{B} \supset_{\mathrm{E}} \right]}{(A \supset B) \supset B} \supset_{\mathrm{I}} \right]}{A \supset (A \supset B) \supset B} \supset_{\mathrm{I}}$$

We stress that we may allow derivations like above as it is always possible to recover the suppressed information mechanically. A complete derivation of intuitionistic propositional logic, in general, must use rules in Prop as well as rules in True.

The above derivation is essentially the same as the following derivation[18] in the conventional natural deduction style. Note, however, that our derivation explicitly indicates scopes of the discharged assumptions and this enhances the readability of the derivation. Also, note that our derivation is an element in E while conventional derivation is an informal figure of the natural deduction system. Because of this difference, while we can easily define the substitution

[17] We will also simply write C for $C : \texttt{True}$.

[18] Gentzen [24] uses numbers as labels for assumptions but recently variables are also used as labels [89].

operation on derivations as we did in §2, the definition of substitution operation on conventional derivations, e.g., in [61], is rather complex.

$$\cfrac{\cfrac{\cfrac{A \supset B^Y \quad A^X}{B}\supset\text{E}}{(A \supset B) \supset B}\supset\text{I}, Y}{A \supset (A \supset B) \supset B}\supset\text{I}, X$$

LEM (Law of the Excluded Middle). We can obtain the classical propositional logic NK by putting

$$\mathsf{NK} :\equiv \mathsf{NJ} \oplus \mathsf{LEM}$$

where the game LEM is defined by the following rule-schema, and in the schema we write $\neg A$ for $A \supset \bot$.

$$\boldsymbol{p} \vee \neg \boldsymbol{p} : \texttt{True} \ \texttt{:-}_{\texttt{LEM}} \ \boldsymbol{p} : \texttt{Prop}.$$

5.2 Second Order Typed Lambda Calculus

The rule-schemata we have considered so far do not use universal judgments. We can express rules that handle quantifications by using universal judgments in premises of rule-schemata. Here, we give examples of such rule-schemata by defining Girard's second order typed lambda calculus F [25].

Type. We define *types* of F by the following rules.

$$\frac{\boldsymbol{A} : \mathsf{Type} \quad \boldsymbol{B} : \mathsf{Type}}{\boldsymbol{A} \to \boldsymbol{B} : \mathsf{Type}} \to \text{F} \qquad \frac{(x)\,[x : \mathsf{Type} \Rightarrow \boldsymbol{A}[\![x]\!] : \mathsf{Type}]}{\Pi(x : \mathsf{Type})\,[\boldsymbol{A}[\![x]\!]] : \mathsf{Type}} \Pi\text{F}$$

Obj. The game Obj defines *objects* of each type. Namely, we define the judgment $\texttt{<}a, A\texttt{>} : \texttt{Obj}$, which means that a is an object of type A. We will write $a \in A$ for $\texttt{<}a, A\texttt{>} : \texttt{Obj}$[19]

$$\frac{\boldsymbol{A} : \mathsf{Type} \quad (x)\,[x \in \boldsymbol{A} \Rightarrow \boldsymbol{b}[\![x]\!] \in \boldsymbol{B}]}{\lambda(x \in \boldsymbol{A})\,[\boldsymbol{b}[\![x]\!]] \in \boldsymbol{A} \to \boldsymbol{B}} \lambda\text{I}(\boldsymbol{b}) \qquad \frac{\boldsymbol{b} \in \boldsymbol{A} \to \boldsymbol{B} \quad \boldsymbol{a} \in \boldsymbol{A}}{\mathsf{app}(\boldsymbol{b}, \boldsymbol{a}) \in \boldsymbol{B}} \lambda\text{E}$$

$$\frac{(x)\,[x : \mathsf{Type} \Rightarrow \boldsymbol{b}[\![x]\!] \in \boldsymbol{B}[\![x]\!]]}{\Lambda(x : \mathsf{Type})\,[\boldsymbol{b}[\![x]\!]] \in \Pi(x : \mathsf{Type})\,[\boldsymbol{B}[\![x]\!]]} \Lambda\text{I}(\boldsymbol{b}, \boldsymbol{B})$$

$$\frac{\boldsymbol{b} \in \Pi(x : \mathsf{Type})\,[\boldsymbol{B}[\![x]\!]] \quad \boldsymbol{A} : \mathsf{Type}}{\mathsf{App}(\boldsymbol{b}, \boldsymbol{A}) \in \boldsymbol{B}[\![\boldsymbol{A}]\!]} \Lambda\text{E}(\boldsymbol{B})$$

[19] Conventional notation for this judgment in type theory is $a : A$, but since ':' has different meaning in our theory, we are using a slightly non-standard notation.

We can now define F by putting:

$$\mathsf{F} :\equiv \mathsf{Type} \oplus \mathsf{Obj}.$$

We give a derivation of the typing judgment for the polymorphic identity function in F.

```
⊢ Λ(A:Type)[λ(x∈A)[x]] ∈ Π(A:Type)[A→A] in F since
Λ(A:Type)[λ(x∈A)[x]] ∈ Π(A:Type)[A→A]
by ΛI((A)[λ(x∈A)[x]],(A)[A→A]) {
    (A)[(X::A:Type)[
        λ(x∈A)[x] ∈ A→A by λI((x)[x]) {
            A:Type by triv(A:Type) {X};
            (x)[(Y::x∈A)[
                x∈A by triv(x∈A) {Y}
            ]]
        }
    ]]
}
```

In the above derivation, we have instantiated application schemata when we applied the rule-schemata λI and ΛI. Let us see in detail, how these rule-schemata are applied. In the case of the application of ΛI above, the rule-schema is applied under the derivation context <>, and the environment ρ for instantiating the rule-schema is such that $\rho(\boldsymbol{b}) \equiv$ (A)[λ(x ∈ A)[x]] and $\rho(\boldsymbol{B}) \equiv$ (A)[A→A]. So, we can compute the instance of the rule-schema by ρ as in the above derivation. The rule-schema λI is applied under the derivation context $A, X :: A : \mathtt{Type}$ and the environment ρ for instantiating the rule-schema is such that $\rho(\boldsymbol{b}) \equiv$ (x)[x].

5.3 Untyped Lambda Calculus

As another example of a calculus that requires application schema in the defining rules, we define the untyped $\lambda\beta$-calculus LambdaBeta as follows.

Term. The λ-terms are defined as follows.

$$\frac{(x)\,[x : \mathtt{Term} \Rightarrow \boldsymbol{M}[\![x]\!] : \mathtt{Term}]}{\lambda(x)\,[\boldsymbol{M}[\![x]\!]] : \mathtt{Term}}\ \lambda\mathrm{F} \qquad \frac{\boldsymbol{M} : \mathtt{Term} \quad \boldsymbol{N} : \mathtt{Term}}{\mathsf{app}(\boldsymbol{M}, \boldsymbol{N}) : \mathtt{Term}}\ \mathrm{appF}$$

An example of a λ-term is given by the following derivation:

```
Y::y:Term ⊢ λ(x)[app(x,y)]:Term in Term since
λ(x)[app(x,y)]:Term by λF {
    (x)[(X::x:Term)[
```

```
            app(x, y):Term by appF {
                x:Term by triv{X};
                y:Term by triv{Y}
            }
        ]]
    }
```

Traditionally, λ-terms are defined as an inductively defined set of words over a fixed alphabet. So, for example, in [5, page 22] , the set of λ-terms Λ is defined as follows:

1. $x \in \Lambda$;
2. $M \in \Lambda \Rightarrow (\lambda x M) \in \Lambda$;
3. $M, N \in \Lambda \Rightarrow (MN) \in \Lambda$;

where x in 1 or 2 is an arbitrary variable.

Fiore et al. [21] reformulates this definition in a *logical* way as follows. We let Γ be a context in the sense of Definition 1.

$$\frac{x \in \Gamma}{\Gamma \vdash x \in \Lambda} \qquad \frac{\Gamma \oplus \texttt{<}x\texttt{>} \vdash M \in \Lambda}{\Gamma \vdash (\lambda x M) \in \Lambda} \qquad \frac{\Gamma \vdash M \quad \Gamma \vdash N}{\Gamma \vdash (MN) \in \Lambda}$$

In this formulation, λ-terms are always viewed under a context which declares all the free variables in the term. Thus, the 1st rule is an assumption rule which says that x is a λ-term under the assumption that x is a λ-term. In our formulation, as we can see from the derivation above, this rule is a built-in rule common to all the derivation games.[20] The 2nd rule shows that the *assumption* x is discharged when we construct $(\lambda x M)$ out of M, and this suggests the *natural* reading of the 2nd clause of Barendregt's [5] definition of Λ. Namely, we can read it as the following judgment:

$$(x)\,[x \in \Lambda \Rightarrow M \in \Lambda] \Rightarrow (\lambda x M) \in \Lambda.$$

Our rule-schema λF simply formalized this judgment as a rule-schema.

EqTerm. We can define the β-equality relation on λ-terms by the game EqTerm which is defined by the following rule-schemata, where we write $M = N$ for $\texttt{<}M, N\texttt{>}$: `EqTerm`.

$$\frac{\lambda(x)\,[\boldsymbol{M}[\![x]\!]] : \texttt{Term} \quad \boldsymbol{N} : \texttt{Term}}{\mathsf{app}(\lambda(x)\,[\boldsymbol{M}[\![x]\!]], \boldsymbol{N}) = \boldsymbol{M}[\![\boldsymbol{N}]\!]}\ \beta(\boldsymbol{M}, \boldsymbol{N})$$

$$\frac{\boldsymbol{M} : \texttt{Term}}{\boldsymbol{M} = \boldsymbol{M}}\ \texttt{refl} \qquad \frac{\boldsymbol{M} = \boldsymbol{N}}{\boldsymbol{N} = \boldsymbol{M}}\ \texttt{sym} \qquad \frac{\boldsymbol{M} = \boldsymbol{N} \quad \boldsymbol{N} = \boldsymbol{L}}{\boldsymbol{M} = \boldsymbol{L}}\ \texttt{trans}$$

[20] See Definition 9 in §3.

$$\frac{M = N \quad Z : \mathtt{Term}}{\mathsf{app}(M, Z) = \mathsf{app}(N, Z)}\ \mathsf{appL} \qquad \frac{Z : \mathtt{Term} \quad M = N}{\mathsf{app}(Z, M) = \mathsf{app}(Z, N)}\ \mathsf{appR}$$

$$\frac{(x)\,[x : \mathtt{Term} \Rightarrow M\,[\![x]\!] = N\,[\![x]\!]]}{\lambda(x)\,[M\,[\![x]\!]] = \lambda(x)\,[N\,[\![x]\!]]}\ \xi(M, N)$$

We can now define the untyped $\lambda\beta$-calculus LambdaBeta by putting:

$$\mathsf{LambdaBeta} :\equiv \mathsf{Term} \oplus \mathsf{EqTerm}.$$

The game LambdaBeta is a completely *formal* definition of the $\lambda\beta$-calculus, while, in [5, pp. 22–23], the $\lambda\beta$-calculus is *informally* defined as the equational theory $\boldsymbol{\lambda}$. For example, in [5, page 23], the rule-schema appL is informally given as:

$$M = N \Rightarrow MZ = NZ.$$

and there the sign '⇒' means informal implication and also the condition that Z is a λ-term is not explicitly mentioned but it is implicitly declared as a variable ranging over λ-terms.[21] In $\boldsymbol{\lambda}$, the β-conversion rule is given as:

$$(\lambda x.\ M)N = M[x := N].$$

Here again the conditions that $\lambda x.\ M$ and N are λ-term are assumed only implicitly.

We give below an example of a formal derivation of a reduction in the $\lambda\beta$-calculus.

```
Y::y:Term ⊢ app(λ(x)[app(x,x)],y) = app(y,y) in LambdaBeta since
app(λ(x)[app(x,x)],y) = app(y,y) by β((x)[app(x,x)],y) {
    λ(x)[app(x,x)]:Term by λF {
        (x)[(X::x:Term)[
            app(x,x):Term by appF {
                x:Term by triv {X};
                x:Term by triv {X}
            }
        ]]
    };
    y:Term by triv {Y};
}
```

In the above derivation, the β-rule-schema is instantiated by the environment ρ such that $\rho(M) \equiv (x)\,[\mathsf{app}(x, x)]$ and $\rho(N) \equiv y$. Hence the schema $M\,[\![N]\!]$ is

[21] In [5, page 22], it is stated that '$M, N, L, \ldots$ denote arbitrary λ-terms' and, by common sense, we understand that the letter Z is embedded in '...'

instantiated as follows:

$$\begin{aligned}\llbracket M \llbracket N \rrbracket \rrbracket_\rho &\equiv [x := \llbracket N \rrbracket_\rho](\mathsf{app}(x,x)) \text{ since } \rho(M) \equiv (x)\,[\mathsf{app}(x,x)] \\ &\equiv [x := \rho(N)](\mathsf{app}(x,x)) \\ &\equiv [x := y](\mathsf{app}(x,x)) \\ &\equiv \mathsf{app}(y,y)\end{aligned}$$

6 Related Works

Our work presented here is strongly influenced by Martin-Löf's philosophical view of mathematics expressed in e.g. [46, 49], and by the design principle of Edinburgh LF [31]. In this section, we will compare our work to these works in detail. We will also compare our work with other related works including the generalized logic calculus GLC of Takeuti [82] which inspired our theory of expressions and schemata.

6.1 Edinburgh LF

We begin with the comparison of NF with Edinburgh LF utilizing a concrete example. In NF, we can define the implicational fragment of intuitionistic propositional logic, which we tentatively call IFIPL, by the following derivation game $\mathsf{IPL}_\supset$:

$$\begin{aligned}\boldsymbol{p}\supset\boldsymbol{q} : \mathtt{Prop} &\;\text{:-}_{\supset\mathrm{F}}\; \boldsymbol{p} : \mathtt{Prop}, \boldsymbol{q} : \mathtt{Prop}. \\ \boldsymbol{p}\supset\boldsymbol{q} : \mathtt{True} &\;\text{:-}_{\supset\mathrm{I}}\; \boldsymbol{p} : \mathtt{Prop},\ \boldsymbol{p} : \mathtt{True} \Rightarrow \boldsymbol{q} : \mathtt{True}. \\ \boldsymbol{q} : \mathtt{True} &\;\text{:-}_{\supset\mathrm{E}}\; \boldsymbol{p}\supset\boldsymbol{q} : \mathtt{True},\ \boldsymbol{p} : \mathtt{True}.\end{aligned}$$

Edinburgh LF employs *judgments-as-types* principle and represents judgements as LF types. LF has three levels of terms: *objects*, *families* and *kinds*, and it has a distinguished kind Type which is used to classify types. In LF, a logical system is defined by a *signature* which is a finite set of declarations of constants specific to the logical system to be defined. So, IFIPL can be defined by the following signature $\Sigma_\supset$ consisting of the following declarations:

$$\begin{aligned}o &: \mathtt{Type}, \\ \supset &: o \to o \to o, \\ \mathtt{True} &: o \to \mathtt{Type}, \\ \supset\mathtt{I} &: \Pi p : o.\Pi q : o.(\mathtt{True}(p) \to \mathtt{True}(q)) \to \mathtt{True}(\supset(p)\,(q)), \\ \supset\mathtt{E} &: \Pi p : o.\Pi q : o.\mathtt{True}(\supset(p)\,(q)) \to \mathtt{True}(p) \to \mathtt{True}(q).\end{aligned}$$

First, the constant o is declared as a type, and it is intended to represent the type of propositions. Then, $\supset$ is introduced as a function that constructs a proposition $\supset(p)\,(q)$ from propositions p and q. For a proposition p, the judgment 'p is true' is represented by the type $\mathtt{True}(p)$ following the judgments-as-types principle, and proofs (derivations) of the judgment is represented as objects having type $\mathtt{True}(p)$. Finally, derivation rules are declared as functions mapping propositions and proofs to proofs. We compare NF and LF using this example.

(1) NF defines the logic as a derivation game $\mathsf{IPL}_{\supset}$ which is a concrete element in S consisting of 3 rule-schemata. LF defines the logic by the signature $\Sigma_{\supset}$ which declares 5 constants. $\Sigma_{\supset}$ has 3 declarations that correspond to the 3 rule-schemata in $\mathsf{IPL}_{\supset}$, and 2 additional declarations. Since LF is a typed calculus, it is necessary to declare the types of o and $\mathtt{True}$, but such explicit declarations are not necessary in NF. The constants o and $\mathtt{True}$ corresponds to two concepts $\mathtt{Prop}$ and $\mathtt{True}$ defined in $\mathsf{IPL}_{\supset}$, and they are *implicitly* declared by the fact that they occur in the concept part of judgment-schemata.

(2) If we reflect on how IFIPL is defined informally[22], we see that it is defined by defining two concepts, namely, the concept *proposition* and the concept *truth.* In fact, if we define IFIPL informally, we would first define *formulae* inductively, and then define *provable formulae* again inductively. Such inductive definitions are given in terms of rules. The following table shows how these concepts are formalized in NF and LF.

Informal judgment	NF	LF
p is a formula	$p : \mathtt{Prop}$	$p : o$
p is provable	$p : \mathtt{True}$	$\mathtt{True}(p)$

In LF only the judgment 'p is provable' is defined as LF type $\mathtt{True}(p)$. We will call such an LF type *judgmental-type.* The following quotation from [31, §4] explains why only one of the two concepts is formalized as an LF judgmental-type.

> The judgments-as-types principle is merely the formal expression of the fact that the assertion of a judgment in a *formal system* is precisely the claim that it has a *formal proof* in that system.

From the above quotation we see that LF's approach is to look at a *formal system* (here, it is IFIPL) and identify the assertions of judgments in the formal system (here, they are $\vdash p$ where p are IFIPL-formulas).

In contrast to LF's approach, NF's approach is based on the following observation about informal mathematics we conduct day by day.

> (*To assert is to prove.*) The assertion of a judgment in mathematics is precisely the claim that it has a proof.[23]

[22] IFIPL is a formal system, but we usually define it informally using informal meta-language like English or Japanese.

[23] The *to-assert-is-to-prove* principle is usually regarded as an intuitionistic principle (see e.g. [46], [6, page 403]). This is probably because the abstract notion of proof has been well developed for constructive mathematics as theories of constructions by Kreisel [43, 44], Scott [74], Martin-Löf [46] and others based on Brouwer-Heyting-Kolmogorov interpretation of logical constants, and less well developed for classical

For example, imagine that you are a teacher of logic and teaching IFIPL to your students, and just defined inductively what is a *formula.* Then you can assert that:

Thm A. *If* p *is a formula, then* $p \supset p$ *is a formula.*

By making the assertion, you are claiming that it has a proof[24]; and you would give a proof of it by saying: "Assume that p is a formula. Then by the implication formation rule in the definition of formulae, we see that $p \supset p$ is a formula." You will then go on to define what it means that a formula is *provable*, and would make a second assertion:

Thm B. *If* p *is a formula, then* $p \supset p$ *is provable.*

Again, you can justify the claim by giving a poof of it.

The derivation game $\mathsf{IPL}_{\supset}$ in NF faithfully reflects such informal practice by translating informal judgments 'p is a formula' ('p is provable', resp.) to $p : \mathtt{Prop}$ ($p : \mathtt{True}$, resp.) and giving appropriate rule-schemata to define them. The above two metatheorems become the following closed judgments in NF:

$$(p)\,\mathtt{[}p : \mathtt{Prop} \Rightarrow p \supset p : \mathtt{Prop]},$$
$$(p)\,\mathtt{[}p : \mathtt{Prop} \Rightarrow p \supset p : \mathtt{True]}.$$

Moreover, we can construct closed $\mathsf{IPL}_{\supset}$-derivations D_{ThmA} and D_{ThmB} such that:

$$\vdash (p)\,\mathtt{[}p : \mathtt{Prop} \Rightarrow p \supset p : \mathtt{Prop]}\ \mathtt{in}\ \mathsf{IPL}_{\supset}\ \mathtt{since}\ D_{\mathrm{ThmA}},$$
$$\vdash (p)\,\mathtt{[}p : \mathtt{Prop} \Rightarrow p \supset p : \mathtt{True]}\ \mathtt{in}\ \mathsf{IPL}_{\supset}\ \mathtt{since}\ D_{\mathrm{ThmB}}.$$

In contrast, in LF with signature $\Sigma_{\supset}$, only Thm B can be represented by the closed LF judgment:

$$\Pi p : o.\ \mathtt{True(}\supset\mathtt{(}p\mathtt{)\,(}p\mathtt{))},$$

and can be justified by constructing a closed LF term M_{ThmB} such that:

$$\vdash_{\Sigma_{\supset}} M_{\mathrm{ThmB}} : \Pi p : o.\ \mathtt{True(}\supset\mathtt{(}p\mathtt{)\,(}p\mathtt{))}.$$

As for ThmA, a naïve translation of it would be:

$$\Pi p : o.\ \supset\mathtt{(}p\mathtt{)\,(}p\mathtt{)} : o.$$

Now, in LF we can derive:

$$p : o \vdash_{\Sigma_{\supset}} \supset\mathtt{(}p\mathtt{)\,(}p\mathtt{)} : o.$$

---

mathematics. However, the to-assert-is-to-prove principle is effective in practice for all modern mathematics, since only assertions with valid proofs are accepted as theorems (see also, Troelstra and van Dalen [88] and Sato [70]).

[24] It is an informal proof that proves a metatheorem (Thm A).

But, from above, we can only conclude that:

$$\vdash_{\Sigma_\supset} \lambda p : o.\supset(p)\,(p) : o \to o$$

and there are no LF typing rules to conclude

$$\vdash_{\Sigma_\supset} M_{\text{ThmA}} : \Pi p : o.\ \supset(p)\,(p) : o$$

for some closed LF term M_{ThmA}. In fact, $\Pi p : o.\ \supset(p)\,(p) : o$ is not even a term acceptable by the LF grammar. In summary, we have the following table.

Informal Thm	NF	LF
Thm A	$(p)\,[p : \texttt{Prop} \Rightarrow p \supset p : \texttt{Prop}]$	—
Thm B	$(p)\,[p : \texttt{Prop} \Rightarrow p \supset p : \texttt{True}]$	$\Pi p : o.\ \texttt{True}(\supset(p)\,(p))$

The above argument relies on a particular choice a signature for defining IFIPL in LF and hence does not exclude the possibility of finding another signature Σ' which can formalize and prove both ThmA and ThmB. But, such a signature would have to be more complex than $\Sigma_\supset$ and would therefore be much more complex than the derivation game $\mathsf{IPL}_\supset$ of NF.

(3) Let us now compare the $\supset$I rule in NF and LF. In NF, it is a rule-schema having two premises, $\boldsymbol{p} : \texttt{Prop}$ and $\boldsymbol{p} : \texttt{True} \Rightarrow \boldsymbol{q} : \texttt{True}$, and the conclusion $\boldsymbol{p} \supset \boldsymbol{q} : \texttt{True}$. So, this rule-schema faithfully translates the informal intuitionistic definition of the logical constant $\supset$ which says that for any expressions p and q, if we can show that p is a proposition and that q is true whenever p is true, then we may conclude that $p \supset q$ is true. In LF, $\supset$I is a constant function which accepts three arguments, and returns a proof of $\texttt{True}(\supset(p)\,(q))$. Of the three arguments, the fist two are p and q each having the type o of propositions and the third argument guarantees that we can have a proof of $\texttt{True}(q)$ whenever we have a proof of $\texttt{True}(p)$. Thus, we see that the implication introduction rule of IFIPL is represented fatithfully both in NF and LF.

There is, however, a minor diffrence in the two representations. Namely, the $\supset$I rule-schema in NF does not have a premise that correspond to the second argument of the function $\supset$I in LF which restricts q to be a proposition. In NF, it is possible to add a premise $\boldsymbol{q} : \texttt{Prop}$, but we did not do that since this will not change the set of derivable judgments[25]. In LF the second argument cannot be removed since the variable q must be declared toghether with its type, so that the rule becomes a closed LF term.

(4) A more essential difference between the $\supset$I rule-schema in NF and the function $\supset$I in LF is the following. The NF rule has a hypothetical judgment

$$\boldsymbol{p} : \texttt{True} \Rightarrow \boldsymbol{q} : \texttt{True}$$

[25] This is because the rule-schemata is designed to guarantee that any true expression is always a proposition. So, in NF, q can be any expression while in LF, it must be a proposition.

in its premise while LF encodes the hypothetical judgment as the type

$$\mathtt{True}(p) \to \mathtt{True}(q).$$

In NF, the operator $\Rightarrow$ constructs a hypothetical judgment, and its meaning is completely determined by the 3rd clause of the Definition 9 and other clauses of G-derivation, where these clauses are defined by purely combinatorial conditions. Therefore, due to the finitistic character of the definition, we can be sure that the meaning of the hypothetical judgment is equally well understood among the readers who might be an intuitionist or not. On the other hand, in LF, $\to$ is a function type constructor in the LF type theory, and the hypothetical judgment in Edinburgh LF is, so to speak, *implemented* by the function type by applying LF's judgments-as-types principle. Now, as is well known, we can assign meanings to such function types in various ways, and this makes the meaning of the original hypothetical judgment unclear. Thus, although, technically speaking, the same (that is the same after the obvious translation) judgments (in IFIPL) are derivable both in NF and LF, LF does not provide an answer to the foundational question: "What is a hypothetical judgment?" We will touch upon this point again in §6.2.

(5) There is yet another difference between NF and LF concerning the extendability and modifiability of the set of rules. In intuitionistic logic, it is possible to give a more liberal formation rule for implication than the $\supset$F rule-schema we gave above. Martin-Löf gives the following implication formation rule in [49]:

$$\frac{A \text{ prop} \quad \begin{array}{c}(A \text{ true})\\ B \text{ prop}\end{array}}{A \supset B \text{ prop}}$$

In NF, we can easily represent this rule by the $\supset$F+ rule-schema as follows:

$$\boldsymbol{p} \supset \boldsymbol{q} : \mathtt{Prop} \ \texttt{:-}_{\supset\mathrm{F+}} \ \boldsymbol{p} : \mathtt{Prop}, \boldsymbol{p} : \mathtt{True} \Rightarrow \boldsymbol{q} : \mathtt{Prop}.$$

It is however not clear how we should modify the LF declaration $\supset : o \to o \to o$, since it is now possible that $p \supset q$ becomes a proposition even if q is not a proposition. In fact, in NF, it is possible to extend $\mathsf{IPL}_{\supset}$ to a theory of rational numbers that has categorical judgments $x : \mathtt{Q}$ meaning that x is a rational number as well as other categorical judgments that represent standard informal relations and operations on rational numbers. Then in such a theory, which is a game in NF, we can show the following judgment:

$$(x : \mathtt{Q})\ [x \neq 0 \supset x \cdot x^{-1} = 1 : \mathtt{True}].$$

By specializing x to 0, we get a true proposition $0 \neq 0 \supset 0 \cdot 0^{-1} = 1$ whose conclusion is not a proposition.

We have thus compared NF with Edinburgh LF using a concrete example. There is one more point we wish compare concerning the 'purity' of the definitional equality relations of the frameworks. We will make a comparison on this point in the next subsection.

6.2 Martin-Löf's Theory of Expressions and His Philosophy

We first compare our theory of expressions and schemata with Martin-Löf's theory of expressions with the notion of arity and then describe the influence of Martin-Löf's philosophy on our work.

Theory of expressions Just like the design of Edinburgh LF was influenced by Martin-Löf's theory of expressions, our framework is also inspired by his theory. According to [55], Martin-Löf's theory of expressions is intended to provide a syntactic framework which can be used to represent various syntactic entities of intuitionistic type theory in a uniform and abstract manner.

In Martin-Löf's theory of expressions, expressions are constructed from given stocks of constants and variables by means of the four operations of (i) forming a *combination* of n expressions out of an ordered list of n expressions, (ii) *selecting* the i-th component of a combination, (iii) forming an *abstract* out of an expression by abstacting away a specified variable, and (iv) *applying* an abstract expression to an expression. If we allow all the expressions freely constructed by these operations, then the definitional equality would become undecidable just as the β-equality on untyped lambda terms is undecidable. So, Martin-Löf restricts such free constructions by assigning an *arity*, which is a notion similar to type[26], to each expression. Thanks to this restriction, the definitional equality on expressions becomes a decidable relation on expressions, and Martin-Löf's theory of expressions provides a framework for developing formal systems that require the notions of abstraction and substitution.

Since the notion of arity is weaker than the notion of dependent type, Martin-Löf's theory of expressions serves as a foundational framework for defining his intuitionistic theory of types on top of it. However, the theory still looked too strong to us because in order to show the decidablity of definitional equality one must verify a number of combinatorial properties of expressions such as the Church-Rosser property and the normalization theorem. Another reason why we felt the theory too strong was that although definitional equality is decidable, deciding the definitional equality of two given expressions is sometimes not entirely evident, and in such a case we would like to have the *evidence* for the equality, and such an evidence should be provided by a derivation like the one we developed in this paper[27]. In this way, we thought that there should be more basic theory of expressions whose defintioal equality is syntactic identity[28]. Our theory of expressions was designed with this motivation.

[26] As pointed out in [4, 55], Martin-Löf's theory of expressions is essentially a simply typed lambda calculus.

[27] As a matter of fact, students and, for that matter, even experts make mistakes in converting an expression to another expression. We can pinpoint such mistakes only if the judgment that two expressions are definitionally equal is asserted together with its derivation.

[28] Then, we can check the equality of two linguistic expressions by comparing each symbol in the expressions one by one. The definitional equality of our expressions is syntactic identity modulo α-equivalence, but it is possible to modify the definition of expressions so that the definitional equality will become syntactic identity.

We have already developed a theory of symbolic expressions in [66, 67]. There, the data structure of symbolic expressions is defined only by means of the *constructors* that construct pairs out of two expressions. The selectors which select left and right component of the pair are introduced only *after* the data structure has been defined. By the same approach, in §2, we defined the data structure E of expressions by means of the two constructors of constructing a pair and constructing an abstraction.

Once the set E of expressions is defined only by means of constructors, we can introduce operators to *destruct* and analyze the data structure. We perform such operations by looking at expressions (which are elements of E) from outside, and therefore we need *meta*-expressions if we wish to have linguistic forms to express such operations. So, in §2, we introduced *schemata* as such meta-expressions, so that we could express the operation of applying an abstract to an expression by a schema. We also saw, without incurring the problems of checking the Church-Rosser property etc., that any such application can be evaluated to a unique expression relative to an environment.

To define a data structure only by means of constructors is a common practice in computer science and we simply followed this practice. For example, as we saw in §4, McCarthy [51] defined symbolic expressions only by means of the pairing operation, and then he introduced meta-expressions, such as *car*[(A . B)], which can be used to analyze symbolic expressions.

Now, concerning the strength of the frameworks, de Bruijn [14] pointed out that:

> Over the years there has been the tendency to strengthen the frameworks by rules that enrich the notion of definitional equality, thus causing impurities in the backbones of those frameworks: the typed lambda calculi.

Then, he made a plea for seeking weaker frameworks by remarking that:

> For several reasons (theoretical, practical, educational and sociological) it may be recommended to keep the frames as simple as possible and as weak as possible.

We cannot agree more with what de Bruijn says in the above two qutoations. However, we do not follow what he then says:

> We restrict the discussions to frames that have typed lambda calculi as their backbone, an idea that is gaining field today, along with the rising popularity of the principle to treat proofs and similar constructions in exactly the same way as the more usual mathematical objects.

Martin-Löf's theory of expressions with arity is 'impure' as a framework since expressions are defined not only by constructors but also by destructors which are unnecessary as we have shown, and other typed lambda calculi (including calculi of AUTOMATH and Edinburgh LF) have the same impurity; and this impurity is the source of the difficulty in verifying the decidability of definitional equality of these calculi. We believe that we have succeeded in designing a framework which is simpler than the frameworks based on typed lambda calculi, and yet can treat proofs and similar constructions in the same way as the more usual mathematical objects.

Martin-Löf's Philosophy. We now turn to the influence of Martin-Löf's philosophy of mathematics on our work.[29] From his articles, e.g., [46, 47, 49], we find that Martin-Löf always wanted to provide a formal framework suitable for developing full-fledged intuitionistic mathematics. Such a framework must be foundationally well-founded and he developed his theory of expressions equipped with decidable definitional equality. He then uses these expressions as a medium to present his intuitionistic theory of types. In developing intuitionistic logic and mathematics, Martin-Löf is very careful (for a natural reason we explain later) about the priority among various notions to be introduced. For example, in [49, page 19], he says:

> Now, the question, What is a judgment? is no small question, because the notion of judgment is just about the first of all the notions of logic, the one that has to be explained before all the others, before even the notions of proposition and truth, for instance.

We completely agree with this remark about the notion of judgment,[30] and we placed the notion of judgment as the most basic notion of our theory.

After explaining judgments in [49], Martin-Löf characterizes an *evident* judgment as a judgment for which a *proof* has been constructed. According to him, a proof is an act of grasping (or knowing) the evidence of the judgment. Therefore, since we conduct the act of proof by applying proof-rules, each proof-rule must be sound in the sense that it sends evident judgments to an evident judgment. To guarantee the soundness of the proof-rule, prior to introduce the rule, one must know the meanings of the categorical judgments involved in the rule. Then, to know the meanings of the categorical judgments properly, the *concepts* involved in the categorical judgments must be introduced in a proper order. Thus, in Martin-Löf's intuitionistic theory of types, *semantical* explanation of each concepts involved is essential in establishing the soundness of the proof-rules of the theory, although the theory itself is a formal theory that manipulates syntactic entities.

In our theory, we introduced the notion of derivation which corresponds to the notion of proof in Martin-Löf's theory. Just like Martin-Löf's proofs, our derivations are attached to judgments and never attached to propositions or formulae. Martin-Löf stressed the character of a proof as an act, but we think that he also accepts the possibility of recording such an act in a linguistic form. So, we think that our deriviations which are elements of E contain proofs in Martin-Löf's sense.

---

[29] What follows is our interpretation of Martin-Löf's philosophy and can be different from what he thinks. See also [6] for Beeson's account of Martin-Löf's philosophy.

[30] Thus, for us, the *judgments-as-types* principle is an unnatural principle since it is an attempt to explain judgments in terms of types.

In this paper, we have introduced the notions of judgments and derivations syntactically without referring to the meanings of categorical judgments. This was made possible by introducing the notion of *derivation game* which we have arrived at by observing the activitiy of mathematicians including ourselves. The result of the observation was that not only an intuitionist mathematician but also a classical mathematician has the right to *assert* a judgment, say the judgment that Fermat's Last Theorem is true, only if he has *proven* it, namely, has carried out the act of proof and has a derivation as the record, or the *trace*[31] in computer science terminology, of his act of making the judgment evident.[32] Thus, the evidence, that is, the derivation, is initially private to the mathematician who proved the judgment, but the derivation can later be used to make the judgment evident to other mathematicians. As we have mentioned briefly in §1, the notion of evidence is not an absolute notion but a notion that depends on the philosophical (epistemological and ontological) standpoint of each mathematician. Thus the law of the excluded middle is evident to classical mathematicians but not so to intuitionist mathematicians. So, evidence is a semantical notion, and we therefore developed our theory without using the notion of evidence, but using the notion of proof-rule (which we call rule-schema) which is a syntactic notion. This is not to say that semantics is unimportant. On the contrary, precisely because we consider semantics to be more important than syntax, we have tried to develop a general theory of judgments and derivations by referring only to syntactic objects, namely, expressions and schemata. In fact, all the notions and theorems we introduced in §§2–3 are explained in a finitistic manner[33] (as advocated by Hilbert in his program) and therefore can be equally well understood both by intuitionist and classical mathematicians. In this way, NF has been designed so as to provide a common framework in which we can formally develop both intuitionistic and classical mathematics.

Another important characterisitic of Martin-Löf's theory of types is its *open-endedness.* (See, e.g., [46, page 3], [49, page 17], [16, 90].) The open-endedness of the theory allows us to introduce new types without affecting the evidence of the judgments already established. This reflects well the common practice of mathematics where mathmaticians extend mathematics by introducing new concepts. NF is also designed to be open-ended, and in NF we can introduce a new concept by adding new rule-schemata that mention the concept to the derivation game we are interested in. For example, consider the mathematical induction rule `ind`

[31] Goto [30] observed that the trace of a computation can be regarded as a derivation.

[32] This is the to-assert-is-to-prove principle upon which our theory of judgments and derivations is based.

[33] There is no consensus as to the exact extent of the finitistically acceptable method. However, it is generally accepted that finitisitic arithmetic contains PRA (primitive recursive arithmetic) (see, e.g., [91, page 303], Takeuti [83, pp. 81 – 85], Tait [81], Simpson [75], Feferman [20, page 196] and Okada [56, §3.2]). Our method is finitistic in the sense that we can carry out all the arguments in §§2–3 in PRA. A natural consequence of this is the primitive-recursive decidability of whether a given expression is a G-derivation or not, and this answers Kresel's dictum in an affirmative way. Thus, we can recognize a proof when we see it.

we showed in §1 again. In traditional presentation of formal logical calculi, the class of *formulae* must be specifed before introducing inference rules. Thus if the calculus is first-order, then the *formula* $P[x]$ must be a first-order formula where the possible terms that may occur in $P[x]$ are also determined by the language of the calculus. So, if one wishes to extend the calculus to, say, a second-order calculus, then one must re-define the language and re-state the `ind`-rule, in spite of the fact that the principle of mathematical induction is valid for any *proposition* on a natural number. By the same reason, recalling the remark in §6.1(5), the IND-rule of Edinburgh LF we cited in §1 is also not open-ended. In contrast, the $\mathtt{ind}(\boldsymbol{P})$ rule of NF is open-ended just as in Martin-Löf's type theory.

6.3 Takeuti's GLC

In 1954, Takeuti [82] introduced the generalized logic calculus (GLC) which is a higher type extension of Gentzen's classical sequent calculus LK. GLC has the notion of *finite type* which is defined by the following grammar:[34]

$$\tau ::= \iota \mid [\tau_1, \ldots, \tau_n].$$

In formulating the basic syntax of GLC, Takeuti introduced the notion of *variety* which he defined simultaneously with the notion of *formula*. Varieties and formulae are therefore related with each other, and among the defining clauses of these notions he has the following clauses.[35]

- If A is a formula and $\varphi_1, \ldots, \varphi_n$ are distinct variables each of type τ_i, then $\{\varphi_1, \ldots, \varphi_n\}A$ is a variety of type $[\tau_1, \ldots, \tau_n]$.
- If φ is a variable of type $[\tau_1, \ldots, \tau_n]$ and $V_1, \ldots, V_n$ are varieties each of type τ_i, then $\varphi[V_1, \ldots, V_n]$ is a formula.

We can clearly see that these clauses are structurally the same as the *abstract* and *application* clauses in Definition 3. Here, we note an important difference between Takeuti's definition of formula and Church's definition of terms of type o in [15]. Namely, while Takeuti does not admit, for example, $\{x, y\}(x = y)[2, 3]$ as a valid formula, Church admits $(\lambda x, y.\ {=}xy)(2)(3)$ as a valid formula, i.e., a valid term of type o. Takeuti, however, admits $\{x, y\}(x = y)[2, 3]$ and similar expressions as meta-expressions which can be calculated at meta-level, rather than at object-level which is the case for Church's simple theory of types. Calculations of such meta-expressions are carried out by substitution in a similar fashion as our computation of application schemata relative to environments. It is to be noted that Takeuti uses the term '*abstract*' instead of 'variety' in his later publications [83, 84].

[34] Takeuti uses the symbol 0 instead of ι, but here we use ι to relate Takeuti's finite type to Church's simple type. Takeuti's $[\tau_1, \ldots, \tau_n]$ roughly corresponds to Church's $\tau_1 \to \cdots \to \tau_n \to o$.

[35] For the sake of simplicity we use a simplified version of the defintion given in [84]. Also, following Gentzen, Takeuti distinguishes free and bound variables syntactically by providing two disjoint sets of variables, but we do not do so.

We may thus say that although Takeuti did not develop a general theory of expressions, he introduced essentially the same mechanism of abstraction and instantiation which we developed in §2.

6.4 Other Related Works

Just as Martin-Löf did in [46, page 3], Katuzi Ono [57] questioned the way the formalistic notions of formula and theorem are presented in standard logic texts, and he developed a unique notion-object theory whose basic concepts are *notion*, *object* and *categorical judgment*, and where a categorical judgment sP (which corresponds to our $s : P$) means that the notion P applies to the object s. The design of NF is considerably influenced by this book[36] of Ono, especially by his attitude to treat the notion of *notion* as more basic than other basic notions such as *set* or *type* as well as by his strong will to *naturally* formalize mathematics as a whole.[37]

Abstract theory of proofs has been developed in order to metamathematically analyze the intuitionistic notion of proof given by Brouwer-Heyting (see for example [34]). There are a lot of studies on this and related research area. Theory of constructions was initially studied by Kreisel [43], Goodman [27], Scott [74] and others (e.g., Aczel [3], chapter XVII of Beeson [6], Kobayashi [40], Sato [69]), and seems to be eventually merged with type theory. Theoretical foundations for de Bruijn's AUTOMATH system [13] and many other proof-checking systems, e.g., [9, 10, 58, 60] so far developed are mostly given by various forms of typed lambda calculi. Kleene's recursive realizability [41] initiated the research in realizability interpretaion and found applications in automatic extraction of programs from proofs of the specifications (see, e.g., Troelstra [87], Hayashi [33], Tatsuta [86]). Gödel's theory T [26] may also be regarded as an abstract theory of proofs, and it was also applied to program extraction by Sato [64] and Goto [28, 29]. Type theory can also be used to develop automatic program extraction systems and its theoretical foundation is given by the Curry-Howard isomorphism (see, e.g., [55]). It is hoped that NF will provide a common framework to implement and compare these different approaches.

In a section on proof theory ([44, pp. 149 – 179]), Kreisel pointed out the possible ambiguities of representation of the derivability relation ('P is a proof of A in the formal system F') within F (via encoding of syntax within F) and asked for a workable framework for presenting formal systems that will admit canonical representation of the derivability relation. Usefulness of pairing structure for providing such frameworks was acknowledged by Feferman [18], Sato [67] and Voda [92, 93]. NF is also intended to provide such a workable framework.

[36] The book is written in Japanese.

[37] Axiomatized set theory ZF, for instance, is not an answer because of its unnaturalness to treat every mathematical entity as a set.

7 Conclusion

We have presented a constructive (finitistic) theory of judgments and derivations, and introduced the Natural Framework (NF) which can be used to implement various forms of computational and logical systems uniformly and naturally as derivation games.

The design goal of NF was to provide a canonical framework for *formally* presenting metamathematical objects such as proofs and theorems, and thereby enable us to treat such objects in the same way as we treat ordinary mathematical objects like numbers and topological spaces. This is also the goal set by de Bruijn[38] and by Kreisel[39]. Our plan was that a framework that attains the goal would be obtained by first observing informal mathematics carefully and then translate it into a framework as *naturally* as possible. The approach we took was very similar to the approach taken by Gentzen [24] who carefully analyzed informal proofs and formulated his natural deduction systems NJ and NK.

In our case, we made three key observations. The first observation was the *to-assert-is-to-prove* principle we described in §6.1. The second observation, which Martin-Löf [49] also made already, was that what we prove in informal mathematics is not a proposition but a *judgment*. The third observation we made about informal mathematics was that the notion of *free variable* which we find in formal mathematics does not exist in informal mathematics. This is because, in informal mathematics, whenever we wish to use a variable, we declare the variable before the usage. Thus, we say 'Let V be a vector space over the field of real numbers' or the like, and then use the declared variables within the scopes of the declarations. This means that any mathematical expression is always considered under a *context.*[40] In formal mathematics, it seems that it is only recently that this practice in working mathematics has started to be respected, notably in type theory and by Fiore et al. [21] in the untyped lambda calculus.

The notions of judgment and derivation have been introduced in this paper in a finitistic way so that these notions are equally well understandable by both intuitionist and classical mathematicians. We could achieve this finitistic character of these fundamental notions by defining these notions without referring to the *meanings* of categorical judgments. By doing so, we could also define the *derivability*, relative to a given game G, of *hypothetical judgments* and *universal judgments* in a finitistic way. It should be noted that our finitistic definition of the *meanings*[41] of these two forms of judgments we gave in Definition 9 is in accordance with the explanations Martin-Löf gave for these forms of judgments in [49]. In [49, page 41], Martin-Löf explains a proof of a hypothetical judgment as follows.

---

[38] See §6.2.

[39] See §6.4.

[40] See also Frege [23, page 659].

[41] Namely, the explanation of what count as the proofs (derivations) of these forms of judgments.

> Now, quite generally, a proof of a hypothetical judgment, or logical consequence, is nothing but a hypothetical proof of the thesis, or consequent, from the hypotheses, or antecedents. The notion of hypothetical proof, in turn, which is a primitive notion, is explained by saying that it is a proof which, when supplemented by proofs of the hypotheses[42], or antecedents, becomes a proof of the thesis, or consequent.

Now, in NF, a *hypothetical proof* of, say, $H \Rightarrow J$ under Γ is nothing but a derivation of J under $\Gamma \oplus$ <X :: H>, and by Theorem 3.4, we see that when substituted by a derivation of H (under Γ) it becomes a proof of J (under Γ). Then, in page 43, he explains a proof of a universal judgments as a *free variable proof* as follows.

> And what is a free variable proof? It is a proof which remains a proof when you substitute anything you want for its free variables, that is, any expressions you want, of the same arities as those variables.

Here, again, Definition 9 and Theorem 3.5 are in complete accordance with the above explanation of a free variable proof. Definition 9 seems to reflect the intrinsically linguistic character of logical reasoning[43], since it defines the meaning of a logical consequence, that is, a hypothetical judgment, by characterizing the linguistic forms of derivations of the logical consequence.

We now turn to the directions we wish to pursue in future. On the practical side, we wish to implement the NF system on a computer. As reported in [73], we have been teaching a computer aided course on computation and logic at Kyoto University. The computer environment for the course is called CAL and through the teaching experience of the course and through the implementation experience of the CAL system we have gradually arrived at the notion of derivation game. The current implementation of CAL does not support the user definition of a derivation game and moreover each derivation game is separately implemented. The NF system, when implemented, will change the situation.

We will then be able to define NF within NF system in a somewhat similar manner as McCarthy 'defined' pure Lisp by writing a meta-circular interpreter [51] of pure Lisp in pure Lisp[44]. The technical device needed is a method to encode S (schemata) into E (expressions) since rule-schemata are elements in S and judgments and derivations are elements in E. McCarthy used the method of quotation in [51] which is much more efficient than Gödel numbering in terms of space ratio between original expressions and encoded expressions. We are planning to use the mechanism of quasi-quotation invented by Quine [62] and used in several dialects of modern Lisp systems, for instance, Scheme [2].

As described in the previous paragraph, we will be able to use the NF system to *define* (or *specify*) computational and logical systems as deriviation games. Then, for each such derivation game G, NF will function as a *proof checker*,

[42] Here he is considering a hypothetical judgment of the form $J\ (H_1, \ldots, H_n)$.

[43] With regard to this claim, the reader is referred to Okada [56] who analyzes Hilbert's and Husserl's philosophy of mathematics.

[44] The interpreter was incorrect and corrected in [52] (see [78]).

and it will check for any expression D whether it is a correct derivation in G. However, demanding the user to manually supply a derivation in a complete form will discourage the use of the system. Therefore, it will be important to provide various means of automating the construction of a derivation. For elementary derivation games, we would be able to fully automate the construction by adding the unification and the backtracking mechanisms to the NF system. Such an NF system can be used as a generic interpreter of the language specified by G. For non-elementary derivation games, addition of the mechanism of higher-order unification (see, Huet [37]) as used, for instance, in the systems described in [39, 53] would become necessary.

As for the theoretical side, we have seen that our theory presented here is more foundational than other frameworks. However, we think that the theory should be made more foundational in at least two points.

The first point is concerned with the fact that we have *not* defined the notion of α-equivalence and substitution on expressions and schemata but relied on the (supposedly) common and standard definitions of these notions. But, as is well known, these notions are very subtle, and, in spite of this, we must choose specific definitions when we actually implement NF on a computer. However, none of existing definitions proposed by others, including, for instance, definitions using de Bruijn indices [12] are satisfactory to us.[45] Since we have our own alternative definitions of substitution and α-equivalence [65–68, 72], we would like to use them to define these subtle notions.

The second point is related to a recent trend in computer science to internalize meta-level syntactic operations and make such operations available at the level of programming languages. Such a trend was geared up by the work of Abadi et al. [1] which introduced the notion of *explicit substitution*. Since then explicit substitution has been extensively studied and other related notions such as *environment* [54, 71] and *context* [8, 11, 32, 50, 59, 63, 72, 85] has been studied. The notion of context here is not the one we used to define expressions, but a context is (intuitively) an expression with holes in it. When the holes are filled with another expression, capturing of variables are not avoided but rather solicited. We can find a perfect example of a context in the mathematical induction rule we considered in §1. The P, which we also write $P[\ ]$, in the rule is such a context, and in the conclusion of the rule, the holes are filled with the variable x and these x's become bound by the quantifier $\forall x$ as expected. As we showed in §1, the induction rule is represented by the rule-schema $\mathtt{ind}(\boldsymbol{P})$ in NF, and when the rule is actually applied, the schematic variable $\boldsymbol{P}$ must be instantiated by an abstract. This means that an abstract expression of NF is nothing but a context. In Edinburgh LF, the context P is realized by a higher-order function ϕ of type $\iota \to o$, and Pitts [59] and Sands [63] used such higher-order functions to represent contexts. We believe that our identification of *context-as-abstract* is more natural than the representations of contexts by higher-order functions. However, although abstracts are expressions in E, we cannot internally instantiate (that is, fill in) the abstracts in the current formulation of NF, since instantiation is a

[45] See also [73].

meta-level operation. To internalize the instantiation operation, we must, again, face the problem of encoding schemata into expressions.

These observations seem to be suggesting us that we will be able to get a more foundational and powerful theory by re-designing the structure of expressions by adding the mechanism of quotation or quasi-quotation. This is, however, only the author's personal speculation. Even for the Natural Framework presented here, there remains a lot to be done, and among which semantical analysis of foundational notions such as *proposition*, *truth*, *type* and *set* by means of derivation games seems to be most interesting and challenging from a theoretical point of view.

Acknowledgements

The author would like to thank John McCarthy and Per Martin-Löf for giving him so much inspiration and philosophical influence through conversations and through their articles.

The author would like to thank Setsuo Arikawa who conceived the concept *discovery science* and led a three year project on discovery science under the sponsorship of the Ministry of Education, Culture, Sports, Science and Technology, Japan. In these three years, the author had the opportunity to organize a group consisting of philosophers, logicians and computer scientists, and discuss freely on foundational issues in discovery science. The focus of attention of the group was the analysis of discovery of concepts by human while other groups mainly studied discovery by machine. The author is grateful to Hiroakira Ono, Keiich Noé, Mitsuhiro Okada, Tetsuo Ida, Masami Hagiya, Yukiyoshi Kameyama and Syun Tutiya who shared many hours of discussions on discovery of concepts with him.

The author benefitted from discussions with many people on matters related to the present work. In particular, the author wishes to thank the following people: Peter Aczel, Michael Beeson, Henk Barendregt, Rod Burstall, Robert Constable, Thierry Coquand, Peter Dybjer, Solomon Feferman, Shigeki Goto, Susumu Hayashi, Rodger Hindley, Gerard Huet, Shigeru Igarashi, Leslie Lamport, Robin Milner, J. Strother Moore, Gopalan Nadathur, Minoru Nagamitsu, Bengt Nordström, Atsushi Ohori, Gordon Plotkin, Randy Pollack, Takafumi Sakurai, Mikio Sato, Andre Scedrov, Dana Scott, Jan Smith, Carolyn Talcott, Masako Takahashi, Satoru Takasu, Gaisi Takeuti, TAKEUTI Izumi, Yoshihito Toyama, Richard Weyhrauch, Yohei Yamasaki, and Mariko Yasugi.

Finally, the author would like to thank Takayasu Ito who carefully read the earlier drafts of the paper and provided useful comments and advices that contributed to improve the paper.

References

1. Abadi, M., Cardelli, L., Curien, P.-L. and Levy, J.-J., Explicit substitutions, pp. 375-416, *Journal of Functional Programming*, **1**, 1991.
2. Abelson, H. et. al, Revised^5 report on the algorithmic language scheme, *Higher-order and symbolic computation*, **11**, pp. 7 – 105, 1998.
3. Aczel, P., Frege structures and the notions of proposition, truth and set, in Barwsise, J. et al. eds., *The Kleene Symposium*, North-Holland, Amsterdam, pp. 31 – 50, 1980.
4. Aczel, P., Carlisle, D.P., Mendler N., Two frameworks of theories and their implementation in Isabelle, pp. 3 – 39, in [38], 1991.
5. Barendregt, H. P., *The Lambda Calculus, Its Syntax and Semantics*, North-Holland, 1981.
6. Beeson, M.J., *Foundations of constructive mathematics*, Springer, 1980.
7. Bell, D., *Frege's Theory of Judgement*, Clarendon Press, Oxford, 1979.
8. Bognar, M., de Vrijer, R., A calculus of lambda calculus contexts, **27**, pp. 29–59, *J. Automated Reasoning*, 2001.
9. Constable R.L. et al., *Implementing Mathematics with the NuPRL Proof Development System*, Prentice-Hall, Englewood Clifs, NJ, 1986.
10. Coquand, T. and Huet, G., The calculus of construction, *Information and Computation*, **76**, pp. 95 – 120, 1988.
11. Dami, L., A lambda-calculus for dynamic binding, pp. 201-231, *Theoretical Computer Science*, **192**, 1998.
12. de Bruijn, D. G., Lambda calculus notation with nameless dummies, a tool for automatic formula manipulation, with application to the Church-Rosser theorem, *Indag. Math.* 34, pp. 381–392, 1972.
13. de Bruijn, D. G., A survey of the project AUTOMATH, in *To H.B. Curry: Essays on Combinatory Logic, Lambda Calculus and Formalism*, pp. 579 – 606, Academic Press, London, 1980.
14. de Bruijn, D. G., A plea for weaker frameworks, pp. 40–67, in [38], 1991.
15. Church, A., A formulation of the simple theory of types, *J. Symbolic Logic*, **5**, pp. 56 – 68, 1940.
16. Dybjer, P., Inductive sets and families in Martin-Löf's type theory and their set theoretic semantics, pp. 280 – 306, in [38], 1991.
17. Feferman, S., A language and axiom for explicit mathematics, in Crossley J.N. ed., *Algebra and Logic*, Lecture Notes in Computer Science, **450**, pp. 87 – 139, 1975.
18. Feferman, S., Indutively presented systems and the formalization of meta-mathemaics, in van Dalen, D. et al. eds., *Logic Colloquim '80*, North-Holland, Amsterdam, 1982.
19. Feferman, S., Finitary inductively presented logics, in Ferro R. et al. eds., *Logic Colloquim '88*, North-Holland, Amsterdam, 1989.
20. Feferman, S., *In the Light of Logic*, Logic and Computation in Philosophy, Oxford University Press, Oxford, 1998.
21. Fiore, M., Plotkin, G., and Turi, D., Abstract syntax and variable binding (extended abstract), in *Proc. 14th Symposium on Logic in Computer Science*, pp. 193-202, 1999.
22. Frege, G., Über Begriff und Gegenstand, *Vierteljahrsschrift für wissenschftliche Philosophie*, **16**, pp. 192–205, 1892.
23. Frege, G., Was ist Funktion?, in *Festschrift Ludwig Boltzmann gewindmet zum sechzigsten Geburstage, 20. Feburar 1904*, Leipzig, pp. 656 – 666, 1904.

24. Genzten, G., Untersuchungen über das logische Schließen, I, Mathematische Zeitschrift, **39**, pp. 175 – 210, 1935, English translation in *The collected papers of Gerhard Gentzen*, Szabo, M.E. ed., pp. 68 – 131, North-Holland, Amsterdam, 1969.
25. Girard, J-Y., Taylor, P., Lafont, Y., *Proofs and Types*, Cambridge University Press, 1989.
26. Gödel, K., Über eine bisher noch nicht benützte Erweiterung des finiten Standpunktes, *Dialectica*, **12**, pp. 280 – 287, 1958.
27. Goodman, N., A theory of constructions equivalent to arithmetic, in *Intuitionism and Proof Theory*, Kino, Myhill and Vesley, eds., North-Holland, pp. 101 – 120, 1970.
28. Goto, S., Program synthesis through Gödel's interpretation, Lecutre Notes in Computer Science, **75**, pp. 302 – 325, 1978.
29. Goto, S., Program synthesis from natural deduction proofs, in *Sixth International Joint Conference on Artificial Intelligence (IJCAI-79)*, pp. 339 – 341, 1979.
30. Goto, S., How to formalize traces of computer programs, *Jumulage Meeting on Typed Lambda Calculi*, Edinbrugh, September, 1989.
31. Harper, R., Honsell, F. and Plotkin, G., A framework for defining logics, *Journal of the Association for Computing Machinery*, **40**, pp. 143–184, 1993.
32. Hashimoto, M., Ohori, A., A typed context calculus, *Theoretical Computer Science*, to appear.
33. Hayashi, S. and Nakano, H., *PX: A Computational Logic*, Foundation of Computing, The MIT Press, Cambridge, 1988.
34. Heyting, A.K., Intuitionism in mathematics, in *Philosophy in the mid-century, a survey*, Klibansky, ed., Florence, pp. 101 – 115, 1958.
35. Hilbert, D., Über das Unendliche, *Mathematische Annalen*, **95**, pp. 161 – 190, English translation in van Heijenoort, J. ed., *From Frege to Gödel: A Source Book in Mathematical Logic*, Harvard University Press, Cambridge, MA, 1967.
36. Howard, W.A., The formulae-as-types notion of construction, in Seldin J.P. and Hindley, J.R. eds., *To H.B. Curry: Essays on Combinatory Logic, Lambda Calculus and Formalism*, pp. 479 – 490, Academic Press, London, 1980.
37. Huet, G., A unification algorithm for typed λ-calculus, *Thoretical Computer Science*, **1**, pp. 27 – 57, 1975.
38. Huet, G., and Plotkin, G. eds., *Logical Frameworks*, Cambridge University Press, 1991.
39. Ida, T., Marin, M., Suzuki, T., Reducing search space in solving higher-order equations, in this volume.
40. Kobayashi, S., Consitency of Beeson's formal system RPS and some related results, in Shinoda, J., Slaman, T.A. and Tugué, T. eds., *Mathematical Logic and Applications, Proceedings of the Logic Meeting held in Kyoto, 1987*, Lecture Notes in Mathematics, **1388**, Springer, pp. 120 – 140, 1989.
41. Kleene, S.C., On the interpretation of intuitionistic number theory, *J. Symbolic Logic*, **10**, pp. 109 – 124, 1945.
42. Kreisel, G., Hilbert's programme, *Dialectica*, **12**, pp. 346–372, 1958.
43. Kreisel, G., Foundations of Intuitionistic Logic, Nagel, Suppes, Tarski eds, *Logic, Methodology and Philosophy of Science*, Stanford University Press, pp. 198 – 210.
44. Kreisel, G., Mathematical Logic, Saaty ed., *Lectures on Modern Mathematics, III*, Wiley, New York, pp. 95 – 195, 1965.
45. Martin-Löf, P., An intuitionistic theory of types: Predicative part, *Logic Colloquium '73*, Rose, H.E. and Shepherdson, J.C., Eds, Studies in Logic and Foundation of Mathematics, **80**, North-Holland, pp. 73 – 118, 1975.

46. Martin-Löf, P., *Intuitionistic Type Theory*, Bibliopolis, 1984.
47. Martin-Löf, P., Truth of a proposition, evidence of a judgment, validity of a proof, *Synthese*, **73**, pp. 407–420, 1987.
48. Martin-Löf, P., Analytic and synthetic judgments in type theory, Parrini, P. ed., *Kant and Contemporary Epistemology*, Kluwer Academic Publishers, pp. 87 – 99, 1994.
49. Martin-Löf, P., On the meanings of the logical constants and the justifications of the logical laws, *Nordic J. of Philosophical Logic*, **1**, pp. 11–60, 1996.
50. Mason, I., Computing with contexts, *Higher-Order and Symbolic Computation*, **12**, pp. 171–201, 1999.
51. McCarthy, J., Recursive functions of symbolic expressions and their computation by machine, Part I, *Comm. ACM*, **3**, pp. 184–195, 1960.
52. McCarthy, J., et. al, LISP 1.5 programmer's manual, MIT Press, 1965.
53. Nadathur, G., Miller, D., Higher-order logic programming, in Gabbay, D.M. et al. eds., *Handbook of Logic in AI and Logic Programming*, **5**, Clarendon Press, Oxford, pp. 499 – 590, 1998.
54. Nishizaki, S., Simply typed lambda calculus with first-class environments, *Publ. RIMS, Kyoto U.*, **30**, pp. 1055 – 1121, 1994.
55. Nordström, B., Petersson,K. and Smith, J.M., *Programming in Martin-Löf's Type Theory*, Oxford University Press, 200 pages, 1990.
56. Okada, M., Ideal concepts, intuitions, and mathematical knowledge acquisitions: Husserl's and Hilbert's cases (a preliminary report), in this volume.
57. Ono, K., *Gainen-Taishou Riron no Kousou to sono Tetugakuteki Haikei* (Eng. *A plan of notion-object theory and its philosophical background*), Nagoya University Press, Nagoya, 1989.
58. Paulson, L., *Logic and Computation*, Cambridge University Press, Cambridge, 1988.
59. Pitts, A.M., Some notes on inductive and co-inductive techniques in the semantics of functional programs, Notes Series BRICS-NS-94-5, Department of Computer Science, University of Aarhus, 1994.
60. Pollack, R., A verified type-checker, in Dezani-Ciancaglini, M. and Plotkin, G. eds., *Proceedings of the Second International Conference on Typed Lambda Calculi and Applications, TLCA '95, Edinburgh*, Springer-Verlag, Lecture Notes in Computer Science **902**, 1995.
61. Prawitz, D., *Natural Deduction: A Proof-Theoretical Study*, Almquist and Wiksell, Stockholm, 1965.
62. Quine, W.V.O., *Mathematical Logic*, Harvard University Press, 1951.
63. Sands, D., Computing with Contexts – a simple approach, in *Proc. Higher-Order Operational Techniques in Semantics, HOOTS II*, 16 pages, *Electronic Notes in Theoretical Computer Science* **10**, 1998.
64. Sato, M., Towards a mathematical theory of program synthesis, in *Sixth International Joint Conference on Artificial Intelligence (IJCAI-79)*, pp. 757 – 762, 1979.
65. Sato, M., and Hagiya, M., Hyperlisp, in de Bakker, van Vliet eds., *Algorithmic Languages*, North-Holland, pp. 251–269, 1981.
66. Sato, M., Theory of symbolic expressions, I, *Theoretical Computer Science*, **22**, pp. 19 – 55, 1983.
67. Sato, M., Theory of symbolic expressions, II, *Publ. of Res. Inst. for Math. Sci.*, Kyoto Univ., **21**, pp. 455–540, 1985.
68. Sato, M., An abstraction mechanism for symbolic expressions, in V. Lifschitz ed., *Artificial Intelligence and Mathematical Theory of Computation (Papers in Honor of John McCarthy)*, Academic Press, pp. 381–391, 1991.

69. Sato, M., Adding proof objects and inductive definition mechanism to Frege structures, in Ito, T. and Meyer, A. eds., *Theoretical Aspects of Computer Software, International Syposium TACS'94 Proceedings*, Lecture Notes in Computer Science, **789**, Springer-Verlag, pp. 179 – 202, 1994.
70. Sato, M., Classical Brouwer-Heyting-Kolmogorov interpretation, in Li, M. Maruoka, A. eds., *Algorithmic Learning Theory, 8th International Workshop, ALT'97, Sendai, Japan, October 1997, Proceedings*, Lecture Notes in Artificial Intelligence, **1316**, Springer-Verlag, pp. 176–196, 1997.
71. Sato, M., Sakurai, T. and Burstall, R., Explicit environments, *Fundamenta Informaticae* 45, pp. 79–115, 2001.
72. Sato, M., Sakurai, T. and Kameyama, Y., A simply typed context calculus with first-class environments, in *Proc. Fifth International Symposium on Functional and Logic Programming (FLOPS)*, Lecture Notes in Computer Science, **2024**, pp. 359–374, 2001.
73. Sato, M., Kameyama, Y., Takeuti, I., CAL: A computer assisted system for learning compuation and logic, in *EuroCAST 2001*, Lecture Notes in Computer Science,**2178**, 2001.
74. Scott, D., Constructive validity, *Symposium on automatic demonstration*, Lecture Notes in Mathematics, **125**, pp. 237 – 275, Springer, Berlin, 1969.
75. Simpson, S.G., Partial realization of Hilbert's program, *J. Symbolic Logic*, **53**, pp. 359 – 394, 1988.
76. Skolem, T., The foundations of elementary arithmetic established by means of the recursive mode of thought, without the use of apparent variables ranging over infinite domains, in [91], pp. 302 – 333.
77. Smorynski, C., The incompleteness theorems, Barwise, J. ed., *Handbook of mathematical logic*, Studies in logic and the foundations of mathematics, **90**, pp. 821 – 865, North-Holland, 1977.
78. Stoyan, H., The influence of the designer on the design – J. McCarthy and LISP, in V. Lifschitz ed., *Artificial Intelligence and Mathematical Theory of Computation (Papers in Honor of John McCarthy)*, Academic Press, pp. 409–426, 1991.
79. Smullyan, R., *Theory of Formal System*, Annals of Mathematics Studies, **47**, Princeton University Press, Princeton, 1961.
80. Sundholm, G., Constructions, proofs and the meaning of logical constants, *J. Phil. Logic*, **12**, pp. 151 – 172, 1983.
81. Tait, W.W., Finitism, *J. of Philosophy*, **78**, pp. 524 – 546, 1981.
82. Takeuti, G., Generalized Logic Calculus, *Journal of Mathematical Society of Japan*, 1954.
83. Takeuti, G., *Proof Theory*, North-Holland, Amsterdam, 1975.
84. Takeuti, G., A conservative extension of Peano arithmetic, *Two Applications of Logic to Mathematics*, Part II, Princeton University Press, and Publications of the Mathematical Society of Japan **13**, 1978.
85. Talcott, C., A Theory of binding structures and applications to rewriting, *Theoretical Computer Science*, **112**, pp. 99-143, 1993.
86. Tatsuta, M., Program synthesis using realizability, *Theoretical Computer Science*, **90**, pp. 309 – 353, 1991.
87. Troelstra, A.S., Realizability and functional interpretation, in Troelstra, A.S. ed., *Metamathematical Investigation of Intuitionistic Arithmetic and Analysis*, Lecture Notes in Mathematics, **344**, Springer-Verlag, 1973.
88. Troelstra, A.S. and van Dalen, D., *Constructivism in Mathematics, An Introduction, I*, North-Holland, Amsterdam, 1988.

89. Troelstra, A.S., Schwichtenberg, H., *Basic Proof Theory*, Cambridge University Press, 1996.
90. Tsukada, Y., Martin-Löf's type theory as an open-ended framework, *International J. of Foundations of Computer Science*, **12**, pp. 31 – 68, 2001.
91. van Heijenoort, J., *From Frege to Gödel, A source Book in Mathematical Logic, 1879 – 1931*, Harvard University Press, Cambridge, MA, 1967.
92. Voda, P., Theory of pairs, part I: provably recursive functions, *Technical Report of Dept. Comput. Science*, UBC, Vancouver, 1984.
93. Voda, P., Computation of full logic programs using one-variable environments, *New Generation Computing*, **4**, pp. 153 – 187, 1986.

Efficient Data Mining from Large Text Databases

Hiroki Arimura[1,2], Hiroshi Sakamoto[1], and Setsuo Arikawa[1]

[1] Department of Informatics, Kyushu Univ., Fukuoka 812-8581, Japan
{arim,hiroshi,arikawa}@i.kyushu-u.ac.jp
[2] PRESTO, Japan Science and Technology Corporation, Japan

Abstract. In this paper, we consider the problem of discovering a simple class of combinatorial patterns from a large collection of unstructured text data. As a framework of data mining, we adopted *optimized pattern discovery* in which a mining algorithm discovers the best patterns that optimize a given statistical measure within a class of hypothesis patterns on a given data set. We present effcient algorithms for the classes of proximity word association patterns and report the experiments on the keyword discovery from Web data.

1 Introduction

The rapid progress of computer and network technologies makes it easy to store large amount of unstructured or semi-structured texts such as Web pages, HTML/XML archives over the Internet. Unfortunately, the current information retrieval and data mining technologies may not be sufficient for huge, heterogeneous, weakly-structured data, and thus, fast and robust text data mining methods are required [8, 18, 28].

Our research goal is to devise an efficient semi-automatic tool that supports human discovery from large text databases. To achieve this goal, we adopt the framework of *optimized pattern discovery* [12, 19, 25], and develop efficient and robust pattern discovery algorithms combining the advanced technologies in string algorithm, computational geometry, and computational learning theory.

As a class of pattern to discover, we introduce the class of *proximity phrase patterns*, which are pairs of a sequence of d phrases and a proximity k such that

$$\pi = (\langle \texttt{attack on} \rangle, \langle \texttt{iranian oil platform} \rangle; 8),$$

which expresses that $\langle$`attacks`$\rangle$ first appears in a text and then $\langle$`iranian oil platform`$\rangle$ follows within 8 words. We also introduced the *unordered version* with unordered set of phrases.

We first devise an efficient algorithm, called *Split-Merge*, that discovers all optimal patterns for the class of *ordered* k-proximity d-phrase association patterns [27]. For nearly random texts of size n, the Split-Merge algorithm quickly runs in almost linear time $O(k^{d-1}(\log n)^{d+1}n)$ using $O(k^{d-1}n)$ space. By experiments on an English text collection of 15MB, the algorithm finds the best

S. Arikawa and A. Shinohara (Eds.): Progress in Discovery Science 2001, LNAI 2281, pp. 123–139.

600 patterns at the entropy measure in a few minutes using a few hundreds mega-bytes of main memory [6].

Then, we also devise another algorithm, called *Levelwise-Scan*, for mining ordered and unordered phrase patterns from large disk-resident text data [11]. Based on the design principle of the Apriori algorithm of Agrawal [3], the Levelwise-Scan algorithm discovers all frequent unordered patterns in time $O(m + k^{d-1}n(\log n)^d)$ and space $O(n \log n + R)$ on nearly random texts of size n with high probability using the random sampling technique, where m is the total length of prefix maximal frequent substrings in the random sample and R is the total size of frequent phrases.

The rest of this paper is organized as follows. In Section 2, we prepare basic definitions and notations on optimized pattern discovery with proximity phrase patterns. In Section 3, we present the Split-Merge algorithm that quickly finds optimal ordered proximity phrase patterns from a collection of binary labeled texts using the suffix array data structure. In Section 4, we present the Levelwise-Scan algorithm for the frequent pattern discovery problem in large disk-resident text data. In Section 5, we report experiments of discovery of important keywords from large Web data. In Section 6, we conclude the results. This paper mainly consists of materials from [6, 11, 17, 27], and Chapter 4 contains some new results.

2 Preliminaries

2.1 Sets and Strings

For a set A, we denote the cardinality of A by $|A|$. We denote by $\mathbf{N}$ the set of the nonnegative integers. For an integer $n \in \mathbf{N}$ and a pair $i \leq j$ of nonnegative integers, we denote by $[n]$ and $[i..j]$ the set $\{1, \ldots, n\}$ and the interval $\{i, \ldots, j\}$, respectively.

Let Σ be a finite alphabet. We assume an appropriate total ordering on the letters in Σ. We denote by ϵ the empty string of length zero. For strings $s, t \in \Sigma$, we denote by $|s|$ the length of s, by $s[i]$ with $1 \leq i \leq |s|$ the ith letter of s, and by $s \cdot t$ or simply by st the concatenation of two strings s and t. For a set $S \subseteq \Sigma^*$ of strings, we denote by $||S|| = \sum_{s \in S} |s|$ the total length of strings in S.

Let t be a string t. If $t = uvw$ for some (possibly empty) strings $u, v, w \in \Sigma^*$ then we say that u, v and w are a *prefix*, a *substring*, and a *suffix* of t, respectively. For a string s, we denote by $s[i..j]$ with a pair $1 \leq i \leq j \leq |s|$ of integers the substring $s[i]s[i+1] \cdots s[j]$. Let s be a substring of $t \in \Sigma^*$ such that $t[i..i + |s| - 1] = s$ for some $1 \leq i \leq |t|$. Then, we say that the index i is an *occurrence of s in t*, or *s occurs in t at position i*. Conversely, we say that the index $j = i + |s| - 1$ is an *end-occurrence of s in t*, or *s occurs in t at end-position j*.

2.2 The Class of Patterns

The class of patterns we consider is the class of proximity phrase association patterns [27], which are association rules over substrings in a given text.

- Monday's **attack on** two Iranian **oil platforms** by American forces in the Gulf.
- ... **attack on** two Iranian **oil platforms** in retaliation for an Iranian attack last Friday on a Kuwaiti ship ...
- ... the action would involve an **attack** on an **oil platform**.
- ... the United States' **attack on** an Iranian **oil platform** on Monday and said it should not worsen the Gulf crisis.
- The **attack on** the **oil platform** was the latest example of a U.S. ...
- ... **attack on** an Iranian **oil platform** in the Gulf on Monday appeared to be ...
- One source said of the **attack on** the **oil platform**: ...
- A top Iranian military official said America's **attack on** an Iranian **oil platform** on Monday had involved the United States in full-scale war ...

Fig. 1. Occurrences of the proximity phrase pattern $(\langle\texttt{attack on}\rangle, \langle\texttt{oil platform}\rangle; 8)$, which were discovered in Reuters-21578 test collection [20] in our experiments (not incliuded in this paper). The alphabet Σ consists of all English words appearing in the collection.

Let Σ be a constant alphabet of letters. For nonnegative integers $d, k \geq 0$, a *k-proximity d-phrase association patterns* ((d, k)-proximity pattern or pattern, for short) is a pair $\pi = (\alpha_1, \cdots, \alpha_d; k)$ of a sequence of d strings $\alpha_1, \cdots, \alpha_d \in \Sigma^*$ and a bounded gap length k, called *proximity*. We refer to each substring α_i of texts as a *phrase*. A phrase pattern can contain arbitrary many but bounded number of phrases as its components. Given $d, k \geq 0$, $\mathcal{P}_k^d$ denotes the class of k-proximity d-phrase association patterns. The class of *followed-by* patterns [13, 23] are special case of proximity patterns with $d = 2$.

For instance, $\pi_1 = (\langle\texttt{tata}\rangle, \langle\texttt{cacag}\rangle, \langle\texttt{caatcag}\rangle; 30)$ and $\pi_2 = (\langle\texttt{attack on}\rangle, \langle\texttt{oil platforms}\rangle; 8)$. are examples of $(3, 30)$-patterns over $\Sigma = \{a, t, c, g\}$ and $(2, 8)$-patterns over alphabet of unique words. In Fig. 1, we also show an example of the occurrences of π_2. In this example, we see that proximity phrase patterns can well represent common patterns hidden in a variety of texts with allowing substrings and proximity gap of variable-length.

Let $\pi = (\alpha_1, \ldots, \alpha_d; k)$ be a (d, k)-pattern. The pattern π *occurs* in a text if the substrings $\alpha_1, \ldots, \alpha_d$ occur in the text in this order and the distance between the occurrences of the consecutive pair of phrases does not exceed k [23]. Formally, an *occurrence* of π in a string $t \in \Sigma^*$ is a tuple $(o_1, \ldots, o_d) \in [1..n]^d$ such that for every $1 \leq j \leq d$, (i) the string α_j appears in t at the position o_j, (ii) o_j satisfies the proximity constraint $0 \leq o_j - o_{j-1} \leq k$. A pattern π *occurs in* t if there exists an occurrence of π in t. If the order of the phrases in a pattern matters as in the above definition then we call it *ordered* [27] and otherwise *unordered* [11].

Another definition of the occurrences as that for patterns with variable-length don't cares (VLDC patterns) is possible. In the definition, π *occurs in a text* $s \in \Sigma^*$ *with inter-string proximity* if there exist some strings $w_0, \ldots, w_d \in \Sigma^*$ such that $s = w_0\alpha_1 w_1 \alpha_2 \cdots w_{d-1}\alpha_n w_d$ and $0 \leq |w_i| \leq k, 1 \leq i \leq d-1$. This is equivalent to the definition of the occurrences where the constraint (ii) is changed to (ii') for every $1 < i \leq d$, o_j satisfies $0 \leq o_j - o_{j-1} - |\alpha_{j-1}| \leq k$.

2.3 Optimized Pattern Discovery

As the framework of data mining, we adopted *optimal pattern discovery* [12, 25], which is also known as *Agnostic PAC learning* [19] in computational learning theory. A *sample* is a pair (S, ξ) of a collection of texts $S = \{s_1, \ldots, s_m\} \subseteq \Sigma^*$ and a binary labeling $\xi : S \to \{0, 1\}$, called *objective function*, which may be a pre-defined or human-specified category.

Let $\mathcal{P}$ be a class of *patterns* and $\psi : [0, 1] \to \mathbf{R}$ be a symmetric, concave, real-valued function called *impurity function* [9, 25] to measure the skewness of the class distribution after the classification by a pattern. The *classification error* $\psi(x) = \min(p, 1-p)$ and the *information entropy* $\psi(x) = -p \log p - (1-p) \log(1-p)$ are examples of ψ [9, 25]. We identify each $\pi \in \mathcal{P}$ to a classification function (matching function) $\pi : \Sigma^* \to \{0, 1\}$ as usual.

Given a sample (S, ξ), a pattern discovery algorithm tries to find a pattern π in the hypothesis space $\mathcal{P}$ that optimizes the *evaluation function*

$$G^{\psi}_{S,\xi}(\pi) = \psi(M_1/N_1)N_1 + \psi(M_0/N_0)N_0.$$

In the above definition, the tuple (M_1, N_1, M_0, N_0) is the *contingency table* determined by π, ξ over S, namely, $M_\alpha = \sum_{x \in S}[\,\pi(x) = \alpha \wedge \xi(x) = 1\,]$ and $N_\alpha = \sum_{x \in S}[\,\pi(x) = \alpha\,]$, where $\alpha \in \{0, 1\}$ is a label indicating the occurrence of π and $[Pred] \in \{0, 1\}$ is the indicator function for a predicate $Pred$. Now, we state our data mining problem [9, 25, 27].

Optimal Pattern Discovery with ψ
Given: A set S of texts and an objective function $\xi : S \to \{0, 1\}$.
Problem: Find a pattern $\pi \in \mathcal{P}$ that minimizes the cost $G^{\psi}_{S,\xi}(\pi)$ within $\mathcal{P}$.

What is good to minimize the cost $G^{\psi}_{S,\xi}(\pi)$? For the case of the classification error ψ used in machine learning, it is known that any algorithm that efficiently solves the above optimization problem can approximate an arbitrary unknown probability distribution over labeled instances and thus can work as a robust learner in a noisy environment [9, 19].

Another data mining problem that we consider is the frequent pattern discovery problem, which is a most popular problem in data mining [2, 3]. Let $S = \{s_1, \ldots, s_m\} \subseteq \Sigma^*$ be a collection of texts. The *frequency* of a pattern $\pi \in \mathcal{P}$ in S, denoted by $freq_S(\pi)$, is the fraction σ of texts that π occurs in S to the number of all texts in S, that is, $freq_S(\pi) = \sum_{x \in S}[\,\pi(x) = 1\,] / \sum_{x \in S}[True] \geq \sigma$. Let $0 < \sigma \leq 1$ be a number,called a *minimum frequency threshold*. Now, we state our data mining problem.

Frequent Pattern Discovery
Given: A set S of texts and a number $0 < \sigma \leq 1$.
Problem: Find all σ-*frequent* patterns $\pi \in \mathcal{P}$ such that $freq_S(\pi) \geq \sigma$.

Agrawal *et al.* [2, 3] give practical algorithms for solving the frequent pattern discovery problem for the class of association rules with Boolean attributes.

2.4 Generalized Suffix Trees

Our data mining algorithms extensively use the *generalized suffix tree* (*GST*, for short) for efficiently storing substrings in a text. Let $T = \{T_1, \ldots, T_m\}$ $(m \geq 0)$ be a collection of texts and \$ be a special delimiter $\$ \notin \Sigma$, which satisfies $\$ \neq \$$. Then, let $\underline{T} = T_1\$ \cdots \$T_m\$$ be the text string obtained by concatenating all texts in a sample $T \subseteq \Sigma^*$ with the delimiters \$. In what follows, we identify $\underline{T}$ to T and let $n = |\underline{T}|$.

The GST for T [22, 28], denoted by $GST(T)$, is the *compacted trie* for all suffices of $s \in T$, which is obtained from the (uncompacted) trie [1], denoted by $UT(T)$, for all suffices of T by removing all internal nodes with a single child and concatenating its labels into a single string. Each node v of a GST represents the string $Word(v)$, called a *branching substring* w.r.t. T, which is obtained by concatenating the labels on the path from the root to v in its order. Each leaf is labeled with the name of the document it belongs to. Each internal node has the *suffix link* which points to the node w such that $Word(v) = c \cdot Word(w)$ for some $c \in \Sigma$. For a node v in $GST(T)$, we write $v \in GST(T)$. In Fig. 4, we show an example of the GST.

$GST(T)$ for T has $O(n)$ nodes, can be represented in $O(n)$ space and can be computed in $O(n)$ time [22]. $GST(T)$ uses $17n$ bytes to store and can be built in $O(n)$ time [22]. The expected height of suffix trees is known to be very small, namely $O(\log n)$, for a random string [10]. Thus, the length of the longest branching substrings is also bounded by $O(\log n)$ in this case.

3 A Fast Text Mining Algorithm with Suffix Arrays

If the maximum number of phrases in a pattern is bounded by a constant d then the frequent pattern problems for both unordered and ordered proximity phrase association patterns are solvable by a naive generate-and-test algorithm in $O(n^{2d+1})$ time and $O(n^{2d})$ scans since there are at most $O(n^2)$ distinct substrings in a given text of length n. A modified algorithm, called *Enumerate-Scan* [28], that uses a suffix tree structure improves to $O(n^{d+1})$ time but requires $O(n^d)$ scans of an input string although it is still too slow to apply real world problems.

In this section, we present an efficient algorithm, called *Split-Merge*, that finds all the optimal patterns for the class of ordered k-proximity d-phrase association patterns for various measures including the classification error and information entropy [7, 27]. The algorithm quickly searches the hypothesis space using dynamic reconstruction of the content index, called a *suffix array* with combining several techniques from computational geometry and string algorithms.

3.1 Outline of the Algorithm

In Fig. 2, we show an outline of the Split-Merge algorithm. The algorithm runs in *almost linear time in average*, more precisely in $O(k^{d-1}N(\log N)^{d+1})$ time using $O(k^{d-1}N)$ space for nearly random texts of size N [27]. The algorithm first *splits* the hypothesis space into smaller subspaces using the content index, called *suffix arrays*, and quickly rebuilds a new content index by *merging* all indexing points that corresponds to the occurrences of the patterns for each subspace. This process is recursively applied to the subspaces and finally results optimized patterns with the sets of their occurrences.

We also show that the problem to find one of the best phrase patterns with arbitrarily many strings is MAX SNP-hard [27]. Thus, we see that there is no efficient approximation algorithm with arbitrary small error for the problem when the number d of phrases is unbounded.

3.2 Canonical Patterns and Suffix Arrays

Let (T, ξ) be a sample, where $T = \{T_1, \ldots, T_m\}$ $(m \geq 0)$. In the following steps, we will work on the text string $\underline{T} = T_1\$ \cdots \$T_m\$$ obtained by concatenating all texts in a sample $T \subseteq \Sigma^*$ with a delimiter $\$ \notin \Sigma$, which satisfies $\$ \neq \$$. Let $n = |\underline{T}|$. The *document function* is a function $\delta : [1..n] \to [1..m]$ such that for every position p in $\underline{T}$, $\delta(p) = i$ if p is an position in the i-th text T_i. In what follows, we identify $\underline{T}$ to T.

Recall that a branching substring is a substring of T that is represented as $Word(v)$ for some $v \in GST(T)$. A (d, k)-proximity pattern $\pi = (\alpha_1, \ldots, \alpha_d; k)$ is said to be *in canonical form* if every substring $\alpha_j, 1 \leq j \leq d$, is a branching substring of T. Let (S, ξ) be a sample. Let $\bot_S \in \Sigma^*$ be an arbitrary string whose length is $\max\{\, |s| \mid s \in S\} + 1$. Clearly, $\bot_S$ occurs no strings in S.

Lemma 1 ([27]). *$G_{S,\xi}(\pi)$ is minimized over all (d, k)-proximity patterns by either a pattern π in canonical form or $\bot_S$.*

We define the *suffix array for T* as follows [24]. For every $1 \leq p \leq n = |T|$, let T_p be the suffix $T[p..n]$ starting at position p. The suffix array is an array $SA[1..n]$ that stores the starting positions of the lexicographically ordered suffixes $T_{p_1} <_{\text{lex}} T_{p_2} <_{\text{lex}} \cdots <_{\text{lex}} T_{p_n}$, where $<_{\text{lex}}$ is the lexicographic order on Σ^*. That is, $SA[i]$ is the starting position of the suffix of T with lexicographic rank i. We also define the *inverse suffix array* $SA^{-1}[1..n]$ that represents the inverse function of $SA[1..n]$ as $SA^{-1}[SA[i]] = i$ for every $1 \leq i \leq n$. The suffix array SA corresponds the list of leaves of $GST(T)$ from left to right order, and built from $GST(T)$ in linear time in n. SA is also directly constructed from T in $O(n \log n)$ time by using string sorting [24].

Let $d, k \geq$ be nonnegative integers. Then, the *diagonal set* $Diag_{d,k} \subseteq [1..n]^d$ is defined as the set of all $(o_1, \ldots, o_d) \in [1..n]^d$ such that (i) for every $1 < j \leq d$, $0 \leq o_j - o_{j-1} \leq k$ holds; (ii) for every $1 \leq j \leq d$, there exists some i such that $\delta(o_j) = i$. The volumn of $Diag_{d,k}$ is at most $k^{d-1}n$. An important observation is that for any branching substring α in T, the occurrences of α in the suffix array SA for T occupies a contiguous subinterval of $[1..n]$, denoted by $Rank(\alpha)$, since the positions in SA are ordered lexicographically. For (d, k)-proximity pattern $\pi = (\alpha_1, \ldots, \alpha_d; k)$, we define the d-dimensional box $Rank(\pi) \subseteq [1..n]^d$ by $Rank(\pi) = Rank(\alpha_1) \times \cdots \times Rank(\alpha_d)$. By the inverse suffix array SA^{-1},

---

Algorithm Split-Merge-with-Array (SMA):
- *Given*: A sample of texts $T \subseteq \Sigma^*$, an objective function $\xi : T \to \{0,1\}$, integers $k, d \geq 0$.
- Initialize $\underline{R} = Create(T)$ by build the suffix array SA, the inverse mapping SA^{-1}, and the height array Hgt. Invoke $Find(R, d, \varepsilon)$ below and report the patterns with the minimal cost.

Find(Q, d, π):
- If $d = 0$, then compute the cost $G^{\psi}_{S,\xi}(\pi)$ from Q, which corresponding to the set of the occurrences of π in T. Record π with its cost.
- Otherwise, $d > 0$.
  Reconstruct $\underline{Q} = \overline{Reconstruct(Rank(T), Q)}$. Then, traverse all branching substrings in Q. For each $\alpha \in \overline{Traverse(\underline{Q})}$, do the followings:
  - (i) $R = Search(\underline{Q}, \alpha)$. This computes the set of all occurrences of α in Q. $P = SA(R)$ and $\overline{P} = Shift(P, 0, k)$. These steps applies inverse transpose to R and shifts the positions to the right by adding the skips $0 \leq x \leq k$.
  - (ii) $R = SA^{-1}(P)$. This transforms positions into the corresponding ranks in R with SA^{-1}.
  - (iii) Finally, invoke $Find(R, d-1, \pi \cdot \alpha)$ recursively.

Fig. 2. The Split-Merge algorithm

we can transform d-tuple $O = (o_1, \ldots, o_d) \in [1..n]^d$ of positions into a d-tuple $Trans(O) = (SA^{-1}[o_1], \ldots, SA^{-1}[o_d])$, called the *transpose* of O. Then, we have the following characterization of the occurrences of patterns.

Lemma 2 ([23, 27]). *Let $d, k \geq 0$ be nonnegative integers. For every (d,k)-proximity pattern $\pi = (\alpha_1, \ldots, \alpha_d; k)$ in canonical form and every $O = (o_1, \ldots, o_d) \in [1..n]^d$, the following equivalence holds: (i) O is an occurrence of π if and only if (ii) $O \in Diag_{d,k}$ and $Trans(O) \in Rank(\pi)$.*

The above lemma tells us that the optimal pattern discovery problem for $\mathcal{P}^d_k$ can reduce to that for d-dimensional boxes in a diagonal set.

3.3 The Algorithm with Suffix Array

In Fig. 2, we show our mining algorithm *Split-Merge-with-Array* (Split-Merge, for short). The input to the Split-Merge algorithm is a tuple $I = (T, \xi, d, k)$, where $T \subseteq \Sigma^*$ is a sample of texts, $\xi : T \to \{0,1\}$ is an objective function, and $d, k \geq 0$ are nonnegative integers. In what follows, we will work on the text $\underline{T} = T_1\$_1 \cdots T_m\$_m$ obtained by concatenating all texts in $T = \{T_1, \ldots, T_m\}$ delimited with unique markers \$. Here, we assume that the document id $\delta(j)$ and the label $\xi(j) = \xi(T_i)$ are attached with each position j. If is is clear from context, we will write T for $\underline{T}$.

A key data structure of our algorithm is a new indexing data structure called a *dynamic reconstructible suffix dictionary* described below, which stores a subset of suffixes of a given input text T. By this data structure, we efficiently implement the idea of reducing the discovery problem for (d,k)-patterns to that for d-dimensional boxes discussed in the previous subsection.

Let $T \in (\Sigma \cup \{\$\})^*$ be an input text of length n. We denote by $Rank(T) = [1..n]$ the set of the lexicographic ranks of all suffixes of T. A *rank set* in T is any subset of $Rank(T)$. Since we identify a suffix and its lexicographic rank, a rank set represents a set of suffixes. A *reconstructible suffix dictionary* for T is data structure for storing a subset of suffixes of T, called the *domain*. The reconstructible suffix dictionary with a rank set $Q \subseteq Rank(T)$ as its domain is denoted by $\underline{Q}$. Then, a reconstructible suffix dictionary is defined as a data structure with the following operations, where $Q \subseteq Rank(T)$ is an arbitrary rank set and $m = |Q|$.

Reconstructible suffix dictionary

- *Create*: Given text T of length n, returns a new global dictionary $Create(T) = \underline{Rank(T)}$ containing all suffixes of T in $O(n \log n)$ time.
- *Reconstruct*: Given dictionary $\underline{Q}$ and a subset $R \subseteq Q$ of its domain, create the dictionary $Reconstruct(\underline{Q}, R) = \underline{R}$ with domain R in $O(|R| \log |R|)$ time.
- *Traverse*: Given dictionary $\underline{Q}$, enumerate all branching substring α occuring in positions in Q in $O(|Q|)$ time. The resulting set of strings is denoted by $Traverse(\underline{Q})$.
- *Search*: Given $\underline{Q}$ and a string α, returns the set $Search(\underline{Q}, \alpha)$ of ranks corresponding to all occurrences of α contained in Q in $O(\alpha \log |Q|)$ time if exists.
- *Domain*: Given $\underline{Q}$, returns its domain $Domain(\underline{Q}) = Q$.

We implemented the above data structure using the suffix array $SA[1..n]$ plus the height array $Hgt[1..n]$ of the given text $T[1..n]$ [24] and the constant-time lcp information [26]. Here, the *height array* is the array that stores the longest common prefix (lcp) length of the adjacent suffixes in SA [24]. For *Reconstruct* operation, we compute the dictionary $\underline{R} = Reconstruct(\underline{Q}, R)$ of size m in $O(m \log m)$ time by using radix sort of ranks in the global dictionary $\underline{Rank(T)}$ and constant-time lcp information [26]. Note that the reconstruction of $\underline{R}$ takes $O(|T| \log |T|)$ time when we use string sorting alone. For *Traverse* operation, we developed an efficient algorithm that simulates the traversal of the suffix tree for T in $O(n)$ time by using SA and Hgt [17], while well-known simulation of the suffix tree by binary search takes $O(n^2)$ time (or $O(n \log n)$ time with lcp info [26]). We also showed that the height array can be directly built in linear time from SA and T [17].

We also require the following operations for manipulating sets of ranks and positions, where R and P denote sets of ranks and positions, respectively.

- *Inverse Transpose*: Given a set R of ranks, returns the set $SA(R) = \{ p \mid p = SA[i], i \in R \}$.
- *Transpose*: Given a set P of positions, returns the set $SA^{-1}(P) = \{ i \mid i = SA^{-1}[p], p \in P \}$.
- *Shift*: Given a set P of positions and a nonnegative integer $k \leq 0$, returns the set $Shift(P, l, h) = \{ p + x \mid p \in P, l \leq x \leq h \}$.

Using these operations, now we can see how the Split-Merge algorithm in Fig. 2 works. Starting from the pair of the universe set of ranks $Rank(T)$ and the empty pattern $(\varepsilon; k)$, the algorithm simultaneously computes the pairs (Q, π) of a set Q of possible occurrences and a possible (d, k)-proximity pattern π with at least one occurrences in T at a position in Q by a devide and conquer manner. To deal with patterns with inter-string proximity, we can change the shift operation $P = Shift(P, 0, k)$ to $P = Shift(P, |\alpha|, |\alpha|+k)$. However, this modification may lose the completeness of the algorithm. By an analysis similar to [27], we have the following theorem in the case of ordinary definition of proximity. Details of the proof will appear in elsewhere.

Theorem 1. *For every d, k, the algorithm Split-Merge-with-Array, given (S, ξ), computes all K (k, d)-proximity patterns that minimize the cost $G^{\psi}_{S,\xi}$ in expected time $O(k^{d-1}n(\log n)^d + K)$ and space $O(dn)$ under the assumption that texts in S are randomly generated from a memoryless source.*

The Split-Merge algorithm with the suffix array is simple, easy to implement, fast in existing computer with memory of moderate size comparing to a structure with suffix trees and lists of occurrences. An implementation of Split-Merge algorithm with suffix array can find thousands of optimal patterns in several seconds ($d = 1$) to several minutes ($d = 4$ and $k = 8$ words) on WS (Ultra SPARC II, 300MHz, 256MB, Solaris 2.6, gcc) for English texts of 15.2MB [6]. This is 10^2–10^3 times speed-up to another algorithm with suffix tree [7].

4 A Scan-Based Algorithm for Ordered and Unordered Patterns

We have to often deal with huge amount of text data that cannot fit into main memory in real applications. In this section, we present a practical algorithm that efficiently finds all frequent (d, k)-proximity patterns in large disk-resident text data [11]. In this section, we consider the frequent pattern discovery problem for proximity patterns with inter-string proximity.

4.1 Outline of the Algorithm

Fig. 3 shows our algorithm Levelwise-Scan for solving the frequent pattern discovery problem, which employs the levelwise search strategy adopted by the *Apriori algorithm* [2, 3]. The algorithm also combines techniques of random sampling, the generalized suffix tree, and the pattern matching automaton to to cope with the huge feature space of text mining.

The input to the Levelwise-Scan algorithm is a tuple $I = (T, d, k, \sigma, \varepsilon, m)$, where $T \subseteq \Sigma^*$ is a sample of texts, $d, k \geq 0$ are nonnegative integers, $0 < \sigma \leq 1$ is a frequency threshold, $0 < \varepsilon < \sigma$ is a small positive number, called *margin*, and $0 \leq m \leq |T|$ is the size of a random sample from T.

Given these inputs, the Levelwise-Scan algorithm works as follows. In Stage 1, the algorithm first builds a random sample $\hat{T} \subseteq T$ by drawing m texts in T with duplicates. Secondly, with a slightly smaller threshold $\hat{\sigma} = \sigma - \varepsilon$ than σ, the algorithm computes the set $\mathcal{S}_{\hat{\sigma}}$ of all $\hat{\sigma}$-frequent (branching) substrings in $\hat{T}$ by computing the GST of $\hat{T}$ in linear time in $m = |\hat{T}|$. Then, we build a deterministic finite automaton $ACm(\mathcal{S})$ for recognizing strings in $\mathcal{S}$, called an AC-machine. Clearly, $\mathcal{L}_1$ is a subset of $\mathcal{S}_{\hat{\sigma}}$ and easily obtained through the construction of $\mathcal{S}_{\hat{\sigma}}$ since $\sigma \geq \hat{\sigma}$.

In Stage 2, the algorithm computes the set $\mathcal{L}_i$ of all σ-frequent d-phrase patterns from $\mathcal{L}_{i-1}$ obtained in the previous stage through the stages $i = 2, \ldots, d_{\max}$. In the following steps, we will work on the text string $\underline{T} = T_1\$ \cdots \$T_m\$$ obtained by concatenating all texts in T with a delimiter $\$ \notin \Sigma$. At every stage $d \geq 2$, the algorithm first build a set $\mathcal{C}_d$ of candidates for frequent (d, k)-patterns from $\mathcal{L}_{d-1}$ and $\mathcal{L}_1$.

Next, the algorithm runs the AC-machine $ACm(\mathcal{S}_{\hat{\sigma}})$ on $\underline{T}$ from left to right and detects the end-positions of $\hat{\sigma}$-frequent substrings in $\mathcal{S}_{\hat{\sigma}}$ by recording such detected end-positions with a bounded buffer B of limited length. At each position $p = 1, \ldots, |\underline{T}|$ in the scan, the algorithm simultaneously detects all (d, k)-patterns in $\mathcal{C}_d$ that appears in T at the current position p as their end-positions and increments their counts. After a scan on $\underline{T}$, the algorithm records all σ-frequent patterns in $\mathcal{C}_d$, and proceeds to the next stage. Iterating these stages, the algorithm computes all frequent patterns in T. In the rest of this section, we will describe the details of the above steps.

4.3 An AC-machine

A key idea of our algorithm is to use a deterministic finite automaton, called an AC-machine, for detecting all substrings in $\mathcal{S}_{\hat{\sigma}}$ appearing in T (Stage 1 (b)). For a set $P \subseteq \Sigma^*$ of strings over Σ, the *Aho-Corasick machine* (AC-machine, for short) for P is a finite state automaton $ACm(P) = (Q, \Sigma, \delta, q_0, Output)$ with ε-transitions [5] that recognizes the language $\Sigma^* P = \{\, \alpha\beta \mid \alpha \in \Sigma^*, \beta \in P \,\}$. In Fig. 5, we show an AC-machine for $S = \{abca\$_1, bbca\$_2\}$. In this figure, we see that this AC-machine is isomorphic to the GST for the same set S in Fig. 4. An AC-machine for P has the output function $Output : Q \to 2^{\Sigma^*}$ that assign to each state $q \in Q$ the set $Output(q)$ of all strings in P that are suffixes of $Word(q)$, the string obtained by concatenating the labels on the path from q_0 to q. We assume that all ε-transitions in $ACm(P)$ are removed by a standard technique [5, 15].

In our algorithm, we build the AC-machine $ACm(\mathcal{S}_{\hat{\sigma}})$ for recognizing the set $\mathcal{S}_{\hat{\sigma}}$ of $\hat{\sigma}$-frequent substrings in $\hat{T}$. Let $\max \mathcal{S}_{\hat{\sigma}}$ be the set of *prefix-maximal* substrings in $\mathcal{S}_{\hat{\sigma}}$ and $m = ||\max \mathcal{S}_{\hat{\sigma}}||$, where a prefix maximal substring in a set $S \subseteq$ is a string that are not a proper prefix of any string in S. By [5], we can show that $ACm(\mathcal{S}_{\hat{\sigma}})$ has $O(m)$ size and can be built from $GST(\hat{T})$ in $O(m)$ time.

Algorithm Levelwise-Scan (LS):

- *Given*: Integers $k, d_{\max} \geq 0$, any numbers $0 < \sigma \leq 1$, $0 < \varepsilon < \sigma$, any integer $\hat{m} \leq m$, and a sample set $T = \{T_1, \ldots, T_m\}$ $(m \geq 0)$ of texts.

1. Preprocessing (stage 1).
   (a) For a random sample $\hat{T} \subseteq T$ of size $\hat{m}$ and $\hat{\sigma} = \sigma - \varepsilon$, compute the set $\mathcal{S}_{\hat{\sigma}}$ of $\hat{\sigma}$-frequent substrings in $\hat{T}$ using $GST(T)$. Let ℓ be the length of the longest branching substring in $\mathcal{S}_{\hat{\sigma}}$.
   (b) Let $\underline{T} = T_1\$ \cdots \$T_m\$$ and let $\mathcal{L}_1 := \mathcal{S}_\sigma$. Build the AC-machine $ACm(\mathcal{S}_{\hat{\sigma}}) = (Q, \Sigma, \delta, q_0, Output)$ for recognizing the strings of $\mathcal{S}_{\hat{\sigma}}$ in $O(||\max(\mathcal{S}_{\hat{\sigma}})||)$ time.
2. For each stage $d = 2, \ldots, d_{\max}$, do:
   (a) Prepare the d-th candidate set $\mathcal{C}_d$ consisting of the d-patterns obtained from $\mathcal{C}_{d-1}$ and $\mathcal{L}_1$. This takes $O(d \cdot |\mathcal{C}_{d-1}| \cdot |\mathcal{L}_1|)$ time.
   (b) Initialize the bounded buffer $B : [1..n] \to \mathcal{S}_{\hat{\sigma}}$ of length $h = (d-1)(k+\ell)$ by $B(i) = \emptyset$ for each $i = 1, \ldots, h$. Initialize the marker $M : \mathcal{C}_d \to [1..n]$ and the counter $C : \mathcal{C}_d \to \mathbf{N}$ as $M(\pi) = -1$ and $C(\pi) = 0$ for each candidate $\pi \in \mathcal{C}_d$. Let $\delta = 1$.
   (c) Scanning T for position $i = 1, \ldots, n = |T|$, do:
      (i) Read the next letter $\underline{T}[i]$. If $\underline{T}[i]$ is the document delimiter \$ then $\delta := \delta + 1$ and go to the next position.
      (ii) Otherwise, make a transition to the next state $q \in Q$ on $ACm(\mathcal{S}_{\hat{\sigma}})$, and execute the substeps (ii)-(iii) below. Record $B[i] := Output(q)$ and discard the oldest entry $B[i-h]$ from the buffer B. This takes $O(\ell)$ time since $|Output(q)| \leq \ell$ holds.
      (iii) Compute the set $Adm(B, i, d, k)$ of all admissible patterns whose end-positions are $o_d = i$ in $O(dk^{d-1}\ell^d)$ time (Lemma 3). For each $\pi = (\alpha_1, \ldots, \alpha_d; k) \in Adm(B, i, d, k)$, do the followings: If $M(\pi) < \delta$ then update the mark by $M(\pi) := \delta$ and increment the counter by $C(\pi) := C(\pi) + 1$.
   (d) Record as σ-frequent patterns all patterns $\pi \in \mathcal{C}_d$ with frequency $C(\pi) \geq m\sigma$. Let $\mathcal{L}_d$ be the set of all σ-frequent d-patterns.
3. Output all members in $\mathcal{L}_1 \cup \cdots \cup \mathcal{L}_{d_{\max}}$.

Fig. 3. Levelwise-scan: The algorithm for finding proximity phrase patterns with an AC-machine

4.4 Counting the Occurrences

Candidate generation. Let $d \geq 2$ be a stage. In step 2(a), the Levelwise-Scan algorithm computes the set $\mathcal{C}_d$ of candidate d-patterns of the form $\pi = (\alpha_1, \ldots, \alpha_{d-1}, \alpha_d; k)$ for all combination of $(\alpha_1, \ldots, \alpha_{d-1}; k) \in \mathcal{L}_{d-1}$ and $\alpha_d \in \mathcal{L}_1$. Clearly, $\mathcal{L}_d \subseteq \mathcal{C}_d$ holds. We store all candidate patterns of $\mathcal{C}_d$ in a hash-based trie [3] on alphabet $\Sigma = \{ \alpha \mid \alpha \in \mathcal{S}_{\hat{\sigma}} \}$ by identifying each substring α as the internal node $v \in GST(\hat{T})$.

Scanning the input text. Next, the algorithm counts the number of occurrences of each candidate pattern of $\mathcal{C}_d$ by scanning a text string $\underline{T} = T_1\$ \cdots \$T_m\$$ of length n on secondary memory. To do this, the algorithm runs the AC-machine

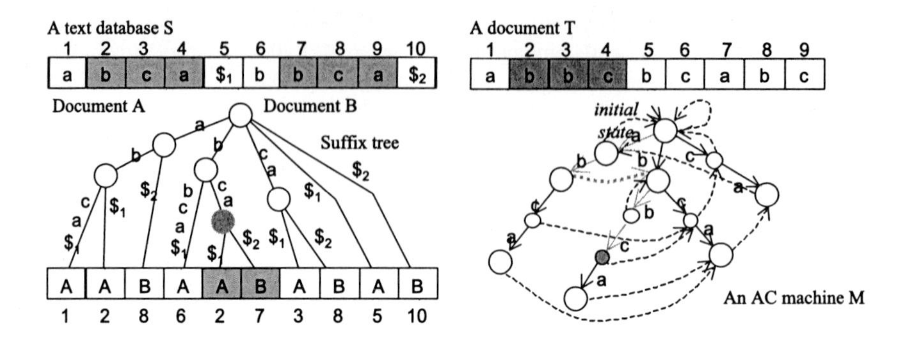

Fig. 4. The generalized suffix tree (GST) ST for $S = \{abca\$_1, bbca\$_2\}$. The shadowed node corresponds to the substring bca and two shadowed leaves correspond to the two occurrences of bca in S. Suffix links are omitted.

Fig. 5. The AC-pattern matching machine M for $P = \{abca, bbca, bca, ca, a\}$, which corresponds to all internal nodes of the GST in Fig. 4. The solid and the dotted lines indicate goto and failure functions, respectively.

$ACm(\mathcal{S}_{\hat{\sigma}})$ on the text $\underline{T}$ which resides in the secondly storage. At every position $i = 1, \ldots, n$, the algorithm reads the next letter $c = \underline{T}[i]$ from the disk, makes a transition from the current state $p \in Q$ to the next state $q \in Q$ on $ACm(\mathcal{S}_{\hat{\sigma}})$ with reading c, and records the output value by $B[i] = Output(i)$ and discards the oldest value $B[i-h]$ in a bounded buffer $B[h-i..i]$ of length h. The precise value of h will be determined later.

After recording the output value at position i, the algorithm detects all patterns in $\mathcal{P}_k^d$ that appear at end-position i at (iii) of step 2(c) as follows. Let $\pi = (\alpha_1, \ldots, \alpha_d; k)$ be a (d, k)-pattern. For a position i, a tuple $(o_1, \ldots, o_d) \in [1..n]^d$ is an *admissible occurrence with end-position* i for π if for every $1 \le j \le d$, (i) the string α_j appears in $\underline{T}$ at the end-position o_j, (ii) o_j satisfies the proximity constraint $0 \le o_j - o_{j-1} - |\alpha_j| \le k$, and (iii) $o_d = i$. A pattern $\pi = (\alpha_1, \ldots, \alpha_d; k)$ is *admissible at* i if there exists an admissible occurrence with end-position i for π. We denote by $Adm(B, i, d, k)$ the set of all admissible patterns with end-position i. Let $\underline{T}$ be a text string. Suppose that for every i, the i-th element $B[i]$ of the bounded buffer contains all strings in $\mathcal{S}_{\hat{\sigma}}$ that appear in $\underline{T}$ at end-position i.

Lemma 3. *Let $d, k \ge 0$ be any integers, ℓ be the length of the longest strings in $\mathcal{S}_{\hat{\sigma}}$, and $h = (d-1)(k+\ell)$ be the length of the buffer B. For every position $1 \le i \le n$, the set $Adm(B, i, d, k)$ has at most $k^{d-1}\ell^d$ members and computable in time $O(dk^{d-1}\ell^d)$.*

Proof. The lemma follows from the following recursive definition: $Adm(i, d, k) = \{\varepsilon\}$ if $d = 0$, and $Adm(B, i, d, k) = \bigcup_{\alpha \in B(i)} \bigcup_{1 \le x \le k} Adm(i - |\alpha| - x, d-1, k) \cdot \{\alpha\}$ if $d > 0$. It is easy to implement this definition to run in the claimed time. □

4.5 The Correctness and the Time Complexity

Now, we see the correctness and the running time of the algorithm. From the arguments in the previous subsections, we can show that the *Levelwise-scan* algorithm computes any σ-frequent (k,d)-proximity pattern π if all strings in π appears in a random sample $\hat{T}$ of the input collection $T \subseteq \Sigma^*$ of texts. Unfortunately, the algorithm may not find all σ-frequent patterns since it uses a random sample $\hat{T}$ to compute the set $\mathcal{S}_{\hat{\sigma}}$.

Now, we have the following theorem which says that for sufficiently large random sample $\hat{T}$ of T, the Levelwise-Scan algorithm correctly computes all σ-frequent patterns with high probability.

Theorem 2. *Let $I = (T,d,k,\sigma,\varepsilon,m)$ be any instance of the frequent pattern discovery problem for $\mathcal{P}^d_k$. Let $h = (d-1)(k+\ell)$ be the length of buffer B, where ℓ is the length of the longest $(\sigma-\varepsilon)$-frequent branching substrings in a random sample $\hat{T}$. The* Levelwise-scan *algorithm of Fig. 3 computes all σ-frequent (d,k)-proximity patterns in T with probability at least $1-\delta$ when the cardinality m of the random sample $\hat{T}$ satisfies:*

$$m \geq \frac{1}{2\sigma\varepsilon^2}(\log\frac{1}{\delta} + \log n + o(\log n)).$$

Furthermore, the algorithm runs in expected time $O(M + dk^{d-1}(\log n)^d n)$ and space $O(M + dk^{d-1}(\log n)^d n)$ on a collection T of randomly generated texts from memoryless source, where M is the total length of the prefix-maximal $\hat{\sigma}$-frequent substrings in a random sample $\hat{T}$.

Proof. For random text collection T of total size n generated by memoryless source, the upper bound ℓ of the length of the longest branching substring in $\mathcal{S}_\sigma$ is expected to be $O(\log n)$ [10]. More precisely, we may take summation of the running time over all values of ℓ with probability [10]. Hence, the time complexity of the algorithm follows from Lemma 3 [11]. The sample complexity of the algorithm is derived by estimating the probability P_{bad} that $\hat{T}$ contains less than $\hat{\sigma}$ texts in which π appears when π is σ-frequent in T using Chernoff bound [16]. □

4.6 A Modified Version for Unordered Proximity Patterns

The algorithm *Levelwise-scan* of Fig. 3 can be modified for discovering unordered versions of σ-frequent proximity patterns. The only difference to the original version is that the algorithm does not distinguish any permutation $(\alpha_{i_1},\ldots,\alpha_{i_d};k)$ from the original pattern $(\alpha_1,\ldots,\alpha_d;k)$, and stores in the hash-based trie $\mathcal{C}_d$ the representative $(\alpha_1,\ldots,\alpha_d;k)$ such that $\alpha_1 < \cdots < \alpha_d$ in some total order $<$ on $\mathcal{S}_\sigma$. Candidate generation is also constrained by this order. Given a sample $T \subseteq \Sigma^*$, the modified algorithm computes all $\hat{T}$-induced σ-frequent unordered (k,d)-proximity patterns in expected time $O(m + d^2 \log dk^{d-1}(\log n)^d n)$ and space $O(m + dk^{d-1}(\log n)^d n)$.

Table 1. Mining the Web. Characterizing communities of Web pages. In each table the first, the second, the third, the fourth columns show, repectively, the rank, the number of occurrences in the positive and the negative documents, and the phrase found. The number of phrases is $d = 1$.

(a) Frequency maximization POS = honda			(b) Entropy minimization POS = honda, NEG = softbank				(c) Entropy minimization POS = honda, NEG = toyota			
1	17893	<the>	1	5477	4	<honda>	1	5477	302	<honda>
2	12377	<and>	2	2125	0	<prelude>	2	2125	9	<prelude>
3	11904	<to>	3	5626	1099	<i>	3	744	5	<vtec>
4	11291	<a>	4	1863	68	<car>	4	732	16	<si>
5	8728	<of>	5	1337	24	<parts>	5	662	23	<bike>
6	8239	<for>	6	1472	92	<engine>	6	629	23	<motorcycle>
7	7278	<in>	7	862	1	<rear>	7	3085	942	<99>
8	5752	<is>	8	744	0	<vtec>	8	376	0	<prelude si>
9	5626	<i>	9	3085	769	<99>	9	448	9	<the honda>
10	5477	<honda>	10	718	0	<exhaust>	10	555	32	<civic>
11	4838	<on>	11	754	16	<miles>	11	396	5	<honda prelude>
12	4584	<s>	12	629	2	<motorcycle>	12	310	0	<valkyrie>
13	4571	<with>	13	662	6	<bike>	13	309	0	<99 time>
14	4545	<you>	14	734	20	<racing>	14	390	12	<honda s>
15	3880	<it>	15	732	29	<si>	15	1208	266	<98>
16	3650	<or>	16	802	45	<black>	16	274	0	<scooters>
17	3447	<this>	17	526	1	<tires>	17	396	17	<rims>
18	3279	<that>	18	586	13	<fuel>	18	271	0	<date 10>
19	3115	<are>	19	555	9	<civic>	19	299	3	<92 96>
20	3085	<99>	20	1307	219	<me>	20	435	29	<accord>

5 Discovery of Important Keywords in Web

In this section, we applied our text mining method to *discovery of important keywords from large Web data* [6]. Based on the Split-Merge algorithm [27] in Section 3, we developed a prototype system on a unix workstation, and run experiments on HTML pages collected from Web.

In the Web search, a user may use a search engine to find a set of Web pages by giving a keyword relevant to an interesting topic. However, for too ambiguous keywords the search engine often returns a large amount of documents as the answers, and thus a postprocessing is required to select the pages that the user is really interested.

5.1 Method

We developed a prototype system on a unix workstation based on the Split-Merge-with-Array algorithm [27] in Section 3 in C++. By experiments on English text data of 15.2MB [20], the prototype system finds the best 600 patterns at the entropy measure in seconds for $d = 2$ and a few minutes for $d = 4$ and with $k = 2$ words using a few hundreds mega-bytes of main memory on Sun Ultra 60 (Ultra SPARC II 300MHz, g++ on Solaris 2.6, 512MB main memory) [6].

In the experiments, the mining system receives two keywords, called the *target* and the *control* as inputs, and computes a pair of collections of Web pages, called the base sets of the target and the control keywords, respectively, as follows. Given a keyword w, the mining system builds the *base set* of w, which is the set of best, say 200, pages obtained by making a query w to a Web search engine together with those pages pointed by the hyperlinks from the first set of pages. It is well known that the base set of a keyword forms a well connected community [18]. Then, the base sets corresponding to the target and the control keywords are given to the text mining system as positive and negative examples, respectively. Then, the system discovers the phrases that characterize the base set of the target keyword relative to the base set of the control keyword by using entropy minimization.

5.2 Data

We used *AltaVista* [4] as an Web search engine and Perl and Java scripts to collect Web pages from Internet. The target keyword is ⟨`honda`⟩, the name of an automobile company, and the control keywords are ⟨`softbank`⟩, the company on Internet business, and ⟨`toyota`⟩, another automobile company. From these keywords ⟨`honda`⟩, ⟨`softbank`⟩, and ⟨`toyota`⟩, we obtained collections of plain text files of size 4.90MB (52,535 sentences), 3.65MB (34,170 sentences), and 5.05MB (40,757 sentences), respectively, after removing all HTML tags. Although the keywords ⟨`honda`⟩ and ⟨`toyota`⟩ can be used as the name of persons, they were used as the names of automobile companies in most pages of high ranks.

5.3 Results

In Table 1, we show the best 20 phrases found in the target set ⟨`honda`⟩ varying the control set. In the table (a), we show the phrases found by traditional frequent pattern mining with the empty control [3]. In the next two tables, we show the phrases found by mining based on entropy minimization, where the target is ⟨`honda`⟩ and the control varies between ⟨`softbank`⟩ and ⟨`toyota`⟩. We see that (b) the mining system found more general phrases, e.g. `cars, parts, engine` in the target set ⟨`honda`⟩ with the control set ⟨`softbank`⟩, while the system found more specific phrases, e.g. ⟨`prelude si`⟩, ⟨`valkyrie`⟩, ⟨`accord`⟩, with the control set ⟨`toyota`⟩. This difference on patterns come from the fact that the target keyword ⟨`honda`⟩ is more similar to the first control keyword ⟨`toyota`⟩ than the second control keyword ⟨`softbank`⟩.

6 Conclusion

In this paper, we investigated efficient text mining based on optimized pattern discovery as a new access method for large text databases. First, we formalized text mining problem as the optimized pattern discovery problem using a statistical measure. Then, we gave fast and robust pattern discovery algorithms, which were applicable for a large collection of unstructured text data. We ran computer experiments on keyword discovery from Web, which showed the effectiveness of our method on real text data.

Acknowledgments

This work is partly supported by a Grant-in-Aid for Scientific Research on Priority Areas "Discovery Science" from the Ministry of Education, Science, Sports, and Culture in Japan. We would like to thank to Junichiro Abe, Hiroki Asaka, Ryouichi Fujino, Shimozono Shinich for fruitful discussions and comments on this issue.

References

1. A. V. Aho, J. E. Hopcroft, and J. Ullman, *The design and Analysis of Computer Algorithms.* Addison-Wesley, 1974.
2. R. Agrawal, H. Mannila, R. Srikant, H. Toivonen, and A. I. Verkamo, Fast discovery of association rules, Advances in Knowledge Discovery and Data Mining, Chap. 12, MIT Press, 307–328, 1996.
3. R. Agrawal, R. Srikant, Fast algorithms for mining association rules, In Proc. VLDB'94, 487–499, 1994.
4. Compaq Computer K.K., AltaVista. `http://www.altavista.com/`, 2000.
5. A. V. Aho, M. J. Corasick, Efficient string matching: An aid to bibliographic search, *Comm. ACM*, 1998
6. H. Arimura, J. Abe, R. Fujino, H. Sakamoto, S. Shimozono, S. Arikawa, Text Data Mining: Discovery of Important Keywords in the Cyberspace, In *Proc. IEEE Kyoto Int'l Conf. Digital Library*, 2001. (to appear)
7. H. Arimura, A. Wataki, R. Fujino, S. Arikawa, A fast algorithm for discovering optimal string patterns in large text databases, In *Proc. the 9th Int. Workshop on Algorithmic Learning Theory* (ALT'98), LNAI 1501, 247–261, 1998.
8. M. Craven, D. DiPasquo, D. Freitag, A. McCallum, T. Mitchell, K. Nigam, and S. Slattery, Learning to construct knowledge bases from the World Wide Web, Artificial Intelligence, 118, 69–114, 2000.
9. L. Devroye, L. Gyorfi, G. Lugosi, *A Probablistic Theory of Pattern Recognition*, Springer-Verlag,1996.
10. L. Devroye, W. Szpankowski, B. Rais, A note on the height of the suffix trees. *SIAM J. Comput.*, 21, 48–53 (1992).
11. R. Fujino, H. Arimura, S. Arikawa, Discovering unordered and ordered phrase association patterns for text mining. Proc. PAKDD2000, LNAI 1805, 281–293, 2000.
12. T. Fukuda, Y. Morimoto, S. Morishita, and T. Tokuyama, Data mining using two-dimensional optimized association rules, In *Proc. SIGMOD'96*, 13–23, 1996.
13. G. Gonnet, R. Baeza-Yates and T. Snider, New indices for text: Pat trees and pat arrays, In William Frakes and Ricardo Baeza-Yates (eds.), *Information Retrieval: Data Structures and Algorithms*, 66–82, 1992.

14. D. Gusfield, Algorithms on Strings, Trees, and Sequences: Computer Science and Computational Biology, Cambridge University Press, New York, 1997.
15. J. E. Hopcroft, and J. Ullman, *Introduction to Automata Theory, Languages and Computation*. Addison-Wesley, 1979.
16. M. J. Kearns and U. V. Vazirani, An Introduction to Computational Learning Theory, MIT Press, 1994.
17. T. Kasai, G. Lee, H. Arimura, S. Arikawa, K. Park, Linear-time longest-common-prefix computation in suffix arrays and its applications, In Proc. CPM'01, LNCS, Springer-Verlag, 2000 (this volumn).
18. J. M. Kleinberg, Authoritative sources in a hyperlinked environment. In Proc. SODA'98, 668–677, 1998.
19. M. J. Kearns, R. E. Shapire, L. M. Sellie, Toward efficient agnostic learning. *Machine Learning*, 17(2–3), 115–141, 1994.
20. D. Lewis, Reuters-21578 text categorization test collection, Distribution 1.0, AT&T Labs-Research, `http://www.research.att.com/~lewis/`, 1997.
21. L. C. K. Lui, Color set size problem with applications to string matching. Proc. the 3rd Annual Symp. Combinatorial Pattern Matching, 1992.
22. E. M. McCreight, A space-economical suffix tree construction algorithm, *JACM*, 23(2):262-272, 1976.
23. U. Manber and R. Baeza-Yates, An algorithm for string matching with a sequence of don't cares. IPL 37, 1991.
24. U. Manber and G. Myers, Suffix arrays: A new method for on-line string searches, *SIAM J. Computing*, 22(5), 935–948 (1993).
25. S. Morishita, On classification and regression, In *Proc. Discovery Science '98*, LNAI 1532, 49–59, 1998.
26. B. Schieber and U. Vishkin, On finding lowest common ancestors: simplifications an parallelization, *SIAM J. Computing*, 17, 1253–1262, 1988.
27. S. Shimozono, H. Arimura, and S. Arikawa, Efficient discovery of optimal word-association patterns in large text databases, *New Generation Computing*, Special issue on Discovery Science, 18, 49–60, 2000.
28. J. T. L. Wang, G. W. Chirn, T. G. Marr, B. Shapiro, D. Shasha and K. Zhang, In *Proc. SIGMOD'94*, 115–125, 1994.

A Computational Model for Children's Language Acquisition Using Inductive Logic Programming

Ikuo Kobayashi[1], Koichi Furukawa[1], Tomonobu Ozaki[2], and Mutsumi Imai[3]

[1] Graduate School of Media and Governance, Keio University.
[2] Keio Research Institute at SFC.
[3] Faculty of Environmental Information, Keio University.
5322 Endo, Fujisawa, Kanagawa,.252-8520, Japan
{ikuokoba,furukawa,tozaki,imai}@sfc.keio.ac.jp
http://bruch.sfc.keio.ac.jp/

Abstract. This paper describes our research activity on developing a computational model for children's word acquisition using inductive logic programming. We incorporate cognitive biases developed recently to explain the efficiency of children's language acquisition. We also design a co-evolution mechanism of acquiring concept definitions for words and developing concept hierarchy. Concept hierarchy plays an important role of defining contexts for later word learning processes. A context switching mechanism is used to select relevant set of attributes for learning a word depending on the category which it belongs to. On the other hand, during acquiring definitions for words, concept hierarchy is developed. We developed an experimental language acquisition system called WISDOM (Word Induction System for Deriving Object Model) and conducted virtual experiments or simulations on acquisition of words in two different categories. The experiments shows feasibility of our approach.

1 Introduction

This paper proposes a computational model for children's language acquisition using inductive logic programming (ILP), based on the observation of the similarity between human learning and concept learning by ILP.

In language acquisition, there are several difficulties related to its efficiency. First, we need to formalize sensory inputs. There are many input data to be potentially used in concept learning. Human beings select only those attributes which are relevant to describe given objects. We assume that humans utilize some kind of context switching mechanism to choose an appropriate set of input stimuli for each learning task. Perception itself may involve learning process to select an appropriate set of attributes for each learning task.

The second problem is to do with further restriction search space in concept learning. In language acquisition, there is a huge number of possibilities for the target of a given label. It is called Quine's paradox [9]. In an attempt to solve this problem, many cognitive psycologists have been working to identify a set of constraints or biases under which children conduct supervised learning of

S. Arikawa and A. Shinohara (Eds.): Progress in Discovery Science 2001, LNAI 2281, pp. 140–155.

concepts description given input sensory stimuli together with labels for target objects [6, 5]. These include Taxonomic Bias, Mutual Exclusivity Bias, Shape Bias and so on. Word learning biases play a central role in reducing search space in concept description. Without such biases, it seems extremly hard for children to learn words concepts from a very small set of examples. The situation is similar in machine learning: biases are essential for learning concepts even for a simple propositional learner [7]. In case of inductive logic programming, there are essentially two kinds of biases: declarative (language) bias and procedural bias. Cognitive biases do not directly correspond to declarative or procedural bias. They correspond to various parts in the ILP procedure. For example, the Shape Bias can be accommodated into inductive logic programming by defining a new evaluation function with heavier weight to shape attributes. On the other hand, the Taxonomic Bias can be realized by introducing different evaluation functions for each task depending on the taxonomic class of the target concept.

One notable characteristics of our model is its co-evolution between *word description* learning ability and *concept hierarchy* building ability. As we arugue later, they are mutually dependent; the word description learning utilizes concept hierarchy and conversely the concept hierarchy building utilizes word description. Also, the Mutual Exclusivity Bias is applied to build a concept hierarchy.

We built a preliminary computational model of children's noun acquisition and conducted virtual experiments by giving a set of attribute-value pairs as input stimuli together with a label to perform supervised learning.

In section 2, constraints theory in cognitive science for acquiring noun is introduced. In section 3, a computational model for children's noun acquisition based on ILP is presented. In section 4, the results of virtual experiments are described. Finally, in section 5, conclusion and future research direction are given.

2 Constraints Theory in Cognitive Science

Identifying a mechanism of human language acquisition has been addressed as an extremely difficult problem.

Induction is deeply tied with learning, so we would like to build noun learning model utilizing induction. However, when we try this, we found difficulty in intensional description of concepts. Intension of a cat is an answer for the question 'What is a cat?'. A cat has furred skin, has four legs, is good at climbing trees, can reach its hind feet to back of its ears, etc. Then, what is the sufficient set of desctiptors? How can children find such a set?

Markman [6] suggested a Constraint Theory on vocabulary acquisition by human children. According to this theory, children's expectation for the questions 'What is named?' and 'How the things to be named should be selected?' is bounded.

We use nouns for different types of intensions. We use 'a cat' or 'a car' to indicate a concrete or an imaginary object from a certain category which we think we know what it is. The words 'you' and 'them' changes their target object/s depending situations. 'Mr. Smith' and 'John' indicates a fixed target in a speaker's community. 'Water' and 'rice' do not have fixed shape.

Although we human beings have such a large number of types of nouns, we learn nouns only from positive examples. The universe of names is complex as we saw above in matured language. Constraint Theory, on the other hand, is a hypothesis on a stepwise vocabulary acquisition by children and can be regarded as defining priority on category types, in genaral.

The Whole Object Bias [6] is a bias Markman herself proposed under the Constraint Theory. This bias states that a child assumes a novel label to refer the whole of a given related object. This means that an unnamed object requires a label for its whole object primarily. We can use such a bias as a constraint for setting the whole object as the one to be referred to by labels.

Other biases we try to assemble in our learning model include: The Taxonomic Bias, the Shape Bias, the Principle of Contrast, and the Mutual Exclusivity Bias.

The Taxonomic Bias [6]: This bias states that a child tends to map a label to a taxonmic category which includes a referred object. This means that children know that the relation between labels and objects is not bijective. In fact, children do generalize a known label to map to objects that are not identical to the one which they saw when they were learning the label.

It is important to consider another aspect of the Taxonomic Bias. According to [1], children recognize animals as analogical beings of human beings or children themselves. Some children say that plants are not living because they do not move; some others say that insects have their heart because they move. These misunderstandings show their knowledge that some objects are alike to human beings while the others are not. This knowldege is referred to as an ability of classifying object taxonomies.

The Shape Bias [5]: This bias states that children think objects are similar when their shapes are similar. Children tend to generalize a label of an object to a concept having similar shape.

The Principle of Contrast: This bias states that children tend to think that different labels can refer to the same object, yet their sets of referring objects cannot be identical.

The Mutual Exclusivity Bias [6]: This bias states that children tend to think different labels do not refer to the same object.

3 A Computational Model for Children's Language Acquisition in ILP

In this paper we propose WISDOM (Word Induction System for Deriving Object Model), our computational model for the childrens' word (noun) acquisition. We use the name 'WISDOM' to refer to our computational model as well as to its implementation. The configuration of WISDOM is shown in Figure 1. The detailed descriptions of their modules are given below. Note that Label Input

Module and Sensory Input Module are not implemented here in WISDOM. They are virtual modules.

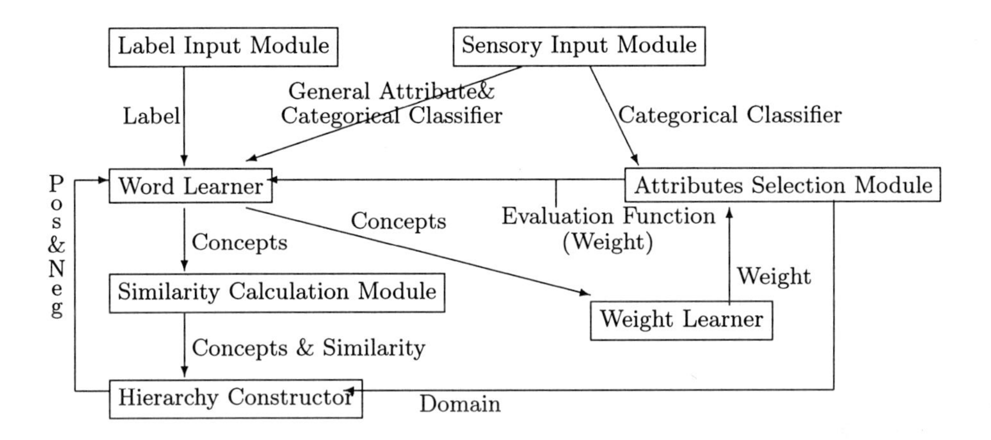

Fig. 1. Configuration of our word acquisition model WISDOM

3.1 Learning Attribute Relevancy (Sensory Input Module, Attributes Selection Module)

As mentioned in section 1, there are many input data to be potentially used in concept learning. Human beings select only those attributes which are relevant to describe given objects. In order to model this feature, we designed a mechanism for selecting an appropriate evaluation function from a set of functions prepared for each category. It is a kind of switch box controlled by categorical information of the object to be learned. Each evaluation function in turn is a linear combination of relevant attributes to the corresponding category. In order to model the learning capability of relevant attributes selection which human perception mechanism realizes, we need to incorporate Weight learning in evaluation functions (Weight Learner). In our first experement, we predefined those weights manually. The automatic weight adjustment is one of our future research targets.

Note that this mechanism should avoid the difficulty of bridging objects perception and words acquisition.

3.2 Inductive Learning of Concept Definition (Word Learner)

ILP [8, 2] is a framework for learning relational concepts when a set of positive and negative examples are given together with background knowledge in the form of horn clauses. More precisely, ILP searches the concept lattice formed by the combinations of background knowledge, and finds the best hypothesis which explains most of the positive and least of the negative examples together with background knowledge. On one hand, the standard ILP systems require at least several positive and negative examples before learning, and those adopt compression gain, or the MDL principle as the evaluation function. On the other hand, the system in our ILP model for the children's vocabulary acquisition continuously repeats generation and revision of the concept description during the life time. It can work even if only one or few examples are provided because a new evaluation function reflecting the cognitive bias is implemented.

We briefly explain how actual word learning corresponds to ILP in our model. We assume that children can correctly identify the object with which a label is associated. Note that the label is given separatedly by, say, their mother as oral signal (Label Input Module). The properties of the object are divided into two kinds, which we call *Categorical Classifier* and *General Attribute*. Categorical Classifiers correspond to children's innate ontology, thus they indicate ontological categories such as 'animate', 'countable', etc. General Attributes such as 'having-four-legs' or 'furred' include all object properties other than those represented as Categorical Classifiers. The General Attributes are further divided into some subtypes such as shapes, colors, textures, etc. Based on these assumptions, we feed both observations (sensory inputs) on an object and a label on it to the system in the form of logical formula which ILP can directly handle.

Since each label presents the category to which the object belongs, the concept to be learned is indicated by the label. The objects are given as examples and their properties are given as background knowledge. When the system takes an object and its label as a pair (we call them current object and current label respectively), the system learns or revises the concept indicated by the current label by generalizing the current object if necessary. At this time, the current object and all objects named by the current label (which are given in the previous learning processes) are used as positive examples. The negative examples are selected from the rest of the examples based on word learning biases.

The concept hierarchy is utilized for providing further positive and negative examples. For example, positive example objects for a subordinate category of the current label are regarded as positive examples for the current label. In contrast, negative examples for a superordinate category are regarded as also negative for the current label.

In our model the concepts are learned in two steps. First, the model determines a *Supercategory*, which is a domain for the label using the Categorical Classifier/s of the current object. Then, relevant General Attributes for the Supercategory are selected for the generalization of a given example.

The obvious benefit of adopting the above two-steps procedure is that one can reduce relevant background knowledge in computing generalization.

Furthermore, it reduces the hypothesis space by restricting negative examples to only meaningful ones. Let us consider the case in which we learn a concept 'cat'. Assume that the given object named cat has 'animate' as its Categorical Classifier and 'having-four-legs' as its General Attribute. Assume also that an object having a label 'desk' has already been given in the previous inference process and it had 'having-four-legs' and 'inanimate' as its General Attribute and Categorical Classifier respectively. In this case, although the object named 'desk' is obviously negative example for the concept 'cat' because the label is different, the model excludes this object from the negative examples for learning because the Categorical Classifiers 'animate' and 'inanimate' are mutually exclusive. If this dynamic selection of negative examples could not be done, the concept 'cat' could not have been inferred because the property 'having-four-legs', which may be crucial for this concept, could not be included in the definition of 'cat'. Furthermore we can reduce the hypothesis space by excluding the candidates having Categorical Classifiers used in selecting negative examples.

3.3 Role of Bias in ILP and Realization of Constraint Theory

Cognitive biases do not correspond directly to declarative or to procedural biases in ILP. We describe how the cognitive biases are implemented.

Let us consider a part of the 'reference problem' of the word acquisition. When a label is associated with an object, what aspect of the object should the label be mapped onto? This is an essential problem for our model because it determines kinds of concept our model can learn. In our current design, the model assumes that every label refers to the entirety of an object according to Whole Object Bias (see section 2). Therefore the model can learn only labels of objects as a whole, while it cannot learn other kinds of concept such as a property or a part of object, an action, an event, and so on. For example, the word 'dog' and 'desk' can be learned, but the word 'ear' (the part of object) and 'running' (the action) cannot be learned.

The first feature of the Taxonomic Bias, the tendency to generalize the label to other like objects, is automatically implemented in our model because ILP by its nature is used for generalization. That is, the model automatically assumes that a label refers to a category. The second feature, that is, the ability of restricting learning domains, is realized by Domain Selection Module.

When two objects referred by two labels are judged to be not similar by the Shape Bias, the model obeys the Mutual Exclusivity Bias and the labels are not allowed to share the same object. On the other hand, when two labels are judged to be similar, the model obeys the Principle of Contrast and the objects are allowed to be hierarchically related.

In our model, the evaluation criterion is introduced so that it reflects the Shape Bias. This allows the model to select an appropriate hypothesis even if only few positive examples are given. Like an ILP system Progol, this evaluation criterion is basically based on the description length of the concept, but some weights are added to each atom consisting of the concept. In our current implementation, the weight of a General Attribute about shape is set 1.5 times heavier than that of an other type General Attributes, and the weight of a Categorical Classifier twice heavier than that of a General Attribulte.

When we learn the concept belonging to a Supercategory or a domain CC_j in the children's innate ontology, the general expression of our evaluation function is defined as:

$$C_{CC_j} = -PE + NE \\ -w_{CC} * |CC_j| - \Sigma_i \ (w_{GA}^{CC_j,i} * GA_i)$$

where PE and NE are the number of positive and negative examples explained by the candidate hypothesis respectively; $|CC_j|$ and w_{CC} are the number of the Categorical Classifiers appearing in the candidate hypothesis and the uniform weight of Categorical Classifiers, respectively. GA_i and $w_{GA}^{CC_j,i}$ are the number of the General Attributes whose subtype is i in the candidate hypothesis and the weight of a General Attribute of that subtype, respectively.

In this evaluation function, the Shape Bias can be realized by placing a heavier weight on the attribute about shape as mentioned earlier. We can also realize some kind of context switching mechanism, *i.e.* the Taxonomic Bias by introducing different sets of evaluation functions and weights for each task depending on the Supercategory of the target concept. By the way, the Shape Bias is not always applied. With the increase of conceptual knowledge, children gradually come to realize that shape similarity is not the most essential factor for determining the membership of an object category. We believe that this bias shift can be simulated by weight learning in the evaluation function.

3.4 Building Concept Hierarchy (Hierarchy Constructor)

As mentioned in section 2, we human beings have knowledge called taxonomy on concepts universe. Materials (like water or rice) are different from objects (like a car or a cat) in the aspects to be referred by labels. In vocabulary acquisiton, it is not the issue to differentiate materials from objects; yet the issue is to differentiate water from rice, and also car from cat. The shape of the whole object makes no sense when we separate rice from water, while the probability that an arbitrary car and an arbitrary cat have the same shape is very low.

So, first of all, children need to distinguish materials from objects before they learn vocabulary. We refer groups like materials or objects as taxonomy class or *Supercategory*. Supercategories which dominate large part of vocabulary universe, such as the ones given above, may be known to children prior to language learning. Experiments involving infants provided some evidence for this [4].

On the other hand, at the lower level of taxonomy, that is, at rather more concrete parts of vocabulary universe, we cannot give a priori categories. For example, artificial tools used in our daily life are classified in various categories. A flower pot is similar to a bowl for dietary use in shape, though the former has hole/s to let water flow out. The former appears in scenes of gardening, while the latter does in meals. We must have learned such categories through the daily life. In other words, such categories are culture-dependent. The flower pot category and the bowl cagegory are both created and developed under the Supercategory of shaped objects. What we refer as concept hierarchy building includes such introducion and following updates of low-lovel, local and cultural categories.

Note that culture-dependent categories include universal ones. As shown in section 2, some children think that plants are non-living, but in any culture plants are separated from non-living things like rocks.

The existance of a higher level Supercategory for the concept to be learned helps a great deal in learning its definition. 'A cat' is learned as a concept within the Supercategory of living animate objects, while 'a bowl' is non-living static objects. Children also know that both of them refer to concepts within the shaped objects' Supercategory. Such knowledge prevents children from analysing irrelevant attributes. When children see a bowl in association with a label 'bowl', they analyse its shape, and possibly, other additional attributes like its material, location and so on. In order to observe and analyse its properties, they may touch it with their hands; they may rub it with their dry and/or wet fingers; or they may taste it with their tongues. However, they do not observe its way of locomotion. They interpret that locomotion is irrelevant as soon as they see the object.

Concept hierarchy is basically united knowledge of hierarchical relations among named concepts. Assume that a child knows the following things: first, a 'dog' is an 'animal'; second, a 'terrier' is a 'dog'. From these propositions, the child can tell that a terrier is an animal. This inference is purely deductive. Note that no sensory device is needed in such inference.

This is an important competence for human beings to understand his or her environments. Our knowledge includes tacit component, like the judging ability whether a given object is a dog, which occupies large parts in our entire knowledge. However, explicit pieces of distributed knowledge such as relationships among labeled concepts are integrated to systematic knowledge by conducting logical inference.

Concept hierarchy enforces vocabulary developing. Recall that children are supported by database which include domain-specific heuristics in learning new names. Concept hierarchy conveys properties of higher concepts to lower concepts. Suppose a child knows what a 'dog' is and an adult comes to tell him 'That is a terrier.' pointing a dog. Unless he adopts the Mutual Exclusivity Bias, he may interpret a 'terrier' is another name of the object, and may think that the category 'terrier' differs from a 'dog', guided by the Principle of Contrast.

The previous story of a dog labeled 'terrier' shows a typical example of hierarchical relationship between two objects. Suppose the learner already knows one of these two concepts, and then sees the other. In the hierarchical case, since one of them includes the other, it is typical that the newly labeled object already has the first label.

We illustrate a representation of one of such cases in our model. On the top of figure 2, a learned concept intension corresponding to a known label '`dog`' is shown. Next three clauses are observations on objects `obj001, obj002, obj003`. Assume that each of these was related with the label '`dog`' when it was observed. Novel category names can be introduced referring to one of these objects. Such a name can be '`cat`', which should be mutually exclusive with '`dog`'; or '`terrier`', which should be subordinate of '`dog`'.

Suppose that one of such labels is newly introduced related only one of these three objects in input, and that the learner then learns a concept named by this label. The learned concepts are also shown in Fig.2 below the list of observations on the objects. Note that we need to identify the case either 'mutually exclusive' or 'subordinate'. For this purpose, we employ a kind of similarlity measure (Similarlity Calculation Module). If the similarity between the concept 'dog' and the newly labeled object is small, the learner would interpret that they are hierarchically related, according to the Principle of Contrast and the Taxonomic Bias. Or, if the similarity is large, he would interpret that they are mutually exclusive based on the Mutual Exculusivity Bias.

(Intension of dog)
```
labeling(dog, A) :-
        tax(A, animation, animate),  attr(A, shape, short_tail).
```

(Observations of objects)
```
tax(obj001, animation, animate).   tax(obj002, animation, animate).
attr(obj001, shape, hanging_ears). attr(obj002, shape, short_tail).
attr(obj001, covering, furred).    attr(obj002, shape, hanging_ears).
tax(obj003, animation, animate).
attr(obj003, shape, short_tail).
attr(obj003, covering, furred).
```

(Derived intensions)
```
labeling(label001, A) :-
        tax(A, animation, animate),
        attr(A, shape, hanging_ears),  attr(A, covering, furred).
labeling(label002, A) :-
        tax(A, animation, animate),
        attr(A, shape, short_tail),  attr(A, shape, hanging_ears).
labeling(label003, A) :-
        tax(A, animation, animate),
        attr(A, shape, short_tail),  attr(A, covering, furred).
```

Fig. 2. List of intension of `dog` and observations for 3 objects

These three objects are different with respect to *typicality* as members of the '`dog`' category. The intension of `label001` relating only with `obj001` seems the least similar to the concept intension of `dog`. In other words, `obj001` seems the *least typical* `dog` of the three, since there appears no General Attribute discriptor `short_tail` in its observation. Of the rest two objects, `obj002` seems a less typical `dog`, because there appears a different discriptor `hanging_ears` from `short_tail` on `shape` dimension. It follows that the last object `obj003` seems the most typical of three. Accordingly, it should lead to the following result: the new label related with `obj001` would get the smallest similarity between the intensions of the `dog` and the one related with `obj003` would get the largest similarity.

We assume that the similarity between the two learned concepts is inversely proportional to the *concept distance* between their intensions. Concept distance between two concept intensions is measured basically by the number of descriptors which occur in one concept intension but do not in the other. Discriptors are weighted based on their importance. First, Categorical Classifiers (represented in the predicate `tax`) are counted 2 times more than basic `attr` predicate descriptors. Additionally, General Attributes (represented in `attr` predicate) can have varying scores by different kind attributes of the descriptor. For example, in the Supercategory `animate`, we put 1.5 points per one difference on `shape` dimension than the other `attr` predicates.

In this case, we count the existence of General Attribute `furred` on `covering` as 1 point. And we give weights for other descriptors as mentioned above: Genral Attributes on `shape` are given 1.5 points, while Categorical Classifiers are given 2 points. Similarities between 'dog' and the newly introduced categories become as follows: 0.25 with `label001`, approximately 0.67 with `label002` and 1.00 with `label003`, respectively.

We let the model judge whether the relation with `dog` is hierarchical for each object by similarities calculated above. Giving a threshold, below which the relation is interpreted mutually exclusive and above which it is interpreted hierarchical, is a simple method. In this case, we set the threshold at $0.17 = 1/6$. This threshold allows us to admit the total occurence of supportive attributes to the target category up to four less than that of non-supportive ones. [1]

It leads that all three introduced categories are interpreted as being hierarchically related with `dog`. `label001` is difficult to judge at current status, but more information and further learning process may be necessary to verify the judgement.

Our computational model WISDOM creates and revises hierarchical and mutual exclusive relations between concepts through the successive learning. The whole of such relations represents knowledge of concept hierarchy.

3.5 Co-evolution of Concept Learning and Concept Hierarchy Building

One of the most characteristic features of our model is the co-evolution of concept learning and concept hierarchy building. Concept learning and concept hierarchy building correspond to two different learning activities: concept learning is achieved by supervised learning whereas concept hierarchy building is achieved by a kind of unsupervised learning. In the concept learning phase, the concept hierarchy serves for identifying the Supercategory which the target object belongs to and therefore selecting an appropriate set of attributes to classify it. On the other hand, in concept hierarchy building, the definitions of related concepts can be used to build a concept hierarchy. One possible trigger to build a concept hierarchy is during concept learning when we apply the Principle of Contrast

[1] Precisely speaking, we sum up the attributes weights. Since 4 times 1.5 (the shape score) is 6, the inverse becomes 1/6.

by assigning either super class category or subordinate category to the labeled object to be learned.

4 Virtual Experiments

In this section, we compare two learning experiments by WISDOM. These two experiments differ from each other in Supercategories that we take example objects from. The domain of the first experiment is silverware. The domain of the other experiment includes animal categories. We intend to find difference between sets of attributes used in intensions of named categories from each other. Each specific domain may be characterized by a set of attributes used in explaining named categories.

In the first experiment, we provide two examples to WISDOM. One is on a fork and the other is on a spoon. These two objects are from the same set. Both are made of stainless steel, have almost the same length (19cm) and have handles in the same design. Then WISDOM tries to make distinction between them by comparing the part having contact to foods.

The input list for the fork given to WISDOM is shown in figure 4. They are the first stimuli for the entirely novice learner. The first clause in the list says that the appearing object is called a `fork`. The third term of this and the next clause shows that this information is the first and the latest for the learner.

The clauses in figure 4 represent perceptual information on the object on the left of figure 3. Assume that the object is put on a table with the handle near to our body. The third clause whose predicate is `tax` says that the object is not animate. The following 4 clauses with the predicate `attr` give information on the whole object. Next 3 clauses with the predicates `subobj` and `connection` state that the object has two parts connected to each other with front-to-back placement. `obj051b`, the handle of the fork, is in contact with the rest part at its backend (`y_plus`). The rest 20 unit clauses of the list represent observations on part of the object.

After giving this input list to WISDOM, we let it start to learn. WISDOM is set to induce the rule of `labeling` given latest. So it induces the condition concluding `labeling(fork, X)`. Selection of the most preferable hypothesis is done based on the hypotheses evaluation function. The result is shown in figure 4.

Although we omit the detailed input data for the spoon here, the reader could easily imagine how they look like. They are similar to those for the fork; the handle is represented in the same formalization and the rest part is represented additionally. We only note that the spoon is given the object identifier `obj052`.

After learning '`fork`', we give the input data for the spoon and let the learner to learn again. The result is shown in figure 5.

After starting the learning, WISDOM constructs a hypothesis. The `obj052` (a spoon) is of course used as a positive example. Note that `obj051` is used as a negative example, because the learner interprets those except explicit positive ones as negative examples by default.

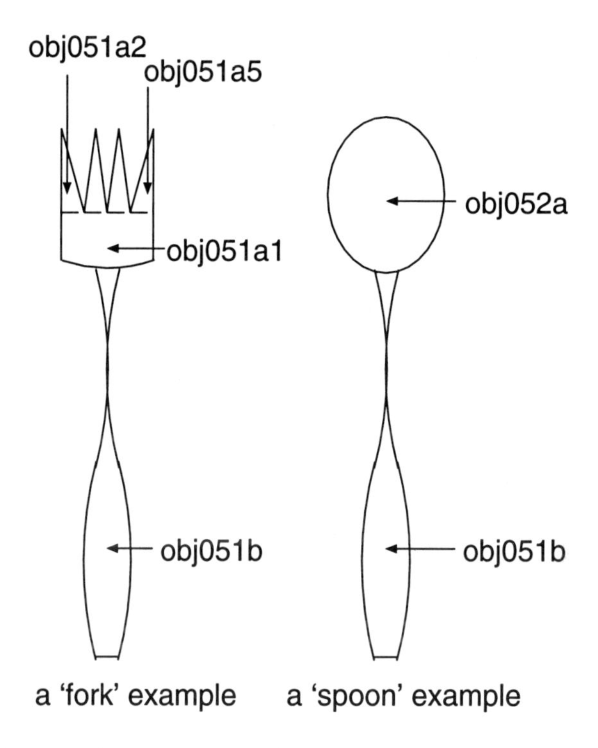

Fig. 3. Example objects `obj051` and `obj052` for the labels `fork` and `spoon`

Then the learner, WISDOM, tries to guess the proper label for the appearing object. This is done by applying all rules whose head predicates are `labeling` to `obj052`. Note that the current rule for the category '`fork`' is strong; it says that every inanimate object is a fork. This suggests that `obj052` is called a '`fork`' as well as a '`spoon`'. However, the learner was given that `obj052` was not a positive example for the category '`fork`', so the learner thinks that the rule on '`fork`' is not correct any longer. It starts learning the category '`fork`' again, with no additional stimuli to obtain more reasonable rule for the fork.

Then we proceed to the second experiment. In this experiment, we provide inputs for two animal instances; a cat, and a dog. This learning is independent from the former experiment. Like the experiment of dining instruments shown above, the learner starts from the status where it does not know any of labelled categories. Figure 6 shows the result of the experiment.

At one time, the information for one object is given, as the experiment shown above. The first input list is for the cat, and the second is for the dog. The given lists are not shown because of the space limitation, but they represent their shapes in detail like the dining instruments experiment, and their colors. In this experiment, an additional attiribute, *i.e.*, the `covering` attribute is also included. The cat is laid on the ground, while the dog stands.

(Input on the fork object obj051)

```
labeling(fork, obj051, 1). newest_labeling(1).
tax(obj051, animation, inanimate).
attr(obj051, color, having_reflection).  attr(obj051, color, shining).
attr(obj051, shape, constant).  attr(obj051, direction, y_axis).
subobj(obj051a, obj051).  subobj(obj051b, obj051).
connection(obj051a, y_minus, obj051b, y_plus).
attr(obj051b, direction, y_axis).  attr(obj051b, shape, board).
attr(obj051b, shape, x_axis/[y_plus, y_minus]).
subobj(obj051a1, obj051a).  subobj(obj051a2, obj051a).
subobj(obj051a3, obj051a).  subobj(obj051a4, obj051a).
subobj(obj051a5, obj051a).
connection(obj051a1, x_minus_y_plus, obj051a2, y_minus).
connection(obj051a1, y_plus, obj051a3, y_minus).
connection(obj051a1, y_plus, obj051a4, y_minus).
connection(obj051a1, x_plus_y_plus, obj051a5, y_minus).
attr(obj051a2, direction, y_axis).  attr(obj051a2, shape, pyramid).
attr(obj051a3, direction, y_axis).  attr(obj051a3, shape, pyramid).
attr(obj051a4, direction, y_axis).  attr(obj051a4, shape, pyramid).
attr(obj051a5, direction, y_axis).  attr(obj051a5, shape, pyramid).
```

(Result of the first induction)

```
labeling(fork, A) :- tax(A, animation, inanimate).
```

Fig. 4. Input and output on the object obj051 associated with the label fork

```
labeling(fork, A) :- tax(A, animation, inanimate).
labeling(spoon, A) :-tax(A, animation, inanimate),
         subobj(B, A), attr(B, shape, oval_semisphere).

obj052 is a fork.
obj052 is a spoon.
Need induction for [fork].

labeling(spoon, A) :- tax(A, animation, inanimate),
         subobj(B, A), attr(B, shape, oval_semisphere).
labeling(fork, A) :- tax(A, animation, inanimate),
          subobj(B, A), subobj(C, B), attr(C, shape, pyramid).

obj052 is a spoon.
Need induction for [].
```

Fig. 5. Result list of learning the label spoon after learning fork

After learning the category cat from the first stimuli set for the cat, WISDOM acquires the knowledge shown in figure 6.

By giving inputs for the dog to the learner having this knowldege, the rule (shown in figure 7) is learned.

```
(Acquired description for cat)
labeling(cat, A) :-tax(A, animation, animate).

(Result of learning dog over the knowing cat)
labeling(cat, A) :- tax(A, animation, animate).
labeling(dog, A) :- tax(A, animation, animate),
         subobj(B, A),attr(B, direction, z_axis).

obj032 is a cat.
obj032 is a dog.
Need induction for [cat].

labeling(dog, A) :- tax(A, animation, animate),subobj(B, A),
         attr(B, direction, z_axis).
labeling(cat, A) :- tax(A, animation, animate), attr(A, shape, barrel).

obj032 is a dog.
Need induction for [].
```

Fig. 6. List on animate categories learning

```
labeling(dog, A) :- tax(A, animation, animate),subobj(B, A),
         attr(B, direction, z_axis).
```

Fig. 7. Result explanation on dog

Both a cat and a dog have their heads, four legs, and tails. In this aspect, they have the same parts of their bodies, while each corresponding part is different.

In this experiment, the difference in posture and figure appears. The direction z_axis is of a leg of the standing dog; the cat is laid and its legs are along x and y axes. Shape barrel is of the trunk of the cat; the dog has no part whose shape is barrel.

If we give more examples in which animals make varied postures, the attribute which represent thier shapes of their heads should play more central role.

In this section, we showed knowledge acquired as results of 2 cases. Let us compare these results. The spoon and the fork has same attribute value on their surfaces, so their difference appears in shapes of parts. The learner picks the fact up that the spoon has a part whose shape is oval_semisphere, while it concerns that the fork has a part whose shape is pyramid. This pyaramid-shaped part is a part of a part of the whole, so difference of connection structures between instruments seems important. In the latter experiment, the model concerns on the standing leg/s of the dog and the shape of the cat's trunk.

The information relating to shape is used in both results. Such information includes shape name attribute, direction of whole or partly objects, connectivity among parts of an object, relative lengths along axes of one object, relative size between different objects, and so on. Detailed analysis down to objects' parts seems effective, especially where the model is given shape-relating input information.

Description like 'object obj001 has four legs' is difficult to use in category induction even the learner knows what a leg is, for two instances of animals can be quadruped having different position, shapes of legs. Such minor difference seems to be important for differentiating categories.

5 Conclusion and Future Work

This paper presented a computational model WISDOM for children's language acquisition using ILP. We proposed a model consisting of two parts; concept learning and concept hierarchy building. In the model, we incorporated cognitive biases such as the Taxonomic Bias, the Shape Bias, the Mutual Exclusive Bias and the Principle of Contrast to reduce a search space in building concept description.

In concept learning, we adopted the learning scheme of Inductive Logic Programming to induce associations between the target concept and various types of knowledge which children have acquired in the past.

In concept hierarchy building, we introduced a kind of similarity measure to judge 1) whether the given target object belongs to some category or not and 2) whether a given label to the target denotes a superordinate or subordinate concept to the conflicting category.

In WISDOM, concept learning and concept hierarchy building co-evolve each other. This feature is very important because it drastically accelarates the learning ability. The usage of taxonomy information in the learning phase introduces the notion of context because each Supercategory defines each context in which concept learning becomes a relatively small and easy task.

We built a learning program (we call the program as WISDOM also) based on our model and simulated language learning tasks for two animals and also for two dining instruments.

In these experiments, we gave shape attributes in detail. We found that different attributes sets were needed for these two tasks. Furthermore we found that common attributes like having-four-legs do not work in differentiating concepts in a domain, but rather subtle attributes such as the shape of trunks are crucial.

We also noted the importance of selecting a relevant domain not only for reducing the entire search space but also for eliminating irrelevant negative examples, which made it possible to learn the right concept efficiently.

We applied a simple similarity measure to judge whether two concept descriptions refer to distinct objects or hierarchicaly related objects. We could not distinguish three cases in our experiments. One reason may be that our similarity measure was too simple. In our future research, we are planning to adopt further sophisticated similarity measure to improve the performance. Another reason may be that experiments of the system is insufficient. It may resolve conflicts in such judgement by further experiments.

Another point that need to be resolved in the future is how a set of weights are assigned to different attributes. There are no strict reasons why we put score 2 the Categorical Classifiers, socre 1.5 for shape attributes and 1 for the rest. We need to find appropriate numbers for these weights to establish the better solutions or better behaviours of the system. It is a weights-adjusting problem and many algorithms are known to perform such tasks. We are particularly interested in adopting a learning algorithm conducted by cerebellum which performs supervised learning using errors directly as teachers signal [3] to implement the weight-adjusting problem.

References

1. Carey, S. *Conceptual Change in Childhood*, MIT Press, 1985.
2. Furukawa, K., Ozaki, T. and Ueno, K. *Inductive Logic Programming*, Kyoritsu-syuppan, 2001 (in Japanese).
3. Kawato, M. *Nou no Keisan Riron* (Computation Theory in Brain) Sangyo Tosyo, 1996 (in Janapese).
4. Gelman, R. First principles orgainize attention to and learning about relevant data: Number and the animate-inanimate distinction as examples, *Cognitive Science*, *14*, pp.79-106, 1990.
5. Haryu, E. and Imai, M. Controlling the application of the mutual exclusivity assumption in the acquisition of lexical hierarchies, *Japanese Psychological research*, Vol.41, No.1, pp.21-34, 1999.
6. Markman, E. M. *Categorization and Naming in Children*, MIT Press series in learning, development and conceptual change MIT Press, 1989.
7. Mitchell, T. M. *MACHINE LEARNING*, WCB/McGraw-Hill, 1997.
8. Muggleton, S. Inverse Entailment and progol, New Generation Computing, 13(4-5), pp.245-286, 1995.
9. Quine, W. V. O. *Word and Object*, MIT Press, 1960.
10. Spelke, E. S., Phillips, A. and Woodward, A. L. Infants' knowledge of object motion and human action. In D. Sperber, D. Premack & A. J. Premack (Eds.), *Causal Cognition: A multidisciplinary debate*, New York: Oxford University Press, 1995.

Some Criterions for Selecting the Best Data Abstractions

Makoto Haraguchi and Yoshimitsu Kudoh

Division of Electronics and Information Engineering,
Hokkaido University, N-13, W-8, Sapporo 060-8628 JAPAN
{makoto,kudo}@db-ei.eng.hokudai.ac.jp

Abstract. This paper presents and summarizes some criterions for selecting the best data abstraction for relations in relational databases. The data abstraction can be understood as a grouping of attribute values whose individual aspects are forgotten and are therefore abstracted to some more abstract value together. Consequently, a relation after the abstraction is a more compact one for which data miners will work efficiently. It is however a major problem that, when an important aspect of data values is neglected in the abstraction, then the quality of extracted knowledge becomes worse. So, it is the central issue to present a criterion under which only an adequate data abstraction is selected so as to keep the important information and to reduce the sizes of relations at the same time. From this viewpoint, we present in this paper three criterions and test them for a task of classifying tuples in a relation given several target classes. All the criterions are derived from a notion of similarities among class distributions, and are formalized based on the standard information theory. We also summarize our experimental results for the classification task, and discuss a future work.

1 Introduction

When the data values processed by data mining algorithms are too concrete or too detailed, the computational complexity needed to find useful knowledge will be generally increased. At the same time, we find some difficulties in reading and analyzing the acquired knowledge, as the knowledge is written in terms of original data values. A notion of data abstraction seems useful to solve the problem of this kind, as we obtain more compact and smaller databases after applying abstractions. The readability of acquired rules will be much improved, since the rules are written in more abstract terms. Here, by the term *data abstraction*, we mean an act of replacing the concrete values, `tokyo`, `osaka` and `new_york` for instance, with a more abstract value `big_city`. In this replacement, the individual aspects of `tokyo`, `osaka` and `new_york` are forgotten except those concerning the bigness of cities. According to the standard terminology [1], the database obtained by such a replacement is called a generalized database by the abstraction.

One of the major problems when we apply data abstractions is that some important information needed to extract useful rules may be lost. In such a case,

S. Arikawa and A. Shinohara (Eds.): Progress in Discovery Science 2001, LNAI 2281, pp. 156–167.

the corresponding precision of the extracted rules will be decreased. Conversely, a data abstraction can be regarded appropriate, provided the important information is preserved even after the abstraction. Needless to say, the problem of what is the important information depends on the purposes of data mining algorithms or the types of data presented to them. So, in this paper, we first discuss this issue in the case of attribute-oriented induction [1] for relational databases, and then present some criterions under which each possible data abstractions are tested for their appropriateness.

All the criterions in this paper are dervied from a notion of *similarities among class distributions*. When the class distributions given attribute values are similar, then we have no particular reason to distinguish them, as every property or characteristic about the classes are captured by the distributions in the attribute-oriented induction. So our first criterion for selecting abstractions is that

> an abstraction that groups attribute values showing similar class distributions should be preferred and selected.

An abstraction satisfying this criterion also has a similar class distribution at abstract level. Consequently, the amount of information after abstraction is preserved at least approximately. From this simple fact, we propose the next criterion.

> An abstraction that almost preserves the amount of information about classes should be regarded as a good candidate.

According to the second principle, we can treat some exceptional distributions, given some attribute values, which are far from the majority of similar distributions. In such a case, any property or characteristic particular to such exceptional distributions has disappeared after the abstraction. In other words, our data abstraction can ignore the exceptional minor class distributions.

Based on these two criterions, the best data abstraction is selected for each attribute independent of another attribute. However in general, the appropriate abstraction for a particular attribute may not be appropriate when we make a condition concerning the other attributes. Taking such a context-dependent aspect of appropriate abstraction into account, we have the third principle.

> Given a condition about several attributes, select the best abstraction for the subrelation specified by the condition.

All the criterions are formally presented and analyzed. A family of data abstraction systems, `ITA` [2], iterative `ITA` [3] and $\mathtt{I^2TA}$ are developed and tested based on the criterions. Our experimental results indicate that the data abstraction is useful not only for the size reduction of relations and acquired knowledge but also for the acquisition of general knowledge covering many particular instances.

2 Data Abstraction

In this section, we introduce our notion of data abstractions, discuss their information loss, and finally present our fundamental criterions to select the appropriate abstraction.abstraction. Let us start with giving some preliminaries.

We consider a relation R with its attributes $A_1, ..., A_m$ with their domains $dom(A_1), ..., dom(A_m)$, respectively. In addition to these attributes, we assume just one target attribute C whose value is called a class. Each element $t \in R$ is therefore represented as a tuple $t = (t[A_1], ..., t[A_m], t[C])$. The j-th component $t[A_j]$ is called an A_j-value of t, and $t[C]$ is the class of t. Our data mining task for this type of data is to find several conditions in terms of attributes $A_1, ..., A_m$ for discriminating or for characterizing the classes in C. Moreover, we suppose a family of learning algoarithms with the entropy-based criterion, e.g. `ID3` and `C4.5`[4]. The underlying probability is given by a uniform distribution over R. Then, every attribute X can be regarded as a random variable. In fact, $\Pr(X = x) = \Pr(\{t \in R | t[X] = x\})$. Moreover, a distribution is represented by a probability vector $(p_1, ..., p_n)$ with $0 \leq p_i \leq 1$ and $\sum_{j=1}^{n} p_j = 1$. Then a property about classes $c_1, ..., c_n$ captured by an attribute value $a \in dom(A)$ is represented by the following (posterior) class distribution given the value a.

$$C_{A=a} = (\Pr(C = c_1 | A = a), ..., \Pr(C = c_n | A = a)) .$$

2.1 Definition of Data Abstractions

Suppose that we have n classes $c_1, ..., c_n$ and a relation R with its attributes $A_1, ..., A_m$. Then, a data abstraction for R is a system of groupings (partitions) of attribute values. In a group of values $a_{i_1}, ..., a_{i_k}$ in $dom(A)$, their individual aspects are forgotten, and only the average properties are considered in the abstraction. As those properties are represented as the class distributions $C_{A=a_{i_j}}$, this means that we neglect their differences and consider only their average which we will call an abstract class distribution.

Definition 1. *(Data Abstraction)*
1. A data abstraction φ_A for an attribute A in $\{A_1, ..., A_m\}$ is defined as a partition $\{g_1, ... g_\ell\}$ of $dom(A)$, where $g_j \subseteq dom(A)$, $g_i \cap g_j = \phi$ whenever $i \neq j$, and $dom(A) = \bigcup_{j=1}^{\ell} g_j$. g_j is called a group of A-values.
2. A data abstraction φ for attributes $A_1, ..., A_m$ is defined as a tuple $(\varphi_{A_1}, ..., \varphi_{A_m})$, where φ_{A_j} is a data abstraction for the attribute A_j.
3. Given a data abstraction $\varphi = (\varphi_{A_1}, ..., \varphi_{A_m})$ and a relation R with the same attributes $A_1, ..., A_m$, we form a generalized relation $\varphi(R)$ which has its attributes $\overline{A_1}, ..., \overline{A_m}$ with their domains $dom(\overline{A_j}) = \varphi_{A_j}$. Thus, each group $g \in \varphi_{A_j}$ is regarded as an abstract attribute value of $\overline{A_j}$. Moreover, a tuple $t = (a_1, ..., a_m, c) \in R$ is called an instance (tuple) of an abstract tuple $\bar{t} = (g_1, ..., g_m, c)$ if $a_j \in g_j$ for any j. In this case, we also say that t is abstracted to $\bar{t}$. Then, $\varphi(R)$ is defined as a set of abstract tuples $\bar{t}$ such that there exists at least one instance in R.

We here illustrate the generalization of relations by a simple example. Suppose that we have a relation with two attributes X and Y (except the target C), shown at the left of Fig. 1, and a data abstraction (φ_X, φ_Y), where $\varphi_X = \{g_{ab} = \{a, b\}, g_{cde} = \{c, d, e\}\}$, and $\varphi_Y = \{under\_10 = \{x|x < 10\}, just\_10 = \{10\}, over\_10 = \{x|10 < x\}\}$. For instance, the expression $g_{ab} = \{a, b\}$ means that $\{a, b\}$ is a group and g_{ab} is its name. Then we see the generalized relation at the right of Fig. 1. For example, $(a, 25)$ is abstracted to $(g_{ab}, over\_10)$, as $a \in g_{ab}$ and $25 \in over\_10$.

X	Y	C	Pr
a	25	c1	1/4
c	15	c2	1/4
d	4	c2	1/4
b	35	c1	1/4

$\overline{X}$	$\overline{Y}$	C	Pr
g_{ab}	over_10	c1	1/2
g_{cde}	over_10	c2	1/4
g_{cde}	under_10	c2	1/4

Fig. 1. Relations before and after a data abstraction

An abstract attribute $\overline{A_j}$ corresponding to A_j can be also considered as a random varialbe. Then, the following fact is often used, as it asserts a fundamental relationship between the original and the abstracted attributes.

Proposition 1. *For a group $g \in \varphi_{A_j} = dom(\overline{A_j})$, the event specified by $\overline{A_j} = g$ is just a disjoint union of all events given by $A_j = a$ with $a \in g$. Therefore, we have* $\Pr(\overline{A_j} = g) = \Pr(A_j \in g) = \sum_{a \in g} \Pr(A_j = a)$.

There may exist many possible data abstractions, so the major problem is to present a criterion for selecting the best one according to which the generalization of relations should be carried out. As we have mentioned in Introduction, we first present in this section a criterion for each attribute. The criterion is concerned with how a class distribution $C_{A=a}$ is changed after data abstractions. The latter class distribution is called an *abstract class distribution.* It is simply defined as the following class distribution (1) given $\overline{A} = g$, and can be represented as a linear combination of original distributions, as shown by the equation (2).

$$C_{\overline{A}=g} = (\Pr(C = c_1|\overline{A} = g), ..., \Pr(C = c_n|\overline{A} = g)) \quad (1)$$

$$= (\sum_{a \in g} \lambda_a \Pr(C = c_1|A = a), ..., \sum_{a \in g} \lambda_a \Pr(C = c_n|A = a))$$

$$= \sum_{a \in g} \lambda_a C_{A=a} \, , \quad \text{where } \lambda_a = \Pr(A = a|\overline{A} = g). \quad (2)$$

2.2 Information Loss Due to Data Abstractions

In this subsection, we analyze some relationships between the appropriateness of our data abstractions and their information loss measured by mutual information.

From the equation (2), an abstract distribution $C_{\overline{A}=g}$ is a mean of original class distributions $C_{A=a}$ such that $a \in g$. Therefore, some information or properties possessed by the individual distributions $C_{A=a}$ may be lost in the abstract distribution. In the worst case, for example, two distributions of two classes $C_{A=a_1} = (1,0)$ and $C_{A=a_2} = (0,1)$ are abstracted to their average $C_{\overline{A}=\{a_1,a_2\}} = (0.5, 0.5)$, provided $\Pr(A = a_1) = \Pr(A = a_2)$. The first two show that if $A = a_j$ then $C = c_j$ with the probability 1, while the abstract distribution is uniform and has no characteristics about the classes. Thus, the actual information about classes has disappeared in the abstraction.

To measure the amount of information loss, we use Shannon's mutual information [5]. It should be noted here that, in the cases of ID3 and C4.5, the mutual information measures a kind of *information gain of attributes for attribute selections*, while in this paper, it does the *information gain of abstracted attributes for abstraction selections.* The latter is equivalent to measure the *information loss for abstraction selections.* More precisely, we compute the subtraction of mutual information before and after the data abstraction.

According to the information theory (see [5], for instance), the following inequalities hold.

$$H(C|A) \leq H(C|\overline{A}) \text{ and } I(C;A) \geq I(C;\overline{A}), \quad \text{where}$$
$$I(C;X) = H(C) - H(C|X), \text{ the mutual information of } C \text{ and } X,$$
$$H(C) = \sum_{c \in dom(C)} L(\Pr(C=c)), \quad L(y) = \text{if } y = 0 \text{ then } 0 \text{ else } -y\log_2 y \ ,$$
$$H(C|X) = \sum_{x \in dom(X)} \Pr(X=x) \left(\sum_{c \in dom(C)} L(\Pr(C=c|X=x)) \right).$$

Therefore, we can define the information loss due to φ_A given by

$$e(\varphi_A) = I(C;A) - I(C;\overline{A}) = H(C|\overline{A}) - H(C|A).$$

Let $e(g;\varphi_A) = H(C|\overline{A} = g) - \sum_{a \in g} \lambda_a H(C|A = a))$, the loss due to each group $g \in \varphi_A$. Then we have $e(\varphi_A) = \sum_{g \in \varphi_A} \Pr(\overline{A} = g) e(g;\varphi_A)$.

To evaluate the loss $e(g;\varphi_A)$ and to establish the basic relationship between the approximation of distributions and the information loss, we need the following definition and proposition.

Definition 2. *(1) For two distributions $D_j = (p_1^j, ..., p_n^j)$ $(j = 1,2)$, the metric between D_1 and D_2 is given by $|D_1 - D_2| = \max_{1 \leq k \leq n} |p_k^1 - p_k^2|$. Moreover, D_1 and D_2 are said δ-similar, denoted by $D_1 \sim_\delta D_2$, if $|D_1 - D_2| < \delta$.*
(2) The diameter of class distributions $\mathcal{C} = \{C_1, ..., C_\ell\}$ is defined by

$$diam(\mathcal{C}) = \max_{i,j} |C_i - C_j|.$$

The following is a direct consequence of the definition and the equation (2) from which we draw our first selection criterion.

Proposition 2. *If* $diam(\{C_{A=a}|a \in g\}) < \delta$, *then* $C_{\overline{A}=g}$ *is* δ*-similar to any* $C_{A=a}$ *in* $\{C_{A=a}|a \in g\}$.

Preservingness of Distributions (PD):
Select a data abstraction $\varphi = (\varphi_{A_1}, ..., \varphi_{A_m})$ such that, for every $g \in \varphi_{A_j}$, any two distributions in $\{C_{A=a}|a \in g\}$ to be abstracted to an abstract distribution $C_{\overline{A}=g}$ are δ-similar with a small δ. Then the distribution $C_{\overline{A}=g}$ after the abstraction is also similar to any distribution in $\{C_{A=a}|a \in g\}$ within the same error δ. As a result, the properties or characteristics about classes captured by the distributions are approximately preserved and are not lost in the abstraction.

As the distributions are thus almost preserved when PD is satisfied, the amount of information before and after the abstraction is also approximately preserved, and the loss of information will be sufficiently small. In fact, Proposition 3 asserts that the information loss, $e(g; \varphi_A)$, is less than $h(\delta)$, which approaches zero whenever the diameter δ of distributions is sufficiently small.

Proposition 3. *There exists a function* $h(\delta)$ *defined on* $[0, 1]$ *such that*

(1) $h(\delta) \to 0$ *as* $\delta \to 0$, *and*
(2) $0 \leq e(g; \varphi_A) \leq h(diam(\{C_a|a \in g\}))$.

On the other hand, when PD is not satisfied, the information loss $e(g : \varphi_A)$ increases in general. However, in some cases, the loss is still small even when PD is not guaranteed. For instance, suppose that we have some minor and exceptional distribution $C_{A=e}$ in the sense that that its probability $\lambda_e = \Pr(A = e|A \in g)$ is very low and that it is far from the majority of similar distributions w.r.t. the metric. The information loss caused by grouping such an exceptional and minor one into the same group can be ignored because of its low probability, while the loss caused by putting the majority of similar distributions into the same one together can be evaluated very small. As a result, the mean $C_{\overline{A}=g}$ is much closer to the average of majority of similar distributions, and the whole information loss is still small. Thus, we can say that our abstraction can capture the majority of distributions, ignoring the exceptional and minor ones. As we consider that abstraction is useful because it can extract general knowledge covering many cases, allowing some exceptions, so we propose to use only the following selection criterion covering PD as its special case. The detail of treating the exceptions is found in [6]

Minimum Information Loss:
Choose a data abstraction whose information loss is the minimum among a class of possible abstractions.

A data abstraction actually chosen by this selection criterion depends on what space of possible abstractions we examine. The most general space is the lattice of all partitions except the trivial one, $\{\{a\}|a \in dom(A)\}$, whose information loss is always 0. As the lattice size is exponential, we will present in

Section 3 its subspace and an algorithm running in it so that finding the best abstraction is computationally tractable. Although the algorithm uses additional parameters and heuristics to improve its performance, the major factor to choose data abstractions is the minimization of information loss.

3 Three Experimental Systems

In this section, we present three experimetal systems, `ITA` , iterative `ITA` and `I`^{2}`TA` . All these are based on Mimimum Information Loss Principle described in Subsection 2.2.

3.1 Conditioning by Paths

In what follows, we evaluate paths in a decision tree in terms of their amount of information. So, before describing the three abstration systems, we here give some necessary notations about them. First, a (concrete level) path is a sequence $p = (A_{i_1} = a_{i_1}, ..., A_{i_k} = a_{i_k})$ of pairs of attribute A_{i_j} and its value $a_{i_j} \in dom(A_{i_j})$. Similarly, an abstract (level) path is defined as a sequence $\overline{p} = (\overline{A_{i_1}} = g_{i_1}, ..., \overline{A_{i_k}} = g_{i_k})$ such that $g_{i_j} \in \varphi_{A_{i_j}} = dom(\overline{A_{i_j}})$. When we examine two or more data abstractions φ_j for an attribute A, the abstract attribute $\overline{A}$ is often denoted by $\overline{A}_{\varphi_j}$ to specify which abstraction is used to form the abstract attribute. When it is clear from the context, we omit φ_j in the expression $\overline{A}_{\varphi_j}$.

Given an attribute X, its value x and a path p, either concerete or abstract, a concatination of path p and $X = x$, meaning an expansion of p to have a longer path in a decision tree, is written $p \cdot (X = x)$. Such a path p defines a conditional probability $\Pr_p(Event) = \Pr(Event|p)$. Particularly, an attribute value distribution for an attribute X (including the target one C) given a path p is defined by $X_p = (\Pr(X = x_1|p), ..., \Pr(X = x_{|dom(X)|}|p))$ whose entropy is also denoted by $H(X_p)$. In the three experimental systems, the mutual information and split information are evaluated for a sub-relation of original R or of its generalized one whose underlying probabilities are given by $\Pr_p$. In fact, we can construct the new sub-relation, R_p, as a set of tuples satisfying the condition specified by p. Then the mutual information $I_p(C; X)$ and the split information $H_p(X)$ for this R_p with the probability $\Pr_p$ are computed by using the following formulas.

$$I_p(C; X) = H_p(C) - H_p(C|X), \quad H_p(X) = \sum_{x \in dom(X)} L(\Pr(X = x|p)) = H(X_p),$$

$$\begin{aligned} H_p(C|X) &= \sum_{x \in dom(X)} \Pr(X = x|p) H_p(C|X = x) \\ &= \sum_{x \in dom(X)} \Pr(X = x|p) H(C_{p \cdot (X=x)}). \end{aligned}$$

```
begin
  for each attribute A do
    select the best abstraction φ_A that maximizes I(C,Ā_{φ_A}) / H(Ā_{φ_A}) ;
  apply the abstractions φ = (φ_{A_1}, ..., φ_{A_m}) to form a generalized relation φ(R) ;
  let the abstract path set P := {ϕ}, where ϕ denotes an empty path ;
  while there exists a path p̄ ∈ P such that
        C_p̄ does not satisfy the required precision do
    begin
      select the best attribute Ā not appearing in p̄, using I_p̄(C,Ā) / H_p̄(Ā)
          for the sub-relation of φ(R) specified by p̄;
      replace p̄ by the set of extended paths {p̄ · (Ā = g)|g ∈ dom(Ā)}, i.e.
          P := P − {p̄} ∪ {p̄ · (Ā = g)|g ∈ dom(Ā)}
    end
end
```

Fig. 2. The ITA algorithm including C4.5

3.2 ITA

The first system, ITA, supposes that the best data abstraction for one attribute can be determined independent to another attribute, according to the principle of minimum informatoin loss. Thus, ITA chooses the best one, φ_A, for each attribute A, and ignores the interrelations among attributes. Moreover, the split information, $H(\overline{A}_{\varphi_A})$, the entropy of attribute after abstraction, is also considered to measure the degree of generalization by φ_A. Formally, ITA calculates the information gain ratio, $\frac{I(C;\overline{A}_\varphi)}{H(\overline{A}_\varphi)}$ of $\overline{A}_\varphi$, to evaluate the amount of information that φ has, and select φ maximizing it. The reason we divide $I(C;\overline{A})$ by the split information $H(\overline{A})$ is the same as C4.5 . In a word, we prefer an abstraction that groups larger number of attribute values together. If a data abstraction φ_1 for A is a refinement of another φ_2, we have $H(\overline{A}_{\varphi_1}) \geq H(\overline{A}_{\varphi_2})$. We prefer φ_2 when the mutual information of $I(C;\overline{A}_{\varphi_1})$ and $I(C;\overline{A}_{\varphi_2})$ are almost equal, because φ_2 puts larger number of attribute values into the same group than φ_1. Thus, we can use $(H(\overline{A}_\varphi))^{-1}$ to measure the degree of generalization.

Such a selection and a generalization of relations are simultaneously performed for every attribute before we apply C4.5 . After that, C4.5 is carried out to produce an abstract level decision tree from the generalized relation, $\varphi(R)$ in Definition 1, with its underlying probability $\Pr_{\varphi(R)}(g_1, ..., g_m, c) = \Pr(A_1 \in g_1, ..., A_m \in g_m, C = c)$. The complete description of ITA embedding C4.5 is described in Fig. 2.

3.3 Iterative ITA

In contrast to ITA, the iterative ITA considers the interrelations among attributes, and selects the best data abstraction, depending on the previously selected attributes and abstractions. As such a history of selections is represented by an abstract path, the best one is consequently choosen according to

begin
  let the abstract path set $\mathcal{P} = \{\phi\}$;
  while there exists a path $\overline{p} \in \mathcal{P}$ such that
      $C_{\overline{p}}$ does not satisfy the required precision do
    begin
      choose the best attribute A in $R_{\overline{p}}$ not yet generalized, using $\frac{I_{\overline{p}}(C,A)}{H_{\overline{p}}(A)}$;
      (*) select the best abstraction φ_A for the chosen A, using $\frac{I_{\overline{p}}(C,\overline{A}_{\varphi_A})}{H_{\overline{p}}(\overline{A}_{\varphi_A})}$;
      $\mathcal{P} := \mathcal{P} - \{\overline{p}\} \cup \{\overline{p} \cdot (\overline{A}_{\varphi_A} = g) | g \in dom(\overline{A}_{\varphi_A})\}$
    end
end

Fig. 3. The iterative ITA algorithm

the conditioning represented by the path. Basically, the iterative ITA follows the behavior of C4.5 at abstract level. So, the selection of data abstraction is inductively defined based on the previously selected attributes $A_{i_1}, ..., A_{i_k}$, their chosen abstractions $\varphi_{A_{i_1}}, ..., \varphi_{A_{i_k}}$ and an abstract level path $\overline{p} = (\overline{A_{i_1}} = g_{i_1}, ..., \overline{A_{i_k}} = g_{i_k})$, where $\overline{A_{i_j}}$ is the abstract attribute formed by $\varphi_{A_{i_j}}$. For our notational convenience, we here suppose without loss of generality that $\overline{p} = (\overline{A_1} = g_1, ..., \overline{A_k} = g_k)$. For the abstract path $\overline{p}$, only the attribute $A_1, ..., A_k$ are generalized to produce a partially generalized relation $R_{\overline{p}}$ defined as a set of tuples $t = (g_1, ..., g_k, a_{k+1}, ..., a_m, c)$ such that, for some $(a_1, ...a_m, c) \in R$, $a_i \in g_i \in \varphi_{A_i}$ $(1 \leq i \leq k)$. The probability for $R_{\overline{p}}$ is the conditional probability

$$\Pr_{\overline{p}}(t) = \Pr(t|\overline{p}) = \frac{\Pr(A_1 \in g_1, ..., A_k \in g_k, A_{k+1} = a_{k+1}, ..., A_m = a_m, C = c)}{\Pr(A_1 \in g_1, ..., A_k \in g_k)}.$$

Note that, in case of $k = 0$, $\overline{p}$ is an empty path, and nothing is generalized. Therefore, $R_\phi = R$ and $\Pr_\phi = \Pr$. Now, C4.5 selects the best attribute A from $\{A_{k+1}, ..., A_m\}$ by the information gatin ratio. After that, we choose the best abstraction φ_A for the selected attribute A, and proceeds with our classification task by expanding the abstract path $\overline{p}$ with its branches $\overline{A}_{\varphi_A} = g \in \varphi_A$. Fig. 3 shows the description of iterative ITA .

3.4 I²TA

The iterative ITA measures the degree of generalizations preformed by data abstractions, using $(H_{\overline{p}}(\overline{A}))^{-1}$. However, according to our experience, a better abstraction φ_1 with greater mutual information is often defeated by another φ_2 with less mutual information, simply because the degree of generalization $(H_{\overline{p}}(\overline{A}_{\varphi_2}))^{-1}$ is much greater than $(H_{\overline{p}}(\overline{A}_{\varphi_1}))^{-1}$. As the degree of generalization is higher, we have more compact relation. We may however miss some important information about classes in the more compact relation. In order to resolve the conflict, we propose the third system, I²TA , which maximizes the mutaul information $I_{\overline{p}}(C; \overline{A}_\varphi)$ among possible data abstractions φ whose split information $H_p(\overline{A}_\varphi)$ is within some range. More precisely, the split change ratio $SA(\varphi) = \frac{H(\overline{A}_\varphi)}{H(A)}$ is calculated. $\mathtt{I^2TA}$ algorithm has two additional parameters, the lower bound s_l and the upperbound s_u of SA, and excutes the same computation as iterative `ITA` , where we choose the best φ_A attaining

$$\max_{s_l \leq SA(\varphi) \leq s_u} I\overline{p}(C; \overline{A}_\varphi).$$

in the step (*) in the algorithm in Fig. 3, instead of maximizing $\frac{I_{\overline{p}}(C,\overline{A}_{\varphi A})}{H_{\overline{p}}(\overline{A}_{\varphi A})}$.

3.5 Exprimental Results

We summarize in this subsection our experimental results of the three systems applied for *Census Database in US Census Bureau* found in the UCI repository[7]. The training data in our experiment consists of 30162 tuples with 11 attributes including *age*, *workclass*, *education*, *occupation*, *relationship*, *race*, *sex*, *hours-per-week*, *native-country* and *salary*. Apart from this training data, we prepare a small census database (called *test data*) consisting of 15060 tuples in order to check the classification accuracy of the constructed decision tree. A class of possible data abstractions for attribute values in the census database is automatically computed from a machine readable dictionary (**MRD**, for short), WordNet[8]. Moreover, we have examined two classes: "$\leq$ \$50K" (NOT more than \$50000) and "$>$ \$50K" (more than \$50000) in the attribute *salary*. Fig. 4 shows the results from which we have the following observations.

MRD and a class of possible abstractions:
We have used a machine readable dictionary, WordNet, to define a class of possible data abstractions. WordNet has various subsumption relationships between words (nouns) which we call primitive views. Regarding these primitive views as partial mappings, we compose them to have a family of composite views relating words at various levels of abstractions. Then the composite view is interpreted as a partition of attribute domain such that two or more values (words) belong to the same group whenever they are mapped to the same value by that view. WordNet classifies words into some semantic categories. So, it suffices to apply the composition in each category. This prevents the composition task from falling into the combinatorial explosion. As a result, the number of all data abstractions thus composed is only 134 in case of $native - country$ attribute. Another attribute has less number of data abstractions.

Error rates:
First of all, the error rates of our three abstraction systems are within the range between 0.15 and 0.18 for both training and test data. So, we cannot observe a significant difference between the test and training data. As the abstraction systems try to extract abstract paths whose distributions have high precision for the abstracted data values, those rules represented by paths tend to have their generality, ignoring the differences between instance tuples and keeping

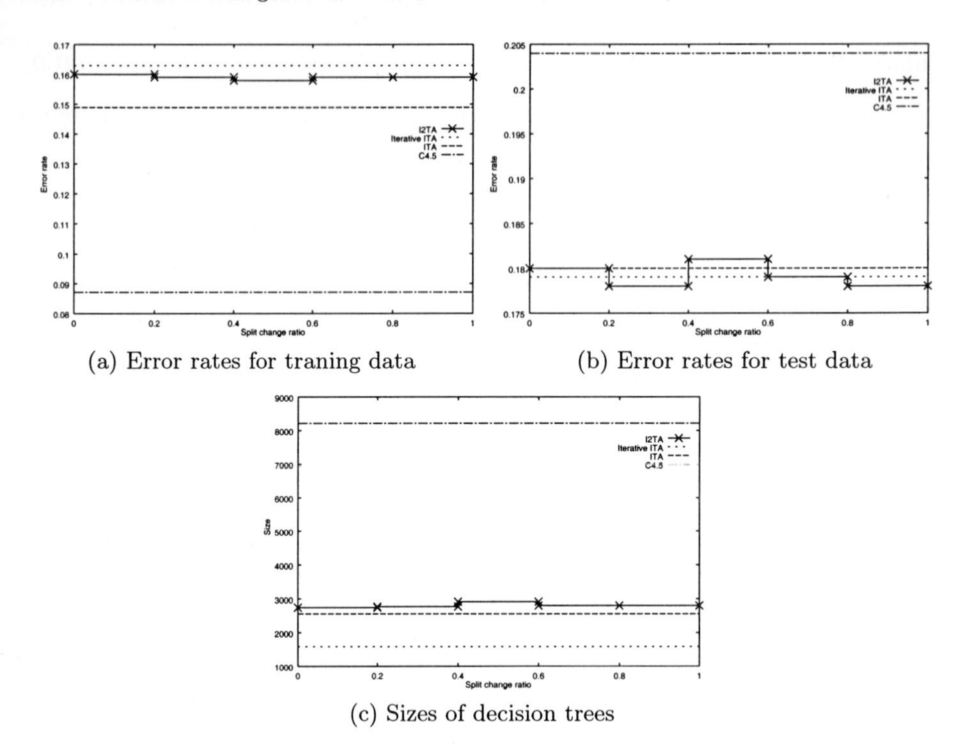

(a) Error rates for traning data

(b) Error rates for test data

(c) Sizes of decision trees

Fig. 4. Error rates and sizes of decision trees constructed by four systems

the precision. On the other hand, the error rate of decision tree constructed by C4.5 increases from 0.088 to 0.204 when we test it for the test data. C4.5 seems much specific to the training data. Even if we apply the error-based pruning to the decision trees, we have observed that the error rates of abstract decision trees are still better than the concrete level decision tree made by C4.5 .

Among the three abstraction systems, $\mathtt{I^2TA}$ achieves the best performance w.r.t. the error rate. In fact, for three intervals of split-change-ratio, we can observe a significant improvement. In the interval, [0.4,0.6], $\mathtt{I^2TA}$ is inferior to the other systems. This implies that MRD does not have adequate views from which we can compose a good abstraction whose split information is within the range.

Sizes:
In contrast to the error rate, the iterative ITA is the best. Since it takes the degree of generalization, $H_{\overline{p}}(\overline{A})^{-1}$, into account at each stage of attribute and abstraction selections, we can say that iterative ITA prefers data abstractions whose degree of generalization is higher. Comparing C4.5 with the three abstraction systems, the size of decision tree is much reduced in all the abstraction systems. This holds even after we apply the error-based pruning technique for C4.5 .

4 Conclusions

From the observations in Subsection 3.5, the error rate of our abstraction systems are not worse than that of C4.5 . Although we have to examine the observations through another experiment for a different kind of data, the experimental result shows that the error rates of abstraction systems become better than C4.5 when we consider the test data. Thus, the abstraction strategy is more stable independent to particular instances. This proves the advantage of using abstraction that is not only the size reduction but also the acquisition of general knowledge not depending on particular cases.

The most important future work is to develop a strategy for synthesizing the class of possible abstractions. We have in fact various approaches for it. For example, some strategy searching for a good abstraction in the lattice of all data abstractions will be a standard one. However, the authors are now investing another new approach related to some knowledge revision techniques. As we have mentioned in Subsection 3.5, I^2TA may be inferior to others w.r.t. the error rate. This can be a trigger of knowledge revision. I^2TA is designed so that it possibly preserves the information about classes, provided an adequate abstraction whose split information is within some range is provided. Conversely, MRD is inadequate, when I^2TA does not work well. Thus, it seems interesting to revise MRD based on the behavior of I^2TA so that its performance is improved.

References

1. Han, J. and Fu, Y.: *Attribute-Oriented Induction in Data Mining.* In Advances in Knowledge Discovery and Data Mining (Fayyad, U.N. et.al. eds.), pp.399-421, 1996.
2. Kudoh, Y. and Haraguchi, M.: An Appropriate Abstration for an Attribute-Oriented Induction Proceeding of The Second International Conference on Discovery Science, LNAI 721, pp.43 - 55, 1999.
3. Kudoh, Y. and Haraguchi, M.: Detecting a Compact Decision Tree Based on an Appropriate Abstraction Proc. of 2nd Intl. Conf. on Intelligent Data Engineering and Automated Learning, LNCS-1983, pp.60–70, 2000.
4. Quinlan, J.R.: *C4.5 - Programs for Machine Learning,* Morgan Kaufmann, 1993.
5. Shannon, C. E.: A Mathematical Theory of Communication, The Bell system technical journal, vol. 27, pp.379-423 (part I), pp.623-656 (part II), 1948.
6. Kudoh, Y., Haraguchi,M. and Okubo,Y.: Data Abstractions for Decision Tree Induction, submitted to an international journal, Jan. 2001.
7. Murphy, P.M. and Aha, D.W.: *UCI Repository of machine learning databases,* http://www.ics.uci.edu/ mlearn/MLRepository.html.
8. Miller, G.A., Beckwith, R., Fellbaum, C., Gross, D. and Miller, K.: *Intorduction to WordNet : An On-line Lexical Database* In : International Journal of lexicography 3 (4), pp.235 - 244, 1990.

Discovery of Chances Underlying Real Data

Yukio Ohsawa

Graduate School of Business Science, The University of Tsukuba,
3-29-1 Otsuka, Bunkyo-ku, Tokyo, 112-0012, Nippon (Japan)
osawa@gssm.otsuka.tsukuba.ac.jp

Abstract. Chance discovery is to notice and explain the significance of a chance, especially if the chance is rare and its significance is unnoticed. Chance discovery is essential for various real requirements in human life. The paper presents the significance and the achieved methods of chance discovery. Fundamental discussions of how to realize chance discovery extracts keys for the progress of chance discovery: communication, imagination, and data mining. As an approach to chance discovery, visualized data mining methods are shown as tools aiding chance discoveries.

1 Introduction to Chance Discovery

Chance Discovery is to discover chances. Here, a "chance" is defined as an event or a situation with significant impact on human decision making, after the "chance" in its second meaning in the Oxford Advanced Learner's Dictionary, i.e., "a suitable time or occasion to do something." In this definition, it is considered that a situation or an event at a certain time is more significant than the time itself, and a decision to do something precedes doing it. A a chance here can be a new event or situation to be conceived either as an opportunity or as a risk. The word "discovery" also has some ambiguity. In discovering a law ruling the nature, sometimes one may "learn" frequent patterns in the observations. On the other hand, I take the general meaning of discovery - to obtain explicit knowledge about the unknown mechanism of an event occurrence. In other words, "discovery" in chance discovery means explaining events which have never been explained on explicit knowledge or hypothesis.

Chance discovery is defined as the awareness on and the explanation of the significance of an event, including rare ones whose significance are unnoticed. By understanding explicitly what actions can put an opportunity into a benefit, desirable effects of opportunities can be actively promoted. Whereas, explicit preventive measures will come to be taken for discovered risks.

2 Chance Discovery versus Prediction

All in all, the stronger-impact situation is the rarer. For example, high-energy states are of low probability according to the Maxwell-Boltzmann distribution. Another example can see in the stock market, where large changes in the stock

S. Arikawa and A. Shinohara (Eds.): Progress in Discovery Science 2001, LNAI 2281, pp. 168–177.

prices occur less frequently than small changes. Because of the rareness, a strong-impact event comes to be the less predictable. Similar problems are with predictive data mining. You can predict the next event by following a frequent *episode* (partial sequence) found in the event-sequence of past data, if no unusual changes in the environment occur [3]. As well, *association rules* discovered from the data of baskets, each of which has a data set of items, can be used to predict future items to occur next, if they were frequent in past baskets [2].

However, an unusual and rare event can make a significant influence onto the future, if its potential impact comes to be realized. Chance discovery aims at explaining how to realize the impact, for creating or surviving the future, based on the awareness on such events.

Recently, the attentions in statistics to *extremals* (rare substances of variables) are motivated by social situations nowadays i.e. rare situations sometimes result in fatal depressions or remarkable prosperity. However, explaining the effects of hidden causes of rare events are out of their research targets.

In the data mining area, the prediction of rare events are coming to be of a novel interest. Methods for learning a high-accuracy rule with a complex condition (A of rule $A \rightarrow B$), having a rare event B in the conclusion-part, have been proposed [4, 5]. However, these methods cannot discover chances. The exceptional rules [5] could not catch Nojima fault (the focus of South-Hyogo Earthquake in 1995) as risky, because a rule was ignored if its *support* or $confidence$ (the probability of the co-occurrence of A and B, and the same co-occurrence under the occurrence of A, respectively) was extremely small. In this case, earthquakes at Nojima fault before 1995 were too rare.

3 Human Roles in Chance Discovery

If there is a commodity with a feature rarely seen, the feature-based prediction, i.e., using the pattern of feature occurrences for predicting the feature of the commodity to be bought next, cannot be applied. In this case, computation methods previously studied cannot see

(1) the significance of new attributes not described in available data, nor
(2) the significance of extremely rare substances of certain attributes.

Because human beings are finely considering (1) and (2) above in real situations, I am directed to answer the question "How can we aid human discovery of chances ?" not only "How can a computer discover chances ?"

4 Three Keys to Chance Discovery

It has been pointed out that a dialogue of a number of people, where chances may be discovered as in Step 3 above, includes the natures below [7]:
Inquiries, where all participants collaborate to answer the question "how can we discover and manage a chance, which may be unknown ?"

Persuasion, where one participant seeks to persuade other(s) to pay attentions to and accept a proposition of a chance management method.
Here, McBurney and Parsons further formalized the communications for chance discovery as a dialogue game formed by a set of formulated locutions. The essential locutions for chance discovery, for example, are $assess(P_i, c, t, p)$ (participant P_i asserts conclusion c can be estimated to mean p, if assessed on criteria t) and $compare(P_i, c_1, c_2, t)$ (P_i compares conclusions c_1 and c_2 on criteria t).

That is, if $assess(P_1, c_1, t, p_1)$ and $assess(P_1, c_2, t, p_2)$ both occur where c_1 and c_2 are conclusions having been presented by P_1 and P_2 respectively, and p_1 and p_2 both mean positive benefits for P_1, then P_1 begins to evaluate P_2 positively - if this occurs to P_2 also, this can be the beginning of a new community of P_1 and P_2 respecting each other. On the other hand, $compare(P_1, c_2, c, t)$ for c, a conclusion from a participant not P_2, is also meaningful in selecting P_2 as a more relevant person than other people. In summary, the introduction of t - meaning a new *value criteria* for evaluating the relevance of an idea to one's benefit or loss - can make a chance for unifying multiple communities.

Note that this study is based on social science dealing with dialogues of people highly concerned with the purpose of the arguments. If t is the value criteria admiring a new product attracting multiple communities, buyers leading each community will be unified first by the mechanism above because they are highly concerned with new products, and lead the communities to be merged to form a new large market on t, as exemplified in Fig.1.

4.1 Scene Information for Aiding Imagination

On the notion of "context shifting," the process of chance perception has been modeled [8]. Because a chance may be extremely rare, average people are not concerned about its meaning. It is essential to shift the mental context of such ignorant people, to be concerned with the situation where the chance comes to be significant. Peripheral information [6] will guide these people efficiently. In the next step, explaining the significance of the chance with central information shifts the subject to clear awareness and understanding of the chance, and more detailed information will shift to decisions and real actions. A concrete action concentrates the attention of the subject, and makes her/him forget other things, and return to a new "ignorance" state where a new *clue event* is desired for recalling relevant knowledge to achieve further progress in problem solving.

Thus, we found three keys to chance discovery:

Key 1: Communications with introducing new value criteria
Key 2: Scenic information for aiding imagination, for shifting user's awareness to the unnoticed context underlying the data.
Key 3: Data mining not discarding rare events.

Key 3 is for catching potentially significant new value criteria in Key 1, which may otherwise be unnoticed forever. The communication in Key 1 includes the

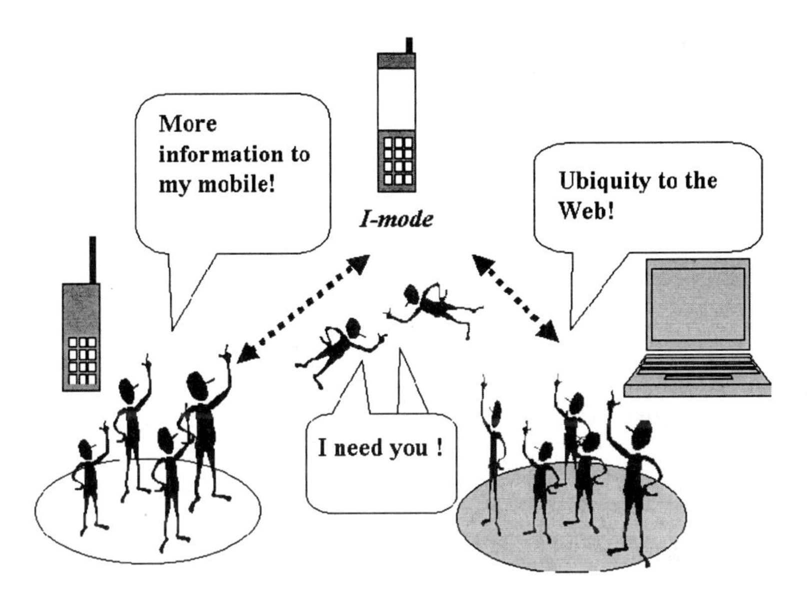

Fig. 1. The diffusion of i-mode, the Internet mobile phone. I-mode made new value criteria in two communities, of mobile phone- and PC-users, and made the communities attract each other. This made a fusion into a big market.

process to externalize and share knowledge, e.g., *design rationale*, essential for appropriate people's catching chances at appropriate moments.

The process shown in Fig.2 clarifies the difference of chance discovery from the KDD (knowledge Discovery from Database) process model [1].

The value of knowledge in the KDD process may be evaluated automatically applying an *objective* measure by matching with test data. Whereas, human is positioned as the major engine to select, interpret and evaluate valuable chances in Fig.2. A number of candidates of chances may be presented by machine, if the interface to human is coordinated to help selecting a sufficient and necessary small number. A chance is here to be evaluated after human "decision/action," with interaction to external information sources. With this interaction, new essential attributes are detected and reflected to "observation" of values for those attributes. The contribution of human to the "selection" step in the KDD process, including the evaluation of rare events which can be chances, is to be entrusted to human interpretation cycles where essential rare events are recognized as "central clues" in each cycle ("selection 2"). Data mining is also desired to urge human concern, by giving "peripheral clues" showing scenic information of the situation where the chances become meaningful.

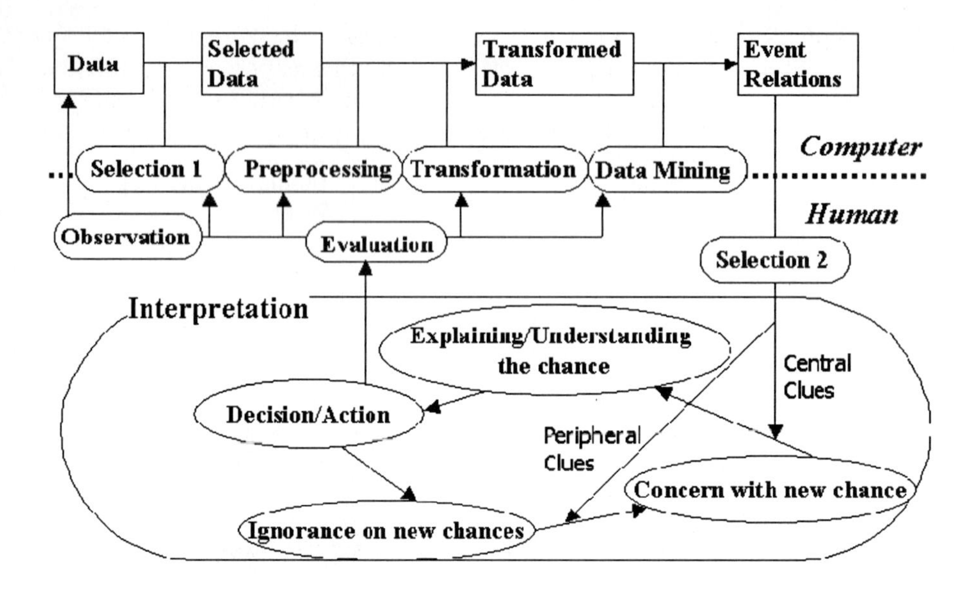

Fig. 2. The process of Chance Discovery

5 Embodying the Keys to Chance Discovery

5.1 Communications Looking at KeyGraph

A marketing method was presented, embodying all the three keys in Section 4 [9]. The method applied a visualized data-mining algorithm KeyGraph [10], to a food consumption data of a number of Japanese families (Key 3). Then the visualized output was presented to a group of housewives and the housewives chatted with looking at the output (Key1). First they imagined the scenes of family-meals, guided by the moderator (Key2), and finally they discovered essential value criteria for rarely consumed foods in breakfasts, lunches, and dinners.

KeyGraph had been proposed for indexing a document and extended to finding risky active faults of earthquakes, on the analogy between a series of words in a document and a time series of active faults where earthquakes occurred [10] (outperformed previous data mining methods). The algorithm can be briefly summarized as follows

The algorithm summary of KeyGraph

KeyGraph-Step 1: Clusters of co-occurring frequent items are obtained as *basic clusters*. That is, items appearing many times in the data (e.g., the word "chance" in this paper) are extracted, and each pair of items occurring often in the same sequence unit (a sentence in a document, a transaction in a sales data, and a breakfast, lunch, or dinner in a food consumption data) are linked to each other, e.g., "chance - discovery" for this document. Each connected graph made of those links and items form one cluster, impling a common cause of included items.

KeyGraph-Step 2: Words not so frequent as ones in clusters but co-occurring with multiple clusters, e.g., "three keys" in this paper, are obtained as chances, i.e., rare items coming to be significant (assertions in a document, or possible future interests of consumers in a consumption data). Such items are significant, e.g., multiple major causes (common to clusters) occur with such an item [10], to make novel value criteria in multiple communities and make a grown market.

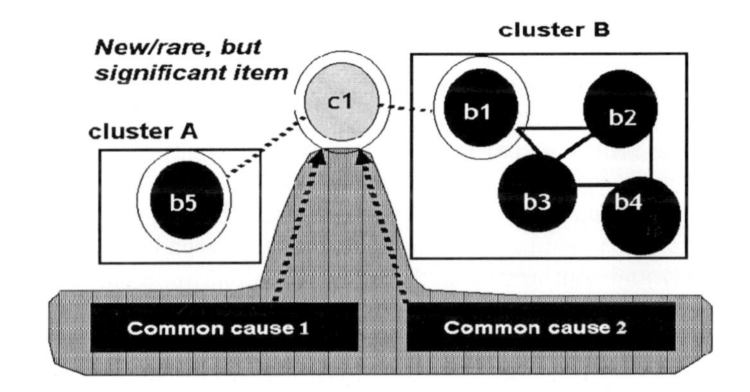

Fig. 3. An example for the output of KeyGraph. The shadowed zone shows hidden common causes of some items' co-occurrences, and their relevance to a novel item $c1$.

The input data D to KeyGraph is given in the form of a "document" below, where each sequence unit in **KeyGraph-Step 1** above is expressed as a *sentence*, i.e., an item-sequence between period('**.**')s.

$$\begin{aligned} &b1, b2, b3, c1, c2, c5, ..., p7, ..., z4. \\ &b2, b3, b4, c0, c3, c4, c5, ..., g13, q3, ..., y0. \\ &b1, b4, ...c1, c3, c6, ... \\ &\vdots \end{aligned} \tag{1}$$

In **KeyGraph-Step 1**, the cluster of four nodes $\{b1, b2, b3, b4\}$ will be obtained as in Fig. 3 from co-occuring pairs $\{b1, b2\}$, $\{b2, b3\}$ etc. A basic cluster of a single node as $b5$ is possible to be made. In **KeyGraph-Step 2**, items co-occuring with items in multiple clusters like $c1$ comes out.

An example use of KeyGraph: Earthquake Risk Discovery [10]

Let us have a data of earthquake sequence, e.g., string sequence D with spaces and periods ("."s), as

$$D = 123\sharp\ 202\sharp\ 84\sharp.\ 1\sharp\ 76\sharp\ 1\sharp.\ 216\sharp\ 202\sharp\ 84\sharp.\ 249\sharp\ 84\sharp.\ 76\sharp\ 249\sharp\ 1\sharp \cdots. \tag{2}$$

Here, "$m\sharp$" means the m th pre-defined active fault, the closest to the epicenters of each earthquake in time-series earthquake data. This example includes three "."s supposing that three big earthquakes (over a threshold magnitude) occurred.

The sources of earthquakes are underground forces, each causing an area of land crust to move, and faults in the area to quake with the land crust movements. Assuming that earthquakes in the same term (i.e., between a pair of nearest "."s) are caused by a common large land crust area, the stress from those areas to the boundary fault tends to cause a big earthquake in a near future. Reflecting this earthquake mechanism, we applied KeyGraph to the data as Equation 2. The outputs were, as expected, 1) co-quaking faults as clusters of nodes, and 2) risky, i.e., stressed faults from an adjacent land crust areas, as "chance" nodes.

Fig.4 is the output of KeyGraph from data of earthquakes in year 1985 in Kansai area. Here active fault No.39, obtained as risky, is Nojima fault which caused the later Southern Hyogo earthquake of $M7.2$ in 1995 (victimized more than 6000 people). From Fig. 4 we can see fault No.39 had been stressed by No.24 and 34 (each node forming one cluster) and the large bold-lined cluster (of faults as No.76, 79, etc.). This means No.39 was stressed between its northern (No.24 and 34) and southern (No.76, 79, etc) neighboring crust areas.(`End of example`)

This algorithm KeyGraph was applied to data of the sequence of foods, during one-year meals of 88 families in the south-east district of Japan [1] [9]. A result from a sequence of one-year breakfasts is in Fig.5. In this case, we find two solid-lined clusters (one in the right and the other in the left of the frame), corresponding to usual Japanese and Western style breakfasts respectively. The dotted links can be regarded as the motivations to eat rare foods (colored red, or the double-circle nodes in the gray-scale printing) in established situations (the two clusters). Thus, nodes "Jelly" "Vitamin Supply" and "Salmon Can" are regarded as essential rare, i.e., latent demands.

We then showed ten of such output figures to gathered seven interviewees, house wives in the ages of forties or fifties (the same ages as subjects in Shoku-MAP data), in the manner of a group interview as in Fig. 5. The moderator guided the interviewees in the following.

1) Asked what the interviewees ate in the morning, and organized a free talk.
2) Explained what is and how to see the KeyGraph's output, in ten minutes.
3) Had interviewees write down what they imagined or thought with looking at the graphical output of KeyGraph, in the balloons as in 5.
4) Had all interviewees talk to each other freely adding lines, nodes, and words to the KeyGraph's output figure, and recorded all the communications.

During the interview, decision proposals for meal-designs including the significance of each rare food appearing as red nodes in the figure was obtained, e.g., "An easy supply of vitamin is essential in the breakfast, for the health care of the family without taking long time for serving the meal."

---

[1] The Shoku-MAP (trademark registered in Japan) data from NTT Data, Inc.

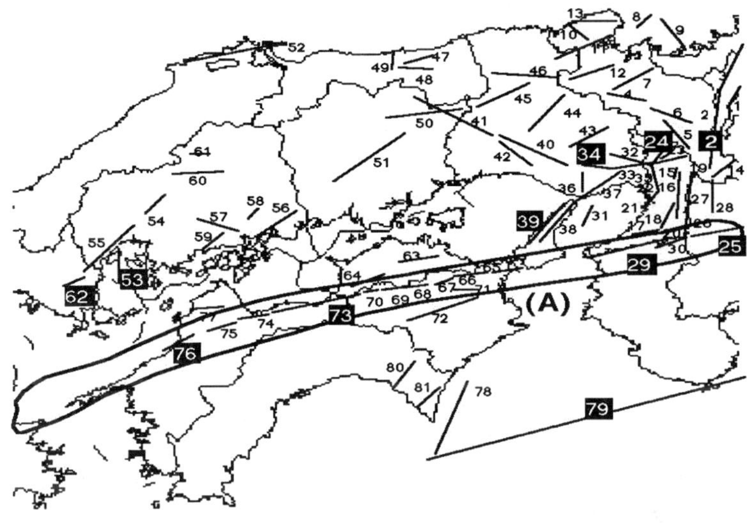

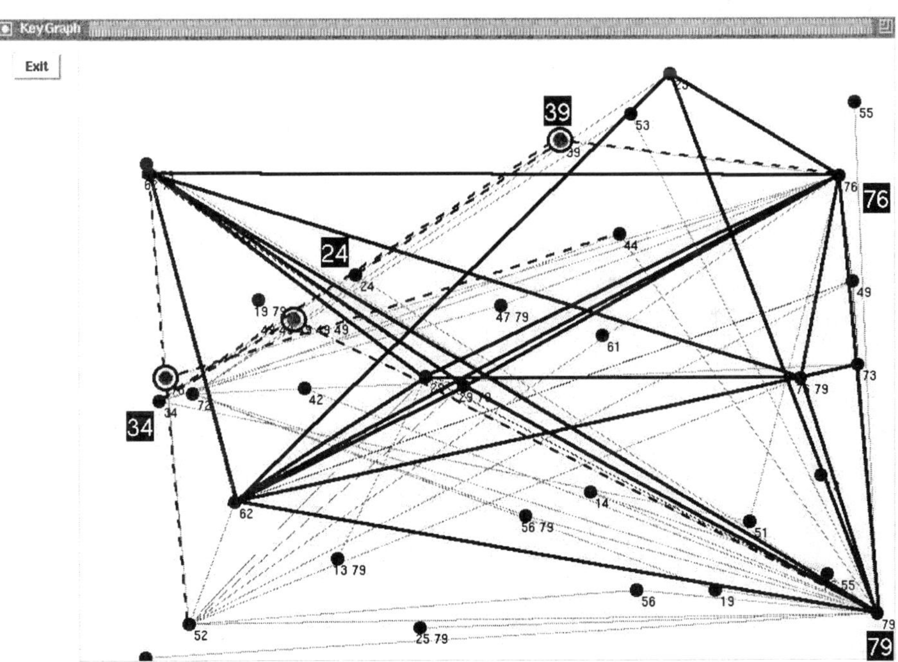

Fig. 4. The result (lower) of F^3 for Kansai area (upper). F^3 obtained the double circle nodes as risky faults. Thick solid lines and black nodes form *clusters* (some clusters include only one node, i.e. one fault), and other lines show *stresses* from clusters to faults. Thick dotted lines are the strongest stresses.

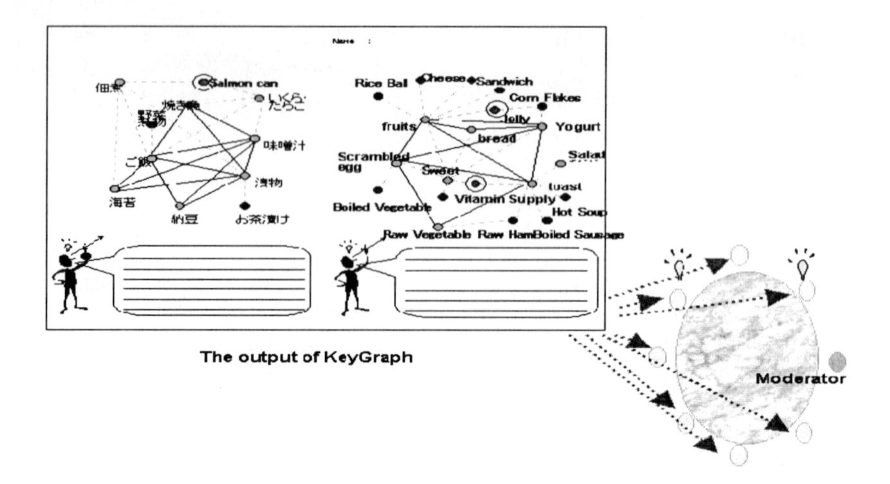

Fig. 5. The design of group interview. The cartoon-like presentation of KeyGraph results helped interviewees imagine the meal scenes and the balloon helped them describe what they imagined.

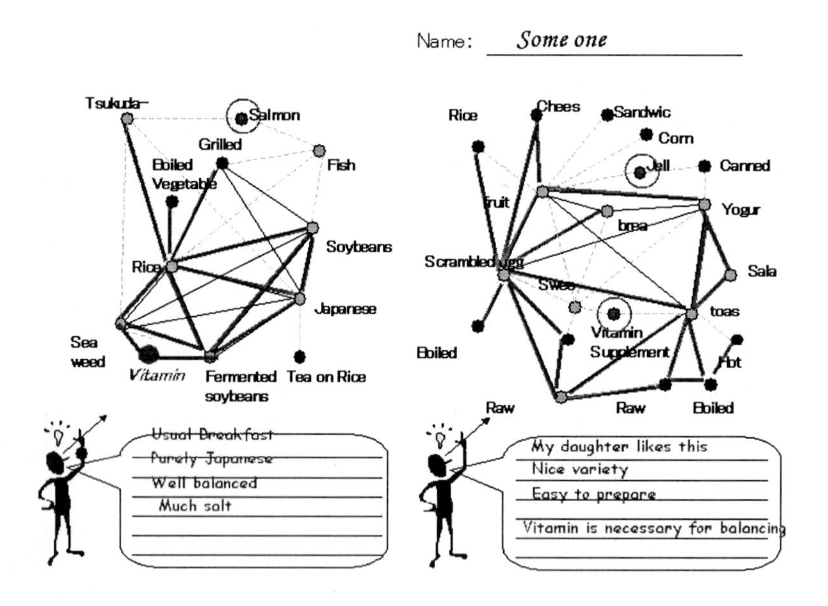

Fig. 6. An interviewee's result of the group interview. The thickest lines, the node "Vitamin" and the words in balloons were added by the interviewee.

The process to reach these opinions was as follows: First, subjective opinions popped out from few interviewees, about the double-circle foods. These opinions were filtered and essential ones became prevalent. Such opinions were accepted in the later half of the two hours interview. "Vitamin" came frequent in the later part of the balloons filled by interviewees (Fig.6).

The principle of KeyGraph is to obtain basic clusters and finding links among them, which can be regarded as community unifications on a new value, reflecting **Key 1**. This principle has been exploited to develop unique methods for the data mining step of chance discoveries [11].

6 Conclusion

Chance discovery intends to support decisions triggered by the awareness on significant events or situations. This paper introduced reader to notice the importance of chance discovery, presented key issues for realizing chance discovery, and some selected examples of realizing chance discovery.

References

1. Fayyad, U., Shapiro, G.P. and Smyth, P., From Data Mining to Knowledge Discovery in Databases, *AI magazine*, Vol.17, No.3, 37–54 (1996)
2. Mining Association Rules between Sets of Items in Large Databases, *Proc. ACM SIGMOD Conf. Management of Data*, 207 – 216 (1993)
3. Mannila, H, et al, *Discovering Frequent Episodes in Event Sequences,* in *Proc. First Conf. on Knowledge Discovery and Data Mining (KDD95)*, (1995)
4. Weiss, G.M. and Hirsh,H. Learning to Predict Rare Events in Event Sequences," in *Proc. of KDD-98* 359-363 (1998)
5. Suzuki, E. and Tsumoto, S., Evaluating Hypothesis-Driven Exception-Rule Discovery with Medical Data Sets, *Knowledge Discovery and Data Mining* (LNAI 1805, Terano, T. et al eds. Springer Verlag) 208–211 (2000)
6. Petty, R.E., and Cacioppo, J.T., *Attitudes and persuasion: Classic and contemporary approaches.* Dubuqute, IA: WC Brown (1981)
7. McBurney, P. and Parsons, S. Chance Discovery Using Dialectical Argumentation in *Proc. CDWS2001* (2001)
8. Ohsawa,Y. and Nara, Y., Decision Trees as a Model of Chance Perception, *Proc. Joint 9th IFSA Congress and 20th NAFIPS International Conference* (2001)
9. Fukuda, H. and Ohsawa, Y., Discovery of Rare Essential Food by Community Navigation with KeyGraph - An introduction to Data-based Community Marketing -, *Proc. KES2001* (IOS press), Osaka (2001)
10. Ohsawa, Y. and Yachida, M., Discover Risky Active Faults by Indexing an Earthquake Sequence, in *Proc. International Conference on Discovery Science* (DS'99)(1999)
11. Ohsawa,Y., Matsumura,N., and Ishizuka, M., Discovering Topics to Enhance Communities' Creation from Links to the Future, *Posters Proc., the World Wide Web Conference (WWW10)* (2001)

Towards the Integration of Inductive and Nonmonotonic Logic Programming

Chiaki Sakama

Department of Computer and Communication Sciences
Wakayama University
Sakaedani, Wakayama 640 8510, Japan
sakama@sys.wakayama-u.ac.jp
http://www.sys.wakayama-u.ac.jp/~sakama

Abstract. Commonsense reasoning and machine learning are two important topics in AI. These techniques are realized in logic programming as *nonmonotonic logic programming* (NMLP) and *inductive logic programming* (ILP), respectively. NMLP and ILP have seemingly different motivations and goals, but they have much in common in the background of problems. This article overviews the author's recent research results for realizing induction from nonmonotonic logic programs.

1 Introduction

Inductive logic programming (ILP) [12, 16] targets the problem of inductive construction of hypothetical knowledge from examples and background theories. The present ILP systems mostly consider *Horn logic programs* as background theories. A Horn logic program is *monotonic* in the sense that adding a clause to the program never leads to the loss of any conclusions previously proved in the program. However, it is known that Horn logic programs are not sufficiently expressive for the representation of *incomplete knowledge*. In the real world, humans perform *commonsense reasoning* when one's knowledge is incomplete. Commonsense reasoning is *nonmonotonic* in its feature, that is, previously concluded facts might be withdrawn by the introduction of new information. *Nonmonotonic logic programming* (NMLP) [2, 4] introduces mechanisms of representing incomplete knowledge and reasoning with commonsense.

ILP and NMLP have seemingly different motivations and goals, however, they have much in common in the background of problems. The process of discovering new knowledge by humans is the iteration of hypotheses generation and revision, which is inherently nonmonotonic. Induction problems assume background knowledge which is incomplete, otherwise there is no need to learn. Therefore, representing and reasoning with incomplete knowledge are vital issues in ILP, though relatively little attention has been paid on them. Comparing ILP and NMLP, ILP performs inductive learning in Horn logic programming, and has limited applications to nonmonotonic theories. By contrast, NMLP realizes commonsense reasoning in logic programming, but a program has no mechanism of learning new knowledge from the input. Thus, both NMLP and ILP

S. Arikawa and A. Shinohara (Eds.): Progress in Discovery Science 2001, LNAI 2281, pp. 178–188.

have limitations in their present frameworks and complement each other. Since both commonsense reasoning and machine learning are indispensable for realizing intelligent information systems, combining techniques of the two fields in the context of *nonmonotonic inductive logic programming* (NMILP) is meaningful and important. Such combination will extend the representation language on the ILP side, while it will introduce a learning mechanism to programs on the NMLP side. Moreover, linking different extensions of logic programming will strengthen the capability of logic programming as a knowledge representation tool in AI. From the practical point of view, the combination will be beneficial for ILP to use well-established techniques in NMLP, and will open new applications of NMLP.

This article gives an overview of the author's recent research results for realizing induction from nonmonotonic logic programs. The rest of this paper is organized as follows. Section 2 reviews frameworks of NMLP and ILP. Section 3 presents techniques for induction in nonmonotonic logic programs. Section 4 presents related work, and Section 5 summarizes the paper.

2 Preliminaries

2.1 Nonmonotonic Logic Programming

Nonmonotonic logic programs considered in this paper are *normal logic programs*, logic programs with negation as failure.

A *normal logic program* (NLP) is a set of *rules* of the form:

$$A \leftarrow B_1, \ldots, B_m, not\, B_{m+1}, \ldots, not\, B_n \tag{1}$$

where each A, B_i $(1 \le i \le n)$ is an atom and *not* presents *negation as failure* (NAF). The left-hand side of $\leftarrow$ is the *head*, and the right-hand side is the *body* of the rule. The conjunction in the body of (1) is identified with the set $\{ B_1, \ldots, B_m, not\, B_{m+1}, \ldots, not\, B_n \}$. The conjunction in the body is often written by the Greek letter Γ. A rule with the empty body $A \leftarrow$ is called a *fact*, which is identified with the atom A. A rule with the empty head $\leftarrow \Gamma$ with $\Gamma \neq \emptyset$ is also called an *integrity constraint*. Throughout the paper a program means a normal logic program unless stated otherwise. A program P is *Horn* if no rule in P contains NAF. A Horn program is *definite* if it contains no integrity constraint. The *Herbrand base* $\mathcal{HB}$ of a program P is the set of all ground atoms in the language of P. Given the Herbrand base $\mathcal{HB}$, we define $\mathcal{HB}^+ = \mathcal{HB} \cup \{ not\, A \mid A \in \mathcal{HB} \}$. Any element in $\mathcal{HB}^+$ is called an *LP-literal*, and an LP-literal of the form *not* A is called an *NAF-literal*. For an LP-literal L, $pred(L)$ denotes the predicate in L and $const(L)$ denotes the set of constants appearing in L. A program, a rule, or an LP-literal is *ground* if it contains no variable. A program/rule containing variables is semantically identified with its ground instantiation, i.e., the set of ground rules obtained from the program/rule by substituting the variables with elements of the Herbrand universe in every possible way.

An *interpretation* is a subset of $\mathcal{HB}$. An interpretation I *satisfies* the ground rule R of the form (1) if $\{B_1, \ldots, B_m\} \subseteq I$ and $\{B_{m+1}, \ldots, B_n\} \cap I = \emptyset$ imply

$A \in I$ (written as $I \models R$). In particular, I satisfies the ground integrity constraint $\leftarrow B_1, \ldots, B_m, not\, B_{m+1}, \ldots, not\, B_n$ if either $\{B_1, \ldots, B_m\} \setminus I \neq \emptyset$ or $\{B_{m+1}, \ldots, B_n\} \cap I \neq \emptyset$. When a rule R contains variables, $I \models R$ means that I satisfies every ground instance of R. An interpretation which satisfies every rule in a program is a *model* of the program. A model M of a program P is *minimal* if there is no model N of P such that $N \subset M$. A Horn logic program has at most one minimal model called the *least model.*

In the presence of negation as failure in a program, a logic program has multiple minimal models in general rather than the single least model in a Horn logic program. In this case, a class of minimal models which provide the intended meaning of a program is selected. In this regard, the *stable model semantics* [7] is one of the most popular semantics of normal logic programs.

Given a program P and an interpretation M, the ground Horn logic program P^M is defined as follows: the rule $A \leftarrow B_1, \ldots, B_m$ is in P^M iff there is a ground rule of the form (1) in the ground instantiation of P such that $\{B_{m+1}, \ldots, B_n\} \cap M = \emptyset$. If the least model of P^M is identical to M, M is called a *stable model* of P. Intuitively, P^M is a program which simplifies P with respect to M by deleting NAF-literals that is true in M and deleting every rule whose condition does not hold in M. Then, M is a stable model if it coincides with the least model of such simplified program P^M.

A program may have none, one, or multiple stable models in general. A program having exactly one stable model is called *categorical* [2]. A stable model is a minimal model and coincides with the least model in a Horn logic program. Given a stable model M, we define $M^+ = M \cup \{\, not\, A \mid A \in \mathcal{HB} \setminus M \,\}$.

Example 2.1. The program

$$p(a) \leftarrow not\, p(b),$$
$$p(b) \leftarrow not\, p(a)$$

has two stable models $M_1 = \{\, p(a) \,\}$ and $M_2 = \{\, p(b) \,\}$. Then, it becomes $M_1^+ = \{\, p(a),\, not\, p(b) \,\}$ and $M_2^+ = \{\, p(b),\, not\, p(a) \,\}$.

A program is *consistent* (under the stable model semantics) if it has a stable model; otherwise a program is *inconsistent.* Throughout the paper, a program is assumed to be consistent unless stated otherwise. If every stable model of a program P satisfies a rule R, it is written as $P \models_s R$. Else if no stable model of a program P satisfies a rule R, it is written as $P \models_s not\, R$. In particular, $P \models_s A$ if a ground atom A is true in every stable model of P; and $P \models_s not\, A$ if A is false in every stable model of P. By contrast, if every model of P satisfies R, it is written as $P \models R$. Note that when P is Horn, the meaning of $\models$ coincides with the classical entailment.

2.2 Inductive Logic Programming

A typical ILP problem is stated as follows. Given a logic program B representing background knowledge and a set E^+ of positive examples and a set E^- of negative examples, find hypotheses H satisfying[1]

1. $B \cup H \models e$ for every $e \in E^+$.
2. $B \cup H \not\models f$ for every $f \in E^-$.
3. $B \cup H$ is consistent.

It is also implicitly assumed that $B \not\models e$ for some $e \in E^+$ or $B \models f$ for some $f \in E^-$, because otherwise there is no need to introduce H. A hypothesis H *covers* (resp. *uncovers*) an example e under B if $B \cup H \models e$ (resp. $B \cup H \not\models e$).

The goal of ILP is then to develop an algorithm which efficiently computes hypotheses that are consistent with the background knowledge and cover every positive example and no negative example. An induction algorithm is *correct* if every hypothesis produced by the algorithm satisfies the above three conditions. An induction algorithm is *complete* if it produces every hypothesis satisfying the conditions. Note that there exist possibly infinite solutions in general and it is impractical to design a complete induction algorithm without any restriction. In order to obtain meaningful hypotheses, additional conditions are usually imposed on possible hypotheses to reduce the search space.[2]

In the field of ILP, most studies consider a Horn logic program as background knowledge and induce Horn clauses as hypotheses. This paper considers an NLP as background knowledge and induces hypothetical rules possibly containing NAF. In our problem setting, the entailment relation $\models$ is replaced by $\models_s$ in the above conditions. In the next section, we give several algorithms which realize this.

3 Induction in Nonmonotonic Logic Programs

3.1 Inverse Resolution

Inverse resolution [13] is based on the idea of inverting the resolution step between clauses. There are two operators that carry out inverse resolution, *absorption* and *identification*, which are called the *V-operators* together. Each operator builds one of the two parent clauses given the other parent clause and the resolvent. Suppose two rules $R_1 : B_1 \leftarrow \Gamma_1$ and $R_2 : A_2 \leftarrow B_2, \Gamma_2$. When $B_1\theta_1 = B_2\theta_2$, the rule $R_3 : A_2\theta_2 \leftarrow \Gamma_1\theta_1, \Gamma_2\theta_2$ is produced by resolving R_2 with R_1. Absorption constructs R_2 from R_1 and R_3, while identification constructs R_1 from R_2 and R_3 (see the figure).

Given a normal logic program P containing the rules R_1 and R_3, absorption produces the program $A(P)$ such that

$$A(P) = (P \setminus \{R_3\}) \cup \{R_2\}\,.$$

[1] When there is no negative example, E^+ is simply written as E.

[2] Such a condition is called an *induction bias*.

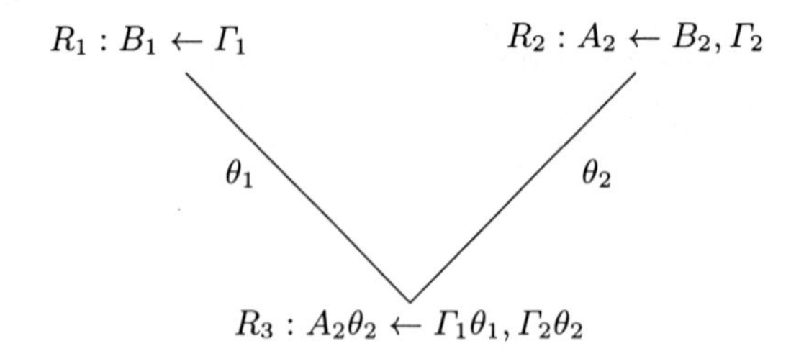

On the other hand, given an NLP P containing the rules R_2 and R_3, identification produces the program $I(P)$ such that

$$I(P) = (P \setminus \{R_3\}) \cup \{R_1\}\,.$$

Note that there are multiple $A(P)$ or $I(P)$ exist in general according to the choice of the input rules in P. We write $V(P)$ to mean either $A(P)$ or $I(P)$.

When P is a Horn logic program, any information implied by P is also implied by $V(P)$, namely

$$V(P) \models P\,.$$

In this regard, the V-operators generalize a Horn logic program. In the presence of negation as failure in a program, however, the V-operators do not work as generalization operations in general.

Example 3.1. Let P be the program:

$$p(x) \leftarrow not\, q(x), \quad q(x) \leftarrow r(x), \quad s(x) \leftarrow r(x), \quad s(a) \leftarrow,$$

which has the stable model $\{\, p(a), s(a)\,\}$. Using the second rule and the third rule, absorption produces $A(P)$:

$$p(x) \leftarrow not\, q(x), \quad q(x) \leftarrow s(x), \quad s(x) \leftarrow r(x), \quad s(a) \leftarrow,$$

which has the stable model $\{\, q(a), s(a)\,\}$. Then, $P \models_s p(a)$ but $A(P) \not\models_s p(a)$.

A counter-example for identification is constructed in a similar manner. The reason is clear, since in nonmonotonic logic programs newly proven facts may block the derivation of other facts which are proven beforehand. As a result, the V-operators may not generalize the original program. Moreover, the next example shows that the V-operators often make a consistent program inconsistent.

Example 3.2. Let P be the program:

$$p(x) \leftarrow q(x), not\, p(x), \quad q(x) \leftarrow r(x), \quad s(x) \leftarrow r(x), \quad s(a) \leftarrow,$$

which has the stable model $\{\, s(a)\,\}$. Using the second rule and the third rule, absorption produces $A(P)$:

$$p(x) \leftarrow q(x), not\, p(x), \quad q(x) \leftarrow s(x), \quad s(x) \leftarrow r(x), \quad s(a) \leftarrow,$$

which has no stable model.

The above example shows that the V-operators have destructive effect on the meaning of programs in general. It is also known that they may destroy the syntactic structure of programs such as acyclicity and local stratification [17].

These observations give us a caution to apply the V-operators to NMLP. A condition for the V-operators to generalize an NLP is as follows.

Theorem 3.1. *(conditions for the V-operators to generalize programs) [17] Let P be an NLP, and R_1, R_2, R_3 be rules at the beginning of this section. Then the following results hold.*[3]

(i) For any NAF-literal $not\, L$ in P, if L does not depend on the head of R_3 in P, then $P \models_s N$ implies $A(P) \models_s N$ for any $N \in \mathcal{HB}$.
(ii) For any NAF-literal $not\, L$ in P, if L does not depend on the atom B_2 of R_2 in P, then $P \models_s N$ implies $I(P) \models_s N$ for any $N \in \mathcal{HB}$.

Example 3.3. Suppose the background program P and a (positive) example E as follows.

$$\begin{aligned} P: \;& flies(x) \leftarrow sparrow(x),\, not\, ab(x), \\ & bird(x) \leftarrow sparrow(x), \\ & sparrow(tweety) \leftarrow, \quad bird(polly) \leftarrow . \\ E: \;& flies(polly). \end{aligned}$$

Using the first rule and the second rule, absorption produces the program $A(P)$ in which the first rule of P is replaced by the next rule:

$$flies(x) \leftarrow bird(x),\, not\, ab(x).$$

Since neither $ab(tweety)$ nor $ab(polly)$ depends on the head of the first rule of P, $P \models_s flies(tweety)$ implies $A(P) \models_s flies(tweety)$. Notice that $P \not\models_s flies(polly)$ but $A(P) \models_s flies(polly)$.

3.2 Inverse Entailment

Suppose an induction problem

$$B \cup \{H\} \models E$$

where B is a Horn logic program and H and E are each single Horn clauses. *Inverse entailment* (IE) [14] is based on the idea that a possible hypothesis H is deductively constructed from B and E by inverting the entailment relation as

$$B \cup \{\neg E\} \models \neg H.$$

[3] Here, *depends on* is a transitive relation defined on atoms as follows: A depends on B if there is a ground rule from P s.t. A appears in the head and B appears in the body of the rule.

When a background theory is a nonmonotonic logic program, however, the IE technique cannot be used. This is because IE is based on the *deduction theorem* in first-order logic, but it is known that the deduction theorem does not hold in nonmonotonic logics in general [21].

To solve the problem, Sakama [18] introduced the *entailment theorem* in normal logic programs. A *nested rule* is defined as

$$A \leftarrow R$$

where A is an atom and R is a rule of the form (1). An interpretation I satisfies a ground nested rule $A \leftarrow R$ if $I \models R$ implies $A \in I$. For an NLP P, $P \models_s (A \leftarrow R)$ if $A \leftarrow R$ is satisfied in every stable model of P.

Theorem 3.2. *(entailment theorem [18]) Let P be an NLP and R a rule such that $P \cup \{R\}$ is consistent. For any ground atom A, $P \cup \{R\} \models_s A$ implies $P \models_s A \leftarrow R$. In converse, $P \models_s A \leftarrow R$ and $P \models_s R$ imply $P \cup \{R\} \models_s A$.*

Note that without the condition $P \models_s R$, $P \models_s A \leftarrow R$ does not imply $P \cup \{R\} \models_s A$ in general.

The entailment theorem corresponds to the deduction theorem and is used for inverting entailment in normal logic programs.

Theorem 3.3. *(IE in normal logic programs [18]) Let P be an NLP and R a rule such that $P \cup \{R\}$ is consistent. For any ground LP-literal L, if $P \cup \{R\} \models_s L$ and $P \models_s \leftarrow L$, then $P \models_s not\, R$.*[4]

Thus, the relation

$$P \models_s not\, R \tag{2}$$

provides a necessary condition for computing a rule R satisfying $P \cup \{R\} \models_s L$ and $P \models_s \leftarrow L$. When L is an atom (resp. NAF-literal), it represents a positive (resp. negative) example. The condition $P \models_s \leftarrow L$ states that the example L is initially false in every stable model of P. To simplify the problem, a program P is assumed to be *function-free* and *categorical* in the rest of this section.

Given two ground LP-literals L_1 and L_2, the relation $L_1 \sim L_2$ is defined if $pred(L_1) = pred(L_2)$ with a predicate of arity ≥ 1 and $const(L_1) = const(L_2)$. Let L be a ground LP-literal and S a set of ground LP-literals. Then, L_1 in S is *relevant* to L if either (i) $L_1 \sim L$ or (ii) L_1 shares a constant with an LP-literal L_2 in S such that L_2 is relevant to L.

Let P be a program with the unique stable model M and A a ground atom representing a positive example. Suppose that the relation $P \cup \{R\} \models_s A$ and $P \models_s \leftarrow A$ hold. By Theorem 3.3, the relation (2) holds, thereby

$$M \not\models R. \tag{3}$$

[4] More precisely, $P \cup \{R\} \models_s L$ implies $P \cup \{\leftarrow L\} \models_s not\, R$ by inverting the entailment. By $P \models_s \leftarrow L$, $P \models_s not\, R$ holds.

Then, we start to find a rule R satisfying the condition (3). Consider the integrity constraint $\leftarrow \Gamma$ where Γ consists of ground LP-literals in M^+ which are relevant to the positive example A.[5] Since M does not satisfy this integrity constraint,

$$M \not\models \leftarrow \Gamma \tag{4}$$

holds. That is, $\leftarrow \Gamma$ is a rule R which satisfies the condition (3).

Next, by $P \models_s \leftarrow A$, it holds that $A \notin M$, thereby $not\, A \in M^+$. Since $not\, A$ is relevant to A, the integrity constraint $\leftarrow \Gamma$ contains $not\, A$ in its body. Then, shifting the atom A to the head produces

$$A \leftarrow \Gamma' \tag{5}$$

where $\Gamma' = \Gamma \setminus \{not\, A\}$.

Finally, the rule (5) is generalized by constructing a rule R^* such that $R^*\theta = A \leftarrow \Gamma'$ for some substitution θ. It is verified that the rule R^* satisfies the condition (2), i.e., $P \models_s not\, R^*$.

The next theorem presents a sufficient condition for the correctness of R^* to induce A.

Theorem 3.4. *(correctness of the IE rule [19]) Let P be a function-free and categorical NLP, A a ground atom, and R^* a rule obtained as above. If $P \cup \{R^*\}$ is consistent and $pred(A)$ does not appear in P, then $P \cup \{R^*\} \models_s A$.*

Example 3.4. Let P be the program

$$bird(x) \leftarrow penguin(x),$$
$$bird(tweety) \leftarrow, \quad penguin(polly) \leftarrow .$$

Given the example $L = flies(tweety)$, it holds that $P \models_s \leftarrow flies(tweety)$. Our goal is then to construct a rule R satisfying $P \cup \{R\} \models_s L$.

First, the set M^+ of LP-literals becomes

$$M^+ = \{\, bird(tweety),\, bird(polly),\, penguin(polly),$$
$$not\, penguin(tweety),\, not\, flies(tweety),\, not\, flies(polly)\, \}.$$

From M^+ picking up LP-literals which are relevant to L, the integrity constraint:

$$\leftarrow bird(tweety),\, not\, penguin(tweety),\, not\, flies(tweety)$$

is constructed. Next, shifting $flies(tweety)$ to the head produces

$$flies(tweety) \leftarrow bird(tweety),\, not\, penguin(tweety)\,.$$

Finally, replacing $tweety$ by a variable x, the rule

$$R^* : \ flies(x) \leftarrow bird(x),\, not\, penguin(x)$$

is obtained, where $P \cup \{R^*\} \models_s L$ holds.

The inverse entailment algorithm is also used for learning programs by negative examples [18].

[5] Since P is function-free, Γ consists of finite LP-literals.

3.3 Learning by Answer Sets

Extended logic programs (ELPs) [8] extend the language of NLPs by introducing explicit representation of negation in addition to NAF. The semantics of ELPs is given by the *answer set semantics* which is an extension of the stable model semantics. Sakama [19] introduces a method of computing inductive hypotheses using answer sets of ELPs. Here, we rephrase the results in the context of NLPs under the stable model semantics.

Theorem 3.5. *[19] Let P be an NLP, R a rule, and A a ground atom. Suppose that $P \cup \{R\}$ is consistent and $P \cup \{R\} \models_s A$. If $P \models_s R$, then $P \models_s A$.*

By this result, $P \cup \{R\} \models_s A$ and $P \not\models_s A$ imply $P \not\models_s R$. This provides a necessary condition for any possible hypothesis R to cover A. Note that the result corresponds to Theorem 3.3, but it is obtained without using the entailment theorem. A candidate hypothesis is then computed in a similar manner to the procedure presented in Section 3.2. That is, firstly computing stable models of P, then constructing a rule which is unsatisfied in a stable model. The method provides the same result as [18] in a much simpler manner. When a program has more than one stable model, different rules are induced by each stable model. In function-free stratified NLPs the algorithm constructs inductive hypotheses in polynomial-time.

4 Related Work

Taylor [22] introduces a new inverse resolution operator called *normal absorption*, which is different from absorption considered in this paper. Normal absorption generalizes normal logic programs, but the effect of the operation on the syntax/semantics of a program is unknown. Muggleton [15] constructs a hypothetical clause using the *enlarged bottom set* which is the least Herbrand model augmented by closed world negation. Our inverse entailment algorithm is close to him in effect, but he does not handle nonmonotonic logic programs.

There are other induction algorithms which learn nonmonotonic logic programs. In [1, 9, 10] monotonic rules satisfying positive examples are firstly constructed and they are subsequently specialized by incorporating NAF literals to the bodies of rules. Dimopoulos and Kakas [5] construct default rules with exceptions in a prioritized hierarchical structure. In [3], candidate hypotheses are given in input to the system, and from those candidates the system selects hypotheses which cover/uncover positive/negative examples. Seitzer [20] firstly computes stable models or the well-founded model of a background NLP and then induces hypotheses using the computed models and examples. Martin and Vrain [11] introduce an algorithm to learn NLPs under the 3-valued semantics. Fogel and Zaverucha [6] propose an algorithm for learning strict and call-consistent NLPs, which effectively searches the hypotheses space using subsumption and iteratively constructed training examples. These algorithms use ordinary induction algorithms for Horn ILP in the process of learning. By contrast, we took an approach to reformulate theories of nonmonotonic ILP and compute induction from nonmonotonic logic programs.

5 Concluding Remark

Techniques in ILP have been centered on clausal logic so far, especially on Horn logic. To enrich the language and to extend the framework, combining ILP and NMLP is an important step towards a better learning tool in AI. However, as nonmonotonic logic programs are different from classical logic, techniques for clausal ILP are not directly applicable to nonmonotonic situations. To this end, we introduced techniques for computing induction in nonmonotonic logic programs. Since inverse resolution and inverse entailment are two basic techniques in Horn ILP, their extensions to nonmonotonic logic programs are meaningful and important. On the other hand, learning by answer sets combines techniques of ILP and NMLP. Such a combination has an important implication that the existing procedures in NMLP are also useful for computing induction.

In contrast to clausal ILP, the field of nonmonotonic ILP is less explored and several issues remain open. Further study is necessary to establish theoretical foundation and to realize efficient induction in nonmonotonic ILP.

Acknowledgements

This research is partly supported by Grand-in-Aid for Scientific Research in the Priority Area "Discovery Science" from the MESSC of Japan. The author thanks Akihiro Yamamoto and Katsumi Inoue for comments on an earlier draft of this paper.

References

1. M. Bain and S. Muggleton. Non-monotonic learning. In: S. Muggleton (ed.), *Inductive Logic Programming*, Academic Press, pp. 145–161, 1992.
2. C. Baral and M. Gelfond. Logic programming and knowledge representation. *Journal of Logic Programming* 19/20:73–148, 1994.
3. F. Bergadano, D. Gunetti, M. Nicosia, and G. Ruffo. Learning logic programs with negation as failure. In: L. De Raedt (ed.), *Advances in Inductive Logic Programming*, IOS Press, pp. 107–123, 1996.
4. G. Brewka and J. Dix. Knowledge representation with logic programs. In: *Proc. 3rd Workshop on Logic Programming and Knowledge Representation, Lecture Notes in Artificial Intelligence* 1471, Springer-Verlag, pp. 1–51, 1997.
5. Y. Dimopoulos and A. Kakas. Learning nonmonotonic logic programs: learning exceptions. In: *Proc. 8th European Conf. on Machine Learning, Lecture Notes in Artificial Intelligence* 912, Springer-Verlag, pp. 122–137, 1995.
6. L. Fogel and G. Zaverucha. Normal programs and multiple predicate learning. In: *Proc. 8th Int'l Workshop on Inductive Logic Programming, Lecture Notes in Artificial Intelligence* 1446, Springer-Verlag, pp. 175–184, 1998.

7. M. Gelfond and V. Lifschitz. The stable model semantics for logic programming. In: *Proc. 5th Int'l Conf. and Symp. on Logic Programming*, MIT Press, pp. 1070–1080, 1988.
8. M. Gelfond and V. Lifschitz. Classical negation in logic programs and disjunctive databases. *New Generation Computing* 9:365–385, 1991.
9. K. Inoue and Y. Kudoh. Learning extended logic programs. In: *Proc. 15th Int'l Joint Conf. on Artificial Intelligence*, Morgan Kaufmann, pp. 176–181, 1997.
10. E. Lamma, F. Riguzzi, and L. M. Pereira. Strategies in combined learning via logic programs. *Machine Learning* 38(1/2), pp. 63–87, 2000.
11. L. Martin and C. Vrain. A three-valued framework for the induction of general logic programs. In: L. De Raedt (ed.), *Advances in Inductive Logic Programming*, IOS Press, pp. 219–235, 1996.
12. S. Muggleton (ed.). *Inductive Logic Programming*, Academic Press, 1992.
13. S. Muggleton and W. Buntine. Machine invention of first-order predicate by inverting resolution. In: [12], pp. 261–280, 1992.
14. S. Muggleton. Inverse entailment and Progol. *New Generation Computing* 13:245–286, 1995.
15. S. Muggleton. Completing inverse entailment. In: *Proc. 8th Int'l Workshop on Inductive Logic Programming, Lecture Notes in Artificial Intelligence* 1446, Springer-Verlag, pp. 245–249, 1998.
16. S.-H. Nienhuys-Cheng and R. de Wolf. *Foundations of inductive logic programming. Lecture Notes in Artificial Intelligence* 1228, Springer-Verlag, 1997.
17. C. Sakama. Some properties of inverse resolution in normal logic programs. In: *Proc. 9th Int'l Workshop on Inductive Logic Programming, Lecture Notes in Artificial Intelligence* 1634, Springer-Verlag, pp. 279–290, 1999.
18. C. Sakama. Inverse entailment in nonmonotonic logic programs. In: *Proc. 10th Int'l Conf. on Inductive Logic Programming, Lecture Notes in Artificial Intelligence* 1866, Springer-Verlag, pp. 209–224, 2000.
19. C. Sakama. Learning by answer sets. In: *Proc. AAAI Spring Symp. on Answer Set Programming*, AAAI Press, pp. 181–187, 2001.
20. J. Seitzer. Stable ILP: exploring the added expressivity of negation in the background knowledge. In: *Proc. IJCAI-95 Workshop on Frontiers of ILP*, 1997.
21. Y. Shoham. Nonmonotonic logics: meaning and utility. In: *Proc. 10th Int'l Joint Conf. on Artificial Intelligence*, Morgan Kaufmann, pp. 388–393, 1987.
22. K. Taylor. Inverse resolution of normal clauses. In: *Proc. 3rd Int'l Workshop on Inductive Logic Programming*, J. Stefan Institute, pp. 165–177, 1993.

EM Learning for Symbolic-Statistical Models in Statistical Abduction

Taisuke Sato

Tokyo Institute of Technology
2-12-1 Ôokayama Meguro-ku Tokyo Japan 152-8552
sato@mi.cs.titech.ac.jp

Abstract. We first review a logical-statistical framework called *statistical abduction* and identify its three computational tasks, one of which is the learning of parameters from observations by ML (maximum likelihood) estimation. Traditionally, in the presence of missing values, the EM algorithm has been used for ML estimation. We report that the *graphical EM algorithm*, a new EM algorithm developed for statistical abduction, achieved the same time complexity as specialized EM algorithms developed in each discipline such as the Inside-Outside algorithm for PCFGs (probabilistic context free grammars). Furthermore, learning experiments using two corpora revealed that it can outperform the Inside-Outside algorithm by orders of magnitude. We then specifically look into a family of extensions of PCFGs that incorporate context sensitiveness into PCFGs. Experiments show that they are learnable by the graphical EM algorithm using at most twice as much time as plain PCFGs even though these extensions have higher time complexity.

1 Introduction

Capturing complex data in an appropriate model to extract information is a fundamental step in knowledge discovery. As data have their own characteristics, it is vital to choose a best model that can abstract essential properties of the given data. Genome data for instance are strings of four (A,T,G and C) letters, and hence it seems reasonable to choose a formal language of some type as their model. Also, taking noise into account, stochastic language models such as HMMs (hidden Markov models) might be better candidates.

In this paper, in search of general yet powerful symbolic modeling tools beyond HMMs, we introduce an abductive framework called *statistical abduction* and report some experimental results with it. The new framework is derived by amalgamating logic programs and ML (maximum likelihood) estimation through abduction, and has the expressive power of predicate calculus and the ability of learning from observations as explained below.

Abduction is one of the three inference modes of human intelligence.[1] It is an inference to the best explanation for an observation. If one finds a fossil of

[1] The other two are deductive inference and inductive inference.

S. Arikawa and A. Shinohara (Eds.): Progress in Discovery Science 2001, LNAI 2281, pp. 189–200.

a trilobite at the top of a mountain, he/she would infer that the place must have been the bottom of the sea 300 million years ago. Logically abduction is formulated as finding an explanation E for the observation O under the background knowledge KB such that $E, KB \vdash O$ where E and KB are consistent. In logic programming literature [8], E is usually taken as a conjunction of primitive hypotheses called *abducibles*.

If abduction is to be applied to the real world however, we obviously must cope with uncertainty. We therefore added to the logical abduction yet another ingredient, a *parameterized probability measure* $P_F(\cdot \mid \boldsymbol{\theta})$ over the set F[2] of abducibles [16] so that abducibles are probabilistically true. The resulting framework is called *statistical abduction* [18, 20]. The introduction of P_F mathematically qualifies each ground atom as a random variable (taking on 1 (true), or 0 (false)), and an explanation E and an observation O have probabilities. What is important is that we not only can determine the most plausible explanation as the one having the highest probability but can learn $\boldsymbol{\theta}$ from observations by ML estimation [16].

We implemented a symbolic-statistical modeling language PRISM [16, 17, 19] as an embodiment of statistical abduction by adopting the simplest form of the probability measure as P_F, i.e. the direct product of countably many generalized Bernoulli trials each having a parameter of success.[3] The user of PRISM can use $\mathtt{msw}(i,n,v)$ atoms as built-in abducibles to represent probabilistic events where i, n and v are ground terms. Our intension is that a set of msw atoms $\{\mathtt{msw}(i,n,v_1),\ldots,\mathtt{msw}(i,n,v_k)\}$ with a common name i and a common trial n collectively represents a discrete random variable $X_{i,n}$ whose values are $\{v_1,\ldots,v_k\}$ in such a way that only one of them exclusively and probabilistically becomes true at each sampling.[4] msw atoms are independent either if names differ or trials differ and the associated parameter $\theta_{(i,v)} = P_F(\mathtt{msw}(i,n,v))$ is learnable from observations by performing ML estimation using the built-in EM learning routine. This P_F looks quite simple, but it is proved [16, 17, 19] that most of existing symbolic-statistical frameworks such as (discrete) Bayesian networks [2], HMMs (hidden Markov models) [15] and PCFGs (probabilistic context free grammars) [13] are expressible as PRISM programs, and on top of that, the parameters associated with a program are efficiently learnable by the *graphical EM algorithm*, a new EM algorithm for ML estimation in statistical abduction[10, 19].[5]

---

[2] Mathematically P_F is a probability measure over the set of Herbrand interpretations of F, but to keep the continuity of finite cases, we sometimes refer to P_F as a kind of a joint distribution (of countably infinite dimension) which generates marginal distributions of finite dimension.

[3] PRISM is available at `http://mi.cs.titech.ac.jp/prism/index.html`.

[4] For instance, a set of ground msw atoms, {msw(gene,once,a), msw(gene,once,b) and msw(gene,once,o)} which are mutually exclusive, collectively represents a probabilistic event that occurs once when he/she inherits one of the ABO blood type genes {a, b, o} from a parent.

[5] The EM algorithm is an iterative algorithm that locally maximizes the likelihood of observations in the presence of missing data [14].

In what follows, after reviewing statistical abduction and time complexity of various EM learning models in Section 2, we report the computational behavior of the graphical EM algorithm applied to PCFGs and their extensions in Section 3 and in Section 4 respectively. Section 5 is a conclusion. The contents of this paper partly overlap with [19, 20] and the reader is assumed to be familiar with logic programming [12], the EM algorithm [14] and stochastic grammars [13].

2 Statistical Abduction and Time Complexity

In statistical abduction, we have a definite clause program $DB = F \cup R$. The fact part F is a set of ground atoms $a_1, a_2, \ldots$ which are supposed to have a probability measure P_F parameterized by $\boldsymbol{\theta}$. F represents primitive probabilistic events such as gene inheritance whereas the rule part R, a set of definite clauses, represents background knowledge such as Mendel's Law. DB models the situation where we know that some ground atoms $\{a_{i_1}, \ldots, a_{i_k}\}$ in F that happens to be true can explain a statistical event G when combined with the background knowledge R, but we do not which of them actually explain it. As $a_1, a_2, \ldots$ play the role of primitive hypotheses, they are called abducibles.[6]

Given a program DB and an atom G representing our observation, we first seek all explanations $E_1, E_2, \ldots, E_n$ for G each of which is a conjunction $a_{i_1} \wedge \ldots \wedge a_{i_k}$ of abducibles such that $\mathrm{comp}(R) \vdash G \leftrightarrow E_1 \vee \cdots \vee E_n$[7] where comp($R$) is the completion of R [12, 7, 6]. The parameterized probability measure $P_F(\cdot \mid \boldsymbol{\theta})$ over the set (of Herbrand interpretations) of abducibles F mathematically induces a parameterized probability measure $P_{DB}(G \mid \boldsymbol{\theta})$, an extension of P_F through the least model semantics (*distribution semantics*)[8] [12, 16, 19] such that we have $P_{DB}(G \mid \boldsymbol{\theta}) = \sum_{i=1}^{n} P_{DB}(E_i \mid \boldsymbol{\theta})$ and in the simplest cases, $P_{DB}(E_i \mid \boldsymbol{\theta}) = \prod_{j=1}^{k_i} P_F(a_{i_j} \mid \boldsymbol{\theta})$ for $E_i = a_{i_1} \wedge \cdots \wedge a_{i_{k_i}}$. PRISM deals only with such simplest cases for the sake of computational efficiency.[9]

By applying the EM algorithm [16, 14] introduced below to $P_{DB}(G \mid \boldsymbol{\theta})$, we are able to obtain the ML estimate $\widehat{\boldsymbol{\theta}} = \mathrm{argmax}_{\boldsymbol{\theta}}\, P_{DB}(G \mid \boldsymbol{\theta})$ and also by computing and comparing the $P_{DB}(E_i \mid \boldsymbol{\theta})$'s, we can select the most likely

[6] Here we use the words "fact" and "rule" from the viewpoint of logic programming in which facts are unit clauses and rules are a generalization of if-then rules. This usage is different from that of abductive logic programming where facts and hypotheses (abducibles) are two disjoint categories such that facts correspond to our rules.

[7] We need all explanations, for instance, to determine the most likely explanation. We assume that there are only finitely many explanations for each observation G.

[8] Suppose we sample true ground atoms F' ($\subset F$) from P_F. Then the least Herbrand model of $F' \cup R$ defines truth value of every ground atom in DB. In other words, every ground atom in DB can be considered as a random variable whose truth value depends on the sample F'. Based on this idea, we can extend P_F to P_{DB}, the probability measure over the set of all Herbrand interpretations for DB.

[9] Theoretically, distribution semantics allows more complex discrete distributions. P_F can be a Boltzmann distribution for instance.

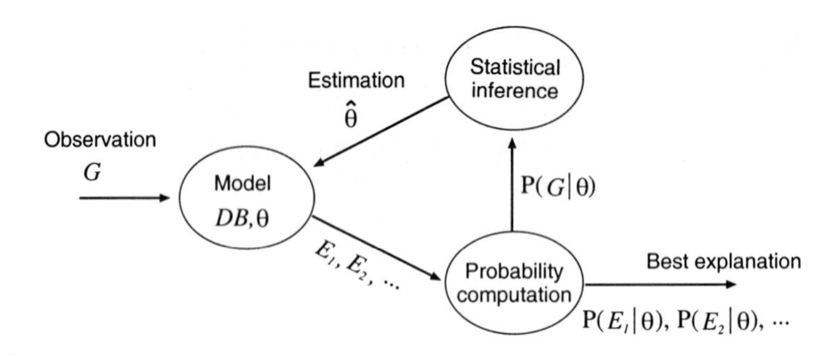

Fig. 1. Statistical abduction

explanation E^* for G. In general, we will have a random sample $G_1, G_2, \dots, G_T$ and $\widehat{\boldsymbol{\theta}}$ is computed as argmax $_{\boldsymbol{\theta}} L(\boldsymbol{\theta})$ where $L(\boldsymbol{\theta})$ is the likelihood function defined by $L(\boldsymbol{\theta}) = \prod_{t=1}^{T} P_{DB}(G_t \mid \boldsymbol{\theta})$.

In the context of statistical abduction, the EM algorithm is introduced as follows. Suppose we have ground atoms $G_1, G_2, \dots, G_T$ representing our observations which our program $DB = F \cup R$ is supposed to explain. First we define a function $Q(\boldsymbol{\theta} \mid \boldsymbol{\theta}')$ by

$$Q(\boldsymbol{\theta} \mid \boldsymbol{\theta}') \stackrel{\text{def}}{=} \sum_{t=1}^{T} \sum_{E \subset F s.t. DB, E \vdash G_t} P_{DB}(E \mid G_t, \boldsymbol{\theta}') \ln P_{DB}(E, G_t \mid \boldsymbol{\theta}) \tag{1}$$

This is called the E(xpectation)-step. We start from an initial value $\boldsymbol{\theta}^{(0)}$ and alternately iterate the E-step and the M(aximization)-step which updates the current value $\boldsymbol{\theta}^{(n)}$ to $\boldsymbol{\theta}^{(n+1)}$ by

$$\boldsymbol{\theta}^{(n+1)} = \text{argmax}_{\boldsymbol{\theta}}\, Q(\boldsymbol{\theta} \mid \boldsymbol{\theta}^{(n)}) \tag{2}$$

Each iteration is guaranteed to increase the likelihood $\prod_{t=1}^{T} P_{DB}(G_t \mid \boldsymbol{\theta})$. We stop iteration when the likelihood saturates. The value of $\boldsymbol{\theta}$ then gives a stationary point of the likelihood. The correctness of this procedure is intuitively understood as follows. In the limit, we have $\boldsymbol{\theta}_\infty = \text{argmax}_{\boldsymbol{\theta}}\, Q(\boldsymbol{\theta} \mid \boldsymbol{\theta}_\infty)$. Assuming $Q(\boldsymbol{\theta} \mid \boldsymbol{\theta}')$ is smooth enough, it follows $\left(\frac{\partial}{\partial \boldsymbol{\theta}}\right) Q(\boldsymbol{\theta} \mid \boldsymbol{\theta}_\infty)\Big|_{\boldsymbol{\theta}_\infty} = 0$. By substituting the definition of $Q(\boldsymbol{\theta} \mid \boldsymbol{\theta}_\infty)$ and after some calculation, we reach $\left(\frac{\partial}{\partial \boldsymbol{\theta}}\right) L(\boldsymbol{\theta})\Big|_{\boldsymbol{\theta}_\infty} = 0$, which means $\boldsymbol{\theta}_\infty$ is an ML estimate. PRISM has a built-in EM routine implementing Equations (1) and (2).

Before stating basic computational tasks of statistical abduction, we give an example of statistical abduction using a small PRISM program for the sake of self-containedness. Note that this program happens to contain neither recursion nor functional terms though they are allowed by distribution semantics.

Fig. 2 is the rule part of PRISM program $DB_{\texttt{abo}}$ that describes how one's phenotype $\texttt{btype}(x)$ (ABO blood type, $x \in \{\texttt{a}, \texttt{b}, \texttt{o}, \texttt{ab}\}$) is determined by

```
btype(X):- pg_table(X,Gf,Gm),genotype(Gf,Gm).

pg_table(a,a,a). pg_table(a,a,o).  pg_table(a,o,a).
pg_table(b,b,b). pg_table(b,b,o).  pg_table(b,o,b).
pg_table(o,o,o). pg_table(ab,a,b). pg_table(ab,b,a).

genotype(Gf,Gm):- gene(father,Gf),gene(mother,Gm).
gene(P,G):-msw(abo,P,G).
```

Fig. 2. The rule part of ABO blood type program DB_{abo}

one's genotype, i.e. a pair of blood type genes ($\in \{\mathsf{a}, \mathsf{b}, \mathsf{o}\}$) inherited from one's parents. The correspondence between one's genotype and one's phenotype is specified by a table $\mathtt{pg\_table}(\cdot,\cdot,\cdot)$. The fact part of DB_{abo}, not explicitly included in the program text but implicitly introduced by the use (and declaration) of msw(abo,P,G), is comprised of abducibles $\{\mathtt{msw(abo,}p\mathtt{,}g\mathtt{)} \mid p \in \{\mathtt{father},\mathtt{mother}\}, g \in \{\mathtt{a,b,o}\}\}$. They have statistical parameters $\boldsymbol{\theta}_{\mathsf{abo}} = \langle \theta_{\mathsf{a}}, \theta_{\mathsf{b}}, \theta_{\mathsf{o}} \rangle$ where $\theta_g = P_{\mathsf{abo}}(\mathtt{msw(abo,\_,}g\mathtt{)})$ ($g \in \{\mathsf{a}, \mathsf{b}, \mathsf{o}\}$) and $\theta_{\mathsf{a}} + \theta_{\mathsf{b}} + \theta_{\mathsf{o}} = 1$. They define a parameterized probability distribution $P_{\mathsf{abo}}(\cdot \mid \boldsymbol{\theta}_{\mathsf{abo}})$ over observable ground atoms $\{\mathtt{btype(}x\mathtt{)} \mid x \in \{\mathsf{a}, \mathsf{b}, \mathsf{o}, \mathsf{ab}\}\}$ through logical equivalences under the completion of DB_{abo}. For example, we have $\mathtt{btype(ab)} \leftrightarrow E_1 \vee E_2$ under the completion of DB_{abo} where $E_1 = \mathtt{msw(abo, father, a)} \wedge \mathtt{msw(abo, mother, b)}$ and $E_2 = \mathtt{msw(abo, father, b)} \wedge \mathtt{msw(abo, mother, a)}$ and hence, $P_{\mathsf{abo}}(\mathtt{btype(ab)}) = 2\theta_{\mathsf{a}}\theta_{\mathsf{b}}$. From random observations such as btype(ab), btype(a), $\cdots$, we can estimate $\boldsymbol{\theta}_{\mathsf{abo}}$ as a (local) maximizer of the likelihood $\prod (2\theta_{\mathsf{a}}\theta_{\mathsf{b}})\ (\theta_{\mathsf{a}}^2 + 2\theta_{\mathsf{a}}\theta_{\mathsf{o}}) \cdots$ by using the built-in EM algorithm in PRISM.

Three basic computational tasks in statistical abduction are stated as follows.

(task-1) computing $P_{DB}(G \mid \boldsymbol{\theta})$, the probability of an atom G representing an observation,
(task-2) finding E^*, the most likely explanation for G, and
(task-3) adjusting the parameters $\boldsymbol{\theta}$ to maximize the likelihood of a given sequence $\mathcal{G} = G_1, G_2, \ldots, G_T$ of observations.

In [10, 18, 19], we have proved that it is possible by introducing *support graphs*, a new compact representation of all explanations for a given goal, to carry out all of the three tasks as efficiently as specialized algorithms that have been developed in each discipline. Especially the graphical EM algorithm for ML estimation [10], a new EM algorithm running on support graphs, has the same time complexity as specialized EM algorithms such as one for singly-connected Bayesian networks [2], the Baum-Welch algorithm for HMMs [15] and the Inside-Outside algorithm for PCFGs [1] as shown below.[10]

[10] For space limitations, we omit the details of the graphical EM algorithm and support graphs though. They are described in [10, 19] and in [21] (URL = http://www.cs.titech.ac.jp/TR/tr01.html).

Model	specialized EM	gEM
HMM	N^2LT (Baum-Welch)	N^2LT
PCFG	M^3L^3T (Inside-Outside)	M^3L^3T
sc-Bnet	$\lvert V\rvert T$ ([2])	$\lvert V\rvert T$

Fig. 3. Time complexity of **task-1** for various probabilistic models

The complexity here is measured by time required for one iteration of the EM algorithm to re-estimate parameters. gEM stands for the graphical EM algorithm, sc-Bnet for singly connected Bayesian networks. N is the number of states in an HMM, M that of non-terminals in a grammar, L the length of a sentence, V the number of nodes in a Bayesian network and T the number of observations.

This table implies that we can realize, at least theoretically, efficient EM learning (complexity-wise) for each model by writing an appropriate PRISM program and running the graphical EM algorithm, if the support graphs are obtainable efficiently with the same complexity as each specialized EM algorithm. We showed in [10, 19] that this is made possible by using a tabulated search technique called OLDT search [23].[11] Have a look at PCFGs (probabilistic context free grammars) for example.[12] We write a probabilistic parser using `msw` atoms for representing the choice of production rules. Observing a sentence s, we seek for all explanations of s, i.e. every possible finite sequence of choice of production rules in the leftmost derivation of s as a conjunction of `msw` atoms while packing these explanations as a support graph by OLDT search in time $O(M^3L^3)$ where L is the length of s. A conjunction of such `msw` atoms has one-to-one correspondence to a parse tree, and it is one of many "hypotheses" that explain s, our observation. After obtaining support graphs for T observed sentences, we run the graphical EM on them to obtain parameters associated with each production rule in the same order $O(M^3L^3T)$ as the Inside-outside algorithm.

We emphasize that what the graphical EM algorithm does *coincide with* these specialized EM algorithms thereby giving the same answers as the specialized ones. The difference is that the graphical EM algorithm uses support graphs, i.e. a general graphical data structure for organizing abducibles. We also remark that statistical abduction requires the search process of all explanations to con-

[11] However the integration of OLDT search and the graphical EM algorithm mentioned in this paper with PRISM is not completed yet. The current version of PRISM uses a tree representation of all explanations and the naive implementation of the EM algorithm.

[12] A PCFG is a stochastic CFG in which each production rule has a probability of application. When there is a nonterminal A having N production rules $\{A \to \alpha_i \mid 1 \leq i \leq N\}$, we associate a probability p_i with each rule $A \to \alpha_i$ $(1 \leq i \leq N)$ $(\sum_{i=1}^{N} p_i = 1)$. The probability of a sentence s is the sum of probabilities of every possible leftmost derivation of s which in turn is the product of probabilities associated with rules used in the derivation.

struct support graphs while these specialized EM algorithms do not contain such process. The effect of the difference is examined in the next section.

3 Parameter Learning of Pure PCFGs

In the previous section, we see that the graphical EM algorithm is competitive with existing specialized EM algorithms in terms of time complexity. In a more realistic setting, it turns out to be more than competitive as far as PCFGs are concerned; learning experiments with existing corpora showed that it runs orders of magnitude faster than the Inside-Outside algorithm, a de facto standard EM algorithm for PCFGs [1]. We state some details of the learning experiments below (see [21] for more details).

In the experiments, we compared the performance of the graphical EM algorithm with that of the Inside-Outside algorithm in terms of updating time per iteration by letting them learn PCFG parameters from a corpus. Two corpora were used. One is a POS (part of speech)-tagged corpus converted from ATR corpus and the other is EDR corpus. Since the purpose of the experiment is speed comparison per iteration of the EM algorithm, we generated support graphs by a Generalized LR parser and did not use PRISM programs to save time for parsing and constructing support graphs. All measurements were made on a 296MHz Sun UltraSPARC-II with Solaris 2.6.

The first corpus, ATR corpus, is a Japanese-English corpus (we used only the Japanese part) developed by ATR (Advanced Telecommunication Research Institute International) [25]. It contains 10,995 sentences (records of conversational sentences) whose minimum length, average length and maximum length are respectively 2, 9.97 and 49. As a skeleton of PCFG, we used a context free grammar G_{atr} comprising 860 rules (172 nonterminals and 441 terminals) manually developed for ATR corpus [24] which generated 958 parses/sentence.

Because the Inside-Outside algorithm only accepts CFGs in Chomsky normal form, we converted G_{atr} into Chomsky normal form G^*_{atr}. G^*_{atr} contains 2,105 rules (196 nonterminals and 441 terminals). We then divided the corpus into subgroups of similar length like $(L = 1,2), (L = 3,4), \ldots, (L = 25,26)$, each containing randomly chosen 100 sentences. After these preparations, we compared the graphical EM algorithm applied to G_{atr} and to G^*_{atr} against the Inside-Outside algorithm applied to G^*_{atr}. The results are shown in Fig. 4.

In the left graph, the Inside-Outside algorithm draws a cubic curve labeled "I-O".[13] The curves by the graphical EM algorithm are not plotted as they coincide with the x axis. The middle graph magnifies the left graph. The curve labeled "gEM (original)" used the original grammar G_{atr} whereas the one labeled "gEM (Chomsky CF)" used G^*_{atr}. Seeing "gEM (original)" at length 10, the average sentence length, we see that the graphical EM algorithm runs about 850 times faster than the Inside-Outside algorithm. The right graph shows linear

[13] An x-axis is the length L of an input sentence, a y-axis time taken by the EM algorithm to re-estimate parameters in one iteration.

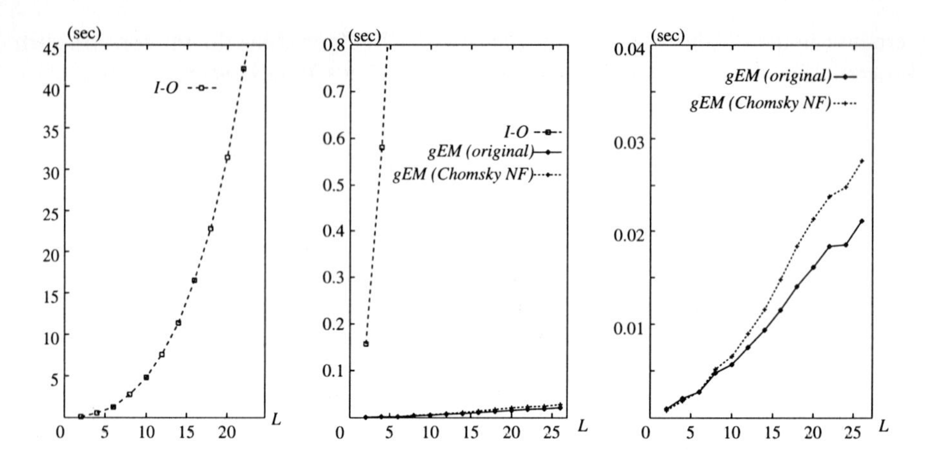

Fig. 4. The Inside-Outside algorithm vs. the graphical EM algorithm

dependency of re-estimation time of the graphical EM algorithm on the sentence length.

When the grammar is more ambiguous than G_{atr}, resulting support graphs tend to be bigger, which is considered advantageous to the Inside-Outside algorithm in speed comparison, and such ratio as 850 would not be expected. We therefore conducted similar experiments using yet another corpus, EDR Japanese corpus [4] which is a Japanese corpus containing 220,000 news article sentences. It is however under the process of re-annotation, and only part of it (9,900 sentences) is available as a labeled corpus at the moment. Compared with the ATR corpus, sentences are much longer (the average length of 9,900 sentences is 20, the maximum length is 63) and the CFG grammar (2,687 original rules and 12,798 rules in Chomsky normal form) developed for it is very ambiguous, having 3.0×10^8 parses/sentence at length 20. By running the graphical EM algorithm and the Inside-Outside algorithm, we found that at average sentence length 20, the former runs 1,300 times faster the latter per iteration (see [21] for details).

This striking difference needs explanation but the reason is simple. When the Inside-Outside algorithm is applied to a specific grammar such as G_{atr}, it tries all possible combinations of CFG rules *in every iteration*. In other words, it parses a sentence anew for each iteration. The graphical EM algorithm on the other hand uses support graphs representing already parsed trees. Although there are other EM algorithms for estimating parameters of PCFGs [5] and [22], they contain redundancies compared to the graphical EM algorithm. [5] employs parsed trees but computes outside probabilities without structure sharing which will make computation time exponential in the length of an input sentence. [22] employs an Earley chart as a data structure for computing probabilities but still redundantly computes inside probabilities.

One might argue that the comparison we made is not fair as what we are comparing is time per iteration of the EM algorithm and the total learning time for the graphical EM algorithm must be calculated by

$$\begin{aligned}&(\mathrm{A} : \text{time for constructing support graphs})\\&\quad + (\mathrm{B} : \text{time per iteration}) \times (\mathrm{C} : \text{the number of iterations})\end{aligned}$$

while for the Inside-Outside algorithm, total learning time is just B×C. We therefore estimated total learning time for both EM algorithms in the case of using the entire ATR corpus as input. B for the Inside-Outside algorithm then was 1215 seconds (average of 20 trials) whereas for the graphical EM algorithm, B was 0.661 second and A was 279 seconds respectively (so the speed ratio per iteration in this case is about 1800). Assuming C = 100 iterations, the ratio of total learning time using the entire corpus is calculated as $(1215 \times 100)/(279 + 0.661 \times 100) = 352$, and hence the graphical EM algorithm will be still orders of magnitude faster than the Inside-Outside algorithm in terms of total learning time.[14] [15]

4 Parameter Learning of Extensions of PCFGs

Now we turn to extensions of PCFGs incorporating context sensitiveness in rule application. We consider four such types. In Pseudo PCSGs (probabilistic context sensitive grammars) [3], the probability of selecting a rule $A \Rightarrow \cdots$ to expand A depends on its parent category (see the leftmost tree in Fig. 5). In the *Bigram* model, it depends on a word just before A as illustrated in the second tree from the left in Fig. 5. The third tree shows the combination of the two, and finally the *Rule Bigram* model [11] takes into account the rule used immediately before the expansion of A (the rightmost tree, boxed rules affect the expansion of A).

We have done parameter learning experiments for these extensions using ATR corpus and the grammar G_{atr} just like the one reported in Section 4. We modified the parser to obtain appropriate support graphs for each grammar model. As Fig. 6 shows, despite higher time complexity ($M^{12}L^3T$) for the rule bigram model for instance), all models draw similar curves to PCFGs. This result suggests that in the case of ATR corpus, support graphs obtained by parsing are relatively small irrespective of models.

[14] The actual number of iterations until convergence by the graphical EM algorithm for the case of the entire ATR corpus is actually 260 (average of 3 trials), which would give more favorable ratio to the graphical EM algorithm.

[15] We also conducted a learning experiment to measure time per iteration for the entire ATR corpus using a faster implementation by Mark Johnson of the Inside-Outside algorithm down-loadable from `http://www.cog.brown.edu/%7Emj/`. This implementation turned out to be twice as fast as our naive implementation, giving B=630 seconds which is still orders of magnitude slower than the graphical EM algorithm.

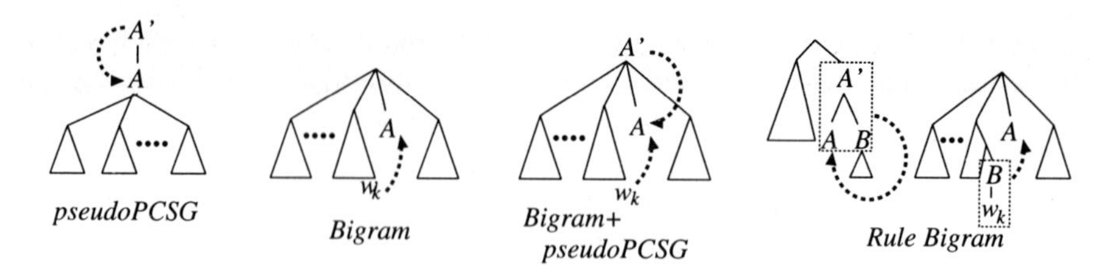

Fig. 5. Various context sensitiveness

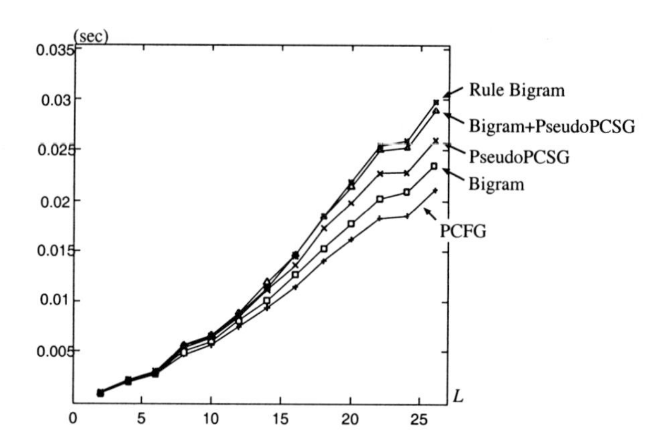

Fig. 6. Learning time per iteration of extended PCFGs by the graphical EM algorithm

5 Conclusion

We have given a brief account of the framework of statistical abduction in which abducibles have a parameterized distribution. Programs in PRISM, an embodiment of statistical abduction as a symbolic-statistical modeling language [17] based on distribution semantics [16], can express most of popular symbolic-statistical systems such as HMMs, PCFGs and Bayesian networks. By introducing a new data structure called support graphs compactly representing hierarchically organized abducibles, it becomes possible to carry out three basic computational tasks, especially EM learning of statistical parameters, required of statistical abduction as efficiently as specialized algorithms developed in each discipline.

We then have compared the performance of the graphical EM algorithm, a new EM algorithm for statistical abduction, and the Inside-Outside algorithm, a standard EM algorithm for PCFGs, by letting them learn parameters of pure PCFGs using two real corpora, ATR corpus and EDR corpus. ATR corpus is comprised of short conversational sentences whose grammar is not very ambiguous whereas EDR corpus contains long sentences from news articles whose grammar is very ambiguous. The experimental results show that in both cases, the graphical EM algorithm runs orders of magnitude faster than the Inside-Outside algorithm per iteration, which seems to apply to the total learning time as well (the measurement of total learning time for EDR corpus was computationally impossible currently). We also have examined the behavior of the graphical EM algorithm applied to extensions of PCFGs that incorporate context sensitivity in one way or another. Contrary to anticipation, i.e. the increase of theoretical time complexity, experimental data suggest that even if much more complex stochastic grammars than PCFGs are employed, parameters are learnable almost as efficiently as PCFGs, a fact which might help in statistical natural language processing where powerful yet efficient unsupervised learning of parameters for sophisticated language models is called for.

References

1. Baker, J. K., Trainable grammars for speech recognition, *Proc. of Spring Conference of the Acoustical Society of America*, pp.547–550, 1979.
2. Castillo, E., Gutierrez, J.M., and Hadi, A.S., *Expert Systems and Probabilistic Network Models*, Springer-Verlag, 1997.
3. Charniak, E. and Carroll,G., Context-sensitive statistics for improved grammatical language models, *Proc. of AAAI'94*, pp.728–733, 1994.
4. EDR Electronic Dictionary Technical Guide (2nd edition), Japan Electronic Dictionary Research Institute, Ltd., Technical report EDR TR–045, `http://www.iijnet.or.jp/edr/E_Struct.html`, 1995.
5. Fujisaki,T. and Jelinek,F. and Cocke,J. and Black,E. and Nishino,T., A Probabilistic Parsing Method for Sentence Disambiguation, *Proc. of International Parsing Workshop '89*, pp.85–94,1989.
6. Fung,T.H. and Kowalski,R., The iff procedure for abductive logic programming, *Journal of Logic Programming*, 33, pp.151–165, 1997.
7. Inoue,K. and Sakama,C., Computing extended abduction through transaction programs, *Annals of Mathematics and Artificial Intelligence*, 25(3,4), pp.339-367, 1999.
8. Kakas, A.C., Kowalski, R.A. and Toni, F., Abductive Logic Programming, *J. Logic Computation*, Vol.2 No.6, pp.719–770, 1992.
9. Kakas, A.C., Kowalski, R.A. and Toni, F., The Role of Abduction in Logic Programming, in *Handbook of Logic in Artificial Intelligence and Logic Programming 5*, D.M.Gabbay, J.J.Hogger and J.A.Robinson eds. Oxford University Press, pp.235–324, 1998.
10. Kameya, Y. and Sato, T., Efficient EM learning for parameterized logic programs, *Proc. of CL2000*, LNAI 1861, pp.269–294, 2000.
11. Kita, K., Morimoto, T., Ohkura, K., Sagayama, S. and Yano, Y., Spoken sentence recognition based on HMM-LR with hybrid language modeling, *IEICE Trans. on Info. & Syst.*, Vol.E77-D, No.2, 1994.
12. Lloyd, J. W., *Foundations of Logic Programming*, Springer-Verlag, 1984.
13. Manning, C. D. and Schütze, H., *Foundations of Statistical Natural Language Processing*, The MIT Press, 1999.
14. McLachlan, G. J. and Krishnan, T., The EM Algorithm and Extensions, Wiley Interscience, 1997.
15. Rabiner, L. R. and Juang, B., *Foundations of Speech Recognition*, Prentice-Hall, 1993.

16. Sato, T., A statistical learning method for logic programs with distribution semantics, *Proc. of ICLP'95*, pp.715-729, 1995.
17. Sato, T. and Kameya, Y., PRISM:A Language for Symbolic-Statistical Modeling, *Proc. of IJCAI'97*, pp.1330–1335, 1997.
18. Sato,T. and Kameya, Y., A Viterbi-like algorithm and EM learning for statistical abduction", *Proc. of UAI2000 Workshop on Fusion of Domain Knowledge with Data for Decision Support*, 2000.
19. Sato,T. and Kameya, Y., Parameter Learning of Logic Programs for Symbolic-statistical Modeling, submitted for publication, 2000.
20. Sato,T., Parameterized Logic Programs where Computing Meets Learning Proc. of FLOPS 2001, LNCS 2024, pp.40-60, 2001.
21. Sato,T., Kameya,Y., Abe,S. and Shirai,K., Fast EM learning of a Family of PCFGs, Titech Technical Report (Dept. of CM) TR01-0006, Tokyo Institute of Technology, 2001.
22. Stolcke, A., An efficient probabilistic context-free parsing algorithm that computes prefix probabilities, *Computational Linguistics*, Vol.21 No.2, pp.165–201, 1995.
23. Tamaki, H. and Sato, T., OLD resolution with tabulation, *Proc. of ICLP'86*, London, LNCS 225, pp.84–98, 1986.
24. Tanaka, H. and Takezawa, T. and Etoh, J., Japanese grammar for speech recognition considering the MSLR method (in Japanese), *Proc. of the meeting of SIG-SLP (Spoken Language Processing)*, 97-SLP-15-25, Information Processing Society of Japan, pp.145–150, 1997.
25. Uratani, N. and Takezawa, T. and Matsuo, H. and Morita, C., ATR Integrated Speech and Language Database (in Japanese), TR-IT-0056, ATR Interpreting Telecommunications Research Laboratories, 1994.

Refutable/Inductive Learning from Neighbor Examples and Its Application to Decision Trees over Patterns

Masako Sato[1], Yasuhito Mukouchi[1], and Mikiharu Terada[2]

[1] Department of Mathematics and Information Sciences
College of Integrated Arts and Sciences
Osaka Prefecture University, Sakai, Osaka 599-8531, Japan
[2] International Buddhist University, Habikino, Osaka 583-8501, Japan

Abstract. The paper develops the theory of refutable/inductive learning as a foundation of discovery science from examples. We consider refutable/inductive language learning from positive examples, some of which may be incorrect. The error or incorrectness we consider is the one described uniformly in terms of a distance over strings. We define a k-neighbor closure of a language L as the collection of strings each of which is at most k distant from some string in L. In ordinary learning paradigm, a target language is assumed to belong to a hypothesis space without any guarantee. In this paper, we allow an inference machine to infer a neighbor closure instead of the original language as an admissible approximation. We formalize such kind of learning, and give some sufficient conditions for a hypothesis space.

As its application to concrete problems, we deal with languages defined by decision trees over patterns. The problem of learning decision trees over patterns has been studied from a viewpoint of knowledge discovery for Genome information processing in the framework of PAC learning from both positive and negative examples. We investigate their learnability in the limit from neighbor examples as well as refutable learnability from complete examples, i.e., from both positive and negative examples. Furthermore, we present some procedures which plays an important role for designing efficient learning algorithms for decision trees over regular patterns.

1 Introduction

The basis of inductive learning for knowledge discovery is the process of generating and refuting hypotheses from examples. Mukouchi&Arikawa [19] proposed the framework with both *refutability* and *learnability* (inferability) of a hypothesis space from examples, as a fundamental component for knowledge discovery. If a target language is a member of the hypothesis space, then the inference machine should identify it *in the limit* (Gold [7]), otherwise it should refute the hypothesis space itself in a finite time.

S. Arikawa and A. Shinohara (Eds.): Progress in Discovery Science 2001, LNAI 2281, pp. 201–213.

In relation to refutable learning, Lange&Watson [13] and Mukouchi [21] proposed learning criteria relaxing the requirements of inference machines, and Jain [10] also dealt with the problem for recursively enumerable languages.

We have so far investigated the following problems within the framework of refutable/inductive learning from a viewpoint of practical applications to knowledge discovery: (1) To propose a new framework of refutable/inductive learning flexibly applicable to concrete problems. (2) To apply the framework to concrete problems and to establish of the fundamental theory which plays an important role for designing efficient learning algorithms in such problems.

Mukouchi et al. [20] focused sequences, called *observations with time passage*, of examples successively generated by a certain kind of system, and dealt with *phenomena* generated by concrete rewriting systems known as $0L$ systems introduced by Lindenmayer [14] and pure grammars due to Gabrielian [6]. And their refutability and learnability from examples with time passage were investigated, and compared with learnability from examples without time passage.

Sato et al. [27] dealt with an efficient learning problem of bounded unions of regular pattern languages discussed by Arimura et al. [4] and gave a characterizing theorem for the class of bounded unions of regular patterns to have *compactness* with respect to containment which plays an important role in the problem. The class of pattern languages was introduced by Angluin [2] as a concrete class learnable from positive examples, and is one of the most basic class in the framework of elementary formal systems due to Smullyan [32]. In practical applications such as Genome informatics, the class is paid much attentions (Arikawa et al. [3]).

In real-world applications, we have to deal with incorrect examples. In order to allow some kind of noisy examples, the present paper proposes a new framework of refutable/inductive learning from *neighbor examples*. There are various approaches to language learning from incorrect examples (cf. e.g. Jain [9], Stephan [31], Case&Jain [5], and Sakakibara&Siromoney [24]). Stephan [31] has formulated a model of noisy data, in which a correct example crops up infinitely often, and an incorrect example only finitely often. There is no connection between incorrect examples considered there and correct examples. Our approach to the noisy examples is to consider that each observed incorrect example has some connection with a certain correct example on a target language to be learned. The incorrect examples we consider here are the ones described uniformly in terms of a *distance* over strings such as the Hamming distance ([8]) and the so-called edit distance ([25]). In connection to the edit distance, Sakakibara&Siromoney [24] discussed a noisy model on learning sets of strings. The noisy model was based on the edit noise and formulated in the framework of PAC learning.

Firstly, we introduce a notion of a recursively generable distance over strings, and define a k-neighbor closure of a language L as the collection of strings each of which is at most k distant from some string in L. Then we define a k-neighbor system as the collection of original languages and their j-neighbor closures with $j \leq k$, and adopt it as a hypothesis space. In ordinary learning paradigm, a target languages, whose examples are fed to an inference machine, is assumed to belong to a hypothesis space without any guarantee. We allow an inference machine to infer a neighbor closure instead of the original target language as an admissible approximation. Roughly speaking, an inference machine M k-neighbor-minimally infers a class $\mathcal{L}$ from positive examples, if for every observed language L, M converges an expression of L's minimal language within a k-neighbor system which is least distant from some language in $\mathcal{L}$. We formalize such kind of learning, and give some sufficient conditions for a hypothesis space. For the particular case of $k = 0$, the above learnability coincides with the paradigm of approximate learning from examples introduced by Mukouchi [18]. In relation to approximate learning, Kobayashi&Yokomori [11] also proposed learning criterion requiring an inference machine to infer an admissible approximate language within the hypothesis space concerned.

Furthermore, we deal with language learning from complete examples under the setting that some positive examples may be presented to the learner as negative examples, and vice versa. We define a k-neighbor system of a language class, and discuss conditions of a language class to be refutable and learnable from complete examples in this framework.

As an application of the above framework to concrete problems, we deal with learning of decision trees over patterns. The problem of learning decision trees over patterns has been studied from a viewpoint of knowledge discovery for Genome information processing in the framework of PAC learning (Arikawa et al. [3] and Miyano [15]). We investigate their refutable/inductive learnability from neighbor examples as well as from positive examples.

We assume the standard definitions in the field of inductive learning (Gold [7] and Angluin [2]). Definitions of other concepts used in this paper may be found as follows: Refutable/inductive inference (Mukouchi et al. [19], Wright [34], and Sato [26]); learning pattern languages (Angluin [1], Shinohara [29, 30], and Arimura et al. [4]).

We omit proofs and so on for the space limitation. Refer to the papers Mukouchi&Sato [22, 23] and Terada et al. [33] for the detailed contents.

2 Refutable/Inductive Learning from Neighbor Examples

In this paragraph, we present a framework for refutable/inductive learning from examples, some of which may be incorrect. The error or incorrectness we deal with is the one described uniformly in terms of a distance over strings.

Definition 1. *Let $N = \{0, 1, 2, \cdots\}$ be the set of all natural numbers, and let Σ be an alphabet. A function $d : \Sigma^+ \times \Sigma^+ \to N \cup \{\infty\}$ is called a* distance *over strings, if it satisfies the following three conditions:*

(1) For every $v, w \in \Sigma^+$, $d(v, w) = 0$ iff $v = w$.

(2) For every $v, w \in \Sigma^+$, $d(v, w) = d(w, v)$.

(3) For every $u, v, w \in \Sigma^+$, $d(u, v) + d(v, w) \geq d(u, w)$.

A distance d is said to be recursive*, if there is an effective procedure that computes $d(v, w)$ for every $v, w \in \Sigma^+$ with $d(v, w) \neq \infty$.*

Assume that the correct example is a string v and the observed example is a string w. In case we are considering the Hamming distance and two strings v and w have the same length but differ just one symbol, then we estimate the incorrectness as their distance of one. In case we are considering the so-called edit distance and w can be obtained from v by deleting just one symbol and inserting one symbol in another place, then we estimate the incorrectness as their distance of two.

Then we define a k-neighbor closure of a language as follows:

Definition 2. *Let $d : \Sigma^+ \times \Sigma^+ \to N \cup \{\infty\}$ be a distance over strings and let $k \in N$.*

The k-neighbor closure $\overline{w}^{(d,k)}$ of a string $w \in \Sigma^+$ w.r.t. d *is the set of all strings each of which is at most k distant from w, that is, we put $\overline{w}^{(d,k)} = \{v \in \Sigma^+ \mid d(v,w) \leq k\}$.*

The k-neighbor closure $\overline{L}^{(d,k)}$ of a language $L \subseteq \Sigma^+$ w.r.t. d *is the set of all strings each of which is at most k distant from some string in L, that is, we put $\overline{L}^{(d,k)} = \bigcup_{w \in L} \overline{w}^{(d,k)} = \{v \in \Sigma^+ \mid \exists w \in L \text{ s.t. } d(v,w) \leq k\}$.*

Definition 3. *A distance d is said to* have finite thickness, *if for every $w \in \Sigma^+$, the set $\overline{w}^{(d,k)}$ is finite.*

A distance d is said to be recursively generable, *if d has finite thickness and there exists an effective procedure that on inputs $k \in N$ and $w \in \Sigma^+$ enumerates all elements in $\overline{w}^{(d,k)}$ and then stops.*

Many of well-known distances such as the Hamming distance and the edit distance are shown to be recursively generable.

Let d be a recursively generable distance and let $k \in N$. Then, for every $v, w \in \Sigma^+$, by checking $v \in \overline{w}^{(d,k)}$, whether $d(v,w) \leq k$ or not is recursively decidable. Therefore d turns to be a recursive distance. Let $L \subseteq \Sigma^+$ be a recursive language. Then, for every $w \in \Sigma^+$, by checking $\overline{w}^{(d,k)} \cap L \neq \phi$, whether $w \in \overline{L}^{(d,k)}$ or not is recursively decidable. Therefore $\overline{L}^{(d,k)}$ is also a recursive language.

Hereafter, we exclusively deal with a *recursively generable distance*, and simply refer it as a *distance* without any notice.

Firstly, we consider the case that the most of positive examples will be presented together with some negative but near examples of a target language. In this case, we force an inference machine to infer a minimal neighbor closure of the original target language as an admissible approximation.

For a class $\mathcal{L}$ and a set S, the set of all minimal languages containing S within $\mathcal{L}$ is denoted by $\mathrm{MIN}(S, \mathcal{L})$.

Definition 4. *Let d be a distance and let $k \in N$.*

For a class $\mathcal{L} = \{L_i\}_{i \in N}$, let us put $\overline{\mathcal{L}}^{(d,k)} = \{\overline{L_i}^{(d,k)}\}_{i \in N}$.

A k-neighbor system $\overline{\mathcal{L}}^{(d,\leq k)}$ of a class $\mathcal{L}$ w.r.t. d *is the collection of languages each of which is a j-neighbor closure w.r.t. d of some language in $\mathcal{L}$ for some $j \leq k$, that is, we put $\overline{\mathcal{L}}^{(d,\leq k)} = \bigcup_{j=0}^{k} \overline{\mathcal{L}}^{(d,j)}$.*

For a nonempty language $L \subseteq \Sigma^+$, *a pair* $(i,j) \in N \times N$ *is said to be a* weak k-neighbor-minimal answer for L, *if* $j \leq k$ *and* $\overline{L_i}^{(d,j)} \in \mathrm{MIN}(L, \overline{\mathcal{L}}^{(d,\leq k)})$.

For a nonempty language $L \subseteq \Sigma^+$, *a pair* $(i,j) \in N \times N$ *is said to be a* k-neighbor-minimal answer for L, *if (1)* (i,j) *is a weak* k*-neighbor-minimal answer for* L *and (2) for every pair* (i',j') *with* $j' < j$, $\overline{L_{i'}}^{(d,j')} \notin \mathrm{MIN}(L, \overline{\mathcal{L}}^{(d,\leq k)})$.

An IIM M *is said to* k-neighbor-minimally *(resp.,* weak k-neighbor-minimally*)* infer a class $\mathcal{L}$ w.r.t. d from positive examples, *if it satisfies the following condition: For every nonempty language* $L \subseteq \Sigma^+$ *and for every positive presentation* σ *of* L, *if* $\mathrm{MIN}(L, \overline{\mathcal{L}}^{(d,\leq k)}) \neq \phi$, *then* M *converges to an integer* $\langle i,j \rangle$ *for* σ *such that* (i,j) *is a* k*-neighbor-minimal (resp., weak* k*-neighbor-minimal) answer for* L, *where* $\langle \cdot, \cdot \rangle$ *represents the Cantor's pairing function.*

A class $\mathcal{L}$ *is said to be* k-neighbor-minimally *(resp.,* weak k-neighbor-minimally*)* learnable w.r.t. d from positive examples, *if there is an IIM which* k*-neighbor-minimally (resp., weak* k*-neighbor-minimally) infers* $\mathcal{L}$ *w.r.t.* d *from positive examples.*

We note that, by the definition, a class $\mathcal{L}$ is (weak) 0-neighbor-minimally learnable w.r.t. d from positive examples, if and only if $\mathcal{L}$ is minimally learnable from positive examples (Mukouchi [18]).

Assume that a class $\mathcal{L}$ is k-neighbor-minimally learnable w.r.t. d from positive examples for some $k \in N$ and for some distance d. For every $L_i \in \mathcal{L}$, $L_i = \overline{L_i}^{(d,0)}$ and $\mathrm{MIN}(L_i, \overline{\mathcal{L}}^{(d,\leq k)}) = \{L_i\}$. Therefore, if (i',j') is a k-neighbor-minimal answer for L_i, then $L_i = L_{i'}$ and $j' = 0$. Thus the class $\mathcal{L}$ is also learnable in the limit from positive examples.

Therefore (weak) k-neighbor-minimal learnability can be regarded as a natural extension of ordinary learnability as well as minimal learnability.

The notion of *finite thickness* for a language class is introduced by Angluin [2] as a sufficient condition for a class to be learnable from positive examples and defined as follows: A class has finite thickness, if for every nonempty set $S \subseteq \Sigma^+$, the set of languages in the class containing S is of finite cardinality.

Here we can show that it also becomes a sufficient condition for k-neighbor-minimal learnability.

Theorem 1. *Let* d *be a distance and let* $k \in N$. *If a class* $\mathcal{L}$ *has finite thickness, then* $\mathcal{L}$ *is* k*-neighbor-minimally learnable w.r.t.* d *from positive examples.*

Two notions of *finite elasticity* and *M-finite thickness* for a language class were introduced by Wright [34] and Moriyama&Sato [16], respectively, as generalized notions of finite thickness. Finite elasticity is a sufficient condition for learnability from positive examples and defined as follows: A language class has finite elasticity, if there is not an infinite sequence of strings $w_0, w_1, \cdots$ and an infinite sequence of languages $L_1, L_2, \cdots$ in $\mathcal{L}$ satisfying $\{w_0, w_1, \cdots, w_{i-1}\} \subseteq L_i$ but $w_i \notin L_i$ for every $i \geq 1$. Any class containing all finite languages does not have finite elasticity.

A language class $\mathcal{L}$ has M-finite thickness, if (1) for every finite set $S \subseteq \Sigma^+$, the set $\mathrm{MIN}(S, \mathcal{L})$ is finite, and (2) if $S \subseteq L$ for some $L \in \mathcal{L}$, then there is a language $L' \in \mathrm{MIN}(S, \mathcal{L})$ such that $L' \subseteq L$. Thus a class containing all finite languages has always M-finite thickness. M-finite thickness is not a sufficient condition for learnability from positive examples, but has a good property as well as finite elasticity as shown below.

Theorem 2 (Wright [34], Moriyama&Sato [16], Sato [26]).
(1) Finite elasticity is a sufficient condition for learnability from positive examples.
(2) A class with M-finite thickness is learnable from positive examples, if and only if for every L in the class, there is a finite tell-tale set $S_L \subseteq L$ for L.
(3) Finite elasticity and M-finite thickness are both preserved under union and intersection operations.

Here a set S_L is a *finite tell-tale set* of L in the class $\mathcal{L}$, if S_L is a finite subset of L, and $L \in \mathrm{MIN}(S_L, \mathcal{L})$ (Angluin [2]).
Note that the above both properties are preserved under another operations such as concatenation, Kleene $*$ and so on (Sato [26]).

Theorem 3. *Let d be a distance and let $k \in N$. If a class $\mathcal{L}$ has M-finite thickness and finite elasticity, then the class is weak k-neighbor-minimally learnable w.r.t. d from positive examples.*

By Theorem 2 and Theorem 3, the next result immediately follows:

Corollary 1. *Let d be a distance, let $k, n \in N$. If a class $\mathcal{L}$ has M-finite thickness and finite elasticity, then the class obtained by at most n times applying union and intersection operations to languages in $\mathcal{L}$ is weak k-neighbor-minimally learnable w.r.t. d from positive examples.*

Furthermore, we also consider inductive learning of languages from complete examples with errors. For a language L, we define a k-neighbor language of L by (1) adding some strings *not* in L each of which is at most k distant from some string in L and by (2) deleting some strings in L each of which is at most k distant from some string *not* in L. Then we force an inference machine to infer the original language L from positive and negative examples of a k-neighbor language of L.
For detailed definitions and results, please refer to Mukouchi&Sato [23].

3 Refutable/Inductive Learning of Decision Trees over Patterns

In this paragraph, we show that the class of languages defined by decision trees with a depth of at most n over patterns (Arikawa et al. [3]) is not learnable from positive examples, if $n \geq 2$. We give a new semantics for decision trees over patterns, and then investigate refutable/inductive learnability of decision trees over patterns from examples. After that, we consider its learnability from neighbor examples introduced in the previous section.

A *pattern* is a finite nonempty string of constant symbols in Σ and variables. The language $L(p)$ defined by a pattern p is the set of constant strings obtained from p by substituting constant nonempty strings to variables in p.

The class $\mathcal{PL}$ of pattern languages was known to have *finite thickness*, and thus it is learnable from positive examples (Angluin [1, 2]).

A *decision tree over patterns* is a binary tree such that the leaves are labeled with 0 or 1 and each internal node is labeled with a pattern. Let us denote by $\mathcal{TP}_n$ and $\mathcal{TP}^n$ the sets of decision trees with a depth of at most n and containing at most n internal nodes, respectively.

For a constant string w, there is a unique path $v_0, v_1, \cdots, v_n$ from the root v_0 to a leaf v_n such that v_{i+1} is the left children of v_i labeled with a pattern p_i if $w \in L(p_i)$, and the right children otherwise, for $i = 0, \cdots, n-1$. A semantics of a decision tree T over patterns is the language $L(T)$ consisting of strings with paths ending at leaves labeled with 1. Thus the language $L(T)$ can be represented by applying three operations of intersection, union and complement to languages defined by patterns appearing in the tree T. For instance, the language $L(T)$ is given as follows for a decision tree T illustrated in Fig. 1:

$$L(T) = (L(xaby) \cap L(axbby)^c) \cup (L(xaby)^c \cap L(xayb)),$$

where $L(p)^c$ is the complement of $L(p)$, i.e., $L(p)^c = \Sigma^+ - L(p)$.

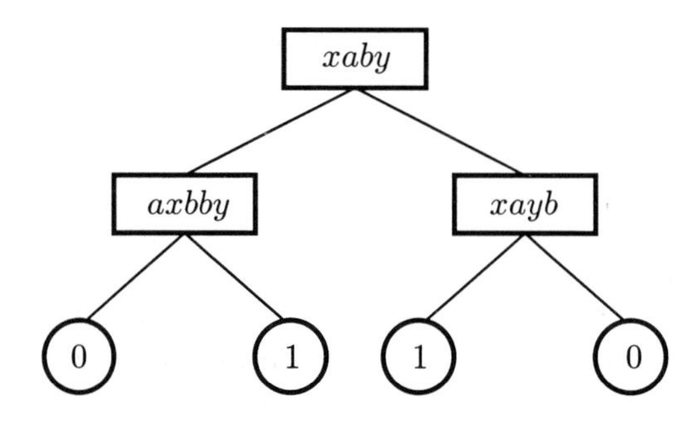

Fig. 1. An example of a decision tree over patterns

We denote by co-$\mathcal{PL}_*$ the class of complements of pattern languages and put $\mathcal{JPL}_* = \mathcal{PL} \cup$ co-$\mathcal{PL}_*$. Then the class $\mathcal{JPL}_*$ is the class of languages defined by decision trees with the depth 1.

Firstly, we investigate the learnability of $\mathcal{JPL}_*$ from positive examples.

The class $\mathcal{PL}$ has finite elasticity as mentioned above, but not co-$\mathcal{PL}_*$. In fact, an infinite sequence of strings $w_0, w_1, \cdots, w_{k-1}, \cdots$ and an infinite sequence of complements $L(w_0)^c, L(w_1)^c, \cdots, L(w_k)^c, \cdots$ satisfy $\{w_0, w_1, \cdots, w_{k-1}\} \subseteq L(w_k)^c$ but $w_k \notin L(w_k)^c$ for each $k \geq 1$ when $w_i \neq w_j$ for $i \neq j$. The class co-$\mathcal{PL}_*$, however, is learnable from positive examples as follows:

Theorem 4 (Shinohara [30]). *The class $\mathcal{PL}$ is learnable from* negative *examples. Thus the class co-$\mathcal{PL}_*$ of complements of pattern languages is learnable from* positive *examples.*

As easily seen, for each $l \geq 1$ and for different variables $x_1, \cdots, x_l$, the complement $L(x_1 x_2 \ldots x_l)^c$ is the finite language, denoted by $\Sigma^{\leq l-1}$, consisting of all strings with length less than l. Hence we can easily show that the class co-$\mathcal{PL}_*$ has M-finite thickness. If two classes $\mathcal{L}$ and $\mathcal{L}'$ have M-finite thickness, the union class $\mathcal{L} \cup \mathcal{L}'$ also has the property (Sato [26]). This implies the following result:

Lemma 1. *The class $\mathcal{JPL}_*$ does not have finite elasticity but has M-finite thickness.*

By the above and Theorem 2, the learnability of the class $\mathcal{JPL}_*$ depends on an existence of a finite tell-tale set for each languages in the class. It can be easily shown that for every p, the set $S_1(p)$ of shortest strings in $L(p)$ is a finite tell-tale set for $L(p)$ and $\Sigma^{\leq |p|} - S_1(p)$ for the complement $L(p)^c$ within the class $\mathcal{JPL}_*$. Hence we obtain the following:

Theorem 5. *The class $\mathcal{JPL}_*$ of languages defined by decision trees with the depth* 1 *is learnable from positive examples.*

Unfortunately the class of languages defined by decision trees with a depth of at most n is not learnable from positive examples, if $n \geq 2$.

Theorem 6. *Let $n \geq 2$. Then the class of languages defined by decision trees with a depth of at most n is not learnable from positive examples.*

Proof. Assume that the class is learnable from positive examples. Then there exists a finite tell-tale set of each language in the class. The language $L(x)$ ($= \Sigma^+$), however, does not have any finite tell-tale set within the class. Indeed, if a finite set $S \subseteq L(x)$ is a finite tell-tale set of $L(x)$, we have $S \subseteq L(x_1 \cdots x_l)^c \cap L(x) \subsetneq L(x)$, where $l = \max\{|w| \mid w \in S\} + 1$. It contradicts the choice of S since the language $L(x_1 \cdots x_l) \cap L(x)$ is defined by a decision tree with the depth 2. □

In order to give a new semantics for a given decision tree, we introduce a particular type of a string p^c called *co-pattern* for each pattern p, and define the semantics $L(p^c)$ of p^c by the subset of the complement $L(p)^c$ consisting of strings with length larger than or equal to $|p|$. That is, $L(p^c) = \{w \in L(p)^c \mid |w| \geq |p|\}$ for every p. Let co-$\mathcal{PL} = \{L(p^c) \mid p \in \mathcal{P}\}$ and $\mathcal{JPL} = \mathcal{PL} \cup$ co-$\mathcal{PL}$.

According to the above semantics, we give a new interpretation $L(T)$ for a decision tree T over patterns as follows: A constant string w is not accepted by the tree if $w \notin L(p) \cup L(p^c)$, i.e., $|w| < |p|$ for some internal node labeled with p in the unique path for w, and the semantics $L(T)$ consists of strings accepted by T and with paths ending at leaves labeled with 1. Using this interpretation instead of complements, the language defined by T illustrated in Fig. 1 is given:

$$L(T) = (L(xaby) \cap L(axbby^c)) \cup (L(xaby^c) \cap L(xayb)).$$

We denote by $\mathcal{TPL}_n$ the class of languages defined by decision trees with a depth of at most n, and $\mathcal{TPL}^n$ similarly.

As easily seen, $w \in L(p^c)$ implies $|w| \geq |p|$ for every $w \in \Sigma^+$ and for every pattern p. Thus the next result immediately follows:

Lemma 2. *The class $\mathcal{JPL}$ $(= \mathcal{TPL}_1)$ has finite thickness.*

By the above result and Theorem 2, for every n the class obtained by at most $n2^{n-1}$ times applying union and intersection operations to languages in $\mathcal{JPL}$ has finite elasticity, and thus it is learnable from positive examples. Since both of $\mathcal{TPL}_n$ and $\mathcal{TPL}^n$ are its subclass, it implies that:

Theorem 7. *Let $n \geq 1$. Then both classes $\mathcal{TPL}_n$ and $\mathcal{TPL}^n$ are learnable from positive examples.*

Note that both classes in the above theorem are shown to be refutably learnable from complete examples as well as the class $\mathcal{PL}$ (Mukouchi&Arikawa [19]).

Now we consider refutable/inductive learning of decision trees from neighbor examples introduced in the previous section. By Theorem 1 and Lemma 2, the next result immediately follows:

Theorem 8. *Let d be a distance and let $k \in N$. The class $\mathcal{JPL}$ of pattern languages and co-pattern languages is k-neighbor-minimally learnable w.r.t. d from positive examples.*

Furthermore, by Theorem 2, Theorem 3 and Lemma 2, the next main result follows:

Theorem 9. *Let d be a distance, let $k \in N$ and let $n \geq 1$. Both classes $\mathcal{TPL}_n$ and $\mathcal{TPL}^n$ are weak k-neighbor-minimally learnable w.r.t. d from positive examples.*

Hereafter, we intend to design an efficient learning algorithm of decision trees over regular pattern from positive examples. A pattern is *regular* if each variable appears at most once in the pattern. As well known, the membership problem of regular pattern languages is polynomial time computable (Shinohara [29]) although that of pattern languages is NP-complete (Angluin [1]).

We denote by co-$\mathcal{RPL}$ the subclass of co-$\mathcal{PL}$ restricted to regular patterns, and $\mathcal{JRPL}$ and $\mathcal{TRPL}_n$ similarly.

By Theorem 7, the subclass $\mathcal{TRPL}_n$ is learnable from positive examples. One of the most important problems to design such an algorithm from positive examples is how to avoid overgeneralizations. In order to solve it, it is an important key to find a decision tree defining a minimal language of a given set of examples within the class. The following procedure computes a co-pattern defining a minimal language of a given set of examples S within the class co-$\mathcal{RPL}$ in time of $O(l_{\max}^2 m)$, where $l_{\max}$ is the length of the longest strings in S and m is the cardinality of the set S.

Procedure co-MINL
input: a set of examples S,
output: a co-regular pattern p^c
begin
let l be the length of the shortest strings in S;
if $\Sigma^l \not\subseteq S$ **then begin**
let $p_1 = a_1 a_2 \cdots a_l$ $(a_i \in \Sigma)$ be an arbitrary string not in S;
for $i = 1$ **to** l **do begin**
let q be the pattern obtained from p_i by replacing the i-th symbol a_i with x_i;
if $S \subseteq L(q^c)$ **then** $p_{i+1} := q$ **else** $p_{i+1} := p_i$
end;
$p := p_{l+1}$
end
else let $p := b_1 b_2 \cdots b_{l-1}$ be an arbitrary string in Σ^{l-1};
output p^c
end

The next procedure computes a regular pattern or co-regular pattern defining a minimal language of a given set of examples S in time of $O(l_{\max}^2 m)$, where **MINL**(S) used in the procedure is the procedure to compute a regular pattern defining a minimal language within the class $\mathcal{RPL}$ which was given by Shinohara [29].

Procedure MINL$_{\mathcal{JRPL}}$
input: a set of examples S,
output: a regular pattern or a co-regular pattern π
begin
let l be the length of the shortest strings in S;
if $\Sigma^l \subseteq S$ **then** $\pi := x_1 x_2 \cdots x_l$
else begin
$\pi :=$ **MINL**(S);
if $\pi = x_1 x_2 \cdots x_l$ **then** $\pi :=$ **co** $-$ **MINL**(S)
end
end

Since the class $\mathcal{JRPL}$ has finite elasticity and the procedure **MINL**$_{\mathcal{JRPL}}$ above is computable in polynomial time, the next result immediately follows (cf. Arimura et al. [4]):

Theorem 10. *The class $\mathcal{JRPL}$ is polynomial update time learnable from positive examples.*

4 Conclusion

We have proposed the framework of refutable/inductive language learning from neighbor examples, and given some sufficient conditions for its learnability using the set-theoretical notions for classes such as finite elasticity, M-finite thickness and so on. Although we omit the detailed definitions for the convenience of

of propositions and $\epsilon < 1$, $\delta < 1$ be arbitrary positive numbers. If $|DNF(f)|$ and $|CNF(f)|$ is small, then we can efficiently discover an approximate critical casebase such that the probability that the classification error rate by the discovered casebase is more than ϵ is at most δ. The sample size of cases is bound in polynomial of $\frac{1}{\epsilon}$, $\frac{1}{\delta}$, $|DNF(f)|$ and $|CNF(f)|$ and necessary number of membership queries is bound in polynomial of n, $|DNF(f)|$ and $|CNF(f)|$.

In [Satoh00b], we have generalized the results in [Satoh00a] to apply our framework to learn a critical casebase for a relation with finite tree-structured values. We show that for every relation $\mathcal{R}$ with a concept tree T, in order to represent $\mathcal{R}$, an upper bound of necessary positive cases is $|DNF(\mathcal{R})|$ and the upper bound of necessary negative cases is $|DNF(\mathcal{R})| \cdot |CNF(\mathcal{R})|$. Then, we give a learning method of a critical casebase and we analyze computational complexity of the method in the PAC learning framework and show that the sample size of cases is at most $(\frac{1}{\epsilon} \ln \frac{1}{\delta}) \cdot |DNF(\mathcal{R})| \cdot (1+|CNF(\mathcal{R})|)$ and necessary number of membership queries is at most $n^2 \cdot width(T) \cdot height(T) \cdot |DNF(\mathcal{R})| \cdot |CNF(\mathcal{R})|$ where $width(T)$ is the number of leaves in T and $height(T)$ is the number of nodes in the longest path from leaves to the root in T.

This paper generalizes these results so that we can apply them to learn a relation with finite semi-lattice structured values. Semi-lattice is a partial-order relation with existence of the least upper bound for every two elements (but not necessary with the greatest lower bound). We show that the results obtained here is almost in parallel with the results [Satoh00b] except the size of negative cases for a critical casebase to express a relation and the complexity of approximating a critical casebase.

2 Case-Based Reasoning for Lattice-Based Relation

Let $\mathcal{A}$ be a finite partial ordered set called a *set of attributes*. We denote a partial order relation over $\mathcal{A}$ as $\preceq$. We furtherly assume that $\mathcal{A}$ forms a *semi-lattice*, in other words, for every two attributes a_1, a_2, there is the least upper bound of $\{a_1, a_2\}$ denoted as $a_1 \vee a_2$. Note that since $\mathcal{A}$ is a finite semi-lattice, there is always the least upper bound of any subset S of $\mathcal{A}$ denoted as $\bigvee S$ and the top element of $\mathcal{A}$ denoted as $\top$. We denote the greatest lower bound of a subset S as $\bigwedge S$ if it exists. We define $min(\mathcal{A})$ as a set of minimal elements in $\mathcal{A}$ w.r.t. $\preceq$ and call such elements as *fundamental elements*. We define $length(\mathcal{A})$ as the number of elements of the longest chain in $\mathcal{A}$ where a chain is a subset of $\mathcal{A}$ such that for every two elements a_1, a_2 of the subset, $a_1 \preceq a_2$ or $a_2 \preceq a_1$.

For a non-top element a of $\mathcal{A}$, we define

$$parent(a) = \{a' | a \prec a' \text{ and there is no } b \text{ s.t. } a \prec b \prec a'\}$$

where $a \prec a'$ means $a \preceq a'$ and $a \neq a'$. We denote $max\{|parent(a)||a \in \mathcal{A}\}$ as $MP(\mathcal{A})$ where $|parent(a)|$ is the number of elements in $parent(a)$.

On the other hand, for a non-minimal element a of $\mathcal{A}$, we define

$$child(a) = \{a'|a' \prec a \text{ and there is no } b \text{ s.t. } a' \prec b \prec a\}.$$

Let a_1, a_2 and a_3 be attributes. We say a_1 *is more or equally similar to* a_2 *than to* a_3 if $a_1 \vee a_2 \preceq a_1 \vee a_3$.

We call an n-ary tuple of fundamental elements *a case*. In other words, a case is in $(min(\mathcal{A}))^n$. We restrict a case as an n-ary tuple of fundamental elements other than any elements since we regard fundamental elements expressing a basic kind of attributes and other elements expressing a derived kind.

Let c be a case. We denote the i-th component of the tuple c as $c[i]$.

We define $c_1 \vee c_2$ as

$$\langle c_1[1] \vee c_2[1], c_1[2] \vee c_2[2], ..., c_1[n] \vee c_2[n] \rangle,$$

and $c_1 \vee c_2 \preceq c_1 \vee c_3$ as for every $i(1 \leq i \leq n)$,

$$c_1[i] \vee c_2[i] \preceq c_1[i] \vee c_3[i].$$

Let c_1, c_2 and c_3 be cases. Then, we say c_1 *is more or equally similar to* c_2 *than to* c_3 if $c_1 \vee c_2 \preceq c_1 \vee c_3$. We have the following important property for $\preceq$.

Proposition 1. *Let c, c_1, c_2 be cases. $c_1 \vee c \preceq c_2 \vee c$ iff $c_1 \vee c_2 \preceq c \vee c_2$.*

We call a subset of $(min(\mathcal{A}))^n$ an *n-ary relation*, or a *relation* for short. In other words, a set of cases is a relation.

Definition 1. *Let $\mathcal{CB}$ be a set of cases which are divided into $\mathcal{CB}^+$ and $\mathcal{CB}^-$. We call $\mathcal{CB}$ a* casebase, *$\mathcal{CB}^+$ a set of* positive cases *and $\mathcal{CB}^-$ a set of* negative cases *respectively.*

We say a case c is positive w.r.t. $\mathcal{CB}$ *if there is a case $c_{ok} \in \mathcal{CB}^+$ such that for every negative case $c_{ng} \in \mathcal{CB}^-$, $c \vee c_{ng} \not\preceq c \vee c_{ok}$. We say* a case c is positive w.r.t. $\mathcal{CB}$ *if c is not positive.*

Note that $c \vee c_{ng} \not\preceq c \vee c_{ok}$ does not imply $c \vee c_{ok} \prec c \vee c_{ng}$ since $\preceq$ is a partial order relation.

In the above definition, "c is positive" means that there is a positive case such that c is not more or equally similar to any negative case than to the positive case. Note also that the definitions of the positiveness and the negativeness are not symmetrical.

Definition 2. *Let $\mathcal{CB}$ be a casebase $\langle \mathcal{CB}^+, \mathcal{CB}^- \rangle$. We say that n-*ary relation *$\mathcal{R}_{\mathcal{CB}}$* is represented by a casebase *$\mathcal{CB}$ if $\mathcal{R}_{\mathcal{CB}} = \{c \in (min(\mathcal{A}))^n | c$ is positive w.r.t. $\mathcal{CB}\}$.*

Conversely, any relation $\mathcal{R}$ can be represented by a casebase $\langle \mathcal{CB}^+, \mathcal{CB}^- \rangle$ where $\mathcal{CB}^+$ is $\mathcal{R}$ itself and $\mathcal{CB}^-$ is $\overline{\mathcal{R}}$ $(=(min(\mathcal{A}))^n - \mathcal{R})$.

From Proposition 1, the following holds.

Proposition 2. *Let $\mathcal{CB}$ be a casebase $\langle \mathcal{CB}^+, \mathcal{CB}^- \rangle$. A case c is positive if and only if there is a case $c_{ok} \in \mathcal{CB}^+$ such that for every case $c_{ng} \in \mathcal{CB}^-$, $c_{ok} \vee c_{ng} \not\preceq c_{ok} \vee c$.*

Note that the above proposition is important if we use a casebase to decide a new case belongs to the given relation. When we decide to fix a casebase, given a new case c to be classified, we need to compute both of $c \vee c_{ng}$ and $c \vee c_{ok}$ in the original definition of representation of relation. However, by the above Lemma, we can "precompile" a fixed casebase for efficient classification, that is, we can compute $c_{ng} \vee c_{ok}$ in advance. Then, we only need to compute $c \vee c_{ok}$ to decide whether c is in the relation.

We can detect redundant negative cases by using the following lemma.

Lemma 1. *Let $\mathcal{CB}$ be a casebase $\langle \mathcal{CB}^+, \mathcal{CB}^- \rangle$. Let $\mathcal{R}_{\mathcal{CB}}$ be a relation represented by $\mathcal{CB}$. Let $C'_{ng} \in \mathcal{CB}^-$ and $\mathcal{CB}' = \langle \mathcal{CB}^+, \mathcal{CB}'^- \rangle$ where $\mathcal{CB}'^- = \mathcal{CB}^- - \{C'_{ng}\}$. Suppose that for all $c_{ok} \in \mathcal{CB}^+$, there exists $c_{ng} \in \mathcal{CB}'^-$ s.t. $c_{ng} \vee c_{ok} \preceq C'_{ng} \vee c_{ok}$. Then $\mathcal{R}_{\mathcal{CB}} = \mathcal{R}_{\mathcal{CB}'}$.*

Definition 3. *Let S be a set of cases and c be a case. We define the nearest cases of S from c, $NN(c, S)$, as follows.*
$NN(c, S) = \{c' \in S | \neg \exists c'' \in S \ s.t. \ c \vee c'' \prec c \vee c'\}$

The following proposition is also related to a reduction of negative cases in a casebase to decide a new case belongs to a given relation.

Proposition 3. *Let $\mathcal{CB}$ be a casebase $\langle \mathcal{CB}^+, \mathcal{CB}^- \rangle$.*
Let $\mathcal{CB}' = \langle \mathcal{CB}^+, \bigcup_{c_{ok} \in \mathcal{CB}^+} NN(c_{ok}, \mathcal{CB}^-) \rangle$. Then, $\mathcal{R}_{\mathcal{CB}} = \mathcal{R}_{\mathcal{CB}'}$.

3 Case-Based Representability

Firstly, we define a language which expresses a relation. This is done to characterize an upper bound of minimal size of casebase to represent a relation. We introduce n variables $x_1, ..., x_n$ which represent the position of arguments in the relation. *An atomic formula* has the one of the following form:

- $x \preceq a$ where x is one of $x_1, ..., x_n$ and a is the name of an attribute in $\mathcal{A}$ which means that x is less or equally general than a.
- a special symbol, **T** which means truth.
- a special symbol, **F** which means falsity.

A formula is the combination of an atomic formula and $\wedge$ and $\vee$ in the usual sense. We denote a set of all formulas as $\mathcal{L}$.

Let us regard an atomic formula as a proposition. Then, $\mathcal{L}$ can be regarded as negation-free propositional language. Then, we can define a disjunctive normal form (DNF) of a formula in $\mathcal{L}$ as a DNF form of the translated propositional language. Similarly, we also define a conjunctive normal form (CNF) of a formula in $\mathcal{L}$ as well.

We can simplify a formula along with the following inference rules (together with usual propositional inference rules):

$$\frac{(x \preceq a_1) \vee ... \vee (x \preceq a_m) \text{ and } child(a) = \{a_1, ..., a_m\}}{x \preceq a}$$

$$\frac{(x \preceq a) \text{ and } child(a) = \{a_1, ..., a_m\}}{(x \preceq a_1) \vee ... \vee (x \preceq a_m)}$$

$$\frac{x \preceq \top}{\mathbf{T}}$$

$$\frac{(x \preceq a_1) \wedge ... \wedge (x \preceq a_m) \text{ and } \bigwedge\{a_1, ..., a_m\} \text{ exists}}{(x \preceq \bigwedge\{a_1, ..., a_m\})}$$

$$\frac{(x \preceq a_1) \wedge ... \wedge (x \preceq a_m) \text{ and } \bigwedge\{a_1, ..., a_m\} \text{ doesn't exist}}{\mathbf{F}}$$

Let F be a formula in $\mathcal{L}$. We define $|DNF(F)|$ as the smallest number of disjuncts in logically equivalent DNF forms to F induced by the above inference rules and we define $|CNF(F)|$ as the smallest number of conjuncts in logically equivalent CNF forms to F as well.

Let c be a case and F be a formula of $\mathcal{L}$. We say that c *satisfies* F denoted as $c \models F$ if one of the following conditions holds.

1. If F is an atomic formula $x_i \preceq a$, then $c[i] \preceq a$.
2. If F is of the form $G \wedge H$, then $c \models G$ and $c \models H$.
3. If F is of the form $G \vee H$, then $c \models G$ or $c \models H$.

We define $\phi(F) = \{c \in \mathcal{A}^n | c \models F\}$.

Next lemma relates with a size of necessary positive cases.

Lemma 2. *Let $\mathcal{R}$ be an n-ary relation and $\mathcal{CB}^+$ be a subset of $\mathcal{R}$ and $D_1 \vee ... \vee D_k$ be a DNF representation of $\mathcal{R}$. Suppose that for every D_i, there exists $c_{ok} \in \mathcal{CB}^+$ such that $c_{ok} \in \phi(D_i)$. Then, $\mathcal{R} = \mathcal{R}_{\mathcal{CB}}$ where $\mathcal{CB} = \langle \mathcal{CB}^+, \overline{\mathcal{R}} \rangle$.*

For the next lemma, we need the definition of $c\downarrow^l_{c'}$ and $PNN(c, \overline{\mathcal{R}})$ defined as follows:

Definition 4. *Let c and c' be cases. We define a set of cases $c\downarrow^l_{c'}$ for $l (1 \leq l \leq n)$ such that $c[l] \neq c'[l]$ as follows. $c'' \in c\downarrow^l_{c'}$ if c'' satisfies the following conditions:*

- $c'[l] \vee c''[l] \in child(c'[l] \vee c[l])$.
- $c'[j] \vee c''[j] = c'[j] \vee c[j]$ *for* $j \neq l (1 \leq j \leq n)$.

$c\downarrow^l_{c'}$ is a set of the nearest cases to c among cases whose least upper bound with c' differs from $c' \vee c$ in the l-th attribute. Note that the number of elements of $c\downarrow^l_{c'}$ for l-th attribute is at most $|min(\mathcal{A})| - 1$.

For the next lemma, we use a set of cases $PNN(c', \overline{\mathcal{R}})$ defined as follows:

$$PNN(c', \overline{\mathcal{R}}) = \{c \in \overline{\mathcal{R}} | \text{for every } l (1 \leq l \leq n) \text{ s.t. } c[l] \neq c'[l], c\downarrow^l_{c'} \subset \mathcal{R}\}$$

$PNN(c', \overline{\mathcal{R}})$ is a set of pseudo nearest neighbor cases from c' with respect to $\overline{\mathcal{R}}$. If $c \in PNN(c', \overline{\mathcal{R}})$, then c is a negative case, and for every c'' such that c'' is nearer to c' with respect to exactly one attribute than c, $c'' \in \mathcal{R}$. It is clear that if $c \in NN(c', \overline{\mathcal{R}})$ then the above property holds, thus $c \in PNN(c', \overline{\mathcal{R}})$.

Lemma 3. *Suppose that $C_1 \wedge ... \wedge C_k$ be a CNF representation for an n-ary relation $\mathcal{R}$ and c be a case. Then, $|PNN(c, \overline{\mathcal{R}})| \leq MP(\mathcal{A})^n \cdot k$.*

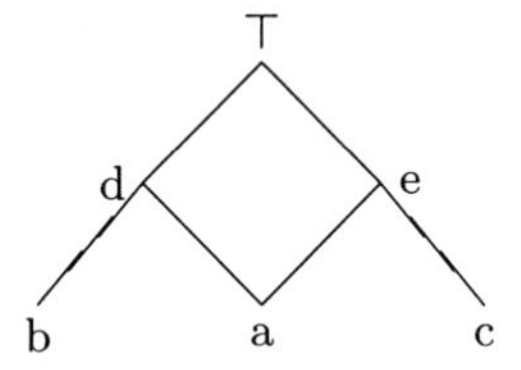

Fig. 1. A lattice whose n-ary relation is difficult to find

Corollary 1. *Let $\mathcal{R}$ be an n-ary relation which has a CNF representation $C_1 \wedge \ldots \wedge C_k$ and c be a case. $|NN(c, \overline{\mathcal{R}})| \leq MP(\mathcal{A})^n \cdot k$. Especially, $|NN(c, \overline{\mathcal{R}})| \leq MP(\mathcal{A})^n \cdot |CNF(\mathcal{R})|$.*

Note the above lemma and corollary are different from [Satoh00b]. To illustrate the difference, consider a semi-lattice in Figure 1 which consists of six elements $\{a, b, c, d, e, \top\}$. Note that $MP(\mathcal{A}) = 2$. Suppose we have an n-ary relation $(x_1 \preceq a) \wedge (x_2 \preceq a) \wedge \ldots \wedge (x_n \preceq a)$. To represent this relation, we need one positive case and 2^n negative cases.

By Lemma 2, Proposition 3 and Corollary 1, we have the following theorem which gives an upper bound of representability of n-ary relations.

Theorem 1. *Let $\mathcal{R}$ be an n-ary relation over attributes. Then, there exists a casebase $\mathcal{CB} = \langle \mathcal{CB}^+, \mathcal{CB}^- \rangle$ such that $\mathcal{R}_{\mathcal{CB}} = \mathcal{R}$, $|\mathcal{CB}^+| \leq |DNF(\mathcal{R})|$, $|\mathcal{CB}^-| \leq MP(\mathcal{A})^n \cdot |DNF(\mathcal{R})| \cdot |CNF(\mathcal{R})|$ and $|\mathcal{CB}| \leq |DNF(\mathcal{R})|(1 + MP(\mathcal{A})^n \cdot |CNF(\mathcal{R})|)$.*

The above theorem means that if we would like to represent a relation over lattice-based attributes, in some cases, we need an exponential number of negative cases. If we restrict attributes as tree-based attributes or boolean-valued attributes, then $MP(\mathcal{A})$ becomes 1 and the number of negative cases depends only on $CNF(\mathcal{R})$ and $DNF(\mathcal{R})$ [Satoh98,Satoh00a].

4 Approximating Critical Casebase

We firstly give a definition of a *critical casebase.*

Definition 5. *Let $\mathcal{R}$ be an n-ary relation and $\mathcal{CB}$ be a casebase $\langle \mathcal{CB}^+, \mathcal{CB}^- \rangle$. $\mathcal{CB}$ is* critical *w.r.t. $\mathcal{R}$ if $\mathcal{CB}$ satisfies the following conditions:*

- $\mathcal{R} = \mathcal{R}_{\mathcal{CB}}$
- *There is no casebase $\mathcal{CB}' = \langle \mathcal{CB}'^+, \mathcal{CB}'^- \rangle$ such that $\mathcal{R} = \mathcal{R}_{\mathcal{CB}'}$ and $\mathcal{CB}'^+ \subseteq \mathcal{CB}^+$ and $\mathcal{CB}'^- \subseteq \mathcal{CB}^-$ and $\mathcal{CB}' \neq \mathcal{CB}$.*

The following results(Theorem 2 and Lemma 4) are related with a minimal set of negative cases and positive cases. These results are actually in parallel with the results in [Satoh00b].

Definition 6. *Let $\mathcal{R}$ be an n-ary relation and $\mathcal{CB}$ be a casebase $\langle \mathcal{CB}^+, \mathcal{CB}^- \rangle$ such that $\mathcal{R}_{\mathcal{CB}} = \mathcal{R}$. $\mathcal{CB}^-$* is a set of minimal negative cases w.r.t. $\mathcal{CB}^+$ and $\mathcal{R}$ *if there is no casebase $\mathcal{CB}' = \langle \mathcal{CB}^+, \mathcal{CB}'^- \rangle$ such that $\mathcal{CB}'^- \subset \mathcal{CB}^-$ and $\mathcal{R}_{\mathcal{CB}'} = \mathcal{R}$.*

The following theorem concerns about necessary and sufficient condition of a set of minimal negative cases given $\mathcal{CB}^+$ and $\mathcal{R}$.

Theorem 2. *Let $\mathcal{R}$ be an n-ary relation and $\mathcal{CB}$ be a casebase $\langle \mathcal{CB}^+, \mathcal{CB}^- \rangle$ such that $\mathcal{R}_{\mathcal{CB}} = \mathcal{R}$. $\mathcal{CB}^-$ is a set of minimal negative cases w.r.t. $\mathcal{CB}^+$ and $\mathcal{R}$ if and only if $\mathcal{CB}^- = \bigcup_{c_{ok} \in \mathcal{CB}^+} NN(c_{ok}, \overline{\mathcal{R}})$.*

The above theorem intuitively means that if $\mathcal{CB}^+$ and a set of negative case $\mathcal{CB}'^-$ represents a relation $\mathcal{R}$, we can reduce $\mathcal{CB}^-$ down to $\bigcup_{c_{ok} \in \mathcal{CB}^+} NN(c_{ok}, \overline{\mathcal{CB}^-})$.

Definition 7. *Let $\mathcal{R}$ be an n-ary relation and $\mathcal{CB}$ be a casebase $\langle \mathcal{CB}^+, \mathcal{CB}^- \rangle$ such that $\mathcal{R}_{\mathcal{CB}} = \mathcal{R}$. $\mathcal{CB}^+$* is a set of minimal positive cases w.r.t. $\mathcal{R}$ *if there is no casebase $\mathcal{CB}' = \langle \mathcal{CB}'^+, \mathcal{CB}'^- \rangle$ such that $\mathcal{CB}'^+ \subset \mathcal{CB}^+$ and $\mathcal{CB}'^-$ is any arbitrary set of negative cases and $\mathcal{R}_{\mathcal{CB}'} = \mathcal{R}$.*

The following lemma shows a sufficient condition on a set of minimal positive cases.

Lemma 4. *Let $\mathcal{R}$ be an n-ary relation and $\mathcal{CB}$ be a casebase $\langle \mathcal{CB}^+, \mathcal{CB}^- \rangle$ such that $\mathcal{R}_{\mathcal{CB}} = \mathcal{R}$. Suppose for every $c_{ok} \in \mathcal{CB}^+$, $c_{ok} \notin \mathcal{R}_{\langle \mathcal{CB}^+ - \{c_{ok}\}, \overline{\mathcal{R}} \rangle}$. Then, $\mathcal{CB}^+$ is a set of minimal positive cases w.r.t. $\mathcal{R}$.*

Now, we propose an approximation method of discovering a critical casebase. In order to do that, we assume that there is a probability distribution $\mathcal{P}$ over $(min(\mathcal{A}))^n$. We would like to have a casebase such that the probability that the casebase produces more errors than we expect is very low.

The algorithm in Fig. 2 performs such an approximation. The algorithm is a simplified version of the algorithm in [Satoh00b]. Intuitively, in the algorithm we try to find counter examples by sampling and if enough sampling is made with no counter examples, we are done. If we find a positive counter example then we add it to $\mathcal{CB}^+$ and if we find a negative counter example then we try to find a "nearest" negative case to a positive case from the found negative counter example.

In the algorithm, $c \in \mathcal{R}?$ expresses a label whether $c \in \mathcal{R}$ or not. If $c \in \mathcal{R}$ then the label is "yes" and otherwise "no".

The following lemma gives an upper bound for a number of positive counter cases.

Lemma 5. *Let $\mathcal{R}$ be an n-ary relation and $D_1 \vee ... \vee D_{|DNF(\mathcal{R})|}$ be a DNF representation with a minimal size $|DNF(\mathcal{R})|$ of $\mathcal{R}$. Suppose that the situation that $c \in \mathcal{R}$ and $c \notin \mathcal{R}_{\mathcal{CB}}$ occurs during the execution of* **FindCCB**(δ, ϵ)*. Then, c does not satisfy any D_k $(1 \leq k \leq |DNF(\mathcal{R})|)$ which some case $c_{ok} \in \mathcal{CB}^+$ satisfies. This situation happens at most $|DNF(\mathcal{R})|$ times.*

The following lemma gives an upper bound for a number of negative counter cases.

Lemma 6. *Let $\mathcal{R}$ be an n-ary relation. Suppose that the situation that $c \notin \mathcal{R}$ and $c \in \mathcal{R}_{\langle \{c_{ok}\}, \mathcal{CB}^- \rangle}$ occurs for some $c_{ok} \in \mathcal{CB}^+$ during the execution of*

FindCCB(δ, ϵ)
begin
$\mathcal{CB}^+ := \emptyset$ and $\mathcal{CB}^- := \emptyset$ and $m := 0$

1. c is taken from $\mathcal{A}^n$ according to the probability distribution $\mathcal{P}$ and get $\langle c, c \in \mathcal{R}? \rangle$ as an oracle.
2. If $c \in \mathcal{R}$ and $c \notin \mathcal{R}_{\langle \mathcal{CB}+, \mathcal{CB}- \rangle}$, then
   (a) $\mathcal{CB}^+ := \mathcal{CB}^+ \cup \{c\}$
   (b) $m := 0$ and Goto 1.
3. If $c \notin \mathcal{R}$ and $c \in \mathcal{R}_{\langle \mathcal{CB}+, \mathcal{CB}- \rangle}$, then for every c_{ok} s.t. $c \in \mathcal{R}_{\langle \{c_{ok}\}, \mathcal{CB}^- \rangle}$,
   (a) $c_{pmin} :=$ **pminNG**(c, c_{ok})
   (b) $\mathcal{CB}^- := \mathcal{CB}^- \cup \{c_{pmin}\}$
   (c) $m := 0$ and Goto 1
4. $m := m + 1$
5. If $m >= \frac{1}{\epsilon} \ln \frac{1}{\delta}$ then
   output $\mathcal{CB}^+$ and $\bigcup_{c_{ok} \in \mathcal{CB}^+} NN(c_{ok}, \mathcal{CB}^-)$
   else Goto 1.

end

pminNG(c, c_{ok})
begin

1. For every $1 \leq l \leq n$ s.t. $c[l] \neq c_{ok}[l]$ and for every $c' \in c\downarrow^l_{c_{ok}}$,
2. Make a membership query for c'.
3. If $c' \notin \mathcal{R}$ then $c := c'$ and Goto 1.
4. output c'. /* $c' \in PNN(c_{ok}, \overline{\mathcal{R}})$ */

end

Fig. 2. Approximating a critical casebase

FindCCB(δ, ϵ). *Then, there exists some* $c' \in PNN(c_{ok}, \overline{\mathcal{R}})$ *such that* $c' \vee c_{ok} \preceq c \vee c_{ok}$ *and* $c' \notin \mathcal{CB}^-$. *This situation happens at most* $MP(\mathcal{A})^n \cdot |CNF(\mathcal{R})|$ *times for each* $c_{ok} \in \mathcal{CB}^+$.

By the above two lemmas, an upper bound for a number of negative counter cases is $|DNF(\mathcal{R})| \cdot MP(\mathcal{A})^n \cdot |CNF(\mathcal{R})|$.

Let $\mathcal{R}_1 \Delta \mathcal{R}_2$ be a difference set between $\mathcal{R}_1$ and $\mathcal{R}_2$ (that is, $(\overline{\mathcal{R}_1} \cap \mathcal{R}_2) \cup (\mathcal{R}_1 \cap \overline{\mathcal{R}_2})$).

The following theorem shows computational complexity of approximating a critical casebase.

Theorem 3. *Let* $\mathcal{R}$ *be an n-ary relation over* $\mathcal{A}$. *The above algorithm stops after taking at most* $(\frac{1}{\epsilon} \ln \frac{1}{\delta}) \cdot |DNF(\mathcal{R})| \cdot (1 + MP(\mathcal{A})^n \cdot |CNF(\mathcal{R})|)$ *cases according to* $\mathcal{P}$ *and asking at most* $n^2 \cdot (|min(\mathcal{A})| - 1) \cdot (length(\mathcal{A}) - 1) \cdot |DNF(\mathcal{R})| \cdot MP(\mathcal{A})^n \cdot |CNF(\mathcal{R})|$ *membership queries and produces* $\mathcal{CB}$ *with the probability at most* δ *such that* $\mathcal{P}(\mathcal{R} \Delta \mathcal{R}_{\mathcal{CB}}) \geq \epsilon$.

The next theorem shows that output from **FindCCB**(δ, ϵ) is an approximation of a critical casebase.

Theorem 4. *Let* $\mathcal{CB}$ *be an output from* **FindCCB**(δ, ϵ). *If* $\mathcal{R}_{\mathcal{CB}} = \mathcal{R}$, $\mathcal{CB}$ *is a critical casebase w.r.t.* $\mathcal{R}$.

For a boolean valued relation, the lattice becomes simplified so that $MP(\mathcal{A}) = 1$ and $min(\mathcal{A}) = 2$ and $length(\mathcal{A}) = 2$. For tree-structure valued relation, the lattice becomes simplified again so that $MP(\mathcal{A}) = 1$. Therefore, we do not need exponentially many cases and membership queries to approximate a give relation in these domains.

5 Conclusion

The contributions of this paper are as follows.

1. We show that for every relation $\mathcal{R}$ for a semi-lattice structure $\mathcal{A}$, in order to represent $\mathcal{R}$, an upper bound of necessary positive cases is $|DNF(\mathcal{R})|$ and the upper bound of necessary negative cases is $MP(\mathcal{A})^n \cdot |DNF(\mathcal{R})| \cdot |CNF(\mathcal{R})|$.
2. We give a learning method of a critical casebase and we analyze computational complexity of the method in the PAC learning framework and show that the sample size of cases is at most $(\frac{1}{\epsilon} \ln \frac{1}{\delta}) \cdot |DNF(\mathcal{R})| \cdot (1 + MP(\mathcal{A})^n \cdot |CNF(\mathcal{R})|)$ and necessary number of membership queries is at most $n^2 \cdot (min(\mathcal{A}) - 1) \cdot (length(\mathcal{A}) - 1) \cdot |DNF(\mathcal{R})| \cdot MP(\mathcal{A})^n \cdot |CNF(\mathcal{R})|$

We would like to pursue the following future work.

1. We would like to extend our language to include negations and extend our method to learn a formula in an extended language.
2. We would like to find a class of relations for lattice-based whose critical casebase can be easily approximated.
3. We would like to apply this method to real domain such as legal reasoning.

References

[Ashley90] Ashley, K. D.: *Modeling Legal Argument: Reasoning with Cases and Hypotheticals* MIT press (1990)

[Ashley94] Ashley, K. D., and Aleven, V.: A Logical Representation for Relevance Criteria. S. Wess, K-D. Althoff and M. Richter (eds.) *Topics in Case-Based Reasoning, LNAI 837* (1994) 338–352

[Bshouty93] Bshouty, N. H.: Exact Learning Boolean Functions via the Monotone Theory. *Information and Computation* **123** (1995) 146–153

[Matuschek97] Matuschek, D., and Jantke, K. P.: Axiomatic Characterizations of Structural Similarity for Case-Based Reasoning. *Proc. of Florida AI Research Symposium (FLAIRS-97)* (1997) 432–436

[Khardon96] Khardon, R., and Roth, D.: Reasoning with Models. *Artificial Intelligence* **87** (1996) 187–213

[Osborne96] Osborne, H. R., and Bridge, D. G.: A Case Base Similarity Framework. *Advances in Case-Based Reasoning, LNAI 1168* (1996) 309–323

[Satoh98] Satoh, K.: Analysis of Case-Based Representability of Boolean Functions by Monotone Theory. *Proceedings of ALT'98* (1998) 179–190

[Satoh00a] Satoh, K., and Nakagawa, R.: Discovering Critical Cases in Case-Based Reasoning (Extended Abstract), *Online Proceedings of 6th Symposium on AI and Math*, http://rutcor.rutgers.edu/ amai/AcceptedCont.htm (2000)

[Satoh00b] Satoh, K.: Learning Taxonomic Relation by Case-based Reasoning. *Proceedings of ALT'00* (2000) 179–193

On Dimension Reduction Mappings for Approximate Retrieval of Multi-dimensional Data

Takeshi Shinohara and Hiroki Ishizaka

Department of Artificial Intelligence, Kyushu Institute of Technology,
Iizuka, 820-8502, Japan
shino@ai.kyutech.ac.jp
ishizaka@ai.kyutech.ac.jp

Abstract. Approximate retrieval of multi-dimensional data, such as documents, digital images, and audio clips, is a method to get objects within some dissimilarity from a given object. We assume a metric space containing objects, where distance is used to measure dissimilarity. In Euclidean metric spaces, approximate retrieval is easily and efficiently realized by a spatial indexing/access method R-tree. First, we consider objects in discrete L_1 (or Manhattan distance) metric space, and present embedding method into Euclidean space for them. Then, we propose a projection mapping H-Map to reduce dimensionality of multi-dimensional data, which can be applied to any metric space such as L_1 or L_∞ metric space, as well as Euclidean space. H-Map does not require coordinates of data unlike K-L transformation. H-Map has an advantage in using spatial indexing such as R-tree because it is a continuous mapping from a metric space to an L_∞ metric space, where a hyper-sphere is a hyper-cube in the usual sense. Finally we show that the distance function itself, which is simpler than H-Map, can be used as a dimension reduction mapping for any metric space.

1 Introduction

Information retrieval can be considered as a process of learning or discovery. In the very early stage of problem solving, we often use past experiences rather than general rules. Case based reasoning seems to be a model based on such intuition. In order to be used as an effective reasoning tool, however, *efficient* retrieval from *large number* of cases is important. For example, the quality or usefulness of document retrieval systems depends on the number of documents, as well as query processing mechanisms. B-tree and its equivalents can supply very efficient indexing for large number of data, as long as a linear order is defined. On the other hand, for multi-dimensional data such as digital images, no useful linear order is defined, and therefore, there exists difficulty in realizing efficient indexing mechanism.

In the present paper, we consider spatial retrieval that extracts objects within given query range in a metric space. Spatial indexing structures, such as R-tree [1, 4], can be used to retrieve multi-dimensional data. R-tree assumes that

S. Arikawa and A. Shinohara (Eds.): Progress in Discovery Science 2001, LNAI 2281, pp. 224–231.

objects are associated with coordinate values. Even if we adopt R-tree, relatively lower dimensionality is required to keep its efficiency [5]. Thus, the method of mapping multi-dimensional data into space of lower dimension is one of the most important subjects in multimedia database, as well as in pattern recognition. In statistical pattern recognition, extracting a feature defined by a combination of data attributes plays a central role and one of the central issues is to find a feature that can extract information as much as possible. In information retrieval of multi-dimensional data, to realize efficient spatial indexing it is necessary to project data into space of lower dimension. To take fully advantage of such an indexing structure as R-tree, it is important to keep information as much as possible in projected data.

Karhunen-Loéve (K-L for short) transformation has been used as dimension reduction mapping in various problems [3]. It is essentially the same as principal component analysis in multivariate analysis. In K-L transformation, data is assumed to be in a Euclidean metric space with coordinate values, and a feature is obtained as a linear combination of coordinate values, which can be considered as an orthogonal projection in the Euclidean space.

However, in applications such as multimedia database, Euclidean distance is not always assumed. In the present paper, we investigate non-Euclidean distance, such as L_1 metric (or Manhattan distance), and edit distance for character strings.

In Section 3, we prove that objects in discrete L_1 (or Manhattan distance) metric space can be embedded into vertices of a unit hyper-cube when the square root of L_1 distance is used as the distance. To take fully advantage of R-tree spatial indexing, we have to project objects into space of relatively lower dimension. We adopt FastMap [2] by Faloutsos and Lin to reduce the dimension of object space. The range corresponding to a query (q, h) for retrieving objects within distance h from an object q is naturally considered as a hyper-sphere even after FastMap projection, which is an orthogonal projection in Euclidean space. However, it is turned out that the query range is contracted into a smaller hyper-box than the hyper-sphere by applying FastMap to objects embedded in the above mentioned way.

In Section 4, we propose a dimension reduction mapping, named H-Map. H-Map is based on a simple mapping defined as $\varphi_{ab}(x) = \frac{D(a,x)-D(b,x)}{2}$, where a and b are arbitrary points in the space, $D(x, a)$ is the distance between x and a. The pair of a and b is called a pivot of the mapping. H-Map can be applied to arbitrary metric space. It requires only distances between data and no coordinate values. We can compare H-Map with K-L transformation for several sample data sets and show its effectiveness as dimension reduction mapping.

FastMap [2] consists of mapping x to $\frac{D(a,b)^2+D(a,x)^2-D(b,x)^2}{2D(a,b)}$ using two points a and b. Taking embedding of L_1 and query range contraction into account, we have the form of H-Map. In Section 5, we derive the simplest mapping $\varphi_a(x) = D(a, x)$ taking one point a as the pivot. This simplest mapping can give a larger variance of images than H-Map.

2 Preliminaries

Here, we give several notions and definitions for approximate retrieval. We assume that objects to be retrieved are given in a metric space X with distance function D. We denote the sets of real numbers by $\mathcal{R}$.

2.1 Metric Space

A *distance function* $D : X \times X \to \mathcal{R}$ of a metric space X satisfies the following *axioms of distance* for any x, y, z in X:

(1) $D(x, y) \geq 0$ ($D(x, y) = 0$ if and only if $x = y$),

(2) $D(x, y) = D(y, x)$, and

(3) $D(x, z) \leq D(x, y) + D(y, z)$.

The condition (3) is also referred as *triangle inequality.*

When $X = \mathcal{R}^n$, a point $x \in X$ is represented by an n-tuple ($x_{(1)}$, $x_{(2)}$, ..., $x_{(n)}$). The following three functions satisfy the axioms of distance:

$$L_2 \text{ (Euclidean) metric: } D(x, y) = \sqrt{\sum_{i=1}^{n} (x_{(i)} - y_{(i)})^2}$$

$$L_1 \text{ metric: } D(x, y) = \sum_{i=1}^{n} \left| x_{(i)} - y_{(i)} \right|$$

$$L_\infty \text{ metric: } D(x, y) = \max_{i=1}^{n} \left| x_{(i)} - y_{(i)} \right|$$

2.2 Approximate Retrieval

Let $S \subset X$ be a finite set of objects. A *query* is given as a pair (q, h) of an object $q \in S$ and a real number h. The *answer* $Ans(q, h)$ to a query (q, h) is the set of objects within distance h from Q, that is,

$$Ans(q, h) = \{x \in S \mid D(q, x) \leq h\}.$$

The above setting of approximate retrieval as $Ans(q, h)$ is very natural and general. When D is Euclidean, most spatial indexing structures are almost directly used to realize approximate retrieval. In many cases, however, unless objects are inherently geometrical like map information, object space is not Euclidean.

2.3 Dimension Reduction Mapping

A *dimension reduction mapping* is a mapping from some space to a space of lower dimension. In some case, such as the set of all the finite strings with edit distance, the dimensionality is not well-defined. Even in such cases, we call a mapping to a space of finite dimension a dimension reduction mapping.

A dimension reduction mapping $f : X \to Y$ is *continuous* if $D'(f(x), f(y)) \leq D(x, y)$ for any x, $y \in X$, where D and D' are distance function of X and Y, respectively.

Let $S \subseteq X$ be a set of objects, $f : X \to Y$ be a continuous dimension reduction Mapping. Then, clearly,

$$Ans(q, h) \subseteq \{x \in S \mid D'(f(q), f(x)) \leq h\}.$$

Thus, we can retrieve all necessary objects by retrieving objects mapped by continuous dimension reduction mapping. The results of approximate retrieval may include irrelevant objects to the query, when a dimension reduction mapping is used. To get exact answer, screening might be needed.

3 Dimension Reduction for Discrete L_1 Metric Space

In this section, we consider discrete L_1 metric (or, Manhattan distance) spaces. Many difference measures might be captured as a discrete L_1 distance. For example, a natural definition of distance between objects consisting of several attribute values may be the sum of the symmetric differences between each attribute values. This definition can be applied to many sort of objects, such as, documents, digital images, and game boards.

We adopt R-tree [1, 4] as a spatial indexing/access method. As Otterman pointed out [5], R-tree can efficiently be used only for relatively low-dimensional objects. Therefore, we have to map high-dimensional objects into a subspace of lower dimension. We can use the FastMap method by Faloutsos and Lin [2] as a dimension reduction mapping. Since FastMap is based on orthogonal projection in Euclidean space, we have to embed objects into a Euclidean space. However, L_1 distance cannot be embedded into any Euclidean space, in general. If we take the square root of L_1 distance as the distance, the objects can be embedded into a Euclidean space. In other words, if we define

$$D^{\frac{1}{2}}(x, y) = \sqrt{D(x, y)},$$

$(S, D^{\frac{1}{2}})$ can be embedded into a Euclidean space. If we appropriately map objects to vertices of unit n_0-cube, then the Euclidean distance between vertices coincides with the square root of the L_1 distance between objects.

Here, we briefly explain the FastMap method. Consider a set S of objects in a Euclidean space, where D is a Euclidean distance function. Let take arbitrarily a pair (a, b) of objects, which is called a *pivot*. The first coordinate $f_{ab}(x)$ of an object x is given by

$$f_{ab}(x) = \overline{aE} = \frac{D(a, b)^2 + D(a, x)^2 - D(b, x)^2}{2D(a, b)}$$

where E is the image of x by the orthogonal projection to the straight line ab. Here, we should note that distances between objects are enough to calculate

$f_{ab}(x)$ and any coordinates of objects are not necessary. Let x' be the image of x by the orthogonal projection to the hyper-plane that is orthogonal to the straight line ab. The distance between x' and y' is given by

$$D(x', y')^2 = D(x, y)^2 - (f_{ab}(x) - f_{ab}(y))^2.$$

Thus, we can repeatedly apply the above projection to get the second and other coordinates of objects. By using k pivots, FastMap gives a projection $p : X \to \mathcal{R}^k$. One of the most important issues in applying FastMap is how to select pivots. Intuitively, the better pivot should provide the more selectivity in retrieval. Details are discussed by Faloutsos and Lin [2].

Let f be an orthogonal projection to $\mathcal{R}^k$ by FastMap, where D is the original distance function and $D^{\frac{1}{2}}$ is used as the distance function in applying FastMap. Since f is an orthogonal projection, and therefore, a continuous dimension reduction mapping, the distance between images of objects in the index space is not larger than the square root of the distance between objects, that is, $d(f(x), f(y)) \le D^{\frac{1}{2}}(x, y)$. For a query (q, h), we have

$$\{x \in S \mid d(f(q), f(x)) \le \sqrt{h}\} \supseteq Ans(q, h).$$

Therefore, we can retrieve all the necessary objects even after reducing dimension by FastMap. Such a retrieval is easily realized by using spatial access method like R-tree.

From the experiments of our method, we observed that the image of the query range by the dimension reduction mapping, which is naturally considered as a k-sphere with radius $h^{\frac{1}{2}}$, is *too large* to get all the necessary objects. Precisely, we can prove

$$\{x \in S \mid |f^{(i)}(q) - f^{(i)}(x)| \le \lambda_i h \text{ for all } i = 1, \ldots, k\} \supseteq Ans(q, h),$$

where $f^{(i)}(x)$ is the i-th coordinate of the image of x by f and λ_i is a constant which is usually much smaller than 1. Thus, the query range of k-box, which is usually smaller than the k-sphere, is enough to retrieve the correct answer. This phenomenon, which is derived from the combination of our object embedding into unit n_0-cube and FastMap, will be theoretically explained as the contraction of query range by FastMap in [7].

More details about the proposed method can be found in [7] with experimental results on approximate retrieval of Japanese Shogi boards.

4 Dimension Reduction Mappings: H-Map

In this section, we introduce a dimension reduction mapping H-Map, which is applied to any metric space.

For any a and b in a space X with distance function D, we define a mapping $\varphi_{ab} : X \to \mathcal{R}$ as follows:

$$\varphi_{ab}(x) = \frac{D(a, x) - D(b, x)}{2}.$$

The pair (a, b) is called a *pivot*. For an k-tuple of pivots $\Pi = ((a_1, b_1), \ldots, (a_k, b_k))$, we define a dimension reduction mapping $\Phi_\Pi : X \to \mathcal{R}^k$ from X to k-dimensional space $\mathcal{R}^k$ by

$$\Phi_\Pi(x) = (\varphi_{a_1 b_1}(x), \varphi_{a_2 b_2}(x), \ldots, \varphi_{a_k b_k}(x)).$$

We call the mapping $\Phi_\Pi : X \to \mathcal{R}^k$ an *H-Map*. We can show that Φ_Π is a continuous mapping from a metric space to an L_∞ metric space, that is,

$$D'(\Phi_\Pi(x), \Phi_\Pi(y)) \leq D(x, y),$$

for any tuple of pivots Π, where D' is the L_∞ distance function in $\mathcal{R}^k$.

H-Map has the following properties:

(1) H-Map can be applied to any metric space such as L_1 or L_∞ metric space, as well as Euclidean metric space. Here, we should note that the continuousness of H-Map depends only on the triangle inequality.
(2) H-Map can be defined only from distances without coordinate values.

A dimension reduction mapping can be considered as a feature extraction function in pattern recognition. By using K-L transformation, we can obtain an optimal linear mapping from $\mathcal{R}^n$ to $\mathcal{R}^k$ which minimizes the square mean error after mapping. The method is essentially the same as the method used in principal component analysis. A mapping obtained from K-L transformation is an orthogonal projection in Euclidean space. Usually, K-L transformation assumes that the metric is Euclidean. From the viewpoint of computation algorithm, K-L transformation requires coordinate values of data, while FastMap and H-Map does not.

Here we should note that minimization of the square mean error after mapping gives maximization of the variance of mapped images. The larger variance of images means that more information of data is kept after mapping and higer efficiency of approximate retrieval via dimension reduction mapping. By using the variance of images as a criteria for good dimension reduction mapping, we can compare methods with each other and observe that H-Map gives larger variances than K-L transformation when the pivot is appropriately chosen. Such comparisons and more details about H-Map, see [6].

5 Simple Dimension Reduction Mapping: S-Map

In the previous section, we introduced a dimension reduction mapping H-Map, which is based on a mapping $\varphi_{ab}(x) = \frac{D(x,a) - D(x,b)}{2}$. In this section we consider its variant with two parameters α, β satisfying $|\alpha| + |\beta| = 1$. We generalize φ_{ab} as

$$\varphi_{ab}(x) = \alpha D(x, a) + \beta D(x, b).$$

Then, we can prove the continuousness of φ_{ab}, that is, we have

$$|\varphi_{ab}(x) - \varphi_{ab}(y)| \leq D(x, y).$$

We can consider an optimization problem for the generalized H-Map. As explained in the previous section, we use the variance of images as the criteria for goodness of dimension reduction. Thus, we have to determine parameters α and β that maximizes the variance of images by φ_{ab}.

Let denote the variances of $\varphi_{ab}(x)$, $D(a,x)$, and $D(b,x)$ by V_φ, V_a, and V_b, respectively, and covariance of $D(a,x)$ and $D(b,x)$ by V_{ab}. Then,

$$V_\varphi = \alpha^2 V_a + 2\alpha\beta V_{ab} + \beta^2 V_b.$$

When V_φ is maximum, $\alpha\beta V_{ab} \geq 0$. Assume $\alpha \geq 0$ and $V_{ab} \geq 0$. In this case, $\beta = 1 - \alpha$. Therefore,

$$V_\varphi = \alpha^2 (V_a - 2V_{ab} + V_b) + 2\alpha(V_{ab} - V_b) + V_b.$$

This can be considered as quadratic expression on α. Since $(V_a - 2V_{ab} + V_b) \geq 0$, V_φ takes the maximum value if $\alpha = 1$ or $\alpha = 0$. We can show the other case in a similar way.

Although the generalized H-Map with parameters α and β is a continuous dimension reduction mapping, we do not need to take two points a and b. Thus, finally, it turns out that a simple mapping using one point as a pivot defined below is enough.

$$\varphi_a(x) = D(x,a).$$

This simplest mapping φ_a can be used as basis for dimension reduction mapping. We call the dimension reduction mapping consisting of φ_a *S-Map*. S-Map inherits almost all the properties of H-Map. Pivots for FastMap and H-Map are pairs of points, while pivot for φ_a is just one point. Naive algorithms of pivot selection for FastMap and H-Map need quadratic running time with respect to the number of objects. Pivot selection for φ_a seems to be much easier than those for FastMap and H-Map.

Here, we give an outline of construction for dimension reduction mappings, FastMap, H-Map, and S-Map. Let S be the set of objects to be mapped and m be the dimension of projected data. First, we select the set C of pivots candidates. Then, starting with the empty set of pivots, we select m pivots that maximize the appropriateness of the mapping. A possible set of the pivot candidates for FastMap and H-Map is the set $S \times S$ of all pairs. On the contrary, the set S it self is enough for the candidates for S-Map. Thus, S-Map seems to be much better than H-Map from the viewpoint of not only dimension reduction but also computation.

6 Concluding Remarks

We started our study on dimension reduction mapping with embedding objects in L_1 metric space into a Euclidean space in order to apply FastMap. From observation on contraction of query range by FastMap, we introduced H-Map that can be applied to any metric space. Finally, we found that distance function itself can be used as a dimension reduction mapping. As one of the most important future subjects should be pivot selection for the S-Map. Since the metric of images of S-Map and H-Map is L_∞, other criteria than variance might be needed.

References

1. Beckmann, N., Kriegal, H.P., Schneider, R. and Seeger, B.: The R*-tree: An Efficient and Robust Access Method for Points and Rectangles. In Proc. ACM SIGMOD International Conference on Management of Data, 19(2):322–331, 1990.
2. Faloutsos, C., Lin, K.I.: FastMap: A Fast Algorithm for Indexing, Data-Mining and Visualization of Traditional and Multimedia Datasets. In Proc. ACM SIGMOD International Conference on Management of Data, **24** (2) (1995) 163–174
3. Fukunaga, K.: Statistical Pattern Recognition. Second Edition, Academic Press (1990)
4. Guttman, A.: R-tree: A Dynamic Index Structure for Spatial Searching. In Proc. ACM SIGMOD, (1984) 47–57
5. Otterman, M.: Approximate Matching with High Dimensionality R-trees. M. Sc. Scholarly paper, Dept. of Computer Science, Univ. of Maryland, 1992
6. Shinohara, T., Chen, J., Ishizaka, H.: H-Map: A Dimension Reduction Mapping for Approximate Retrieval of Multi-Dimensional Data. In Proc. The Second International Conference on Discovery Science, LNAI 1721 (1999) 299–305
7. Shinohara, T., An, J., Ishizaka, H.: Approximate Retrieval of High-Dimensional Data with L_1 metric by Spatial Indexing. New Generation Computing, **18** (2000) 49–47

Rule Discovery from fMRI Brain Images by Logical Regression Analysis

Hiroshi Tsukimoto[1], Mitsuru Kakimoto[2], Chie Morita[2], and Yoshiaki Kikuchi[3]

[1] Tokyo Denki University, 2-2, Kanda-Nishiki-cho, Chiyoda-ku, Tokyo 101-8457 Japan
tsukimoto@c.dendai.ac.jp
[2] Corporate Research & Development Center, Toshiba Corporation,1, Komukai Toshiba-cho, Saiwai-ku, Kawasaki 212-8582 Japan
{mitsuru.kakimoto, chie.morita}@toshiba.co.jp
[3] Tokyo Metropolitan University of Health Sciences, Higashi-ogu 7-2-10, Arakawa-ku, Tokyo 116-8551 Japan
ykikuchi@post.metro-hs.ac.jp

Abstract. This paper presents rule discovery from fMRI brain images. The algorithm for the discovery is the Logical Regression Analysis, which consists of two steps. The first step is regression analysis. The second step is rule extraction from the regression formula obtained by the regression analysis. In this paper, we use nonparametric regression analysis as a regression analysis, since there are not sufficient data in rule discovery from fMRI brain images. The algorithm was applied to several experimental tasks such as finger tapping and calculation. This paper reports the experiment of calculation, which has rediscovered well-known facts and discovered new facts.

1 Introduction

Analysis of brain functions using functional magnetic resonance imaging(f-MRI), positron emission tomography(PET), magnetoencephalography(MEG) and so on is called non-invasive analysis of brain functions[4]. As a result of the ongoing development of non-invasive analysis of brain function, detailed functional brain images can be obtained, from which the relations between brain areas and brain functions can be understood, for example, the relation between a subarea and another subarea in the motor area and a finger movement.

Several brain areas are responsible for a brain function. Some of them are connected in series, and others are connected in parallel. Brain areas connected in series are described by "AND" and brain areas connected in parallel are described by "OR". Therefore, the relations between brain areas and brain functions are described by rules.

Researchers are trying to heuristically discover the rules from functional brain images. Several statistical methods, for example, principal component analysis, have been developed. However, the statistical methods can only present some principal areas for a brain function. They cannot discover rules. This paper presents an algorithm for the discovery of rules from fMRI brain images.

S. Arikawa and A. Shinohara (Eds.): Progress in Discovery Science 2001, LNAI 2281, pp. 232–245.

fMRI brain images can be dealt with by supervised inductive learning. However, the conventional inductive learning algorithms[5] do not work well for fMRI brain images, because there are strong correlations between attributes(pixels) and a small number of samples.

There are two solutions for the above two problems. The first one is the modification of the conventional inductive learning algorithms. The other one is nonparametric regression. The modification of the conventional inductive learning algorithms would require a lot of effort. On the other hand, nonparametric regression has been developed for the above two problems. We use nonparametric regression for the rule discovery from fMRI brain images. The outputs of nonparametric regression are linear formulas, which are not rules. However, we have already developed a rule extraction algorithm from regression formulas[9],[10],[12].

The algorithm for rule discovery from fMRI brain images consists of two steps. The first step is nonparametric regression. The second step is rule extraction from the linear formula obtained by the nonparametric regression. The method is a Logical Regression Analysis(LRA), that is, a rule discovery algorithm consisting of regression analysis[1] and rule extraction from the regression formulas.

Since brains are three dimensional, three dimensional LRA is appropriate. However, the three dimensional LRA needs a huge computation time, for example, many years. Therefore, we applied two dimensional LRA to f-MRI images as the first step.

We applied the algorithm to artificial data, and we confirmed that the algorithm works well for artificial data[11]. We applied the algorithm to several experimental tasks such as finger tappings and calculations. In the experiments of finger tapping, we compared the results of LRA with z-score[6], which is the typical conventional method. In the experiments, LRA could rediscover a little complicated relation, but z-score could not rediscover the relation. As the result, we confirmed that LRA works better than z-score, which cannot be described in this paper due to space limitations. In the experiments of calculations, we confirmed that LRA worked well, that is, rediscovered well-known facts regarding calculations, and discovered new facts regarding calculations. This paper reports the experiments of calculations.

Section 2 overviews the rule discovery from fMRI images by Logical Regression Analysis. Section 3 briefly explains nonparametric regression analysis. Section 4 briefly explains rule extraction. Section 5 describes experiments.

2 The Outline of Rule Discovery from fMRI Brain Images by Logical Regression Analysis

The brain is 3-dimensional. In fMRI brain images, a set of 2-dimensional images(slices) represents a brain. See Fig. 1. 5 slices are obtained in Fig. 1. Fig. 2

[1] The regression analysis includes the nonlinear regression analysis using neural networks

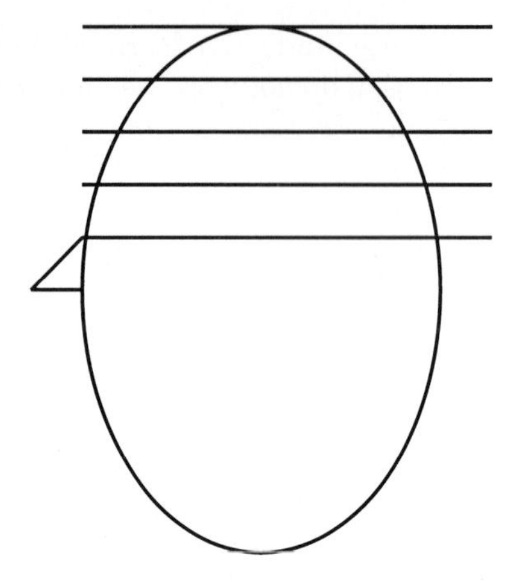

Fig. 1. fMRI images(3 dimension)

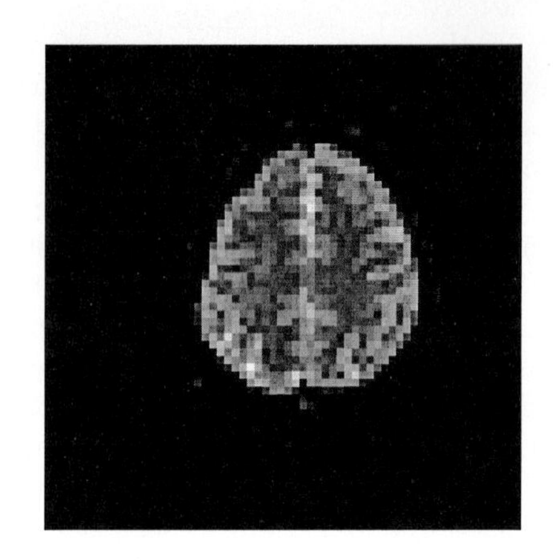

Fig. 2. A fMRI image

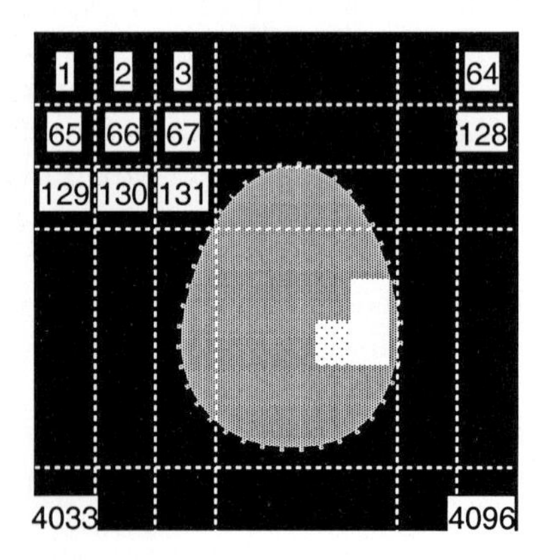

Fig. 3. An example of fMRI image

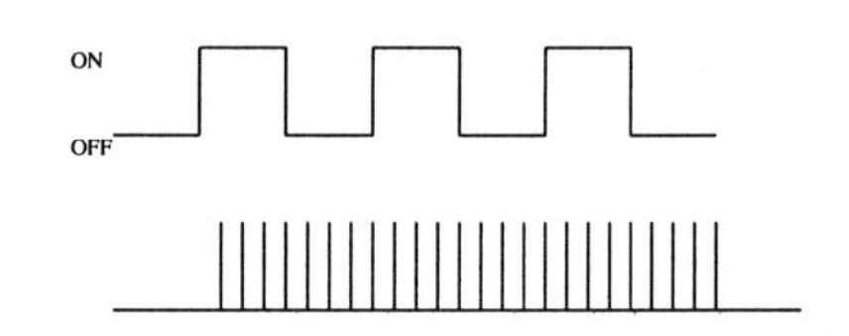

Fig. 4. Measurement

shows a real fMRI brain image. When an image consists of $64 \times 64 (= 4096)$ pixels, Fig. 2 can be represented as Fig. 3. In Fig. 3, white pixels mean activations and dot pixels mean inactivations. Each pixel has the value of the activation.

An experiment consists of several measurements. Fig. 4 means that a subject repeats a task(for example, finger tapping) three times. "ON" in the upper part of the figure means that a subject executes the task and "OFF" means that the subject does not executes the task, which is called rest. Bars in the lower part of the figure means measurements. The figure means 24 measurements. When 24 images(samples) have been obtained, the data of a slice can be represented as Table 1.

Table 1. fMRI Data

	1	2	...	4096	class
S1	10	20	...	11	Y
S2	21	16	...	49	N
...	...	...	...	...	...
S24	16	39	...	98	N

Y(N) in the class stand for on(off) of an experimental task. From Table 1, machine learning algorithms can be applied to fMRI brain images. In the case of Table 1, the attributes are continuous and the class is discrete.

Attributes(pixels) in image data have strong correlations between adjacent pixels. Moreover it is very difficult or impossible to obtain sufficient fMRI brain samples, and so there are few samples compared with the number of attributes (pixels). Therefore, the conventional supervised inductive learning algorithms such as C4.5[5] do not work well, which was confirmed in [11].

Nonparametric regression works well for strong correlations between attributes and a small number of samples. The rule extraction algorithm can be applied to the linear formula obtained by nonparametric regression. The algorithm for the discovery of rules from fMRI brain images consists of nonparametric regression and rule extraction.

Fig. 5 shows the processing flow. First, the data is mapped to the standard brain[7]. Second, the brain parts of fMRI brain images are extracted using Standard Parametric Mapping:SPM(a software for brain images analysis[6]) Finally, LRA is applied to each slice.

3 Nonparametric Regression

First, for simplification, the 1-dimensional case is explained[2].

3.1 1-Dimensional Nonparametric Regression

Nonparametric regression is as follows: Let y stand for a dependent variable and t stand for an independent variable and let $t_j (j = 1, .., m)$ stand for measured values of t. Then, the regression formula is as follows:

$$y = \sum a_j t_j + e (j = 1, .., m),$$

where a_j are real numbers and e is a zero-mean random variable. When there are n measured values of y

$$y_i = \sum a_j t_{ij} + e_i (i = 1, .., n)$$

For example, In the case of Table 1, $m = 4096$, $n = 24$, $t_{11} = 10, t_{12} = 20, t_{1\ 4096} = 11$, and $y_1 = Y$.

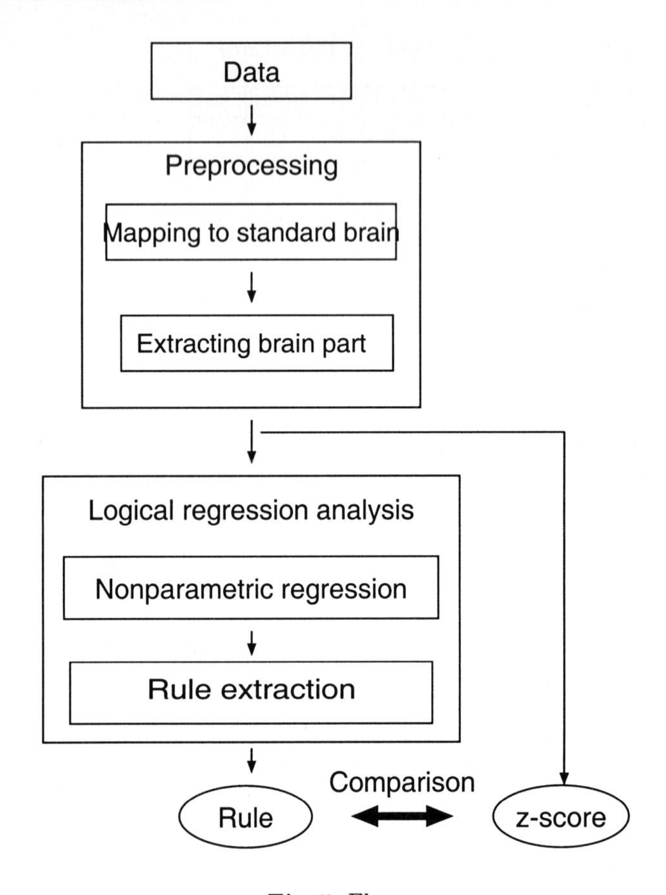

Fig. 5. Flow

In usual linear regression, error is minimized, while, in nonparametric regression, error plus continuity or smoothness is minimized. When continuity is added to error, the evaluation value is as follows:

$$1/n \sum_{i=1}^{n} (y_i - \hat{y}_i)^2 + \lambda \sum_{i=1}^{n} (\hat{y}_{i+1} - \hat{y}_i)^2,$$

where $\hat{y}$ is an estimated value. The second term in the above formula is the difference of first order between the adjacent dependent variables, that is, the continuity of the dependent variable. λ is the coefficient of continuity. When λ is 0, the evaluation value consists of only the first term, that is, error, which means the usual regression. When λ is very big, the evaluation value consists of only the second term, that is, continuity, which means that the error is ignored and the solution $\hat{y}$ is a constant.

The above evaluation value is effective when the dependent variable has continuity, for example, the measured values of the dependent variable are adjacent in space or in time. Otherwise, the above evaluation value is not effective. When the dependent variable does not have continuity, the continuity of coefficients a_js is effective, which means that adjacent measured values of the independent variable have continuity in the influence over the dependent variables. The evaluation value is as follows:

$$1/n \sum_{i=1}^{n} (y_i - \hat{y_i})^2 + \lambda \sum_{j=1}^{m} (a_{j+1} - a_j)^2$$

When λ is fixed, the above formula is the function of a_i($\hat{y_i}$ is the function of a_i). Therefore, a_is are determined by minimizing the evaluation value, and the optimal value of λ is determined by cross validation.

3.2 Calculation

Let $\mathbf{X}$ stand for $n \times m$ matrix. Let t_{ij} be an element of $\mathbf{X}$. Let $\mathbf{y}$ stand for a vector consisting of y_i. $m \times m$ matrix $\mathbf{C}$ is as follows:

$$\mathbf{C} = \begin{pmatrix} 1 & -1 & & & \\ -1 & 2 & -1 & & \\ & -1 & 2 & -1 & \\ & & & & \cdots \end{pmatrix}$$

Cross validation CV is as follows:

$$CV = n\tilde{\mathbf{y}}^{\mathbf{t}}\tilde{\mathbf{y}}$$

$$\tilde{\mathbf{y}} = \mathbf{Diag}(\mathbf{I} - \mathbf{A})^{-1}(\mathbf{I} - \mathbf{A})\mathbf{y}$$

$$\mathbf{A} = \mathbf{X}(\mathbf{X}^{\mathbf{t}}\mathbf{X} + (\mathbf{n} - \mathbf{1})\lambda\mathbf{C})^{-1}\mathbf{X}^{\mathbf{t}},$$

where $\mathbf{Diag}(\mathbf{A})$ is a diagonal matrix whose diagonal components are $\mathbf{A}$'s diagonal components. The coefficients $\hat{\mathbf{a}}$ are as follows:

$$\hat{\mathbf{a}} = (\mathbf{X}^{\mathbf{t}}\mathbf{X} + \mathbf{n}\lambda_{\mathbf{o}}\mathbf{C})^{-1}\mathbf{X}^{\mathbf{t}}\mathbf{y},$$

where λ_o is the optimal λ determined by cross validation.

3.3 2-Dimensional Nonparametric Regression

In 2-dimensional nonparametric regression, the evaluation value for the continuity of coefficients a_{ij} is modified. In 1 dimension, there are two adjacent measured values, while, in 2 dimensions, there are four adjacent measured values. The evaluation value for the continuity of coefficients is not

$$(a_{i+1} - a_i)^2,$$

but the differences of first order between a pixel and the four adjacent pixels in the image. For example, in the case of pixel 66 in Fig. 3, the adjacent pixels are pixel 2, pixel 65, pixel 67, and pixel 130, and the evaluation value is as follows:

$$(a_2 - a_{66})^2 + (a_{65} - a_{66})^2 + (a_{67} - a_{66})^2 + (a_{130} - a_{66})^2.$$

Consequently, in the case of 2 dimensions, **C** is modified in the way described above. When continuity is evaluated, four adjacent pixels are considered, while smoothness is evaluated, eight adjacent pixels are considered, for example, in the case of pixel 6 in Fig. 3, the adjacent pixels are pixel 1, pixel 2, pixel 3, pixel 65,pixel 67, pixel 129, pixel 130 and pixel 131.

4 Rule Extraction

This section briefly explains rule extraction. Details can be found in [9], [10], [12].

4.1 The Basic Algorithm of the Approximation Method

The basic algorithm approximates a linear function by a Boolean function. Let $f(x_1, ..., x_n)$ stand for a linear function, and $(f_i)(i = 1, ..., 2^n)$ be the values of the linear function. Note that the values of the linear function is normalized into [0,1] in advance by a certain method. Let the values be the interval [0,1]. Let $g(x_1, ..., x_n)$ stand for a Boolean function, and $(g_i)(g_i = 0 \text{ or } 1, i = 1, ..., 2^n)$ be the values of the Boolean function.

The basic algorithm is as follows:

$$g_i = \begin{cases} 1 \ (f_i \geq 0.5), \\ 0 \ (f_i < 0.5). \end{cases}$$

The Euclidean distance between the linear function and the Boolean function is described as

$$\sum_{i=1}^{2^n} (f_i - g_i)^2,$$

each term can be minimized independently, and $g_i = 1$ or 0. Therefore, the above algorithm minimizes Euclidean distance.

The Boolean function is represented as follows:

$$g(x_1, ..., x_n) = \sum_{i=1}^{2^n} g_i \phi_i,$$

where $\sum$ is disjunction, g_i is calculated by the above algorithm and ϕ_i stands for an atom[1].

Example Fig. 6 shows a case of two variables. Crosses stand for the values of a linear function and dots stand for the values of the Boolean function. 00, 01, 10 and 11 stand for the domains, for example, 00 stands for $x = 0, y = 0$. In this case, there are four domains as follows:

$$(0, 0), (0, 1), (1, 0), (1, 1)$$

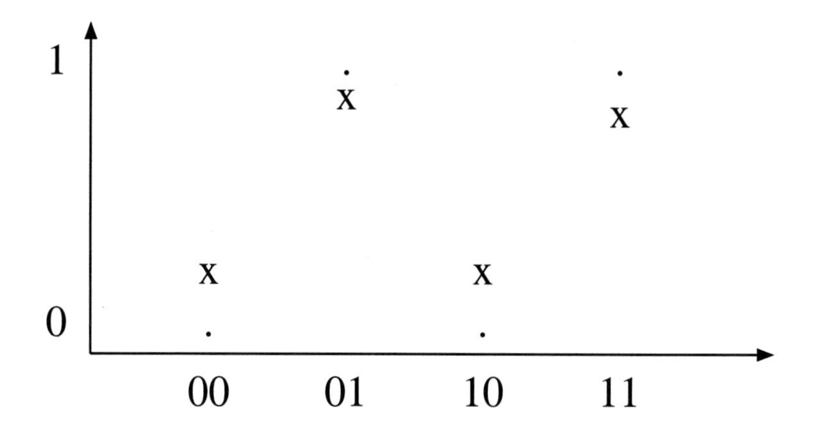

Fig. 6. Approximation

The atoms corresponding to the domains are as follows:

$$\begin{aligned}(0,0) &\Leftrightarrow \bar{x}\bar{y},\\(0,1) &\Leftrightarrow \bar{x}y,\\(1,0) &\Leftrightarrow x\bar{y},\\(1,1) &\Leftrightarrow xy.\end{aligned}$$

The values of the Boolean function $g(x, y)$ are as follows:

$$g(0,0) = 0, g(0,1) = 1, g(1,0) = 0, g(1,1) = 1.$$

Therefore, in the case of Fig. 6, the Boolean function is represented as follows:

$$g(x, y) = g(0,0)\bar{x}\bar{y} \vee g(0,1)\bar{x}y \vee g(1,0)x\bar{y} \vee g(1,1)xy.$$

The Boolean function is reduced as follows:

$$\begin{aligned}g(x, y) &= g(0,0)\bar{x}\bar{y} \vee g(0,1)\bar{x}y \vee g(1,0)x\bar{y} \vee g(1,1)xy,\\g(x, y) &= 0\bar{x}\bar{y} \vee 1\bar{x}y \vee 0x\bar{y} \vee 1xy,\\g(x, y) &= \bar{x}y \vee xy,\\g(x, y) &= y.\end{aligned}$$

The basic algorithm is exponential in computational complexity, and so the method is unrealistic when variables are many. A polynomial algorithm was presented[9],[10].

4.2 Extension to the Continuous Domain

Continuous domains can be normalized to [0,1] domains by some normalization method. So only [0,1] domains have to be discussed. First, we have to present a system of qualitative expressions corresponding to Boolean functions, in the

[0,1] domain. The expression system is generated by direct proportion, inverse proportion, conjunction and disjunction. The direct proportion is $y = x$. The inverse proportion is $y = 1 - x$, which is a little different from the conventional one ($y = -x$), because $y = 1 - x$ is the natural extension of the negation in Boolean functions. The conjunction and disjunction will be also obtained by a natural extension. The functions generated by direct proportion, reverse proportion, conjunction and disjunction are called continuous Boolean functions, because they satisfy the axioms of Boolean algebra. For details, refer to [8]. In the domain [0,1], linear formulas are approximated by continuous Boolean functions. The algorithm is the same as in the domain $\{0,1\}$.

Rule extraction is applied to the results of nonparametric analysis. That is, the linear functions obtained by nonparametric regression analysis are approximated to continuous Boolean functions, Therefore, the domains of the linear functions should be [0,1], and so the values of pixels should be normalized to [0,1]. See Table 1.

5 Experiments

LRA was applied to several experimental tasks such as finger tapping and calculation. In the experiments of finger tapping, we compared the results of LRA with z-score[6], which is the typical conventional method. In the experiments, LRA could rediscover a little complicated relation, but z-score could not rediscover the relation. As the result, we confirmed that LRA works better than z-score, which cannot be described in this paper due to space limitations. In the experiments of calculations, we confirmed that LRA worked well, that is, rediscovered well-known facts regarding calculations, and discovered new facts regarding calculations. The experiment of calculation is described. In the experiment, a subject adds a number repeatedly in the brain. The experimental conditions follow:

Magnetic field : 1.5tesla
Pixel number : 64×64
Subject number : 8
Task sample number : 34
Rest sample number : 36

Table 2 shows the errors of nonparametric regression analysis. Slice 0 is the image of the bottom of brain and slice 31 is the image of the top of brain. We focus on the slices whose errors are small, that is, the slices related to calculation. 133, ...,336 in the table are the ID numbers of subjects.

Table 3 summarizes the results of LRA. Numbers in parenthesis mean slice numbers. Fig. 7,.., Fig. 12 show the main results

White means high activity, dark gray means low activity, and black means non-brain parts. White and dark gray ares are connected by conjunction. For example, let A stand for the white area in Fig. 7 and B stand for the dark gray area in Fig. 7. Then, Fig. 7 is interpreted as $A \wedge \bar{B}$, which means area A is activated and area B is inactivated.

Table 2. Results of nonparametric regression analysis(calculation)

slice	133	135	312	317	321	331	332	336
0	0.924	0.882	0.444	0.547	0.0039	0.870	0.455	0.306
1	0.418	0.030	0.546	0.587	0.298	0.814	0.028	0.946
2	0.375	0.538	0.337	0.435	0.278	0.723	0.381	0.798
3	0.016	0.510	0.585	0.430	0.282	0.743	0.402	0.798
4	0.456	0.437	0.519	0.446	0.157	0.636	0.419	0.058
5	0.120	0.469	0.473	0.376	0.265	0.698	0.385	0.366
6	0.965	0.434	0.602	0.138	0.380	0.475	0.420	0.541
7	1.001	0.230	0.430	0.309	0.119	0.175	0.482	0.547
8	1.001	0.388	0.434	0.222	0.478	0.246	0.387	0.704
9	0.968	0.473	0.362	0.281	0.390	0.409	0.193	0.913
10	1.001	0.008	0.447	0.357	0.341	0.358	0.227	0.908
11	1.001	0.066	0.383	0.380	0.167	0.275	0.115	0.914
12	1.001	0.736	0.302	0.312	0.397	0.021	0.181	0.909
13	0.828	0.793	0.525	0.222	0.455	0.845	0.204	0.733
14	0.550	0.822	0.349	0.523	0.023	0.229	0.130	0.474
15	0.528	0.805	0.298	0.569	0.107	0.439	0.338	0.374
16	0.571	0.778	0.494	0.509	0.008	0.354	0.377	0.493
17	0.009	0.007	0.159	0.615	0.238	0.159	0.561	0.774
18	0.089	0.060	0.663	0.010	0.011	0.033	0.519	0.711
19	0.642	0.238	0.573	0.405	0.185	0.426	0.470	0.689
20	0.887	0.514	0.383	0.376	0.149	0.177	0.214	0.430
21	0.282	0.532	0.256	0.028	0.018	0.219	0.303	0.548
22	0.281	0.415	0.613	0.167	0.045	0.213	0.352	0.528
23	0.521	0.422	0.229	0.227	0.048	0.306	0.050	0.450
24	0.814	0.270	0.401	0.439	0.013	0.212	0.350	0.570
25	0.336	0.394	0.411	0.195	0.469	0.148	0.414	0.689
26	0.603	0.008	0.390	0.180	0.477	0.107	0.358	0.541
27	0.535	0.062	0.324	0.191	0.308	0.279	0.455	0.413
28	0.719	0.010	0.371	0.271	0.167	0.436	0.237	0.649
29	0.942	0.310	0.400	0.257	0.169	0.353	0.023	0.775
30	0.898	0.360	0.547	0.283	0.209	0.467	0.464	0.157
31	0.746	0.026	0.023	0.445	0.187	0.197	0.084	0.195

The extracted rules are represented by conjunctions, disjunctions and negations of areas. The conjunction of areas means the co-occurrent activation of the areas. The disjunction of areas means the parallel activation of the areas. The negation of an area means a negative correlation.

LRA can generate rules including disjunctions. However, the rules including disjunctions are too complicated to be interpreted by human experts in brain science, because they have paid little attention to the phenomena. Therefore, the rules including disjunctions are not generated in the experiments.

Researchers in brain science have paid attention to positive correlations, and have not paid attention to negative correlations. LRA can detect negative correlations, and so is expected to discover new facts.

Table 3. Results of LRA

NO.		Left	Right
133	cingulate gyrus(17)	Cerebellum(3) superior frontal gyrus(17) inferior frontal gyrus(17) superior temporal plane(17,18) middle frontal gyrus(18,21) angular gyrus(21,22)	Cerebellum(0,5) middle frontal gyrus(25)
135		inferior frontal gyrus(17,18) superior temporal plane(17,18) precuneus(26,28)	superior frontal gyrus(26) superior parietal gyrus(26,28)
312	cingulate gyrus(21)	inferior frontal gyrus(15) angular gyrus(21) supramarginal gyrus(18,21) middle frontal gyrus(23)	angular gyrus(17)
317	cingulate gyrus(27)	inferior frontal gyrus(18,21) cuneus(21,22,26,27)	angular gyrus(26)
321	cingulate gyrus(16,22,24)	inferior frontal gyrus(14) postcentral gyrus(16,18) cuneus(21) parieto-occipital sulcus(22) supramarginal gyrus(24)	cuneus(16,18) parieto-occipital sulcus(21) supramarginal gyrus(22,24)
331	cingulate gyrus(25,26)	inferior frontal gyrus(12,17,18) angular gyrus(17,18,26) supramarginal gyrus(17,18)	angular gyrus(17,18,26) middle temporal gyrus(7,12,17)
332		inferior temporal gyrus(1) Cerebellum(1) postcentral gyrus(33) middle temporal gyrus(12) pre-,post-central gyrus(14) angular gyrus(23) middle frontal gyrus(29)	inferior temporal gyrus(1) Cerebellum(1) middle frontal gyrus(29) superior parietal gyrus(29,43)
336		Cerebellum(0,5) middle temporal gyrus(4,5) middle frontal gyrus(30,31) precentral gyrus(31) superior frontal gyrus(32)	Cerebellum(0,5) superior parietal gyrus(30) occipital gyrus(11)

Figures are taken from feet, and so the left side in the figures means the right side of the brain, and the right side in the figures means the left side of the brain. The upper side in the figures means the front of the brain, and the lower side in the figures means the rear of the brain.

Activation in the left angular gyrus and supramarginal gyrus was observed in 4 and 3 cases, respectively, and that in the right angular gyrus and supramarginal gyrus was observed in 3 cases and 1 case, respectively. Clinical observations show that damage to the left angular and supramarginal gyrii causes acalculia which is defined as an impairment of the ability to calculate. Despite the strong association of acalculia and left posterior parietal lesions, there are certain characteristics of acalculia that have led to the suggestion of a right-hemispheric

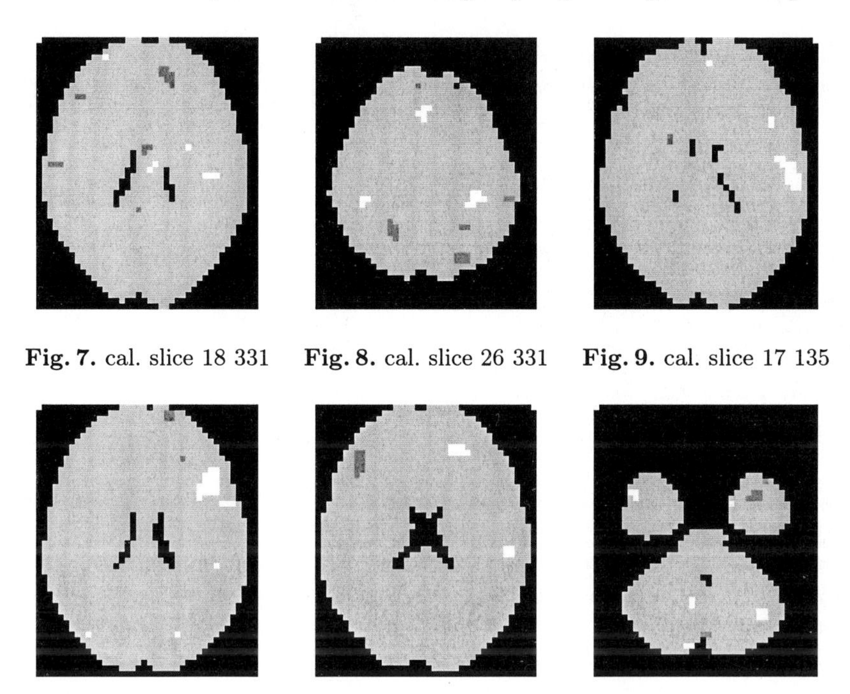

Fig. 7. cal. slice 18 331 **Fig. 8.** cal. slice 26 331 **Fig. 9.** cal. slice 17 135

Fig. 10. cal. slice 18 317 **Fig. 11.** cal. slice 17 133 **Fig. 12.** cal. slice 1 332

contribution. Clinical observations also suggest that acalculia is caused by lesions not only in the left parietal region and frontal cortex but also in the right parietal region. Fig. 7 shows slice 18 of subject 331 and Fig. 8 shows slice 26 of subject 331.

Significant activation was observed in the left inferior frontal gyrus in 6 out of 8 cases. On the other hand, none was observed in the right inferior frontal gyrus. The result suggests that the left inferior frontal gyrus including Broca's area is activated in most subjects in connection with implicit verbal processes required for the present calculation task. Furthermore, significant activation in frontal region including middle and superior frontal regions was found in 8 cases (100%) in the left hemisphere and in 3 cases in the right hemisphere. The left dorsolateral prefrontal cortex may play an important role as a working memory for calculation. Fig. 9 shows slice 17 of 135 and Fig. 10 shows slice 18 of subject 317.

In addition to these activated regions, activation in cingulate gyrus, cerebellum, central regions and occipital regions was found. The activated regions depended on individuals, suggesting different individual strategies. Occipital regions are related to spatial processing, and cingulate gyrus is related to intensive attention. Central regions and cerebellum are related to motor imagery. 5 out of 8 subjects use cingulate gyrus, which means that they are intensively attentive. Fig. 11 shows slice 17 of 133. 3 out of 8 subjects use cerebellum, which is thought not to be related to calculation. Fig. 12 shows slice 1 of 332. The above two results are very interesting discoveries that have never been experimentally confirmed so far. The problem of whether these regions are specifically related to mental calculation or not is to be investigated in further research with many subjects.

LRA has generated rules consisting of regions by conjunction and negation. As for conjunctions and negations, the results showed the inactivated regions simultaneously occurred with the activated regions. In the present experiment, inactivation in the brain region contralateral to the activated region was observed, suggesting inhibitory processes through corpus callosum. LRA has the possibility of providing new evidence in brain hemodynamics.

6 Conclusions

This paper has presented an LRA algorithm for the discovery of rules from fMRI brain images. The LRA algorithm consists of nonparametric regression and rule extraction from the linear formula obtained by the nonparametric regression. The LRA algorithm works better than z-score and has discovered new relations respecting brain functions.

In the experiments, LRA was applied to slices, that is, 2-dimensional fMRI brain images. However, complicated tasks such as calculation are related to at least a few areas, and so the application of LRA to a set of a few slices is necessary for a fruitful rule discovery from fMRI brain images. Moreover, 2-dimensional nonparametric regression analysis can be regarded as rule discovery after attribute selection. Therefore, it is desired that LRA be applied to 3-dimensional fMRI brain images. However, the nonparametric regression analysis of 3-dimensional fMRI brain images needs a huge computational time. Therefore, the computational time should be reduced, which is included in future work.

References

1. Birkhoff,G. and Bartee,T.C.: *Modern Applied Algebra*, McGraw–Hill, 1970.
2. Eubank,R.L.: *Spline Smoothing and Nonparametric Regression*, Marcel Dekker,Newyork, 1988.
3. Miwa,T.:private communication, 1998.
4. Posner,M.I. and Raichle,M.E.: *Images of Mind*, W H Freeman & Co, 1997.
5. Quinlan,J.R.: Induction of decision tree. *Machine Learning* 1, pp.81-106, 1986.
6. http://www.fil.ion.ucl.ac.uk/spm/
7. Talairach,J. and Tournoux,P.: *Coplanar Streoaxic atlas of the human brain*, New York: Thieme Medica. 1988.
8. Tsukimoto,H.: On continuously valued logical functions satisfying all axioms of classical logic, *Systems and Computers in Japan*,Vol.25, 12, 33-41, SCRIPTA TECHNICA, INC., 1994.
9. Tsukimoto,H. and Morita,C.: Efficient algorithms for inductive learning-An application of multi-linear functions to inductive learning, *Machine Intelligence 14*, pp.427-449, Oxford University Press, 1995.

10. Tsukimoto,H., Morita,C., Shimogori,N.: An Inductive Learning Algorithm Based on Regression Analysis. *Systems and Computers in Japan*, Vol.28, No.3. pp.62-70, 1997.
11. Tsukimoto,H. and Morita,C.: The discovery of rules from brain images,*The First International Conference on Discovery Science*, pp.198-209, 1998.
12. Tsukimoto,H: Extracting Rules from Trained Neural Networks, *IEEE Transactions on Neural Networks*, pp.377-389, 2000.

A Theory of Hypothesis Finding in Clausal Logic

Akihiro Yamamoto[1,2] and Bertram Fronhöfer[3]

[1] Division of Electronics and Information Engineering and MemeMedia Laboratory
Hokkaido University, Sapporo 060-8628, JAPAN
yamamoto@meme.hokudai.ac.jp
[2] "Information and Human Activity", PRESTO
Japan Science and Technology Corporation (JST)
[3] Institut für Informatik, TU München, D–80290 München, GERMANY
fronhoef@informatik.tu-muenchen.de

Abstract. Hypothesis finding constitutes a basic technique for fields of inference related to Discovery Science, like inductive inference and abductive inference. In this paper we explain that various hypothesis finding methods in clausal logic can be put on one general ground by using the combination of the upward refinement and residue hypotheses. Their combination also gives a natural extension of the relative subsumption relation. We also explain that definite bottom clauses, a special type of residue hypotheses, can be found by extending SLD-resolution.

1 Introduction

Discovering new knowledge from data is a fundamental activity of scientists. In Computer Science and Artificial Intelligence mechanizing this activity has been investigated in specific fields related to Discovery Science, such as inductive inference, abductive inference, and machine learning. Deductive logic, in particular first-order clausal logic, has often been used as a foundation in such research. Plotkin [17] and Shapiro [20] applied clausal logic to inductive inference with the aim to establishing a new logic of discovery. Their work has been followed by many researchers in the area of inductive logic programming ([3] is a survey of this area). Poole et al. [18] gave a formalization of abduction.

As in Plotkin's formalization, all of the logic-based work listed above has used deductive inference in order to derive (candidate) hypotheses. The aim of our present research is to handle all these types of inference in a unified way, *hypothesis finding*, based on the resolution principle. Hypothesis finding is formalized as follows: For given logical formulae B and E such that $B \not\models E$, hypothesis finding means the generation of a formula H such that $B \wedge H \models E$. The formulae B, E, and H are intended to represent a *background theory*, a *positive example*, and a *correct hypothesis* respectively. For example, the deductive part of abduction [18] and its improvement [9] may be regarded as hypothesis finding. Developing and analyzing hypothesis finding methods is one of the main subjects in inductive logic programming. Examples of methods developed there include *bottom method* (or *bottom generalization method*) [22, 24, 25], *inverse entailment* [15], and *saturation* [19].

S. Arikawa and A. Shinohara (Eds.): Progress in Discovery Science 2001, LNAI 2281, pp. 246–257.

We put hypothesis finding on general grounds, by using upward refinement and residue hypotheses. Upward refinement is the inverse of the logical derivation of clausal theories, which is called downward refinement [12] in inductive inference. It is also called generalization. Our general method for solving hypothesis finding problems is as follows: If a background theory B is empty, every correct hypothesis H can be obtained by applying upward refinement to a given example E. If B is non-empty, upward refinement is not applied directly to E, but to residue hypotheses which are generated from B and E. In the case where both B and E are clausal theories such that $B \not\models E$, a clausal theory H satisfies $B \wedge H \models E$ if and only if H is obtained by the general method. We show that the hypothesis finding methods listed above can be embedded into the relative subsumption of clausal theories.

This survey is organized as follows: In the next section we define hypothesis finding in clausal theories. In Section 3 we introduce upward refinement and residue hypotheses, and show that their combination derives correct hypotheses only and can derive all correct hypotheses. In Section 4 we show that each of the previous hypothesis finding methods listed above can be regarded as a restricted version of the combination of upward refinement and residue hypotheses. As a corollary of the result it is shown that relative subsumption between clausal theories is an extension of relative subsumption between clauses. In the last section we give some remarks on our results.

2 Hypothesis Finding Problem

Let $\mathcal{L}$ be a first-order language. We assume that each variable starts with a capital letter. We also assume that, for each variable X, we can prepare a new constant symbol c_X called the *Skolem constant* of X. We let $\mathcal{L}^s$ denote the language whose alphabet is obtained by adding all the Skolem constants to the alphabet of $\mathcal{L}$.

A *clausal theory* is a finite set of clauses *without any tautological clauses*, which represents the conjunction of these clauses. Let S and T be clausal theories. We write $S \models T$ if T is a logical consequence of S. For a clausal theory S the union of its ground instances in $\mathcal{L}$ is denoted by $ground(S)$.

We define a hypothesis finding problem in clausal logic as follows:

Definition 1. An instance of the *hypothesis finding problem* (*HFP*, for short) in clausal logic is defined as a pair (B, E) of satisfiable clausal theories such that $B \not\models E$. The instance is denoted by HFP(B, E). The theory B is called a *background theory*, and each clause in E is called an *example*. A solution to HFP(B, E) is given by any satisfiable clausal theory H such that $B \cup H \models E$. Such a solution is also called a *correct hypothesis*.

In inductive inference and machine learning, two types of examples, positive and negative, are generally used. Every example in this paper is a *positive example*. We call a correct hypothesis simple a *hypothesis*, if this causes no ambiguity.

In the original definition by Plotkin [17] it was assumed that each example e_i in E should be a definite clause $f_i \leftarrow v_i$ such that f_i represents a fact obtained by observation and v_i represents evidence relevant to f_i. In our definition E can be an arbitrary, but consistent clausal theory.

Example 1 ([3]). Let us consider a case when we construct a deductive database of families. We prepare predicate symbols p, m, f, and gf which respectively represent relations of is_a_parent_of, is_male, is_the_father_of, and is_a_grandfather_of. Assume that the following logic programs are currently given:

$$\begin{aligned} DB_{parent} &= \{p(a,b) \leftarrow, \quad p(b,c) \leftarrow, \quad p(s,t) \leftarrow, \quad p(t,u) \leftarrow\}, \\ DB_{gender} &= \{m(a) \leftarrow, \quad m(b) \leftarrow\}, \\ DB_{gf} &= \{gf(a,c) \leftarrow, \quad gf(s,u) \leftarrow\}, \\ R_{gf} &= \{gf(X,Y) \leftarrow f(X,Z), p(Z,Y)\}. \end{aligned}$$

This program is lacking the definition of the predicate f. Let $B_1 = DB_{parent} \cup DB_{gender} \cup R_{gf}$ and $E_1 = DB_{gf}$. Then

$$H_{11} = \{f(X,Y) \leftarrow p(X,Y), m(X)\} \text{ and } H_{12} = \{f(X,Y) \leftarrow \ \}$$

are examples of correct hypotheses for HFP(B_1, E_1). If a predicate symbol q is in the alphabet of $\mathcal{L}$, then

$$H_{13} = \{f(X,Y) \leftarrow q(X), \quad q(X) \leftarrow p(X,Y), m(X)\}$$

is also a correct hypothesis.

3 A General Theory of Hypothesis Finding

3.1 Upward Refinement

At first we treat HFP in the case where $B = \emptyset$ and E is an arbitrary clausal theory. Since each solution of HFP$(\emptyset, E)$ is a clausal theory H which satisfies $H \models E$, solving HFP$(\emptyset, E)$ seems to coincide with inverting a logical derivation of E from H. Therefore we start our discussion with formalizing logical derivation of clausal theories.

It is well-known that inference in clausal logic is based on deriving factors of a clause, deriving clauses subsumed by a clause, and deriving resolvents of two clauses. A clause C *subsumes* a clause D if there is a substitution θ such that every literal in $C\theta$ occurs in D. Note that a factor of a clause C is subsumed by C. Note also that a resolvent of clauses C and D is obtained by applying the *excluding middle* rule to some clauses C' and D' which are subsumed by C and D, respectively. We represent these operations in the form of inference rules which derive a clausal theory from another clausal theory.

Definition 2. A clausal theory T is *directly derivable* from another clausal theory S, if one of the following three conditions holds:

1. (Weakening) $T = S - \{C\}$, where $C \in S$.
2. (Excluding Middle) $T = S \cup \{C\}$, where C is such a clause that both $C \vee A$ and $C \vee \neg A$ are in S for some atom A.
3. (Subsumption) $T = S \cup \{D\}$, where D is a clause subsumed by some clause C in S.

We say that T is derivable from S and write $S \vdash T$ if there is a sequence of S_0, S_1, ..., S_n such that $S = S_0$, $S_n = T$, and S_{i+1} is directly derivable from S_i for every $i = 0, 1, \ldots, n-1$. The sequence is called a *derivation* of T from S.

The deductive completeness of the inference rules is represented as follows:

Theorem 1 ([12, 13]). *Let S be a clausal theory and T be a clausal theory. Then T is a logical consequence of S iff $S \vdash T$.*

Now we consider inverting the inference rules in order to solve HFP for the case where the background theory is empty.

Definition 3. Let S and T be clausal theories. If $S \vdash T$, we say that S is derived by *upward refinement* from T, and also that S is a *generalization* of T.

When we implement upward refinement in the form of an effective procedure, we have to give a method with which we recover the atoms deleted by applications of the resolution rule. This implies that the condition $H \models E$ is too weak as the specification of H. In previous research on inductive inference and abductive inference, giving such a method was often avoided by disallowing the application of the resolution rule in the derivation of E from H. Such restricted derivability is called a *subsumption relation* for clausal theories. The relation is usually defined as follows:

Definition 4. Let S and T be clausal theories. We define $S \sqsupseteq T$ iff, for every clause D in T, there is a clause C in S such that C subsumes D.

3.2 Residue Hypotheses

A clausal theory H is a solution of HFP(B, E) iff $H \models \neg(B \wedge \neg E)$. Under the condition that $B \neq \emptyset$, the formula $\neg(B \wedge \neg E)$ is not always a clausal theory. In order to use the upward refinement strategy for this case, we replace the formula $\neg(B \wedge \neg E)$ with some clausal theories called residue hypotheses. A residue is defined for a clausal theory S and then a residue hypothesis is defined for HFP(B, E).

Definition 5. Let $S = \{C_1, C_2, \ldots, C_m\}$ be a ground clausal theory, and $C_i = L_{i,1} \vee L_{i,2} \vee \ldots \vee L_{i,n_i}$ for $i = 1, 2, \ldots, m$. We define the *complement* of S as

$$\overline{S} = \{\neg L_{1,j_1} \vee \neg L_{2,j_2} \vee \ldots \vee \neg L_{m,j_m} \mid 1 \leq j_1 \leq n_1, 1 \leq j_2 \leq n_2, \ldots, 1 \leq j_m \leq n_m\}.$$

Note that we distinguish the logical negation $\neg S$ and the complement $\overline{S}$, though they are logically equivalent.

The complement $\overline{S}$ may contain tautological clauses. We define another clausal theory which contains no tautology, but is still logically equivalent to $\neg S$.

Definition 6. For a ground clausal theory S we define the *residue* of S as the clausal theory which is obtained by deleting all tautological clauses from $\overline{S}$. The residue is denoted by $\mathrm{Res}(S)$.

$\mathrm{Res}(S)$ can be computed in practice by using the fact that a ground clause is a tautology iff it contains any pair of complementary literals.

Proposition 1. *In the case where both B and E are ground clausal theories such that $B \not\models E$, a clausal theory H is a solution of* $\mathrm{HFP}(B, E)$ *iff* $H \vdash \mathrm{Res}(B \cup \mathrm{Res}(E))$.

This means that all ground solutions of $\mathrm{HFP}(B, E)$ are obtained by applying upward refinement to $\mathrm{Res}(B \cup \mathrm{Res}(E))$.

Now we define residue hypotheses for $\mathrm{HFP}(B, E)$ when at least one of B and E contains variables. We call such a pair *non-ground.*

Definition 7 ([7, 8, 26]). Let B and E be clausal theories. A clausal theory H is a *residue hypothesis* for $\mathrm{HFP}(B, E)$, if $H = \mathrm{Res}(F)$ for some $F \subseteq ground(B) \cup \mathrm{Res}(E\sigma_E)$, where σ_E is a substitution replacing each variable in E with its Skolem constant.

Note that $ground(B)$ may be infinite and that residue hypotheses for $\mathrm{HFP}(B, E)$ are not unique in general for a non-ground pair (B, E). We have to revise Proposition 1 so that we may derive solutions of $\mathrm{HFP}(B, E)$ even if (B, E) is non-ground. For this purpose we use Herbrand's Theorem (see e.g. [4, 6, 14] [1]).

Theorem 2 (Herbrand). *A finite set S of clauses is unsatisfiable if and only if there is a finite and unsatisfiable subset of $ground(S)$.*

The following is the revised version of Proposition 1.

Theorem 3. *Let B and E be clausal theories such that $B \not\models E$. A clausal theory H is a solution of* $\mathrm{HFP}(B, E)$ *if and only if there is a residue hypothesis K for* $\mathrm{HFP}(B, E)$ *such that $H \vdash K$.*

However, as mentioned in Section 3.1, the inverse of $\sqsupseteq$, which does not rely on the excluding middle rule, is preferred to that of $\vdash$. So we give the following definition.

Definition 8. Let H, E, and B be clausal theories. Then we say that H *subsumes E relative to B* iff H subsumes a residue hypothesis of $\mathrm{HFP}(B, E)$.

In Section 4 we will show that some methods developed in the past for induction and abduction derive only hypotheses which subsume E relative to B.

Example 2. Let $B_2 = R_{gf}$ in Example 1 and $E_2 = \{gf(a, c) \leftarrow\}$. We solve $\mathrm{HFP}(B_2, E_2)$. By using a ground instance $G_2 = \{gf(a, c) \leftarrow f(a, b), p(b, c)\}$ of B_2, we get a residue hypothesis

$$K_2 = \left\{ \begin{array}{l} f(a, b), gf(a, c) \leftarrow, \\ p(b, c), gf(a, c) \leftarrow \end{array} \right\}.$$

[1] Of the two versions given by Chang and Lee [6], we adopt "Herbrand's Theorem, Version II".

A clausal theory $H_2 = \{f(X,Y) \leftarrow,\ p(X,Y) \leftarrow\}$ is a correct hypotheses satisfying $H_2 \sqsupseteq K_2$, and therefore H_2 subsumes E_2 relative to B_2.

Example 3 ([26]). Consider a background theory $B_3 = \{even(0) \leftarrow,\ even(s(X)) \leftarrow odd(X)\}$ and an example $E_3 = \{odd(s^5(0)) \leftarrow\}$. The predicates *even* and *odd* are respectively intended to represent even numbers and odd numbers. The constant 0 means zero, and the function s is the successor function for natural numbers. The term which is an n-time application of s to 0 is written as $s^n(0)$. Then an expected solution of HFP(B_3, E_3) is $H_3 = \{odd(s(X)) \leftarrow even(X)\}$. We show that H_3 subsumes E_3 relative to B_3. At first we choose a clausal theory $G_3 = \{even(0) \leftarrow,\ even(s^2(0)) \leftarrow odd(s(0)),\ even(s^4(0)) \leftarrow odd(s^3(0))\}$, which is a subset of $ground(B_3)$. Then a residue hypothesis using G_3 is

$$\mathrm{Res}(G_3 \cup \mathrm{Res}(E_3\sigma_{E_3})) = \left\{ \begin{array}{l} odd(s^5(0)) \leftarrow even(s^4(0)), even(s^2(0)), even(0) \\ odd(s^5(0)), odd(s(0)) \leftarrow even(s^4(0)), even(0) \\ odd(s^5(0)), odd(s^3(0)) \leftarrow even(s^2(0)), even(0) \\ odd(s^5(0)), odd(s^3(0)), odd(s(0)) \leftarrow even(0) \end{array} \right\},$$

which is subsumed by H_3.

The relative subsumption relation for clausal theories defined above is an extension of that for clauses defined by Plotkin [17]. We show this fact in the next section with a property of the bottom method.

4 Pruning the Search Space for Upward Refinement

In this section we show that applying downward refinement to $B \cup \mathrm{Res}(E\sigma_E)$ prunes the search space of upward refinement of $\mathrm{Res}(B \cup \mathrm{Res}(E\sigma_E))$. From now on $\mathrm{CT}(\mathcal{L})$ ($\mathrm{CT}(\mathcal{L}^s)$) denotes the set of all clausal theories in $\mathcal{L}$ ($\mathcal{L}^s$, resp.).

4.1 Using Downward Refinement

At first we give a schema of hypothesis finding procedures.

Definition 9 ([26]). A *base enumerator* Λ is a procedure which takes a clausal theory S in $\mathrm{CT}(\mathcal{L})$ as its input and enumerates elements of a set of ground clausal theories in $\mathrm{CT}(\mathcal{L}^s)$. The set of clausal theories enumerated with this procedure is denoted by $\Lambda(S)$ and called a *base*. A *generalizer* Γ takes a ground clausal theory K in $\mathrm{CT}(\mathcal{L}^s)$ and generates clausal theories in $\mathrm{CT}(\mathcal{L})$. The set of clauses generated by Γ is denoted by $\Gamma(K)$.

Let us consider the following schema of procedures which call a base enumerator Λ and a generalizer Γ.

Procedure $\mathrm{FIT}_{\Lambda,\Gamma}(B, E)$
/* $B \in \mathrm{CT}(\mathcal{L})$ for a background theory and $E \in \mathrm{CT}(\mathcal{L})$ for an example */
1. Choose non-deterministically a ground clausal theory K from $\Lambda(B \cup \mathrm{Res}(E\sigma_E))$.

2. Return non-deterministically clausal theories H in $\Gamma(K)$.

If either of the sets $\Lambda(B \cup \mathrm{Res}(E\sigma_E))$ and $\Gamma(K)$ is infinite, and if we enumerate all H such that $H \in \Gamma(K)$ for some $K \in \Lambda(B \cup \mathrm{Res}(E\sigma_E))$, we need to use some dovetailing method in order to enumerate all elements in these sets.

According to the discussion in the previous section, we assume a base enumerator GT and two generalizers AE and AS satisfying

$$\mathrm{GT}(S) = \{K \in \mathrm{CT}(\mathcal{L}^s) \mid K = \mathrm{Res}(T) \text{ for some } T \text{ such that } T \subseteq \mathit{ground}(S)\ \},$$
$$\mathrm{AE}(K) = \{H \in \mathrm{CT}(\mathcal{L}) \mid H \vdash K\}, \text{ and } \mathrm{AS}(K) = \{H \in \mathrm{CT}(\mathcal{L}) \mid H \sqsupseteq K\}.$$

Using terminology of deductive logic, Theorem 3 shows that $\mathrm{FIT}_{\mathrm{GT,AE}}$ is sound and complete as a generator of solutions of $\mathrm{HFP}(B, E)$. The procedure $\mathrm{FIT}_{\mathrm{GT,AS}}$ is sound but not complete in general. It follows directly from the definition of relative subsumption that a clausal theory H is generated by $\mathrm{FIT}_{\mathrm{GT,AS}}$ iff H subsumes E relative to B.

Theorem 4 ([26]). *Let S be a clausal theory and T be a ground clausal theory. If $S \vdash T$, then there is a finite subset U of $\mathit{ground}(S)$ such that* $\mathrm{Res}(T) \sqsupseteq \mathrm{Res}(U)$.

4.2 Abduction and Hypothesis Finding

The method for abduction developed by Poole et al. [18] is $\mathrm{FIT}_{\mathrm{AB,AS}}$, where the base enumerator AB satisfies

$$\mathrm{AB}(S) = \{\mathrm{Res}(K) \mid K \text{ consists of one ground clause and is derivable from } S\}.$$

The abductive method $\mathrm{FIT}_{\mathrm{AB,AS}}$ is a restricted version of $\mathrm{FIT}_{\mathrm{GT,AS}}$. This result is given by the following corollaries of Theorem 4.

Corollary 1. *Let B and E be clausal theories satisfying that $B \not\models E$. Then any clausal theory H generated by* $\mathrm{FIT}_{\mathrm{AB,AS}}(B, E)$ *can be generated by* $\mathrm{FIT}_{\mathrm{GT,AS}}(B, E)$.

Example 4. The clausal theory H_{12} in Example 1 is generated by $\mathrm{FIT}_{\mathrm{AB,AS}}(B_1, E_1)$.

4.3 Finding Clauses

Let $E = \{e_1, e_2, \ldots, e_n\}$, and consider the following strategy :

1. For each $i = 1, 2, \ldots, n$, find a solution of $\mathrm{HFP}(B, \{e_i\})$ which consists of only one clause h_i.
2. Then generate H by generalizing the clausal theory $\{h_1, h_2, \ldots, h_n\}$.

Though, as is explained below, this strategy might be incomplete for solving $\mathrm{HFP}(B, E)$, it is very often employed not only in the practical construction of ILP systems [10] but also in the theoretical analysis of learning algorithms [1, 2].

For the first step, we define the bottoms of $\mathrm{HFP}(B, E)$[2].

[2] The bottom method was not well distinguished from inverse entailment in previous work, namely [21, 22].

Definition 10. Let S be a satisfiable clausal theory. A ground clause k is a *bottom* of S if k is a ground clause such that $\neg L$ is derivable from S for every literal L in k. We also define $\mathrm{BT}(S) = \{\{k\} \mid k \text{ is a bottom of } S\}$.

It is easy to see that the set of outputs of $\mathrm{FIT}_{\mathrm{BT,AS}}$ coincides with that of $\mathrm{FIT}_{\mathrm{BT,AI}}$ where $\mathrm{AI}(K) = \{H \in \mathrm{CT}(\mathcal{L}) \mid H\theta = K \text{ for some substitution } \theta\ \}$.

The bottom method $\mathrm{FIT}_{\mathrm{BT,AS}}$ is a restricted version of $\mathrm{FIT}_{\mathrm{GT,AS}}$.

Corollary 2. *Let B and E be clausal theories satisfying that $B \not\models E$. Then any clausal theory H generated by $\mathrm{FIT}_{\mathrm{BT,AS}}(B, E)$ can be generated by $\mathrm{FIT}_{\mathrm{GT,AS}}(B, E)$.*

Example 5. The clausal theory H_2 in Example 2 is generated by $\mathrm{FIT}_{\mathrm{BT,AI}}(B_2, E_2)$.

Now we discuss the semantics of bottoms. We use the original relative subsumption defined by Plotkin as a relation between two clauses, not between two clausal theories.

Definition 11 ([17]). Let c and d be clauses and B a clausal theory. We say that *c subsumes d relative to B* iff $\forall(c\theta \to d)$ is a logical consequence of B for some substitution θ.

This original definition of relative subsumption looks somehow artificial, but we can show that it is an extension of our definition in Section 3.2 by considering the procedural aspect of the original relative subsumption [22].

Theorem 5 ([22]). *Let c is a non-empty clause. Let d be a clause and B a clausal theory such that $B \not\models \{d\}$. Then $\{c\}$ is generated by $\mathrm{FIT}_{\mathrm{BT,AI}}(B \cup \mathrm{Res}(\{d\}))$ iff c is a clause subsuming d relative to B in Plotkin's sense.*

This theorem shows that the bottom method is complete for deriving clauses c subsuming d relative to B in Plotkin's sense. Together with Corollary 2 we get the following corollary.

Corollary 3. *Let c is a non-empty clause. Let d be a clause and B a clausal theory such that $B \not\models \{d\}$. Then $\{c\}$ subsumes $\{d\}$ relative to B if c subsumes d relative to B in Plotkin's sense.*

Example 6. Let us consider $\mathrm{HFP}(B_1, E_1)$ in Example 1. A ground definite clause $f(a, b) \leftarrow p(a, b), m(a)$ is a bottom of $B_1 \cup \mathrm{Res}(E_1)$. The bottom is subsumed by the definite clause in H_{11}.

It is clear that the bottom method cannot derive any clausal theory consisting of more than one clause, like H_2 in Example 2. We showed in [23, 24] that the hypothesis H_3 in Example 3 cannot be derived with the bottom method.

Theorem 4 shows which hypotheses may be missed by the bottom method. Let k be a bottom of S and T its negation. By the repeated use of the theorem, we get such a subset U of S and a derivation $U_0 = U, U_1, \ldots, U_m = T$ such that $\mathrm{Res}(U_i) \sqsupseteq \mathrm{Res}(U_{i-1})$ for every $i = 1, 2, \ldots, m$. Then $\mathrm{FIT}_{\mathrm{BT,AS}}(B, E)$ may not contain a hypothesis H such that $H \sqsupseteq U_i$ for some $i = 0, 1, \ldots, m-1$ and $H \not\sqsupseteq T$. The hypothesis H_3 in Example 3 is such a hypothesis, and therefore it is missed by the bottom method.

5 Finding Definite Clauses

When we restrict every hypothesis to be a singleton set of definite clauses, we can represent the base enumerators $\mathrm{AB}(S)$ and $\mathrm{BT}(S)$ with an extension of SLD-derivation. For a definite clause $c = A_0 \leftarrow A_1, \ldots, A_n$, we write the atom A_0 as $hd(c)$ and the sequence $A_1, \ldots, A_n$ as $bd(c)$.

The base enumerator $\mathrm{AB}(S)$ must generate clauses which are logical consequences of S. For this purpose we introduce the consequence finding technique invented by Inoue [9].

Definition 12 ([21, 25]). Let P be a definite program, g be a goal clause, and R be a computation rule. An *SOLD-derivation* of $P \cup \{g\}$ is a finite sequence Δ of quadruples $\langle g_i, f_i, \theta_i, c_i\rangle$ $(i = 0, 1, \ldots, n)$ which satisfies the following conditions:

1. g_i, and f_i are goal clauses, θ_i is a substitution, and c_i is a variant of a definite clause in P the variables of which are standardized apart by renaming.
2. $g_0 = g$ and $f_0 = \Box$.
3. $g_n = \Box$.
4. For every $i = 0, \ldots, n-1$, if $g_i = \leftarrow A_1, \ldots, A_k$, $f_i = \leftarrow B_1, \ldots, B_h$, and A_m is the atom selected from g_i by R, then either
   (a) $g_{i+1} = \leftarrow A_1, \ldots, A_{m-1}, A_{m+1}, \ldots, A_k$, $f_{i+1} = \leftarrow B_1, \ldots, B_h, A_m$, and both θ_i and c_i can be arbitrary, or
   (b) $g_{i+1} = (\leftarrow A_1, \ldots, A_{m-1}, bd(c_i), A_{m+1}, \ldots, A_k)\theta_i$ where θ_i is the mgu of A_m and $hd(c_i)$.

The goal f_n is called the *consequence* of Δ.

Proposition 2. *Let g and f be goal clauses and P a definite program. Then $f \in \mathrm{AB}(P \cup \{g\})$ iff there is an SOLD-derivation of $P \cup \{g\}$ whose consequence subsumes f.*

Example 7. Let us take B_1 from Example 1 and E_2 from Example 2. A goal clause $\leftarrow f(a,b)$ is the consequence of an SOLD-derivation of $B_1 \cup \mathrm{Res}(E_2)$,

Now we show how to generate bottoms with SOLD-derivation.

Proposition 3. *Let B be a definite program and e be a definite clause such that $B \not\models \{e\}$. Then a ground definite clause k is a bottom of e under B iff $\leftarrow hd(k)$ is subsumed by the consequence f of some SOLD-derivation of $B \cup \mathrm{Res}(\{e\sigma_e\})$ and there is an SLD-refutation of $B \cup \mathrm{Res}(\{\leftarrow bd(e\sigma_e)\}) \cup \{\leftarrow bd(k)\}$.*

Note that $\mathrm{Res}(\{\leftarrow bd(e\sigma_e)\}) \cup \{\leftarrow bd(k)\}$ is obtained by replacing $\leftarrow hd(e\sigma_e)$ in $\mathrm{Res}(\{e\})$ with $\leftarrow bd(k)$. This proposition shows that Theorem 5 still holds even if we assume that B is a definite logic program and both of c and d are definite.

Example 8. Consider B_1 and E_1 from Example 1, and let e_1 be the clause $gf(a,c) \leftarrow$ in E_1. A ground definite clause $f(a,b) \leftarrow p(a,b), m(a)$ is a bottom of $B_1 \cup \mathrm{Res}(E_1)$, because $\leftarrow f(a,b)$ is a consequence of an SOLD-derivation of $B_1 \cup \{e_1\}$, and there is an SLD-refutation of $B \cup \mathrm{Res}(\{\leftarrow bd(e_1)\}) \cup \{\leftarrow p(a,b), m(a)\}$.

Saturants have very often been used for finding hypotheses in ILP.

Definition 13. A ground definite clause k is a *saturant* of a definite clause e under B if $hd(k)$ is a ground instance of $hd(e)$ and there is an SLD-refutation of $B \cup \text{Res}(\{\leftarrow bd(e\sigma_e)\}) \cup \{\leftarrow bd(k)\})$. The set of all saturants of e under B is denoted by $\text{SATU}(B, e)$. For a definite program E such that $B \not\models E$, a *saturant* k of E under B is defined as a definite clause generated by $\text{FIT}_{\text{SATU,AI}}$ where the base enumerator SATU satisfies that $\text{SATU}(B \cup \text{Res}(E\sigma_E)) = \bigcup_{e \in E} \text{SATU}(B, e)$.

A bottom k of $B \cup \text{Res}(E\sigma_E)$ is a saturant of E under B if k is used in the first step of an SLD-refutation of $B \cup \text{Res}(\{e\sigma_e\})$ for some $e \in E$. The name of saturants comes from that this process tries to saturate the evidence for the fact $f = hd(e)$ with the support of B.

Saturants are characterized with Buntine's extended subsumption.

Definition 14. Let A be a ground atom and I be an Herbrand interpretation. A definite clause C *covers* A in I if there is a substitution θ such that $hd(C)\theta = A$ and $B\theta$ is true in I for every B in $bd(C)$.

Definition 15 ([5]). Let B be a definite program and h and e be two definite clauses. Then h *subsumes* e *w.r.t.* B if, for any Herbrand model M of B and for any ground atom A, h covers A in M whenever e covers A.

Theorem 6 ([11]). *Let B and E be definite programs such that $B \not\models E$. Then a definite clause k is in* $\text{FIT}_{\text{SATU,AS}}(B \cup \text{Res}(\{E\sigma_E\}))$ *iff k subsumes some e in E* w.r.t. B.

Example 9. Let us consider Example 1 and put $B_4 = DB_{parent} \cup DB_{gender} \cup \{f(X,Y) \leftarrow p(X,Y), m(X)\}$ and $E_4 = \{gf(a,c) \leftarrow\}$. The definite clause in R_{gf} is a generalization of a saturant $gf(a,c) \leftarrow f(a,b), p(b,c)$ of E_4 under B_4.

6 Concluding Remarks

In this paper we have shown that the combination of deriving residue hypotheses and of deriving their upward refinements is a complete method for solving any hypotheses finding problems in clausal logic. In the papers [7, 8] we initially defined residue hypotheses on the basis of the terminology of the Connection Method, which is a special method for theorem proving [4]. Using this terminology, the complement of S corresponds to *the set of negated paths* in the matrix representation of S.

Considering complexity, the derivation of $\text{Res}(S)$ from S is equivalent to the enumeration of all satisfiable interpretations of S. A similar problem, counting such interpretations, is denoted by ♯SAT and is in the class ♯P (see [16] for standard definitions on complexity). Therefore the complexity of deriving $\text{Res}(S)$ is quite high. If S has m clauses with at most n literals, $\text{Res}(S)$ has n^m clauses at most.

This complexity analysis might explain why the abductive hypothesis finding method and the bottom method were discovered earlier than our method.

Assuming severe restrictions on hypotheses, the earlier methods derive clausal theories whose residue hypotheses are easily computed. In fact, the abductive method generates theories consisting of a clause $L_1 \vee \ldots \vee L_n$ and the bottom method derives theories of the form $L_1 \wedge \ldots \wedge L_n$. In both cases the residue hypotheses of derived theories are computed in linear time. But the discussion in Section 4 shows that the achieved efficiency is paid for by missing hypotheses which might be important.

In this paper we did not treat the problem of how to recover the clauses deleted by applications of the weakening rule. In order to tackle this problem we have proposed restricting this rule either to the inverse of resolution or to that of subsumption. This is related to relevant logic [7, 8]. The relation between HFP and relevant logic will be studied more precisely in near future.

Acknowledgments

The first author thanks Prof. Taisuke Sato, Prof. Katsumi Inoue and Prof. Koichi Hirata for their helpful comments. The authors thank Prof. Hiroki Arimura for fruitful discussions with him.

References

1. D. Angluin, M. Frazier, and L. Pitt. Learning Conjunctions of Horn Clauses. *Machine Learning*, 9:147–164, 1992.
2. H. Arimura. Learning Acyclic First-order Horn Sentences From Implication. In *Proceedings of the 8th International Workshop on Algorithmic Learning Theory (LNAI 1316)*, pages 432–445, 1997.
3. H. Arimura and A. Yamamoto. Inductive Logic Programming : From Logic of Discovery to Machine Learning. *IEICE Transactions on Information and Systems*, E83-D(1):10–18, 2000.
4. W. Bibel. *Deduction: Automated Logic*. Academic Press, 1993.
5. W. Buntine. Generalized Subsumption and its Applications to Induction and Redundancy. *Artificial Intelligence*, 36:149–176, 1988.
6. C.-L. Chang and R. C.-T. Lee. *Symbolic Logic and Mechanical Theorm Proving*. Academic Press, 1973.
7. B. Fronhöfer and A. Yamamoto. Relevant Hypotheses as a Generalization of the Bottom Method. In *Proceedings of the Joint Workshop of SIG-FAI and SIG-KBS*, SIG-FAI/KBS-9902, pages 89–96. JSAI, 1999.
8. B. Fronhöfer and A. Yamamoto. Hypothesis Finding with Proof Theoretical Appropriateness Criteria. Submitted to the AI journal, 2000.
9. K. Inoue. Linear Resolution for Consequence Finding. *Artificial Intelligence*, 56:301–353, 1992.
10. K. Ito and A. Yamamoto. Finding Hypotheses from Examples by Computing the Least Generalization of Bottom Clauses. In , *Proceedings of the First International Conference on Discovery Science (Lecture Notes in Artificial Intelligence 1532)*, pages 303–314. Springer, 1998.
11. B. Jung. On Inverting Generality Relations. In *Proceedings of the 3rd International Workshop on Inductive Logic Programming*, pages 87–101, 1993.

12. P. D. Laird. *Learning from Good and Bad Data.* Kluwer Academic Publishers, 1988.
13. R.C.T. Lee. *A Completeness Theorem and Computer Program for Finding Theorems Derivable from Given Axioms.* PhD thesis, University of California, Berkeley, 1967.
14. A. Leitsch. *The Resolution Calculus.* Springer, 1997.
15. S. Muggleton. Inverse Entailment and Progol. *New Generation Computing*, 13:245–286, 1995.
16. C. H. Papadimitriou. *Computational Complexity.* Addison Wesley, 1993.
17. G. D. Plotkin. A Further Note on Inductive Generalization. In *Machine Intelligence 6*, pages 101–124. Edinburgh University Press, 1971.
18. D. Poole, R. Goebel, and R. Aleliunas. Theorist: A Logical Reasoning System for Defaults and Diagnosis. In N. Cercone and G. McCalla, editors, *The Knowledge Frontier: Essays in the Representation of Knowledge*, pages 331–352. Springer-Verlag, 1987.
19. C. Rouveirol. Extentions of Inversion of Resolution Applied to Theory Completion. In S. Muggleton, editor, *Inductive Logic Programming*, pages 63–92. Academic Press, 1992.
20. E. Y. Shapiro. Inductive Inference of Theories From Facts, 1981. Also in: *Computational Logic* (Lassez, J.-L. and Plotkin, G. eds.), The MIT Press, Cambridge, MA, pp.199–254, 1991. .
21. A. Yamamoto. Representing Inductive Inference with SOLD-Resolution. In *Proceedings of the IJCAI'97 Workshop on Abduction and Induction in AI*, pages 59 – 63, 1997.
22. A. Yamamoto. Which Hypotheses Can Be Found with Inverse Entailment? In *Proceedings of the Seventh International Workshop on Inductive Logic Programming (LNAI 1297)*, pages 296 – 308, 1997. The extended abstract is in *Proceedings of the IJCAI'97 Workshop on Frontiers of Inductive Logic Programming, pp.19–23 (1997).*
23. A. Yamamoto. Logical Aspects of Several Bottom-up Fittings. To appear in Proceedings of the 9th International Workshop on Algorithmic LEarning Theory, 1998.
24. A. Yamamoto. An Inference Method for the Complete Inverse of Relative Subsumption. *New Generation Computing*, 17(1):99–117, 1999.
25. A. Yamamoto. Using abduction for induction based on bottom generalization. In A. Kakas and P. Flach, editor, *Abductive and Inductive Reasoning: Essays on their Relation and Integration*, pages 267–280. Kluwer-Academic Press, 2000.
26. A. Yamamoto and B. Fronhöfer. Hypothesis Finding via Residue Hypotheses with the Resolution Principle. In *Proceedings of the the 11th International Workshop on Algorithmic Learning Theory (LNAI 1968)*, pages 156–165, 2000.

Efficient Data Mining by Active Learning

Hiroshi Mamitsuka[1] and Naoki Abe[2,*]

[1] Computational Engineering Technology Group
Computer and Communications Research
NEC Corporation
mami@ccm.cl.nec.co.jp
[2] Mathematical Sciences Department
IBM Thomas J. Watson Research Center
nabe@us.ibm.com

Abstract. An important issue in data mining and knowledge discovery is the issue of data scalability. We propose an approach to this problem by applying active learning as a method for data selection. In particular, we propose and evaluate a selective sampling method that belongs to the general category of 'uncertainty sampling,' by adopting and extending the 'query by bagging' method, proposed earlier by the authors as a query learning method. We empirically evaluate the effectiveness of the proposed method by comparing its performance against Breiman's Ivotes, a representative sampling method for scaling up inductive algorithms. Our results show that the performance of the proposed method compares favorably against that of Ivotes, both in terms of the predictive accuracy achieved using a fixed amount of computation time, and the final accuracy achieved. This is found to be especially the case when the data size approaches a million, a typical data size encountered in real world data mining applications. We have also examined the effect of noise in the data and found that the advantage of the proposed method becomes more pronounced for larger noise levels.

1 Introduction

One of the most important issues in data mining and knowledge discovery is that of data scalability. With the Internet becoming a popular means of communication and transactions, there has been a tremendous increase in the amount of data accumulated each day in enterprises. Similarly, more and more scientific data are becoming globally available today, also helped by the establishment of the Internet as the method of exchange of ideas as well as data in sciences. As a consequence of such developments, methods are in demand that can efficiently analyze massive amout of data and discover meaningful rules in them, be it from scientific or business data. As a possible solution to this problem, we propose

* Supported in part by a Grant-in-Aid for Scientific Research on Priority Areas "Discovery Science" from the Ministry of Education, Science, Sports and Culture of Japan. This work was carried out while this author was with NEC and Tokyo Institute of Technology.

S. Arikawa and A. Shinohara (Eds.): Progress in Discovery Science 2001, LNAI 2281, pp. 258–267.

to apply active learning as a method of selective sampling from very large data sets, and show that it is useful for designing efficient and effective data mining algorithms.

Provost and Kolluri [PK99] give a comprehensive survey of methods for 'scaling up' inductive algorithms, for the purpose of mining large data sets. Of the approaches surveyed in this article, we are concerned with that of 'data partitioning', and 'sequential multi-subset learning with a model-guided instance selection,' in particular. In this approach, relatively small subsets of data are sequentially sampled, using a model guided instance selection strategy, and the successive models are combined to give the final resulting model. A number of methods that have been proposed to date in the literature belong to this category, including Windowing [Q83], Integrative Windowing [F98], boosting [FS97], and Ivotes [Breiman99]. One thing that is common among all of these methods is that they employ a sampling method that makes use of the label information in the candidate instances. For example, Ivotes uses a sampling method called 'importance sampling', which chooses examples on which the current hypothesis makes a mistake and (with high probability) discards those on which the current hypothesis predicts correctly. It has been reported, however, that some of these methods do not work well, in the presence of abundant noise in the training data (c.f. [C91,PK99]).

As an approach to this problem, we propose to use what is generically known as 'uncertainty sampling', which samples those examples that cannot be reliably predicted at that point. Note that, in uncertainty sampling, the label information in the candidate instances is *not* used in making selections, and thus such a method can be interpreted as a 'query learning method' that queries for the labels of selected instances. The purpose of the present paper is to examine how well this approach works in the current context of efficient mining from large data sets, and to characterize under what conditions it works better than importance sampling, in particular. The particular sampling method we employ is based on the idea of 'query by bagging' proposed in [AM98], which was in turn obtained by combining ideas of query by committee [SOS92] and 'bagging' [Breiman96]. The basic idea is that query points are chosen by picking points on which the predictions made by the hypotheses resulting from applying the component inductive algorithm to sub-samples obtained via re-sampling from the original data set, are most evenly spread. This method is like Breiman's Ivotes, except committee-based uncertainty sampling is used in place of the importance sampling employed in Ivotes.

We empirically evaluated the performance of this method, using a number of different types of data sets. In our first experiments, we used synthetic data sets of size one million each, generated from the 'generator' functions of [AIS93], used often as benchmark data for evaluating data mining methods. We found that the performance of QbagS was favorable as compared to that of Ivotes, both in terms of the computation time required to reach the same predictive accuracy, and the final accuracy attained.

In order to better understand the conditions under which QbagS performs well, we varied a parameter called the 'perturbation factor,' which controls the noise level of the 'generator' functions. It was found that for larger perturbation factors, the significance level by which QbagS out-performed Ivotes became larger. This result confirms the thesis that uncertainty sampling is more desirable than sampling methods that concentrate on those instances on which prediction errors are made, when the data is noisy. This thesis is further supported by the results of our experimentation on a real world data set, which is inevitably noisy. Specifically, we compared the two methods using a data set in the area of database marketing (internet provider churn data) of size roughly a million. Here we found that the predictive accuracy of QbagS was significantly better than Ivotes.

2 The Mining/Learning Methods

2.1 Proposed Method

In this section, we describe the mining/learning method we propose and evaluate in this paper, which we call QbagS, standing for 'Query by bagging with a single loop.' This procedure provides a sampling strategy that uses an arbitrary component learning algorithm as a subroutine, and works roughly as follows. (See the pseudocode shown below.) At each iteration, it randomly samples a relatively large number of candidate examples (R, say 10,000) from the database (line 1). It then selects a small enough (D, say 1,000) subset of this set and applies the component learning algorithm to it to obtain a new hypothesis (line 3). When making this selection, it uses the hypotheses from the past iterations to predict the labels of the candidate examples, and then pick those on which the predicted values are split most evenly. More precisely, it calculates the 'margin' of each candidate instance, that is, the difference between the number of votes by the past hypotheses for the most 'popular' label, and that for the second most popular label (line 2). Then, D instances having the least values of margin are selected from the candidates (line 3). The final hypothesis is defined by the majority vote over all the hypotheses obtained in the above process.

Algorithm: Query-by-Bagging:Single (QbagS)
Input: Number of iterations: M
  Component learning algorithm: A
  Number of candidates at each iteration: R
  Number of selected examples at each iteration: D
  Candidates at the i-th iteration: C_i
  Selected (training) examples at the i-th iteration: S_i
Initialization: 1. Randomly sample initial sample
$S_1 = \langle (x_1, y_1), \cdots, (x_D, y_D) \rangle$ from the database.
2. Run A on S_1 and obtain hypothesis h_1.
For $i = 1, ..., M$
  1. Randomly sample R examples C_i from database.

2. For all $x \in C_i$, calculate 'margin' $m(x)$ using past hypotheses $h_1, \cdots, h_i$
   $m(x) = \max_y |\{t \leq i : h_t(x) = y\}|$
   $\quad - \max_{y \neq y_{\max}(x)} |\{t \leq i : h_t(x) = y\}|$
   where $y_{\max}(x) = \arg\max_y |\{t \leq i : h_t(x) = y\}|$
3. Select D examples $\langle(x_1^*, y_1^*), \cdots, (x_D^*, y_D^*)\rangle$ from C_i having the smallest $m(x)$ $(x \in C_i)$ and let $S_{i+1} = \langle(x_1^*, y_1^*), \cdots, (x_D^*, y_D^*)\rangle$.
4. Run A on S_{i+1} and obtain hypothesis h_{i+1}.

End For
Output: Output final hypothesis given by:
$h_{fin}(x) = \arg\max_{y \in Y} |\{t \leq M : h_t(x) = y\}|$

Notice that, in QbagS, re-sampling is done directly from the database, and past hypotheses are used to judge what examples to sample next. In fact, this is significantly simplified as compared to the original procedure of query by bagging(Qbag) of [AM98]: As can be seen in the pseudocode presented below,[1] in the original procedure the selected examples are accumulated to form the training data set (line 5). Then at each iteration, re-sampling is done from this set of training data (line 1), and the resulting hypotheses are used to judge what examples to select next, using a committee-based uncertainty sampling (lines 2 to 5). Since re-sampling is done at each iteration, using embedded looping, this procedure is computationally more demanding than the new version QbagS, but may be more data efficient. This is consistent with the fact that the original Qbag was designed as a *data efficient* query learning method, whereas QbagS is meant primarily as a *computationally efficient* method for mining from large databases.

Algorithm: Query-by-Bagging (Qbag)
Input: Number of stages: M
Component learning algorithm: A
Number of re-sampling at each iteration: T
Number of candidates at each iteration: R
Number of selected examples at each iteration: D
Candidates at the i-th iteration: C_i
Selected (training) examples at the i-th iteration: S_i
Initialization:
Randomly sample initial sample
$S_1 = \langle(x_1, y_1), \cdots, (x_D, y_D)\rangle$ from the database.
For $i = 1, ..., M$

1. By re-sampling from S_i with uniform distribution, obtain sub-samples $S'_1, .., S'_T$ (of same size as S_i).
2. Run A on $S'_1, .., S'_T$ to obtain hypotheses $h_1, ..., h_T$.

[1] There is one modification from the way Qbag was presented in [AM98]: Now it is presented for multi-valued prediction, while the original version was assuming a binary-valued prediction.

3. Randomly select R examples C_i from database.
4. For all $x \in C_i$, compute the margin $m(x)$ by:
$m(x) = \max_y |\{t \leq T : h_t(x) = y\}|$
$- \max_{y \neq y_{\max}(x)} |\{t \leq T : h_t(x) = y\}|$
where $y_{\max}(x) = \arg\max_y |\{t \leq T : h_t(x) = y\}|$
5. Select D examples $\langle (x_1^*, y_1^*), \cdots, (x_D^*, y_D^*) \rangle$ from C_i having the smallest values of $m(x)$ $(x \in C_i)$ and update the training data as follows.
$S_{i+1} = append(S_i, \langle (x_1^*, y_1^*), \cdots, (x_D^*, y_D^*) \rangle)$

End For

Output: Output final hypothesis given as follows.
$h_{fin}(x) = \arg\max_{y \in Y} |\{t \leq T : h_t(x) = y\}|$
where h_t $(t = 1, \cdots, T)$ are the hypotheses of the final (M-th) stage.

2.2 Breiman's Ivotes

We briefly review 'Ivotes' (importance sampling) [Breiman99], with which we compare the performance of our method. Like QbagS, Ivotes takes a sample from the database and applies the component learning algorithm at each iteration, discards the data and keeps just the hypotheses. When sampling, it uses what is called 'importance sampling.' In this sampling method, if the label of an example is wrongly predicted by its current combined hypothesis (out-of-bag prediction), it is automatically chosen. If the prediction on an example is correct, then it is selected with probability $e/(1-e)$, where e is the error probability (measured using a separately reserved test set) of the current hypothesis. Here, the out-of-bag prediction is done by majority vote over those hypotheses trained on sub-samples *not* containing the current example. Breiman [Breiman99] claims that this feature contributes greatly to improving the performance of Ivotes, and proposes a particular implementation method for computing out-of-bag predictions: It keeps records of how many times each label of each example in the database has been predicted. We follow this implementation exactly.

3 Empirical Evaluation

We empirically evaluated the performance of the proposed method and that of Ivotes, using a series of large scale synthetic data sets, generically referred to as Generator in [AIS93], often used as benchmark data for evaluating data mining methods.

In our experiments, we used both C4.5 [Q93] and CART[2] as the component algorithm. Our evaluation was mainly done in terms of the total computation time to achieve a given prediction accuracy (on separate test data), including disk access time. We also compare the 'final accuracy' attained by each of the

[2] To be precise, we used a version of CART included in the IND package due to Wray Buntine.

methods. By 'final accuracy,' we mean the accuracy level reached for data sizes large enough that the predictive performance appears to be saturating.[3] For the evaluation on the real world data set, we also used the measures of precision and recall, standard performance measures in the field of information retrieval. Note that 'recall' is defined as the probability of correct prediction given that the actual label is 1 (or whatever label of interest), and 'precision' is defined as the probability of correct prediction given that the predicted label is 1.

In our experiments, the evaluation was done by 5-fold cross validation. That is, we split the data set into 5 blocks of roughly equal size, and at each trial 4 out of these 5 blocks were used as training data, and the last block was reserved as test data. The results (prediction accuracy, learning curves, precision and recall) were then averaged over the five runs. Since the statlog data come with pre-specified test data, the average was taken over five randomized runs. All of our experiments were run on an Alpha server 4100 (466MHz, 512 MB).

It is said in general (See [PK99]) that large scale data sets are those with sizes in the order of a million. It is, however, difficult to find publically available real world data sets having data sizes in the order of a million.[4] We thus used a series of synthetic data introduced in [AIS93] called Generator. These data sets have been used often as benchmark data in evaluating the scalability issues of data mining methods (e.g. see [GGRL99,RS98]).

Generator contains 10 distinct functions for generating the labels, taking a number of different forms. Here we chose 5 out of the 10 (Function 2, 5, 8, 9 and 10) on which the predictive performance of ID3, as reported by Agrawal et al. [AIS93] was relatively poor. Generator also has a parameter called 'perturbation factor,' which is used to randomly perturb the continuous valued attributes. In our first set of experiments, we chose to set the perturbation factor to be either 0.2 or 0.6. For each of the five functions, we generated a data set of size one million, and performed 5-fold cross-validation.[5]

We note that there is a parameter to be tuned in Ivotes, namely the number of examples to be selected at each iteration. As a test, we observed the predictive performance of Ivotes on one of the five data sets (Function 2), varying this parameter. It was observed that the value of 1,000 was doing about the best, so we set this parameter to be 1,000 in all of our experiments. This is consistent with the observation in the experiments reported in [Breiman99] that the performance of Ivotes improves up to 800 but appears to be almost saturating. We remark that we did not make extensive effort in optimizing the parameters in the other methods.

[3] To be sure, we performed the 'mean difference significance test' (See Section 3.1 for the definition) for the predictive accuracy attained at the end of each run and that reached 1,000 seconds prior to that point. We found that the significance level was typically around 0.2 and at most 0.5, and thus the difference was insignificant.

[4] The largest data sets we found were those in statlog. The largest ones in UCI ML Repository we could find (mushroom and thyroid0387), for example, contain less than 10,000 data.

[5] We randomly picked 10,000 out of the test data to evaluate the predictive performance.

We show part of the results of the above experimentation (with perturbation factor 0.2 and CART) in the form of learning curves in Figure 1. Note that, in these curves, the average predictive accuracy (on test data) is plotted against the total computation time.

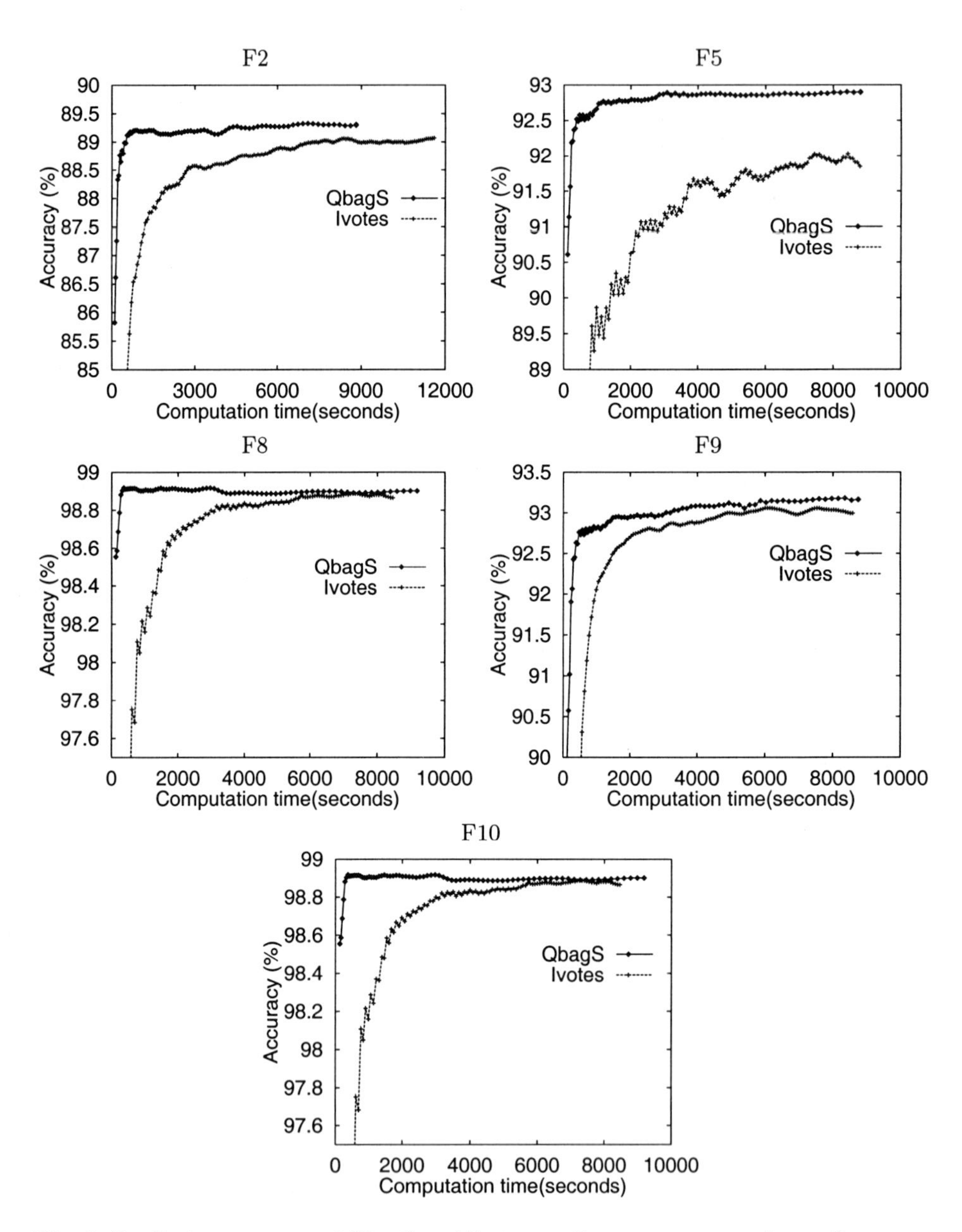

Fig. 1. Prediction accuracy of QbagS and Ivotes on Generator, averaged over five runs, plotted as function of computation time.

From these results, it is seen that, in terms of the speed for reaching the same level of accuracy, QbagS is favored over Ivotes in all cases (when using CART). The speed-up factor achieved by these methods over Ivotes, as measured by the amount of computation time they take to reach a given performance level, is anywhere from 2 to 30.

In terms of the 'final prediction accuracy', that is, the prediction accuracy reached for data sizes large enough that the performance appears to be saturating, the results were also favorable for QbagS. More concretely, we examined the final accuracies reached by the two methods for the five functions and the 'Z' values of the mean difference significance test for the respective cases are exhibited. Here, the Z values are calculated using the following well-known formula (e.g. see [WI98]):

$$Z = \frac{acc(A) - acc(B)}{\sqrt{\frac{var(A)}{n_A} - \frac{var(B)}{n_B}}}$$

where we let, in general, $acc(A)$ denote the accuracy estimate for method A, and $var(A)$ the variance of this estimate, and n_A the data size used for this estimate (5 in our case). For example, if Z is greater than 2 then it is more than 95 per cent significant that A achieves higher accuracy than B. For perturbation factor 0.2 and using C4.5 as the component algorithm, QbagS did significantly better than Ivotes in 4 out of the 5 cases, and Ivotes did significantly better in the other case. For perturbation factor 0.2 and using CART, QbagS did significantly better than Ivotes in 1 out of the 5 cases, slightly (insignificantly) better in 3 cases, and Ivotes did slightly better in the other case. When perturbation factor is set to 0.6 and when using CART, QbagS did better than Ivotes in all 5 cases, 4 out of them being statistically significant. These results are exhibited in the two graphs in Figure 2; one plots how the Z values of the mean difference significance test vary as the perturbation factor is changed from 0.2 to 0.6 for the five functions, and the other plots how the computation time ratios change. We can see, in

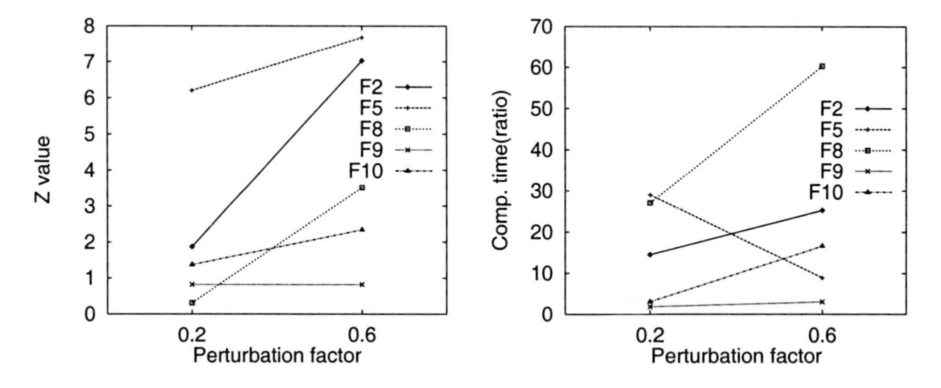

Fig. 2. Significance test values Z (Left) and computation time ratios (Right) between QbagS and Ivotes on Generator, plotted as functions of perturbation factor.

these results, that for higher values of perturbation factor, the significance of the difference in the performance of the two methods becomes more pronounced.

4 Concluding Remarks

We have proposed a sampling method for data mining that is especially effective for efficient mining from *large, noisy* data sets. The key property of our method which contributes to this advantage is its sampling strategy that does *not* make use of the label information in the candidate instances. Interesting future work is to characterize more systematically the conditions (noise level and data size) under which the proposed method works well, and better than importance sampling in particular. It would also be interesting to investigate the relationship and possibly combinations between uncertainty sampling and importance sampling and/or boosting, in the context of efficient mining from large data sets.

Acknowledgement

The authors would like to thank Osamu Watanabe and Carlos Domingo for invaluable discussions on the topics of this paper.

References

[AM98] N. Abe and H. Mamitsuka. Query Learning Strategies Using Boosting and Bagging Proceedings of Fifteenth International Conference on Machine Learning, 1–9, 1998.

[AIS93] R. Agrawal and T. Imielinski and A. Swami. Database Mining: A Performance Perspective *IEEE Transactions on Knowledge and Data Engineering*, 5(6):914–925, 1993.

[Breiman96] L. Breiman. Bagging Predictors *Machine Learning* 24:123–140, 1996.

[Breiman99] L. Breiman. Pasting Small Votes for Classification in Large Databases and on-line *Machine Learning* 36:85–103, 1999.

[C91] J. Catlett. Megainduction: A test flight Proceedings of Eighth International Workshop on Machine Learning, 596–599, 1991.

[FS97] Y. Freund and R. Schapire. A decision-theoretic generalization of on-line learning and an application to boosting *Journal of Computer and System Sciences* 55(1), 119–139, 1997.

[F98] J. Furnkranz Integrative windowing Journal of Artificial Intelligence Research 8:129-164, 1998.

[GGRL99] J. Gehrke and V. Ganti and R. Ramakrishnan and W-Y. Loh BOAT – Optimistic Decision Tree Construction Proceedings of the ACM SIGMOD International Conference on Management of Data, 169–180, 1999.

[MST94] D. Michie and D. Spiegelhalter and C. Taylor (Editors). *Machine Learning, Neural and Statistical Classification*, Ellis Horwood, London, 1994.

[PK99] F. Provost and V. Kolluri. A Survey of Methods for Scaling Up Inductive Algorithms *Knowledge Discovery and Data Mining* 3(2):131-169, 1999.

[Q83] J. R. Quinlan. Learning efficient classification procedures and their applications to chess endgames *Machine Learning: An artificial intelligence approach*, R. S. Michalski and J. G. Carbonell and T. M. Mitchell (Editors), San Francisco, Morgan Kaufmann, 1983.

[Q93] J. R. Quinlan C4.5: Programs for Machine Learning, San Francisco, Morgan Kaufmann, 1993.

[RS98] R. Rastogi and K. Shim Public: A Decision Tree Classifier that integrates building and pruning Proceedings of 24th International Conference on Very Large Data Bases, New York, Morgan Kaufmann, 404-415, 1998.

[SOS92] H. S. Seung and M. Opper and H. Sompolinsky. Query by committee Proceedings of 5th Annual Workshop on Computational Learning Theory, 287–294, New York, ACM Press, 1992.

[WI98] S. M. Weiss and N. Indurkhya *Predictive Data Mining*, Morgan Kaufmann, San Francisco, 1998.

Data Compression Method Combining Properties of PPM and CTW

Takumi Okazaki[1], Kunihiko Sadakane[2], and Hiroshi Imai[1]

[1] University of Tokyo
[2] Tohoku University

Abstract. Universal compression and leaning has been interacting with each other. This paper combines two compression schemes, PPM (Prediction by Partial Match) and CTW (Context Tree Weighting), to a scheme which can predict the multi-alphabet probabilities to attain better compression ratio.

1 Introduction

Compression and leaning has strong connection. For example, a principle of 'All things being equal, the simplest explanation tends to be right,' called Occam's razor, is often used in learning theory. Lossless compression maintains the completely same information as small as possible. Lossy compression represents the original information as much as possible with reducing the complexity very much by retaining important factors of the source information. Thus, compression is intrinsically a kind of learning. Universal compression does not assume any *a priori* knowledge of the information source, and learns the probability structure of the source while reading a sequence of symbols produced by the source on-line. However, it does not output its learned results explicitly, since the aim of compression schemes is to produce compressed objects. Or, the loss function to minimize may be said to be different in compression and learning algorithms. Now, dictionary-based universal compression schemes originally proposed by Lempel and Ziv are widely used as `gzip`, `compress`, etc. In connection with Occam's razor, the minimum description length (MDL) principle, which was originally proposed in coding and compression theory, now becomes a standard technique to avoid overlearning.

In this paper, we extend the Context Tree Weighting (CTW for short) method [15] by incorporating the Prediction by Partial Match (PPM) method [5] to predict the multi-alphabet probabilities better. Both PPM and CTW can be called the "context-based" compressions, and should predict the encoding probabilities from the substrings just before the encoding character. These probabilities should be learned how frequently each character occurs after each context in the data which must be compressed. The CTW method for binary alphabets has sound theory in the context of universal compression based on the KT-estimator in predicting the probability of the next contexts, but the case for multi-alphabets has been left unsolved somehow. We consider utilizing nice

S. Arikawa and A. Shinohara (Eds.): Progress in Discovery Science 2001, LNAI 2281, pp. 268–283.

properties of PPM in this combination and learning a key parameter value in the CTW method. Through computational experiments, good features of our schemes, together with some findings of DNA sequences, are revealed.

2 CTW: Context Tree Weighting

CTW [15] is the encoding method for the tree sources, especially for the FSMX information sources, which is outlined in this section. Originally, CTW is designed for binary alphabet sequences, and in this section its basic encoding method [15] is explained, together with some implementation schemes.

First, we introduce a context tree. Let $\mathcal{A}$ be an alphabet which organizes information source. Denote by $\mathcal{A}^* = \bigcup_{l=0}^{\infty} \mathcal{A}^l$ the set of all finite sequences from $\mathcal{A}$. Let $\lambda \in \mathcal{A}^*$ be an empty sequence. Denote by x_i^j $(i \leq j)$ a finite sequence $x_i \cdots x_j$. A context tree $T \subset \mathcal{A}^*$ is defined as a tree which satisfies any postfix of any $s \in T$ is an element of T. A context tree is also called a postfix tree. Figure 1 shows a example of a context tree and a set of leaves. In the context tree, each node express a context, and each edge express one character of context. When $T = \{\lambda, 0\}$, the set of leaves for T can be defined as $\partial T = \{00, 10, 1\}$.

Any context tree can be applied to CTW. Practically, the maximum depth of the context tree must be restricted to the fixed value D. In implementing the context tree for CTW, each context (node) s holds frequencies of character which occurred after s until that time. For binary data, the context s preserves the count a_s of 0 and the count b_s of 1. Then, children of the parent node s are set to $0s$ and $1s$. From the definition of frequency parameters, counts of characters at children satisfies $a_{0s} + a_{1s} = a_s, b_{0s} + b_{1s} = b_s$.

In CTW, on the basis of the information of frequencies in the context tree, estimated probabilities P_e^s of a specific sequence which consists of a_s zeros and b_s ones are calculated. For this estimation, Krichevsky-Trofimov(KT)-estimator, a special case of Dirichlet distribution, is widely used. Computationally, the KT-estimator may be defined as follows. When CTW algorithm updates P_e^s, initially we set $P_e^s(0,0) = 1$, and whenever one character must be encoding (or decoding) the following recursion formula

$$P_e^s(a_s+1, b_s) = \frac{a_s + \frac{1}{2}}{a_s + b_s + 1} P_e^s(a_s, b_s), \quad P_e^s(a_s, b_s+1) = \frac{b_s + \frac{1}{2}}{a_s + b_s + 1} P_e^s(a_s, b_s)$$

are applied. This formula makes the update of P_e^s easily. Here, in order to avoid, e.g., estimating the probability of 0 for a sequence of all 1's as zero, $\frac{1}{2}$ is necessary. As a merit of the KT-estimator, it is well-known that the redundancy for avoiding this problem of probability 0 is always bounded to $\frac{1}{2}\log(a_s + b_s) + 1$. Then, we calculate probabilities of P_w^s, the weighting sum of some values of P_e, which can be calculated by the following recursion formula:

$$P_w^s(x_1^t) = \begin{cases} \gamma P_e^s(x_1^t) + (1-\gamma) P_w^{0s}(x_1^t) P_w^{1s}(x_1^t) & \text{(node } s \text{ is not a leaf)} \\ P_e^s(x_1^t) & \text{(node } s \text{ is a leaf)} \end{cases}$$

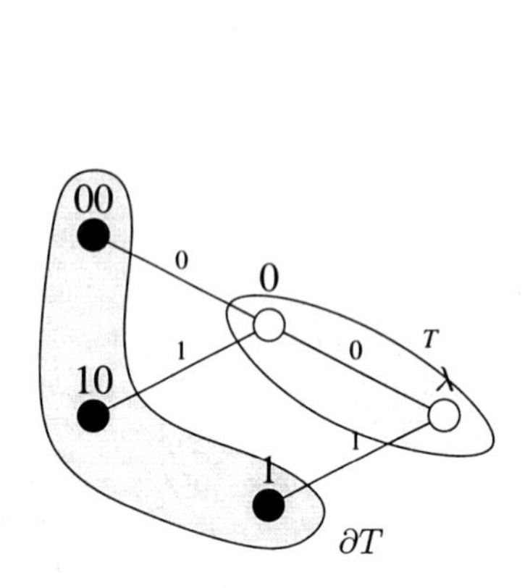

Fig. 1. Context tree and set of leaves when $T = \{\lambda, 0\}$

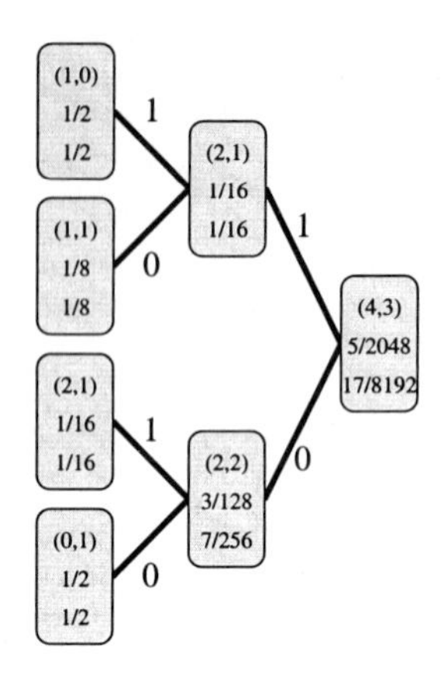

Fig. 2. Context tree with weighting of depth 2 for $x_1^T = 0110100$ and $x_{1-D}^0 = 10$

where γ is a real number which satisfies $0 < \gamma < 1$, and $P_e^s(x_1^t)$ is the KT-estimator $P_e(a,b)$ where a is the frequency of 0 just behind the context s in sequence x_1^t, and b is the frequency of 1 in the similar condition. The parameter γ is a value which determines the ratio how much information is weighted from some context and information from the context which is longer by one character than the above-mentioned context. The encoding code of sequence x_1^t is obtained by encoding the probability $P_w^\lambda(x_1^t)$ of context λ (namely, the probability of the root in the context tree) to the arithmetic code.

It is probable to misunderstand the multiplication term $P_w^{0s}(x_1^t)P_w^{1s}(x_1^t)$ to be the addition term. This misunderstanding is generated by mistaking a sequence probability $P_w^s(x_1^t)$ for one character probability. For the sequence x_1^t, it can be considered that the frequency information of context s consists of the frequency of context $0s$ **and** the one of context 1_s. Therefore, the multiplication term $P_w^{0s}(x_1^t)P_w^{1s}(x_1^t)$ is involved to the recursion formula.

Figure 2 shows the example of context tree as mentioned above. Numbers in each node mean, in descending order, (a_s, b_s), P_e^s, and P_w^s. (a_s, b_s) means that 0 occurs a_s times and 1 does b_s times just after the context s. P_e^s, which is the second number from the top in each node, means KT-estimator. It can be calculated only from the value of a_s and b_s. An edge denotes a character of a context. Leftmost and second node(leaf) from the top indicate the information after the context "01". In the leftmost column, P_w^s is defined to the same value as P_e^s, because a node in the leftmost column is a leaf. In the center column, P_w^s is defined by the recursion formula from P_e^s in the same column as P_w^s and from P_w^{0s}, P_w^{1s} in the left column. For example, 7/256 in the bottom node means P_w^0 and it can be calculated by the formula;

$$P_w^0 = \frac{1}{2}P_e^0 + \frac{1}{2}P_w^{00}P_w^{10} = \frac{1}{2} \cdot \frac{3}{128} + \frac{1}{2} \cdot \frac{1}{16} \cdot \frac{1}{2} = \frac{7}{256}.$$

Thus calculating P_w in each node, we compute the value of P_w at the rightmost as the probability assigned to the encoder of the arithmetic coding.

Willems et al. [15] suggested an algorithm of coding block probabilities each of which consists of T characters to the arithmetic code. This has been extended to more general cases, say in [11] .

When CTW is implemented, as far as the memory and the time of computation are finite, the size of the context tree must be restricted. A relatively simple method of this restriction is to define the maximum depth of the context tree. CTW can achieve stable compression performance by using weighting skillfully without dependence on the maximum depth. If the maximum depth is too short, the compression performance drops naturally because information of the longer context cannot be investigated. However, when the maximum depth is long, deep contexts are weighted with little ratio because of the system of weighting computation. This property cannot be seen in PPM as below.

3 PPM: Prediction by Partial Match

PPM [5] is the compression method for the multi-alphabet data such as English texts. It defines a probability of an encoding symbol by contexts which mean sequences directly before this symbol. An important peculiarity of PPM is to predict probabilities by imaginary *escape symbols.*

While PPM encodes data, it checks all characters that have followed k-length subsequence for $k = 0, 1, 2, \cdots$ in the given input, and counts the number of times of occurrence of each character. Practically, k is set to some fixed value. Then, it encodes probabilities based on that information of frequencies. The arithmetic coding is used for encoding probabilities. A problem occurs when count of a symbol which must be encoded is zero at firstly seen context. For example, when 'a' occurs four times and 'b' does five times after some context, the probability that next symbol is 'a' for the same context cannot be predicted as 4/9 because probabilities assigned characters except 'a' and 'b' become zero, which makes it impossible to encode such characters. To solve this problem, an 'escape' symbol and its probability is defined, in order to encode unseen characters. When an unseen character occurs, first, PPM encodes an escape symbol. Next, it shortens the watched-context by one character. Then it tries to encode from the frequency information of this shortened context. Owing to decrease of context-length, an universe of frequency information becomes larger. Therefore, the possibility of encoding jumps up. If the count of symbol which must be encoded is still zero for the shorten context, PPM outputs an escape symbol and shorten the context again. PPM repeats these until the watched context has the prediction of the encoding symbol.

Figure 3 shows the PPM model after processing the string "ababcab". "order" means the length of watched-context. This example displays the PPM model whose maximum order is defined as two (PPM of order 2). The root node involves the information of the context λ, an empty sequence. This information shows frequency of each character regardless of contexts. Namely, because the

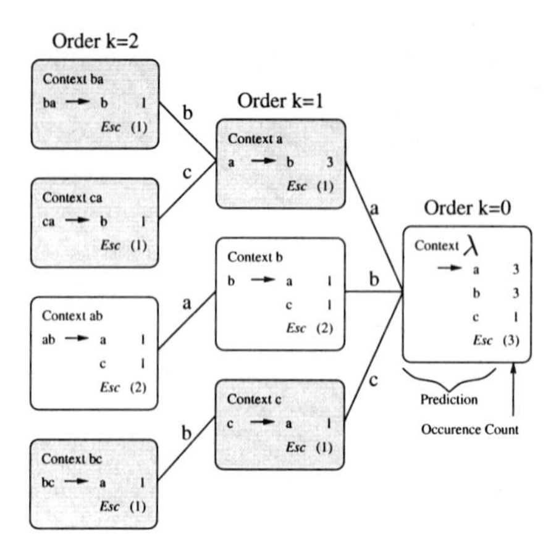

Fig. 3. Tree of PPM model for "ababcab"

sequence "ababcab" involves three 'a's, three 'b's and one 'c', these data are described at the root. Moreover, at the same place, the escape frequency is written in parentheses, since it is not like other frequency counts but means the number of prediction patters, which is thought to be better than the real counts (For example, at the root, escape probability should have weight of "three" because the root has three predictions, a, b and c). Here, for the ease of explanation, we regard the weight of three as the integer count of three. Namely, when 'a' is encoded at the root, the encoding probability of 'a' should be $\frac{3}{3+3+1+3} = \frac{3}{10}$. This definition of an escape frequency is called Method C[8]. As another definition, Method D (and its variation D+ [13]) is well-known and will be used in the experiments of this paper.

Then, from the condition of modeling "ababcab" as stated above, we consider the following symbol to be encoded. For the following symbol, the longest context of the PPM model is "ab" because the maximum order of PPM model is limited to two. Therefore, we start encoding from the context "ab". The behavior of encoding is different in compliance with the following symbol.

Case 'c' The context "ab" is checked to check 'c' occurs after "ab", and the encoding probability is set by the data of frequencies as $\frac{1}{1+1+2} = \frac{1}{4}$.

Case 'b' The context "ab" is checked, but there is no prediction that 'b' occurs after "ab". An escape symbol is then encoded with the probability of $\frac{2}{1+1+2} = \frac{1}{2}$. The context is shortened by one character and the context "b" is checked (in the tree, this corresponds to moving to the parent node). At the parent node, there is again no prediction 'b', and an escape symbol is encoded by the probability of $\frac{2}{1+1+2} = \frac{1}{2}$. Then at the shortened context λ, the prediction that 'b' occurs can be found, and its encoding probability is set to $\frac{3}{3+3+1+3} = \frac{3}{10}$.

Case 'z' Until the model of Order $k = 0$, similar for 'b'. In this case, the order-0 model cannot encode 'z' because it has no prediction of 'z'. An escape occurs again, and the probability is calculated as $\frac{3}{3+3+1+3} = \frac{3}{10}$. Then, the context should be further shortened. For that reason, the model of order $k = -1$

must be defined. This model encodes every alphabet which has a possibility of occurrence equally. This equal probability is defined for the alphabet $\mathcal{A}$ as $1/|\mathcal{A}|$.

This encoding method is a basic one, which may have the following deficiency in estimating probability. When an escape occurs, predictions which cannot be used are takes into account at the short context, which make a part of estimated probabilities unreliable. With the exclusion technique, these can be overcome. This is illustrated by using the running example for the cases of 'b' and 'z'.

Case 'b' There is no problem until an escape occurs at the context "ab". At the context "b", the predictions of occurrence of 'a' and 'c' might be thought to exist. However, these predictions cannot exist actually, because if 'a' or 'c' is encoded, the character has been encoded at the context "ab" which has the predictions of occurrence of 'a' and 'c'. Hence, as far as moving from the context "ab" to the context "b", the escape probability at the context "b" should be considered by deleting the probabilities of 'a' and 'c'. This mechanism as deleting probabilities in this way is called *exclusion*. For this example, an escape occurs by the probability of 1 at the context 'b'. Then, at the context λ, the probabilities of 'a' and 'b' need not be considered for the same reason. Consequently, only the probabilities of 'b' and escape symbol must be considered.

Case 'z' Similar until the context λ. At the context λ, an escape probability is encoded by the probability of $\frac{3}{3+3} = \frac{1}{2}$. Then, exclusion can be also considered at the model of order $k = -1$. Because 'a', 'b' and 'c' cannot occur for the similar reason, 'z' is encoded by the probability at the model of order $k = -1$ as $\frac{1}{|\mathcal{A}|-3}$.

Even in this small running example, the exclusion is seen to work quite efficiently. Especially, for encoding 'b', the second step of encoding is considerably different between probabilities with exclusion and without exclusion. This difference causes the difference of the code length by about three bits.

In concluding this section, we mention inherent learning problems concerning PPM. One of the drawbacks with PPM is that it performs relatively poorly at the beginning of texts. To overcome this, one of the simple schemes is to use training text. This scheme has a problem of which and how much training text to use. Variations on PPM may be mainly regarded as learning/setting appropriate escape probabilities. Also, there is a problem of overlearning. The experiment shows that, even if the order is made much large, the compression rate may become worse. Certainly, when the maximum context length increases, the prediction of the PPM model becomes specific. However, the chance also increases that any prediction cannot apply, that is, the time of escape events is enlarged at the same time. The increase of escape events has a harmful influence on the compression performance. This is related to overlearning problem. CTW does not have this defect, and, at this point, CTW is superior to PPM.

4 Multi-alphabet CTW Using PPM

CTW is known as one of compression schemes attaining the best compression performance in many cases. However, because it is originally for binary alphabet, it becomes less effective when it encodes multi-alphabet data. We here show a new method of applying the prediction of probabilities by PPM which is fit to encode multi-alphabet data to CTW.

4.1 A CTW Method Incorporating PPM

CTW displays strong power for binary data. However, for data of texts, etc., contexts depend on the unit of character (for example, the unit of 8 bits) rather than the the unit of bit. Hence, if CTW can correspond to multi-alphabet data, a method suitable for data of texts can be realized.

The recursion formula of CTW for binary alphabet

$$P_w^s(x_1^t) = \begin{cases} \gamma P_e^s(x_1^t) + (1-\gamma)P_w^{0s}(x_1^t)P_w^{1s}(x_1^t) & \text{(node } s \text{ is not a leaf)} \\ P_e^s(x_1^t) \quad \text{(node } s \text{ is a leaf)} \end{cases}$$

can be expanded directly to the multi-alphabet. For binary data, a probability estimated by longer context is calculated from the multiplication of P_w. For multi-alphabet, a similar technique applies. Denote the set of all the alphabets by $\mathcal{A}$. Then multi-alphabet CTW can be represented by

$$P_w^s(x_1^t) = \begin{cases} \gamma P_e^s(x_1^t) + (1-\gamma) \prod_{a \in \mathcal{A}} P_w^{as}(x_1^t) & \text{(node } s \text{ is not a leaf)} \\ P_e^s(x_1^t) \quad \text{(node } s \text{ a leaf)} \end{cases} .$$

Since methods like KT-estimator,which assigns each character some probability, cannot deal with characters successfully whose frequency of occurrence is extremely small, predicting values of P_e becomes a problem. Specially, in the case of texts where definite character always occurs after specific context, it is not a wise method to treat frequent characters and unseen ones equally.

To solve this problem, we describe a method of computing P_e by PPM mechanism. This idea was proposed in [1], and have been incorporated in the method of CTW implementation using PPM [11]. PPM introduces the notion of the escape symbol. When an unseen character for the observed context is encoded, an escape symbol is encoded and the information for the shorter context is checked. Determining escape frequency by the appropriate method realizes the assignment of probabilities for unseen characters. An idea of escape is unifying unseen characters to one symbol, whose probability is computed on the same level as the characters which has occurred already. By the introduction of escape, unseen characters also can be predicted to the probability being extremely precise. Originally, it is known that PPM itself has higher compression performance than dictionary-based compression methods such as `gzip`. Therefore, PPM can predict probabilities more precisely. Setting probabilities predicted by PPM to P_e^s can achieve higher compression performance.

However, for CTW, the counts of computing P_e^s is extremely large in order that P_e^s must be calculated at each node and at each character. When PPM computes every P_e^s independently, and with this method these computations for CTW take enormous time totally. Hence, P_e^s must be computed efficiently.

To solve this problem, we notice the following fact. Whenever CTW encodes one character, it updates the information of frequency. This updated information is only at the one path of context tree for CTW algorithm. This fact helps the efficient computation of P_e^s.

First, we compute probabilities of characters collectively at each node to decrease the number of times to check the context tree. By the collective computation of probabilities, every P_e^s needed by encoding of one character can be computed along only one path of the context tree.

This method still has a problem. For PPM, the consideration of exclusion produces the improvement of compression performance. However, the above algorithm cannot realize the exclusion. Therefore, predicted probabilities by this algorithm involve some waste.

We can resolve this problem by using the fact that encoding one character for CTW updates only one path of the context tree. Originally, exclusion can be considered when the scope is shifted from the longer context to the shorter one for PPM. Therefore, we can say it is natural that there is a relationship between the exclusion mechanism and the recursion of weighting for CTW.

Computation of P_e^s is executed many times. However, this computation is on the path of context tree. Hence, exclusion can be considered at the definite place and probabilities with exclusion do not change at the same context because information of shorter contexts always involves that of longer contexts. Namely, when exclusion can be considered at some context, exclusion occurs always for the same characters. This occurrence has no connection with the depth of context where PPM starts the prediction.

Exclusion must be considered from the deeper context. However, calculating P_e^s must be consider from the shallower (shorter) context. The reason is that PPM model predicts probabilities by the escape mechanism which uses prediction of shorter context. When probabilities for longer context of PPM are considered, information of probabilities for shorter context is needed.

Accordingly, the path of context must be traced twice in order to compute P_e^s. At the first trace, the probability of each character on the path is checked, and escape probability is encoded from the longer context to handle exclusion. At this time, both probabilities with exclusion and without exclusion are computed. Probabilities without exclusion are applied to probabilities of PPM from that context. Probabilities with exclusion are used in the case of computing probabilities of the longer context by one character. At the second trace, from these probabilities, each P_e^s is computed from the shorter context because of escape mechanism. Then, each P_w must be computed from the longer context because of its definition. As a whole, the context tree must be traced three times for encoding one character. This algorithm can be described as follows;

1. Check the deepest node of the context tree to find its length m.
2. For $d = m, \cdots, 0$ as the length of context, compute probabilities of escape and every character which can be predicted at the watched context. At this time, compute both probabilities with exclusion and without exclusion.

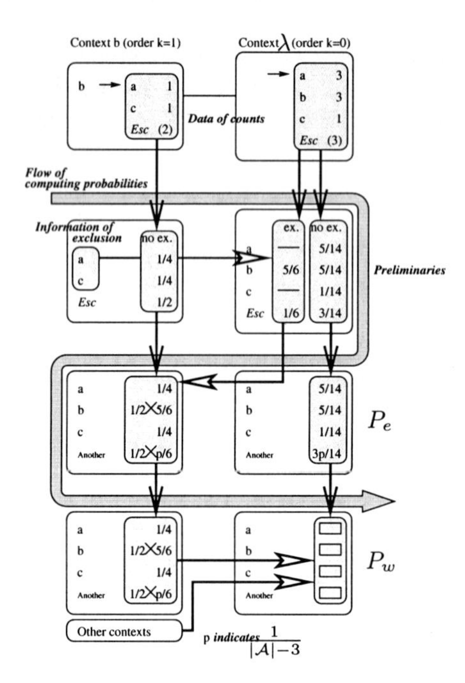

Fig. 4. Implementing multi-alphabet CTW for "ababcab"

3. For $d = 0, \cdots, m$ as the length of context, find $P_e[d][a]$ of each character a by using probabilities computed as above.
4. For $d = m, \cdots, 0$ as the length of context, find $P_w[d][a]$ of each character a.
5. Encode/Decode probabilities according to $P_w[0]$.

Figure 4 depicts the above-mentioned algorithm for encoding the next character to the sequence "ababcab". In Figure 4, the topmost nodes show the data of counts and inner numbers of them mean the counts of characters. Numbers except at the topmost nodes indicate probabilities. Numbers at the second nodes from the top mean values of P_e, and at the bottommost nodes show values of P_w. Arrows in the figure mean that end-data of them is calculated on the basis of base-data. White arrows indicate important ones because they create the flow of this algorithm critically.

At the step of preliminaries, on the basis of the given information of frequency, probabilities of characters which can be computed directly (without escape mechanism) are calculated. These calculations are based on PPMD+. At this time, both probabilities with exclusion and without exclusion must be computed. Therefore, the computation at this step must be done from the deeper context in order to investigate exclusion.

Then, P_e is computed. P_e at each order corresponds to the probability of the character without exclusion. However, the information of probabilities with exclusion at the shorter context becomes needed when a probability of a character which causes escape is investigated.

Lastly, P_w is computed. at the leaves, the value of P_w is simply the same as the value of P_e. However, at their parent nodes, the information of P_w at the children nodes becomes needed because of the recursive definition of P_w. Accordingly, the computation of P_w must be done from the deeper context.

4.2 Experimental Results

We use the workstation Sun Ultra60 (2048MB memory). As sample data of multi-alphabet, Calgary corpus [2] is adopted. It is referred widely to papers of data compression in comparing compression performance.

Table 1. Compression performance of multi-alphabet CTW using PPM

	`gzip -9`	PPMD+	binary CTW			multi-alphabet CTW		
title\\γ			0.5	0.2	0.05	0.5	0.2	0.05
bib	2.509	1.862	2.131	1.926	1.878	1.8030	**1.7945**	1.8123
book1	3.250	2.303	2.266	2.181	2.179	2.2080	**2.2131**	2.2277
book2	2.700	1.963	2.052	1.929	1.916	1.8728	**1.8747**	1.8931
geo	5.345	4.733	4.370	4.366	4.394	4.4543	**4.4569**	4.4619
news	3.063	2.355	2.649	2.450	2.426	2.2952	**2.2895**	2.3043
obj1	3.839	3.728	4.164	3.898	3.925	3.6719	**3.6682**	3.6797
obj2	2.628	2.378	2.842	2.589	2.555	2.2674	**2.2510**	2.2610
paper1	2.790	2.330	2.624	2.412	2.406	2.2773	**2.2751**	2.2991
paper2	2.887	2.315	2.473	2.300	2.293	**2.2355**	2.2378	2.2593
pic	0.817	0.795	0.777	0.772	0.776	**0.7688**	0.7690	0.7705
progc	2.678	2.363	2.746	2.485	2.471	2.3040	**2.2988**	2.3210
progl	1.805	1.677	2.006	1.770	1.706	1.5744	**1.5600**	1.5794
progp	1.812	1.696	2.057	1.832	1.795	1.5824	**1.5616**	1.5746
trans	1.611	1.467	1.909	1.630	1.564	1.3695	**1.3495**	1.3725

Table 1 shows the difference of compression performance between CTW for binary and CTW for multi-alphabet which uses predictions of PPM. Numbers in this table indicate the compression performance, whose units are determined as bpc which means "bit per character". For the implementation of CTW, we must define the maximum depth of the context tree. We define this depth as $D = 64$. Besides, weighting parameters of CTW, γ, are fixed to 0.5, 0.2 and 0.05. As seen from this table, PPM and CTW achieve much higher compression performance than `gzip` with the option of maximizing compression performance. This fact shows the usefulness of PPM and CTW for compression performance. PPMD+ is as good as CTW for binary. CTW cannot achieve the best performance when some model based on multi-alphabet is applied directly to CTW for binary. However, the CTW method for the multi-alphabet case shows higher compression performance than PPMD+. The method using predictions of PPM for CTW can compress data well.

5 Optimizing CTW Parameters

CTW is the method which weights and sums the mutual model of predicting probabilities. How to define these weighting parameters has scarcely been discussed theoretically. Here, we propose the idea of these definition and the method of them. Then, we describe the experimental results based on this method, which achieves the stable compression performance.

5.1 Learning Weighting Parameters

We consider the meaning of the recursion formula of weighting.

$$\gamma P_e^s(x_1^t) + (1-\gamma) P_w^{0s}(x_1^t) P_w^{1s}(x_1^t).$$

This formula means that it weights the probability predicted by only the watched context and the probability by the longer context by one character. Then, it collects these probabilities by using the parameters γ. If the value of this γ is large, the shorter context is regarded as important. Conversely, if the value of γ is small, longer contexts become important. The optimum values of γ ought to be different in different encoding data. Therefore, if we can learn an optimum value for γ adaptively, the prediction of probabilities of characters becomes more precise, which achieves the higher compression performance.

Willems, Shtarkov and Tjalkens, who established the basis of CTW algorithm [15] showed the example that every value of γ is $\frac{1}{2}$ and described the possibility of values of γ. The case of $\gamma = \frac{1}{2}$ has connection with Rissanen's universal compression scheme. However, they do not state clearly what are the optimum values of γ.

We proposed the method in [11] as follows. Whenever one character is encoded, each of predicted probabilities of occurrence of characters are computed for each of several numbers as γ, and the probability with the minimum entropy is adapted to γ. The "candidates" of the numbers as γ are calculated as $\gamma = 0.1(0.8)^i$ for $i = 0, 1, 2, 3, 4$.

This method can achieve better compression performance than that of fixing γ to a certain value. However, this method needs the computation of counts of candidates of γ. When naively implemented, this method takes several times as large as the counts of candidates. Besides, the theoretical basis of the numbers of candidates is weak. Moreover, these methods commonly use the same value of γ at every context when a character is encoded. However, in the case that only the specific character occurs at the specific context, it is more natural that the optimum value of γ is assumed to be different at contexts. Accordingly, if we can regard γ as the function of s, we can predict probabilities for CTW more precisely.

Now, we consider which context we weight greatly in order to bring γ close to the optimum. Because of the argument as mentioned above, we should weight a context greatly where definite characters occur and do a context lightly where characters occur randomly.

We can adopt the entropy as a method of realizing this condition, because the entropy corresponds to predicted code-length of binary bits. When the entropy of a context is large, characters occur randomly at this context. Conversely, when the entropy is small, some definite rules may exist. Accordingly, if we define appropriate rules that a context with large entropy is weighted greatly and vice versa, we obtain the precise predictor for CTW.

However, for the problem of PPM, it is claimed in [3] that considering the most probable symbol's probability is simpler to compute and better in compression performance than considering entropies. This strategy will be called MPS-P hereafter. This reason is not well understood. However, both MPS-P and entropy can assign probabilities as a context with prejudiced probability distribution is weighted greatly.

Both MPS-P and entropy can be guides of teaching the "self-confidence" of the context. If MPS-P of some context is large, or entropy is small, this context has great self-confidence. Then, we consider adapting the ratio of each parameter of MPS-P or entropy to the ratio of γ directly.

However, the shape of the recursion formula of CTW becomes a problem. Because predicted probabilities at the longer context has the the repetitive multiplication of γ, they are comparatively smaller than probabilities at the shorter context. Therefore, it is undesirable that values which is simply normalized as the sum of the above-mentioned parameters equals to one is assigned to γ.

For the solution of this problem, we use the following [10]. When, we watch the shape of the recursion formula. we find that the multiplication of γ for the shortest context P_e^λ is only once. This γ can be the value of the direct ratio of the parameter corresponding to the shortest context to context of every length which must be considered. In other words, γ at context λ can be computed by the parameter at context λ over the sum total of the parameters.

Next, we remove the information at the shortest context λ. Because of the shape of the recursion formula, the probability at the shortest context in resultant information is multiplied by γ only once. Hence, this γ is assigned to the value normalized in the information without that of the shortest context λ. By repeating this computation, we can find γ at context of every length.

5.2 Experimental Results

We again make experiments for Calgary corpus [2]. Tables 2, 3, and 4 show the results of these experiments. Numbers in each table mean compression ratios in bit per character. "mix" in Table 2 means the method [11] that each entropy for five fixed candidates of γ is computed and a certain γ corresponding to the minimum entropy is used in encoding each character, as mentioned before. Although, this method is expected to achieve high compression performance, the computation time is naturally slow.

Table 2 shows the result of adapting our method for the simple CTW method [12]. This method involves the technique of binary decomposition [14] for multi-alphabet data at the viewpoint of CTW.

Table 2. Compression performance compared with each fixed value of γ

title\\γ	0.5	0.2	0.05	mix	Entropy
bib	2.131	1.926	1.878	**1.860**	1.884
book1	2.266	2.181	2.179	2.164	**2.163**
book2	2.052	1.929	1.916	**1.899**	1.907
geo	4.370	4.366	4.394	4.384	**4.327**
news	2.649	2.450	2.426	**2.397**	2.412
obj1	4.164	3.898	3.925	**3.860**	3.941
obj2	2.842	2.589	2.555	**2.523**	2.530
paper1	2.624	2.412	2.406	**2.361**	2.369
paper2	2.473	2.300	2.293	**2.260**	**2.260**
pic	0.777	0.772	0.776	0.772	**0.766**
progc	2.746	2.485	2.471	**2.426**	2.448
progl	2.006	1.770	1.706	**1.688**	1.729
progp	2.057	1.832	1.795	**1.765**	1.780
trans	1.909	1.630	1.564	**1.540**	1.579

Table 3. Compression performance for CTW using PPM

title\\γ	0.5	0.3	0.2	0.1	MPS-P
bib	1.803	1.794	1.795	1.802	**1.789**
book1	2.208	2.210	2.213	2.220	**2.205**
book2	1.873	1.872	1.875	1.883	**1.866**
geo	4.454	4.456	4.457	4.459	**4.452**
news	2.295	2.289	2.290	2.296	**2.277**
obj1	3.672	3.668	3.668	3.673	**3.663**
obj2	2.267	2.254	2.251	2.254	**2.250**
paper1	2.277	2.272	2.275	2.286	**2.267**
paper2	2.236	2.234	2.238	2.248	**2.227**
pic	0.769	0.769	0.769	0.770	**0.762**
progc	2.304	2.297	2.299	2.308	**2.289**
progl	1.574	1.562	1.560	1.567	**1.560**
progp	1.582	1.566	1.562	1.565	**1.561**
trans	1.370	1.353	**1.350**	1.357	1.357

In the case of fixed γ, the compression performance is biased according to γ. Moreover, the optimum value of γ is different according to sample data. If we adopt entropy as the scale of distributing γ, more stable compression performance is achieved, and it is unrelated what is the optimum value of γ. For example, the optimum is 0.2 for `geo` and 0.05 for `book1`. However, for these example, our method achieves higher performance than the method fixing γ.

Besides, for a few example, our method is better than the method of mix. Our method and mix commonly try to decrease entropy. As a merit of our method, compression performance to a certain extent can be achieved without use of the method like mix which spends much time.

Table 3 displays the compression performance for multi-alphabet CTW introducing prediction of probabilities by PPM. In this table, the usefulness of our method using MPS-P as scales of γ is shown. Because multi-alphabet can be treated owing to the computation of PPM, numbers of alphabet decreases which express context. CTW tends to assign probabilities equally from the shortest context to the longest context. Accordingly, a long context creates the scattering of weight, which causes imprecise prediction. CTW using PPM avoid this situation because of the shortness of contexts.

Table 4 shows the comparison of compression performance between entropy and MPS-P in determining the distribution of γ. Bloom's claim [3] that MPS-P is better as scales for PPM is observed to fit to the CTW introducing PPM in this table. However, we cannot state that his claim is suitable also to the independent CTW. The length of context may be related to this reason. If the length of context is small, like PPM, because probabilities are distributed in the neighborhood the adoption of MPS-P related to PPM technique is useful. However, if the longer context is treated, the adoption of MPS-P is useless. Concerning the time of computation, the time of the algorithm controlling γ is

Table 4. Difference of compression performance based on scales of defining γ

title	CTW		CTW of PPM	
	Entropy	MPS-P	Entropy	MPS-P
bib	**1.884**	1.896	1.811	**1.7888**
book1	**2.163**	2.170	2.2128	**2.2053**
book2	**1.907**	1.915	1.8806	**1.8660**
geo	**4.327**	4.331	4.4525	**4.4523**
news	**2.412**	2.423	2.2919	**2.2772**
obj1	3.941	**3.929**	3.6715	**3.6626**
obj2	**2.530**	2.544	2.2731	**2.2503**
paper1	**2.369**	2.388	2.2875	**2.2674**
paper2	**2.260**	2.274	2.2407	**2.2269**
pic	**0.766**	**0.766**	0.7639	**0.7621**
progc	**2.448**	2.468	2.3129	**2.2891**
progl	**1.729**	1.744	1.5908	**1.5599**
progp	**1.780**	1.804	1.5837	**1.5607**
trans	**1.579**	1.601	1.3874	**1.3574**

much slighter than the other computation, such as calculating P_e. The loss of time for our method is little.

Next, we make experiment for DNA sequences. Table 5 displays the experimental result for DNA sequences. Each algorithm exists the preliminary computation as mentioned above. The DNA sequences consist of 256 alphabets. This table shows that controlling γ by MPS-P obtains higher compression performance than by entropy because character of DNA sequence depends on comparatively short context. However, other experiment shows that controlling γ cannot work well when DNA sequences consist of 4 alphabets, probably because entropy or MPS-P shows the similar value for DNA sequences.

Table 6 shows the evidence of this property of DNA sequence. In this table, CTW algorithm is for binary data. For some DNA sequences, PPMD+ with the

Table 5. Difference of compression performance for DNA sequences

DNA sequences	0.5	0.2	0.05	0.02	0.005	Entropy	MPS-P
CHMPXX	1.8400	**1.8384**	1.8390	1.8400	1.8406	1.8388	1.8388
CHNTXX	1.9341	**1.9331**	1.9336	1.9342	1.9343	1.9335	1.9334
HEHCMVCG	1.9612	**1.9596**	1.9603	1.9613	1.9620	1.9604	1.9603
HUMDYSTROP	1.9202	**1.9196**	1.9205	1.9215	1.9223	1.9200	1.9202
HUMGHCSA	1.9247	1.8034	1.5981	**1.5706**	**1.5706**	1.6423	1.6380
HUMHBB	1.9188	**1.9176**	1.9178	1.9184	1.9191	1.9179	1.9180
HUMHDABCD	1.9346	1.9132	1.8854	**1.8824**	1.8876	1.8992	1.8971
HUMHPRTB	1.9258	1.9138	1.8931	**1.8907**	1.8963	1.9037	1.9027
MPOMTCG	1.9641	**1.9631**	1.9634	1.9640	1.9643	1.9633	1.9632
PANMTPACGA	1.8685	1.8658	1.8655	1.8666	1.8680	**1.8654**	**1.8654**
SCCHRIII	1.9487	**1.9477**	1.9480	1.9485	1.9486	1.9484	1.9481
VACCG	1.9042	1.8954	**1.8942**	1.8968	1.8992	1.8980	1.8965

Table 6. Compression ratio of DNA sequences for each maximum length of PPM

DNA sequence	maximum length of context					
	5	4	3	2	1	0
PANMTPACGA	1.930	1.884	**1.869**	**1.869**	1.871	1.879
MPOMTCG	2.009	1.974	**1.964**	**1.964**	1.967	1.981
SCCHRIII	1.980	1.955	**1.948**	1.950	1.952	1.959
VACCG	1.945	1.917	**1.908**	1.909	1.914	1.918
HEHCMVCG	1.996	1.966	**1.959**	1.969	1.973	1.979

preliminary computation for alphabet of two bits is applied as each maximum length of context. Moreover, because alphabet of DNA sequence is limited as four symbols, PPMD+ is remodeled as if every alphabet occurs, probability of escape symbol becomes zero. This table shows that PPMD+ whose maximum length of contexts is three has the best compression performance. Accordingly, DNA sequence depends on shorter context than English texts, which causes the great efficiency of MPS-P. Besides, the experiment [6, 7] indicates the dependency of DNA sequences except HUMGHCSA on short contexts. Only HUMGHCSA can achieve the best compression performance in investigating about length-11 context according to the experimentation [6]. This sequence is known as the repetition of long string frequently occurs.

Then, we return to Table 5. CTW controlling γ by MPS-P does not have the best compression performance. However, the performance near the best can be achieved. For example, for CHMPXX, only fixing γ to 0.2 is better than MPS-P. On the other hand, for HUMGHCSA, MPS-P tries to shift the performance to that of $\gamma = 0.05, 0, 02, 0.005$.

Consequently, our method by entropy or MPS-P is adaptive for a kind of data. Generally, our method provides the improvement and the stability of compression performance.

6 Conclusion

We have adopted two lossless compression methods, PPM and CTW both of them predict the occurrence probabilities of characters from the relationship of dependence on contexts, thus to learn the probabilities is a key for compression. We have shown the relationship between PPM and CTW and to improve their efficiency and interpretations from this relationship.

There are some algorithms which uses EM-algorithm to learn such probabilities, but has less compression performance. Our results really attains almost the best compression ratio without *a priori* knowledge for DNA sequences.

Acknowledgment

The work of the authors has been supported by the Grant-in-Aid for Priority Areas ‘Discover Science’ of the Ministry of Education, Culture, Sports, Science and Technology of Japan.

References

1. J. Åberg and Y. M. Shtarkov. Text compression by context tree weighting. In *IEEE Data Compression Conference*, pages 377–386, March 1997.
2. The Calgary corpus. `http://corpus.canterbury.ac.nz/`.
3. C. Bloom. Solving the problems of context modeling, March 1998. `http://www.cco.caltech.edu/~bloom/papers/ppmz.zip`.
4. J. G. Cleary and W. J. Teahan. Experiments on the zero frequency problem. In *IEEE Data Compression Conference*, page 480, April 1995.
5. J. G. Cleary and I. H. Witten. Data compression using adaptive coding and partial string matching. *IEEE Trans. Communications*, COM-32(4):396–402, 1984.
6. T. Matsumoto. DNA sequence compression algorithms. Bachelor's thesis, University of Tokyo, Department of Information Science, Faculty of Science, 2000.
7. T. Matsumoto, K. Sadakane, H. Imai, and T. Okazaki. Can general-purpose compression schemes really compress DNA sequences? In *Currents in Computational Molecular Biology*, pages 76–77, 2000. Poster at RECOMB 2000.
8. A. Moffat. Implementing the PPM compression scheme. *IEEE Transactions on Communications*, COM-38(11):1917–1921, November 1990.
9. T. Okazaki. *Data Compression Method Combining Properties of PPM and CTW.* Master's thesis, University of Tokyo, March 2001.
10. T. Okazaki and H. Imai. Optimization of weighting parameters for CTW data compression. IPSJ SIG Note SIGAL-75-9, IPSJ, November 2000. (in Japanese).
11. K. Sadakane, T. Okazaki, and H. Imai. Implementing the context tree weighting method for text compression. In *IEEE Data Compression Conference*, March 2000.
12. K. Sadakane, T. Okazaki, T. Matsumoto, and H. Imai. Implementing the context tree weighting method by using conditional probabilities. In *The 22nd Symposium on Information Theory and Its Applications.* SITA, December 1999. (in Japanese).
13. W. J. Teahan. PPMD+. Program. `http://www.cs.waikato.ac.nz/~wjt/software/ppm.tar.gz`.
14. T. J. Tjalkens, P. A. J. Volf, and F. M. J. Willems. A context-tree weighting method for text generating sources. In *IEEE Data Compression Conference*, page 472, March 1997. Posters at `http://ei1.ei.ele.tue.nl/~paul/p_and_w/dcc97p.ps.gz`.
15. F. M. J. Willems, Y. M. Shtarkov, and T. J. Tjalkens. The context tree weighting method: Basic properties. *IEEE Trans. Inform. Theory*, IT-41(3):653–664, 1995.

Discovery of Definition Patterns by Compressing Dictionary Sentences

Masatoshi Tsuchiya[1], Sadao Kurohashi[2], and Satoshi Sato[1]

[1] Graduate School of Informatics, Kyoto University
tsuchiya@pine.kuee.kyoto-u.ac.jp
sato@i.kyoto-u.ac.jp
[2] Graduate School of Information Science and Technology, Tokyo University
kuro@kc.t.u-tokyo.ac.jp

Abstract. This paper proposes an automatic method to discover definition patterns from an ordinary dictionary. There are frequent patterns to describe words and concepts in a ordinary dictionary. Each definition pattern gives a set of similar words and can be used as a template to clarify distinctions among them. To discover these definition patterns, we convert definition sentences into tree structures, and compress them using the MDL principle. The experiment on a Japanese children dictionary is reported, showing the effectiveness of our method.

1 Introduction

How to handle meaning of words is the first, crucial problem to realize intelligent natural language processing. A thesaurus is one way of representing meaning of words in which their hyponym-hypernym relations are described.

A thesaurus, however, lacks information about distinctions among synonyms. When 'a swimming pool' and 'an athletics stadium' are classified into the same group 'a stadium', this classification clearly shows similarity among them, but gives no information of their distinctions. Therefore, we cannot describe the rule to distinguish between the natural sentence "A boy swims in a swimming pool" and the unnatural sentence "A boy swims in an athletics stadium". When a thesaurus which keeps distinctions among synonyms is available, it is easy to solve this difficulty.

This paper proposes an automatic method to discover sets of similar words and to find distinctions among them from an ordinary dictionary. This is surely a step to build a thesaurus which can distinguish among synonyms. There are frequent sub-sentential patterns in a dictionary because its definition sentences are written carefully to describe words and concepts. To discover such sub-sentential patterns, we employ data compression using the MDL principle, which is widely used to discover common sub-structures in given data. Our approach consists of two parts: 1) parsing definition sentences to convert them into graphs, and 2) compressing graphs using the MDL principle to extract those patterns. Each extracted sub-sentential pattern gives a set of similar words and can be used as a template to clarify distinctions among them.

S. Arikawa and A. Shinohara (Eds.): Progress in Discovery Science 2001, LNAI 2281, pp. 284–295.

We now proceed as follows. In Section 2, we discuss about frequent sub-sentential patterns in dictionary sentences. The next section (Section 3) explains the formulation of the description length of the dictionary and shows the algorithm to compress it. The experiment on a Japanese children dictionary is reported in Section 4, showing the effectiveness of our method in Section 5. We finish this paper with a conclusion section (Section 6).

2 Definition Sentences and Definition Patterns

2.1 Definition Patterns

Reading a dictionary, we find that some expressions are used frequently to describe words and concepts. For example, the expression 'comes into flowers' is used 68 times in Reikai Shogaku Kokugojiten [1]. Three of them are:

aburana a type of plant which comes into yellow flowers in spring.
magnolia a type of plant which comes into white flowers in summer.
nemunoko 'silk tree' a type of plant which comes into red flowers in summer, forming a cluster.

The common expression, 'comes into flowers', indicates that these words are plants which bear flowers. More careful observation for these sentences makes us to find that these sentences include words which represent color of flowers and words which denote seasons. Using the variable < color > which denotes an adjective of color, and using the variable <season> which denotes a noun representing a season, the common sub-sentential pattern among these sentences can be unified into the same format as follows:

comes into <color> flowers in <season>.

In this paper, we will call such common sub-sentential patterns *definition patterns*, and call their fixed parts *common features* and call their variable parts *common slots.*

We think that 'aburana', 'magnolia', and 'nemunoki' are similar because they belong the set of words which are described by this definition pattern. In other words, it is able to get a set of similar words when a definition pattern is discovered. Distinctions among them are also extracted by different values of common slots.

2.2 Arbitrariness of Definition Patterns

We have focused the definition pattern 'comes into < color > flowers in < season >' in the previous section, and there are many possible definition patterns of different sizes, such as 'comes into <color> flower', 'comes into flowers' and so on. In the similar way, there are possible patterns of different ranges of values. We can suppose the common slot <noun> is the space which can hold any noun and the definition pattern 'comes into flowers in <noun>'.

To select an appropriate pattern from many possible patterns, we employ the Minimum Description Length (MDL) principle, proposed by Rissanen [2]. In the MDL principle, models are evaluated on the sum of the description length of the model and the description length of the given data based on it, and the model which gives the minimum sum is the best model to describe the given data. There are the preceding study [3] which employs the MDL principle to discovery common sub-structures such as benzene from molecular structures, and the others [4, 5] discovering characters of language from a large corpora based on the MDL principle.

3 Dictionary Description Length

This section explains the definition of the description length of the dictionary to compress its definition sentences using the MDL principle. In our approach, definition sentences are converted into tree structures, then definition patterns are mapped into sub-trees among them. We will define the description length of the dictionary, considering compression using common sub-trees.

3.1 Tree Representation of Definition Sentences

At first, each definition sentence of a dictionary is converted into a tree. Each of its vertices has a label corresponding to a content word, and each of its edges has a connection type representing a modifier-head relation between content words.

To do such a conversion, we need language-dependent analysis. In our experiments, a Japanese morphological analyzer JUMAN, and a parser KNP were employed [6]. Both of them can analyze definition sentences of a Japanese dictionary with a fairly satisfactory accuracy.

After the conversion of raw sentences into trees, we believe our method is language-independent. Although our experiment was done on a Japanese dictionary, in this paper we use their English translations for the explanation.

As a matter of notational convenience, we put a formal definition of tree. A tree t which have n vertices is defined as a 4-tuple,

$$t = (n, V_t, E_t, \Phi_t), \tag{1}$$

where V_t denotes its vertex set, E_t denotes its edge set, and Φ_t denotes a mapping from E_t to a set of ordered pairs of instances of V_t.

3.2 Tree Representation of Definition Patterns

We focus attention on tree representation of definition patterns included in definition sentences.

It is illustrated clearly in Figure 1–a that the expression of the common feature, ‘comes into flowers’, is converted into the common sub-tree which is discovered from both trees.

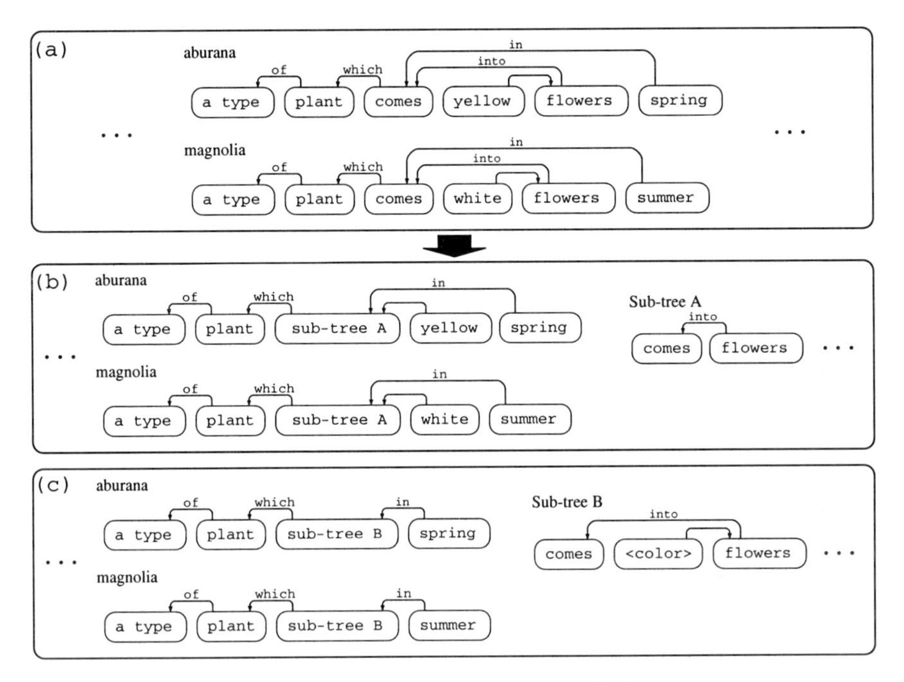

Fig. 1. Compression examples of a dictionary.

Expressions of common slots, such as 'yellow' and 'white', are mapped to vertices which have an edge connecting to the common sub-tree. To represent a vertex corresponding to a common slot, we suppose a *semantic class* which is a set of labels of vertices corresponding to common slots. In our approach, a label is a content word and a semantic class is defined as a set of content words. Suppose the semantic class $<\text{color}>=\{\text{yellow}, \text{white}, \ldots\}$, and the definition pattern 'comes into $<$color$>$ flowers' can be extracted as the sub-tree B which include a semantic class vertex in Figure 1–c.

We also introduce *connection classes* which enable more flexible expressions of definition patterns. A connection class is defined as a set of connection types, considering a hierarchy of relations between content words based on syntactic knowledge.

3.3 Formal Definition of Description Length of a Dictionary

In this section, we give a formal definition of the description length of a dictionary.

Class of a Target Tree Set. To employ the MDL principle, it is necessary that a class of a target data comes a finite set or a countably infinite set. We introduce a label set Σ and a connection type set Γ to express a class of a target tree set which represents a dictionary. A semantic class is defined as a set of content words and a label set Σ satisfies

$$\Sigma \subseteq 2^{\Sigma_0},$$

where Σ_0 denotes a set of all content words in a dictionary. For example, the minimum label set Σ_{min} to describe all trees and sub-trees in Figure 1–c is defined as

$$\Sigma_{\text{min}} = \left\{ \begin{array}{l} \text{a type, plant, comes, flowers, spring,} \\ \text{summer, white, yellow, <color>} \end{array} \right\}.$$

Because a connection class is defined as a set of connection types, a connection type set Γ satisfies

$$\Gamma \subseteq 2^{\Gamma_0},$$

where Σ_0 denotes a set of all modifier-head relations between content words in a dictionary. Finally, considering that both of a label set and a connection type set come a finite set, and that an element of the target set is a tree which is defined like Formula 1, they leads that the class of the target tree set comes a countably infinite set.

Description Length of a Simple Tree. Let us first consider how to calculate a simple tree description length $L(t|\Sigma_0, \Gamma_0)$ where t denotes a discrete tree which meets these restrictions: 1) all vertices are labeled by instances of the label set Σ_0 which includes no semantic classes, 2) all edges are typed by instances of the connection type set Γ_0 which includes no type classes. On the given label set Σ_0 and the given type set Γ_0, we need

$$L(t|\Sigma_0, \Gamma_0) = -\log P(t|\Sigma_0, \Gamma_0), \tag{2}$$

to encode t using a non-redundant code. When all probability distributions of size of tree, vertices, edges, and graph structure are independent, the probability of realizing t is defined as a joint probability distribution:

$$\begin{aligned} P(t|\Sigma_0, \Gamma_0) &= P(n, V_t, E_t, \Phi_t|\Sigma_0, \Gamma_0) \\ &= P(n)P(V_t|\Sigma_0)P(E_t|\Gamma_0)P(\Phi_t|V_t, E_t). \end{aligned} \tag{3}$$

If we assume that each occurrence of vertices is independent, the conditional probability $P(V_t|\Sigma_0)$ is estimated as

$$P(V_t|\Sigma_0) = \prod_{v \in V_t} P(v|\Sigma_0). \tag{4}$$

When the probability distribution over Σ_0 is supposed as a uniform distribution, the conditional probability $P(v|\Sigma_0)$ is estimated as

$$P(v|\Sigma_0) = \frac{1}{|\Sigma_0|}. \tag{5}$$

Suppose that each occurrence of edges is independent, the conditional probability $P(E_t|\Gamma_0)$ of mapping from each edge of E_t to a instance of Γ_0 is estimated as

$$P(E_t|\Gamma_0) = \prod_{e \in E_t} P(e|\Gamma_0). \tag{6}$$

When the probability distribution over Γ_0 is supposed as a uniform distribution, the conditional probability $P(e|\Gamma_0)$ is estimated as

$$P(e|\Gamma_0) = \frac{1}{\Gamma_0}. \tag{7}$$

In Japanese, all vertices except the last one of the sentence have an edge connecting to a following vertex respectively. Then, the number of possible tree structures is equal to the factorial of $(|V_t|-1)$. When the probability distribution of realizing t over V_t and E_t is a uniform distribution, we estimate

$$P(\Phi_t|V_t, E_t) = \frac{1}{(|V_g| - 1)!}. \tag{8}$$

The probability $P(n)$ must satisfy two restrictions: 1) it is greater than zero over all positive numbers, and 2) it satisfies $\sum_n^\infty P(n) \leq 1$. As such probability distribution, we employ the probability P_{cer} that gives universal encoding of positive numbers.

$$P(n) = P_{cer}(n) \tag{9}$$

Finally, the description length of a tree t is derived as follows:

$$L(t|\Sigma_0, \Gamma_0) = |V_t| \log |\Sigma_0| + |E_g| \log |\Gamma_0| + \sum_{i=1}^{|V_t|} \log i - \log P_{cer}(n). \tag{10}$$

Description Length of a Sub-tree. The description length of a sub-tree s is derived in the similar way with Formula 10 as

$$L(s|\Sigma, \Gamma) = |V_s| \log |\Sigma| + |E_s| \log |\Gamma| + \sum_{i=1}^{|V_s|} \log i, \tag{11}$$

where V_s denotes a vertex set of s, E_s denotes an edge set of s.

Description Length of a Condensed Tree. We focus attention on the description length of a tree t which was condensed by using sub-trees included in the sub-graph set Ω. Sub-trees may include semantic class vertices and connection class edges, then we require more bits to describe their actual labels and their actual types. Some bits are also required to describe possibilities of connections inside and outside of sub-trees.

Let us consider bits which is required to describe actual labels of semantic class vertices. The number of possible mappings from each vertex of the vertex set V_s to its actual label belongs Σ_0 is computed as

$$X(s) = \prod_{v \in V_s} |C(v)|, \tag{12}$$

where $C(v)$ is the function of v gives its semantic class. As a matter of notational convenience, we put that $C(v)$ returns $\{v\}$ when the label of the vertex v is not semantic class. If we assume that the distribution of the occurrences of actual labels is uniform, Formula 5 is replaced by this expression,

$$P'(v|\Sigma) = \frac{1}{|\Sigma|} \cdot \frac{1}{X(v)}. \tag{13}$$

We will discuss bits required to describe actual types of connection class edges in the similar way. On the sub-tree s, the number of possible mappings from each edge of the edge set E_s to its actual label belongs Γ_0 is computed as

$$Y(s) = \prod_{e \in V_s} |C(e)|, \tag{14}$$

where $C(e)$ is the function of e gives its connection class. The function $C(e)$ returns $\{e\}$ when the type of the edge e is not connection class. If we assume that the distribution of the occurrences of actual types is uniform, Formula 7 is replaced by this expression,

$$P'(e|\Gamma) = \frac{1}{|\Gamma|} \cdot \frac{1}{Y(e)}. \tag{15}$$

When a vertex is labeled as a sub-tree and some edges connect it, to distinguish their inside end, some bits are also required. Therefore, Formula 8 is replaced by this expression,

$$P'(\Phi_t|V_t, E_t) = \frac{1}{(|V_t| - 1)! \cdot \prod_{v \in V_t} |V_v|^{|E_v|}}, \tag{16}$$

where E_v denotes a set of edges connecting to the vertex v, V_v denotes a vertex set of the sub-tree v. When v is not labeled as a sub-tree, we put that $V_v = \{v\}$.

Finally, the description length of a tree t which was condensed by using sub-trees is defined as

$$\begin{aligned} L(t|\Sigma + \Omega, \Gamma) = {} & |V_t| \log |\Sigma + \Omega| + \sum_{v \in V_t} \log |C(v)| \\ & + |E_t| \log |\Gamma| + \sum_{e \in E_t} \log |C(e)| \\ & + \sum_{i=1}^{|V_t|} \log i + \sum_{v \in V_t} |E_v| \log |V_v| - \log P_{cer}(n). \end{aligned} \tag{17}$$

Description Length of a Dictionary. Consequently, the description length of a set of trees D is derived as

$$L(D) = \sum_{t \in D} L(t|\Sigma + \Omega, \Gamma) + \sum_{s \in \Omega} L(s|\Sigma, \Gamma) + L(\Sigma) + L(\Gamma), \quad (18)$$

where $L(\Sigma)$ denotes the description length of Σ, $L(\Gamma)$ denotes the description length of Γ.

Since there is no clear ground to select a good Σ, we assume that $L(\Sigma)$ is equal to any Σ. The same may be said of $L(\Gamma)$. Then, the third item and the fourth item of Formula 18 can be neglected, and we obtain

$$L'(D) = \sum_{t \in D} L(t|\Sigma + \Omega, \Gamma) + \sum_{s \in \Omega} L(s|\Sigma, \Gamma) \quad (19)$$

as the objective function of the dictionary compression based on the MDL principle.

3.4 Search Algorithm

As we mentioned above, the goal of the MDL based compression is to find a label set Σ, a type set Γ and a set of sub-graphs Ω which minimize the value of the objective function. The possible combination of Σ, Γ and Ω, however, becomes so large that it is intractable to find the optimal solution by considering all possible cases.

Furthermore, since their changes have relation to each other, neither divide and conquer method nor dynamic programming method can handle this problem. Therefore, we give up to find the optimal solution, and take an iterative improvement method based on heuristics.

First of all, we set Σ as s set of all content words in a dictionary and all semantic classes in a handmade thesaurus to reduce search space. Secondly, a fixed type set Γ is prepared based on syntactic knowledge of the target language. Having discussed that it is impossible to get the optimal solution of a sub-graph set Ω, we design a search procedure based on an iterative improvement method for Ω and a beam search for each element of Ω. Our search procedure consists of the following steps:

1. A list of all pairs of vertices is made which exist in the target graph set D.
2. If a sub-graph in the list contains a word vertex or a simple typed edge, new sub-graphs are added to the list whose word vertex or simple typed edge is replaced with any semantic classes or type classes.
3. All sub-graphs are sorted depending on its frequency. For the top-n sub-graphs, their scores are calculated and the rest sub-graphs are deleted from the list. The score of a sub-graph s is equal to the value of the objective function when s is added to Ω.
4. For the best sub-graph which give the minimum score, if its score is worse than L'_{best}, go to the step 7.

5. The best sub-graph is substituted for s_{best}, and its score for L'_{best}.
6. All sub-graphs are sorted depending on its score. New sub-graphs consists of a sub-graph of the top-m sub-graphs in the list, and a vertex connecting to it are added to the new list. The list is replaced with the new one, and back to the step 2.
7. The best subgraph s_{best} is added to Ω, and each of its occurrence in D is replaced to a condensed vertex.
8. This iterative procedure will be continued while the last replacement improves the value of the objective function.

4 Experiment

We applied the method described so far into Reikai Shogaku Kokugojiten, a Japanese dictionary for children [1]. Table 1 shows statistics of the dictionary. Definition sentences were extracted from the dictionary by a simple filter, and converted to graph structures by JUMAN and KNP.

It was impossible to make an experiment by using a whole dictionary because the memory of our computer was limited (Our computer is Sun Enterprise–3500 with 4GB memory). Therefore, we reduced by half the number of the head words and eliminated sentences which included words which have no category on the thesaurus.

Table 1. Statistics of the dictionary.

	whole	target
# of head words	28015	10087
# of definition sentences	52919	13486
# of vertices	179597	53212

As a thesaurus to reduce the search space of Σ, we employed Bunruigoihyou, a Japanese thesaurus [7]. In case of some words which have several categories on the thesaurus, one of them was selected randomly and the other were ignored (Words of this thesaurus have 1.2 categories on average). Because our heuristic search procedure depending on frequency is confused by general semantic classes, such classes which were close to the root of the thesaurus were deleted. Next, we defined the fixed type set Γ based on dependency-types between content words assigned by KNP.

We employed the discussed algorithm on the fixed sets Σ and Γ, and discovered 1409 sub-graphs. Through this search process, the value of the objective function decreased from 1153455.5 bit to 993121.6 bit. Figure 2 shows the trace of $L'(D)$. It was equivalent to 13.9% compression.

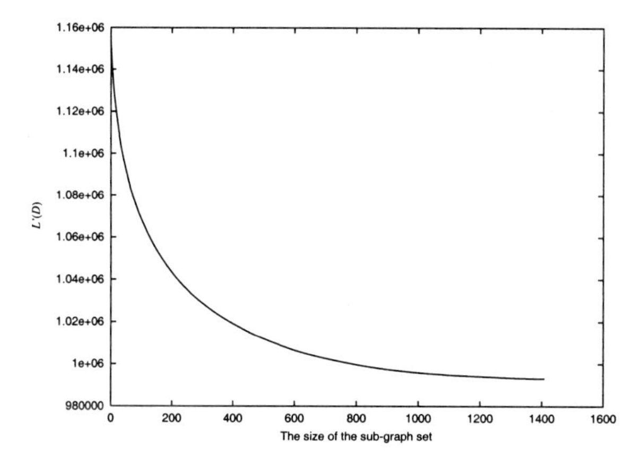

Fig. 2. Description length of the dictionary.

5 Discussion

Figure 3 shows some compressed definition sentences which are translated into English. Most features detected by our method are reasonable based on our linguistic intuition.

The common sub-graph "to < arrange > neatly" is discovered from three definition sentences of *tidying*, *arrangement* and *appearance*. Definition sentences of other head words such as *hairdressing* also include the similar sub-graph "to < arrange> ", but the these three head words are the most similar among them. We checked there is no suitable word in 22 head words which include the expression, "neatly".

The common sub-graph "is <cook> to eat" is discovered from the definition sentences of the five head words such as *kidney beans* and *spaghetti.* All these five head words are foodstuffs and it is interesting that proper semantic class <cook> is discovered. The definition sentences of *asari*, *eel* and *shijimi* include the other sub-graph "is made into <dish> to eat", which includes the common part "to eat", and they also concerns foodstuffs. There is few foodstuff head words except these eight head words in the dictionary.

We discover the common sub-graph "comes into < color > flowers" from the definition sentences whose head words are *gardenia*, *magnolia* and so on. These head words are hyponyms of plants which bloom flowers. These examples shows the effectiveness of our method which can extract the hyponym-hypernym relation and the attribute <color> which is shared by these hyponyms.

6 Conclusion

In this paper, we proposed an automatic method to discover sets of similar words and distinctions among them. Observing definition sentences in a real dictionary

haircut
to shape the hair. <arrange>

tidying
to arrange neatly. <arrange>

arrangement
to tidy things neatly. <arrange>

appearance
to straighten clothes neatly. <arrange>

precise
neatly and correct in regard to the smallest details.

kidney beans
their green pods which are boiled <cook> to eat, and their beans which are made into anko.

spaghetti
an Italian food which is boiled <cook> to eat and served with some kind of sauce.

soumen 'vermicelli'
a Japanese food which is boiled <cook> to eat.

tanishi 'mud snail'
food snail which is poached <cook> to eat.

hijiki 'algae'
food algae which is boiled <cook> to eat.

asari 'clam'
food clam which is made into miso soup <dish> to eat.

eel
food fish which is made into kabayaki <dish> to eat.

shijimi 'clam'
food clam which is made into miso soup <dish> to eat.

gardenia
a type of plant which comes into aromatic white <color> flowers in summer

magnolia
a type of plant which comes into big white <color> flowers in early summer

Fig. 3. Samples of compressed definition sentences.

written in natural language, we found that there are definition patterns which are used frequently to describe words and concepts. These definition patterns consist of common features and common slots, and give sets of similar words and distinctions among them. To discover these patterns, an automatic method was designed. Its first step is parsing dictionary sentences to convert them into graphs, and its second step is to compress graphs using the MDL principle. We reported an experiment to compress a Japanese children's dictionary, and showed several interesting results of discovery of definition patterns among definition sentences, indicating the effectiveness of our method.

The target of our future research is to generate a thesaurus which includes descriptions of distinctions among synonyms.

References

1. Junichi Tadika, editor. *Reikai Shogaku Kokkugojiten.* Sansei-do Co., 1997.
2. J. Rissanen. *Stochastic Complexity in Stochastic Inquiry.* World Scientific Publishing Company, 1989.
3. Diane J. Cook and Laerence B. Holder. Substructure discovery using minimum description length and background knowledge. *Journal of Artificial Intelligence Research*, 1:231–255, 1994.
4. Hang Li. Generalizing case frames using a thesaurus and the mdl principle. *Computational Linguistics*, 24(2):217–244, 1998.
5. H. Bunt and R. Muskens, editors. *Minimum Description Length and Compositionality*, volume 1, pages 113–128. Kluwer, 1999.
6. Sadao Kurohashi and Makoto Nagao. A syntactic analysis method of long Japanese sentences based on the detection of conjunctive structures. *Computational Linguistics*, 20(4), 1994.
7. The National Language Research Institute. *Bunruigoihyou*, 1993.

On-Line Algorithm to Predict Nearly as Well as the Best Pruning of a Decision Tree

Akira Maruoka and Eiji Takimoto

Graduate School of Information Sciences, Tohoku University
Sendai, 980-8579, Japan
{maruoka, t2}@ecei.tohoku.ac.jp

Abstract. We review underlying mechanisms of the multiplicative weight-update prediction algorithms which somehow combine experts' predictions to obtain its own prediction that is almost as good as the best expert's prediction. Looking into the mechanisms we show how such an algorithm with the experts arranged on one layer can be naturally generalized to the one with the experts laid on nodes of trees. Consequently we give an on-line prediction algorithm that, when given a decision tree, produces predictions not much worse than the predictions made by the best pruning of the given decision tree.

1 Introduction

Multiplicative weight-update prediction algorithms for predicting the classification of an instance from environment have been intensively investigated (see [1, 2, 5, 7, 8]). In particular the computational mechanism behind such algorithms has been applied to seemingly different topics such as the boosting algorithm, game theory and pruning of decision trees. In this paper we present the underlying computational mechanisms in most applicable forms and show how the off-line dynamic programming scheme can be transformed into a corresponding on-line algorithm so that we can obtain a corresponding on-line algorithm that, given a decision tree, performs nearly as well as the best pruning of the decision tree. Compared to a similar algorithm due to Helmbold and Schapire which is supposed to work only for the absolute loss function, ours is so simple and general that it works for a wide class of loss functions. We can also give an on-line prediction algorithm that is competitive not only with the best pruning but also with the best prediction values for the nodes in the decision tree.

2 On-Line Prediction Model

In the on-line prediction model a *prediction algorithm* produces its prediction combining somehow predictions given by *experts*. The pool of the experts is denoted by $\mathcal{E} = \{\mathcal{E}_1, \ldots, \mathcal{E}_N\}$. Let the *instance space* be denoted by X, the *outcome space* be denoted by Y, and the *prediction space* be denoted by $\hat{Y}$. At each trial $t = 1, 2, \cdots$ the i th expert $\mathcal{E}_i$ receives an instance $x^t \in X$, generates its prediction

S. Arikawa and A. Shinohara (Eds.): Progress in Discovery Science 2001, LNAI 2281, pp. 296–306.

$\xi_i^t \in \hat{Y}$, and sends a prediction to the *master algorithm.* The master algorithm makes its prediction $\hat{y}^t \in \hat{Y}$ by combining somehow the experts' predictions $\xi_1^t \cdots, \xi_N^t$. After these predictions being made, the correct outcome $y^t \in Y$ for instance x^t is observed. The master algorithm and the experts suffer loss which is given in terms of a loss function $\lambda : Y \times \hat{Y} \to [0, \infty]$: The master algorithm suffers loss $\lambda(y^t, \hat{y}^t)$ and the ith expert suffers loss $\lambda(y^t, \xi_i^t)$. Various loss functions are introduced to measure loss incurred by the discrepancy between the predictions and the actual outcomes. Typical examples of the outcome space and the prediction space are $\{0, 1\}$ and $[0, 1]$, respectively.

In the on-line prediction model, the prediction algorithm maintains a weight $w_i^t \in [0, 1]$ for each expert $\mathcal{E}_i$ that reflects the actual performance of the expert $\mathcal{E}_i$ up to time t. For simplicity, we assume that the weight is set to $w_i^1 = 1/N$ for $1 \leq i \leq N$ at time $t = 1$. In order to specify the update rule we introduce the parameter $\beta \in (0, 1)$ of the algorithm called the *exponential learning rate.* After receiving the correct outcome $y_t \in Y$ at the tth trial, the algorithm updates the weights of the experts according to the rule

$$w_i^{t+1} = w_i^t \beta^{\lambda(y_t, \xi_i^t)}$$

for $1 \leq i \leq N$ and $1 \leq t \leq T$. Because $0 < \beta < 1$, the larger the expert $\mathcal{E}_i$'s loss is, the more its weight decreases. The *cumulative loss* of a prediction algorithm A and that of the ith expert for outcome sequence $y = (y^1, \ldots, y^T)$ are given by

$$L_A(y) = \sum_{t=1}^{T} \lambda(y^t, \hat{y}^t)$$

and

$$L_i(y) = \sum_{t=1}^{T} \lambda(y^t, \xi_i^t),$$

respectively. The goal of the prediction algorithm A is to minimize the cumulative loss $L_A(y) = \sum_{t=1}^{T} \lambda(y^t, \hat{y}^t)$ for outcome sequence $y = (y^1, \ldots, y^T)$ arbitrarily given. The cumulative loss is simply called the loss of A.

What is mentioned above is a typical framework for prediction algorithms. To design a particular prediction algorithm we need to specify what the outcome and prediction spaces are and how to determine a prediction at each trial and so on. A simple and yet good way of making a prediction is to take the inner product of the normalized weight vector and the vector consisting of experts' predictions:

$$\begin{aligned} \bar{w}^t \cdot \xi^t &= (\bar{w}_1^t, \cdots, \bar{w}_N^t) \cdot (\xi_1^t, \cdots, \xi_N^t) \\ &= \sum_{i=1}^{N} \bar{w}_i^t \xi_i^t, \end{aligned}$$

where $\bar{w}_i^t$'s are the normalized weights given as follows:

$$\bar{w}_i^t = w_i^t \Big/ \sum_{i=1}^{N} w_i^t.$$

Suppose that

$$\beta^{\lambda(y^t,\sum_i \bar{w}_i^t \xi_i^t)} \geq \sum_i \bar{w}_i^t \beta^{\lambda(y^t,\xi_i^t)}$$

holds for any $1 \leq t \leq T$. In another words, putting

$$f_y(x) = \beta^{\lambda(y,x)},$$

suppose that the Jensen's inequality

$$f_y(E(X)) \geq E(f_y(X))$$

hold for any $y \in Y$, where $E(\ )$ denotes the expectation under the distribution $\{\bar{w}_i^t\}$. [4] verified that the loss of the prediction algorithm A that produces predictions given by the inner product of the normalized weight vector and the experts' prediction vector is upper bounded by

$$L_A(y) \leq \min_{1 \leq i \leq N} \{L_i(y) + \frac{\ln N}{\ln(1/\beta)}\}.$$

This is a typical statement concerning loss of a prediction algorithm saying that performance of the prediction algorithm is not "much worse" than the predictions made by the best expert. The upper bound above is valid for any loss function as long as the loss function satisfies Jensen's inequality. In this paper we deal with an algorithm that works for an arbitrary loss function and drive an upper bound on the loss of the algorithm. The algorithm we deal with is called an aggregating algorithm (AA for short).

In section 3, we give an algorithm for predicting nearly as well as the best pruning of a decision tree. The algorithm is based on the observation that finding the best pruning can be efficiently solved by a dynamic programming in the "batch" setting. In order to implement the idea in the "on-line" setting we extend the notion of the Aggregating Algorithm to have the Aggregating Pseudo-Algorithm (APA for short), which gives a "first approximation" to the Aggregating Algorithm by combining and generating not the actual predictions but what we call the pseudopredictions. When we want distinguish actual predictions from pseudopredictions, we refer to the predictions given by elements of $\hat{Y}$ as *genuine predictions*. In what follows we use the same symbol ξ to represent not only a genuine prediction but also a pseudoprediction. A genuine prediction $\xi \in \hat{Y}$ on each round can also be treated as a pseudoprediction $\xi : Y \to [0, \infty]$: For any $y \in Y$ $\xi(y)$ is equal to the loss $\lambda(y^t, \xi)$, where y^t is the correct outcome. Here again we use ξ to denote the two different objects.

More precisely, a pseudoprediction is a function from Y to $[0, \infty]$. In particular, the pseudoprediction, denoted $r(y)$, that is obtained by combining genuine predictions $\xi_1^t, \cdots, \xi_N^t$ produced by N experts at the t th trial is given by

$$\beta^{r(y)} = \sum_{i=1}^{N} \beta^{\lambda(y,\xi_i^t)} \bar{w}_i^t,$$

where $\bar{w}_i^t$ is the normalized weight of the i th expert at the t th trial. Similarly the pseudoprediction $r(y)$ that is obtained by combining pseudopredictions $\xi_1^t, \cdots, \xi_N^t$ produced by N experts is given by

$$\beta^{r(y)} = \sum_{i=1}^{N} \beta^{\xi_i^t(y)} \bar{w}_i^t.$$

So the pseudoprediction actually gives the weighted average of losses of N experts for each outcome in Y. Note that the pseudoprediction is rewritten as

$$r(y) = \log_\beta(\sum_{i=1}^{N} \beta^{\lambda(y,\xi_i^t)} \bar{w}_i^t)$$

as in the expressions in the following algorithms.

We first explain the algorithm APA. In APA not only the master algorithm but also the experts are supposed to produce pseudopredictions. The weight-update rule for the algorithm is given by

$$w_i^{t+1} := w_i^t \beta^{\xi_i^t(y^t)}$$

for $1 \leq i \leq N$ and $1 \leq t \leq T$, where the algorithm receives y^t just before the update of the weight. The complete description of the algorithm APA due to Vovk [8] is given in Figure 1.

```
for i ∈ {1,...,N} do
    w_i^1 := 1
for t := 1,2,... do
    receive (ξ_1^t,...,ξ_N^t)
    for y ∈ Y do
        r^t(y) := log_β Σ_i β^{ξ_i^t(y)} w̄_i^t
    output r^t
    receive y^t
    for i ∈ {1,...,N} do
        w_i^{t+1} := w_i^t β^{ξ_i^t(y_t)}
```

Fig. 1. Algorithm APA(β)

The loss of APA(β) and that of expert $\mathcal{E}_i$ for $y = (y^1, \ldots, y^T)$ are given by

$$L_{\mathrm{APA}(\beta)}(y) = \sum_{t=1}^{T} r^t(y^t)$$

and

$$L_i(y) = \sum_{t=1}^{T} \xi_i^t(y^t),$$

respectively. Then we have the following theorem.

Theorem 1 ([8]). *Let $0 < \beta < 1$. Then, for any $N \geq 1$, any N experts $\mathcal{E}_i$'s and for any $y \in Y^*$*

$$L_{\mathrm{APA}(\beta)}(y) = \sum_{t=1}^{T} r^t(y^t) \leq \min_{1 \leq i \leq N} \left(L_i(y) + \frac{\ln N}{\ln(1/\beta)} \right).$$

When the prediction algorithm and the experts are required to produce genuine predictions we need to take into account loss incurred in terms of the loss function. So the loss $\xi_i^t(y)$ in APA is replaced with $\lambda(y, \xi_i^t)$ in AA, and AA outputs genuine prediction $\hat{y}^t$ which is somehow determined relying on pseudoprediction r. As shown in AA the $\hat{y}^t$ is given in terms of the substitution function, denoted $\sum_\beta$. That is, $\hat{y}^t = \sum_\beta(r)$. We do not have enough space for the argument on the definition of the substitution function $\sum_\beta$. But it turns out that in order to obtain an upper bound, like the one given by [4], on the cumulative loss of AA, it suffices to show that the following inequality holds for *any* $y \in Y$ for *appropriately chosen* $\hat{y} \in \hat{Y}$ and a constant c

$$\lambda(y, \hat{y}) \leq c r^t(y),$$

where r^t is the pseudoprediction at the tth trial. Let us define a *β-mixture* as the function $r : Y \to [0, \infty]$ given by

$$\beta^{r(y)} = \sum_{\xi \in \hat{Y}} \beta^{\lambda(y,\xi)} P(\xi)$$

for a probability distribution P over $\hat{Y}$. A *β-mixture* can be viewed as a possible pseudoprediction. Let R_β denote the set of all β-mixture. When we try to drive an upper bound on the loss of the aggregating algorithm, $\sum_{t=1}^{T} \lambda(y^t, \hat{y}^t)$, it is crucial to make the inequality

$$\lambda(y^t, \hat{y}^t) \leq c r^t(y^t)$$

holds on each round. Together with the inequality $\sum_{t=1}^{T} r^t(y^t) \leq \min_{1 \leq i \leq N}(L_i(y) + (\ln N)/\ln(1/\beta))$ given by Theorem 1, the inequality above gives an upper bound on the loss $\sum_{t=1}^{T} \lambda(y^t, \hat{y}^t)$, which will be described as in Theorem 2. Since, the smaller gets the coefficient c, the stronger becomes the upper bound thus obtained, we'll take, as constant c, the smallest number that satisfies the inequality. The consideration above leads to the following definition.

Definition 1. The *mixability curve*, denoted $c(\beta)$, is defined as the function that maps $\beta \in (0, 1)$ to the real value given by

$$c(\beta) = \inf\{c \mid \forall r \in R_\beta \exists \hat{y} \in \hat{Y} \forall y \in Y : \lambda(y, \hat{y}) \leq c r(y)\}$$

with $\inf \phi = \infty$. Moreover, a function $\sum_\beta : R_\beta \to \hat{Y}$ that maps β-mixture r to a prediction $\hat{y} = \sum_\beta(r)$ such that $\lambda(y, \hat{y}) \leq c(\beta) r(y)$ is called a *β-substitution function*. (Throughout the paper we assume that such $\hat{y}$ always exists.)

The constant $c(\beta)$ and the substitution function $\Sigma_\beta(r)$ have been obtained for popular loss functions such as the absolute loss, the square loss and the log loss functions [8]. In particular, when we consider the absolute loss function $\lambda(y,\hat{y}) = |y - \hat{y}|$ for $y \in Y = \{0,1\}$, $\hat{y} \in \hat{Y} = [0,1]$, it was shown [8] that

$$c(\beta) = \frac{\ln(1/\beta)}{2\ln(2/(1+\beta))},$$

and

$$\sum\nolimits_\beta(r) = \mathbf{I}_0^1\left(\frac{1}{2} + c(\beta)\frac{r(0)-r(1)}{2}\right),$$

where

$$\mathbf{I}_0^1(t) = \begin{cases} 1, & \text{if } t > 1, \\ 0, & \text{if } t < 0, \\ t, & \text{otherwise.} \end{cases}$$

Note that the absolute loss $|y-\hat{y}|$ is exactly the probability of the probabilistically predicted bit differing from the true outcome y. As shown in Theorem 2 below, when we assume the absolute loss the upper bound on the loss of AA is given in terms of the constant $c(\beta)$ for the absolute loss. This contrasts with the fact that, since the absolute loss does not satisfies the condition for Jensen's inequality, the bound for the weighted average of experts' losses can no longer be apply for the absolute loss. Thus the mixability curve $c(\beta)$ and the β-substitution function $\sum_\beta(r)$ in Definition 1 can be viewed as the ones for giving the optimal upper bounds on the cumulative loss whatever loss function we may assume.

We show algorithm $AA(\beta)$ in Figure 2. It should be noticed that the algorithm does not necessarily fully specify how the prediction is computed, e.g., it is not described how to compute the function $\sum_\beta(r)$ in the algorithm. As we can see above, what is left unspecified in the algorithm is given when the loss function in question is specified.

for $i \in \{1,\ldots,N\}$ **do**
  $w_i^1 := 1/N$
for $t := 1,2,\ldots$ **do**
  receive $(\xi_1^t,\ldots,\xi_N^t)$
  for $y \in Y$ **do**
    $r^t(y) := \log_\beta \sum_i \beta^{\lambda(y,\xi_i^t)} \overline{w}_i^t$
  output $\hat{y}_t := \Sigma_\beta(r^t)$
  receive y^t
  for $i \in \{1,\ldots,N\}$ **do**
    $w_i^{t+1} := w_i^t \beta^{\lambda(y^t,\xi_i^t)}$

Fig. 2. Algorithm AA(β)

The loss of the aggregating algorithm, denoted $L_{\mathrm{AA}(\beta)}(y)$, is defined as

$$L_{\mathrm{AA}(\beta)}(y) = \sum_{t=1}^{T} \lambda\left(y^t, \Sigma_\beta(r^t)\right).$$

By the definitions of the mixability curve $c(\beta)$ and the substitution function Σ_β, we have an upper bound on the loss of the aggregating algorithm as follows.

Theorem 2 ([8]). *Let $0 < \beta < 1$, and let $c(\beta)$ give the value of the mixability curve at β. Then, for any $N \geq 1$, any N experts $\mathcal{E}$ and for any $y \in Y^*$,*

$$\begin{aligned} L_{\mathrm{AA}(\beta)}(y) &= \sum_{t=1}^{T} \lambda(y^t, \Sigma_\beta(r^t)) \\ &\leq \min_{1 \leq i \leq N} \left(c(\beta) L_i(y) + \frac{c(\beta) \ln N}{\ln(1/\beta)} \right). \end{aligned}$$

3 Applying the Aggregating Algorithm to Prune Decision Trees

Suppose that a decision tree $\mathcal{T}$ over an alphabet $\sum$ is given whose leaves, as well as its inner nodes, are associated with elements from the prediction space $\hat{Y}$. A *sample* is a sequence of pairs from $\sum^* \times Y$, which is denoted by $S = ((x^1, y^1), \cdots, (x^T, y^T))$. A pruned tree of $\mathcal{T}$ is naturally thought of as a predictor for a sample, where y_t is considered as the correct classification of instance x^t for the t th trial. Loss for a pruned tree can be defined similarly as a cumulative loss in terms of a loss function as in the previous section. In this section we consider the following problem: Given a decision tree $\mathcal{T}$ and a sample S, find the pruned tree $\mathcal{P}$ of $\mathcal{T}$ that minimizes the cumulative loss on sample S. To solve the problem in the on-line setting one may enumerate all the pruned trees of a given decision tree and apply the AA with the pruned trees being taken as experts so as to achieve performance that is "not much worse" than the best pruning of the given decision tree. Unfortunately this naive approach has a fatal drawback that we have to deal with the exponential number of the prunings of $\mathcal{T}$.

A template tree $\mathcal{T}$ over Σ is a rooted, $|\Sigma|$-ary tree. Thus we can identify each node of $\mathcal{T}$ with the sequence of symbols in Σ that forms a path from the root to that node. In particular, if a node u of $\mathcal{T}$ is represented by $x \in \Sigma^*$ (or a prefix of x), then we will say that x *reaches* the node u. The leaf l that x reaches is denoted by $l = \mathrm{leaf}_{\mathcal{T}}(x)$. Given a tree $\mathcal{T}$, the set of its nodes and that of its leaves are denoted by nodes($\mathcal{T}$) and leaves($\mathcal{T}$), respectively. A string in Σ^* that reaches any of the leaves of $\mathcal{T}$ is called an *instance*.

As in the usual setting, an instance induces a path from the root to a leaf according to the outcomes of classification tests done at the internal nodes on the path. So, without loss of generality, we can identify an instance with the path it induces and thus we do not need to explicitly specify classification rules at the internal nodes of $\mathcal{T}$. A label function V for template tree $\mathcal{T}$ is a function that maps the set of nodes of $\mathcal{T}$ to the prediction space $\hat{Y}$. A pruning $\mathcal{P}$ of the template tree $\mathcal{T}$ is a tree obtained by replacing zero or more of the internal nodes (and associated subtrees) of $\mathcal{T}$ by leaves. Note that $\mathcal{T}$ itself is a pruning of $\mathcal{T}$ as well. The pair $(\mathcal{P}, V)$ induces a pruned decision tree that makes its prediction $V(\text{leaf}_{\mathcal{P}}(x))$ for instance x . The set of all pruning of $\mathcal{T}$ is denoted by $\text{PRUN}(\mathcal{T})$.

Elaborating a data structure, Helmbold and Schapire constructed an efficient prediction algorithm for the absolute loss that works in a manner equivalent to the naive algorithm without enumerating prunings. The performance of their algorithm is given by Theorem 3. This will be generalized to the case where a loss function is given arbitrarily.

Theorem 3 (Helmbold and Schapire [3]). *In the case of the absolute loss, there exists a prediction algorithm A such that for any $\mathcal{T}$, V and $y \in \{0,1\}^*$, when given $\mathcal{T}$ and V as input, A makes predictions for y so that the loss is at most*

$$L_A(y) \le \min_{\mathcal{P} \in \text{PRUN}(\mathcal{T})} \frac{L_{\mathcal{P},V}(y) \ln(1/\beta) + |\mathcal{P}| \ln 2}{2 \ln(2/(1+\beta))},$$

where $|\mathcal{P}|$ denotes the number of nodes of $\mathcal{P}$ minus $|\text{leaves}(\mathcal{T}) \cap \text{leaves}(\mathcal{P})|$. A generates a prediction at each trial t in time $O(|x^t|)$. Moreover, the label function V may depend on t.

We shall explore how to use the APA in the previous section to construct an algorithm that seeks for a weighted combination of the mini-experts located at the inner nodes of a given decision tree so that the weighted combination of the mini-experts performs nearly as well as the best pruning of the decision tree.

Our algorithm is in some sense a quite straightforward implementation of the dynamic programming. To explain how it works, we still need some more notations. For node $u \in \Sigma^*$, let $\mathcal{T}_u$ denote the subtree of $\mathcal{T}$ rooted at u. For an outcome sequence $y \in Y^*$, the loss suffered at u, denoted $L_u(y)$, is defined as follows:

$$L_u(y) = \sum_{t : x^t \text{ reaches } u} \lambda(y^t, V(u)).$$

Then, for any pruning $\mathcal{P}_u$ of $\mathcal{T}_u$, the loss suffered by $\mathcal{P}_u$, denoted $L_{\mathcal{P}_u}(y)$, can be represented by the sum of $L_l(y)$ for all leaves l of $\mathcal{P}_u$. In other words, we can write $L_{\mathcal{P}_u}(y) = L_u(y)$ if $\mathcal{P}_u$ consists of a single leaf u and $L_{\mathcal{P}_u(y)} = \Sigma_{a \in \Sigma} L_{\mathcal{P}_{ua}}(y)$ otherwise. Here $\mathcal{P}_{ua}$ is the subtree of $\mathcal{P}_u$ rooted at ua. We have for any internal node u of $\mathcal{T}$

$$\min_{\mathcal{P}_u \in \text{PRUN}(\mathcal{T}_u)} L_{\mathcal{P}_u}(y) = \min \left\{ L_u(y), \sum_{a \in \Sigma} \min_{\mathcal{P}_{ua} \in \text{PRUN}(\mathcal{T}_{ua})} L_{\mathcal{P}_{ua}}(y) \right\}.$$

Since dynamic programming can be applied to solve the minimization problem of this type, we can efficiently compute $\mathcal{P}_\epsilon \in \text{PRUN}(\mathcal{T}_\epsilon)$ that minimize $L_{\mathcal{P}_\epsilon}(y)$, which is the best pruning of $\mathcal{T}$. But if we try to solve the minimization problem based on the formula above in a straightforward way, we have to have the sequence of outcomes $y = (y^1, \ldots, y^T)$ ahead of time.

In the rest of this section we try to construct an algorithm that solves the minimization problem in an on-line fashion by applying aggregating pseudo-algorithm recursively on the decision trees. We associate two mini-experts $\mathcal{E}_u = \{\mathcal{E}_{u\perp}, \mathcal{E}_{u\downarrow}\}$ with each internal node u, one $\mathcal{E}_{u\perp}$ treating the node u as a leaf and the other $\mathcal{E}_{u\downarrow}$ treating the node u as an internal node. To combine these experts we apply the APA recursively, which is placed on each inner node of $\mathcal{T}$: the APA at an inner node u, denoted $\mathrm{APA}_u(\beta)$, combines the pseudopredictions of the experts $\mathcal{E}_{u\perp}$ and $\mathcal{E}_{u\downarrow}$ to obtain its own pseudoprediction r^t_u, and pass it to the APA at the parent node of u. More precisely, when given an instance x_t that goes through u and ua, the first expert $\mathcal{E}_{u\perp}$ generates $V(u)$ and the second expert $\mathcal{E}_{u\downarrow}$ generates r^t_{ua}, i.e., the pseudoprediction made by the APA at node ua, denoted $\mathrm{APA}_{ua}(\beta)$. Then, taking the weighted average of these pseudopredictions $V(u)$ and r^t_{ua} (recall that the genuine prediction $V(u)$ is regarded as a pseudoprediction), $\mathrm{APA}_u(\beta)$ obtains the pseudoprediction r^t_u at u. To obtain the genuine prediction $\hat{y}_t$, our algorithm applies the β-substitution function to the pseudoprediction only at the root during every trial, that is, $\hat{y}_t = \Sigma_\beta(r^t_\epsilon)$: in the internal nodes we combine not genuine predictions but pseudopredictions using the APA.

In Figure 3 and Figure 4 we present below the prediction algorithm constructed this way, which we call *Structured Weight-based Prediction algorithm* (SWP(β) for short). Here, $\mathrm{path}(x_t)$ denotes the set of the nodes of $\mathcal{T}$ that x_t reaches. In other words, $\mathrm{path}(x_t)$ is the set of the prefixes of x_t. For node u, $|u|$ denotes the depth of u, i.e, the length of the path from the root to u.

procedure PSEUDOPRED(u, x_t)
  if $u \in \mathrm{leaf}(\mathcal{T})$ **then**
    for $y \in Y$ **do**
      $r^t_u(y) := \lambda(y, V(u))$
  else
    choose $a \in \Sigma$ such that $ua \in \mathrm{path}(x^t)$
    $r^t_{ua} :=$ PSEUDOPRED(ua, x_t)
    for $y \in Y$ **do**
      $r^t_u(y) := \log_\beta \left(\overline{w}^t_{u\perp} \beta^{\lambda(y, V(u))} + \overline{w}^t_{u\downarrow} \beta^{r^t_{ua}(y)} \right)$
  return r^t_u

procedure UPDATE(u, y^t)
  if $u \in \mathrm{leaf}(\mathcal{T})$ **then**
    return
  else
    choose $a \in \Sigma$ such that $ua \in \mathrm{path}(x^t)$
    $w^{t+1}_{u\perp} := w^t_{u\perp} \beta^{\lambda(y^t, V(u))}$
    $w^{t+1}_{u\downarrow} := w^t_{u\downarrow} \beta^{r^t_{ua}(y^t)}$
    return

Fig. 3. Algorithm $\mathrm{APA}_u(\beta)$

```
for u ∈ nodes(T)\leaves(T) do
    w^1_{u⊥} := 1/2
    w^1_{u↓} := 1/2
for t = 1, 2, ... do
    receive x^t
    r^t_ε := PSEUDOPRED(ε, x^t)
    ŷ^t := Σ_β(r^t_ε)
    output ŷ^t
    receive y^t
    for u ∈ path(x^t) do
        UPDATE(u, y^t)
```

Fig. 4. Algorithm SWP(β)

Let the loss suffered by $\mathrm{APA}_u(\beta)$ be denoted by $\hat{L}_u(y)$. That is,

$$\hat{L}_u(y) = \sum_{t:u\in\mathrm{path}(x_t)} r_u^t(y^t).$$

Since the first expert $\mathcal{E}_{u\perp}$ suffers the loss $L_u(y)$ and the second expert $\mathcal{E}_{u\downarrow}$ suffers the loss $\Sigma_{a\in\Sigma}\hat{L}_{ua}(y)$, Theorem 1 says that for any internal node u of $\mathcal{T}$,

$$\hat{L}_u(y) \le \min\left\{L_u(y), \sum_{a\in\Sigma}\hat{L}_{ua}(y)\right\} + (\ln 2)/(\ln(1/\beta)).$$

By the similarity between the inequality above and the equation

$$\min_{\mathcal{P}_u\in\mathrm{PRUN}(\mathcal{T}_u)} L_{\mathcal{P}_u}(y) = \min\left\{L_u(y), \sum_{a\in\Sigma}\min_{\mathcal{P}_{ua}\in\mathrm{PRUN}(\mathcal{T}_{ua})} L_{\mathcal{P}_{ua}}(y)\right\}$$

for the minimization problem, we can roughly say that $\hat{L}_u(y)$ is not much larger than the loss of the best pruning of $\mathcal{T}_u$. We have the following theorem.

Theorem 4 ([7], cf. [6]). *There exists a prediction algorithm A such that for any $\mathcal{T}$, V and $y \in Y^*$, when $\mathcal{T}$ and V are given as input, A makes predictions for y so that loss is at most*

$$L_A(y) \le \min_{\mathcal{P}\in\mathrm{PRUN}(\mathcal{T})}\left(c(\beta)L_{\mathcal{P},V}(y) + \frac{c(\beta)\ln 2}{\ln(1/\beta)}|\mathcal{P}|\right).$$

4 Concluding Remarks

Based on dynamic programming, we gave a simple on-line prediction algorithm that performs nearly as well as the best pruning $\mathcal{P}$ of a given tree $\mathcal{T}$ for a wide class of loss functions. It was shown that its loss bound is as good as that of

Helmbold and Schapire's algorithm in the case of the absolute loss game. We expect that our algorithm could be applied to many on-line optimization problems solved by dynamic programming. Furthermore, the time spent by procedure PSEUDOPRED to make a pseudoprediction in each trial might be greatly saved if it is made to be randomized. More precisely, instead of taking weighted average of the predictions by two experts, just choose either of predictions randomly with probability according to their weights. Then, the expected loss of the modified algorithm turns out to be almost the same but the expected time spent by PSEUDOPRED might be much smaller. In particular, we conjecture that the expected time is proportional to the depth of the best pruning rather than the depth of $\mathcal{T}$.

Based on the idea that the predictions can be made independently at each leaf, we can also give for the absolute loss game an on-line prediction algorithm using a given template tree $\mathcal{T}$ that performs nearly as well as the best pruning with the best labelings. The property of independence comes from the distinctive feature of the tree. More precisely, the tree $\mathcal{T}$ can be thought of as specifying how to partition the instance space Σ^* into subclasses and how to assign a prediction value to each subclass. Because our algorithm does not use the internal structure of $\mathcal{T}$ but only uses the subclasses defined by $\mathcal{T}$, it can easily be generalized for any given rule that partitions the instance space into subclasses.

Acknowledgment. We are grateful to the referees for their helpful comments to improve the presentation of the paper.

References

1. N. Cesa-Bianchi, Y. Freund, D. Haussler, D. P. Helmbold, R. E. Schapire, and M. K. Warmuth. How to use expert advice. *Journal of the ACM*, 44(3):427–485, 1997.
2. Y. Freund and R. E. Schapire. A decision-theoretic generalization of on-line learning and an application to boosting. *J. Comput. Syst. Sci.*, 55(1):119–139, Aug. 1997.
3. D. P. Helmbold and R. E. Schapire. Predicting nearly as well as the best pruning of a decision tree. *Machine Learning*, 27(1):51–68, 1997.
4. J. Kivinen and M. K. Warmuth. Averaging expert predictions. In *Proc. 4th European Conference on Computational Learning Theory*, volume 1572 of *Lecture Notes in Artificial Intelligence*, pages 153–167. Springer-Verlag, 1999.
5. N. Littlestone and M. K. Warmuth. The weighted majority algorithm. *Inform. Comput.*, 108(2):212–261, 1994.
6. E. Takimoto, K. Hirai, and A. Maruoka. A simple algorithm for predicting nearly as well as the best pruning labeled with the best prediction values of a decision tree. In *Algorithmic Learning Theory: ALT '97*, volume 1316 of *Lecture Notes in Artificial Intelligence*, pages 385–400. Springer-Verlag, 1997.
7. E. Takimoto, A. Maruoka, and V. Vovk. Predicting nearly as well as the best pruning of a decision tree through dynamic programming scheme. *to appear in Theoretical Computer Science.*
8. V. G. Vovk. A game of prediction with expert advice. In *Proc. 8th Annu. Conf. on Comput. Learning Theory*, pages 51–60. ACM Press, New York, NY, 1995.

Finding Best Patterns Practically

Ayumi Shinohara, Masayuki Takeda, Setsuo Arikawa,
Masahiro Hirao, Hiromasa Hoshino, and Shunsuke Inenaga

Department of Informatics, Kyushu University 33, Fukuoka 812-8581, JAPAN
{ayumi, hirao, hoshino, s-ine, takeda, arikawa}@i.kyushu-u.ac.jp

Abstract. Finding a pattern which separates two sets is a critical task in discovery. Given two sets of strings, consider the problem to find a subsequence that is common to one set but never appears in the other set. The problem is known to be NP-complete. Episode pattern is a generalized concept of subsequence pattern where the length of substring containing the subsequence is bounded. We generalize these problems to optimization problems, and give practical algorithms to solve them exactly. Our algorithms utilize some pruning heuristics based on the combinatorial properties of strings, and efficient data structures which recognize subsequence and episode patterns.

1 Introduction

In these days, a lot of text data or sequential data are available, and it is quite important to discover useful rules from these data. Finding a *good rule* to separate two given sets, often referred as *positive examples* and *negative examples*, is a critical task in Discovery Science as well as Machine Learning. String is one of the most fundamental structure to express and reserve information. In this paper, we review our recent work [7, 8] which find a best pattern practically.

First we remind our motivations. Shimozono et al. [13] developed a machine discovery system BONSAI that produces a decision tree over regular patterns with alphabet indexing, from given positive set and negative set of strings. The core part of the system is to generate a decision tree which classifies positive examples and negative examples as correctly as possible. For that purpose, we have to find a *pattern* that maximizes the goodness according to the entropy information gain measure, recursively at each node of trees. In the current implementation, a pattern associated with each node is restricted to a *substring pattern*, due to the limit of computation time. One of our motivations of this study was to extend the BONSAI system to allow *subsequence patterns* as well as substring patterns at nodes, and accelerate the computation time.

However, there is a large gap between the complexity of finding the best *substring pattern* and *subsequence pattern*. Theoretically, the former problem can be solved in linear time, while the latter is NP-hard. In [7], we introduced a practical algorithm to find a best subsequence pattern that separates positive examples from negative examples, and showed some experimental results.

A drawback of subsequence patterns is that they are not suitable for classifying *long* strings over *small* alphabet, since a short subsequence pattern matches

S. Arikawa and A. Shinohara (Eds.): Progress in Discovery Science 2001, LNAI 2281, pp. 307–317.

with almost all long strings. Based on this observation, in [8] we considered *episode patterns*, which were originally introduced by Mannila et al. [10]. An episode pattern $\langle v, k\rangle$, where v is a string and k is an integer, *matches* with a string t if v is a subsequence for some substring u of t with $|u| \leq k$. Episode pattern is a generalization of subsequence pattern since subsequence pattern v is equivalent to episode pattern $\langle v, \infty\rangle$. We gave a practical solution to find a best episode pattern which separates a given set of strings from the other set of strings.

In this paper, we summarize our practical implementations of exact search algorithms that practically avoids exhaustive search. Since these problems are NP-hard, essentially we are forced to examine exponentially many candidate patterns in the worst case. Basically, for each pattern w, we have to count the number of strings that contain w as a subsequence in each of two sets. We call the task of counting the numbers as *answering subsequence query*. The computational cost to find the best subsequence pattern mainly comes from the total amount of time to answer these subsequence queries, since it is relatively heavy task if the sets are large, and many queries will be needed. In order to reduce the time, we have to either (1) ask queries as few as possible, or (2) speed up to answer queries. We attack the problem from both these two directions.

At first, we reduce the search space by appropriately pruning redundant branches that are guaranteed not to contain the best pattern. We use a heuristics inspired by Morishita and Sese [12], combined with some properties on the subsequence languages, and episode pattern languages.

Next, we accelerate answering for subsequence queries and episode pattern queries. Since the sets of strings are fixed in finding the best pattern, it is reasonable to preprocess the sets so that answering query for any pattern will be fast. We take an approach based on a deterministic finite automaton that accepts all subsequences of a string. Actually, we use subsequence automata for sets of strings, developed in [9] for subsequence query, and episode pattern recognizer for episode pattern query. These automata can answer quickly for subsequence query, at the cost of preprocessing time and space requirement to construct them.

2 Preliminaries

Let $\mathcal{N}$ be the set of integers. Let Σ be a finite *alphabet*, and let Σ^* be the set of all *strings* over Σ. For a string w, we denote by $|w|$ the length of w. For a set $S \subseteq \Sigma^*$ of strings, we denote by $|S|$ the number of strings in S, and by $||S||$ the total length of strings in S.

We say that a string v is a *prefix* (*substring*, *suffix*, resp.) of w if $w = vy$ ($w = xvy$, $w = xv$, resp.) for some strings $x, y \in \Sigma^*$. We say that a string v is a *subsequence* of a string w if v can be obtained by removing zero or more characters from w, and say that w is a *supersequence* of v. We denote by $v \preceq_{\mathrm{str}} w$ that v is a substring of w, and by $v \preceq_{\mathrm{seq}} w$ that v is a subsequence of w. For a string v, we define the *substring language* $L^{\mathrm{str}}(v)$ and *subsequence language*

$L^{\rm seq}(v)$ as follows:

$$L^{\rm str}(v) = \{w \in \Sigma^* \mid v \preceq_{\rm str} w\}, \text{ and}$$
$$L^{\rm seq}(v) = \{w \in \Sigma^* \mid v \preceq_{\rm seq} w\}, \text{ respectively.}$$

An *episode pattern* is a pair of a string v and an integer k, and we define the *episode language* $L^{\rm eps}(\langle v,k\rangle)$ by

$$L^{\rm eps}(\langle v,k\rangle) = \{w \in \Sigma^* \mid {}^{\exists}u \preceq_{\rm str} w \text{ such that } v \preceq_{\rm seq} u \text{ and } |u| \le k\}.$$

The following lemma is obvious from the definitions.

Lemma 1 ([7]). *For any strings $v, w \in \Sigma^*$,*

(1) if v is a prefix of w, then $v \preceq_{str} w$,
(2) if v is a suffix of w, then $v \preceq_{str} w$,
(3) if $v \preceq_{str} w$ then $v \preceq_{seq} w$,
(4) $v \preceq_{str} w$ if and only if $L^{str}(v) \supseteq L^{str}(w)$,
(5) $v \preceq_{seq} w$ if and only if $L^{seq}(v) \supseteq L^{seq}(w)$.

We formulate the problem by following our previous paper [7]. Readers should refer to [7] for basic idea behind this formulation. We say that a function f from $[0, x_{\max}] \times [0, y_{\max}]$ to real numbers is *conic* if

- for any $0 \le y \le y_{\max}$, there exists an x_1 such that
  - $f(x,y) \ge f(x',y)$ for any $0 \le x < x' \le x_1$, and
  - $f(x,y) \le f(x',y)$ for any $x_1 \le x < x' \le x_{\max}$.
- for any $0 \le x \le x_{\max}$, there exists a y_1 such that
  - $f(x,y) \ge f(x,y')$ for any $0 \le y < y' \le y_1$, and
  - $f(x,y) \le f(x,y')$ for any $y_1 \le y < y' \le y_{\max}$.

We assume that f is conic and can be evaluated in constant time in the sequel.

The following are the optimization problems to be tackled.

Definition 1 (Finding the best substring pattern according to f).
Input *Two sets $S, T \subseteq \Sigma^*$ of strings.*
Output *A string v that maximizes the value $f(x_v, y_v)$, where $x_v = |S \cap L^{str}(s)|$ and $y_s = |T \cap L^{str}(s)|$.*

Definition 2 (Finding the best subsequence pattern according to f).
Input *Two sets $S, T \subseteq \Sigma^*$ of strings.*
Output *A string v that maximizes the value $f(x_v, y_v)$, where $x_v = |S \cap L^{seq}(v)|$ and $y_v = |T \cap L^{seq}(v)|$.*

Definition 3 (Finding the best episode pattern according to f). .
Input *Two sets $S, T \subseteq \Sigma^*$ of strings.*
Output *A episode pattern $\langle v,k\rangle$ that maximizes the value $f(x_{\langle v,k\rangle}, y_{\langle v,k\rangle})$, where $x_{\langle v,k\rangle} = |S \cap L^{eps}(\langle v,k\rangle)|$ and $y_{\langle v,k\rangle} = |T \cap L^{eps}(\langle v,k\rangle)|$.*

```
1  pattern FindMaxPattern(StringSet S, T)
2      maxVal = −∞;
3      for all possible pattern π do
4          x = |S ∩ L(π)|;
5          y = |T ∩ L(π)|;
6          val = f(x, y);
7          if val > maxVal then
8              maxVal = val;
9              maxPat = π;
10     return maxPat;
```

Fig. 1. Exhaustive search algorithm.

We remind that the first problem can be solved in linear time [7], while the latter two problems are NP-hard.

We review the basic idea of our algorithms. Fig. 1 shows a naive algorithm which exhaustively examines and evaluate all possible patterns one by one, and returns the best pattern that gives the maximum value. The most time consuming part is obviously the lines 4 and 5, and in order to reduce the search time, we should (1) reduce the possible patterns in line 3 *dynamically* by using some appropriate pruning method, and (2) speed up to compute $|S \cap L(\pi)|$ and $|T \cap L(\pi)|$ for each π. In Section 3, we deal with (1), and in Section 4, we treat (2).

3 Pruning Heuristics

In this section, we introduce some pruning heuristics, inspired by Morishita and Sese [12].

For a function $f(x, y)$, we denote $F(x, y) = \max\{f(x,y), f(x,0), f(0,y), f(0,0)\}$. From the definition of conic function, we can prove the following lemma.

Lemma 2. *For any patterns v and w with $L(v) \supseteq L(w)$, we have*

$$f(x_w, y_w) \leq F(x_v, y_v).$$

3.1 Subsequence Patterns

We consider finding subsequence pattern in this subsection. By Lemma 1 (5) and Lemma 2, we have the following lemma.

Lemma 3 ([7]). *For any strings $v, w \in \Sigma^*$ with $v \preceq_{seq} w$, we have*

$$f(x_w, y_w) \leq F(x_v, y_v).$$

In Fig. 2, we show our algorithm to find the best subsequence pattern from given two sets of strings, according to the function f. Optionally, we can specify the maximum length of subsequences. We use the following data structures in the algorithm.

```
1   string FindMaxSubsequence(StringSet S, T, int maxLength = ∞)
2       string prefix, seq, maxSeq;
3       double upperBound = ∞, maxVal = −∞, val;
4       int x, y;
5       PriorityQueue queue;    /* Best First Search*/
6       queue.push("", ∞);
7       while not queue.empty() do
8           (prefix, upperBound) = queue.pop();
9           if upperBound < maxVal then break;
10          foreach c ∈ Σ do
11              seq= prefix+ c;    /* string concatenation */
12              x = S.numOfSubseq(seq);
13              y = T.numOfSubseq(seq);
14              val = f(x, y);
15              if val > maxVal then
16                  maxVal = val;
17                  maxSeq = seq;
18*             upperBound = F(x, y);
19              if |seq| < maxLength then
20                  queue.push(seq, upperBound);
21      return maxSeq;
```

Fig. 2. Algorithm *FindMaxSubsequence*.

StringSet Maintain a set S of strings.

- **int** *numOfSubseq*(**string** *seq*) : return the cardinality of the set $\{w \in S \mid seq \preceq_{\text{seq}} w\}$.

PriorityQueue Maintain strings with their priorities.

- **bool** *empty*() : return **true** if the queue is empty.
- **void** *push*(**string** *w*, **double** *priority*) : push a string w into the queue with priority *priority.*
- **(string, double)** *pop*() : pop and return a pair (*string*, *priority*), where *priority* is the highest in the queue.

The next theorem guarantees the completeness of the algorithm.

Theorem 1 ([7]). *Let S and T be sets of strings, and ℓ be a positive integer. The algorithm FindMaxSubsequence(S, T, ℓ) will return a string w that maximizes the value $f(x_v, y_v)$ among the strings of length at most ℓ, where $x_v = |S \cap L^{str}(s)|$ and $y_s = |T \cap L^{str}(s)|$.*

Proof. We first consider the case that the lines 18 is removed. Since the value of *upperBound* is unchanged, **PriorityQueue** is actually equivalent to a simple queue. Then, the algorithm performs the exhaustive search in a breadth first manner. Thus the algorithm will compute the value $f(x_v, y_v)$ for *all* strings of length at most *maxLength*, in increasing order of the length, and it can find the best pattern trivially.

We now focus on the line 9, by assuming the condition $upperBound < maxVal$ holds. Since the *queue* is a priority queue, we have $F(x_v, y_v) \leq upperBound$ for any string v in the queue. By Lemma 3, $f(x_v, y_v) \leq F(x_v, y_v)$, which implies $f(x_v, y_v) < maxVal$. Thus no string in the queue can be the best subsequence and we jump out of the loop immediately.

Next, we consider the lines 18. Let v be the string currently represented by the variable *seq*. At lines 12 and 13, x_v and y_v are computed. At line 18, $upperBound = F(x_v, y_v)$ is evaluated, and if *upperBound* is less than the current maximum value *maxVal*, v is not pushed into *queue*. It means that any string w of which v is a prefix will not be evaluated. We can show that such a string w can never be the best subsequence as follows. Since v is a prefix of w, we know v is a subsequence of w, by Lemma 1 (1) and (3). By Lemma 3, we know $f(x_w, y_w) \leq F(x_v, y_v)$, and since $F(x_v, y_v) < maxVal$, the string w can never be the maximum. □

3.2 Episode Pattern

We now show a practical algorithm to find the best episode patterns. We should remark that the search space of episode patterns is $\Sigma^* \times \mathcal{N}$, while the search space of subsequence patterns was Σ^*. A straight-forward approach based on the last subsection might be as follows. First we observe that the algorithm *FindMaxSubsequence* in Fig. 2 can be easily modified to find the best episode pattern $\langle v, k \rangle$ *for any fixed threshold* k: we have only to replace the lines 12 and 13 so that they compute the numbers of strings in S and T that match with the episode pattern $\langle seq, k \rangle$, respectively. Thus, for each possible threshold value k, repeat his algorithm, and get the maximum. A short consideration reveals that we have only to consider the threshold values up to l, that is the length of the longest string in given S and T.

However, here we give a more efficient solution. Let us consider the following problem, that is a subproblem of *finding the best episode pattern* in Definition 3.

Definition 4 (Finding the best threshold value).
Input *Two sets* $S, T \subseteq \Sigma^*$ *of strings, and a string* $v \in \Sigma^*$.
Output *Integer* k *that maximizes the value* $f(x_{\langle v,k \rangle}, y_{\langle v,k \rangle})$, *where* $x_{\langle v,k \rangle} = |S \cap L^{eps}(\langle v, k \rangle)|$ *and* $y_{\langle v,k \rangle} = |T \cap L^{eps}(\langle v, k \rangle)|$.

The next lemma give a basic containment of episode pattern languages.

Lemma 4 ([8]). *For any two episode patterns* $\langle v, l \rangle$ *and* $\langle w, k \rangle$, *if* $v \preceq_{seq} w$ *and* $l \geq k$ *then* $L^{eps}(\langle v, l \rangle) \supseteq L^{eps}(\langle w, k \rangle)$.

By Lemma 2 and 4, we have the next lemma.

Lemma 5 ([8]). *For any two episode patterns* $\langle v, l \rangle$ *and* $\langle w, k \rangle$, *if* $v \preceq_{seq} w$ *and* $l \geq k$ *then* $f(x_{\langle w,k \rangle}, y_{\langle w,k \rangle}) \leq F(x_{\langle v,l \rangle}, y_{\langle v,l \rangle})$.

For strings $v, s \in \Sigma^*$, we define the *threshold value* θ of v for s by $\theta = min\{k \in \mathcal{N} \mid s \in L^{eps}(\langle v, k \rangle)\}$. If no such value, let $\theta = \infty$. Note that $s \notin L^{eps}(\langle v, k \rangle)$ for any $k < \theta$, and $s \in L^{eps}(\langle v, k \rangle)$ for any $k \geq \theta$. For a set S of strings and a string v, let us denote by $\Theta_{S,v}$ the set of threshold values of v for some $s \in S$.

A key observation is that a best threshold value for given $S, T \subseteq \Sigma^*$ and a string $v \in \Sigma^*$ can be found in $\Theta_{S,v} \cup \Theta_{T,v}$ without loss of generality. Thus we can restrict the search space of the best threshold values to $\Theta_{S,v} \cup \Theta_{T,v}$.

From now on, we consider the numerical sequence $\{x_{\langle v,k \rangle}\}_{k=0}^{\infty}$. (We will treat $\{y_{\langle v,k \rangle}\}_{k=0}^{\infty}$ in the same way.) It clearly follows from Lemma 4 that the sequence is non-decreasing. Remark that $0 \leq x_{\langle v,k \rangle} \leq |S|$ for any k. Moreover, $x_{\langle v,l \rangle} = x_{\langle v,l+1 \rangle} = x_{\langle v,l+2 \rangle} = \cdots$, where l is the length of the longest string in S. Hence, we can represent $\{x_{\langle v,k \rangle}\}_{k=0}^{\infty}$ with a list having at most $\min\{|S|, l\}$ elements. We call this list *a compact representation of the sequence* $\{x_{\langle v,k \rangle}\}_{k=0}^{\infty}$ (*CRS*, for short).

We show how to compute CRS for each v and a fixed S. Observe that $x_{\langle v,k \rangle}$ increases only at the threshold values in $\Theta_{S,v}$. By computing a sorted list of all threshold values in $\Theta_{S,v}$, we can construct the CRS of $\{x_{\langle v,k \rangle}\}_{k=0}^{\infty}$. If using the counting sort, we can compute the CRS for any $v \in \Sigma^*$ in $O(|S|ml + |S|) = O(||S||m)$ time, where $m = |v|$.

We emphasize that the time complexity of computing the CRS of $\{x_{\langle v,k \rangle}\}_{k=0}^{\infty}$ is the same as that of computing $x_{\langle v,k \rangle}$ for a single k ($0 \leq k \leq \infty$), by our method.

After constructing CRSs $\bar{x}$ of $\{x_{\langle v,k \rangle}\}_{k=0}^{\infty}$ and $\bar{y}$ of $\{y_{\langle v,k \rangle}\}_{k=0}^{\infty}$, we can compute the best threshold value in $O(|\bar{x}| + |\bar{y}|)$ time. Thus we have the following, which gives an efficient solution to the finding the best threshold value problem.

Lemma 6. *Given $S, T \subseteq \Sigma^*$ and $v \in \Sigma^*$, we can find the best threshold value in $O(\,(||S|| + ||T||) \cdot |v|\,)$ time.*

By substituting this procedure into the algorithm *FindMaxSubsequence*, we get an algorithm to find a best episode pattern from given two sets of strings, according to the function f, shown in Fig. 3. We add a method *crs*(v) to the data structure **StringSet** that returns CRS of $\{x_{\langle v,k \rangle}\}_{k=0}^{\infty}$, as mentioned above.

By Lemma 5, we can use the value *upperBound* $= F(x_{v,\infty}, y_{v,\infty})$ to prune branches in the search tree computed at line 20 marked by (*). We emphasize that the value $F(x_{\langle v,k \rangle}, y_{\langle v,k \rangle})$ is insufficient as *upperBound*. Note also that $x_{\langle v,\infty \rangle}$ and $y_{\langle v,\infty \rangle}$ can be extracted from $\bar{x}$ and $\bar{y}$ in constant time, respectively. The next theorem guarantees the completeness of the algorithm.

Theorem 2 ([8]). *Let S and T be sets of strings, and ℓ be a positive integer. The algorithm FindBestEpisode(S, T, ℓ) will return an episode pattern that maximizes $f(x_{\langle v,k \rangle}, y_{\langle v,k \rangle})$, with $x_{\langle v,k \rangle} = |S \cap L^{eps}(\langle v, k \rangle)|$ and $y_{\langle v,k \rangle} = |T \cap L^{eps}(\langle v, k \rangle)|$, where v varies any string of length at most ℓ and k varies any integer.*

```
1       string FindBestEpisode(StringSet S, T, int ℓ)
2           string prefix, v;
3           episodePattern maxSeq; /* pair of string and int */
4           double upperBound = ∞, maxVal = −∞, val;
5           int k';
6           CompactRepr x̄, ȳ; /* CRS */
7           PriorityQueue queue;    /* Best First Search*/
8           queue.push("", ∞);
9           while not queue.empty() do
10              (prefix, upperBound) = queue.pop();
11              if upperBound < maxVal then break;
12              foreach c ∈ Σ do
13                  v = prefix+ c;    /* string concatenation */
14                  x̄ = S.crs(v);
15                  ȳ = T.crs(v);
16                  k' = argmax_k{f(x_⟨v,k⟩, y_⟨v,k⟩)} and val = f(x_⟨v,k'⟩, y_⟨v,k'⟩);
17                  if val > maxVal then
18                      maxVal = val;
19                      maxEpisode = ⟨v, k'⟩;
20(*)                upperBound = F(x_⟨v,∞⟩, y_⟨v,∞⟩);
21                  if upperBound > maxVal and |v| < ℓ then
22                      queue.push(v, upperBound);
23          return maxEpisode;
```

Fig. 3. Algorithm *FindBestEpisode*.

4 Using Efficient Data Structures

We introduces some efficient data structures to speed up answering the queries.

4.1 Subsequence Automata

First we pay our attention to the following problem.

Definition 5 (Counting the matched strings).
Input *A finite set* $S \subseteq \Sigma^*$ *of strings.*
Query *A string* $seq \in \Sigma^*$.
Answer *The cardinality of the set* $S \cap L^{seq}(seq)$.

Of course, the answer to the query should be very fast, since many queries will arise. Thus, we should preprocess the input in order to answer the query quickly. On the other hand, the preprocessing time is also a critical factor in some applications. In this paper, we utilize automata that accept subsequences of strings.

In [9], we considered a subsequence automaton as a deterministic complete finite automaton that recognizes all possible subsequences of a set of strings, that is essentially the same as the directed acyclic subsequence graph (DASG) introduced by Baeza-Yates [2]. We showed an online construction of subsequence

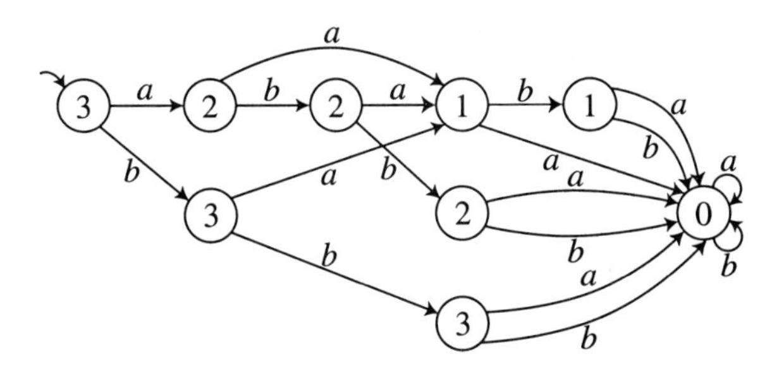

Fig. 4. Subsequence automaton for $S = \{abab, abb, bb\}$, where $\Sigma = \{a, b\}$. Each number on a state denotes the number of matched strings. For example, by traverse the states according to a string ab, we reach the state whose number is 2. It corresponds to the cardinality $|L^{\mathrm{seq}}(ab) \cap S| = 2$, since $ab \preceq_{\mathrm{seq}} abab$, $ab \preceq_{\mathrm{seq}} abb$ and $ab \not\preceq_{\mathrm{seq}} bb$.

automaton for a set of strings. Our algorithm runs in $O(|\Sigma|(m + k) + N)$ time using $O(|\Sigma|m)$ space, where $|\Sigma|$ is the size of alphabet, N is the total length of strings, and m is the number of states of the resulting subsequence automaton. We can extend the automaton so that it answers the above *Counting the matched strings* problem in a natural way (see Fig. 4).

Although the construction time is linear to the size m of automaton to be built, unfortunately $m = O(n^k)$ in general, where we assume that the set S consists of k strings of length n. (The lower bound of m is only known for the case $k = 2$, as $m = \Omega(n^2)$ [4].) Thus, when the construction time is also a critical factor, as in our application, it may not be a good idea to construct subsequence automaton for the set S itself. Here, for a specified parameter $mode > 0$, we partition the set S into $d = k/mode$ subsets $S_1, S_2, \ldots, S_d$ of at most $mode$ strings, and construct d subsequence automata for each S_i. When asking a query seq, we have only to traverse all automata simultaneously, and return the sum of the answers. In this way, we can balance the preprocessing time with the total time to answer (possibly many) queries. In [7], we experimentally evaluated the optimal value of the parameter $mode$.

4.2 Episode Directed Acyclic Subsequence Graphs

We now analyze the complexity of *episode pattern matching*. Given an episode pattern $\langle v, k\rangle$ and a string t, determine whether $t \in L^{\mathrm{eps}}(\langle v, k\rangle)$ or not. This problem can be answered by filling up the edit distance table between v and t, where only insertion operation with cost one is allowed. It takes $\Theta(mn)$ time and space using a standard dynamic programming method, where $m = |v|$ and $n = |t|$. For a fixed string, automata-based approach is useful. We use the Episode Directed Acyclic Subsequence Graph (EDASG) for string t, which was recently introduced by Troíček in [14]. Hereafter, let $EDASG(t)$ denote the EDASG for t. With the use of $EDASG(t)$, episode pattern matching can be answered quickly in practice, although the worst case behavior is still $O(mn)$. EDASG(t) is also useful to compute the threshold value θ of given v for t quickly in practice. As an

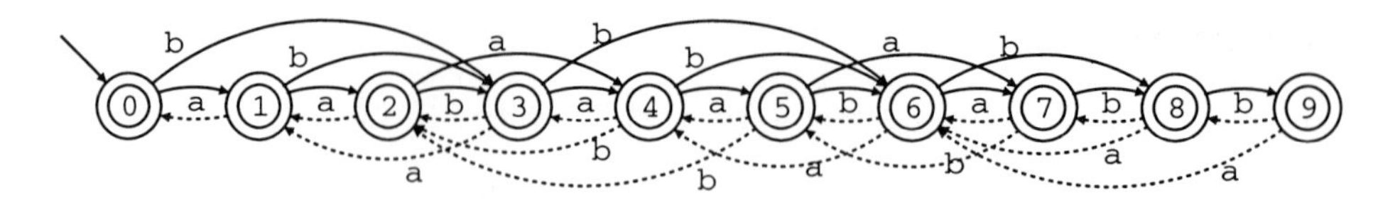

Fig. 5. *EDASG*(t), where $t = aabaababb$. Solid arrows denote the forward edges, and broken arrows denote the backward edges. The number in each circle denotes the state number.

example, *EDASG*($aabaababb$) is shown in Fig. 5. When examining if an episode pattern $\langle abb, 4\rangle$ matches with t or not, we start from the initial state 0 and arrive at state 6, by traversing the forward edges spelling abb. It means that the shortest prefix of t that contains abb as a subsequences is $t[0:6] = aabaab$, where $t[i:j]$ denotes the substring $t_{i+1}\dots t_j$ of t. Moreover, the difference between the state numbers 6 and 0 corresponds to the length of matched substring $aabaab$ of t, that is, $6-0 = |aabaab|$. Since it exceeds the threshold 4, we move backwards spelling bba and reach state 1. It means that the shortest suffix of $t[0:6]$ that contains abb as a subsequence is $t[1:6] = abaab$. Since $6-1 > 4$, we have to examine other possibilities. It is not hard to see that we have only to consider the string $t[2:*]$. Thus we continue the same traversal started from state 2, that is the next state of state 1. By forward traversal spelling abb, we reach state 8, and then backward traversal spelling bba bring us to state 4. In this time, we found the matched substring $t[4:8] = abab$ which contains the subsequence abb, and the length $8-4=4$ satisfies the threshold. Therefore we report the occurrence and terminate the procedure.

It is not difficult to see that the EDASGs are useful to compute the threshold value of v for a fixed t. We have only to repeat the above forward and backward traversal up to the end, and return the minimum length of the matched substrings. Although the time complexity is still $\Theta(mn)$, practical behavior is usually better than using standard dynamic programming method.

5 Conclusion

In this paper, we focused on *finding the best* subsequence pattern and episode patterns. However, we can easily extend our algorithm to *enumerate all strings* whose values of the objective function exceed the given threshold, since essentially we examine all strings, with effective pruning heuristics. Enumeration may be more preferable in the context of *text data mining* [3, 5, 15].

It is challenging to apply our approach to find the best *pattern* in the sense of *pattern languages* introduced by Angluin [1], where the related consistency problems are shown to be very hard [11]. Hamuro *et al.* [?] implemented our algorithm for finding best subsequences, and reported a quite successful experiments on business data. We are now in the process of installing our algorithms into the core of the decision tree generator in the BONSAI system [13].

References

1. D. Angluin. Finding patterns common to a set of strings. *J. Comput. Syst. Sci.*, 21(1):46–62, Aug. 1980.
2. R. A. Baeza-Yates. Searching subsequences. *Theoretical Computer Science*, 78(2):363–376, Jan. 1991.
3. A. Califano. SPLASH: Structural pattern localization analysis by sequential histograms. *Bioinformatics*, Feb. 1999.
4. M. Crochemore and Z. Troníček. Directed acyclic subsequence graph for multiple texts. Technical Report IGM-99-13, Institut Gaspard-Monge, June 1999.
5. R. Feldman, Y. Aumann, A. Amir, A. Zilberstein, and W. Klosgen. Maximal association rules: A new tool for mining for keyword co-occurrences in document collections. In *Proc. of the 3rd International Conference on Knowledge Discovery and Data Mining*, pages 167–170. AAAI Press, Aug. 1997.
6. R. Fujino, H. Arimura, and S. Arikawa. Discovering unordered and ordered phrase association patterns for text mining. In *Proc. of the 4th Pacific-Asia Conference on Knowledge Discovery and Data Mining*, volume 1805 of *Lecture Notes in Artificial Intelligence*. Springer-Verlag, Apr. 2000.
7. M. Hirao, H. Hoshino, A. Shinohara, M. Takeda, and S. Arikawa. A practical algorithm to find the best subsequence patterns. In *Proc. of The Third International Conference on Discovery Science*, volume 1967 of *Lecture Notes in Artificial Intelligence*, pages 141–154. Springer-Verlag, Dec. 2000.
8. M. Hirao, S. Inenaga, A. Shinohara, M. Takeda, and S. Arikawa. A practical algorithm to find the best episode patterns. In *Proc. of The Fourth International Conference on Discovery Science*, Lecture Notes in Artificial Intelligence. Springer-Verlag, Nov. 2001.
9. H. Hoshino, A. Shinohara, M. Takeda, and S. Arikawa. Online construction of subsequence automata for multiple texts. In *Proc. of 7th International Symposium on String Processing and Information Retrieval*. IEEE Computer Society, Sept. 2000. (to appear).
10. H. Mannila, H. Toivonen, and A. I. Verkamo. Discovering frequent episode in sequences. In U. M. Fayyad and R. Uthurusamy, editors, *Proc. of the 1st International Conference on Knowledge Discovery and Data Mining*, pages 210–215. AAAI Press, Aug. 1995.
11. S. Miyano, A. Shinohara, and T. Shinohara. Polynomial-time learning of elementary formal systems. *New Generation Computing*, 18:217–242, 2000.
12. S. Morishita and J. Sese. Traversing itemset lattices with statistical metric pruning. In *Proc. of the 19th ACM SIGACT-SIGMOD-SIGART Symposium on Principles of Database Systems*, pages 226–236. ACM Press, May 2000.
13. S. Shimozono, A. Shinohara, T. Shinohara, S. Miyano, S. Kuhara, and S. Arikawa. Knowledge acquisition from amino acid sequences by machine learning system BONSAI. *Transactions of Information Processing Society of Japan*, 35(10):2009–2018, Oct. 1994.
14. Z. Troníček. Episode matching. In *Proc. of 12th Annual Symposium on Combinatorial Pattern Matching*, Lecture Notes in Computer Science. Springer-Verlag, July 2001. (to appear).
15. J. T. L. Wang, G.-W. Chirn, T. G. Marr, B. A. Shapiro, D. Shasha, and K. Zhang. Combinatorial pattern discovery for scientific data: Some preliminary results. In *Proc. of the 1994 ACM SIGMOD International Conference on Management of Data*, pages 115–125. ACM Press, May 1994.

Classification of Object Sequences Using Syntactical Structure

Atsuhiro Takasu

National Institute of Informatics
2-1-2 Hitotsubashi, Chiyoda-ku, Tokyo 101-8430, Japan
takasu@nii.ac.jp

Abstract. When classifying a sequence of objects, in an ordinary classification, where objects are assumed to be independently drawn from identical information sources, each object is classified independently. This assumption often causes deterioration in the accuracy of classification. In this paper, we consider a method to classify objects in a sequence by taking account of the context of the sequence. We define this problem as component classification and present a dynamic programming algorithm where a hidden Markov model is used to describe the probability distribution of the object sequences. We show the effectiveness of the component classification experimentally, using musical structure analysis.

1 Introduction

Classification is one of the most fundamental problems in pattern recognition and machine learning. In classification, an object to be classified is represented by a feature vector. When classifying a sequence of objects, in an ordinary classification, each object is classified independently, i.e., objects are assumed to be independently drawn from an identical information source. This assumption often causes a deterioration in the accuracy of classification.

Syntactical pattern analysis [3] is one way to classify an object, by taking account of the context of the object sequence. In syntactical pattern analysis, for a feature space X, a composite object is observed as a feature vector sequence $\boldsymbol{x} \in X^l$, where X^l represents a set of feature vector sequences of length l. One typical task of syntactical pattern analysis is the conversion of a feature vector sequence to a string in a certain alphabet A. Let possible strings be $\boldsymbol{A} \subseteq A^*$. Then, statistically, this problem is solved by finding the optimal string $\boldsymbol{a} \in \boldsymbol{A}$ for a given feature vector sequence $\boldsymbol{x}$, such that

$$\arg\max_{\boldsymbol{a}} P(\boldsymbol{a}|\boldsymbol{x}) = \arg\max_{\boldsymbol{a}} \frac{P(\boldsymbol{a})P(\boldsymbol{x}|\boldsymbol{a})}{P(\boldsymbol{x})} = \arg\max_{\boldsymbol{a}} P(\boldsymbol{a})P(\boldsymbol{x}|\boldsymbol{a}) \qquad (1)$$

where $P(\boldsymbol{A})$ and $P(\boldsymbol{X})$ represent probability distributions of strings and feature vector sequences, respectively, while $P(\boldsymbol{X}|\boldsymbol{A})$ and $P(\boldsymbol{A}|\boldsymbol{X})$ respectively represent conditional probability distributions (densities) of a feature vector sequence conditioned by a string and a string conditioned by a feature vector sequence.

S. Arikawa and A. Shinohara (Eds.): Progress in Discovery Science 2001, LNAI 2281, pp. 318–326.

Syntactical pattern analysis is applied to various problems. For example, in speech recognition [6], a feature vector sequence $\boldsymbol{x}$ and a string $\boldsymbol{a}$, respectively, correspond to acoustic data and a word sequence, where $P(\boldsymbol{A})$ and $A(\boldsymbol{X}|\boldsymbol{A})$ represent a language model and an acoustic model, respectively. Speech is recognized by finding the most likely word sequence from the acoustic data in the sense of (1). Fuzzy fulltext searching [7] is another example. This problem involves finding a string in a text database that contains errors caused by recognizers such as Optical Character Recognition (OCR). In this case, $\boldsymbol{x}$ and $\boldsymbol{a}$ correspond to a recognized erroneous string and an original clean string, where $P(\boldsymbol{A})$ and $P(\boldsymbol{X}|\boldsymbol{A})$ correspond to a language model and an error model of a recognizer.

Another problem is an error-correcting parsing [3] of composite objects. The purpose of this problem is to extract the structure of a composite object from an observed feature vector sequence $\boldsymbol{x}$. In this problem, an alphabet corresponds to a set of components that constitute complex objects. Although (1) is used to find the optimal string, the main concern of this problem is to find a structure in $\boldsymbol{a}$ such as the syntactical analysis of sentences. Usually, a grammar is used to define the probability $P(\boldsymbol{A})$, and a parsing tree consisting of an alphabet is generated from a feature vector sequence. This task is applied to a composite object analysis such as document layout analysis [5] where a document's logical structure is extracted from the scanned document images.

In order to apply (1), we need to know the probabilities $P(\boldsymbol{A})$ and $P(\boldsymbol{X}|\boldsymbol{A})$. Usually, there exist an infinite number of vector sequences, and very many strings, and therefore certain models are required to estimate the probabilities $P(\boldsymbol{A})$ and $P(\boldsymbol{X}|\boldsymbol{A})$. For example, in speech recognition, a word trigram language model is often used as a language model $P(\boldsymbol{A})$, and a hidden Markov model (HMM) is used for the acoustic model $P(\boldsymbol{X}|\boldsymbol{A})$ [6]. In fuzzy retrieval, an n-gram language model is often used for $P(\boldsymbol{A})$, and several models have been proposed, such as the confusion matrix, probabilistic automata, and HMM for $P(\boldsymbol{X}|\boldsymbol{A})$.

In syntactical analysis, the target of the classification is itself a composite object. However, we are often interested in classifying the components of a composite object. For example, let us consider the determination of the types of phrases in a song and choose a representative phrase of the song. In this case, we do not need to determine the whole phrase structure of the song, but determine a type of each phrase in a song, i.e., whether the phrase is representative or not. This situation enables us to use a simpler model of phrase structure, which eases the parameter estimation of the model, and a more accurate model may be estimated. We discuss this *component classification in composite object* problem, which is referred to as *component classification* throughout this paper. We first define the problem formally and give an efficient algorithm in Section 2, and then, apply the proposed method to musical analysis for information retrieval in Section 3. In this paper, we assume that a composite object is represented by a sequence of components.

Table 1. Example of Conditional Probabilities

P(AA|xx)	
	xx
aa	0.1
ab	0.5
ba	0.1
bb	0.3

2 Component Classification in an Object Sequence

Component classification converts a given feature vector sequence $\boldsymbol{x}(\equiv x_1 x_2 \cdots x_l)$, to a class sequence $\boldsymbol{a}$ $(\equiv a_1 a_2 \cdots a_l) \in A^l$, where A is a set of classes. This is the same as the syntactical pattern analysis described in the previous section. However, this problem is different from the syntactical pattern analysis in finding the optimal class a_i for each feature vector x_i. In this sense, this problem can be seen as an ordinary classification problem when the object is not drawn independently from certain information sources.

Let us formally define the problem. Suppose that an observed feature vector sequence is $\boldsymbol{x} = x_1 x_2 \cdots x_l$. Let us consider the probability, denoted as $P_k(a|\boldsymbol{x})$, that the k-th component of $\boldsymbol{x}$ is classified to a. This probability is obtained by summing the conditional probabilities $P(\boldsymbol{a}|\boldsymbol{x})$ for a string $\boldsymbol{a}$ whose k-th element is a. Then, the optimal k-th class a_k^* for a feature vector sequence $\boldsymbol{x}$ is

$$\begin{aligned} a_k^* &\equiv \arg\max_{a \in A} P_k(a|\boldsymbol{x}) \\ &= \arg\max_{a \in A} \sum_{\{\boldsymbol{a} \mid \boldsymbol{a}[k]=a\}} P(\boldsymbol{a}|\boldsymbol{x}) \\ &= \arg\max_{a \in A} \sum_{\{\boldsymbol{a} \mid \boldsymbol{a}[k]=a\}} \frac{P(\boldsymbol{x}|\boldsymbol{a})P(\boldsymbol{a})}{P(\boldsymbol{x})} \\ &= \arg\max_{a \in A} \sum_{\{\boldsymbol{a} \mid \boldsymbol{a}[k]=a\}} P(\boldsymbol{x}|\boldsymbol{a})P(\boldsymbol{a}) \end{aligned} \tag{2}$$

where $\boldsymbol{a}[k]$ represents the k-th element of a sequence $\boldsymbol{a}$. For a given feature vector sequence $\boldsymbol{x}$, the component classification problem finds $a_1^* a_2^* \cdots a_l^*$. For a given feature vector sequence, the criterion (1) of the syntactical pattern analysis means to obtain the most likely class sequence, whereas the criterion (2) of the component classification means to obtain the most likely class for each position in $\boldsymbol{x}$.

In order to compare an ordinary classification and a component classification, let us assume that a set of class is $\{a, b\}$ and a given feature vector sequence is xx, i.e., the first and the second feature vectors are identical. Suppose a conditional probability distribution $P(AA|xx)$ is given in Table 1. Then, a feature vector sequence xx is classified into ab according to the first line of (2), whereas

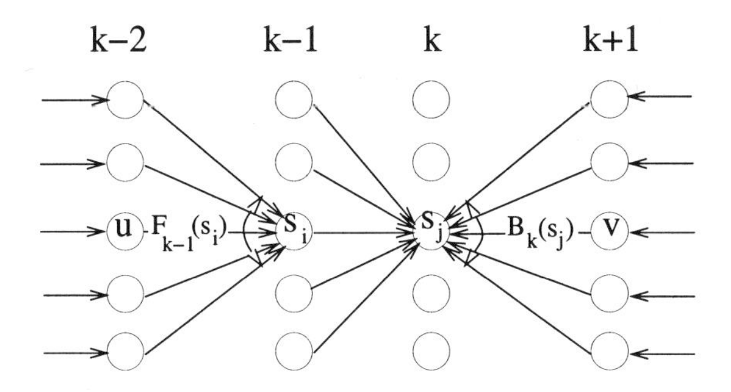

Fig. 1. Forward and Backward Joint Probabilities

both feature vectors are classified into the same class b in an ordinary classification. The distortion of the conditional probability over positions in the sequence demonstrates the difference between component classification and ordinary classification.

We use an HMM for determining $P(\boldsymbol{A})$. HMM is represented as a four-tuple (S, A, t, o), where S and A are a set of states and an output alphabet, respectively. Function t is a state transition function and $t(s_i, s_j)$ represents the transition probability from a state s_i to a state s_j. Let $t(0, s_i)$ denote an initial probability of a state s_i. Function o is an output function and $o(s_i, a)$ represents the probability that the output is a at state s_i. On the other hand, we assume that conditional probability $P(x_i|a_i)$ is independent, that is,

$$P(\boldsymbol{X} = x_1 x_2 \cdots x_l | \boldsymbol{A} = a_1 a_2 \cdots a_k) = \prod_i P(x_i|a_i)$$

holds. If this assumption does not hold, we can use an HMM for joint probability $P(\boldsymbol{X}, \boldsymbol{A})$. However, we need to estimate probability density $P(X|A)$ for each state, and consequently, we require very large training data for this case.

In order to solve the component classification, we can apply the forward and backward procedures used in an HMM learning algorithm[2]. When given a feature vector sequence $\boldsymbol{x} \equiv x_1 x_2 \cdots x_l$, forward and backward probabilities are defined for each state and each time, just as in HMM learning. For a state s and time k, the forward probability $F_k(s)$ is the probability that after $k-1$ transitions from any initial state,

- the HMM produces the output sequence $\boldsymbol{a} = a_1 a_2 \cdots a_k$,
- we reach state s, and
- $\boldsymbol{a}$ is represented with feature vector sequence $x_1 x_2 \cdots x_k$.

Let T_s^k be a set of state transitions of length k that start from any initial state and end at state s. Then, the forward probability $F_k(s)$ is expressed by

$$\sum_{\boldsymbol{a} \in A^k} \sum_{\boldsymbol{s} \in T_s^k} P(\boldsymbol{x}_1^k | \boldsymbol{a}) P(\boldsymbol{a}, \boldsymbol{s}) \tag{3}$$

where $\boldsymbol{x}_1^k$ represents a partial sequence $x_1 x_2 \cdots x_k$ of the given feature vector sequence $\boldsymbol{x}$, and $P(\boldsymbol{a}, \boldsymbol{s})$ represents the joint probability that the state transition is $\boldsymbol{s}$ and the corresponding output sequence is $\boldsymbol{a}$.

The forward probability (3) is represented by the following recursive form:

$$F_0(s) = t(0, s)$$
$$F_k(s) = \sum_{u \in S} t(s, u) F_{k-1}(u) \sum_{a \in A} o(s, a) P(x_k | a) \ . \tag{4}$$

Similarly, for a state s and time k, the backward probability $B_k(s)$ is the probability that, starting from the state s,

- the HMM produces output sequence $\boldsymbol{a} = a_{k+1} \cdots a_l$, and
- $\boldsymbol{a}$ is represented with feature vector sequence $x_{k+1} \cdots x_l$.

The recursive forms of backward probabilities are given as

$$B_l(s) = 1$$
$$B_k(s) = \sum_{u \in S} t(s, u) B_{k+1}(u) \sum_{a \in A} o(u, a) P(x_{k+1} | a) \ . \tag{5}$$

Using forward and backward probabilities, the probability of (2) is calculated as follows:

$$\sum_{\{\boldsymbol{a} \mid \boldsymbol{a}[k]=a\}} P(\boldsymbol{x}|\boldsymbol{a}) P(\boldsymbol{a}) = \sum_{s_i \in S} \sum_{s_j \in S} F_{k-1}(s_i) t(s_i, s_j) P(x_k | a) o(s_j, a) B_k(s_j) \ . \tag{6}$$

The trellis [6] in Figure 1 illustrates the sequence of operations required to calculate (6) when the state is s_j at time k. By applying this sequence of operations to all states, we derive (6).

For a feature vector sequence $x_1 x_2 \cdots x_l$, $a_1^* a_2^* \cdots a_l^*$ is obtained by the following steps:

1. calculate $F_k(s)$ for each $1 \leq k \leq l$ and $s \in S$ by (3),
2. calculate $B_k(s)$ for each $1 \leq k \leq l$ and $s \in S$ by (5),
3. calculate a_k^* for each $1 \leq k \leq l$ by (6).

In order to apply component classification, we need to construct an HMM model for the probability distribution $P(\boldsymbol{A})$, and estimate the conditional probability density $P(\boldsymbol{X}|\boldsymbol{A})$. For an HMM, there are well-known parameter estimation algorithms such as the Baum-Welch algorithm [2]. In order to estimate $P(\boldsymbol{X}|\boldsymbol{A})$, we can use a classifier and estimate the class conditional probabilities. For example, let us consider a decision tree as a classifier. Suppose an induced decision tree has leaves $l_1, l_2, \cdots, l_n$. Each leaf of a decision tree corresponds to a subspace of feature space. For a class a, suppose that there exists n_a objects in the training data and $n_{a,i}$ objects are assigned to a leaf l_i. Then, an estimation of the probability $P(l_i|a)$ is $n_{a,i}/n_a$. Let the volume of the subspace corresponding to l_i be v_i. Then, for a feature vector x in a subspace l_i, the estimated value of the class conditional probability $P(x|a)$ is $n_{a,i}/(v_i n_a)$.

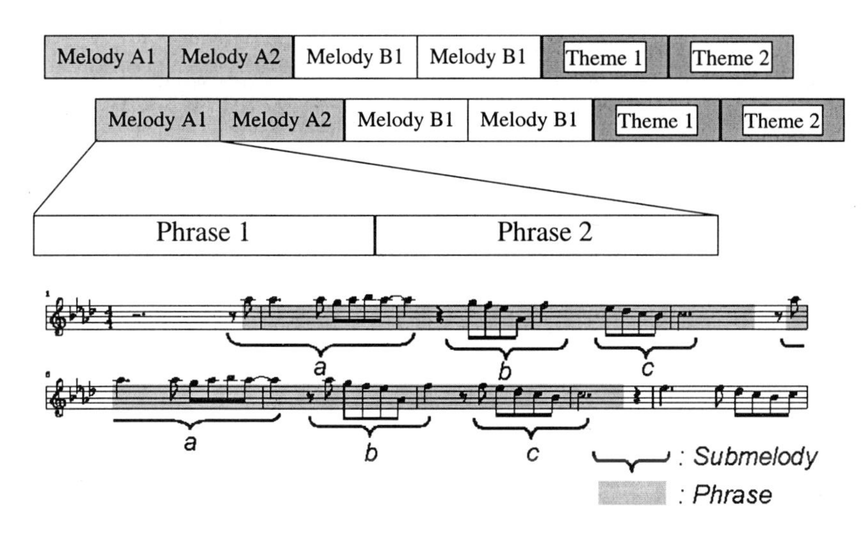

Fig. 2. An Example of Music Structure

3 Application to Musical Structure Analysis

Component classification can be applied to various error-tolerant pattern analyses and to pattern matchings such as fuzzy retrieval of erroneous text [9]. This section describes an application of component classification to musical information analysis for music retrieval. Figure 2 shows an example of a tonal structure of a Japanese popular song. As shown in Figure 2, a song structure is hierarchical, i.e., a song consists of a sequence of melodies, a melody consists of a sequence of phrases, and finally, a phrase consists of a sequence of notes. A phrase is believed to be a fundamental structural element of musical information and it can be used in various ways. Phrases play a central role in theories of musical structure, such as the Generative Theory of Tonal Music [8]. From a practical point of view, it improves the performance of music retrieval [12]. In music retrieval, a user usually uses a phrase, or a few phrases, as a query. By segmenting songs in a database into phrases, we can handle the local features of songs and consequently achieve high accuracy in matching a query to a song. As another practical use of phrases, we can use them as a concise representation of the music retrieval result, which is important for multimedia information retrieval. Retrieval results are usually presented to users as a ranked list. When judging whether the resultant music is the one intended or not, users need to listen to each piece of music. In this step, we need concise data on music. In video retrieval, skimming is used (e.g., [11]), whereas a theme phrase is adequate for music retrieval.

In this section, we discuss a theme phrase extraction using component classification. For this purpose, we need to classify phrases into two classes: a theme phrase and a non-theme phrase. Let us denote a theme phrase and a non-theme

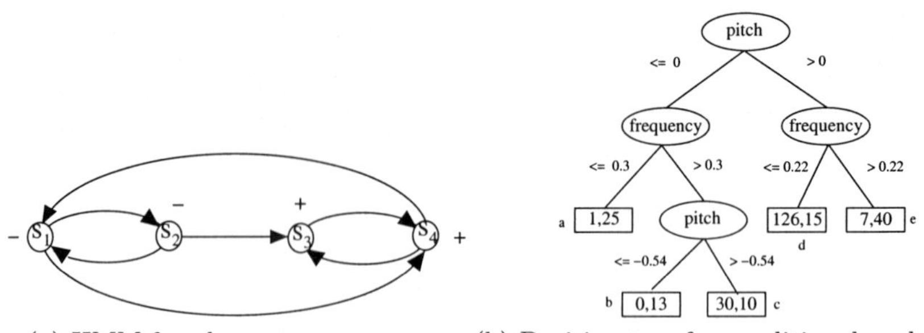

(a) HMM for phrase-type sequence (b) Decision tree for conditional probability

Fig. 3. Hidden Markov Model and Decision Tree for Theme Phrase Classification

phrase as + and −, respectively, in the following discussion. Theme phrases are expected to have certain characteristics, such as appearing frequently in a song and having high pitch and a high degree of loudness. From these considerations, we use four features to construct a feature vector of a phrase: pitch, duration, loudness, and frequency of a phrase. As shown in Figure 2, a phrase consists of a sequence of notes. Pitch and duration are obtained directly from its note sequence. Although loudness is not specified explicitly in a score, it can be obtained from a digital form of musical information such as Musical Instrument Digital Interface (MIDI). The frequency of the phrase in a song is obtained by clustering phrases and counting the size of the cluster to which the phrase belongs. See [12] for details of phrase extraction from MIDI data and feature measurement.

Figure 3 (a) shows an HMM for extracting theme phrases. States s_1 and s_2 correspond to non-theme phrases and produce − with a probability of 1.0, whereas states s_3 and s_4 correspond to theme phrases and produce + with a probability of 1.0. States s_1 and s_3 are entry points to sequences of non-theme and theme phrases, respectively. This HMM is used to estimate $P(\boldsymbol{A})$ in Section 2.

Figure 3 (b) shows a simplified decision tree for estimating $P(X|A)$ as described in Section 2. The purpose of the decision tree is to estimate the class conditional probability, rather than to determine a specific class. We therefore assign numbers n_+ and n_- to leaves instead of classes, where n_+ (resp. n_-) represents the number of theme (resp. non-theme) phrases that meet the condition of the leaf. For example, the leaf e in Figure 3 (b) indicates that a total of 47 phrases meet the condition of e, and seven of these are themes. Let $P(+|e)$ denote the conditional probability that a phrase meeting the condition of e is a theme. Then, the estimated value of $P(+|e)$ is $7/137 = 0.051$, since there exists a total of 137 theme phrases.

We carried out experiments of theme phrase extraction from Japanese popular songs by component classification. The database consisted of 94 Japanese popular songs in MIDI format. We decomposed the note sequences of these songs into phrases and obtained a total of 886 phrases. We then calculated the features of each phrase and classified them into theme/non-theme phrases manually.

Table 2. Experimental Results

pitch	velocity	duration	frequency
0.828	0.730	0.696	0.870

(a) Effectiveness of features

size (songs)	50	55	60	65	70
accuracy	0.843	0.873	0.855	0.893	0.912

(b) Accuracy w.r.t. size of training data

Songs were separated to training data and test data. Training data were used to estimate HMM parameters and to induce a decision tree.

First we measured the effectiveness of each feature with a five-fold cross validation. Table 2 (a) shows the resultant average accuracy for each feature. As shown in the table, pitch and frequency are effective features for theme phrase extraction. From this result, we can see that:

- theme phrases tend to have high pitch
- theme phrases are repeated more than non-theme phrases

as we expected.

Second, we measured the accuracy of component classification with respect to the size of the training data. Table 2 (b) shows the resultant accuracy of classification. The table shows that the accuracy increases as training data size increases.

Finally, we compared phrase classification using two settings: one is by a decision tree only, and the other is by our method. The average accuracy of the five-fold cross validation of the decision tree was 0.852, whereas the average accuracy of our method was 0.912. That is, we could improve the accuracy by about 6% by using syntactical information as well as the features of the phrases.

4 Conclusions

In this paper, we discussed the component classification problem. The conventional classification method classifies objects independently, while the component classification classifies objects by taking the context of the sequence into account to improve the classification accuracy. We showed that a dynamic programming technique can be applied to component classification. We applied the component classification to musical information analysis and showed experimentally that it improves the accuracy compared with ordinary classification.

In this paper, we assume that the feature vector length and class sequence length are equivalent. However, in real world problems, they are not always the same. In order to handle such cases, we need another model for class conditional probability $P(\boldsymbol{X}|\boldsymbol{A})$, such as an HMM that produces variable length outputs [4].

In order to estimate class conditional probability $P(\boldsymbol{X}|\boldsymbol{A})$, we used a decision tree. The optimal decision tree for ordinary classification and the optimal tree for component classification are not always the same. In order to obtain the optimal decision tree, an information criterion, such as Akaike's information criterion [1] and Rissanen's stochastic complexity (e.g., [10]), will be effective. In practical terms, we need to study whether these information criteria work well for modeling. This will be the basis of another future study concerning component classification.

In order to estimate $P(\boldsymbol{A})$, we must define a graph structure of HMM. In current syntactical pattern analysis, parameter estimation is generally performed. However, other parts of the model, such as the states and transitions in HMM, must be determined manually, depending on the application. For our method to be a powerful tool for discovery, we need to develop a method to acquire all parts of a model, including the parameter estimation.

References

1. H. Akaike. A New Look at the Statistical Model Identification. *IEEE Transactions on Automatic Control*, AC-19:716–723, 1974.
2. L. E. Baum. An Inequality and Associated Maximization Technique in Statistical Estimation for Probabilistic Functions of a markov Process. *Inequalities*, 3:1 – 8, 1972.
3. H. Bunke and A. Sanfeliu, editors. *Syntactic and Structural Pattern Recognition, Theory and Applications.* World Scientific, 1990.
4. S. Deligne and F. Bimbot. Language modeling by variable length sequences: Theoretical formulation and evaluation of multigrams. In *Proc. of IEEE International Conference on Acoustics, Speech, and Signal Processing*, pages 169–172, 1995.
5. D. Dori, D. Doermann, C. Shin, R. Haralick, I. Phi llips, M. Buchman, and D. Ross. The Representation of Document Structure: A Generic Object-Pro cess Analysis. In E. Bunke and P.S.P. Wang, editors, *Handbook of Character Recognition and Document Image Analysis*, pages 421 – 456. World Scientific, 1997.
6. Frederick Jelinek. *Statistical Methods for Speech Recognition.* The MIT Press, 1997.
7. Karen Kukich. "Techniques for Automtically Correcting Words in Text". *ACM Computing Surveys*, 24(4):377–439, 1992.
8. F. Lerdahl and R Jackendoff. *A Generative Theory of Tonal Music.* The MIT Press, 1983.
9. M. Ohta, A. Takasu, and J. Adachi. "Probabilistic Automaton Model for Fuzzy English-text Retriev al". In *Lecture Notes in Computer Science 1923*, pages 35–44, 2000.
10. Jorma Rissanen. *Stochastic Complexity in Statiscal Inquiry.* World Scientific, 1989.
11. M. Smith and T. Kanade. Video Skimming and Characterization through the Combination of Image and Language Understanding. Technical report, CMU School of Computer Science, 1996.
12. T. Yanase, A. Takasu, and J. Adachi. Phrase Based Feature Extraction for Musical Information Retrieval. In *Proc. of IEEE Pacific Rim Conference on Communications, Computers and Signal Processing (PACRIM'99)*, pages 396–399, 1999.

Top-Down Decision Tree Boosting and Its Applications

Eiji Takimoto and Akira Maruoka

Graduate School of Information Sciences, Tohoku University
Sendai, 980-8579, Japan. {t2, maruoka}@maruoka.ecei.tohoku.ac.jp

Abstract. Top-down algorithms such as C4.5 and CART for constructing decision trees are known to perform boosting, with the procedure of choosing classification rules at internal nodes regarded as the base learner. In this work, by introducing a notion of pseudo-entropy functions for measuring the loss of hypotheses, we give a new insight into this boosting scheme from an information-theoretic viewpoint: Whenever the base learner produces hypotheses with non-zero mutual information, the top-down algorithm reduces the conditional entropy (uncertainty) about the target function as the tree grows. Although its theoretical guarantee on its performance is worse than other popular boosting algorithms such as AdaBoost, the top-down algorithms can naturally treat multiclass classification problems. Furthermore we propose a base learner LIN that produces linear classification functions and carry out some experiments to examine the performance of the top-down algorithm with LIN as the base learner. The results show that the algorithm can sometimes perform as well as or better than AdaBoost.

1 Introduction

Boosting is a technique for finding a hypothesis with high accuracy by combining many *weak* hypotheses that are only moderately accurate. The procedure of boosting is described as in the following general scheme: take any learning algorithm as a base learner (sometimes called a weak learner), rerun it many times with different distributions on the given sample to get many "weak" hypotheses, and combine them somehow to form a master hypothesis. The master hypothesis is hopefully a better classifier than any of the weak hypotheses produced by the base learner. So far many boosting algorithms have been proposed and extensively studied both in practice and theory [12, 4–6, 8, 13]. Most of them are essentially developed for binary classification problems and require weak hypotheses to have classification error less than 1/2. This requirement seems necessary because the error of 1/2 can be achieved by random guessing, from which we cannot extract any information about the target. Surprisingly, however, we can do boosting in some cases where hypotheses have error just 1/2. In particular, Natarajan's algorithm [10] works on such a base learner that produces hypotheses with error just 1/2 but being *one-sided*, i.e., mistakes occurring on positive (or negative) examples only. It turns out that a hypothesis with one-sided error has positive *mutual information* about the target function whenever

S. Arikawa and A. Shinohara (Eds.): Progress in Discovery Science 2001, LNAI 2281, pp. 327–337.

the error is below 1 (so including the case of 1/2) [17]. This observation suggests that it would be reasonable to measure the loss of a hypothesis in terms of the amount of information it brings, rather than its classification error.

In this work, we investigate *information-based* boosting where algorithms boost the amount of information about the target function. The first result on information-based boosting is the work due to Kearns and Mansour [8] although it was not explicitly stated as such. They analyzed the performance of top-down algorithms for growing decision trees such as C4.5 [11] and CART [3] from the viewpoint of error-based boosting as usual. A top-down algorithm begins with a single leaf and repeats the following procedure: replace a leaf of the current tree with an internal node, label it with some classification rule, and create child leaves accordingly. To decide which classification rule (hypothesis) should be chosen, the top-down algorithm uses a function G called the *splitting criterion*. For instance, C4.5 uses the Shannon entropy function ($G(p) = -p\log p - (1-p)\log(1-p)$) and CART uses the Gini Index ($G(p) = p(1-p)$). The claim stated by Kearns and Mansour was that top-down algorithms perform error-based boosting when we consider the procedure of choosing classification rules at internal nodes as the base learner. We give an information-theoretic interpretation of this result and restate the claim as follows: Whenever the base learner produces hypotheses with non-zero mutual information, the top-down algorithm grows tree T so that the conditional entropy (uncertainty) about the target given tree T becomes smaller [15]. Here the splitting criterion G plays the role of the entropy function as well. So we sometimes call the function G a *pseudo-entropy function*, or, more specifically, the G-entropy function.

For the following three reasons, this information-theoretic criterion is very useful for considering multiclass classification problems, where target functions and hypotheses may take more than two values. Firstly, the mutual information can be defined between any two multiclass functions, even when the ranges of the two functions may be different. Secondly, a weak hypothesis is naturally defined as the hypothesis that attains non-zero mutual information. Thirdly, the boosting property can be shown to hold for multiclass classification problems. These contrast with the fact that, for the error-based criterion, it is unclear to define "error" of multiclass functions and we may need the problems somehow reduced to the binary classification problems [6, 14].

Recently, Aslam modified the most popular boosting algorithm AdaBoost and gave a partly information-based boosting algorithm called InfoBoost [2]. InfoBoost also combines weak hypotheses with non-zero mutual information to form a master hypothesis with small classification error. Although InfoBoost seems much more effective than decision tree boosting, the analysis is still error-based and it is unclear how InfoBoost can be applied to multiclass classification problems without reduction. Friedman, Hastie and Tibshirani proposed another variant of AdaBoost, called LogitBoost, based on a statistical analysis of AdaBoost [7]. LogitBoost can naturally handle multiclass classification problems, but its performance on boosting remains unclear.

On the other hand, top-down algorithms are shown to realize information-based boosting in the multiclass classification setting in general. Unfortunately, the theoretical guarantee on its performance of top-down algorithms is worse than other popular boosting algorithms such as AdaBoost. To examine the empirical performance of the top-down algorithm, we apply it on some data sets from UCI Machine Learning Repository. In particular, we propose a new base learner called LIN that intends to produce linear classifiers with large mutual information so that it fits the top-down algorithm. Note that the master hypothesis produced by the top-down algorithm with LIN as the base learner is of the form of a decision tree where its internal nodes are labeled with linear classifiers. This representation is interesting in the sense that this is the "reverse" of the form of a (thresholded) linear combination of decision trees, which is the same form as the master hypothesis produced by a typical application of AdaBoost, namely, AdaBoost on top of C4.5. The experimental results show that the top-down algorithm with LIN seems to be comparable to AdaBoost applied on C4.5, in its performance.

2 Information-Theoretic Criterion

Let X denote an instance space and Y a finite set of labels. We assume $Y = \{1, \ldots, N\}$ with $N \geq 2$. In what follows we fix a target function $f : X \to Y$. Let D be a probability distribution over X. We consider the loss of a function $h : X \to Z$ w.r.t. D, where Z is a finite set but possibly different from Y. In the special case where $Y = Z$, we typically define the loss of h to be the probability of misclassification, i.e., $\Pr_D(h(x) \neq f(x))$. We call this measure the *error* of h. In this work, we define the loss of h from an information-theoretic view point. To do so, we extend the notion of entropy and introduce a function $G : [0,1]^N \to [0,1]$ having the following three properties:

1. For any $(q_1, \ldots, q_N) \in [0,1]^N$ with $\sum_i q_i = 1$,

$$\min_{1 \leq i \leq N} (1 - q_i) \leq G(q_1, \ldots, q_N).$$

2. For any $(q_1, \ldots, q_N) \in [0,1]^N$ with $\sum_i q_i = 1$,

$$G(q_1, \ldots, q_N) = 0 \quad \Leftrightarrow \quad q_i = 1 \text{ for some } 1 \leq i \leq N.$$

3. G is concave.

We call such a function G a *pseudo-entropy function.* Note that the Shannon entropy function $-\frac{1}{\log N} \sum_{i=1}^N q_i \log q_i$ and the Gini index $1 - \sum_{i=1}^N q_i^2$ are also pseudo-entropy functions. In the following, G is assumed to be an arbitrary pseudo-entropy function.

Interpreting $G(q_1, \ldots, q_N)$ as uncertainty of the label with respect to the probability distribution $(q_1, \ldots, q_N)$ over Y, we can define the entropy of f, where f is considered to be a random variable that takes value i with probability q_i.

Definition 1. *For* $1 \leq i \leq N$, *let* $q_i = \Pr_D(f(x) = i)$. *Then the* G-entropy of f w.r.t. D, *denoted* $H_D^G(f)$, *is defined as*

$$H_D^G(f) = G(q_1, \ldots, q_N).$$

Similarly we can define the notion of conditional G-entropy. For hypothesis h, the conditional G-entropy of f given h can be interpreted as uncertainty of f remaining after receiving the value of h. So we can think of this as the loss of h.

Definition 2. *For* $1 \leq i \leq N$ *and* $z \in Z$, *let* $q_{i|z} = \Pr_D(f(x) = i \mid h(x) = z)$. *Then the* conditional G-entropy of f given h w.r.t. D, *denoted* $H_D^G(f|h)$, *is defined as*

$$H_D^G(f|h) = \sum_{z \in Z} \Pr_D(h(x) = z) G(q_{1|z}, \ldots, q_{N|z}).$$

Finally the mutual G-information between f and h w.r.t. D, *denoted* $I_D^G(f;h)$, *is defined as*

$$I_D^G(f;h) = H_D^G(f) - H_D^G(f|h).$$

In what follows, we omit the superscript G and simply write $H_D(f)$, $H_D(f|h)$ and $I_D(f;h)$ to denote the G-entropy, the conditional G-entropy and the mutual G-information, respectively.

3 The Top-Down Algorithm

In this section we give the top-down algorithm TopDown that grows decision trees from samples. Assume that TopDown is given as input a sample $S = \{(x_1, y_1), \ldots, (x_m, y_m)\} \subseteq X \times Y$ and a probability distribution D over S, where $y_j = f(x_j)$ for $1 \leq j \leq m$. Usually the distribution D is assumed to be uniform, but in this report we let TopDown work for any input distribution. TopDown is also given access to a base learner L whose hypothesis class is $H \subseteq \{h : X \to Z\}$, where Z is a finite set. TopDown produces a decision tree $\mathcal{T}$ with each internal node labeled with a classification rule in H, and the edges labeled with members of Z. Note that $\mathcal{T}$ is a $|Z|$-ary tree. The goal of TopDown is to produce $\mathcal{T}$ such that the conditional G-entropy $H_D(f|\mathcal{T})$ is small.

Instance $x \in X$ induces a path from the root to a leaf in the obvious way. So we can think of the tree $\mathcal{T}$ as a function from X to the set of leaves of $\mathcal{T}$ denoted by leaves($\mathcal{T}$). For leaf $\ell \in$ leaves($\mathcal{T}$), let S_ℓ denote the set of examples in S that reach ℓ. Then the conditional G-entropy of f given $\mathcal{T}$ w.r.t. D can be written as

$$H_D(f|\mathcal{T}) = \sum_{\ell \in \text{leaves}(\mathcal{T})} w_\ell G(q_{1|\ell}, \ldots, q_{N|\ell}),$$

where $w_\ell = \Pr_D(\mathcal{T}(x_j) = \ell)$ and $q_{i|\ell} = \Pr_D(y_j = i \mid \mathcal{T}(x_j) = \ell)$. Tree $\mathcal{T}$ also represents a function from X to Y in the following way: when instance x reaches leaf ℓ, then $\mathcal{T}$ takes as the value on x the majority of the labels in the sample S_ℓ, i.e., $\mathcal{T}(x) = \arg\max_{1 \leq i \leq N} q_{i|\ell}$. It is easy to see that the error of $\mathcal{T}$ is upper bounded by the conditional G-entropy given $\mathcal{T}$, i.e., $\Pr_D(\mathcal{T}(x) \neq f(x)) \leq H_D(f|\mathcal{T})$. Therefore, growing tree $\mathcal{T}$ so as to reduce the conditional G-entropy $H_D(f|\mathcal{T})$ tends to reduce the error of $\mathcal{T}$ as well.

Now we are ready to describe the algorithm TopDown. The algorithm starts with a single leaf and repeats the following procedure until it meets some terminating condition. For leaf ℓ, let D_ℓ denote the distribution obtained by restricting the original distribution D to S_ℓ. In each round, TopDown runs the base learner L fed with S_ℓ together with the restricted distribution D_ℓ over S_ℓ to get a hypothesis $h_\ell \in H$ for all leaves $\ell \in$ leaves($\mathcal{T}$). Then it chooses the leaf ℓ such that the weighted mutual G-information $w_\ell I_{D_\ell}(f; h_\ell)$ is maximum, and replace the leaf ℓ by the internal node with $|Z|$ child leaves, where the internal node is labeled with h_ℓ. The tree obtained in this way is denoted by $\mathcal{T}(\ell, h_\ell)$. In Figure 1 we give algorithm TopDown in more detail. It should be noticed that using a heap for storing $w_\ell I_{D_\ell}(f; h_\ell)$ for all $\ell \in$ leaves($\mathcal{T}$), we can choose in each round the leaf to be replaced more efficiently.

```
TopDown(S)
begin
    Let T be the single leaf tree.
    while some terminating condition is met do
        begin
            for all ℓ ∈ leaves(T) do
                h_ℓ := L(S_ℓ, D_ℓ);
            ℓ := arg max_{ℓ∈leaves(T)} w_ℓ I_{D_ℓ}(f; h_ℓ);
            T := T(ℓ, h_ℓ);
        end
    output T;
end.
```

Fig. 1. Algorithm TopDown

The criterion for choosing the leaf is simply the greedy strategy that reduces the conditional G-entropy $H_D(f|\mathcal{T})$ most. To see this, observe the next equality.

$$\begin{aligned} H_D(f|\mathcal{T}) - H_D(f|\mathcal{T}(\ell, h_\ell)) &= w_\ell(H_{D_\ell}(f) - H_{D_\ell}(f|h_\ell)) \\ &= w_\ell I_{D_\ell}(Y; h_\ell). \end{aligned}$$

From this we can see that as long as the hypothesis h_ℓ gives non-zero mutual information locally at leaf ℓ, the conditional G-entropy $H_D(f|\mathcal{T})$ decreases as $\mathcal{T}$ grows. This implies that TopDown is an information-based boosting algorithm. The next theorem describes the performance of TopDown more precisely.

Theorem 1 ([15]). *Let L be a base learner that, when given any sample S and any distribution D' over S, produces a hypothesis h that satisfies*

$$H_{D'}(f|h) \leq (1 - \gamma) H_{D'}(f) \tag{1}$$

for some positive constant $\gamma > 0$. Then, TopDown given access to L produces a decision tree $\mathcal{T}$ such that $H_D(f|\mathcal{T}) \leq \epsilon$ when the number of internal nodes of $\mathcal{T}$ reaches

$$\left(\frac{1}{\epsilon}\right)^{\frac{\log |Z|}{\gamma}}.$$

4 A Pseudo-Entropy Function

Theorem 1 requires the base learner to produce a hypothesis h that satisfies condition (1) for *any* distribution D'. We call such a base learner a *weak learner*. That is, it seems that we need a distribution-free weak learner. However, a particular choice of G significantly relaxes the condition of weak learning. In particular, we propose the following pseudo-entropy function

$$G(q_1, \ldots, q_N) = \frac{1}{\sqrt{N-1}} \sum_{1 \leq i \leq N} \sqrt{q_i(1-q_i)}, \tag{2}$$

where the coefficient $1/\sqrt{N-1}$ is a normalization factor. We show in Theorem 2 below that our choice of G helps us design a weak learner which is required to work only for balanced distributions.

Definition 3 (Balanced distributions). *Let D be a probability distribution over X and $q_i = \Pr_D(f(x) = i)$ for $i \in Y = \{1, \ldots, N\}$. The distribution D is* balanced *if there exists a subset $Y' \subseteq Y$ such that $q_i = 1/|Y'|$ if $i \in Y'$ and $q_i = 0$ otherwise.*

Theorem 2 ([15]). *Assume that we choose G given by (2). If there exists a weak learner with respect to balanced distributions, then there exists a distribution-free weak learner.*

In the binary case where $N = 2$, the pseudo-entropy function is $G(q_1, q_2) = 2\sqrt{q_1(1-q_1)}$, which is exactly the same as the splitting criterion that Kearns and Mansour proposed to show that TopDown is an efficient error-based boosting algorithm. It is not hard to see that any error-based weak hypothesis h with $\Pr_D(f(x) \neq h(x)) \leq 1/2 - \gamma$ immediately implies an information-based weak hypothesis with $H_D(f|h) \leq (1 - 2\gamma^2) H_D(f)$ with respect to any balanced distribution. Therefore, Theorem 2 says that any error-based weak learning algorithm can be transformed to an information-based weak learning algorithm in the distribution-free setting. Because we have an information-based boosting algorithm, the overall transformations involved gives an error-based boosting algorithm. Here we have a much simpler proof of the main result of Kearns and Mansour.

5 Base Learner LIN

In this section, we consider the case where the hypothesis class for the base learner H consists of linear classifiers. A linear classifier h is represented by

$$h(x) = \text{sign}(v \cdot x + b) = \text{sign}\left(\sum_{i=1}^{n} v_i x_i + b\right)$$

for some vector $v = (v_1, \ldots, v_n) \in \mathbf{R}^n$ and a real number $b \in \mathbf{R}$, where $\text{sign}(z) = 1$ if $z \geq 0$ and $\text{sign}(z) = 0$ if $z < 0$. So the range of the hypotheses is $Z = \{0, 1\}$, which may be different from the range $Y = \{1, \ldots, N\}$ of the target. Recall that we can take binary classifiers as weak hypotheses even when we treat a multiclass classification problem. We choose G given by (2) as the pseudo-entropy function.

We assume that the input distribution D is uniform. In this case, the distribution that TopDown feeds to the base learner is always uniform over a sample (a subset of S). As we see in Theorem 1, the larger γ is, the more efficiently TopDown performs. In other words, a reasonable base learner would be to produce the linear classifier that minimizes the conditional entropy $H_U(f|h)$, where U is the uniform distribution over a sample. Unfortunately, minimizing $H_U(f|h)$ is most likely NP-hard because a similar related problem is known to be NP-hard [1]. So we approximate the conditional entropy by some continuous function w.r.t. v and b, and employ the gradient decsent search to find a pair (v, b) that gives a minimal point of the approximate function. Below we describe this base learner called LIN more precisely.

Let $S = \{(x_1, y_1), \ldots, (x_m, y_m)\}$ and U be the uniform distribution over S. Then we have $\Pr_U(h(x_j) = z) = |S_z|/|S|$, where $S_z = \{(x_j, y_j) \in S \mid h(x_j) = z\}$ for $z \in \{0, 1\}$. Plugging these equalities with

$$q_{i|0} = \Pr_U(y_j = i \mid h(x_j) = 0) = \left(\sum_{j:y_j=i} (1 - h(x_j))\right) / |S_0|,$$

$$q_{i|1} = \Pr_U(y_j = i \mid h(x_j) = 1) = \left(\sum_{j:y_j=i} h(x_j)\right) / |S_1|$$

into $H_U(f|h) = \Pr_U(h(x_j) = 0) G(q_{1|0}, \ldots, q_{N|0}) + \Pr_U(h(x_j) = 1) G(q_{1|1}, \ldots, q_{N|1})$, we have

$$\sqrt{N-1}|S| H_U(f|h) = \left[\left(\sum_{j:y_j=i} (1 - h(x_j))\right)\left(\sum_{j:y_j \neq i} (1 - h(x_j))\right)\right]^{1/2} + \left[\left(\sum_{j:y_j=i} h(x_j)\right)\left(\sum_{j:y_j \neq i} h(x_j)\right)\right]^{1/2}.$$

Now we replace sign by the sigmoid function $\sigma(z) = 1/(1 + e^{-\mu z})$ so that the function obtained becomes continuous with respect to v and b. Here $\mu > 0$ is a parameter that represents the steepness of $\sigma(z)$ around $z = 0$. More specifically, our approximate function is

$$F(v,b) = \left[\left(\sum_{j:y_j=i}(1-\sigma(v\cdot x_j+b))\right)\left(\sum_{j:y_j\neq i}(1-\sigma(v\cdot x_j+b))\right)\right]^{1/2}$$
$$+\left[\left(\sum_{j:y_j=i}\sigma(v\cdot x_j+b)\right)\left(\sum_{j:y_j\neq i}\sigma(v\cdot x_j+b)\right)\right]^{1/2}.$$

Our base learner LIN seeks for the pair (v, b) that minimizes $F(v, b)$. To do so LIN employs the gradient descent search. That is, LIN updates a pair (v, b) until it converges according to the following formulas:

$$v_i := v_i - \eta\frac{\partial F(v,b)}{\partial v_i}$$

for all $1 \leq i \leq n$ and

$$b := b - \eta\frac{\partial F(v,b)}{\partial b},$$

where $\eta > 0$ is a learning parameter.

6 Experiments

We evaluated the performance of algorithm TopDown with LIN as the base learner, denoted TopDown+LIN, on four data sets, car-evaluation, tic-tac-toe, abalone, balance-scale-weight from the UCI Machine Learning Repository. To examine the effectiveness of the proposed base learner LIN, we compare our algorithm with TopDown with the brute force base learner which produces the decision stump that minimizes the conditional G-entropy. A decision stump is a function of the form of $\text{sign}(x_i > b)$ for some attribute x_i and real number b. This algorithm is simply called TopDown. Note that a decision stump can be seen as a linear classifier (v, b) where only one component of v is non-zero. In our experiments, we chose as the initial hypothesis of LIN the linear classifier (v, b) induced by the decision stump that the brute force base learner produces. Furthermore we chose the parameters $\mu = 2.0$ and $\eta = 5.0$. For both TopDown+LIN and TopDown, we grew decision trees $\mathcal{T}$ until the conditional G-entropy $H_U(f|\mathcal{T})$ converged to 0, that is, the final tree $\mathcal{T}$ had no training error.

Figure 2 shows how the training and test errors change as the tree $\mathcal{T}$ grows for both algorithms. Here we randomly picked 25% of the available data as the training set and the rest as the test data. From these we can see that TopDown+LIN converges very quickly and performs better than TopDown.

We also compared the performance of these algorithms with AdaBoost using TopDown as the base learner with the following terminating condition: TopDown must stop growing trees before the depth exceeds d. We call this algorithm

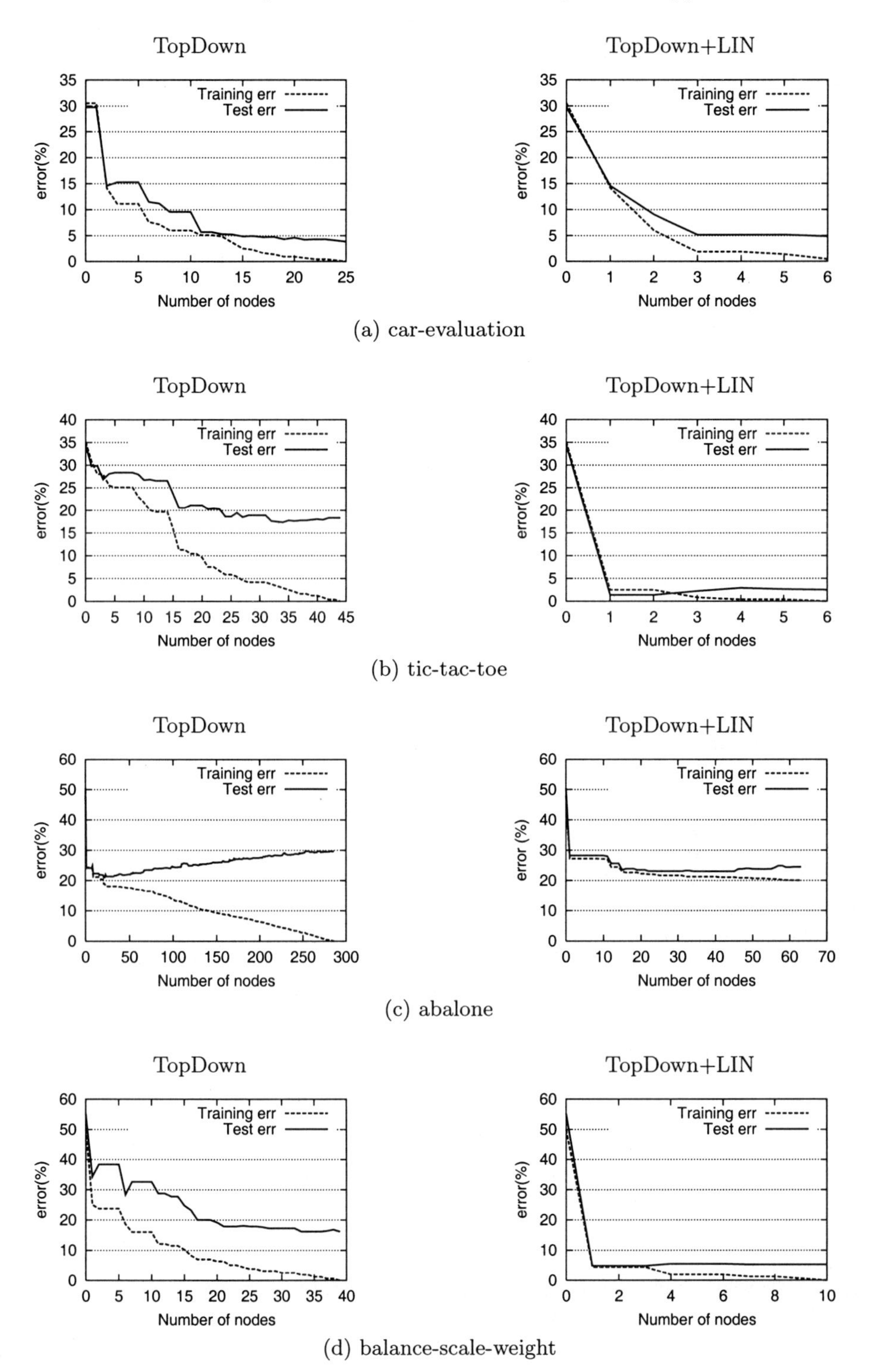

Fig. 2. Learning curves for algorithms TopDown and TopDown+LIN

AdaBoost+TopDown(d). Furthermore, we terminate AdaBoost after 200 rounds. This means that AdaBoost+TopDown produces a linear combination of 200 decision trees.

For each data set, we performed 4-fold cross validation with 25% of the data used as the training data and the rest reserved as the test data. We reran it 5 times for a total of 20 runs of each algorithm. Table 1 shows the average test errors of the final hypotheses of the algorithms. This shows that TopDown+LIN could be comparable to AdaBoost+TopDown. A nice property of TopDown+LIN is that in most cases it produces hypotheses of size much smaller than those that AdaBoost+TopDown produces.

Table 1. Average test error (%) of the algorithms

	car	tic	abalone	balance
TopDown+LIN	4.81	3.10	24.56	6.66
TopDown	4.20	16.00	28.98	17.28
AdaBoost+TopDown(1)	5.78	4.13	24.56	16.26
AdaBoost+TopDown(3)	2.37	5.20	23.62	9.96
AdaBoost+TopDown(5)	2.72	5.68	24.78	13.88

7 Concluding Remarks

We introduced the notion of pseudo-entropy function and showed that for any pseudo-entropy function G, the conditional G-entropy is useful for measuring the loss of hypotheses especially for the multiclass classification problems. We showed that the top-down decision tree learning algorithm with G as its splitting criterion realizes information-based boosting with respect to G. Unfortunately, however, the performance of boosting of the top-down algorithm is much worse than other popular boosting algorithms such as AdaBoost. In particular, the sample complexity of our algorithm is of the form of $(1/\epsilon)^{1/\gamma}$ (see Theorem 1), whereas that of AdaBoost is only $poly((1/\gamma)\log(1/\epsilon))$. So we cannot allow γ to be as small as $1/poly(n, s)$, where n and s are the domain and the target complexities, respectively. This inefficiency comes from the fact that the top-down algorithm divides the domain into small pieces and the local modification at a small piece contributes little to the performance of the whole tree. To overcome this difficulty, Mansour and MacAllester recently develop an algorithm [9] that is based on the top-down algorithm but now some leaves are sometimes merged into one, producing a final hypothesis as a branching program. It is worth analyzing this algorithm using the conditional G-entropy. In spite of the relatively moderate theoretical performance of TopDown as a boosting algorithm, the experiments shows that TopDown sometimes outperforms AdaBoost+TopDown, which is one of the most popular applications of boosting.

Another issue with our algorithm is the question of what pseudo-entropy function G is the best one to use. The function G we proposed seems to help the base learner because it is required to work well only for balanced distributions. However, in order to show that this really helps, we need to give instances of natural and non-trivial weak learning algorithms.

Other promising approaches to designing information-based boosting is to analyze InfoBoost [2] and LogitBoost [7] in terms of the conditional G-entropy and show that these algorithms have a boosting property even when they are applied to multiclass classification problems.

References

1. M. Anthony and P. Bartlett. *Neural Network Learning: Theoretical Foundations.* University Press, Cambridge, 1999.
2. J. A. Aslam. Improving algorithms for boosting. In *13th COLT*, pages 200–207, 2000.
3. L. Breiman, J. H. Friedman, R. A. Olshen, and C. J. Stone. *Classification and Regression Trees.* Wadsworth International Group, 1984.
4. Y. Freund. Boosting a weak learning algorithm by majority. *Inform. Comput.*, 121(2):256–285, Sept. 1995. Also appeared in COLT90.
5. Y. Freund and R. E. Schapire. Game theory, on-line prediction and boosting. In *Proc. 9th Annu. Conf. on Comput. Learning Theory*, pages 325–332. 1996.
6. Y. Freund and R. E. Schapire. A decision-theoretic generalization of on-line learning and an application to boosting. *J. Comput. Syst. Sci.*, 55(1):119–139, 1997.
7. J. Friedman, T. Hastie, and R. Tibshirani. Additive logistic regression: a statistical view of boosting. Technical report, Stanford University, 1998.
8. M. Kearns and Y. Mansour. On the boosting ability of top-down decision tree learning algorithms. *J. of Comput. Syst. Sci.*, 58(1):109–128, 1999.
9. Y. Mansour and D. McAllester. Boosting using branching probrams. In *13th COLT*, pages 220–224, 2000.
10. B. K. Natarajan. *Machine Learning: A Theoretical Approach.* Morgan Kaufmann, San Mateo, CA, 1991.
11. J. R. Quinlan. *C4.5: Programs for machine learning.* Morgan Kaufmann, 1993.
12. R. E. Schapire. The strength of weak learnability. *Machine Learning*, 5(2):197–227, 1990.
13. R. E. Schapire, Y. Freund, P. Bartlett, and W. S. Lee. Boosting the margin: a new explanation for the effectiveness of voting methods. In *Proc. 14th International Conference on Machine Learning*, pages 322–330. Morgan Kaufmann, 1997.
14. R. E. Schapire and Y. Singer. Improved boosting algorithms using confidence-rated predictions. In *Proc. 11th Annu. Conf. on Comput. learning Theory*, 1998.
15. E. Takimoto and A. Maruoka. Top-down decision tree learning as information based boosting. *to appear in Theoretical Computer Science.* Earlier version in [16].
16. E. Takimoto and A. Maruoka. On the boosting algorithm for multiclass functions based on information-theoretic criterion for approximation. In *Proc. 1st International Conference on Discovery Science*, volume 1532 of *Lecture Notes in Artificial Intelligence*, pages 256–267. Springer-Verlag, 1998.
17. E. Takimoto, I. Tajika, and A. Maruoka. Mutual information gaining algorithm and its relation to PAC-learning algorithm. In *Proc. 5th Int. Workshop on Algorithmic Learning Theory*, volume 872 of *Lecture Notes in Artificial Intelligence*, pages 547–559. Springer-Verlag, 1994.

Extraction of Primitive Motion and Discovery of Association Rules from Human Motion Data

Kuniaki Uehara and Mitsuomi Shimada

Department of Computer and Systems Engineering
Kobe University
{uehara, shimada}@ai.cs.kobe-u.ac.jp
http://www.ai.cs.kobe-u.ac.jp

Abstract. In past several years, more and more digital multimedia data in the forms of image, video and audio have been captured and archived. This kind of new resource is exiting, yet the sheer volume of data makes any retrieval task overwhelming and its efficient usage impossible. In order to deal with the deficiency, tagging method is required so as to browse the content of multimedia data almost instantly.
In this paper, we will focus on tagging human motion data. The motion data have the following features: movements of some body parts have influence on other body parts. We call this dependency motion association rule. Thus, the task of tagging motion data is equal to the task of expressing motion by using motion association rules. Association rules consist of symbols, which uniquely represent basic patterns. We call these basic patterns primitive motions. Primitive motions are extracted from the motion data by using segmentation and clustering processes. Finally, we will discuss some experiments to discover association rules from multi-stream of the motion data.

1 Introduction

Human motion data are practically used in some domains such as SFX movies, CGs, PC games and so on. Motion data possess spatial information so that it enables us to replay natural human movement on computer displays. Creators use the motion data to produce exciting and dangerous scenes without real human actors.

Human motion data include dependencies between body parts. That is, motion is the combination of postures. For example, we swing both arms to keep walking straight. In this experiment, stepping left foot forward causes swing of right arm. We must pre-define the set of postures to represent the motions data. Furthermore, since the human motion data have the dependencies, the motion data should be treated as multi-stream [5].

Multi-stream includes unexpectedly frequent or infrequent co-occurrences among different streams. This means one stream is related to other events of other streams, which seems to be independent of the former event. We call this dependency motion association rule. It is an expensive task to discover motion association rules because of the huge amount of the motion data.

S. Arikawa and A. Shinohara (Eds.): Progress in Discovery Science 2001, LNAI 2281, pp. 338–348.

In order to find the motion association rule, we define the primitive motions and we give an unique symbol to each of them. Every motion is represented as the combination of primitive motions. For example, "taking 2 steps" and "taking 20 steps" is composed of the same motion "stepping". Furthermore "stepping" is composed of primitive motions, such as "moving the left foot forward" and "moving the right foot forward". Next, we represent the motion data as a sequence of the defined symbols. However, there are two problems to find primitive motions. Firstly, we must consider the ambiguity of the borders between primitive motions. The definition of borders for movement is unclear. It is necessary to define borders in consecutive movement. Secondly, we must consider the reconstructivity of movement by using primitive motions.

2 Human Motion Data as Multi-stream

Figure 1 shows the body parts used in motion capture system. The motion-captured data are mass of information of the body parts. It means that captured motion data can be described by the 3 dimensional time series streams, which represent positions of major body joints. We can get the data for the following body parts: the upper torso, root of neck, head, roots of collar bones, shoulders, elbows, wrists, hips, knees and ankles.

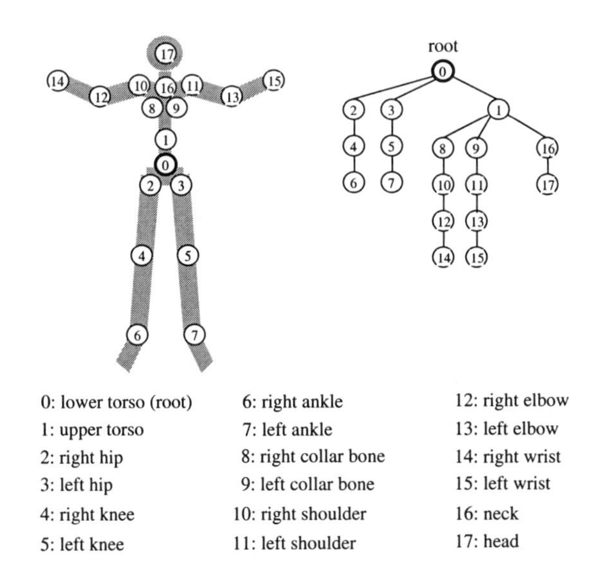

Fig. 1. Body parts used for the experiment.

We use an optical motion capture system to get time series of 3-D motion data. This system consists of six infra-red cameras, a personal computer (PC) and a workstation (WS). The procedure to get time series of motion data is as follows: An actor puts on 18 markers which reflect infra-red ray, and performs some actions with the markers being surrounded with 6 cameras. The cameras record actor's actions as video images, and all the video images are gathered to the PC and WS, which process the images and calculate 3-D locations of the markers. We set the sampling rate to 120 ($times/sec$) to capture the motion data.

Figure 2 shows an example of the motion data, which represents "after one finished raising one's right hand, one starts lowering the left hand". The example has unfixed temporal intervals between actions on each stream (T_1 and T_2). In Figure 2, "raising the right hand" and "lowering the left hand" occur twice in each stream, however, the temporal intervals between "raising the right hand" and "lowering the left hand" are different from each other ($T_1 \neq T_2$), which can be explained by the personal characteristics of the actor. Because everyone cannot perform the same motion precisely. Furthermore, human activity is composed of a continuous flow of actions. This continuity makes it difficult to define borders of actions. Therefore, we will deal with the human motion data qualitatively. That is, real values of human motion data are described in terms of their ordinal relations with a finite set of symbolic landmark values, rather than in terms of real numbers.

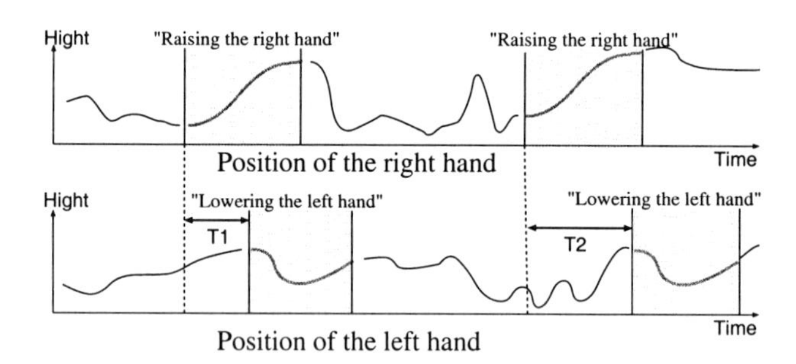

Fig. 2. An example of correlation in multi-stream of human motion.

3 Content-Based Automatic Symbolization by Using Curve Simplification

Human recognizes movement of the object by perceiving the change of its velocity. Generally, the human motion consists of 3 states: occurrence, transition and termination. Occurrence represents a change of speed from zero to a positive value, transition represents increase or decrease of the speed and termination represents a change of the speed from a positive value to zero. Thus, each change in the speed of a body part can be considered as a breakpoint. We will divide motion data into segments [6] at those breakpoints where velocity changes. Those detected breakpoints are candidates for segmentation as shown in Figure 4.

Techniques for the segmentation of patterns have received considerable attention when these patterns are planar [4]. The goal of this technique is to simplify the general shape of the curve to a few perceptually significant breakpoints. However the variation of the curves for motion data between these breakpoints is small and includes noise made by unconscious movement, which is not related to the occurrence of the changes of contents.

The unconscious movement is mainly caused by the vibration of the body parts. The noise should be removed by considering the 3-D distances between breakpoints. Removal of noise brings semantic summarization of the high dimensional patterns of motion. For this reason, the algorithm calculates the variance among all the candidate breakpoints. If the variance is small, then it should be removed from the set of candidate breakpoints.

To remove the noise, we compare the height of the candidate breakpoints as shown in Figure 3.

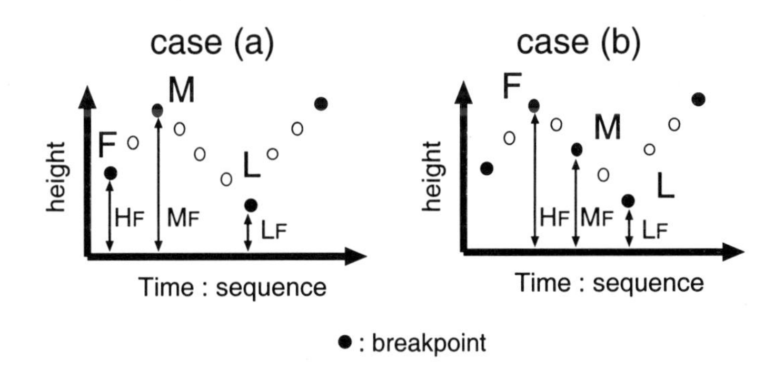

Fig. 3. Discernment of breakpoints.

The motion data represented in Figure 3 (a) include ten candidate breakpoints, where four of them are proper breakpoints (black circles). Figure 3 (b) represents the noise which were detected by a proper breakpoint detection algorithm. One of the candidate breakpoint ("M" in Figure 3 (b)) was removed because the change of velocity does not occur at the candidate.

To explain the proper breakpoint detection algorithm, let us consider the sequence of three candidate breakpoints (F, M and L). Firstly, we calculate respectively the height for F (H_F), (H_M) for M and (H_L) for L as shown in Figure 3 (a). Secondly, we compare H_F, H_M, H_L. If ($H_F < H_M < H_L$) or ($H_L < H_M < H_F$) then M should be consider as a noise. But if ($H_M > H_F > H_L$) or ($H_M > H_L > H_F$) then M should be considered as a proper breakpoint.

Figure 4 (a) shows an original motion data. Black and white circles in Figure 4 (b) represent candidate breakpoints that would be remained for the segmentation process. After checking the variance of the original data (see Figure 4 (a)), only the candidate breakpoints remains (see Figure 4 (b)), where the white circles represent a noise and the black points represent proper breakpoints. For example, four white circles in Figure 4 (b) are detected as noise. Thus, they should not be considered as proper breakpoints.

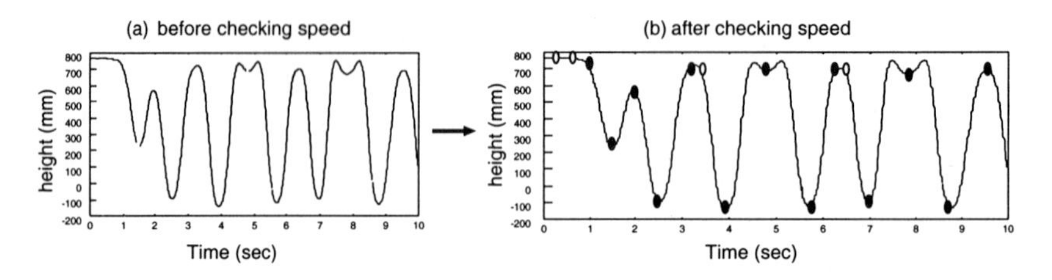

Fig. 4. An example of checking speed.

4 DTW as an Evaluation Measure for Motion Data

Human voice has a fixed number of phonemes, but human motion does not have such pre-defined patterns. That is, the number of primitive motions is unknown. For this purpose, we employ a simple and powerful unsupervised learning algorithm, Nearest Neighbor (NN) algorithm [1] for clustering primitive motions. It can be represented by the following rules: classify an unknown pattern, and choose the class of the nearest example in the training set as measured by distant metric.

Then we give the features of a motion data to NN classifier. The features are multiple dimensional data: time series position data of 3-D space. To evaluate multiple features of the motion data, we employ Dynamic Time Warping (DTW) method [2] that was developed in speech recognition domain. In our case, we apply this method to the motion data. DTW calculates the similarity between patterns reducing the influence of variation. When we apply DTW to our motion data, the best correspondence between two time series can be found.

We will explain how DTW calculates similarity between two time series. Assume two discrete time series A and B are given as follows:

$$\begin{aligned} A &= a_1, \ldots, a_i, \ldots, a_M \\ B &= b_1, \ldots, b_j, \ldots, b_N \end{aligned} \tag{1}$$

where a_i and b_j are the i-th and j-th factors of A and B respectively. The similarity between A and B is given by the following formulas (2) and (3).

$$d_{i,j} = \sqrt{(a_i - b_j)^2} \tag{2}$$

$$\begin{aligned} S_{i,j} &= min \begin{cases} S_{i-1,j-2} + 2d_{i,j-1} + d_{i,j} \\ S_{i-1,j-1} + 2d_{i,j} \\ S_{i-2,j-1} + 2d_{i-1,j} + d_{i,j} \end{cases} \\ S_{1,1} &= d_{1,1} \\ D(A,B) &= S_{M,N} \end{aligned} \tag{3}$$

Motion data are the time series of spatial position. They are extended discrete time series of voice wave, which are one-dimensional discrete patterns. Then we calculate Euclid distance for motion data by using the following formula (4) instead of formula (2),

$$X = x_{a_i} - x_{b_j},\ Y = y_{a_i} - y_{b_j},\ Z = z_{a_i} - z_{b_j}$$
$$d_{i,j} = \sqrt{X^2 + Y^2 + Z^2} \tag{4}$$

where x_{a_i} and x_{b_j} are the projections of a_i and b_j on the x-axis, y_{a_i} and y_{b_j} are the projections of a_i and b_j on the y-axis, z_{a_i} and z_{b_j} are the projections of a_i and b_j on the z-axis respectively. The similarity between A and B can also be calculated by a dynamic programming shown as formula (3).

By using DTW as evaluation measure for the motion data, nearest neighbor based algorithm takes the following phases: The algorithm calculates distances between the new data and the already existing clusters. Then the algorithm finds the nearest distance, and compares that distance to the given threshold. If it is smaller than the threshold, the algorithm classifies the new data to the cluster that gave the smallest distance. If not, it makes a new cluster for the new data.

Figure 5 shows the process of clustering A, B, C, and D.

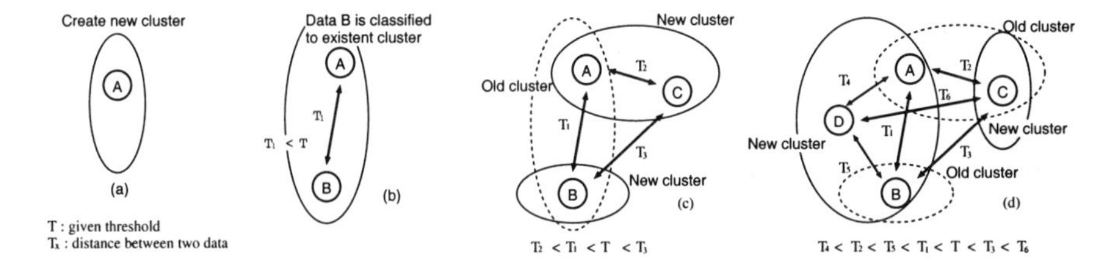

Fig. 5. Clustering algorithm

1. For a given data A,
   the algorithm creates a new cluster for the data A (see phase (a)).
2. For the next data B,
   the algorithm calculates T_1, which is the distance between the new data B and the already existing cluster (A). In this case, T_1 is smaller than the given threshold (T), then the algorithm classifies B in the pre-existed cluster (see phase (b)).
3. For the next data C,
   the algorithm calculates T_1, T_2 and T_3 which are respectively the distances between (A, B), (A, C) and (B, C). In this case, we suppose that T_2 is smaller than T, and T_3 is larger than T. Moreover, T_1 is larger than T_2, then the algorithm classifies A and C in the same cluster and makes a new cluster for B. (see phase (c))

This is a simple case, but when a new data D is introduced, it becomes a more complicated case (see phase (d)).

4. When D is introduced, we calculate the distances (D, A), (D, B) and (D, C) and we note them respectively by T_4, T_5 and T_6.
   T_4 is smaller than T_2, and T_1, T_2. T_5 are smaller than T. Thus A, B and D can be classified in a new cluster. Whereas T_3 and T_6 are larger than the threshold T, so only C must be classified to a new cluster.

As a result, there will be two clusters, one includes A, B and D, and the other includes only C.

5 Discovery of Motion Association Rules in Symbol Multi-stream

After segmenting and clustering processes, human motion data are expressed as a multiple sequences of symbols. We call this sequence symbol multi-stream. Each symbol represents the segment of certain cluster, so the length of zones with the same symbol is usually different. In addition, segmentation process is executed on streams of the body parts. Therefore, borders of zones on the streams rarely appear at the same point on the time axis. Thus, the symbol multi-stream is synchronized at the points where the changes of the content occur as shown in Figure 6 (a).

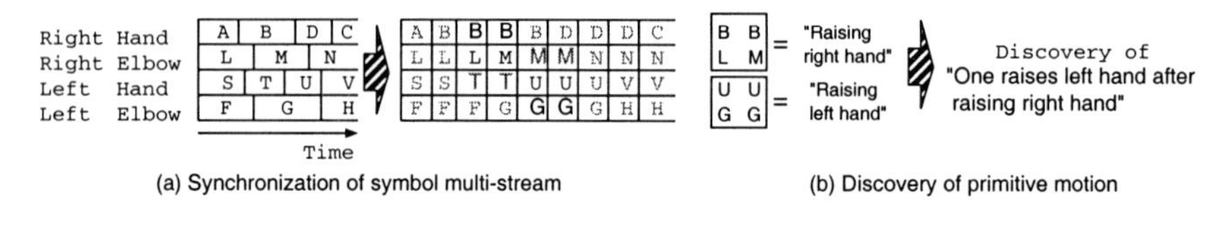

Fig. 6. The process of multi-stream.

We call motions that have influence on the others "active motion" and we call "passive motions" the actions influenced by the "active motions." In order to find active motions and passive motions, we set 2 windows W_a (*a*ctive) and W_p (*p*assive) with fixed interval. By sliding the 2 windows on synchronized symbol multi-stream, we can find motion association rules. Then combinations of symbol patterns in W_a and W_p represent motion association rules.

We define the strength of motion association rules by using the frequency of occurrence for two symbol patterns. The frequency is calculated by the following function,

$$f(co) = \frac{t(P_a \wedge P_p)}{t(P_a)}$$

where *co* is the identifier of a motion association rule, P_a and P_p are patterns which appears in 2 windows W_a and W_p, $t(P_a)$ is the number of occurrences of P_a and $t(P_a \wedge P_p)$ is the number of simultaneous occurrences of P_a and P_p. That means the system calculates the probability of occurrence of a pattern B which occurs after a pattern A in certain blocks of interval.

Our algorithm finds the pattern A which includes symbol "B", "B", "L", "M" in streams of right hand and elbow. After two blocks of symbols from the beginning of A, the algorithm finds the pattern B which consists of "U", "U", "G", "G", in streams of left hand and elbow as shown in Figure 6 (a). Assume that the combination of A and B occurs frequently in symbol multi-stream, then we can understand that "raising the left hand" occurs after "raising the right hand" at the higher probability as shown in Figure 6 (b).

In order to decrease the influence of the variation, we set the size of windows W_a and W_p to W_{asize} and W_{psize}, and the interval of W_a and W_p to int. The values of W_{asize}, W_{psize} and int may be determined flexibly. This flexibility allows the system to find P_a and P_p with the same interval, even if those sizes are set $W_{asize} = 2$, $W_{psize} = 2$ and $int = 2$, or, $W_{asize} = 3$, $W_{psize} = 3$ and $int = 1$. This flexibility decreases the influence of the variation.

The system discovers many motion association rules from motion multi-stream and each of them has a degree of strength to characterize the motion. Probability is an important parameter to define the strength and to select motion association rules as primitive motions for the motion. However, it still requires a deep consideration to define it.

6 Experimental Results

We prepared experimental data from 2 types of motion data. One is rumba (RM), known as moody and passionate dance from Middle America. The other is Japanese traditional physical exercise (JTPE). Rumba is usually performed by a couple, but in this experiment we prepared the data performed only by a man. We prepared basic and frequently used steps of RM. Arms are allowed to have wide variation of expression in dance, but for our data, the performer keeps arms quiet. The performer tried not to play extreme acts. The step does not include jumping. On the other hand, JTPE does not necessarily involve the whole body but may be performed with a set of body parts only. Multiple actions can be performed in parallel way if they use non-intersecting sets of body parts. Some motions include bending knees and jumping.

Table 1 shows the details of motion data for RM and JTPE.

Table 1. Details of motion data.

	RM	JTPE
Amount	23	25
Kinds	6	12
Length(sec))	6 - 12	10 - 20

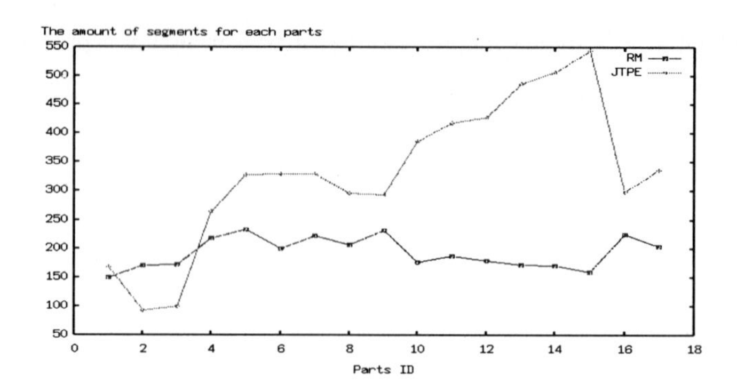

Fig. 7. Details of segmentation.

Figure 7 shows the number of segments for each body part. Parts ID in Figure 7 corresponds to the number in Figure 1.

Table 2 shows clustering result of JTPE. Indicated are the number of clusters generated by our algorithm, the number of incorrect clusters, and the predictive accuracy. The predictive accuracy is shown in terms of percentage of clusters correctly clustered in Table 2. The percentage of correct clusters gives a good indication for the quality of DTW as a similarity measure. For example, predictive accuracy of wrists are higher than those of elbows, because movement of wrist is larger than those of elbows and characteristics of motion appeared more at wrists than at elbows.

Table 2. Clustering result.

Parts name	N_C	N_E	Predictive accuracy
Left wrist	203	27	86.7
Right wrist	187	23	87.7
Left elbow	100	19	81.0
Right elbow	107	18	82.4
Left knee	30	12	60.0
Right knee	25	9	64.0
Left ankle	50	20	60.0
Right ankle	50	17	64.0

$Performance\ accuracy = 100 \times (N_C - N_E)/N_C$
N_C : The number of clusters
N_E : The number of incorrect clusters

We heuristically set $W_{asize} = 5$, $W_{psize} = 5$ and $int = 0$. $W_{asize} = 5$ (or $W_{psize} = 5$) is about 0.5 second long, and we can extract motion association rules in the scope of $0.5 + 0.5 + 0 = 1.0$ second long for this setting. 1.0 second is proper to find motion association rules from RM and JTPE.

The following examples are discovered motion association rules for RM.

$(RM\ a)$ $if\ 'LAnkle\ 19'\ \&\ 'LHip\ 17'$ $then\ 'RElbow\ 18'$
$(RM\ b)$ $if\ 'LHip\ 16'$ $then\ 'RWrist\ 2'$
$(RM\ c)$ $if\ 'LAnkle\ 19'$ $then\ 'RShoulder\ 13'$

$(RM\ a)$ is one of the motion association rules found for RM. The association rule can be illustrated in Figure 8 using our skeleton model. It represents that after either ankle gets to higher position than usual and the hip on the same side draws the arc (Figure 8 (a)), the arm on the opposite side spreads wide (Figure 8 (b)).

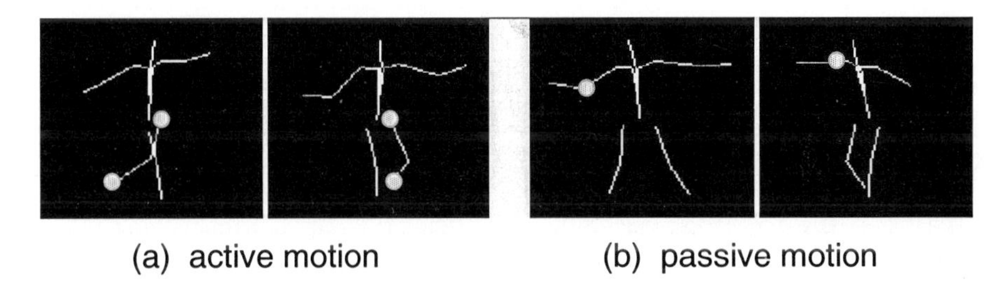

Fig. 8. Visualized motion association rule.

This rule is very important for dancers to keep exactly their weight on their foot and dancers create action from their foot through hips. $(RM\ a)$, $(RM\ b)$ and $(RM\ c)$ show that left side of the body parts affects right side of the body parts. This motion association rule is a specific characteristic for RM. For example, a motion for RM is the step that the performer moves to both sides of woman by turns with spreading his arms.

The following examples show discovered motion association rules of JTPE.

$(JTPE\ a)$ $if\ 'LAnkle\ 259'\ \&\ 'RAnkle\ 260'\ then\ 'LShoulder\ 377'\ \&\ 'RElbow\ 350'$
$(JTPE\ b)$ $if\ 'LAnkle\ 252'\ \&\ 'RAnkle\ 259'\ then\ 'LWrist\ 381'\ \&\ 'RWrist\ 259'$
$(JTPE\ c)$ $if\ 'LAnkle\ 252'\ \&\ 'RAnkle\ 259'\ then\ 'LShoulder\ 382'\ \&\ 'RElbow\ 362'$

$(JTPE\ a)$ is one of the motion association rules of JTPE: when ankles are at the peak or on the way to the peak of a jump, shoulder and elbow is also at or on the way to the peak. This motion association rule can be extracted without performer's consciousness. Legs push body up by kicking the floor and the action causes shoulder and elbow to get to higher position than the normal position.

7 Conclusion

We proposed the method to discover motion association rules from human motion data. Motion association rules appear many times in multi-stream and motion data are represented by the combination of motion association rules. Motion association rules can be tags for motion retrieval and recognition. It is difficult to express features of motion exactly only by words, however, those discovered motion association rules are visually understandable tags. This means our method is effective to discover tags for multi-stream of motion.

There still remains the problem to decide which motion association rules are better to be given to motion as tags. The occurrence probability is one of effective parameters, however, we should consider the peculiarity of the motion association rules: motion association rules that are frequently discovered in some categories of motion cannot be specific tags to certain category of motion.

Many researchers have focused on the extraction of human motion information from time series of 2-D images [3]. The extracted information does not possess correlation and they do not process the information as multi-stream. From this point of view, our method is more effective to recognize and retrieve human motion data.

References

1. S. D. Bay: Combining nearest neighbor classifiers through multiple feature subsets. In *Proc. of 15th International Conference on Machine Learning*, pages 37–45, 1998.
2. D. J. Berndt and J. Clifford: Finding patterns in time series: A dynamic programming approach. In U. M. Fayyad, G. Piatetsky-Shapiro, P. Smyth, and R. Uthurusamy, editors, *Advances in Knowledge Discovery and Data Mining*, pages 229–248, AAAI Press, 1996.
3. T. Darrell and A. Pentland: Space–time gestures. In *Proc. of IEEE Computer Society Conference on Computer Vision and Pattern Recognition*, pages 335–340, 1993.
4. D. DeMenthon, V. Kobla, and D. Doermann: Video summarization by curve simplification. In *Proc. of the sixth ACM international conference on Multimedia*, pages 211–217, 1998.
5. T. Oates and P. R. Cohen: Searching for structure in multiple stream of data. In *Proc. of 13th International Conference on Machine Learning*, pages 346–354, 1996.
6. T. Pavlidis and S. L. Horowitz: Segmentation of plane curves. *IEEE Transactions on Computers*, 23(8):860–870, 1974.
7. Ryuta Osaki, Mitsuomi Shimada and Kuniaki Uehara: Extraction of Primitive Motion for Human Motion Recognition. In *Proc. of the Second International Conference on Discovery Science (DS1999)*, pages 351–352, 1999.
8. Mitsuomi Shimada and Kuniaki Uehara: Discovery of Correlation from Multi-stream of Human Motion. In *Proc. of the Third International Conference on Discovery Science (DS2000)*, pages 290–294, 2000.
9. Ryuta Osaki, Mitsuomi Shimada and Kuniaki Uehara: A Motion Recognition Method by Using Primitive Motions. In *Proc. of 5th IFIP 2.6 Working Conference on Visual Database Systems*, pages 117–128, 2000.

Algorithmic Aspects of Boosting

Osamu Watanabe

Dept. of Mathematical and Computing Sciences, Tokyo Institute of Technology
watanabe@is.titech.ac.jp

Abstract. We discuss algorithmic aspects of boosting techniques, such as Majority Vote Boosting [Fre95], AdaBoost [FS97], and MadaBoost [DW00a]. Considering a situation where we are given a huge amount of examples and asked to find some rule for explaining these example data, we show some reasonable algorithmic approaches for dealing with such a huge dataset by boosting techniques. Through this example, we explain how to use and how to implement "adaptivity" for scaling-up existing algorithms.

1 Introduction

During the *Discovery Science Project*, the author and his colleagues, Carlos Domingo and Ricard Gavaldà, have developed simple and easy-to-use sampling techniques that would help us scaling up algorithms designed for machine learning and data mining. (For studying our sampling techniques, see, for example, technical papers [DGW01]. Also see survey papers [Wat00a,Wat00b] for adaptive sampling techniques in general.)

A common feature of these sampling techniques is "adaptivity". Tuning up algorithms is important for making them applicable to heavy computational tasks. It often needs, however, experts' knowledge on algorithms and tasks, and it is sometimes very difficult even to experts. A typical example is a case where the performance of an algorithm A depends on some parameters and we need to set these parameters appropriately. An expert may be able to set the parameters appropriately (using some complicated formulas), but even an expert may not be able to do so because the choice depends on some *hidden* parameter x that is specific to the task that A is supposed to solve. (In the case of boosting, for example, this is the quality (more specifically, the advantage) of an obtained weak hypothesis.) But as it sometimes occurs in data mining applications, such a hidden parameter x can be detected on the course of solving the task. Then one might hope for an algorithm that infers x and set up the parameters required for A appropriately; that is, an algorithm that *adapts* it to the current circumstances and achieves nearly best performance. We have developed sampling techniques with such adaptivity.

Here we explain how to use and how to implement this "adaptivity" for scaling-up existing algorithms, by using boosting algorithms as example. (In fact, being inspired by AdaBoost [FS97], a boosting technique with some adaptive property, we have studied sampling techniques for developing a boosting

S. Arikawa and A. Shinohara (Eds.): Progress in Discovery Science 2001, LNCS 2281, pp. 349–359.

based fully adaptive classifier.) We discuss algorithmic aspects of boosting techniques; we propose some ways to design scalable algorithms based on boosting techniques. Through this explanation, we show simple but useful techniques for adaptive algorithms.

Boosting, or more precisely, a boosting technique, is a way to design a strong PAC learning algorithm by using an assumed weak PAC learning algorithm. The first boosting technique has been introduced by Schapire [Sch90], for showing the equivalence between the strong and weak PAC-learnability notions. Since then, in particular, due to the success of AdaBoost, several boosting techniques have been proposed; see, e.g., the Proceedings of *the Thirteenth Annual Conference on Computational Learning Theory* (2000). Various important investigations have been also made concerning boosting properties; see, e.g., [FHT98,CSS00]. While these statistical investigations are essential for boosting techniques, we could also consider some algorithmic aspects of boosting techniques, such as actual methods to design *efficient* learning algorithms based on boosting techniques, which has not been addressed so much. Here we focus on three boosting techniques — Majority Vote Boosting (which we abbreviate MajBoost) [Fre95], AdaBoost [FS97], and MadaBoost [DW00a] — and discuss how to apply these techniques to design classification algorithms that could be applicable for huge datasets.

2 Boosting Techniques and Our Goal

In this paper, we consider some simple "data mining task" for our example and explain how to design reasonable algorithms based on boosting techniques. Here we specify our example problem; we also review basics of boosting techniques.

Our example problem is simply to find a binary classification rule explaining a given dataset, i.e., a set X of examples. More specifically, X is a set of labeled examples, i.e., a pair $(x, l_*(x))$, where x is an actual attribute data of some *instance* and $l_*(x)$ is the *label* of the instance. The form of data x is not important for our discussion; we may intuitively regard it as a vector consisting of values of a certain set of attributes. On the other hand, it is important for us that $l_*(x)$ takes only two values, either $+1$ or -1. That is, we consider the binary classification problem.

An important point (we assume here) is that X is huge, and it is not practical even just going through the whole examples in X. We also assume that X reflects the real world quite well without any serious errors; thus, our task is simply to find a good classification rule for X under the uniform distribution on X.

In summary, our task is to find, with the confidence $> 1 - \delta_0$, some binary classification rule for a huge set X of labeled examples whose error on X under the uniform distribution is bounded by ε_0. Let us call it "our learning goal". In order to simplify our discussion, we will assume that δ_0 is some constant and ignore the factor for the confidence (which is $\log(1/\delta_0)$) in the following analysis.

Next we review boosting techniques. A boosting technique is a way to design a "strong" learning algorithm by using an assumed "weak" learning algorithm **WeakLearn**. (Throughout this paper, we use **WeakLearn** to denote a weak

Procedure Generic Boost;
Given: A set S of labeled examples and a distribution D on S;
begin
$t \leftarrow 1$; $D_1 \leftarrow D$;
repeat
train **WeakLearn** on S under D_t
and obtain a weak hypothesis h_t with advantage γ_t;
compute a weight α_t of h_t;
modify D_t and obtain a new distribution D_{t+1};
$t \leftarrow t+1$;
until $t > T$, or the accuracy of a combined hypothesis f_t (see below) from weak hypotheses $h_1, ..., h_t$ obtained so far reaches to a certain level;
output the following hypothesis f_T

$$f_T(x) = \text{sign}\left(\sum_{1 \le i \le T} \alpha_t h_i(x)\right),$$

where T is the total number of obtained weak hypotheses;
end-procedure.

Fig. 1. A Genetic Boosting Algorithm

learning algorithm.) Precisely, **WeakLearn** is an algorithm that asks for a certain amount of examples under a fixed distribution (that is unknown to the algorithm) and outputs a weak hypothesis with error probability $P_{\rm err}$ strictly smaller than 1/2 under the distribution from which the examples were generated. The quantity $1/2 - P_{\rm err}$ is called the *advantage* of the obtained weak hypothesis.

Almost all boosting techniques follow some common outline. In particular, three boosting techniques we will consider here share the same outline that is stated as a procedure of Figure 1. This boosting procedure runs **WeakLearn** several times, say T times, under distributions $D_1, ..., D_T$ that are slightly modified from the given distribution D and collects weak hypotheses $h_1, ..., h_T$. A final hypothesis is built by combining these weak hypotheses. Here the key idea is to put more weight, when making a new weak hypothesis, to "problematic instances" for which the previous weak hypotheses perform poorly. That is, at the point when $h_1, ..., h_t$ have been obtained, the boosting algorithm computes a new distribution D_{t+1} that puts more weight on those instances that have been misclassified by most of $h_1, ..., h_t$. Then a new hypothesis h_{t+1} produced by **WeakLearn** on this distribution D_{t+1} should be strong on those problematic instances, thereby improving the performance of the combined hypothesis built from $h_1, ..., h_{t+1}$. (In the following, we call each execution of **WeakLearn** a *boosting round* or *boosting step*.)

Boosting techniques differ typically on (i) a way weak hypotheses are combined, and (ii) a way to define modified distributions. Here we specify these points to define three boosting techniques MajBoost, AdaBoost, and MadaBoost.

The first point (i) is specified by simply showing how to define the weight α_t used in the above generic procedure.

$$\begin{array}{lll} \underline{\text{MajBoost:}}\ \alpha_t &=& 1. \\ \underline{\text{AdaBoost:}}\ \alpha_t &=& \log \beta_t^{-1}. \\ \underline{\text{MadaBoost:}}\ \alpha_t &=& \log \beta_t^{-1}. \end{array} \quad \left(\text{Where } \beta_t = \sqrt{\frac{1-2\gamma_t}{1+2\gamma_t}}. \right)$$

For the second point (ii), we define a weight $w_{t-1}(x)$ of each instance x in S. Then $D_t(x)$ is defined by $D_t(x) = w_{t-1}(x)/W_{t-1}$, where $W_{t-1} = \sum_{x \in S} w_{t-1}(x)$. (Below β_i is defined as above, and $\text{cons}(h_i, x)$ is 1 if $h_i(x) = l_*(x)$, and -1 otherwise.)

<u>MajBoost:</u>

$$w_{t-1}(x) = D(x) \times \binom{T-t}{T/2-r} \left(\frac{1}{2}+\gamma\right)^{T/2-r} \left(\frac{1}{2}-\gamma\right)^{T/2-t+r},$$

where γ is a lower bound of $\gamma_1, ..., \gamma_T$, and
r is the number of hypotheses $h_1, ..., h_{t-1}$ misclassifying x.

<u>AdaBoost:</u>

$$w_{t-1}(x) = D(x) \times \prod_{1 \le i \le t-1} \beta_i^{\text{cons}(h_i,x)}.$$

<u>MadaBoost:</u>

$$w_{t-1}(x) = \begin{cases} D(x) \times \prod_{1 \le i \le t-1} \beta_i^{\text{cons}(h_i,x)}, & \text{if } \prod_{1 \le i \le t-1} \beta_i^{\text{cons}(h_i,x)} < 1, \text{ and} \\ D(x), & \text{otherwise.} \end{cases}$$

Notice that MajBoost uses the total number T of boosting rounds and a lower bound of advantages γ to compute weights; thus, these values should be determined in advance. (Since there is a formula (for MajBoost) to compute T from γ, what we need to know in advance is only γ.)

Now by applying α_t and D_t defined above in the procedure of Figure 1, we can define MajBoost, AdaBoost, and MadaBoost. The following theorem states the efficiency of these techniques.

Theorem 1. *In the execution of each boosting technique, suppose that the advantage of obtained weak hypotheses are $\gamma_1, \gamma_2, ...$ and that γ_{lb} is their lower bound. Then for any ε, if T, the total number of boosting rounds, satisfies the following inequality, then the error probability of combined hypothesis f_T on S under the distribution D is less than ε.*

$$\begin{array}{ll} \text{MajBoost:} & T \ge \dfrac{1}{2\gamma_{lb}^2} \ln \dfrac{1}{2\varepsilon} \quad (\text{if } \gamma_{lb} \text{ is used for } \gamma), \\ \text{AdaBoost:} & \sum_{1 \le i \le T} \gamma_i^2 \ge \dfrac{1}{2} \ln \dfrac{1}{\varepsilon} \left(\text{or roughly, } T \ge \dfrac{1}{2\gamma_{lb}^2} \ln \dfrac{1}{\varepsilon} \right) \\ \text{MadaBoost:} & \sum_{1 \le i \le T} \gamma_i^2 \ge \dfrac{1}{2} \cdot \dfrac{1}{\varepsilon} \left(\text{or roughly, } T \ge \dfrac{1}{2\gamma_{lb}^2} \cdot \dfrac{1}{\varepsilon} \right) \end{array}$$

Remark. *These bounds are: for MajBoost from Corollary 2.3 of [Fre95], for AdaBoost from (21) and (22) of [FS97], and for MadaBoost from Theorem 2 of [DW00a]. Theorem 2 of [DW00a] proves the bound for a slightly modified version of MadaBoost, but here we ignore this difference.*

Notice that the convergence time of MadaBoost is exponentially larger than that of MajBoost and AdaBoost. Although we have not been able to prove a better bound, our experiments show that there is not so much difference in convergence speed between MadaBoost and AdaBoost, and we believe that MadaBoost has more or less similar convergence speed. In fact, we can guarantee, for MadaBoost, a similar convergence speed for the first few, say, ten, steps. We can also argue that a similar convergence speed can be achieved if we ignore examples that are misclassified many times by previously obtained weak hypotheses.

As a final remark on boosting, we mention methods or frameworks for implementing boosting techniques. There are two frameworks:- *boosting by sub-sampling* and *boosting by filtering* [Fre95] [1]. The difference between these two frameworks is simply a way to define the "training set", the set S used in the procedure of Figure 1. In the sub-sampling framework, S is a set of a certain number m of examples chosen from X by using the original example generator, which we denote by $\mathbf{EX}_X$. On the other hand, in the filtering framework, S is simply the original set X. The distribution D that is given first is defined accordingly. That is, in the sub-sampling framework, D is the uniform distribution on S, while it is the original distribution D_X assumed on X in the filtering framework. (In our example task, the assumed distribution D_X is just the uniform distribution on X.)

Now we have gone through (though quickly) basics of boosting techniques. Are we ready to write a program for solving our example task? Maybe not. What is missing in the above explanation? First of all, we have not explained how to design the weak learner **WeakLearn**. But for weak learners, several weak learning algorithms have been proposed, from a simple decision stump selection to an advanced decision tree constructor such as C4.5. Here we leave this issue to other papers (see, e.g., [Qui96,Die98]) and assume that some reasonable weak learner is available. Then are we ready now? Still not. We still have to determine some parameters that are important for the performance of an obtained algorithm, and setting these parameters appropriately is not a trivial task at all!

Take MajBoost as our first choice, and consider the implementation of MajBoost in the sub-sampling framework. Then we will soon face the parameter setting problem; we have to determine the parameter γ, which should be a lower bound of advantages $\gamma_1, ..., \gamma_T$ of obtained weak hypotheses; otherwise, the boosting mechanism does not seem to work (at least, there is no theoretical justification). On the other hand, since T is roughly proportional to $1/\gamma^2$, if we chose γ too small, the total boosting rounds would become unnecessarily large. This problem is solved in the other two boosting techniques, i.e., AdaBoost and MadaBoost. In these techniques, weights are determined appropriately depending on the advantages of (so far) obtained hypotheses. That is, they enable us to design an algorithm that is adaptive to the advantages of weak hypotheses.

Note, however, that we have not yet completely solved the parameter setting problem. When we apply, e.g., AdaBoost to design an algorithm in the sub-

[1] These frameworks are also sometimes referred to as *boosting by re-weighting* and *boosting by re-sampling.*

sampling framework, we still have to determine the total number T of boosting rounds and the sample size m. It seems that setting these parameters is still at the art level. It would be nice if we can solve such parameter setting problem *algorithmically*. In the following sections, we will show some simple approaches solving this problem.

3 First Approach: Doubling and Cross Validation

As pointed out in the previous section, we would like to determine the total boosting round T and the size m of the training set S. Note first that we do not have to fix T in advance if we terminate the boosting round based on the quality of the obtained combined hypothesis f_t. In fact, as we will see below, it is enough if the error probability of f_t on S under D becomes less than $\varepsilon_0/2$ (provided m is large enough). Thus, let us use this as our stopping condition. Let **AB** denote the algorithm implementing AdaBoost in the sub-sampling framework with this stopping condition. (We assume that **WeakLearn** is appropriately designed. On the other hand, the parameter m is not still fixed in **AB**.) It follows from the bound in Theorem 1, the total boosting steps, which we denote by T_*, is bounded by $\mathcal{O}((1/\gamma_{lb}^2)\ln(1/\varepsilon_0))$.

For the choice of m, we, of course, have some theoretical formula, which is derived by the standard Occam razor type argument [FS97] based on the estimation of the descriptive complexity of composed hypothesis f_{T_*}. We recall it below. (In the following, we use ℓ_{wk} to denote the descriptive complexity of weak hypotheses that **WeakLearn** produces, which is, more concretely, defined as $\log|H_{\mathrm{wk}}|$, where H_{wk} is the set of all possible weak hypotheses.)

Theorem 2. **AB** *yields a hypothesis satisfying our learning goal if m is chosen to satisfy $m \geq (c/\varepsilon_0)\cdot(\ell_{\mathrm{wk}}T_*\log T_*)$, for some constant c.*
Remark. *This bound is based on the generalization error bound derived from the estimation given in Theorem 8 of [FS97] by using the Chernoff bound. In the literature, on the other hand, a similar generalization error bound is given that is derived by using the Hoeffding bound (which is also known as the additive Chernoff bound). But the sample size estimated by using the latter bound becomes $\mathcal{O}((1/\varepsilon_0^2)\cdot(\ell_{\mathrm{wk}}T_*\log T_*))$.*

Thus, we roughly have $m = \widetilde{\mathcal{O}}(\ell_{\mathrm{wk}}T/\varepsilon_0)$. (By $\widetilde{\mathcal{O}}$ notation, we mean to ignore some logarithmic factors.) Then from the bound for T_*, we could estimate the sample size $m = \widetilde{\mathcal{O}}(\ell_{\mathrm{wk}}/(\varepsilon_0\gamma_{lb}^2))$. But this sample size is essentially the same as the one that MajBoost needs in the sub-sampling framework! The advantage of "adaptivity" of AdaBoost is lost at this point.

We can get around this problem by using some simple algorithmic tricks[2]. Instead of fixing m in advance, we execute **AB** several times by changing m;

[2] Since both "doubling" and "cross validation" techniques are now standard, the idea explained here may not be new. In fact, in the paper [FS97] introducing AdaBoost, the "cross validation" technique is suggested.

starting from a certain small value, e.g., $m = 20$, we double m at each execution. The crucial point of using this "doubling trick" is the existence of some reliable termination test. That is, it should be possible to determine whether m exceeds the size that is necessary and sufficient for **AB** to produce a desired hypothesis. Here we use the standard "cross validation" technique. That is, we use $\mathbf{EX}_X$ to generate examples of X and check the generalization error of a hypothesis that **AB** yields under the current choice of m. By using the Chernoff bound, we can show that $\widetilde{\mathcal{O}}(1/\varepsilon_0)$ examples are enough for this cross validation test. (Some small logarithmic factor is necessary to bound the *total* error probability. In the following, the $\widetilde{\mathcal{O}}$ notation will be introduced time to time, due to the same reason.)

Now let $\mathbf{AB}_{\mathrm{CV}}$ be an algorithm implementing this idea. Also let m_* denote the sample size that is necessary and sufficient for our task. Then we can show the following bound for $\mathbf{AB}_{\mathrm{CV}}$.

Theorem 3. *With high probability,* $\mathbf{AB}_{\mathrm{CV}}$ *terminates and yields a hypothesis satisfying our learning goal by executing* **AB** *for* $\lceil \log m_* \rceil$ *times. In this case, the number of examples needed is at most* $4m_*$ *(for the executions of* **AB***) plus* $\widetilde{\mathcal{O}}(1/\varepsilon_0)$ *(for the cross validation).*

So far, we have discussed about the sample complexity. But it is easy to see that in the execution of **AB**, the time needed for each boosting round is roughly $\widetilde{\mathcal{O}}(m)$ except for the time needed for the weak learner **WeakLearn**. Thus, if the computation time of **WeakLearn** is also linear (or, almost linear) to the sample size m, we may assume that the actual computation time of **AB** and $\mathbf{AB}_{\mathrm{CV}}$ is proportional to the number of examples used.

4 Second Approach: Filtering Framework

A good news is that in the filtering framework, we do not have to worry about the parameters m and T. In the filtering framework, S is just X, and, as we will see below, the termination (of boosting rounds) can be tested in terms of the quality of obtained combined hypothesis on X. In a sense, we can incorporate the cross validation into the boosting process. Unfortunately, however, AdaBoost cannot be implemented in the filtering framework, and this was why we proposed MadaBoost in [DW00a].

In the filtering framework, examples are taken from X directly, and it is necessary to generate examples under modified distributions D_1, D_2, ... But we are provided only $\mathbf{EX}_X$, the procedure generating examples of X under the original (which is uniform in our case) distribution D_X. Thus, we need some procedure generating examples from X under a given distribution D_t. Such a procedure is given by Freund in [Fre95]. Let us first recall this procedure. Figure 2 states the outline of the procedure, which is denoted as $\mathbf{FiltEX}_{D_t}$, that generates examples under the distribution D_t that is used at the tth boosting round. The function w_{t-1} is the one defined in Section 2 for each boosting technique. Recall that D in the definition of w_{t-1} is D_X in the filtering framework.

```
Procedure FiltEX_{D_t};
begin
   repeat
      use EX_X to generate an example (x, b);
      p(x) ← w_{t-1}(x)/D_X(x);
      accept (x, b) with probability p(x);
   until some example (x, b) is accepted;
   return (x, b);
end-procedure.
```

Fig. 2. An example generator $\mathbf{FiltEX}_{D_t}$ for the filtering framework.

The probability that each (x, b) is generated at one repeat-iteration of $\mathbf{FiltEX}_{D_t}$ is $p(x)D_X(x) = w_{t-1}(x)$. Hence, the probability that some $x \in X$ is generated at one repeat-iteration is $W_{t-1} = \sum_{x \in X} w_{t-1}(x)$. Thus, for generating one example by $\mathbf{FiltEX}_{D_t}$, its repeat-iteration (hence, $\mathbf{EX}_X$) is executed *on average* $1/W_{t-1}$ times. Then if W_{t-1} is small, the cost of generating one example by $\mathbf{FiltEX}_{D_t}$ becomes large. Fortunately, however, from the following relation for MajBoost and MadaBoost, we can prove that W_{t-1} cannot be so small.

Theorem 4. *Under the definition of w_{t-1} of MajBoost (resp., MadaBoost), it holds that $D_X(\{ f_{t-1}(x) \neq l_*(x) \}) \leq W_{t-1}$.*
Remark. *For AdaBoost, this bound does not hold, which is due to its exponentially increasing weight function.*

It follows from this theorem that if the current composed hypothesis f_{t-1} is not accurate enough (more specifically, if its error probability is larger than ε_0), then the probability that some example is generated at one repeat-iteration of $\mathbf{FiltEX}_{D_t}$ is at least ε_0; hence, with high probability, $\mathbf{FiltEX}_{D_t}$ generates one example by executing $\mathbf{EX}_X$ at most c/ε_0 times. Therefore, if $\mathbf{FiltEX}_{D_t}$ does not terminate after executing $\mathbf{EX}_X$ $\widetilde{\mathcal{O}}(1/\varepsilon_0)$ times, the boosting algorithm can safely yield the current composed hypothesis f_{t-1} as an answer. That is, the termination test of the whole boosting procedure is included in this generation procedure.

There is one more point that we need to consider for the filtering framework. We have assumed that the weak learner **WeakLearn** *somehow* produces, at each tth boosting round, some weak hypothesis h_t with advantage γ_t. But we need to implement such a weak learner, and in many situations, it is said that $\mathcal{O}(1/\gamma_t^2)$ examples is necessary and sufficient to guarantee that the obtained h_t indeed has advantage approximately γ_t. If we know that some lower bound γ_{lb} of γ_t, then there is an easy implementation. Just generate $\mathcal{O}(1/\gamma_{lb}^2)$ examples by using $\mathbf{FiltEX}_{D_t}$, and pass them to the weak learner. But we would like to obtain such a hypothesis without knowing γ_{lb} in advance. This can be achieved by our adaptive sampling algorithms [DGW01].

The basic idea is simple and in fact the same as before. We use "doubling" and "cross validation". To be concrete, let us consider here the problem of estimating the advantage γ_* of a given weak hypothesis h (that is obtained at some boosting round). We estimate the advantage of h by computing its advantage on generated examples, where the number of examples is determined by the current guess γ of γ_*. We start with $\gamma = 1/2$, that is, guessing h has the maximum advantage, and reduce γ by half each time the estimation fails. At each estimation, we generate examples and estimate γ_* on these examples; let $\widehat{\gamma}$ be the obtained estimation. The number of examples is large enough so that γ_* is in the range $[\widehat{\gamma} - \gamma/4, \widehat{\gamma} + \gamma/4]$ with high probability. (By using the Hoeffding bound, we can show that $\mathcal{O}(1/\gamma^2)$ examples are enough.) This estimation fails if $\widehat{\gamma} < \gamma/2$, because then our current guess γ of γ_* is too large. Otherwise, we can conclude that $\widehat{\gamma} - \gamma/4$ is a reasonable estimation of γ_*. It is easy to see that the estimation computed in this way is within the range $[\gamma_*/2, \gamma_*]$ with high probability. Also the total number of examples is $\mathcal{O}(1/\gamma_*^2)$, which is of the same order compared with the case when we *know* γ_* in advance.

The approach is essentially the same for computing a weak hypothesis by using any reasonable weak learning algorithm. Following this approach, we can obtain a nearly best weak hypothesis h_t at the tth boosting round by using $\widetilde{\mathcal{O}}(1/\gamma_t^2)$ examples. In [DGW01], a slightly more sophisticated algorithm for a simple hypothesis selection is presented and its performance is analyzed more carefully.

Now, following the ideas explained above, we can define two algorithms $\mathbf{MjB}_{\mathrm{Filt}}$ and $\mathbf{MaB}_{\mathrm{Filt}}$ that implement respectively MajBoost and MadaBoost in the filtering framework. (For implementing MajBoost, we assume that γ_{lb} is given in advance.)

Finally, let us estimate the efficiency of these algorithms $\mathbf{MjB}_{\mathrm{Filt}}$ and $\mathbf{MaB}_{\mathrm{Filt}}$. Here we estimate the number m_* of executions of $\mathbf{EX}_X$, because the total computation time is almost proportional to m_* (provided **WeakLearn** is almost linear time). As before, we use T_* to denote the total number of boosting rounds executed by each boosting algorithm. (Thus, T_* takes different values for $\mathbf{MjB}_{\mathrm{Filt}}$ and $\mathbf{MaB}_{\mathrm{Filt}}$.) By summarizing our discussion, we can show the following bound.

Theorem 5. *With high probability,* $\mathbf{MjB}_{\mathrm{Filt}}$ *and* $\mathbf{MaB}_{\mathrm{Filt}}$ *respectively terminates and yields a hypothesis satisfying our learning goal by executing* $\mathbf{EX}_X$ *at most* $\widetilde{\mathcal{O}}(1/\varepsilon_0) \cdot \widetilde{\mathcal{O}}(\sum_{1 \le i \le T_*} 1/\gamma_i^2)$ *times. This bound is, roughly,* $\widetilde{\mathcal{O}}(T_*/(\varepsilon_0 \gamma_{lb}^2))$.

5 Concluding Remarks

We have shown two algorithmic approaches for implementing boosting techniques to solve a simple classification problem on a huge dataset. Since these implementations are adaptive, it is hard to compare their efficiency analytically. Here we *very roughly*[3] assume that the boosting steps T_* is more or less the same

[3] In fact, this is a quite rough comparison. It seems that $\mathbf{MjB}_{\mathrm{Filt}}$ that requires to set the parameter γ by an appropriate γ_{lb} is slow, though again we may use the doubling trick. On the other hand, while it seems that MadaBoost has convergence speed similar to AdaBoost, it has not been proved yet.

for any of three boosting algorithms, and compare their running time. Then provided that **WeakLearn** runs in almost linear time, we have the following time complexity: $\mathbf{AB}_{\rm CV}$ $\widetilde{\mathcal{O}}(T_*/\varepsilon_0)$, $\mathbf{MjB}_{\rm Filt}$ and $\mathbf{MaB}_{\rm Filt}$ $\widetilde{\mathcal{O}}((1/\varepsilon_0)\cdot(\sum_{1\le i\le T_*} 1/\gamma_i^2))$, which is roughly $\widetilde{\mathcal{O}}((T_*/\varepsilon_0)\cdot(1/\gamma_{lb}^2))$. Thus, the implementation by the filtering framework is slower by the factor of $1/\gamma_{lb}^2$. The situation, however, would change if we want to use a bit more complicated weak learner for **WeakLearn**. Usually a weak learning algorithm, in the sub-sampling framework, needs to use the whole m_* examples in S to obtain a weak hypothesis. Thus, if it has a quadratic running time, then the running time of $\mathbf{AB}_{\rm CV}$ becomes $\widetilde{\mathcal{O}}(m_*^2)$, which is $\widetilde{\mathcal{O}}(T_*^2/\varepsilon_0^2)$. On the other hand, the number of examples passed to **WeakLearn** in $\mathbf{MjB}_{\rm Filt}$ and $\mathbf{MaB}_{\rm Filt}$ is $\widetilde{\mathcal{O}}(1/\gamma_t^2)$ at the tth boosting round. Thus, the computation time becomes $\widetilde{\mathcal{O}}((1/\varepsilon_0)\cdot(\sum_{1\le i\le T_*} 1/\gamma_i^4))$, and $\mathbf{MaB}_{\rm Filt}$ may be faster than $\mathbf{AB}_{\rm CV}$ in this situation. As pointed out, e.g., by Morishita [Mor01], a stronger weak learner sometimes yields a better learning algorithm, and in general the filtering framework is better for using costly weak learning algorithms.

We have explained some algorithmic techniques to apply boosting for huge data classification problems. But in practice, almost all example sets have some amount of errors, and the proportion of error may become even larger if a dataset becomes large. On the other hand, it has been criticized that boosting techniques, e.g., AdaBoost, are too sensitive to erroneous examples. One interesting idea is to ignore examples that are misclassified by previous weak hypotheses more than a certain number of times. (Motivated this, Freund [Fre99] proposed a new boosting algorithm — BrownBoost.) But then we will have another parameter setting problem; that is, determining this threshold. An algorithmic approach to this parameter setting problem would be an important and interesting future problem.

Acknowledgments

This paper is based on a series of joint works [DGW01,DW00a,DW00b] with Carlos Domingo and Ricard Gavaldà. I would like to thank these coauthors for their collaborations. I would like to thank anonymous referees for their comments on an earlier draft that helped me to improve the presentation of the paper.

References

[CSS00] M. Collins, R.E. Schapire, and Y. Singer, Logistic regression, AdaBoost and Bregman Distance, in *Proc. of the Thirteenth Annual ACM Workshop on Computational Learning Theory* (COLT'00), ACM, 158–169, 2000.

[Die98] T.G. Dietterich, An experimental comparison of three methods for constructing ensembles of decision trees: bagging, boosting and randomization, *Machine Learning* 32, 1–22, 1998.

[DGW01] C. Domingo, R. Gavaldà, and O. Watanabe, Adaptive sampling methods for scaling up knowledge discovery algorithms, *Data Mining and Knowledge Discovery* (special issue edited by H. Liu and H. Motoda), 2001, to appear.

[DW00a] C. Domingo and O. Watanabe, MadaBoost: A modification of AdaBoost, in *Proc. of the Thirteenth Annual ACM Workshop on Computational Learning Theory* (COLT'00), ACM, 180–189, 2000.

[DW00b] C. Domingo and O. Watanabe, Scaling up a boosting-based learner via adaptive sampling, in *Proc. of Knowledge Discovery and Data Mining* (PAKDD'00), Lecture Notes in AI 1805, 317–328, 2000.

[Fre95] Y. Freund, Boosting a weak learning algorithm by majority, *Information and Computation*, 121(2):256–285, 1995.

[Fre99] Y. Freund, An adaptive version of the boost by majority algorithm, in *Proc. of the Twelfth Annual Conference on Computational Learning Theory* (COLT'99), ACM, 102–113, 1999.

[FHT98] J. Friedman, T. Hastie, and R. Tibshirani, Additive logistic regression: a statistical view of boosting, Technical report, 1998.

[FS97] Y. Freund and R.E. Schapire, A decision-theoretic generalization of on-line learning and an application to boosting, *J. Comput. Syst. Sci.*, 55(1):119–139, 1997.

[Mor01] S. Morishita, Computing optimal hypotheses efficiently for boosting, in this issue.

[Qui96] J.R. Quinlan, Bagging, boosting, and C4.5, in *Proc. of the 13th National Conference on Artificial Intelligence*, 725–730, 1996.

[Sch90] R.E. Schapire, The strength of weak learnability, *Machine Learning*, 5(2):197–227, 1990.

[Wat00a] O. Watanabe, Simple sampling techniques for discovery science, *IEICE Transactions on Information and Systems*, E83-D(1), 19–26, 2000.

[Wat00b] O. Watanabe, Sequential sampling techniques for algorithmic learning theory, in *Proc. 11th Int'l Conference on Algorithmic Learning Theory* (ALT'00), Lecture Notes in AI 1968, 27–40, 2000.

Automatic Detection of Geomagnetic Jerks by Applying a Statistical Time Series Model to Geomagnetic Monthly Means

Hiromichi Nagao[1], Tomoyuki Higuchi[2], Toshihiko Iyemori[3], and Tohru Araki[1]

[1] Department of Geophysics, Graduate School of Science, Kyoto University, Kitashirakawa Oiwake-cho, Sakyo-ku, Kyoto 606-8502, Japan
{nagao, araki}@kugi.kyoto-u.ac.jp
http://www-step.kugi.kyoto-u.ac.jp/
[2] The Institute of Statistical Mathematics, 4-6-7, Minami-Azabu, Minato-ku, Tokyo 106-8569, Japan
higuchi@ism.ac.jp
[3] Data Analysis Center for Geomagnetism and Space Magnetism, Graduate School of Science, Kyoto University, Kitashirakawa Oiwake-cho, Sakyo-ku, Kyoto 606-8502, Japan
iyemori@kugi.kyoto-u.ac.jp

Abstract. A geomagnetic jerk is defined as a sudden change in the trend of the time derivative of geomagnetic secular variation. A statistical time series model is applied to monthly means of geomagnetic eastward component obtained at 124 geomagnetic observatories to detect geomagnetic jerks objectively. The trend component in the model is expressed by a second order spline function with variable knots. The optimum parameter values of the model including positions of knots are estimated by the maximum likelihood method, and the optimum number of parameters is determined based on the Akaike Information Criterion. The geomagnetic jerks are detected objectively and automatically by regarding the determined positions of knots as the occurrence epochs. This analysis reveals that the geomagnetic jerk in 1991 is a local phenomenon while the 1969 and 1978 jerks are confirmed to be global phenomena.

1 Introduction

A geomagnetic secular variation observed on the ground is believed to have several origins. One is the dynamo action in the outer core, which is the liquid metallic layer in the interior of the earth, and the others are the currents flowing external of the earth such as the magnetopause currents, the ionospheric currents, the ring current, the tail current, and the field-aligned currents [17]. The geomagnetic field of the core origin is much larger than that of the external origins [8].

It has been reported that the trends of the time derivative of geomagnetic secular variations in the eastward component changed suddenly at Europe around 1969 [6], 1978 [18], and 1991 [5] [15] as shown in Fig. 1. This phenomenon is

S. Arikawa and A. Shinohara (Eds.): Progress in Discovery Science 2001, LNAI 2281, pp. 360–371.

termed "geomagnetic jerk" [16], which mathematically means that the secular variation has an impulse in its third order time derivative. It is discussed for a couple of decades whether the origin of the jerks is internal or external of the earth and whether their distribution is worldwide or local [3] [7] [16].

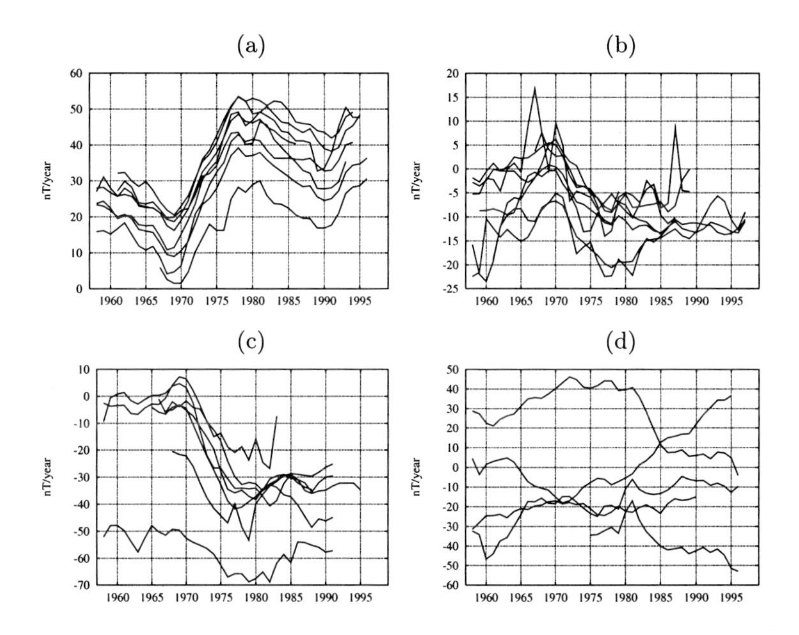

Fig. 1. The time derivative of annual means of the geomagnetic eastward component in (a) Europe, (b) Eastern Asia, (c) North America, and (d) the Southern Hemisphere. The trends sometimes change suddenly in a few years around 1969, 1978, and 1991. These phenomena are called geomagnetic jerks.

In analyzing the geomagnetic jerks, most of the papers assume that the jerks occurred around 1969, 1978, and 1991, then fit several lines to the time derivative of the secular variation, and obtain a jerk amplitude as the difference of the trends of two lines connected at the occurrence epoch of the jerk. Their methods are subjective in the determination of occurrence epochs [3]. Two objective methods using the optimal piecewise regression analysis [24] and the wavelet analysis [1] [2] were developed for determination of the jerk occurrence epochs. The former method was applied to the geomagnetic annual means, but it is better to adopt monthly means rather than annual means for jerk analyses because it takes only a few years to change the trends of the time derivative of geomagnetic secular variation as seen in Fig. 1, and the temporal resolution of annual means is not sufficient to analyze the jerk. The latter method, on the other hand, was applied to geomagnetic monthly means, but this method still has room for improvement because seasonal adjustment is not taken into account. In this paper, a statistical time series model, in which the adjustments of both seasonal and short time scale variations are taken into account, is applied to geomagnetic monthly means to estimate the trend component of geomagnetic secular variation. The geomagnetic time series is decomposed into the trend, the seasonal, the stationary autoregressive (AR), and the observational noise components. The model parameters are estimated by the maximum likelihood method and the best model is selected based on the Akaike Information Criterion (AIC). Finally, the occurrence epochs of the jerks are determined automatically.

The data description shall be given in Section 2 and the method of our analysis is developed in Section 3. A result of the decomposition, the occurrence index of jerks, and the global distribution of jerk amplitudes are shown in Section 4.

2 Geomagnetic Data

We use monthly means of geomagnetic eastward component obtained at observatories distributed worldwide. The data are collected through the World Data Center system (e.g., see `http://swdcdb.kugi.kyoto-u.ac.jp/`). The time series at an observatory should be continuously maintained for more than ten years, since the geomagnetic jerks are the sudden changes in the trends of geomagnetic decadal changes. The time series of 124 geomagnetic observatories are selected according to this criterion. The distribution of these observatories is shown in Fig. 2. Since the worldwide coverage of the observatories in operation is not enough before International Geophysical Year (IGY) 1957, we use the time series from 1957 to 1999. Obvious artificial spikes and baseline jumps in the data are corrected manually before the analysis. The influences of small errors in the data at an observatory, which are difficult to be corrected manually, to result of the analysis are considered to be identified by comparing with results obtained from the data at other observatories as shown in Section 4.

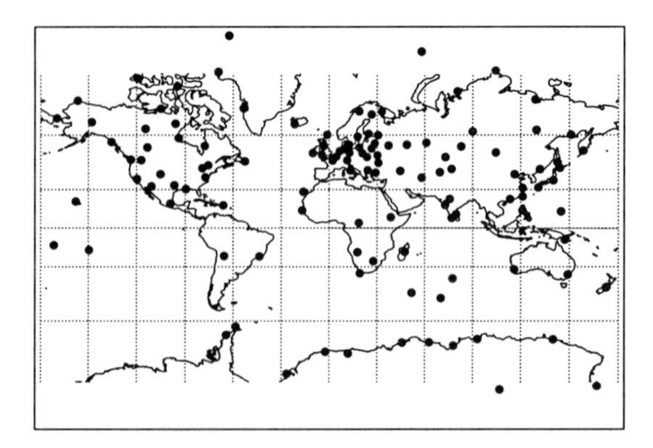

Fig. 2. The distribution of the 124 geomagnetic observatories used in this paper.

3 Method of Analysis

3.1 Statistical Time Series Model

A geomagnetic time series of monthly means in the eastward component obtained at an observatory is described by the following statistical time series model:

$$Y_n = t_n + s_n + p_n + w_n \ , \tag{1}$$

where Y_n is a monthly mean value at the consecutive month n starting from January 1957, t_n is the trend component, s_n is the seasonal component, p_n is the stationary AR component, and w_n is the observational noise component.

Trend Component. The time derivative of decadal geomagnetic time series of the eastward component is approximated by a piecewise linear curve (first order spline) as seen in Fig. 1. The trend component t_n of the geomagnetic time series, therefore, is expressed by the second order spline function with knots $\xi_1, \xi_2, \cdots, \xi_K$, where K is the number of knots. Here $\xi_1, \xi_2, \cdots, \xi_K$ are integer values, that is, a knot locates at a data point. The position of a knot corresponds to the occurrence epoch of a geomagnetic jerk.

To derive models for the trend component, we present the continuous time model:

$$f(t) = f(t - \Delta t) + f'(t - \Delta t)\Delta t + \frac{1}{2} f''(t - \Delta t)(\Delta t)^2 + O\left((\Delta t)^3\right) \tag{2}$$
$$f'(t) = f'(t - \Delta t) + f''(t - \Delta t)\Delta t + O\left((\Delta t)^2\right) \tag{3}$$
$$f''(t) = f''(t - \Delta t) + O(\Delta t) \ . \tag{4}$$

The models for the trend component t_n, the first time differential component δt_n, and the second time differential component $\delta^2 t_n$ can be derived by discretizing the equations (2), (3), and (4), respectively:

$$t_n = t_{n-1} + \delta t_{n-1} + \frac{1}{2}\delta^2 t_{n-1} \tag{5}$$
$$\delta t_n = \delta t_{n-1} + \delta^2 t_{n-1} \tag{6}$$
$$\delta^2 t_n = \delta^2 t_{n-1} + v_{n1} \ , \tag{7}$$

where v_{n1} is the system noise sequence which obeys the normal distribution function with mean zero and variance τ_1^2, i.e., $v_{n1} \sim N(0, \tau_1^2)$. The system noise sequences for the trend component model (5) and the first time differential component model (6) are assumed to be zero. Here each interval between successive knots is assumed to be more than five years to avoid too many knots concentrating in a short period [24], and an initial and end knots are kept to be separated from $n = 1$ and $n = N$ by more than three years, respectively. Similar related approaches to a trend component model can be found in [9] and [20]. τ_1^2 has a large value at every knot where the second order time derivative of the spline function is not continuous while the value of τ_1^2 is zero elsewhere.

The values of τ_1^2 at the knots $\xi_1, \xi_2, \cdots, \xi_K$ are written as $\tau_{11}^2, \tau_{12}^2, \cdots, \tau_{1K}^2$, respectively. The parameters which should be estimated in the trend component are the number and the positions of the knots and the variances of the system noise at each knot. An alternative way to realize an abrupt change in $\delta^2 t_n$ is to replace $N(0, \tau_{1k}^2)(k = 1, 2, \cdots, K)$ by the heavy-tailed non-normal distribution [10] [12]. Namely, we employ the non-normal linear trend model to detect an abrupt change automatically and objectively. Meanwhile this approach mitigates a difficulty of estimating the parameter values $(\xi_k, \tau_{1k}^2)(k = 1, 2, \cdots, K)$, an identification of jerks is not straightforward from the estimated $\delta^2 t_n$ obtained by applying the non-normal trend model. Therefore we did not adopt this approach in this study. If the jerks occurred simultaneously in the global extent, it is better to consider a time series model to treat all of observations simultaneously instead of analyzing data from each observatory separately. However we did not adopt this generalized model because the jerks were reported to occur, in the southern hemisphere, a few years after the occurrence in the northern hemisphere [2] [7].

Seasonal Component. The seasonal component s_n, which represents the annual variation in the data, should have a twelve months periodicity, i.e., $s_n \approx s_{n-12}$. This condition can be rewritten as

$$\sum_{i=0}^{11} s_{n-i} = v_{n2} , \tag{8}$$

where v_{n2} is the system noise sequence which obeys $v_{n2} \sim N(0, \tau_2^2)$ [12]. τ_2^2 is a parameter to be estimated.

Stationary Autoregressive Component. Unless the stationary AR component is included in the model (1), the best model tends to have more knots than expected as shown in the next section. This component may represent short time scale variations less than one year, such as the solar effects, or responses to the external field which reflect the mantle conductivity structures. The stationary AR component is expressed as

$$p_n = \sum_{i=1}^{m} a_i p_{n-i} + v_{n3} , \tag{9}$$

where m is the AR order and v_{n3} is the system noise sequence which obeys $v_{n3} \sim N(0, \tau_3^2)$. τ_3^2 is a parameter to be estimated.

Observational Noise. The observational noise w_n is assumed to be a white noise sequence which obeys the normal distribution function with mean zero and variance σ^2, i.e., $w_n \sim N(0, \sigma^2)$, where σ^2 is a parameter to be estimated.

3.2 Parameter Estimation and Model Identification

When a parameter vector involved in the equations (1)-(9)

$$\boldsymbol{\theta} = (\xi_1, \xi_2, \cdots, \xi_K, a_1, a_2, \cdots, a_m, \tau_{11}^2, \tau_{12}^2, \cdots, \tau_{1K}^2, \tau_2^2, \tau_3^2, \sigma^2) \tag{10}$$

is given, the equations (1)-(9) can be represented by the state space model [14]

$$\boldsymbol{x}_n = F\boldsymbol{x}_{n-1} + G\boldsymbol{v}_n \tag{11}$$

$$Y_n = H\boldsymbol{x}_n + w_n \tag{12}$$

with the state space vector $\boldsymbol{x}_n$ and the system noise vector $\boldsymbol{v}_n$

$$\boldsymbol{x}_n = (\, t_n \quad \delta t_n \quad \delta^2 t_n \quad s_n \quad \cdots \quad s_{n-10} \quad p_n \quad \cdots \quad p_{n-m+1}\,)^t \tag{13}$$

$$\boldsymbol{v}_n = (\, v_{n1} \quad v_{n2} \quad v_{n3}\,)^t \,, \tag{14}$$

where the superscript t denotes the transpose operation, and the matrices F, G, and H are

$$F = \left(\begin{array}{ccc|ccccc|cccc}
1 & 1 & 1/2 & & & & & & & & & \\
0 & 1 & 1 & & & & & & & & & \\
0 & 0 & 1 & & & & & & & & & \\
\hline
 & & & -1 & \cdots & -1 & -1 & & & & & \\
 & & & 1 & & & & & & & & \\
 & & & & \ddots & & & & & & & \\
 & & & & & 1 & & & & & & \\
\hline
 & & & & & & & & a_1 & \cdots & a_{m-1} & a_m \\
 & & & & & & & & 1 & & & \\
 & & & & & & & & & \ddots & & \\
 & & & & & & & & & & 1 &
\end{array}\right) \tag{15}$$

$$G = \left(\begin{array}{c|c|c}
0 & & \\
0 & & \\
1 & & \\
\hline
 & 1 & \\
 & 0 & \\
 & \vdots & \\
 & 0 & \\
\hline
 & & 1 \\
 & & 0 \\
 & & \vdots \\
 & & 0
\end{array}\right) \tag{16}$$

$$H = (\underbrace{1 \;\; 0 \;\; 0}_{3} \mid \underbrace{1 \;\; 0 \;\; \cdots \;\; 0}_{11} \mid \underbrace{1 \;\; 0 \;\; \cdots \;\; 0}_{m}) \,. \tag{17}$$

The log-likelihood function of this model as a function of $\boldsymbol{\theta}$ is

$$\ell(\boldsymbol{\theta}) = -\frac{1}{2}\left[N \log 2\pi + \sum_{n=1}^{N} \left\{ \log d_{n|n-1} + \frac{(Y_n - Y_{n|n-1})^2}{d_{n|n-1}} \right\} \right] \,, \tag{18}$$

where N is the number of observational data, $Y_{n|n-1}$ is the mean of the predictive distribution for Y_n given observational data $Y_i(i = 1, 2, \cdots, n-1)$, and $d_{n|n-1}$ is the variance of that distribution [12]. The quantities of $Y_{n|n-1}$ and $d_{n|n-1}$ in the equation (18) can be obtained by the Kalman filter algorithm recursively:

$$\begin{aligned} \boldsymbol{x}_{n|n-1} &= F\boldsymbol{x}_{n-1|n-1} && (19)\\ V_{n|n-1} &= FV_{n-1|n-1}F^t + GQG^t && (20)\\ K_n &= V_{n|n-1}H^t(HV_{n|n-1}H^t + \sigma^2)^{-1} && (21)\\ \boldsymbol{x}_{n|n} &= \boldsymbol{x}_{n|n-1} + K_n(Y_n - H\boldsymbol{x}_{n|n-1}) && (22)\\ V_{n|n} &= (I - K_nH)V_{n|n-1}\ , && (23) \end{aligned}$$

where $\boldsymbol{x}_{n|j}$ is the mean vector and $V_{n|j}$ is the variance covariance matrix of the state vector $\boldsymbol{x}_n$ given observational data $Y_i(i = 1, 2, \cdots, j)$, I is the unit matrix, and Q is the variance covariance matrix of the system noise:

$$Q = \begin{pmatrix} \tau_1^2 & 0 & 0 \\ 0 & \tau_2^2 & 0 \\ 0 & 0 & \tau_3^2 \end{pmatrix} . \tag{24}$$

The positions of knots $\xi_i(i = 1, 2, \cdots, K)$ move to other data points in the course of this iterative estimation. Other parameters in the parameter vector (10) is iteratively searched by employing the quasi-Newton method [19] which makes the log-likelihood (18) larger. This procedure is iterated until the parameter vector $\boldsymbol{\theta}$ makes the log-likelihood maximum. The best model among the models with different number of the parameters is selected by minimizing the AIC:

$$\begin{aligned} \mathrm{AIC} &= -2\ell(\hat{\boldsymbol{\theta}}) + 2\dim\hat{\boldsymbol{\theta}} \\ &= \begin{cases} -2\ell(\hat{\boldsymbol{\theta}}) + 2(2K+2) & (m = 0) \\ -2\ell(\hat{\boldsymbol{\theta}}) + 2(2K+m+3) & (m \neq 0) \end{cases} . \end{aligned} \tag{25}$$

The exceptional handling of $m = 0$ is necessary to avoid the identification problem between innovation variance τ_3^2 of the 0-th AR component and variance σ^2 of the observational noise. See [11] for an application of the AIC to determination of the number and positions of the knots. There are various kinds of information criteria besides the AIC [13]. Minimum description of length (MDL) for the selection of the orders of the models is discussed in [21]. Modeling and simplicity in a field of scientific discovery and artificial intelligence are discussed in [22] and [23].

The final estimate of the state vector variable for the best model is obtained by the fixed interval smoother algorithm [12]:

$$\begin{aligned} A_n &= V_{n|n}F_{n+1}^tV_{n+1|n}^{-1} && (26)\\ \boldsymbol{x}_{n|N} &= \boldsymbol{x}_{n|n} + A_n(\boldsymbol{x}_{n+1|N} - \boldsymbol{x}_{n+1|n}) && (27)\\ V_{n|N} &= V_{n|n} + A_n(V_{n+1|N} - V_{n+1|n})A_n^t\ . && (28) \end{aligned}$$

The estimated positions of the knots are regarded as the occurrence epochs of jerks. The jerk amplitude $\delta^3 Y_n$ at a knot ξ_i is defined as:

$$\delta^3 Y_{\xi_i} = \delta^2 t_{\xi_i} - \delta^2 t_{\xi_i - 1} \, . \tag{29}$$

The unit of this jerk amplitude is not nT(nano tesla)/year^3 but nT/year^2 because it is defined as a simple difference between successive second time differential components. The definition of the jerk amplitude (29) follows that of the previous papers (e.g., [7]).

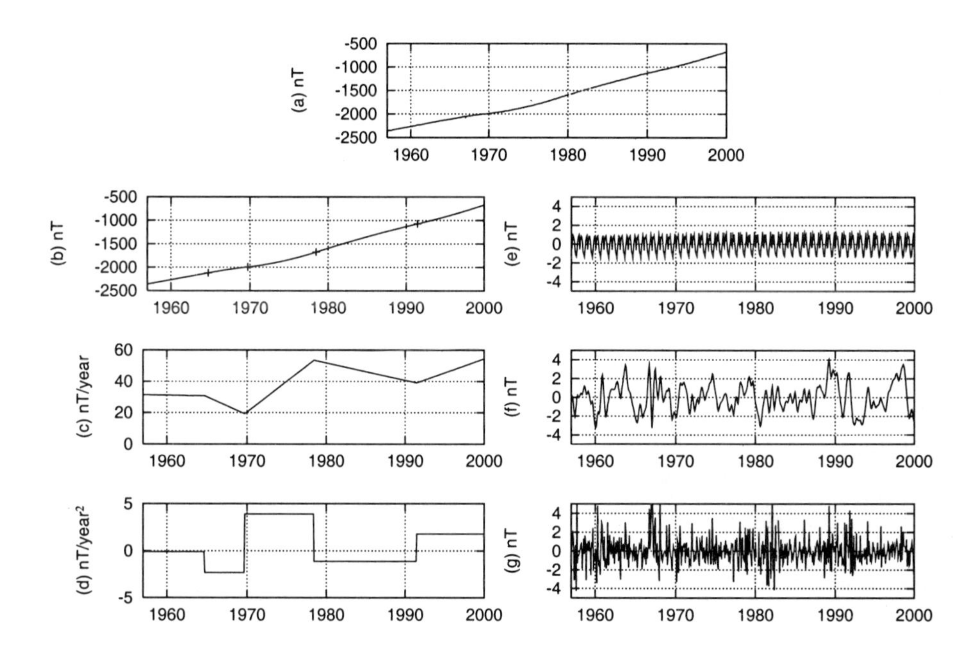

Fig. 3. (a) The time series of monthly means of geomagnetic eastward component Y_n at Chambon-la-Foret, France, (b) the trend component t_n, (c) the first time differential component δt_n, (d) the second time differential component $\delta^2 t_n$, (e) the seasonal component s_n, (f) the stationary AR component p_n, and (g) the observational noise w_n. The trend component is expressed by a second order spline function, and estimated positions of the knots (denoted by the plus symbols, i.e., September 1964, September 1969, June 1978, and June 1991) are regarded as the occurrence epochs of the geomagnetic jerks.

4 Results and Discussion

An example of the decomposition of geomagnetic time series by the method mentioned in Section 3 is illustrated in Fig. 3. The data used in Fig. 3 are the geomagnetic monthly means of the eastward component at Chambon-la-Foret (48.02°N, 2.27°E), France. Although there are no data from March to April 1980 and in October 1981, values of state vector (11) in the missing periods can be estimated by the Kalman filter and smoother algorithms [4]. The number of the knots for the best model is $\hat{K} = 4$ and they locate at $\hat{\xi}_1$ = September 1964, $\hat{\xi}_2$ = September 1969, $\hat{\xi}_3$ = June 1978, and $\hat{\xi}_4$ = June 1991; our method can identify the jerks which are believed to have occurred except for 1964. Jerks are detected by our method around 1969, 1978, and 1991 from geomagnetic eastward component at most of the European observatories. On the other hand, the ratio of the number of observatories where a jerk is detected in 1964 to the number of observatories where a jerk is not detected in that year is much lesser than that in the years when the jerks are believed to have occurred. That is, the detected 1964 jerk may be a local phenomenon at Chambon-la-Foret. The number of the knots becomes 5 and the AIC increases from 2461.06 to 2558.86 unless the stationary AR component p_n is included in the model (1). This result indicates that an inclusion of the stationary AR component is reasonable. The same tendency is seen in the results obtained from the data at most of the observatories.

Fig. 4 shows the occurrence index of geomagnetic jerks detected by our method. The occurrence index in the year y is defined by

$$R_y = \frac{1}{n_y} \sum_i \sum_{\xi_j \in y} |\delta^3 Y_{\xi_j}(O_i)| \,, \tag{30}$$

where $\delta^3 Y_{\xi_j}(O_i)$ is the jerk amplitude obtained at the i-th observatory O_i and n_y is the number of observatories available in the year y. It can be confirmed from Fig. 4 that R_y has clear peaks in 1969 and 1977 when the jerks are believed to have occurred [6] [18], which is consistent with the result obtained by [2], and also has small peaks in 1989 and 1994. However there is no peak in 1991, when a jerk is also believed to have occurred [5] [15]. Fig. 5 shows the global distributions of the jerk amplitudes $\delta^3 Y_n$ defined by the equation (29) in the years when the geomagnetic jerks believed to have occurred (i.e., 1969, 1978, and 1991). The circles indicate positive jerk amplitudes and the triangles indicate negative ones. It should be noted from these figures that the 1991 jerk is not a worldwide phenomenon in comparison with the 1969 and 1978 ones. This result is inconsistent with the conclusion by the previous paper [7].

5 Conclusions

We developed a method to analyze the time series of geomagnetic monthly means for identifying the geomagnetic jerks. Each geomagnetic time series is decomposed into the trend, the seasonal, the stationary AR, and the observational noise components by applying a statistical time series model. The trend component is expressed by a second order spline function because a jerk is an impulse in the third order time derivative of the geomagnetic time series. The model parameters including the positions of the knots of the spline function are estimated by the maximum likelihood method and the number of the knots and

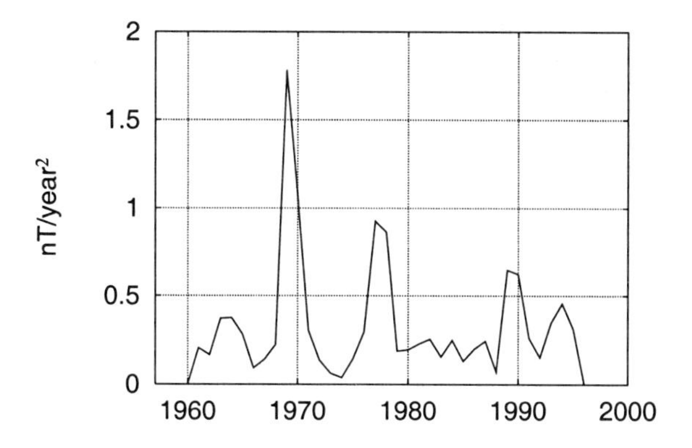

Fig. 4. The occurrence index of geomagnetic jerks defined by the equation (30).

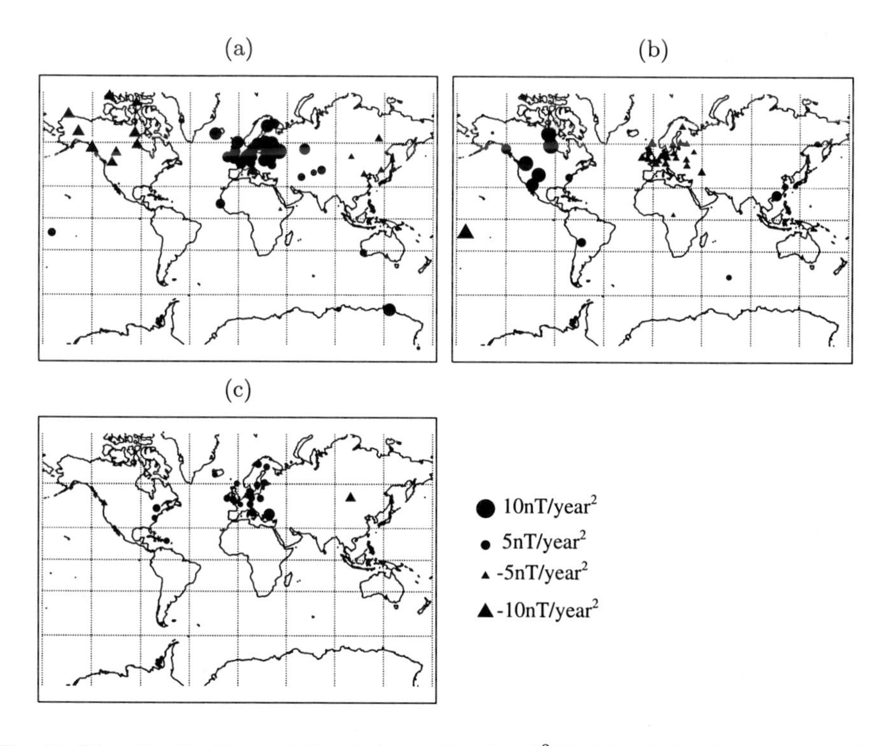

Fig. 5. The distributions of the jerk amplitudes $\delta^3 Y$ defined by the equation (29) around (a) 1969, (b) 1978, and (c) 1991. The circles show positive jerk amplitudes and the triangles show negative ones. The magnitude of a jerk amplitude is indicated by the radius of the symbol.

the AR order is selected based on the AIC. Distributions of jerk amplitudes are obtained by regarding the estimated positions of the knots as the occurrence epochs of the jerks and by defining a jerk amplitude at an optimum knot as the difference between the successive second time differential components in the trend component connected at the knot. The 1969 and 1978 jerks are confirmed to be worldwide phenomena while the 1991 jerk is found to be localized. Our method developed in this paper gives a valuable contribution to discovery in the applied domain.

Acknowledgments

We are grateful to staff of Data Analysis Center for Geomagnetism and Space Magnetism, Graduate School of Science, Kyoto University for supplying us the high quality geomagnetic data, and to Yukiko Yokoyama, Shin'ichi Ohtani, Akinori Saito, and Genta Ueno for valuable discussions. We also thank Makio Ishiguro and another anonymous referee for reviewing this paper and giving us constructive comments. This research was supported by the Ministry of Education, Science, Sports and Culture (MESSC) of Japan, Grant-in-Aid for Scientific Research on Priority Area "Discovery Science," 10143106, 1998-2001. Work at the Institute of Statistical Mathematics was in part carried out under the ISM Cooperative Research Program (H12-ISM.CRP-2026). The computation in this work has been done using SGI2800 of the Institute of Statistical Mathematics.

References

1. Alexandrescu, M., D. Gibert, G. Hulot, J. L. Le Mouël, G. Saracco: Detection of geomagnetic jerks using wavelet analysis. *J. Geophys. Res.*, **100** (1995) 12557–12572
2. Alexandrescu, M., D. Gibert, G. Hulot, J. L. Le Mouël, G. Saracco: Worldwide wavelet analysis of geomagnetic jerks. *J. Geophys. Res.*, **101** (1996) 21975–21994
3. Alldredge, L. R.: A discussion of impulses and jerks in the geomagnetic field. *J. Geophys. Res.*, **89** (1984) 4403–4412
4. Anderson, B. D. O., J. B. Moore: Optimal Filtering. Prentice-Hall (1979)
5. Cafarella, L., A. Meloni: Evidence for a geomagnetic jerk in 1990 across Europe. *Annali di Geofisica*, XXXVIII (1995) 451–455
6. Courtillot, V., J. Ducruix, J. L. Le Mouël: Sur une accélération récente de la variation séculaire du champ magnétique terrestre. *C. R. Acad.*, Série D (1978) 1095–1098
7. De Michelis, P, L. Cafarella, A. Meloni: Worldwide character of the 1991 geomagnetic jerk. *Geophys. Res. Lett.*, **25** (1998) 377–380
8. Gauss, C. F.: Allgemeine Theorie des Erdmagnetismus. *Resultate aus den Beobachtungen des Magnetischen Vereins im Jahre 1838*, Leipzig (1839) 1–57
9. Harvey, A. C.: Forecasting, structural time series models and the Kalman filter. *Cambridge University Press* (1989)
10. Higuchi, T., G. Kitagawa: Knowledge discovery and self-organizing state space model. *Ieice Trans. Inf. & Syst.*, **E83-D** (2000) 36–43

11. Higuchi, T., S. Ohtani: Automatic identification of large-scale field-aligned current structures. *J. Geophys. Res.*, **105** (2000) 25305–25315
12. Kitagawa, G., W. Gersch: Smoothness priors analysis of time series. *Lecture Notes in Statistics*, **116**, Springer-Verlag, New York (1996)
13. Kitagawa, G., T. Higuchi: Automatic Transaction of Signal via Statistical Modeling. *The proceedings of The First International Conference on Discovery Science*, Springer-Verlag, New York (1998) 375–386
14. Kitagawa, G., T. Higuchi, F. N. Kondo: Smoothness prior approach to explore the mean structure in large time series data. *The proceedings of The Second International Conference on Discovery Science*, Springer-Verlag, New York (1999) 230–241
15. Macmillan, S.: A geomagnetic jerk for the early 1990's. *Earth Planet. Sci. Lett.*, **137** (1996) 189–192
16. Malin, S. R. C., B. M. Hodder: Was the 1970 geomagnetic jerk of internal or external origin? *Nature*, **296** (1982) 726–728
17. Merrill, R. T., M. W. McElhinny, P. L. McFadden: The magnetic field of the earth. *International Geophysics Series*, **63**, Academic Press, San Diego (1998)
18. Nevanlinna, H., C. Sucksdorff: Impulse in global geomagnetic "secular variation" 1977–1979. *J. Geophys.*, **50** (1981) 68–69
19. Press, W. H., B. P. Flannery, S. A. Teukolsky, W. T. Vetterling: Numerical Recipes: The Art of Scientific Computing. *Cambridge University Press* (1986)
20. Rahman, M. M., M. Ishiguro: Estimating linear relationships among trends of nonstationary time series. *The proceedings of The 2nd International Symposium on Frontiers of Time Series Modeling, Nonparametric Approach to Knowledge Discovery* (2000) 244–245
21. Rissanen, J.: Modeling by shortest data description. *Automatica*, **14** (1978) 465–471
22. Simon, H. A.: Models of Discovery. *D. Reidel Publishing Company* (1977) 38–40
23. Simon, H. A., R. E. Veldes-Perez, D. H. Sleeman: Scientific discovery and simplicity of method. *Artificial Intelligence*, **91** (1997) 177–181
24. Stewart, D. N., K. A. Whaler: Optimal piecewise regression analysis and its application to geomagnetic time series. *Geophys. J. Int.*, **121** (1995) 710–724

Application of Multivariate Maxwellian Mixture Model to Plasma Velocity Distribution

Genta Ueno[1,3], Nagatomo Nakamura[2], Tomoyuki Higuchi[3], Takashi Tsuchiya[3], Shinobu Machida[1], and Tohru Araki[1]

[1] Department of Geophysics, Graduate School of Science, Kyoto University, Kyoto 606-8502, Japan
genta@kugi.kyoto-u.ac.jp
[2] Sapporo Gakuin University, Hokkaido 069-8555, Japan
[3] The Institute of Statistical Mathematics, Tokyo 106-8569, Japan

Abstract. Recent space plasma observations have provided us with three-dimensional velocity distributions having multiple peaks. We propose a method for analyzing such velocity distributions via a multivariate Maxwellian mixture model where each component of the model represents each of the multiple peaks. The parameters of the model are determined through the Expectation-Maximization (EM) algorithm. For the automatic judgment of the preferable number of components in the mixture model, we introduce a method of examining the number of extrema of a resulting mixture model. We show applications of our method to velocity distributions observed in the Earth's magnetotail.

1 Introduction

From direct measurements of space plasma by spacecraft we have obtained macroscopic physical quantities by calculating velocity moments of plasma particle velocity distributions (e.g., number density, bulk velocity and temperature). This macroscopic description assumes that the plasma is in a state of local thermal equilibrium. Under this assumption a particle velocity distribution is given as a normal distribution which is called the Maxwellian distribution in a scientific domain, plasma physics. The Maxwellian distribution is given by

$$g\left(\boldsymbol{v}|\,\boldsymbol{V},\mathrm{T}\right)=\left(\frac{m}{2\pi\mathrm{T}}\right)^{3/2}\exp\left[-\frac{m\left|\boldsymbol{v}-\boldsymbol{V}\right|^{2}}{2\mathrm{T}}\right], \tag{1}$$

where m [kg] is the mass of the particle, $\boldsymbol{V}$ [m/s] is the bulk velocity vector and T [J] is the temperature.

Observational techniques are progressing notably today, and making it possible to measure detailed shapes of velocity distributions in the three-dimensional velocity space. These observations have revealed that there frequently happen cases in which space plasmas are not in a state of thermal equilibrium and their velocity distributions are not the Maxwellian but consist of multiple peaks.

S. Arikawa and A. Shinohara (Eds.): Progress in Discovery Science 2001, LNAI 2281, pp. 372–383.

This is because space plasmas are basically collisionless with large mean-free-path. Therefore we have to be aware that they may give the same velocity moments even if the shapes of distributions differ. For instance, when a plasma has two beam components whose velocity vectors are sunward and anti-sunward, and each component has the same numbers of particles, the bulk velocity becomes zero because their velocity vectors cancel out. On the other hand, when a stagnant plasma is observed, the bulk velocity also becomes zero. When we deal with two-beam distributions, we should separate the distributions into two beams and calculate the velocity moments for each beam. Such non-equilibrium multi-component distribution have been reported many times, and a kinetic description of space plasmas that accounts the shape of the velocity distribution have come to be required.

It has been difficult, however, to evaluate the shape of the velocity distribution. Some of the previous researchers have calculated the velocity moments of each component separated by their visual inspections; it would be expected to take time and have limitation. Moreover, resultant moment values will not be estimated accurately when more than one components partially overlap each other.

In this paper we develop a method of representing a three-dimensional distribution by a multivariate Maxwellian mixture model [7, 6] in which the parameter values are obtained by the Expectation-Maximization (EM) algorithm [7, 3, 5]. This method enables us to express the shape of the distribution and to find a feasible way to conduct a statistical analysis for many multi-component cases. The organization of this paper is the following. In Sect. 2, we describe the data of plasma velocity distribution. A fitting method with multivariate Maxwellian mixture model is described in Sect. 3, followed by considerations on how to judge the preferable number of components in the mixture model in Sect. 4. Two applications are demonstrated in Sect. 5. In Sect. 6, we discuss a problem of model selection. We conclude this paper in Sect. 7.

2 Data

We used ion velocity distributions obtained by an electrostatic analyzer named LEP-EA on board the Geotail spacecraft. LEP-EA measured three-dimensional velocity distributions by classifying the velocity space into 32 for the magnitude of the velocity, 7 for elevation angles and 16 for azimuthal sectors (Fig. 1). Let us assume that LEP-EA detected the ion count $C(\boldsymbol{v}_{pqr})$ [#] in a sampling time τ [s], where $\boldsymbol{v}_{pqr}$ [m/s] is the ion velocity. Subscription p, q and r are indicators of the magnitude of the velocity, elevation angle and azimuthal sector, and they take integers $p = 1, \cdots, 32$; $q = 1, \cdots, 7$; and $r = 1, \cdots, 16$. Thus we obtain the total ion count N [#]:

$$N = \sum_{p,q,r} C(\boldsymbol{v}_{pqr}) . \tag{2}$$

Under the assumption that the incident differential ion flux is uniform within the energy and angular responses of the analyzer, the velocity distribution $f_0(\boldsymbol{v}_{pqr})$

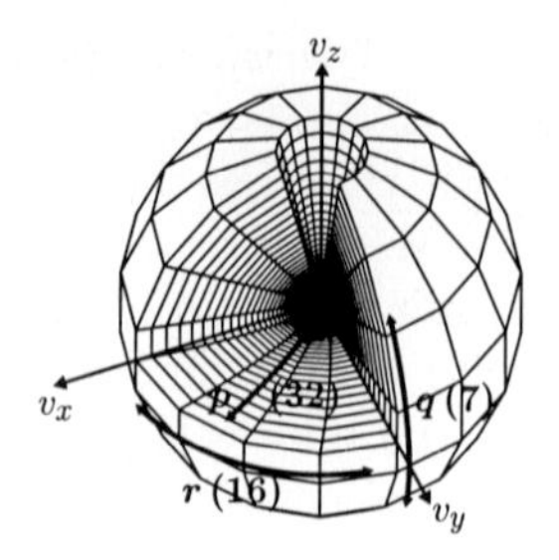

Fig. 1. Classes for observation of a velocity distribution with LEP-EA. Three orthogonal axes v_x, v_y, and v_z are taken in accordance with the spacecraft coordinates

[$\mathrm{s}^3/\mathrm{m}^6$] is given by

$$f_0\left(\boldsymbol{v}_{pqr}\right) = 2 \times 10^4 \frac{1}{\tau \varepsilon_q \mathcal{G}_q} \frac{C\left(\boldsymbol{v}_{pqr}\right)}{\left(\boldsymbol{v}_{pqr}^T \boldsymbol{v}_{pqr}\right)^2}, \tag{3}$$

where ε_q is the detection efficiency and $\mathcal{G}_q$ [$\mathrm{cm}^2\,\mathrm{sr}\,\mathrm{eV/eV}$] is the g-factor. Integrating $f_0\left(\boldsymbol{v}_{pqr}\right)$ over the velocity space, we obtain the number density n [$\#/\mathrm{m}^3$]:

$$n = \sum_{p,q,r} f_0\left(\boldsymbol{v}_{pqr}\right) d\boldsymbol{v}_{pqr}, \tag{4}$$

where $d\boldsymbol{v}_{pqr}$ is the class interval whose class mark is $\boldsymbol{v}_{pqr}$.

For convenience of statistical modeling, we deal with a probability function $f\left(\boldsymbol{v}_{pqr}\right)$ instead of $f_0\left(\boldsymbol{v}_{pqr}\right)$ defined by:

$$f\left(\boldsymbol{v}_{pqr}\right) = \frac{f_0\left(\boldsymbol{v}_{pqr}\right) d\boldsymbol{v}_{pqr}}{n}, \tag{5}$$

so that $\sum_{p,q,r} f\left(\boldsymbol{v}_{pqr}\right) = 1$. Since $f\left(\boldsymbol{v}_{pqr}\right)$ is a function of discrete variables, it is necessary to consider the mixture of probability functions, but we approximate it by the mixture of the Maxwellian that is a probability *density* function.

3 Method

3.1 Multivariate Maxwellian Mixture Model

To deal with multiple components seen in the probability function (Eq. (5), we fit Eq. (5) by the mixture model composed of the sum of s multivariate Maxwellian distributions:

$$f\left(\boldsymbol{v}_{pqr}\right) \simeq \sum_{i=1}^{s} n_i g_i\left(\boldsymbol{v}_{pqr} | \boldsymbol{V}_i, \mathbf{T}_i\right), \tag{6}$$

where n_i is the mixing proportion of the Maxwellians ($\sum_{i=1}^{s} n_i = 1,\ 0 < n_i < 1$). Each multivariate Maxwellian g_i is written as

$$g_i\left(\boldsymbol{v}_{pqr}\middle|\,\boldsymbol{V}_i,\ \mathbf{T}_i\right) = \left(\frac{m}{2\pi}\right)^{3/2} \frac{1}{\sqrt{|\mathbf{T}_i|}} \exp\left[-\frac{m}{2}\left(\boldsymbol{v}_{pqr} - \boldsymbol{V}_i\right)^T \mathbf{T}_i{}^{-1}\left(\boldsymbol{v}_{pqr} - \boldsymbol{V}_i\right)\right], \tag{7}$$

where $\boldsymbol{V}_i$ [m/s] is the bulk velocity vector and $\mathbf{T}_i$ [J] is the temperature matrix of i-th multivariate Maxwellian, and superscript T denotes transposition operation. The multivariate Maxwellian (7) is a generalized form of the Maxwellian (1) as a thermal equilibrium velocity distribution, and can also deal with anisotropic temperature. Plasma with anisotropic temperature is often observed in space because of its collisionless property; it takes a long time for non-equilibrium plasma to relax into a state of thermal equilibrium due to rare particle interactions.

3.2 Parameter Estimation Procedure

The log-likelihood of this mixture model (Eq. (6)) becomes

$$l(\theta) = N \sum_{p,q,r} f\left(\boldsymbol{v}_{pqr}\right) \log \sum_{i=1}^{s} n_i g_i\left(\boldsymbol{v}_{pqr}|\boldsymbol{V}_i,\ \mathbf{T}_i\right) \tag{8}$$

where $\theta = (n_1,\ n_2,\ \cdots,\ n_{s-1},\ \boldsymbol{V}_1,\ \boldsymbol{V}_2,\ \cdots,\ \boldsymbol{V}_s,\ \mathbf{T}_1,\ \mathbf{T}_2,\ \cdots,\ \mathbf{T}_s)$ denotes the all unknown parameters. These parameters directly correspond to the velocity momemts for each component.

Partially differentiate (8) with respect to $\boldsymbol{V}_i,\ \mathbf{T}_i^{-1}\ (i = 1,\ 2,\cdots,\ s)$ and put them equal to zero, maximum likelihood estimators (denoted by $\hat{}$) of the mixing proportion, the bulk velocity vector and the temperature matrix of each Maxwellian are given by

$$\hat{n}_i = \sum_{p,q,r} f\left(\boldsymbol{v}_{pqr}\right) \hat{P}_i\left(\boldsymbol{v}_{pqr}\right), \tag{9}$$

$$\hat{\boldsymbol{V}}_i = \frac{1}{\hat{n}_i} \sum_{p,q,r} f\left(\boldsymbol{v}_{pqr}\right) \hat{P}_i\left(\boldsymbol{v}_{pqr}\right) \boldsymbol{v}_{pqr}, \tag{10}$$

$$\hat{\mathbf{T}}_i = \frac{1}{\hat{n}_i} \sum_{p,q,r} f\left(\boldsymbol{v}_{pqr}\right) \hat{P}_i\left(\boldsymbol{v}_{pqr}\right) \frac{1}{2} m \left(\boldsymbol{v}_{pqr} - \hat{\boldsymbol{V}}_i\right)\left(\boldsymbol{v}_{pqr} - \hat{\boldsymbol{V}}_i\right)^T, \tag{11}$$

where

$$\hat{P}_i\left(\boldsymbol{v}_{pqr}\right) = \frac{\hat{n}_i g_i\left(\boldsymbol{v}_{pqr}\middle|\,\hat{\boldsymbol{V}}_i,\ \hat{\mathbf{T}}_i\right)}{\sum_{j=1}^{s} \hat{n}_j g_j\left(\boldsymbol{v}_{pqr}\middle|\,\hat{\boldsymbol{V}}_j,\ \hat{\mathbf{T}}_j\right)} \tag{12}$$

is an estimated posterior probability.

We should note that when applying a single-Maxwellian model ($s = 1$), we will obtain the parameters identical to the usual velocity moments. This is because the number density is obtained by the usual moment calculation (Eq. (4)), and because the bulk velocity and the temperature matrix as maximum likelihood estimators prove to be identical with those obtained from the moment calculation [8].

On the basis of Eqs. (9)–(12), we estimate the unknown parameters by the EM algorithm [7]. In the following procedure, t denotes an iteration counter of the EM algorithm. Suppose that superscript (t) denotes the current values of the parameters after t cycles of the algorithm for $t = 0, 1, 2, \cdots$.

Setting Initial Value: $t = 0$. We classify the observed velocity space in s groups (G_i; $i = 1, 2, \cdots, s$) using the k-means algorithm, and set the initial value of the posterior probability as

$$P_i^{(0)}(\boldsymbol{v}_{pqr}) = \begin{cases} 1 & (\boldsymbol{v}_{pqr} \in G_i) \\ 0 & (\boldsymbol{v}_{pqr} \notin G_i) \end{cases}, \tag{13}$$

where $i = 1, 2, \cdots, s$. With $P_i^{(0)}(\boldsymbol{v}_{pqr})$, we calculate $\hat{n}_i^{(0)}$, $\hat{\boldsymbol{V}}_i^{(0)}$, and $\hat{\mathbf{T}}_i^{(0)}$ by Eqs. (9), (10) and (11).

Parameter Estimation by EM Algorithm: $t \geq 1$. On the t-th iteration ($t \geq 1$), we compute $n_i^{(t)}$ and $P_i^{(t)}(\boldsymbol{v}_{pqr})$ by Eqs. (9) and (12) as the E-step. At the M-step, we choose $\boldsymbol{V}_i^{(t)}$ and $\mathbf{T}_i^{(t)}$ as maximum likelihood estimators by Eqs. (10) and (11).

Judgment of Convergence. We finish the iteration if

$$\left| l\left(\hat{\theta}^{(t)}\right) - l\left(\hat{\theta}^{(t-1)}\right) \right| < \epsilon \quad \text{and} \quad \left\| \hat{\theta}^{(t)} - \hat{\theta}^{(t-1)} \right\| < \delta, \tag{14}$$

where ϵ and δ are sufficiently small positive number. If the above convergence condition is not satisfied, return to the E-step with replacing t by $t + 1$.

4 Preferable Number of Components

When we fit a velocity distribution by the Maxwellian mixture model, we should examine how reasonable the fit is. That is, for instance, it is inappropriate to approximate a unimodal observation by a multi-Maxwellian mixture model. Here we introduce a method of judging which of two models, i.e., a single-Maxwellian model or a two-Maxwellian mixture model, is preferable for each observation. We adopt the following principle. If a two-Maxwellian mixture model which resulted from an observation shows two peaks, the observation will also have two peaks. Hence we conclude that the two-Maxwellian mixture model is reasonable to use.

On the other hand, if the resulting two-Maxwellian mixture model has only one peak, the observation will be a unimodal distribution: We should use a usual single-Maxwellian fitting.

To judge whether the fitting result is reasonable or not, we enumerate the number of peaks of the resulting fitted model. Actually, to count the number of peaks, we enumerate the number of extrema of the model. Let us consider when we fit some data $f(\boldsymbol{v})$ by a two-Maxwellian mixture model and the fitting result is computed as

$$f(\boldsymbol{v}) \simeq n_1 g_1 (\boldsymbol{v}|\boldsymbol{V}_1, \mathbf{T}_1) + n_2 g_2 (\boldsymbol{v}|\boldsymbol{V}_2, \mathbf{T}_2). \tag{15}$$

To count the number of peaks, we need to count the number of $\boldsymbol{v}$ satisfying

$$\frac{d}{d\boldsymbol{v}} [n_1 g_1 (\boldsymbol{v}|\boldsymbol{V}_1, \mathbf{T}_1) + n_2 g_2 (\boldsymbol{v}|\boldsymbol{V}_2, \mathbf{T}_2)] = 0. \tag{16}$$

It is difficult, however, to treat the three-dimensional variable $\boldsymbol{v}$. We then reduce this three-dimensional problem to one set of simultaneous equations of one-dimensional variables (details are given in Ref.[11, 12]):

$$\eta(\xi) = \xi, \tag{17}$$

$$\eta(\xi) = \frac{n_1 g_1 (\boldsymbol{w}(\xi)|0, \mathbf{I})}{n_2 g_2 (\boldsymbol{w}(\xi)|\boldsymbol{W}, \mathbf{M}^{-1})}, \tag{18}$$

whose number of solutions is equivalent to the number of $\boldsymbol{v}$ satisfying Eq. (16). Here we put $\boldsymbol{w}(\xi) = (\mu_1 W_1/(\xi+\mu_1), \mu_2 W_2/(\xi+\mu_2), \mu_3 W_3/(\xi+\mu_3))$, $\boldsymbol{W} = (\boldsymbol{y}_1, \boldsymbol{y}_2, \boldsymbol{y}_3)^{-1} \mathbf{L}^T (\boldsymbol{V}_2 - \boldsymbol{V}_1)$, where $\mathbf{L}$ is a matrix that satisfies $\mathbf{L}\mathbf{L}^T = \mathbf{T}_1^{-1}$, and μ_1, μ_2 and μ_3 are the eigenvalues of $(\mathbf{L}^T \mathbf{T}_2 \mathbf{L})^{-1}$ whose corresponding unit eigenvectors are $\boldsymbol{y}_1$, $\boldsymbol{y}_2$ and $\boldsymbol{y}_3$. Consequently, we need to count the nodes of the line (17) and the curve (18) in the ξ-η plane.

Utilizing Eqs. (17) and (18), we can also evaluate a mixture model with three or more Maxwellians. That is, we first pick up all the combinations of two Maxwellians from the multiple Maxwellians, and then apply Eqs. (17) and (18) for each combination.

5 Application

The left-hand panel of Fig. 2(a) show an ion velocity distribution observed in the plasma sheet boundary layer of the Earth's magnetotail. Displayed distribution is a slice by the v_x-v_y plane, whose values are black-to-white-coded as shown in the bar. Used coordinate system is taken in accordance with the spacecraft coordinate system: The v_z axis is parallel to the spacecraft spin axis and positive northward, the v_x axis is parallel to the projection of the spacecraft-Sun line on the plane whose normal vector is the v_z axis and is positive sunward, and the v_y axis completes a right-hand orthogonal system. We can observe a hot component and a cold component whose bulk velocities are $(v_x, v_y) \simeq (1000, 0)$ km/s and $(v_x, v_y) \simeq (-200, -500)$ km/s, respectively.

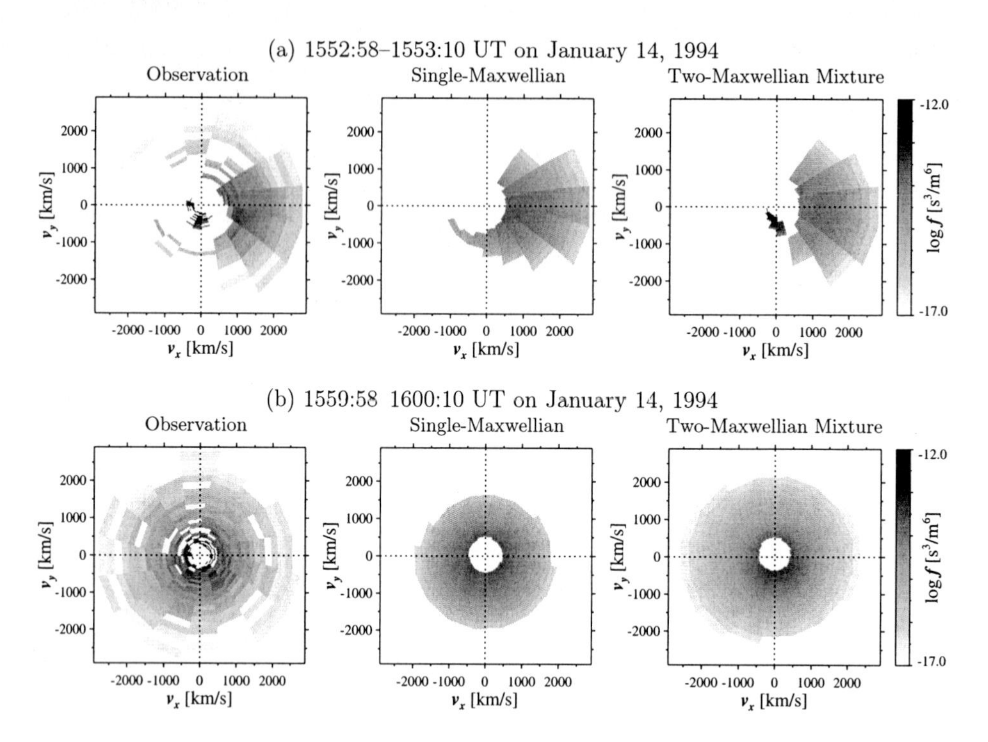

Fig. 2. (a) Observation of an ion velocity distribution on the v_x-v_y plane in the time interval 1552:58–1553:10 UT on January 14, 1994 (left), fitting functions by a single-Maxwellian model (center) and a two-Maxwellian mixture model (right). (b) For the observation between 1559:58–1600:10 UT on the same day

When we fit this data with a single-Maxwellian, we obtain the estimated parameters given in the second row of Table 1(a). A single-Maxwellian model with these parameters produces the distribution shown in the central panel of Fig. 2(a). This corresponds to what we deal with by the usual velocity moments, and it is hot and has a shifted bulk velocity compared with the observation.

This problem, however, is easily resolved by applying a two-Maxwellian mixture model. Similarly, we give the estimated parameters in the third and fourth rows of Table 1(a), and display the produced distribution in the right-hand panel of Fig. 2(a). It can be seen that both the hot and cold components existing in the observed distribution were properly reproduced.

For this example, we find that the fitting by the two-Maxwellian model is preferable to the single-Maxwellian model by counting the number of solutions with simultaneous equations (17) and (18). The two-Maxwellian fitting is therefore justified, which agrees with our inspection of the observed distribution.

The other example is a distribution observed in the central plasma sheet. As can be seen in the left-hand panel of Fig. 2(b), it is appropriate to expect that this consists of a single hot component whose bulk velocity is located near

Table 1. Estimated parameters for single-Maxwellian and two-Maxwellian mixture models. The first column n is the number density, that is, the mixing proportion multiplied by the number density

(a) 1552:58–1553:10 UT on January 14, 1994

	n [/cc]	V_x [km/s]	V_y	V_z	T_{xx} [eV]	T_{xy}	T_{xz}	T_{yy}	T_{yz}	T_{zz}
1	0.040	793	−104	6	5515	327	−216	2177	−129	2089
1	0.012	−101	−260	47	120	−138	23	326	−31	95
2	0.028	1167	−38	−11	2815	−347	−90	2800	−130	2912

(b) 1559:58–1600:10 UT on January 14, 1994

	n [/cc]	V_x [km/s]	V_y	V_z	T_{xx} [eV]	T_{xy}	T_{xz}	T_{yy}	T_{yz}	T_{zz}
1	0.087	−5	−121	−135	3203	16	44	2936	−174	3185
1	0.029	−94	−46	−27	6495	192	187	5676	−275	6359
2	0.058	39	−158	−189	1512	−20	48	1536	−187	1523

the origin of the velocity space. Hence, when we fit the data, we should adopt a single-Maxwellian model rather than a two-Maxwellian mixture model.

In the central panel of Fig. 2(b), we show the calculated distribution with the single-Maxwellian model. The parameters used are given in the second row of Table 1(b). In this case the single-Maxwellian fitting appears to be sufficient. Furthermore, we display the result with the two-Maxwellian mixture model. The right-hand panel shows the calculated distribution with the estimated parameters given in the third and fourth rows of Table 1(b).

By examining the number of solutions of the simultaneous equations for ξ and η, we find that they have only one solution. We therefore adopt the usual velocity moments obtained by the single-Maxwellian fitting.

6 Discussion

To select the preferable number of components, we adopt in this study an empirical approach; we first fit the data with a two-Maxwellian mixture model, then examine whether there is a saddle point on the segment between the bulk velocities of the model. Generally, AIC (Akaike Information Criterion [1]) defined by

$$\mathrm{AIC} = -2 \max l(\theta) + 2 \dim \theta, \tag{19}$$

has been frequently employed for this problem [6]. However, AIC selected a mixture model having a larger number of components compared with our intuition. According to AIC, the best number of components was found to be six or more for both distributions shown in Fig. 2.

This is expected to be due to following three reasons. First, it is not so appropriate in our data set to adopt the Maxwellian distribution as a component distribution of a mixture model. While the Maxwellian distribution can well represent the observed distribution near the peak, it cannot follow the distribution in the large-velocity range. Since the observed distribution has a heavy tail in the large-velocity range, it is necessary to have many components for fitting such a tail accurately. In fact, the two-Maxwellian mixture model shown in Fig. 2(b) consists of two components which present a peak and a heavy tail, respectively. This problem would be solved when using a mixture model which consists of heavy tail distributions instead of the Maxwellian distributions.

One of the heavy tail distributions is the κ distribution defined by

$$g_i\left(\boldsymbol{v}_{pqr} \middle| \boldsymbol{V}_i, \mathbf{T}_i, \kappa_i\right) = \left(\frac{m}{2\pi}\right)^{3/2} \frac{\Gamma\left(\kappa_i\right)}{\Gamma\left(\kappa_i - 3/2\right)} \frac{1}{\sqrt{|\mathbf{T}_i|}} \cdot \left[1 + \frac{m}{2}\left(\boldsymbol{v}_{pqr} - \boldsymbol{V}_i\right)^T \mathbf{T}_i^{-1}\left(\boldsymbol{v}_{pqr} - \boldsymbol{V}_i\right)\right]^{-\kappa_i} . \quad (20)$$

This converges to the Maxwellian distribution in the limit of $\kappa \to \infty$, so it can give a more comprehensive treatment of the data. When we select the κ distribution as a component distribution, the algorithm presented in Sect. 3 can work by including the κ renewing step. However, our experiments found that the estimated κ value was the order of 10^1, which means that the resultant distribution is practically the Maxwellian.

The second reason is that the observation has a large total ion count ($N = 2925$ and 4762 for the Figs. 2 (a) and (b), respectively). The log-likelihood $l(\theta)$ is multiplied by N as defined in Eq. (8), and $\max l(\theta)$ was on the order of 10^4–10^5 in the two examples. On the other hand, the dimension of free parameters θ was on the order of 10^0–10^1. AIC was determined practically by $\max l(\theta)$ and was not affected by $\dim\theta$ as a penalty term.

To take large N into account, we evaluated the competing models by BIC (Bayesian Information Criterion [9]) instead of AIC. BIC is an information criterion such that posterior probability is maximized and defined as

$$\mathrm{BIC} = -2 \max l(\theta) + \log N \dim \theta . \quad (21)$$

BIC, however, yielded the same result as with AIC in our cases.

Finally, we should notice that there exist some classes of $\boldsymbol{v}_{pqr}$ around $\boldsymbol{v} = 0$ such that $f(\boldsymbol{v}_{pqr}) = 0$, which is identified as a white region around $v_x = v_y = 0$ in the two observations displayed in Fig. 2. This is due to the instrument, an electrostatic analyzer. That is, when we observe the ambient velocity distribution as shown in the left-hand panel of Fig. 3 by the electrostatic analyzer, we obtain the count $C(\boldsymbol{v}_{pqr})$ as in the central panel. Since $C(\boldsymbol{v}_{pqr})$ is a count data, it becomes zero if it is less than unity (under the one count level presented by the dotted line). The zero count is then converted to zero probability $f(\boldsymbol{v}_{pqr})$ through Eqs. (3) and (5). Namely, probability below the one-count level curve becomes zero in the observation (see the right-hand panel). This cut off effect with quantization occurs especially around $\boldsymbol{v} = 0$, which produces the observed distribution, $f(\boldsymbol{v}_{pqr})$, having a "hole" around $\boldsymbol{v} = 0$ as in the right-hand panel.

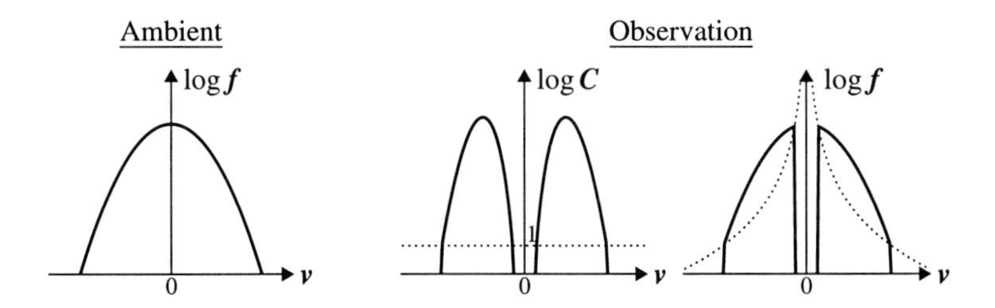

Fig. 3. Creation of a "hole" in the observed velocity distribution near the origin. When the ambient velocity distribution (left) is observed, the obtained count and velocity distribution become the ones as shown in the central and right-hand panels. The dotted line and curve present one count level

AIC is expected to choose the model with multi-components to present the edge of the "hole" precisely.

This "hole" effect will be reduced if we give a probabilistic description to the ion count $C(\boldsymbol{v}_{pqr})$. We assume that the ion count observation is generated from a multinominal distribution in which a detection probability of an ion of velocity $\boldsymbol{v}_{pqr}$ is $h(\boldsymbol{v}_{pqr})$. When an ion whose velocity is $\boldsymbol{v}_{pqr}$ is detected $C(\boldsymbol{v}_{pqr})$ times in N trials, the likelihood function becomes

$$\lambda\left(C\left(\boldsymbol{v}_{pqr}\right)|h\left(\boldsymbol{v}_{pqr}\right)\right) = N! \prod_{p,q,r} \frac{h\left(\boldsymbol{v}_{pqr}\right)^{C(\boldsymbol{v}_{pqr})}}{C\left(\boldsymbol{v}_{pqr}\right)!}. \tag{22}$$

An expected count of the ion of velocity $\boldsymbol{v}_{pqr}$ is $Nh(\boldsymbol{v}_{pqr})$. We also assume that the velocity distribution as a function of $h(\boldsymbol{v}_{pqr})$ can be approximated by a mixture model. This approximation can be realized by considering the following prior distribution for $h(\boldsymbol{v}_{pqr})$ [4, 10]

$$\pi\left(h\left(\boldsymbol{v}_{pqr}\right)|\theta\right) = \prod_{p,q,r} \left[\sum_{i=1}^{s} n_i g_i\left(\boldsymbol{v}_{pqr}|\boldsymbol{V}_i, \mathbf{T}_i\right)\right]^{Nf'(\boldsymbol{v}_{pqr}, h(\boldsymbol{v}_{pqr}))}, \tag{23}$$

where we set

$$f'\left(\boldsymbol{v}_{pqr}, h\left(\boldsymbol{v}_{pqr}\right)\right) = \frac{f_0'\left(\boldsymbol{v}_{pqr}, h\left(\boldsymbol{v}_{pqr}\right)\right) d\boldsymbol{v}_{pqr}}{\sum_{p,q,r} f_0'\left(\boldsymbol{v}_{pqr}, h\left(\boldsymbol{v}_{pqr}\right)\right) d\boldsymbol{v}_{pqr}}, \tag{24}$$

$$f_0'\left(\boldsymbol{v}_{pqr}, h\left(\boldsymbol{v}_{pqr}\right)\right) = 2 \times 10^4 \frac{1}{\tau \varepsilon_q \mathcal{G}_q} \frac{Nh\left(\boldsymbol{v}_{pqr}\right)}{\left(\boldsymbol{v}_{pqr}^T \boldsymbol{v}_{pqr}\right)^2}. \tag{25}$$

With Eqs. (22) and (23), the log-likelihood function is defined by

$$l'\left(C\left(\boldsymbol{v}_{pqr}\right)|\theta\right) = \log \int \lambda\left(C\left(\boldsymbol{v}_{pqr}\right)|h\left(\boldsymbol{v}_{pqr}\right)\right) \pi\left(h\left(\boldsymbol{v}_{pqr}\right)|\theta\right) dh\left(\boldsymbol{v}_{pqr}\right). \tag{26}$$

An optimal θ can be also obtained by maximizing $l'(C(\boldsymbol{v}_{pqr})|\theta)$. An evaluation of the competing models is again carried out by comparing AIC which is defined by

$$\mathrm{AIC} = -2\max l'\left(C\left(\boldsymbol{v}_{pqr}\right)|\theta\right) + 2\dim\theta. \tag{27}$$

In this framework, AIC defined by Eq. (27) is sometimes called ABIC (Akaike Bayesian Information Criterion [2]). In this approach, an evaluation of model is conducted by considering two distribution of $C\left(\boldsymbol{v}_{pqr}\right)$ and $f\left(\boldsymbol{v}_{pqr}\right)$, namely a Bayesian approach. The low count around $\boldsymbol{v} = 0$ yields a small "weight" so that the fitting of $f\left(\boldsymbol{v}_{pqr}\right)$ is allowed to be done less precisely around $\boldsymbol{v} = 0$. Moreover, a Bayesian approach will treat a problem of a heavy tail in the high-velocity range. Since the count of ions of high-velocity is also small, modeling and model comparison with a smaller effect of the heavy tail will be tractable.

7 Conclusion

We have proposed a method for analyzing a multi-peaked velocity distribution via a multivariate Maxwellian mixture model. With the fitting of this model, we can extract the velocity moments for each component of the multiple peaks automatically. The parameters of the model are determined through the EM algorithm. For the automatic judgment of the preferable number of components of the mixture model, we introduced a method of examining the number of extrema of the resulting mixture model. When we use the method, we can adopt an appropriate fitting result and obtain a tool for a kinetic description of the plasma dynamics; this is especially effective for dealing with a large data set. Application of our method to observations confirmed that the method works well as shown in Fig. 2. More scientific application is presented in our recent work [12].

Acknowledgment

We would like to thank T. Mukai for providing us with Geotail/LEP data. This work was carried out under the auspices of JSPS Research Fellowships for Young Scientists.

References

1. Akaike, H.: A New Look at the Statistical Model Identification. IEEE Trans. Autom. Control **AC-19** (1974) 716–723
2. Akaike, H.: Likelihood and the Bayes procedure. In: Bernardo, J. M., De Groot, M. H., Lindley, D. V., Smith, A. F. M. (eds.): Bayesian Statistics. University Press, Valencia Spain (1980) 1–13
3. Dempster, A. P., Laird, N. M., Rubin, D. B.: Maximum Likelihood from Incomplete Data via the *EM* Algorithm. J. Roy. Statist. Soc. B **39** (1977) 1–38
4. Ishiguro, M., Sakamoto, Y.: A Bayesian Approach to Binary Response Curve Estimation. Ann. Inst. Statist. Math. **35** (1983) 115–137

5. McLachlan, G. J., Krishnan, T.: The EM Algorithm and Extensions. Wiley Series in Probability and Statistics. John Wiley and Sons, New York (1997)
6. McLachlan, G. J., Peel, D.: Finite Mixture Models. Wiley Series in Probability and Statistics. John Wiley and Sons, New York (2000)
7. Nakamura, N., Konishi, S., Ohsumi, N.: Classification of Remotely Sensed Images via Finite Mixture Distribution Models (in Japanese with English abstract). Proceedings of the Institute of Statistical Mathematics **41** (1993) 149–167
8. Sakamoto, Y., Ishiguro, M., Kitagawa G.: Akaike Information Criterion Statistics. KTK Scientific Publishers. D. Reidel, Tokyo (2000)
9. Schwarz, G.: Estimating the Dimension of a Model. Ann. Statist. **6** (1978) 461–464
10. Tanabe, K., Sagae, M.: An Exact Cholesky Decomposition and the Generalized Inverse of the Variance-Covariance Matrix of the Multinominal Distribution, with Applications. J. R. Statist. Soc. B. **54** (1992) 211–219
11. Ueno, G., Nakamura, N., Higuchi, T., Tsuchiya, T., Machida, S., Araki, T: Application of Multivariate Maxwellian Mixture Model to Plasma Velocity Distribution Function. In: Arikawa, S., Morishita, S. (eds.): Discovery Science. Lecture Notes in Computer Science, Vol. 1967. Springer-Verlag, Berlin Heidelberg New York (2000) 197–211
12. Ueno, G., Nakamura, N., Higuchi, T., Tsuchiya, T., Machida, S., Araki, T., Saito, Y., Mukai, T.: Application of Multivariate Maxwellian Mixture Model to Plasma Velocity Distribution Function. to appear in J. Geophys. Res. (2001)

Inductive Thermodynamics from Time Series Data Analysis

Hiroshi H. Hasegawa[1,⋆], Takashi Washio[2], and Yukari Ishimiya[1]

[1] Department of Mathematical Sciences, Ibaraki University,
2-1-1 Bunkyo, Mito 310-8512, Japan
hiroshih@mito.ipc.ibaraki.ac.jp
[2] I.S.I.R., Osaka University,
8-1 Mihogaoka, Ibaragi, Ohsaka 567-0047, Japan

Abstract. We propose an inductive thermodynamics from time series data analysis. Using some modern techniques developed in Statistical Science and Artificial Intelligence, we construct a mathematical model from time series data. We introduce an effective potential for a steady state distribution and construct thermodynamics following the recent work by Sekimoto–Sasa. We apply our idea to neutron noise data from a test nuclear reactor. We interpret the slow transformation of the control bar as work. The response to the transformation appears as excess heat production in accordance with the second law.

1 Introduction

Nowadays computers in research labs and companies keep huge amounts of data. It is an important problem to obtain useful knowledge from the database using modern techniques of artificial intelligence and statistical science. Recently scientists have tried to discover scientific rules using these techniques (Discovery Science) [1]. For example, many scientists to find a rule relating some special sequences in the genome and a specific disease. In this paper we try to find some statistical rules instead of deterministic ones. We propose an inductive thermodynamics from time series data analysis.

We believe that the concept of effective theory plays an important role in Discovery Science. Physicists understand the diversity of laws as a result of a separation of hierarchies and the coexistence of effective theories. An effective theory is a closed theory within a hierarchy. A typical example is thermodynamics. It is a closed theory using macro scale variables such as pressure and temperature. The effect of a single molecule is negligible because of the huge scale separation. When a scientist finds a simple rule we expect the existence of an effective theory. We will inductively construct an effective theory from time series data to find some statistical laws.

⋆ Center for Statistical Mechanics, University of Texas, Austin,TX78712, USA.
Tel 1-512-471-7253, Fax 1-512-471-9612, e-mail: hiroshi@physics.utexas.edu

S. Arikawa and A. Shinohara (Eds.): Progress in Discovery Science 2001, LNAI 2281, pp. 384–394.

Inductive thermodynamics is applicable for a system which has a steady state distribution. It describes the response to a slow external transformation. Inductive thermodynamics has the following three parts:

(1) Mathematical models from time series data

We construct mathematical models using the modern techniques in artificial intelligence and statistical science. For instance, we can mention some linear models such as Auto-Regressive Moving Average (ARMA) model and state vector model and also some nonlinear models such as neuron networks and return map for 1/f noise. In this paper we will consider the ARMA model identified by Akaike Information Criteria (AIC)[8].

(2) An effective potential from steady state distribution

We assume that the system has a steady state when we do not operate externally. We calculate the steady state distribution using the mathematical model. We may directly estimate it from actual data. From the steady state distribution $P_s(\mathbf{x}, \alpha)$, we define the effective potential $U(\mathbf{x}, \alpha)$ and te free potential $F(\alpha)$ by analogy to the Boltzmann distribution:

$$P_s(\mathbf{x}, \alpha) = \exp[F(\alpha) - U(\mathbf{x}, \alpha)] \tag{1}$$

where we may choose the zero of the effective potential for convenient coordinates $\mathbf{x}$ and parameters α. Ordinarily we choose the minimum value of the effective potential as zero. This ambiguity is same as for the energy.

We can introduce an effective potential for both economic and biological systems. It will play the same role as the energy in thermodynamics [1] [4].

(3) Thermodynamics following Sekimoto–Sasa

Recently Sekimoto and Sasa reconstructed the thermodynamics of a system governed by the Langevin equation [2]. They considered the Langevin equation as a balance of forces equation. So they immediately derived the first law. Then they interpreted the heat as work by the reaction force to the heatbath. They derived the second law using a perturbation expansion. We consider the ARMA model as a Langevin equation in discrete time. We obtain the laws of thermodynamics by applying their argument to the ARMA model.

We hope to apply inductive thermodynamics to both economic and biological systems. Although the fluctuations in these systems are not thermal ones, we can formally construct a thermodynamics in which the effective potential plays the same role as energy in ordinary thermodynamics. In the steady state distributions of such systems, detailed balance may be broken and irreversible circulation of fluctuations may appear [6]. In the next section, we will extend the argument by Sekimoto–Sasa to the case of irreversible circulation of fluctuations.

[1] We defined the effective potential dimensionless. The temperature is already renormalized in it. It is convenient to define dimensionless potential for both economic and biological systems.

2 Sekimoto–Sasa Theory for Irreversible Circulation of Fluctuations

In this section, we will derive thermodynamics for a system whose steady state distribution has irreversible circulation of fluctuations. Our argument is completely parallel with the original Sekimoto–Sasa theory except that the effective potential includes house-keeping heat caused by the irreversible circulation of fluctuations.

We start with the following two-dimensional linear Langevin equation,

$$\dot{\mathbf{x}} = -\mathcal{A}\mathbf{x} + \mathbf{e}(t), \tag{2}$$

where the coordinate vector $\mathbf{x} = (x_1, x_2)^{\mathrm{T}}$, $\mathcal{A}$ is an asymmetric 2×2 matrix, and the normalized Gaussian white noise vector $\mathbf{e}(t) = (e_1(t), e_2(t))^{\mathrm{T}}$ satisfies

$$< e_i(t)e_j(t') >= 2\Theta\delta_{ij}\ \delta(t-t'). \tag{3}$$

This normalization condition defines the temperature as Θ. For convenience, we choose $\Theta = 1$ in this paper.

The Fokker–Plank equation is derived as

$$\frac{\partial P(\mathbf{x},t)}{\partial t} = \frac{\partial}{\partial \mathbf{x}} \cdot \mathbf{J}(\mathbf{x},t). \tag{4}$$

Here the current $\mathbf{J}(\mathbf{x},t)$ is defined as

$$\mathbf{J}(\mathbf{x},t) = [\frac{\partial}{\partial \mathbf{x}} + \mathcal{A}\mathbf{x}]P(\mathbf{x},t). \tag{5}$$

We parameterize the matrix $\mathcal{A}$ as

$$\frac{1}{2}\mathcal{A} = \begin{pmatrix} \alpha_1 & \beta-\gamma \\ \beta+\gamma & \alpha_2 \end{pmatrix}, \tag{6}$$

where $2\gamma = -(\mathcal{A}_{12} - \mathcal{A}_{21})/2$ is the parameter of asymmetry.

We separate the righthand side of the Fokker–Plank equation into a symmetric part and an asymmetric part:

$$\frac{\partial P(\mathbf{x},t)}{\partial t} = \frac{\partial}{\partial \mathbf{x}} \cdot [\frac{\partial}{\partial \mathbf{x}} + \frac{\partial V(\mathbf{x})}{\partial \mathbf{x}}]P(\mathbf{x},t) + 2\gamma\mathbf{x} \times \frac{\partial}{\partial \mathbf{x}}P(\mathbf{x},t), \tag{7}$$

where
$V(\mathbf{x}) = \mathbf{x} \cdot \mathcal{A}\mathbf{x}/2$ and
$\mathbf{x} \times \partial/\partial\mathbf{x} = x_1\partial/\partial x_2 - x_2\partial/\partial x_1$.
Notice that the term proportional to the asymmetric part is just the rotation operator. As we will show later, the steady state is the equilibrium state for the symmetric case. On the other hand, the current exists and detailed balance is broken for the asymmetric case.

When the parameters $\alpha = (\alpha_1, \alpha_2, \beta, \gamma)$ are constant, the asymptotic distribution becomes the steady state. Because of the linearity of the Langevin equation, the steady state distribution becomes Gaussian:
$P_s(\mathbf{x}, \alpha) = \exp[F(\alpha) - U(\mathbf{x}, \alpha)]$,
$F(\alpha) = -\log[\det(2\pi R_s(\alpha))]/2$, and
$U(\mathbf{x}, \alpha) = \mathbf{x} \cdot R_s^{-1}(\alpha)\mathbf{x}/2$.

If the matrix $\mathcal{A}$ is symmetric, the potential $U(\mathbf{x}, \alpha) = V(\mathbf{x}, \alpha)$. The steady state is the equilibrium state,

$$P_{\rm eq}(\mathbf{x}, \alpha) = \exp[F(\alpha) - V(\mathbf{x}, \alpha)]. \tag{8}$$

Then the current $\mathbf{J} = \mathbf{0}$.

If the matrix $\mathcal{A}$ is asymmetric, the steady state distribution is derived by solving the Fokker–Plank equation. The effective potential becomes

$$U(\mathbf{x}, \alpha) = \frac{V(\mathbf{x}, \alpha)}{1+\tilde{\gamma}^2} + \frac{\tilde{\gamma}\mathbf{x}}{1+\tilde{\gamma}^2} \cdot \begin{pmatrix} \gamma+\beta & (\alpha_2-\alpha_1)/2 \\ (\alpha_2-\alpha_1)/2 & \gamma-\beta \end{pmatrix} \mathbf{x}, \tag{9}$$

where $\tilde{\gamma} = 2\gamma/(\alpha_1 + \alpha_2)$. In this case, the steady state is nonequilibrium and the current $\mathbf{J} \neq \mathbf{0}$. This nonequilibrium steady state is known as the irreversible circulation of fluctuations. Tomita–Tomita first studied it[6]. Kishida and his collaborators applied it to nuclear-reactor observations [7].

Now we consider an external transformation. We slowly change the parameters $\alpha(t)$ from $\alpha(0) = \alpha_0$ to $\alpha(T) = \alpha_T$. Then the first law is given as

$$\Delta U = W + D, \tag{10}$$

where the change of potential ΔU, the work W, and the dissipation D are respectively defined as follows:

$$\Delta U \equiv < U(\alpha_T) >_T - < U(\alpha_0) >_0, \tag{11}$$

$$W \equiv \int_0^T dt \int d\mathbf{x} \frac{d\alpha(t)}{dt} \cdot \frac{\partial U(\alpha(t))}{\partial \alpha(t)} P(\mathbf{x}, t), and \tag{12}$$

$$D \equiv \int_0^T dt \int d\mathbf{x} U(\alpha(t)) \frac{\partial P(\mathbf{x}, t)}{\partial t}, \tag{13}$$

where $< \cdot >_t \equiv \int d\mathbf{x} \cdot P(\mathbf{x}, t)$. $P(\mathbf{x}, t)$ can be obtained by averaging over ensemble of data [2].

(1)Work in quasi-static processes

In a quasi-static process, the probability distribution $P(\mathbf{x}, t)$ can be approximated as the steady state $P_s(\mathbf{x}, \alpha(t))$. Then

$$W_{\rm QS} = F(\alpha_T) - F(\alpha_0). \tag{14}$$

[2] In this paper we define the work W based on the effective potential $U(\mathbf{x}, \alpha)$. This work W is different from the one defined using the potential $V(\mathbf{x})$ which appeared in the Fokker–Plank equation. Because of the irreversible circulation of fluctuations, the steady state deviates from equilibrium. The difference is call house-keeping heat [3].

The work does not depend on its path and is given as the difference between the initial and final free potential. The derivation is given in Appendix A.
(2)Work in near-quasi-static processes
We can estimate the work in a near-quasi-static process W_{NQS} under the assumption of the validity of the Fokker–Plank equation during the transformation. The work in a near-quasi-static process is greater than in the quasi-static case,i.e.

$$\delta W = W_{\mathrm{NQS}} - W_{\mathrm{QS}} \geq 0. \tag{15}$$

The excess heat production $\delta W \geq 0$ means that thermodynamic entropy increasing: the second law holds. We can obtain the inequality in the second order perturbation expansion with respect to the time derivative of the parameters $(\alpha_T - \alpha_0)/T$. See Appendix B.

In this section, we have derived the first and second laws of thermodynamics for the system displaying the irreversible circulation of fluctuations. The original argument by Sekimoto–Sasa works using the effective potential including housekeeping heat [3]. Following Oono–Paniconi, Hatano–Sasa and Tasaki–Sasa have developed steady state thermodynamics in a more general and elegant way [5].

In the next section, we will apply our arguments to actual time series data from a test nuclear reactor.

2.1 Kinetic Model of Zero-Power Point Reactor

The kinetic model of a zero-power point reactor [9] is governed by the following coupled Langevin type equations:

$$dn/dt = (\rho - \tilde{\beta})/\Lambda * n + \sum_i \lambda_i C_i + \mathcal{S} + e_0(t) \tag{16}$$

$$dC_i/dt = \tilde{\beta}_i/\Lambda * n - \lambda_i C_i + e_i(t) \quad i = 1, 2, \cdots \tag{17}$$

where n is the concentration of neutrons, C_i is the concentration of the i-th precursor nuclear, $\beta = \sum \beta_i$ is the rate of total delayed neutrons, Λ is the lifetime of prompt neutrons, λ_i is the decay constant of the i-th precursor nuclear, S: external neutron current, ρ is the reactivity, e_i is the white noise.

We assume that the inverse of the lifetime of the prompt neutron is much greater than the decay constant of the advanced nuclei, $\beta/\Lambda \gg \lambda_i$ and that the ratio of total delayed neutron is much greater than the reactivity, $\beta \gg \rho$.

Without the noise $e_i(t) = 0$, the time evolution of system is a superposition of the following eigenmodes:
(1) One extremely fast exponentially decaying eigenmode, the eigenvalue $\sim O(\beta\lambda/\Lambda) * \lambda$.
(2) Five or six exponentially decaying eigenmodes with eigenvalues $\sim O(1) * \lambda$.
(3) One extremely slow exponentially decaying or growing eigenmode with eigenvalue $\sim O(\rho/\beta) * \lambda$.
When the external noise is turned on, the system shows irreversible circulation of fluctuations [7].

Although the reactivity ρ is negative, the external neutron source $\mathcal{S}$ makes the system to achieve a steady state. By adjusting the reactivity ρ, we can control the power of the reactor. It is straightforward to apply our previous argument to the kinetic model of the zero point reactor, since the system is governed by coupled Langevin equations. The external transformation corresponds to the change of reactivity ρ realized by movement of the control bar.

2.2 ARMA Model

From time series data of neutron noise $\{x(t)\}$ we identify the ARMA model [8],

$$x(t) + a_1 x(t-1) + \cdots + a_{n_a} x(t - n_a) \tag{18}$$

$$= c_0 e(t) + c_1 e(t-1) + \cdots + c_{n_c} e(t - n_c) \tag{19}$$

where $e(t)$ is Gaussian white noise at time t.

We have analyzed the experimental data obtained by Professor Yamada's group from test nuclear reactor UTR-KINKI at Kinki University in November, 1991. For steady state, they sampled data with a sampling frequency 48kHz after lowpass filtering at 500 Hz. They resampled the analog data to digital. In nonstationary processes, they sampled data after additional highpass filtering at 0.1 Hz. We refiltered the original data using sharp lowpass filter.

There are steady state data for the reactivity $\rho = -2.19\$, -1.25\$, -0.729\$,$ $-\ 0.588\$, -0.289\$$ [9]. There are also nonstationary data from $\rho = -2.19\$$ to $-1.25\$$ and $-0.588\$$ to $-0.289\$$. We divide the total data into 50 groups. Each group has 3000 data-points and we identified ARMA model using each 3000 data-points. After statistical treatment of AIC[8], we judged that ARMA(2,2) is the best except $\rho = -0.289\$$. The coefficients of the ARMA(2,2) model are as follows,

ρ	−2.19	−1.25	−0.729	−0.588	−0.289
a_1	0.2688	0.1653	0.0319	0.0519	−0.0485
a_2	−0.3975	−0.5097	−0.5274	−0.5586	−0.5546
$\bar{c}_1$	0.7013	0.7419	0.6710	0.7090	0.6559
$\bar{c}_2$	−0.1242	−0.1080	−0.1216	−0.1165	−0.1133

(20)

where $\bar{c}_1 = c_1/c_0$ and c_0 was directly determined from the variances of the data.

2.3 Thermodynamics of the ARMA Model

We will now consider the thermodynamics of ARMA(2,2) model.

(1)Steady state distribution

We rewrite the ARMA(2,2) model as a two-dimensional vector model,

$$\mathbf{x}(t) = A\mathbf{x}(t-2) + C_0\mathbf{e}(t) + C_1\mathbf{e}(t-2) \tag{21}$$

where $\mathbf{x}^{\mathrm{T}} = (x(t), x(t-1))$ and the matrices A, C_0, C_1 are given as follows:

$$A = \begin{pmatrix} -a_2 + a_1^2 & a_1 a_2 \\ -a_1 & -a_2 \end{pmatrix} \tag{22}$$

$$C_0 = \begin{pmatrix} c_0 & c_1 - a_1 c_0 \\ 0 & c_0 \end{pmatrix} \quad C_1 = \begin{pmatrix} c_2 - a_1 c_1 & -a_1 c_2 \\ c_1 & c_2 \end{pmatrix} \tag{23}$$

The two eigenmodes of the matrix A correspond to the two slow decay modes in the kinetic model. From the arguments in the previous section, we can construct thermodynamics for the kinetic model of the zero point reactor. On the other hand, we can construct thermodynamics from time series data through ARMA(2,2) model. The two thermodynamics are related to each other. Some quick decay modes associated with precursor nuclei are coarse-grained in the measurement and the filtering. In the thermodynamics of ARMA(2,2), these decay modes are renormalized into the heatbath.

Now we consider the time evolution of the probability distribution in the ARMA(2,2) model. From linearity, we assume the Gaussian distribution

$$P(\mathbf{x}, t) = \frac{1}{\sqrt{\det(2\pi R(t))}} \exp[-\frac{1}{2}\mathbf{x} \cdot R(t)^{-1}\mathbf{x}]. \tag{24}$$

Then the time evolution of the covariance matrix gives us information about the distribution:

$$R(t) = AR(t-2)A^{\mathrm{T}} + C_0 C_0^{\mathrm{T}} + C_1 C_1^{\mathrm{T}} + C_1 C_0^{\mathrm{T}} A^{\mathrm{T}} + AC_0 C_1^{\mathrm{T}}. \tag{25}$$

When the parameters $\mathbf{a} = (a_1, a_2)$ and $\mathbf{c} = (c_0, c_1, c_2)$ are constant, the model has steady state distribution. We can obtain the covariance matrix by solving the following equation:

$$R_{\mathrm{s}} - AR_{\mathrm{s}}A^{\mathrm{T}} = C_0 C_0^{\mathrm{T}} + C_1 C_1^{\mathrm{T}} + C_1 C_0^{\mathrm{T}} A^{\mathrm{T}} + AC_0 C_1^{\mathrm{T}} \tag{26}$$

We will explicitly write it below.

(2) Thermodynamics

From the steady state distribution, the effective potential $U(\mathbf{x}, \mathbf{a}, \mathbf{c})$ and the free potential $F(\mathbf{a}, \mathbf{c})$ are defined as follows:

$$P_{\mathrm{s}}(\mathbf{x}, \mathbf{a}, \mathbf{c}) = \exp[F(\mathbf{a}, \mathbf{c}) - U(\mathbf{x}, \mathbf{a}, \mathbf{c})], \tag{27}$$

where $\mathbf{a} = (a_1, a_2)$, $\mathbf{c} = (c_0, c_1, c_2)$. Using the inverse of the covariance matrix R_{s}^{-1},

$$F(\mathbf{a}, \mathbf{c}) = -\log[\det(2\pi R_{\mathrm{s}}(\mathbf{a}, \mathbf{c})]/2,$$

$$U(\mathbf{x}, \mathbf{a}, \mathbf{c}) = \mathbf{x} \cdot R_{\mathrm{s}}^{-1}\mathbf{x}/2$$

where

$$\begin{aligned}
R_{\mathrm{s}11}^{-1} = R_{\mathrm{s}22}^{-1} &= (1 - a2)((1 + a2)(\mathbf{c} \cdot \mathbf{c}) - 2a1c1(c0 + c2) \\
&\quad + 2(a1^2 - a2 - a2^2)c0c2)/D, \\
R_{\mathrm{s}12}^{-1} = R_{\mathrm{s}21}^{-1} &= (1 - a2)(a1(\mathbf{c} \cdot \mathbf{c} - 2a2c0c2) \\
&\quad + (1 + a1^2 - a2^2)(-c0c1 + a1c0c2 - c1c2))/D, \\
D = (\mathbf{c} \cdot \mathbf{c} &- a2c0c2 + (1 + a1 - a2)(-c0c1 + a1c0c2 - c1c2)) \\
(\mathbf{c} \cdot \mathbf{c} &- a2c0c2 + (-1 + a1 + a2)(-c0c1 + a1c0c2 - c1c2)).
\end{aligned}$$

The first law

The first law is given as

$$\Delta U = W + D, \tag{28}$$

where the work W and the dissipation D are defined as integrations over the continuous time. The data and ARMA model are defined on the discrete time. We need to approximate the integration over the continuous time as a sum on the discrete time. We assume analyticity of the distribution functions with respect to time. Then we can approximate them using the Lagrange interpolation formula. In the second order approximation, the integrations are given using the trapezoid formula:

$$W \sim \sum_{t=0}^{T-1} \int d\mathbf{x}[U(t+1) - U(t)][P(\mathbf{x}, t+1) + P(\mathbf{x}, t)]/2, \tag{29}$$

$$D \sim \sum_{0}^{T-1} \int d\mathbf{x}[U(t+1) + U(t)][P(\mathbf{x}, t+1) - P(\mathbf{x}, t)]/2. \tag{30}$$

We may use the Simpson formula for a higher-order approximation.

The second law

In a quasi-static process, the distribution can be approximated as steady state for the parameters at time t. The work is given as the difference between the initial and final free potential,

$$W_{\mathrm{QS}} = F(T) - F(0). \tag{31}$$

We can estimate the work in a near-quasi-static process W_{ARMA} under the assumption of the validity of the ARMA model during the transformation. We have shown that the work in a near-quasi-static process is greater than in the quasi-static case,i.e.

$$\Delta W_{\mathrm{ARMA}} = W_{\mathrm{ARMA}} - W_{\mathrm{QS}} \geq 0. \tag{32}$$

See Appendix C.

In the nonstationary processes in the experiment, the simple ARMA model is not valid. Many other degrees of freedom are excited, so we expect more excess heat,

$$\Delta W_{\mathrm{EX}} = W_{\mathrm{EX}} - W_{\mathrm{QS}} \geq \Delta W_{\mathrm{ARMA}}. \tag{33}$$

Estimation of work

In the estimation of the work W_{EX} using the experimental data, we assumed: (1) the coefficients approximately depend on time linearly; and (2)the ensemble average can be approximated by a long time one, since the change of reactivity $\rho(t)$ is so slow. Then we can estimate the work as

$$W_{\mathrm{EX}} \sim \sum_{t} [\frac{dR_{\mathrm{s11}}^{-1}(t)}{dt} x(t)^2 + \frac{dR_{\mathrm{s12}}^{-1}(t)}{dt} x(t)x(t-1). \tag{34}$$

Similarly we can estimate W_{ARMA} using data from numerical simulation.

By analyzing the experimental and simulation data, we have obtained the following results.

ρ	$-2.19\$ \to -1.25\$$	$-0.588\$ \to -0.289\$$
W_{QS}	-0.8587	-0.8616
ΔW_{ARMA}	0.00007	0.00015
ΔW_{EX}	0.1035	0.1622
T	41556	14667

. (35)

These show our expected relation $\Delta W_{\text{EX}} \geq \Delta W_{\text{ARMA}} \geq 0$. Since ΔW_{ARMA} is so small, we estimated it using the scaling argument: $\Delta W_{\text{ARMA}} \sim 1/T$. ΔW_{EX} is about 100 times greater than ΔW_{ARMA}. Roughly speaking, this means that there exists 100 times longer time correlation in the nonstationary process.

3 Conclusion and Remarks

We consider thermodynamics as a general theory describing how a system responds to slow external transformations. We expect such a theory in biological or economical systems, in which random fluctuations are not thermal.

In this paper we have proposed an inductive thermodynamics from time series data analysis. It is based on two recent advances. One is the development of mathematical modeling from data analysis in statistical science and artificial intelligence. The other is the development of effective theories in physics for explaining the diversity of the nature. We have combined the auto-regressive moving average model and Sekimoto–Sasa theory.

We have applied our arguments to this actual nuclear reactor noise. Since detailed balance is broken in this system, we have extended Sekimoto-Sasa theory for the circulation of fluctuations. From the steady state distribution, we define the effective potential, which includes the house-keeping heat. We have confirmed the excess heat production or the second law from the data.

We show that the effective potential, determined from steady state distribution, plays a role of the energy in thermodynamics. We can introduce the concept of potential in biological or economical systems. But it is not so clear how important or useful these potentials are. It is interesting to construct the inductive thermodynamics of a system, in which real thermodynamics exists.

We are trying to construct it from time series of chemical data. Because of the development of technology we can directly measure position of a microscopic size bead or a DNA molecule in liquid [10]. It is important which scale we observe. It is an interesting question how inductive thermodynamics related with macro scale chemical thermodynamics.

Acknowledgement. We thank Professor S. Yamada for providing us the time series data of nuclear reactor noise. This work is partly supported by a Grant -in-Aid for Scientific Research on Priority Areas "Discovery Science" from the Ministry of Education, Science and Culture, Japan. We also deeply appreciate

Professor S. Miyano and Ms. A. Suzuki for taking care of us in the project of "Discovery Science". We thank K. Nelson for his useful comments.

A. Work in a Quasi-static Process

$$\begin{aligned} W_{\mathrm{QS}} &= \int_0^T dt \int d\mathbf{x} \frac{1}{2}\mathbf{x} \cdot \frac{dR_{\mathrm{s}}^{-1}(\alpha(t))}{dt}\mathbf{x} P_{\mathrm{s}}(\mathbf{x}, \alpha(t)) \\ &= \int_0^T dt\ \frac{1}{2}\mathrm{Tr}[\frac{dR_{\mathrm{s}}^{-1}(\alpha(t))}{dt} R_{\mathrm{s}}(\alpha(t))] \\ &= \int_0^T dt\ \frac{1}{2}\frac{d}{dt}\mathrm{Tr}[\log\{R_{\mathrm{s}}^{-1}(\alpha(t))\}] \\ &= \int_0^T dt\ \frac{1}{2}\frac{d}{dt}\log[\det\{R_{\mathrm{s}}^{-1}(\alpha(t))\}] \\ &= F(\alpha_T) - F(\alpha_0) \end{aligned}$$

B. Excess Heat Production in the System of the Langevin Equation

For $\Delta W = \int dt \delta W(t)$ and $\delta R(t) \equiv R(t) - R_{\mathrm{s}}(\alpha(t))$,

$$\delta W(t) = \frac{1}{2}\mathrm{Tr}[\frac{dR_{\mathrm{s}}^{-1}(\alpha(t))}{dt}\delta R(t)].$$

From the Fokker–Plank equation, we can derive

$$\frac{dR_{\mathrm{s}}^{-1}(\alpha(t))}{dt} = \delta R(t)^{-1}\{A - 2R_{\mathrm{s}}^{-1}(\alpha(t))\} + \{A^{\mathrm{T}} - 2R_{\mathrm{s}}^{-1}(\alpha(t))\}\delta R(t)^{-1}.$$

By substituting it into the first equation in Appendix B, we obtain

$$\delta W(t) = \frac{1}{2}\mathrm{Tr}[R_{\mathrm{s}}(\alpha(t))\delta R^{-1}(t)\{2R_{\mathrm{s}}^{-1}(\alpha(t)) - A\}R_{\mathrm{s}}(\alpha(t))\delta R^{-1}(t)].$$

Here we introduce four dimensional vector representation $\mathbf{R}^{\mathrm{T}} = (R_0, R_1, R_2, R_3)$ where

$$R_{\mathrm{s}}\delta R^{-1} = \begin{pmatrix} R_0 + R_1 & R_2 + R_3 \\ R_2 - R_3 & R_0 - R_1 \end{pmatrix}.$$

We can rewrite $\delta W(t)$ using 4×4 matrix representation of $2R_{\mathrm{s}}^{-1} - A$, M_{2R-A},

$$\delta W(t) = 2\mathbf{R} \cdot M_{2R-A}\mathbf{R}.$$

The four eigenvalues of the matrix M_{2R-A} are given as two degenerated ones, $\mathrm{Tr}A$ and $\mathrm{Tr}A \pm \sqrt{\mathrm{Tr}A^2 - \det A}$. Since $\mathrm{Tr}A > 0$ and $\det A > 0$, $\delta W(t) \geq 0$.

C. Excess Heat Production in the ARMA Model

The argument is almost same as in Appendix B. We can derive the following eqation from Eq.(25),

$$\delta R(t) \sim -\frac{dR_s(t)}{dt} + A(t-2)\delta R(t-2)A^{\mathrm{T}}(t-2).$$

By neglecting the time dependence of dR_s/dt and A within the correlation time,

$$\delta R(t) \sim -\sum_{n=0}^{\infty} A^n(t)\frac{dR_s(t)}{dt}(A^n(t))^{\mathrm{T}}.$$

By substituting it into the first equation in Appendix B, we obtain

$$\delta W(t) \sim \frac{1}{2}\sum_{n=0}^{\infty} \mathrm{Tr}[\frac{dR_s(t)}{dt}R_s^{-1}(t)A^n(t)\frac{dR_s(t)}{dt}(R_s^{-1}(t)A^n(t))^{\mathrm{T}}].$$

Since $dR_s(t)/dt$ has real eigenvalues and $\det A > 0$, $\delta W(t) \geq 0$.

References

1. Proceedings of the First International Conference on Discovery Science, eds. S. Arikawa and H. Motoda, LNAI **1532**, Springer-Verlag (1998) and references therein.
2. K. Sekimoto: J. Phys. Soc. Japan, **66** (1997) 1234;
K. Sekimoto and S. Sasa: ibid. 3326.
3. Y. Oono and M. Paniconi: Prog. Theor. Phys. Suppl, **130** (1998) 29;
K. Sekimoto: ibid. 17.
4. H.H. Hasegawa, T. Washio and Y. Ishimiya: LNAI **1721** Springer-Verlag (1999) 326; LNAI **1967** Springer-Verlag (2000) 304.
5. T. Hatano and S. Sasa: Phys. Rev. Lett., **86** (2001) 3463;
H. Tasaki and S. Sasa: Meeting (Kiken Kenkyukai) (2001).
6. K. Tomita and H. Tomita: Prog. Theor. Phys. **51** (1974) 1731;
K. Tomita,T. Ohta and H. Tomita: ibid. **52** (1974) 1744.
7. K. Kishida, S. Kanemoto, T. Sekiya and K. Tomita:
J. Nuclear Sci. Tech. **13** (1976) 161;
K. Kishida, N. Yamada and T. Sekiya:
Progress Nuclear Energy **1** (1977) 247.
8. H. Akaike: A New Look at the Statistical Model Identification, IEEE Transaction on Automatic Control **AC 19** (1974) 716;
H. Akaike: Prediction and Entropy, A Celebration of Statistics, eds. A.C. Atkinson and S.E. Fienberg, Springer-Verlag, New York (1985) 1;
G. Kitagawa: Changing Spectrum Estimation, Journal of Sound and Vibration **89** (1983) 433.
9. Tohn R. Lamarsh: Introduction Nuclear Reactor Theory, Addison–Wesley Publishing Company INC, (1974).
10. K. Yoshikawa, H. Mayama and K. Magome: in private communications.

Mining of Topographic Feature from Heterogeneous Imagery and Its Application to Lunar Craters

Rie Honda[1], Yuichi Iijima[2], and Osamu Konishi[1]

[1] Department of Mathematics and Information Science,
Kochi University, Akebono-cyo 2-5-1 Kochi, 780-8520, JAPAN
{honda, konishi}@is.kochi-u.ac.jp
http://www.is.kochi-u.ac.jp
[2] Institute of Space and Astronautical Science,
Yoshino-dai 3-1-1, Sagamihara Kanagawa, 229-8510, JAPAN
iijima@planeta.sci.isas.ac.jp

Abstract. In this study, a crater detection system for a large-scale image database is proposed. The original images are grouped according to spatial frequency patterns and both optimized parameter sets and noise reduction techniques used to identify candidate craters. False candidates are excluded using a self-organizing map (SOM) approach. The results show that despite the fact that a accurate classification is achievable using the proposed technique, future improvements in detection process of the system are needed.

1 Introduction

Recent advances in sensors and telemetry systems have increased the amount and quality of imagery available for researchers in fields such as astronomy, earth observation, and planetary exploration. However such advances have also increased the need for a large-scale database of scientific imagery and associated data mining techniques. [1][2][4][8][14][13].

Smyth et al.[13] and Burl et al. [1] developed a trainable software system that learns to recognize Venusian volcanos in a large set of synthetic aperture radar imagery taken by the spacecraft Magellan. A machine leaning approach was adopted because it is easier for geologists to identify feature examples rather than describe feature constraints. Experimental results showed that the system was able to successfully identify volcanos in similar imagery but performance deteriorated when significantly different scenes were used. Burl et al. also proposed an automated feature detection system for planetary imagery named Diamond Eye[2] which was applied to crater detection and showed a good performance, however, a difficulty similar with the previous study was expected.

Hamada et al.[6] reported on the automated construction of image processing techniques based on misclassification rate and an expert system composed of a large set of image processing modules.

S. Arikawa and A. Shinohara (Eds.): Progress in Discovery Science 2001, LNAI 2281, pp. 395–407.

In this paper, attention is focused on two difficulties in feature detection in optically observed image databases. The first is heterogeneity of image quality due to differences in illumination and surface conditions that affect the parameters included in the detection process. The second is the wide range of target feature sizes. For example, the diameter of lunar craters ranges from 1000 km to just 100 m (approximately equal to the size of several pixels in the object space).

A feature detection system for a large database of scientific imagery is proposed particularly focusing on detecting features with a wide range of sizes from large scale imagery of various quality at the best performance. The technique is applied to the detection of craters in lunar optical imagery.

2 System Overview

Craters are hollow features of varying size and shape and are frequently observed on solid planetary surfaces. Most craters were formed as a result of meteoroid impact. Their number and size distributions provide significant information about meteoroid activity in the past, the age and rheological properties of the planetary surface. Crater analysis has relied on human visual interpretation because of the difficulties in implementing efficient and accurate automation techniques.

In optical imagery, craters are generally recognized by shadows around the rim and represented according to the illumination conditions. Furthermore, image quality varies due to albedo, surface roughness, and illumination conditions, which further complicates the detection process.

Considering these difficulties, the following detection process is proposed: edge detection filtering, binarization, and circular pattern detection using Hough transforms or a genetic algorithm (GA). Concentrating on edge patterns reduces difficulties caused by changing illumination conditions. However, additional parameters such as the binarization threshold are introduced into the detection process and optimization of these parameters should be considered.

Thus the proposed crater detection system is descried as follows.

1. Clustering of original images.
2. Selection of representative image for each cluster and generation of teacher images by manually extracting features.
3. Optimization of detection process for each representative image by comparison with the result of 2.
4. Learning of candidate pattern for solution screening.
5. Detection of feature candidates and screening of unknown images using information obtained in 1 - 4.
6. Storage of extracted feature information in secondary database.
7. High level spatial pattern mining.

A schematic overview of processes 1 to 5 is shown is Figure 1. In this study, processes 1, 2, 3 and 4 are examined in detail and the effectiveness of integrated process evaluated by application to new imagery.

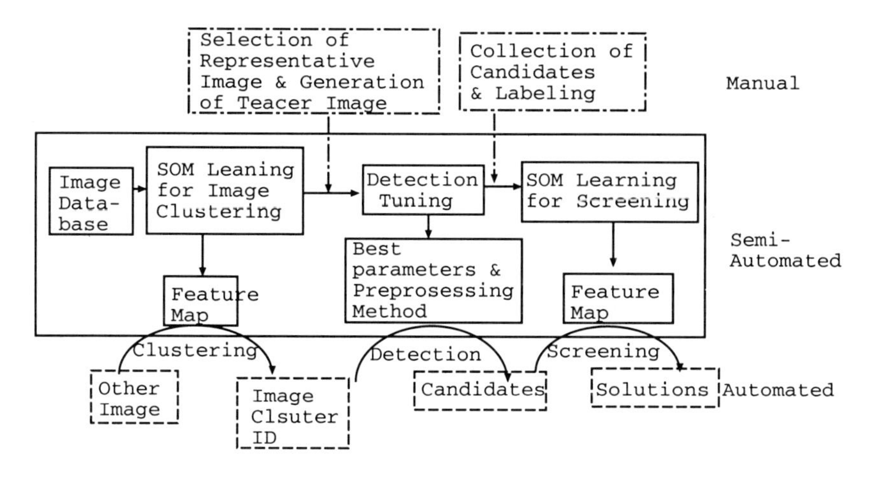

Fig. 1. System overview.

3 Candidate Detection

3.1 Crater Detection Method

In this section, the use of Hough transforms and genetic algorithm is shown as possible crater detection modules. The details of these techniques are described in the following.

Combinational Hough Transform. Hough transforms are used for the extraction of geometrically simple parametric figures from binary images[10]. For crater detection, the target parameters are the center and the radius of the crater rim. Firstly, the parameter space is divided into cells (bins). Probable parameter values (or trace) are calculated for each signal (white pixel) in a binary image assuming that the signal is a part of the figure, and the count of the corresponding cell is increased by one. After all signals are counted in the parameter space, parameter sets of the figures that exist in the binary image are obtained by extracting parameter cells whose count number exceeds a threshold.

Watanabe and Shibata [15] proposed combinational Hough Transform (CHT) that uses a pair of signals in a restricted region and multiresolution images to simplify projection into a parameter space. The results showed the use of a CHT reduced computation time and significantly improved the solution accuracy. Therefore, a CHT with additional noise reduction and other minor processes to improve accuracy is proposed for crater detection[9].

The algorithm of crater detection based on CHT are summarized as follows.

1. The original binary image are preprocessed by using some of the following methods: isolated noise reduction, expansion and shrinking, thinning by Hilditch's algorithm, pyramid-like signal reduction.
2. The image is degraded using the $W \times W$ pixel filter matrix.
3. The image is divided into the $L \times L$ pixels blocks.

4. The radius of the target circle is set to be $r = L/4$. The following process from 5 to 8 are proceeded increasing r by 1 while $r \leq L/2$.
5. The processes of 6 and 7 are performed for all blocks.
6. Among pairs of white pixels in the block extended by 50% ($2L \times 2L$ pixels), $P_{i1} = (x_{i1}, y_{i1})$ and $P_{i2} = (x_{i2}, y_{i2})$, the pairs that satisfy $r \leq |P_{i1}P_{i2}| < 2r$ are selected as signal candidates.
7. The center of the circle (x_{ic}, y_{ic}) is calculated for each pair assuming they exist on a circle rim with radius of r.
8. The count of the (x_{ic}, y_{ic}, r) cell in the parameter space is increased by 1.
9. The cells are sorted concerned with number of count. If the count is larger than 0, a circle of (x_{ic}, y_{ic}, r) is projected on the image, and the normalized count and the matching ratio are calculated. The definition of both values are given by

$$NC = count(x, y, r)/np^2, \tag{1}$$

$$M = np_w/np, \tag{2}$$

where NC is the normalized count, $count(x, y, r)$ is the count of the cell (x, y, r), np is the number of pixels on the rim of projected circle, M is the matching ratio, np_w is the number of white pixels of the rim of projected circle.
Furthermore, to exclude the false solutions caused by random noises, the internal noise ratio IN within the circle with the radius of hr is introduced, where $0 < h < 1$ (typically $h = 0.6$).
10. The cells satisfying $(NC > NC_{th}) \cap (M > M_{th}) \cap (IN < IN_{th})$ are extracted as the solutions, where NC_{th}, M_{th} and IN_{th} are the thresholds for NC, M, and IN, respectively.

Since the radius of circle is restricted by L, we utilize the multiresolution image of the original grayscale image to detect the circle with the radius larger than $L/2$. It should be noted that appropriate three threshold values and noise reduction methods must be chosen to optimize the performance.

Genetic Algorithm. Genetic algorithms are frequently used to obtain a single solution in optimization problems[5]. In order to implement such an algorithm for circular object detection, based on [12], a gene is set as a binary string that sequentially expresses a parameter set of (x_i, y_i, r_i), where (x_i, y_i) and r_i are the center and radius of the circle represented by the i-th gene, respectively. The fitness of the i-th gene, g_i, is calculated by projecting the circle represented by the i-th gene onto the binary image and checking its overlapping ratio, $g_i = n_i/N_i$, where n_i is the number of white pixels on the circle and N_i is the total number of pixels on the circle. In order to avoid random noise being incorporated into the solution, we modified g_i as follows:

$$g_i' = g_i - g_{i,r=fr_i}, \tag{3}$$

where $g_{i,r=fr_i}$ is the ratio of the white pixels on a circle with a radius of fr_i and $0 < f < 1.0$ (typically $f = 0.3$).

A variety of genes are then randomly produced and evolved through selection, crossing, and mutation. After iteration, genes with a fitness higher than the threshold are extracted as solutions.

Since it is possible to have many solutions (craters) in a single image, a process to unify similar genes and delete detected circles from the original images is introduced to improve the system's ability to detect multiple solutions[9]. After removal of solution circles, genes are newly generated and the process is iterated.
The algorithm of crater detection by GA is summarized as follows.

1. The original image is degraded using $W \times W$ pixel filter matrix.
2. Initial populations of genes are generated.
3. The following process from 4 to 6 are iterated for a given number of generations.
4. The fitness of genes, g'_i, are calculated.
5. The genes are selected, crossed, and mutated.
6. The genes with the same attributes are unified.
7. The genes with $g'_i > g'_{th}$ are detected as solutions, where g'_{th} is the threshold of g'.
8. The solutions are projected as circle rims on the image. The intensity of pixels on the projected circle rims are changed to 0 (black).
9. The processes from 2 to 8 are iterated for a given number of times.

The proposed algorithm also includes several parameters (e. g. g'_{th} or mutation rate) that affect solution accuracy and optimization of these parameters is dependent on image quality.

3.2 Optimization of the Detection Process

Preliminary results of tests using the above method have indicate that noise and signal gaps have a significant deterioration effect on detection accuracy[9]. The optimization of parameters such as the binarization or count threshold is effective technique for improving accuracy, however, the optimized values are dependent on image quality.

The following optimization process is suggested: (1) cluster source images, (2) select representative image from each group, (3) produce teacher image by manual visual recognition, (4) optimize crater detection process by comparing results from teacher images and the result from the corresponding original image. The details of these sub-processes are provided in the following section.

Clustering of Frame Images. It is suggested that the rough grouping of images with respect to image quality is an effective technique for simplifying optimization of the detection process. In this study, the clustering of original images using Kohonen's self-organizing maps (SOM) [11] was examined.

SOM is an unsupervised learning algorithm that uses a two-layer network of input layer and competition layers, both of which are composed of units with n-th dimensional vectors. SOM effectively maps the similar pattern of the input layer on the competitive layer. In the SOM algorithm, the distance (usually Euclidean) between the input vector and each unit vector of the competition layer is calculated and the input vector is placed into the winner unit, which has the smallest distance. At the same time, the unit vectors in the cells adjacent to the winner cell (defined by the neighborhood distance) are modified so that they move closer to the input vector. As a result of this iterative projection and learning, the competitive layer learns to reflect variation of the input vectors and can obtain adequate clustering of the input vectors. Presently, SOM is widely used for the clustering, visualization and abstraction of unknown data sets.

Selection and preprocessing of input vectors is crucial to improve SOM accuracy. In order to group lunar images according to roughness or contrast, the FFT power spectrum of normalized images is adopted as the input vector.

After clustering, a representative image, which has the largest similarity with the unit vector and also includes many craters, is selected for each unit cell (cluster). Then craters are marked in each representative image and binary teacher images generated (see Figure 2(a) and 2(b)).

Method of Optimization. The detection process is divided into three parts for optimization purposes: binarization, preprocessing including noise reduction, and circle detection. These processes are optimized sequentially using teacher images.

Firstly, edge detection filtering is carried out on the original images. Then, optimal binarization threshold that produce a binary image most similar to the teacher image is identified for each cluster. Based on [6], the evaluation function is defined by

$$\begin{aligned} E_k = \; & P(T_k(i,j) = S_k(i,j) \mid T_k(i,j) = 1) \\ & -\alpha P(T_k(i,j) \neq S_k(i,j) \mid T_k(i,j) = 0), \end{aligned} \quad (4)$$

where k is the cluster ID, $T_k(i,j)$ is the intensity of the (i,j) pixel of teacher binary image, α is a weight parameter (typically $\alpha = 0.3$), and $S_k(i,j)$ is the intensity of (i,j) pixel of the final binary image defined by

$$S_k(i,j) = \begin{cases} 1 & \text{for } Q_k(i,j) > Q_{th,k} \\ 0 & \text{for } Q_k(i,j) \leq Q_{th,k}, \end{cases} \quad (5)$$

where $Q_k(i,j)$ is image intensity after edge detection filtering and $Q_{th,k}$ is the binarization threshold. The value of $Q_{th,k}$ is searched greedily to maximize E_k.

Next the combination of preprocessing methods that maximizes true positive detection rate (ratio of detected true solutions to all possible true solutions) of craters defined by $Pr_k = N_k/Nt_k$ is identified, where N_k and Nt_k are the numbers of craters detected from the binary image and the teacher binary image for cluster k, respectively.

Finally, the circle detection parameters that maximize Pr_k for the preprocessed image using selected methods is identified. Figure 2 shows a schematic view of the optimization process.

As shown in Figure 2(d), extracted solutions may include many false solutions, which will be excluded in the post-processing stage described in the following section.

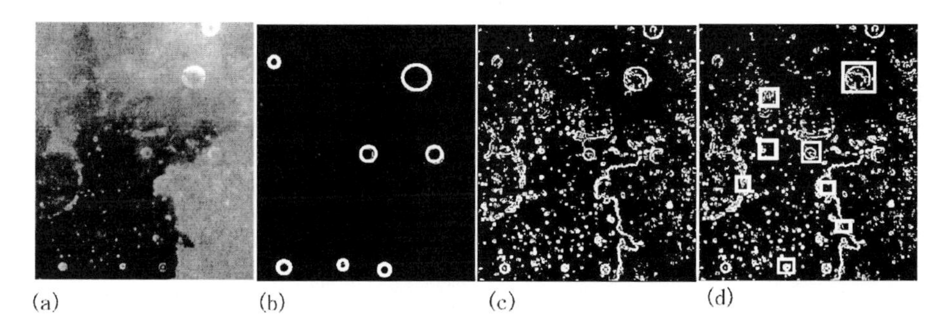

Fig. 2. Schematic view of optimization process. (a), (b), (c), and (d) show the original image, teacher image, tuned binary image, and results of detection, respectively. White squares in (d) indicate the extracted candidates.

3.3 Screening of Solutions

A solution screening process is used in the post-processing stage to exclude false solutions. Candidate crater images are cut out, normalized with respect to its size and intensity, and visually labeled true or false. The candidate pattern is learned by SOM taking the normalized intensity vectors or FFT power spectrum as the input vectors. Each unit in the competition layer is labeled either true or false by evaluating the ratio of candidates in it. If we assume that the properties of the new data set are similar to those of the studied data set, the class (true or false) of the newly detected candidate is decided by projecting it onto the SOM feature map.

4 Experiments

4.1 Description of Data Set

A total of 984 medium browse images from Lunar Digital Image Model (LDIM), which had been mosaiced by the U. S. Geological Survey based on the lunar global images obtained by the U. S. Clementine spacecraft. The images were between 322 and 510 pixels in width and 480 pixels in height, and resampled at a space resolution of approximately 500 m/pixel using a sinusoidal projection. Images in the polar regions were not used to avoid distortions due to the map projection. The radius of target craters ranged from 2 to 18 pixels.

For clustering of original images, an area of 256×256 pixels was extracted from the center of the normalized images, and the FFT power spectrum calculated as the input vectors of the SOM. The size of the SOM competition layer was defined as 4×4 units because only a rough grouping was needed. One hundred images were sampled and the SOM leaning process iterated 100000 times. All images were then projected onto the competition layer, and the unit cell vectors adjusted using K-means method[3]. No images with extremely large distance were identified in this process.

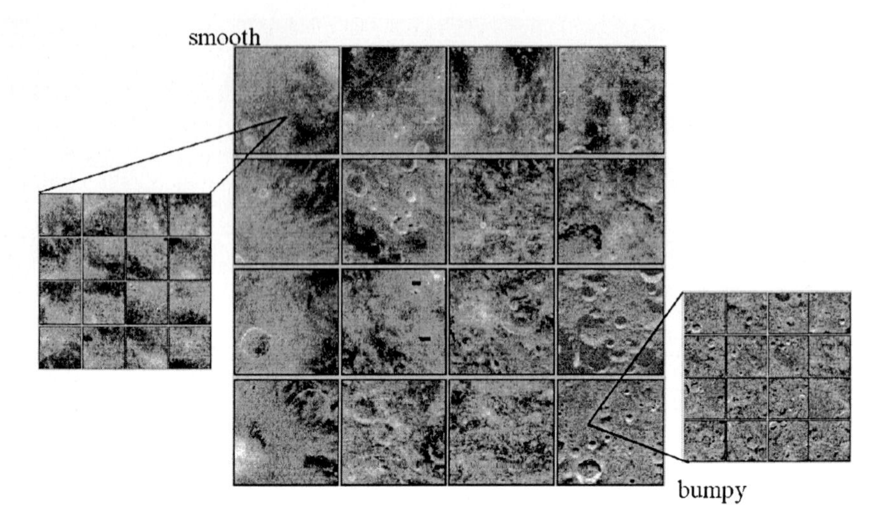

Fig. 3. SOM feature map for clustering of original images. Cluster ID 0, 1, 2, ..., and 15 are for each cell from the upper left corner to the lower right corner in raster-scan order.

4.2 Result of Image Clustering

Figure 3 shows the resulting competition layer, hereafter denoted the feature map. In this map, the image with the smallest distance with each unit vector is displayed in each cell to visualize the clustering result. It can be seen that relatively smooth images including the Mare recognized by a dark region, are clustered on the left side, and the rugged terrains called the Highland with many clearly identifiable craters are clustered in the lower right corner. This result indicates that learning by SOM successfully distinguishes between variations in image quality and groups them effectively. Based on the clustering result, a representative image was manually selected and a teacher binary image produced for each cluster.

4.3 Result of Detection Optimization

Table 1 shows an example of optimized parameters for CHT case, in which the result of noise reduction method are common to both CHT and GA.

The binarization threshold ranged from 30 to 125 and binary images that approximated the teacher images were produced automatically. It is suggested that a single threshold value for the entire image will not be adequate in some cases because of spatial variations within the image, and that this problem should be solved at the pre-processing stage.

For the optimization of the noise reduction processes, 12 combinations of four noise reduction methods was examined: thinning by Hilditch's algorithm[7], pyramid-like signal reduction, isolated noise reduction, and expansion and shrinking. Result of experiments showed that combination of thinning and isolated

Table 1. Teacher Image list together with optimized parameters for CHT case. Abbreviations TH, PY, IS and ES represent thinning by Hilditch's algorithm[7], pyramid-like signal reduction, isolated noise reduction, and expansion and shrinking, respectively.

ClusterID	ImageID	No. of target crater	Q_{th}	Noise reduction method	NC_{th}	M_{th}	IN_{th}
0	bi03n051	9	30	IS+TH	0.55	0.30	0.47
1	bi10n093	10	70	IS+TH	0.45	0.20	0.47
2	bi45n065	11	50	TH+PY	0.55	0.25	0.80
3	bi31n063	12	55	IS+TH	0.55	0.60	0.47
4	bi17n039	6	25	TH+PY	0.55	0.60	0.80
5	bi38s335	35	55	ES	0.45	0.45	0.80
6	bi17s225	19	85	IS+TH	0.45	0.60	0.80
7	bi38s045	44	95	IS+TH	0.35	0.60	0.80
8	bi17s027	12	95	IS+TH	0.55	0.60	0.47
9	bi03n285	15	85	IS+TH	0.45	0.50	0.80
10	bi31s039	20	125	IS+TH	0.55	0.40	0.47
11	bi80s015	36	70	IS+PY	0.45	0.20	0.80
12	bi03s039	10	80	IS+TH	0.55	0.20	0.80
13	bi38s155	13	90	IS+TH	0.55	0.35	0.30
14	bi45s335	22	105	IS+TH	0.45	0.25	0.80
15	bi59s052	59	95	IS+TH	0.35	0.20	0.30

noise reduction scored the highest positive solution detection rate for most cases (12 cases in 16). Thus we applied combination of thinning and isolated noise reduction to images of all clusters in the following process for simplicity.

Figure 4 summarizes the results of the optimization of the detection process for both CHT and GA techniques as performance curves represented by true positive detection rate as functions of the false solution number. In general, a decrease in threshold leads to an increase in both the true positive detection rate and the number of false solutions. Figure 4 shows that true positive detection rate increases with number of false solutions when the number of false solutions is relatively small, however, it remains constant for lager numbers. Thus the initial point of the flat portion of the performance curve is considered to be the optimum performance condition. In most cases, this coincides with the point that minimizes the false number and maximizes the positive rate.

Figure 4 also shows that CHT performs significantly better than GA. This is mainly caused by the fact that GA's are used to obtain a single solution. Although the GA was modified to obtain multiple solutions, the results show that the GA can acquire only a few solutions per trial even for teacher binary images that include clearly identifiable circles, and requires many iterations to acquire multiple solutions. Thus, CHT was selected as the circle detection module suitable for crater detection.

After optimization, the CHT positive solution detection rate increased significantly up to values ranging from 0.2 to 0.6. However, since information was lost

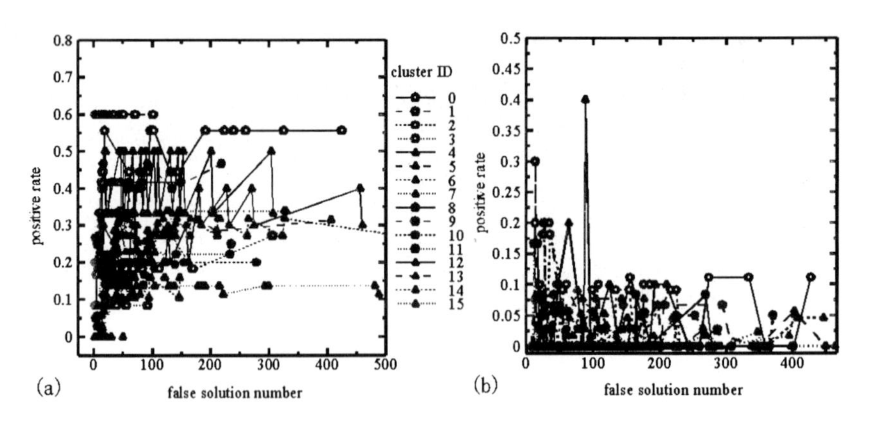

Fig. 4. Optimization result for CHT(a) and GA(b).

during the binarization process, it will be necessary to consider other detection methods for some clusters to further improve detection performance.

4.4 Result of Screening of Detected Candidates

SOM clustering of crater candidates extracted in the previous process was performed for screening purposes to produce the candidate classifier. A total of 646 candidates were visually labeled either true or false. Half of the candidates were randomly sampled for learning and the remainder were used for examination purpose. The percentage of true candidates for both groups was 25.4% and 27.2%, respectively. All images were rotated such that the direction of sunlight incidence was equal, and normalized with respect to intensity and size. Two types of input vectors were examined: image vectors represented by pixel intensities aligned in a raster-scan order, and the FFT power spectrum. The size of the SOM competition layer was set to 6×6 units by trial and error and 323000 iteration were performed. The neighborhood distance at iteration t is given by $2(1 - t/323000)$.

Figure 5 shows an example feature map obtained after SOM learning. Cells enclosed by thick frames contain more than 50% true solutions and hence were labeled true candidate cells. The remainder are labeled false cells. Figure 5 shows a cluster of true candidate cells in the upper right corner. To examine SOM classification ability, the precisions for true cells and false cells (ratio of actual true solutions to the solutions predicted true and vice versa), and accuracy (ratio of solutions predicted correctly to all data) were calculated from 5 trials.

Table 2 summarizes the results of both learning from the study data and clustering for the test data using the map with the best precision for true. The result shows that learning using image vectors was more accurate than that using the FFT power spectrum and classified candidates with an accuracy of 89.7%, which is much higher compared with the value of 78.4% for FFT power spectrum.

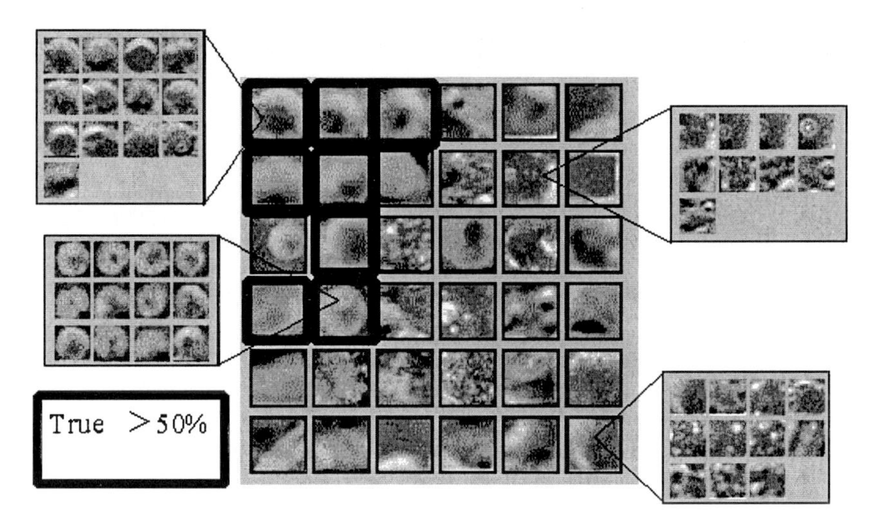

Fig. 5. Example of SOM feature map for classification of crater candidates. Input vectors are set to be image vectors.

Table 2. Result of SOM learning for crater candidate screening.

Input data	Case	Precision(true)	Precision(false)	Accuracy
	study	0.812±0.037	0.931±0.009	0.897±0.016
Image Vector	best/study	0.855	0.946	0.929
	test	0.776	0.891	0.864
	study	0.691±0.037	0.803±0.096	0.784±0.013
FFT power spectrum	best/study	0.733	0.788	0.783
	test	0.605	0.768	0.755

The most accurate map classified the unknown data with an accuracy of 86%. This indicates that the utilization of SOM feature map learned from image vectors is an effective technique for the classification of solution candidates. It should also be noted that selecting the most suitable map from the trials is important to improve classification accuracy because performance varied significantly according to the initial conditions.

5 Application to Other Imagery

The effectiveness of the proposed technique for crater detection was examined using imagery that had not been used in the optimization process. In addition, multiresolution images were used to handle craters with a wide range of sizes. By setting the radius of target crater ranging from 9 to 18 pixels, it was possible to detect craters with a radius up to 72 pixels using the multiresolution images of three levels.

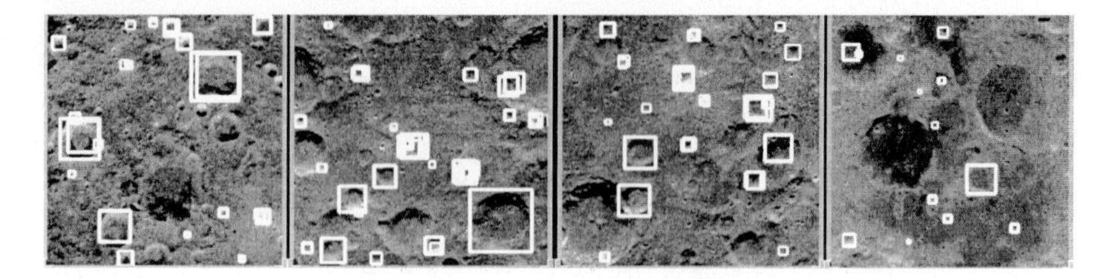

Fig. 6. Example of crater detection for other general images. The images are 480 pixels in height.

Figure 6 shows examples of detection and screening results for four images. It can be seen that detection ability is improved significantly even without manual operations. Unfortunately, the achieved detection rate is not for scientific analysis (e. g. detection rate required to be much more than 60%), thus other detection methods should be considered for some groups and the selection of circle detection modules should also be included in future work. However, our methods is considered to be sufficient for some applications such as autonomous spacecraft tracking in that high precision and moderate detection rate are required.

6 Conclusions

A technique for mining features from sets of large scale of optical imagery of various quality has been proposed. The original images were grouped according to spatial frequency patterns, and optimized parameter sets and noise reduction methods were used in the detection process. Furthermore, to improve solution accuracy, false solutions were excluded using SOM feature map that learned true and false solution patterns from a large number of crater candidates. Application of the extracted information to new imagery verified effectiveness of this approach.

The detection rate achieved in this study, however, is not sufficient in comparison with the requirements for scientific analysis and it is necessary to include other detection methods in the future work. However, we believe combining automated abstraction and summarization processes with the accurate manual techniques is crucial for the development of an accurate scientific data mining system. The proposed technique is applicable to various applications in which specific features need to be extracted from large-scale of imagery databases.

Acknowledgements

The authors are grateful to Syuta Yamanaka and Hisashi Yokogawa for their past contribution amd anonymous reviewers for helpful comments. This research is supported by a grant-in-aid for intensive research (A) (1) (Project 1130215) from the Ministry of Education, Culture, Sports, Science and Technology of Japan.

References

1. Burl, M.C., Asker, L., Smyth, P., Fayyad, U.M., Perona, P., Crumpler, L., Aubele, J.: Learning to recognize volcanos on Venus. Machine Learning, **Vol. 30**, (2/3) (1998) 165–195
2. Burl, M. C. et al., Mining for Image Content, Systems, Cybernetics, and Informatics / Information Systems: Analysis and Synthesis, (Orlando, FL) (1999)
3. Duda, R. O. and Hart, P. E., Pattern Classification and Scene Analysis, John Willey and Sons, New York (1973)
4. Fayyad, U.M., Djorgovski, S.G., Weir, N.: Automatic the analysis and cataloging of sky surveys. Advances in Knowledge Discovery and Data Mining, AAAI Press/MIT Press (1996) 471–493
5. Goldberg, D. E., Genetic Algorithm in search optimization and machine learning, Addison Wesley, Reading (1989)
6. Hamada, T., Shimizu, A., Hasegawa, J., Toriwaki, J.: A Method fo Automated Construction of Image Processing Procedure Based on Misclassification Rate Condition and Vision Expert System IMPRESS-Pro (in Japanese), Transaction of Information Processing Society of Japan, **Vol. 41**, No. 7 (2000) 1937–1947
7. Hilditch, C. J. Linear skeletons from square cupboards, Machine Intelligence 4, Edinburgh Univ. Press, Edinburgh, (1969) 403-420
8. Honda. R., Takimoto, H, Konishi, O.: Semantic Indexing and Temporal Rule Discovery for Time-series Satellite Images, Proceedings of the International Workshop on Multimedia Data Mining in conjunction with ACM-SIGKDD Conference, Boston, MA (2000) 82–90
9. Honda, R., Konishi, O, Azuma, R, Yokogawa, H, Yamanaka, S, Iijima, Y : Data Mining System for Planetary Images - Crater Detection and Categorization -, Proceedings of the International Workshop on Machine Learning of Spatial Knowledge in conjunction with ICML, Stanford, CA (2000) 103–108
10. Hough, P. V. C., Method and means for recognizing complex patterns, U. S. Patent, 069654 (1962)
11. Kohonen, T.: Self-Organizing Maps, 2nd eds., Springer (1997)
12. Nagase, T. and Agui, T. and Nagahashi, H., Pattern matching of binary shapes using a genetic algorithm (in Japanese), Transaction of IEICE, **Vol. J76-D-II**, No. 3, (1993) 557-565
13. Smyth, P., Burl, M.C., Fayyad, U.M.: Modeling subjective uncertainty in image annotation. In Advances in Knowledge Discovery and Data Mining, AAAI Press/MIT Press (1996) 517–539
14. Szalay, A., Kunszt, P. Thakar, A., Gray, J. Slutz, D. and Brunner, R.J. : Designing and Mining Multi-terabyte Astronomy Archives: The Sloan Digital Sky Survey, Proceeding ACM-SIGMOD International Conference on Management of Data, Dallas TX (2000) 451–462
15. Watanabe, T. & Shibata, T., Detection of broken ellipse by the Hough transforms and Multiresolutional Images (in Japanese), Transaction of IEICE, **Vol. J83-D-2**, No. 2, (1990) 159–166

Application of Neural Network Technique to Combustion Spray Dynamics Analysis

Yuji Ikeda and Dariusz Mazurkiewicz

Kobe University, Rokkodai, Nada, Kobe 657-8501, Japan
ikeda@mech.kobe-u.ac.jp
http://www.ms-5.mech.kobe-u.ac.jp

Abstract. This paper presents results from several analytical and empirical combustion process investigations using data mining tools and techniques. An artificial neural network was used to analyze the performance of data in phase Doppler anemometry (PDA) and particle image velocimetry (PIV) which can measure droplet size and velocity in combustion spray. The dataset used for the analysis was obtained from measurements in a practical combustion burner. The preliminary results are discussed, and improvements to the neural network architecture are suggested. The inclusion of additional input variables and modified data pre-processing improved the results of the classification process, providing a higher level of accuracy and narrower ranges of classified droplet sizes.

1 Introduction

Mankind has used combustion since the beginning of recorded history, and it is now used as a light source, for heating and power generation, and in engines [Glassman, 1987]. Even though the combustion process itself remains unchanged, our knowledge and understanding of its uses have improved, from the primitive ideas of times past to the highly specialized, advanced technological applications of today. Unfortunately, our knowledge and understanding of the process is not yet complete: owing to its complexity, several problems and topics remain unsolved, even using modern research tools.

Typical combustion research is focused on the physics of the process and on specific applications, such as diesel engines, particulate matter, aeroplane engines, power generation, incinerating plants, and the reduction of combustion-generated pollutants (for example, dioxin and CO_2). Combustion research is necessary, not only for the sake of science itself, but also to solve practical engineering problems so that the results can be applied to benefit society.

Current combustion research uses sophisticated laser diagnostic techniques. While the large amounts of data collected can be used to infer new knowledge about combustion, the sheer volume of data is also a handicap, as some information about the combustion process remains hidden inside huge databases. We tried to apply suitable analytical processes and research methods, with the purpose of describing their advantages and limitations. In addition, we started

S. Arikawa and A. Shinohara (Eds.): Progress in Discovery Science 2001, LNAI 2281, pp. 408–425.

to search for new solutions. Unfortunately, the number of tools and methods presently available was insufficient to fully analyse such data. Therefore, data mining (DM) techniques were proposed, to detect, for example, large-scale structures from gasoline injector spray data. Several other typical sub-topics of combustion research are also described in the present work, and an algorithm for analysing combustion processes using DM is proposed. Finally, artificial neural networks were used to classify and predict combustion data, as a possible method for future process control, monitoring, and optimisation.

2 Typical Limitations of Combustion Experiments

Combustion processes are extremely complicated and complex phenomena. A better understanding of these processes is necessary to meet strong industrial and social needs, and this requires new experimental research methods in sub-topics such as fluid mechanics, heat and mass transfer, chemical kinetics, turbulence, and thermodynamics [Kuo, 1986; Lefebvre, 1989]. The newest and most sophisticated research methods being used for this research include laser diagnostic techniques (Fig. 1) and computer tomography.

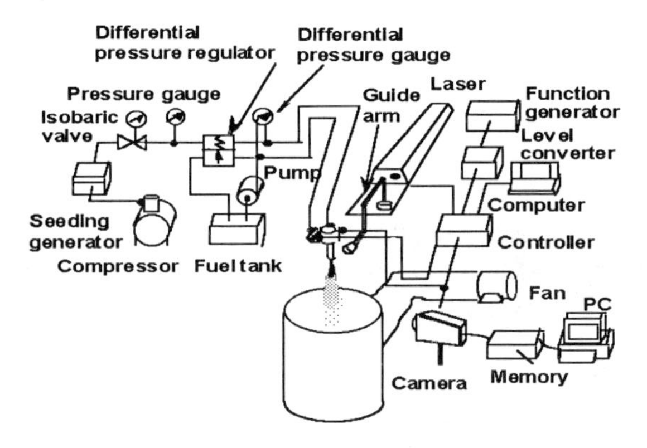

Fig. 1. Example of typical experimental apparatus for combustion researches with laser techniques

2.1 Laser Diagnostic Techniques

The most popular laser diagnostics techniques are phase Doppler anemometry (PDA) and particle image velocimetry (PIV) [Lauterborn and Vogel, 1994; Roy G., 1998]. These optical techniques are widely applied in fluid mechanics and combustion research to obtain measurements of velocity and density fields. PDA can be classified as a flow-visualization technique, which makes use of light scattered by particles in the fluid. Using laser Doppler anemometry (LDA), fluid velocity can be measured as a function of time with a degree of accuracy, but unfortunately it can be measured only at a single point in the fluid at any given time. On the other hand, PIV is an experimental technique that was developed to measure instantaneous fields of scalars and vectors [Adrian, 1994; Chigier, 1991; Glassman, 1987]. It can also be described as a velocity-measuring technique that measures the motion of small, marked regions of a fluid. In PIV, a pulsed sheet of light illuminates the particles in a fluid. PIV and PDA provide different information about the measured process; both provide detailed information about droplet velocities. PDA measures the arrival and transit time of each single droplet, as well as its diameter, but spatially resolved information is not available. In addition to measuring velocities, PIV also provides flow visualization data, but droplet size cannot be determined owing to the planar measurement technique. Both PIV and PDA can be used to obtain a great deal of valuable information about combustion processes; however, because of the imperfections of PIV and PDA, additional measurement techniques should also be employed.

Laser-based measurement techniques have been used previously to investigate spray dynamics and combustion characteristics for various flow fields (Fig. 2) [Ikeda et al., 1996, 1997, 1999, 2000]; for example, we measured a gasoline injector spray using PIV. This improved our understanding of instantaneous spatial spray structures, which are important in automotive applications such as gasoline direct injection (GDI) engine development, where it is necessary to know how small the spray droplets are and how they should be distributed. PIV measurements can demonstrate differences in droplets' response to turbulent motion,

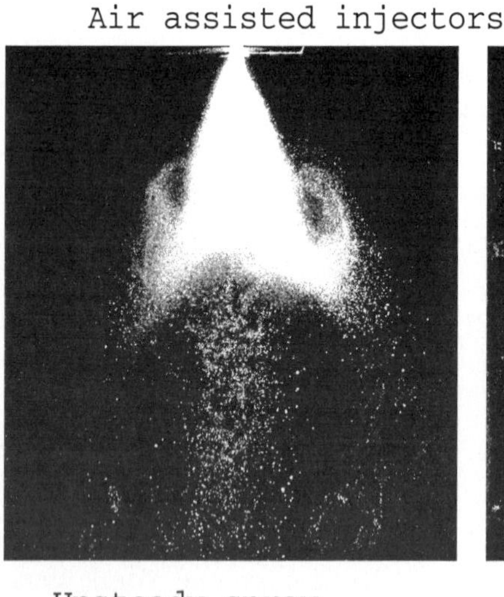

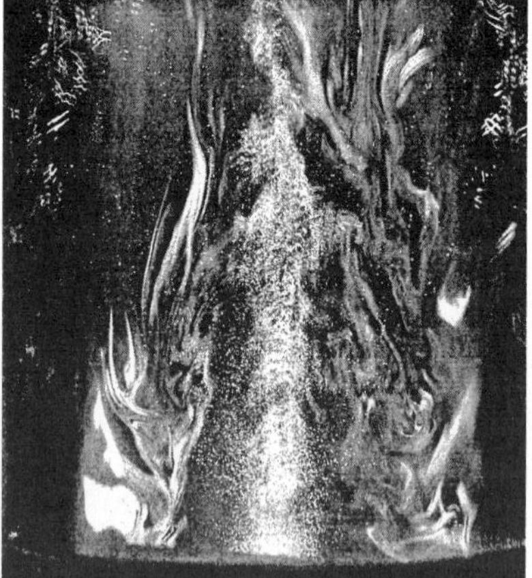

Fig. 2. Examples of combustion researches with the use of laser diagnostic techniques

and provide information about groups of droplets that penetrate the flow field. Spray combustion is used in most practical internal combustion engines, and combustion research has led to higher thermal efficiencies and lower emissions. Laser diagnostic techniques are widely used to investigate the behavior and interaction between droplets and air. Point measurement methods have been used to investigate droplet velocities and diameters at very high data acquisition rates. For this reason, planar measurement (PIV) is used to obtain instantaneous spatial velocity distributions. One target of the present research is to develop a PIV droplet size classification system that is based on the intensity distribution of the source image. This specialized aspect of this research is also an example of how a source image can be separated into multiple images.

2.2 Problems Using Laser Techniques for Combustion Research

There are many unanswered questions in combustion research. For example, how does one observe the flame and the fluid flow? How does one construct proper measurement tools? How does one collect data? How does one analyze the data and control the process in real-time? And finally, how can combustion systems be optimized? Some of these are general research problems that are similar to those found in other experimental disciplines, where some solutions already exist that can be implemented without any special problems. However, in the case of combustion processes, there are additional, more complicated problems from an experimental point of view (Fig. 3), each involving different limitations:

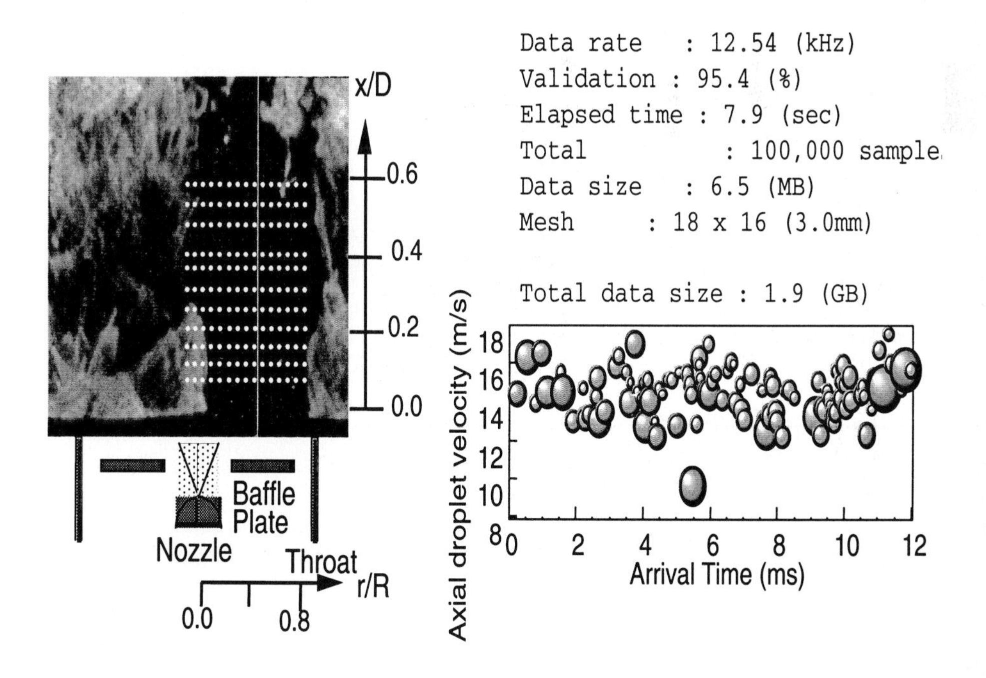

Fig. 3. Raw data and their source

- The combustion process is unsteady, and is characterized by different physical and chemical time scales.
- Combustion experiments are limited by the resolution of the measuring equipment; thus, the area of data acquisition is limited.
- Reactions during the process are extremely fast.
- Measured data are discrete (figs. 4 and 5).
- The experimental databases are very large.
- Combustion itself is a complex phenomenon that includes several different sub-topics, such as fluid mechanics, heat and mass transfer, thermodynamics, etc.
- The combustion flow is in random and irregular motion.
- The combustion flow is a mixture of gases, solids, and liquids, the properties of which must also be considered.
- The time required to process data is still too slow for on-line monitoring, control, or optimization.
- The investigation of combustion processes is complex and involves interactions between many different scientific disciplines.

This list includes only the basic problems and their related limitations; such a list can be extended almost to infinity. Most of these problems have not yet been solved. Therefore, combustion research is complicated by the number of parameters that must be considered, such as differences in time scales and the discrete nature of the data. These problems make it necessary to search for new solutions beyond typical combustion research methods. The time and expense spent on experimental data acquisition using laser techniques, coupled with unsatisfactory data analysis based on statistical methods, forced us to investigate and apply new modeling tools and algorithms. The results of the present research suggest that solutions can be found using data mining (DM), knowledge discovery in databases (KDD), and other similar computer science (CS) tools.

When searching for the most appropriate methods and tools, we considered DM, and especially artificial neural networks, as particularly suitable for analyzing data obtained from laser diagnostics. Previous applications, including comparisons of experimental data and the results predicted by neural networks, confirmed the accuracy of our neural network approach to experimental data analysis [Ikeda and Mazurkiewicz, 2000, 2001].

3 Typical Examples of DM Tools and Methods Used for Combustion Research

Data mining (DM), which can be defined as the process of exploiting and analyzing large quantities of data by automatic or semi-automatic means [Berry and Gordon, 1997], has applications in several different disciplines. There are many technologies and tools available for data mining, some of which can be used, for example, to improve marketing campaigns or operational procedures, while others aid with examining medical records or process monitoring. Until now, DM has been used mostly in business, but it has applications in other fields, too, including engineering.

DM is one of several terms, including knowledge extraction, information harvesting, data archaeology, and data dredging, which describe the concept of knowledge discovery in huge databases [Westphal and Blaxton, 1998]. DM is of interest to many researchers, in fields such as pattern recognition, artificial neural networks (artificial intelligence), machine learning, database analysis, statistics, expert systems, and data visualization. Many DM techniques can also be successfully applied to combustion research problems, particularly those related to experiments that use laser techniques and require further data analysis, process optimisation, monitoring, or control.

DM tools, especially artificial neural networks (ANN), have proved particularly suitable for analyzing combustion data and modeling empirical processes. For example, an algorithm has been used to detect laminar and turbulent flows, providing both qualitative and quantitative information about the flow fields [Nogawa et al., 1997]. DM has also been used in research on three-dimensional vortex extraction in turbulent fluid flows, in which classification of the fluid motions as well as vortex structure detection in the boundary layers of turbulent flows were performed [Zhong et al., 1998]. ANN has been used to predict and describe combustion processes, with the aim of providing a temporal evolution model of a reduced combustion chemical system [Blasco et al., 1998]. Homma and Chen [2000] optimized the post-flame combustion process with genetic algorithms, achieving a reduction in NO_2 emissions. Neural networks have been used for laser-induced thermal acoustic data analysis [Schlamp et al., 2000] and for a combustion control system for utility boiler applications [Allen et al., 1993]. Wei [2001] proposed a discrete singular convolution (DSC) algorithm for spatial discretization of incompressible fluid flow, and Wu et al. [2001] developed an intelligent identification method for the regime of oil-gas-water multiphase flows, using ANN. All of these applications prove that DM tools are very promising solutions for several combustion data analysis and modeling problems.

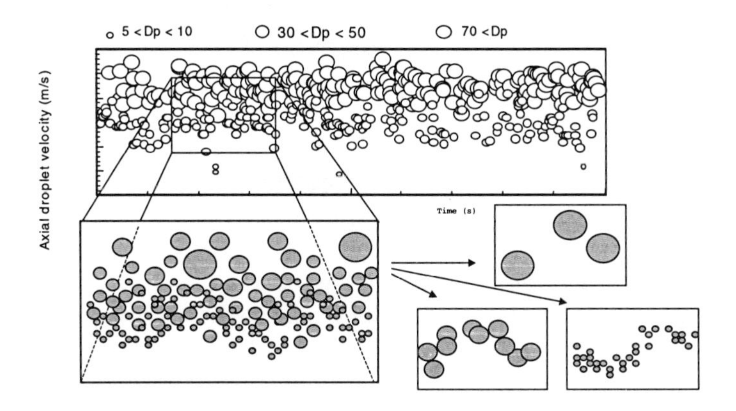

Fig. 4. Examples of spray droplets behavior passing at the measurement point

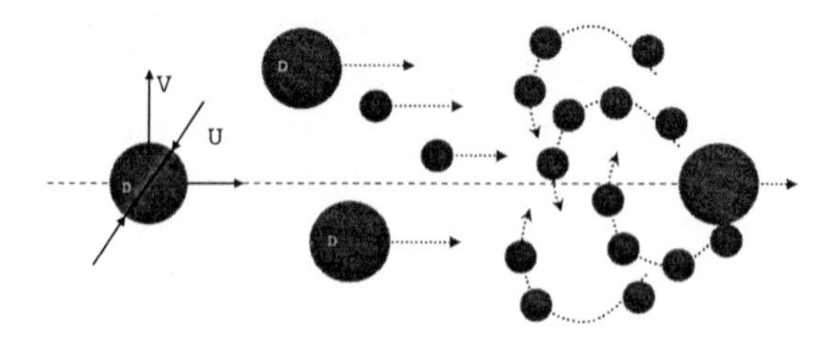

Fig. 5. Examples of droplets behavior in a fluid (simplified model)

4 DM Tools and Methods for PIV-PDA Experiments

DM techniques have previously been used as a knowledge-based method for experimental data analysis in droplet size detection [Ikeda and Mazurkiewicz, 2000, 2001]. A special algorithm for combustion process analysis was used to describe the main steps that must be followed (Fig. 5). The aim was to obtain more sophisticated information about the combustion process from PIV and PDA measurements, for possible future application in combustion process control, monitoring, and optimisation. A neural network was used, either to classify droplets into reasonable size ranges or to predict their future appearance, from PDA data. The same network was then used to analyse PIV data, where droplet size is unavailable. In this manner, additional information about droplet size was obtained from the PIV data, after using the PDA dataset to train the network [Ikeda and Mazurkiewicz, 2000]. This is explained in more detail below. The possibility of using classification trees to perform the same analysis was also discussed, to find solutions to combustion measurement problems by using the most effective DM tools. The results obtained using ANN were improved by increasing the number of input variables and modifying the data pre-processing step during further investigations and analysis.

4.1 Model Development

Information obtained previously from PDA experiments was used to classify or predict the size of droplets, using PIV data (Fig. 6). There is a choice of PDA input variables, such as U_{vel}, V_{vel}, the arrival time, the travel time, and the droplet size (Table 1). Initially, PIV predictions based on PDA input information were analysed. The output variable was D, the droplet diameter (Fig. 7). If the network is trained to predict droplet diameters based on proper PDA inputs, then the same network can be used for predictions from PIV data, which do not contain droplet size information. The aim of such modeling is to create a neural network and then check its ability to classify PIV droplet sizes based on a PDA training dataset. Previous analysis of these datasets identified some empirical relations between the output (droplet diameter) and the inputs (velocity) (Fig. 8). For effective modeling of PIV/PDA droplet size detection using ANN, the following basic questions must be answered:

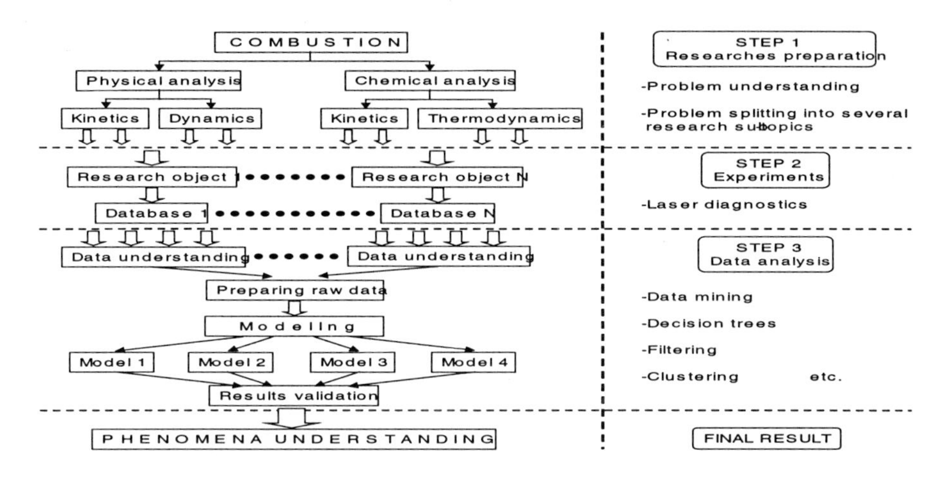

Fig. 6. Droplet size classification by multi-intensity-layer

- What type of network architecture should be used?
- What type of training method should be employed?
- What is the acceptable network error?
- What droplet size-classification rate should be used?
- Which methods and activities should be included in the data pre-processing step?

Table 1. Example element of PDA raw data sheet

PT	$A.T.$(sec)	$T.T.$(msec)	U_{vel}(m/s)	V_{vel}(m/s)	$D(\mu$m)
0	0.000257	0.002600	-21.447094	26.161095	9.223717
1	0.000326	0.007800	-17.935289	-10.095972	20.292175
2	0.000368	0.003900	-33.688240	-6.755212	3.689487
3	0.000385	0.006500	-24.958899	-11.275064	11.068460
4	0.000453	0.005200	-31.330315	-3.217937	6.764059
5	0.000647	0.020800	-8.152405	0.908883	21.522005
6	0.000719	0.016900	-4.590432	2.284490	33.820293
7	0.000773	0.003900	-4.941612	26.848898	6.149144
8	0.000792	0.003900	-6.898189	33.383030	6.149144
9	0.000798	0.003900	-6.547709	24.785488	0.614914
10	0.000828	0.011700	-7.550382	23.704653	3.074572
11	0.000837	0.005200	-0.125423	30.287914	4.304401
12	0.000939	0.003900	1.580310	24.785488	3.689487
13	0.000940	0.005200	-5.593805	37.657238	3.689487
14	0.000949	0.003900	-6.898189	30.287914	1.844743
15	0.000952	0.001300	-3.888071	29.944014	11.683374

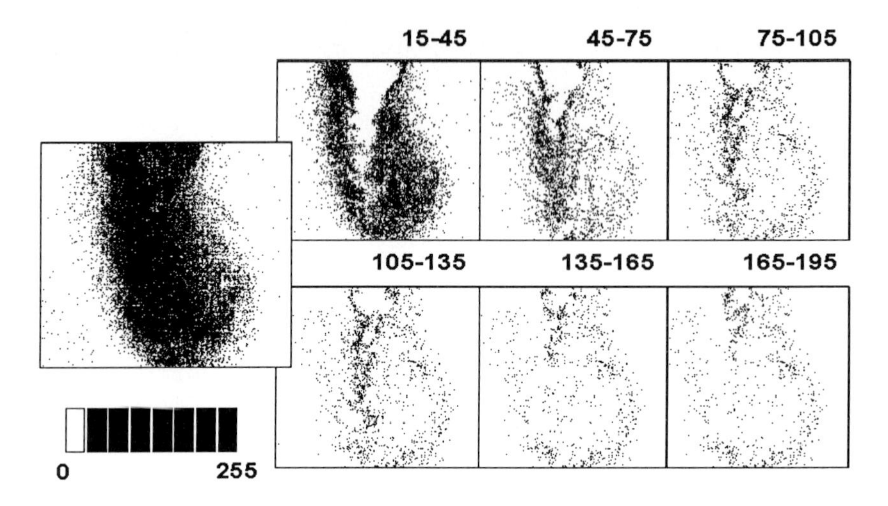

Fig. 7. Droplet size classification by multi-inteisty-layer

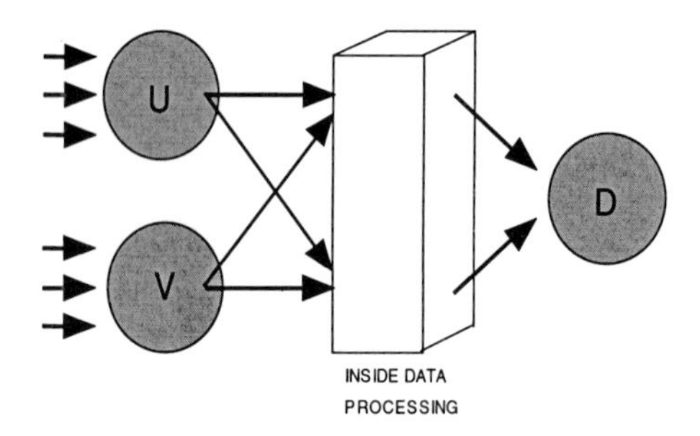

Fig. 8. Simplified scheme of neural network architecture for droplet size classification

Only very basic network architectures were used to obtain results in the present case, without spending a lot of time on sophisticated training methods that have higher degrees of accuracy. The aim was not to perform a time-consuming analysis, but simply to verify the basic ability of the technique to provide the information necessary for further investigation of combustion.

4.2 Data Pre-processing

To obtain satisfactory results, a representative training dataset that includes information about typical relations between the input and output variables must be prepared in a suitable format. If a supervised training process is used, the network must be shown data from which it can find correct relations that properly describe combustion phenomena. The data cannot be random, since further training or classification predictions largely depend on finding the correct relations among the data. Therefore, once a research problem has been defined, regions of data must be identified with patterns that have the proper values for further modeling. In other words, the data must be selected to include the necessary information related to the defined problem. Proper training datasets must be prepared: raw experimental datasets cannot be used directly for neural network training.

Since combustion itself is too complicated to use as a basis for training datasets, it must be divided into several well-defined modeling sub-problems (Table 2). This step is complicated and time consuming, requiring a good knowledge of combustion and its experimental datasets, as well as the format of the data. Pre-processing also includes PDA raw data extraction. During a single PDA experiment used in this study, 5 variables were collected at 264 points 10,000 times. A single dataset therefore included 13,200,000 values, from which typical training data had to be selected. An example of a data flowchart for such a case is shown in Fig. 9.

Table 2. Definition of basic modeling targets for ANN modeling

Target	Input variables	Output variables
To classify droplet appearance basing on information about previous droplets	TT, AT, U_{vel}, V_{vel}	D
To predict droplet appearance basing on information about previous droplets	TT_{-n}, AT_{-n}, U_{vel-n}, $V_{vel-n} \cdots$	D
To predict droplet size basing on previous droplet diameters	$D_{-n}, \cdots$	D

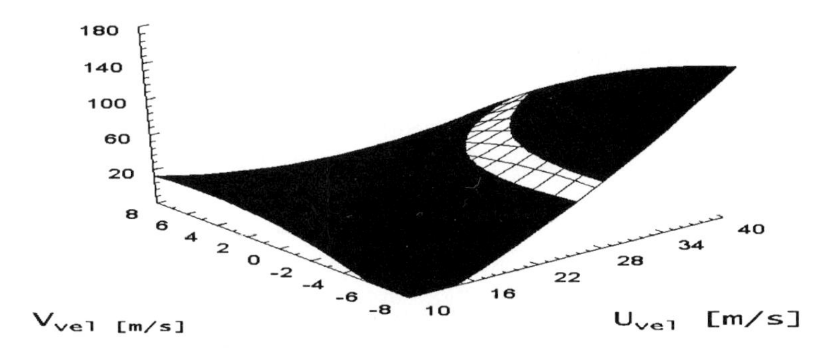

Fig. 9. Surface plot of droplet size, as a function of its velocities

A raw laser diagnostic experimental dataset is too large to be used for ANN modeling, and the training process takes too long. Raw datasets also have improper formatting that does not meet ANN requirements. Some information about droplet behavior is stored more than once in each dataset, and most of the droplets in each dataset have diameters that fall within only one range. Thus, the raw dataset includes irrelevant data that should be eliminated, according to the problem definition to avoid complications or errors in classification.

For neural network training, the same number of representative droplets should be selected from each range of droplet size. This selection must be performed separately for each previously defined modeling problem. Thus, target datasets must be created by selecting the proper data from the entire database. If the appropriate data are used, the training process will be more effective. However, the proper elements must be included in each target dataset. If the data are to be analysed as a time series, then the same number of input and output variables should be prepared for the same type of information. If the available PIV input information consists only of U_{vel} and V_{vel}, then irrelevant PDA data that are also available need not be used for training (such as the travel and arrival time of each droplet). The input and output variables for both the PDA training set and the PIV classifications or predictions should have the same format. In the case of neural network modeling, the output variable can have both categorical and quantitative formats. When a neural network is used, the data format depends on the defined modeling problem and on the results expected.

4.3 ANN Modeling Results

The ANN was trained with the examples available using pre-processed data to reveal and represent some complex relationships within the data. Various combinations of a number of hidden layers and neurons were tested for different architectures, and several sets of independent data were used for training. The main goal was to minimize the sum of the squares of the output value errors. The Multi-layer Perceptron (MLP) with the Levenberg-Marquardt algorithm performed the best for supervised training. The network architecture was relatively simple, including only two hidden layers with 4 and 3 neurons, respectively. Relatively good correspondence was obtained between the ANN predictions and actual data. In most tests, the network correctly identified 65-75% of the objects analysed (Fig. 10). This network proved to be a very effective method for droplet size classification.

MLP networks are probably the best type of feed forward networks to use in this type of research problem. Each neuron in the hidden layer sums up its input signals (x_i) after first weighting them according to the strength of the respective connections (w_{ji}) from the input layer. It then computes the output (y_j) as a function of the sum:

$$y_j = f(\sum w_{ji}x_i) \tag{1}$$

In this case, f was sigmoid function,

$$f(s) = 1/(1 + \exp(-s)) \tag{2}$$

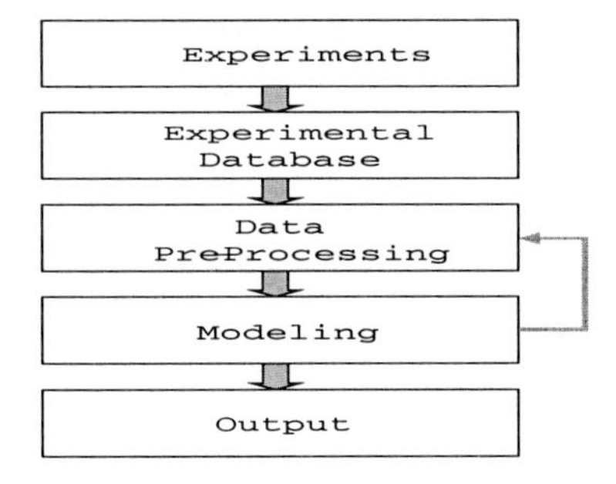

Fig. 10. Example of data flow from experiment to required output

The performance of the ANN improved when the Levenberg-Marquardt optimization method was used rather than back-propagation. This method is an approximation of Newton's method:

$$\triangle W = (\boldsymbol{J}^T \boldsymbol{J} + \mu I)^{-1} \boldsymbol{J}^T e \tag{3}$$

where $\boldsymbol{J}$ is the Jacobian matrix of the derivatives of each error to each weight, is a scalar, and e is an error vector. If the scalar is very large, the above expression approximates gradient descent. When it is small, the expression becomes the Gauss-Newton method [Demuth et al., 1996].The Levenberg-Marquardt algorithm for supervised training is more effective than conjugate gradient descent for this type of problem, especially when networks with single outputs are used. It is well known that feed forward networks with one or two hidden layers and a sufficient number of neurons that use sigmoid activation functions (or, in some cases, threshold functions) are universal approximators [Hornik et al., 1999].

Several other applications for this type of modeling are possible in combustion research. It is expected that in this manner a set of new algorithms will be created for data analysis, which will extract previously unknown information from databases. This will help to solve several of the present research problems. The results show that better definition of research problems and pre-processing of data can increase the number of correct classifications found.

4.4 Improvement of Modeled Results

The theory of artificial neural networks and previous experience in combustion research suggest that the results obtained can be improved. For the present network, results are 10% closer to actual values than results obtained using the preliminary approach described above (Fig. 11). The new results were obtained by changing the number of input (independent) variables, modifying the network architecture, and improving the data pre-processing.

Modification of Network Architecture. For future applications of DM, the most important parameters are training data and network architecture, and their

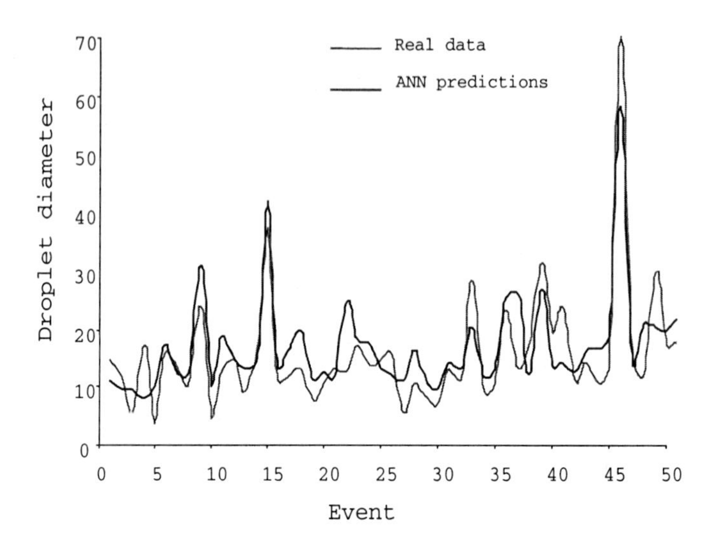

Fig. 11. Droplet size real data (red line) and their ANN predictions (blue line)

selection. The size of the input layer was very important in this investigation. The output layer always consisted of just one neuron, owing to the goal of the investigation (droplet size classification or prediction), but the number of input neurons could be changed, depending on the information desired and on the data available when training the network. Thus, the size of the input layer depends on the amount of information available about the process from the independent variables measured during the experiment. For PDA experiments, the available parameters include the U and V velocities, the arrival time of each droplet (ΔT), and the travel time of each droplet (T). The fluid dynamics can be described from this information. In the preliminary tests, the neural network consisted of only two neurons in the input layer, which represented the two main independent variables - the U and V velocities. This produced good results with simple network architecture and a short training time. Different network architectures using different numbers of hidden layers and units could be tested in a relatively short time. Even though the network considered was quite simple, the correspondence observed between its predictions and actual data was relatively good.

Our knowledge of the combustion process and experience of several experiments using laser techniques prompted us to find new solutions for neural network design. The number of independent variables need not be restricted to two. If several parameters are measured with PDA, why not use all of them for ANN modeling? What would the result be if we showed the network all the available information about the process, increasing the number of input layer neurons? To answer these questions, additional independent parameters were added to the input layer of our network. The output layer remained unchanged as a single output neuron (Fig. 12).

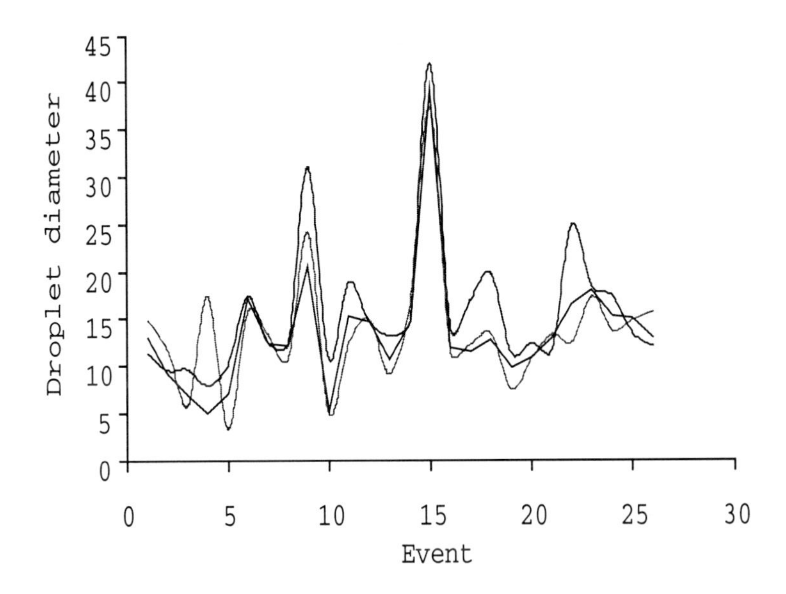

Fig. 12. Droplet size real PDA data (red line), and their ANN predictions: with previous network (blue solid line), with new improved network (blue dot line)

The number of input variables that could be used depended on the definitions of basic combustion modeling problems [Ikeda and Mazurkiewicz, 2000]. There were three possibilities:

- To obtain the best droplet size classification or prediction results, the basic mechanical properties (U, V, ΔT, TT, and L) could be used.
- To predict droplet appearance, including size information, a group of previously measured droplets (U_{-n}, V_{-n}, TT_{-n}, ΔT_{-n}, D_{-n}, $\cdots$ U_{-1}, V_{-1}, TT_{-1}, ΔT_{-1}, D_{-1}) could be used. The influence of previous droplets could be examined in this manner.
- To predict droplet size based on the diameters of previous droplets, D_{-n}, $\cdots$ D_{-2}, D_{-1} could be used.

These three possibilities were verified using further ANN modeling. When the input layer variables described in the second and third possibilities were included in the modeling, the training time became extremely long, resulting in complications and poor performance. In addition, it was difficult to determine how many previously measured droplets should be considered. Thus, the number of previous droplets that influenced the present droplet could not be calculated using this method of modeling. The network classifications and predictions were more effective for the first type of analysis. The only problem was with the last input variable, the measurement point location. When this was included in the input layer, the training process became very long and the results were unsatisfactory. Therefore, this parameter was excluded from the input layer and considered during data pre-processing instead. The best results were obtained when U, V,

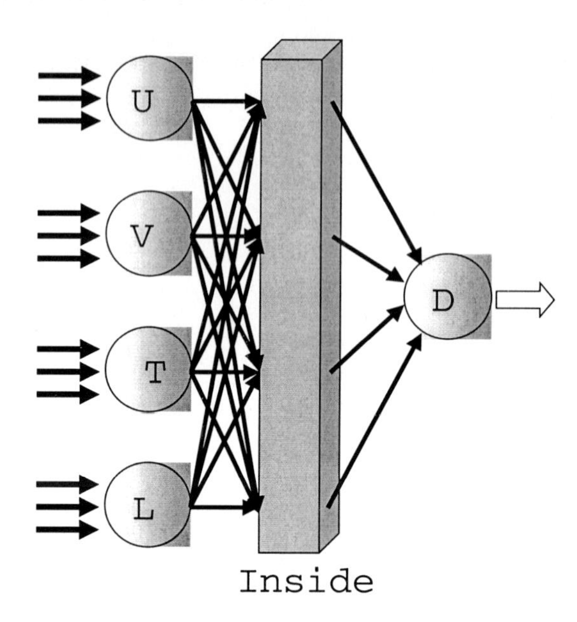

Fig. 13. Example of improved neural network architecture for PDA droplet size classification

and ΔT were used as the input-independent variables. Such a network included three input neurons and one output neuron. Once this improvement was made, it was noted that the droplet appearance time intervals greatly affected the final results.

Modification of Data Pre-processing. Artificial neural networks are a good choice for most classification and prediction tasks [Fayyad et al., 1996; Groth, 2000], but they do not work well when there are too many input features. A large number of features makes it more difficult for the network to find patterns, and can result in extremely long training phases that never converge in a good solution [Berry and Linoff, 2000; Groth, 2000]. This was observed in the previous section. Unfortunately, data were collected from 264 points during our typical PDA measurements, and each point had its own characteristics that influenced the final results of the modeling. In addition, the sheer volume of input information caused several problems. The effectiveness of the training process was improved by choosing a group of representative points from the 264, selected according to their location within the most typical and important phenomena of the combustion process. Such choices must be based on experience and knowledge of the process under consideration. Separate networks are required for each selected measurement point, which greatly increases the number of networks used to describe a single problem. However, these additional networks will have simple architectures and short learning times compared to preparing a single, huge network to analyze the entire problem. It is better to analyze a defined problem step by step, obtaining partial output information, and then slowly increase the complexity of the problem. This also provides valuable information for future multiple modeling studies.

The number of points selected and the resulting neural networks depends on how deeply the process must be analysed, and on how many steps should be included in the investigation.

There is also one other advantage to modifying the data pre-processing in such a manner. The modeling method and the number of points used for analysis can be chosen according to the output required. Thus, in this case the value of the analysis to an understanding of the combustion process could be increased by providing additional analysis of droplet behaviour at different locations, without making use of more complicated neural networks.

5 Conclusions

Combustion data analysis is extremely difficult, and very often introduces problems that are impossible to solve using conventional analytical methods. Numerical simulations based on algorithms of combustion data analysis are very important. Artificial neural networks can predict or classify the outcomes of complex problems accurately, and can therefore be used for modeling combustion processes, especially when analyses of experimental data or predictions are required. When the algorithms are trained using suitably selected data, good results are obtained 60-65% of the time. The use of neural networks in combustion research requires knowledge about the actual combustion process, combined with some practical experience. In our analysis, artificial neural networks were successfully applied to obtain PIV/PDA droplet size classifications and predictions. The preliminary models and architectures were later modified to obtain better results: the size of the input layer was changed, and new data pre-processing was introduced based on experience from previous investigations. The new results were 10% closer to actual values than neural networks tested previously. This type of analysis both improved network performance and solved several combustion data analysis problems.

New information was obtained about the combustion process, and the analysis became both easier and more effective. Models of neural networks can also be applied in combustion research to improve the real-time systems for control, monitoring, or optimisation that will be required by industry in the 21st century.

References

1. Adrian R.J., 1994 - Particle-Imaging Techniques for Experimental Fluid Mechanics. Selected Papers on Particle Image Velocimetry edited by Grant I., SPIE Optical Engineering Press, Washington, pp. 43-64.
2. Allen M. G., Butler C. T., Johnson S. A., Lo E. Y. and Russo R., 1993 - "An Imaging Neural Network Combustion Control System for Utility Boiler Applications". Combustion and Flame, vol. 94 (1993), pp.: 205-214.

3. Beronov K.N., Ishihara T. and Kaneda Y., 1998 - Numerical Studies of Large Scale Turbulent Flows in Two and Three Dimensions. Oxford Kobe Seminars, The Second Meeting on Systems Engineering and Applied Mathematics. St. Catherine's College - University of Oxford Kobe Institute, 1998, pp. 145-151.
4. Berry M.J.A., Linoff G.S., 1997 - Data Mining Techniques for Marketing, Sales and Customer Support. John Wiley & Sons Publishing, New York, 1997.
5. Berry M.J.A., Linoff G.S., 2000 - Mastering Data Mining - The Art and Science of Customer Relationship Management. Wiley Computer Publishing, New York, 2000.
6. Blasco J. A., Fueyo N., Dopazo C., Ballester J., 1998 - modeling the Temporal Evolution of a Reduced Combustion Chemical System With an Artificial Neural Networks". Combustion and Flame, Nr. 113 (1998), pp.: 38-52.
7. Chigier N. (Ed.), 1991 - "Combustion Measurement". Hemisphere Publishing Corporation, New York 1991.
8. Demuth H., Beale M., 1996 - "Neural Network Toolbox For Use with MATLAB". The MathWorks, Inc. 1996.
9. Fayyad U. M., Piatetsky-Shapiro G., Smyth P., Uthurusamy R. (Editors), 1996 - "Advances in Knowledge Discovery and Data Mining". MIT Press 1996.
10. Glassman I., 1987 - "Combustion". Academic Press, San Diego 1987.
11. Groth R., 2000 - Data Mining - Building Competitive Advantage. Prince Hall PTR, New Jersey, 2000.
12. Homma R., Chen J.-Y., 2000 - "Combustion Process Optimization by Genetic Algorithms: Reduction of NO2 Emission via Optimal Post-flame Process". The Twenty Eight International Symposium on Combustion, Edinburgh, Scotland, August 2000.
13. Hronik K. M., Stinchcombe M., White M., 1999 - "Multi-layer feedforward networks are universal approximators". Neural Networks 1999, 2 (5), pp.: 356-366.
14. Ikeda Y., Mazurkiewicz D., 2000 - "Combustion Experimental Data Preparation For Data Mining". International Conference of the Japanese Society for Artificial Intelligence (JSAI) - A04 and A05 Group Meeting. Senri Life Science Center, Osaka (Japan), 26-28 October 2000, pp.: 79-84.
15. Ikeda Y., Mazurkiewicz D., 2000 - "The First Step for Combustion Diagnostic - Problem Understanding and Viewing". International Conference of the Japanese Society for Artificial Intelligence (JSAI) - A04 Group Meeting. Kochi, Shikoku Island (Japan), 13-15 July 2000, pp.: 57-62.
16. Ikeda Y., Mazurkiewicz D., 2000 - "Data Mining Aided Analysis In Laser Diagnostic Experiments". Discovery Science 2000, Progress Report Edited by S. Arikawa, M. Sato, T. Sato, A. Marouka, S. Miyano, Y. Kanada. The 3rd International Conference DS'2000, Kyoto (Japan), 4-6 December 2000, pp.: 88-90.
17. Ikeda Y., Mazurkiewicz D., 2000 - "The First Step for Combustion Diagnostic - Problem Understanding and Viewing". International Conference of the Japanese Artificial Intelligence Society - A04 Group Meeting. Kochi, Shikoku Island (Japan), 13-15 July 2000, pp.: 57-62.
18. Ikeda Y., Mazurkiewicz D., 2001- "Neural Network for PDA Droplet Size Classification". Japanese Society for Artificial Intelligence (JSAI A04 Group). March 2001, pp.: 215-221.
19. Ikeda Y., Nakajima T., Kurihara N., 1996 - "Spray formation and dispersion of size-classified fuel droplet of air-assist injector". Eighth International Symposium on Applications of Laser Techniques to Fluid Mechanics. Lisbon, Portugal, July 8-11 1996.

20. Ikeda Y., Nakajima T., Kurihara N., 1997 - "Size-Classified Droplet Dynamics and its Slip Velocity Variation of Air-Assist Injector Spray". SAE Technical Paper Series 1997 (paper no. 970632).
21. Ikeda Y., Yamada N., Nakajima T., 2000 - "Multi-intensity-layer particle-image velocimetry for spray measurement". Measurement Science and Technology, Vol. 11, Nr. 6, June 2000.
22. Kawahara N., Ikeda Y., Hirohata T., Nakajima T., 1996 - "Size-Classified Droplets Dynamics of Combustion Spray in 0.1 MW Oil Furnace". Eighth International Symposium on Applications of Laser Techniques to Fluid Mechanics. Lisbon, Portugal, July 8-11 1996, pp. 10.5.1-10.5.12.
23. Kuo K. K., 1986 - "Principles of Combustion". A Wiley-Interscience Publication, New York 1986.
24. Lauterborn W., Vogel A., 1994 - Modern Optical Techniques in Fluid Mechanics. Selected Papers on Particle Image Velocimetry edited by Grant I., SPIE Optical Engineering Press, Washington, pp. 3-13.
25. Lefebvre A. H., 1989 - "Atomization and Sprays". Hemisphere Publishing Corporation, New York 1989.
26. Mazurkiewicz D., 1999 - "Data Mining - new concept in data processing". Maintenance and Reliability - Journal of the Polish Academy of Science Branch in Lublin and Polish Maintenance Society. Nr 2/1999, pp.: 24-30.
27. Nogawa H., Nakajima Y., Sato Y., Tamura S., 1997 - "Acquisition of Symbolic Description from Flow Fields: A New Approach Based on a Fluid Model". IEEE Transactions on Pattern Analysis and Machine Intelligence, Vol. 19, No. 1, January 1997, pp.: 58-62.
28. Roy G. (Ed.), 1998 - "Propulsion Combustion. Fuels to Emissions". Combustion: An International Series. Taylor and Francis, New York 1998.
29. Wei G.W., 2001 - A new algorithm for solving some mechanical problems. Computer Methods Appl. Mech. Engrg. 190 (2001), pp. 2017-2030.
30. Westphal Ch., Blaxton T, 1998 - Data Mining Solutions - Methods and Tools for Solving Real-world Problems. Wiley Computer Publishing, New York, 1998.
31. Wu H., Zhou F., Wu Y., 2001 - Intelligent identification system of flow regime of oil-gas-water multiphase flow. International Journal of Multiphase Flow 27 (2001), pp. 459-475.
32. Zhong J., Huang T. S., Adrian R. J, 1998. - "Extracting 3D Vortices in Turbulent Fluid Flow". IEEE Transactions on Pattern Analysis and Machine Intelligence, Vol. 20, No. 2, February 1998, pp.: 193-199.

Computational Analysis of Plasma Waves and Particles in the Auroral Region Observed by Scientific Satellite

Yoshiya Kasahara, Ryotaro Niitsu, and Toru Sato

Graduate School of Informatics, Kyoto University
Yoshida, Sakyo-ku, Kyoto 606-8501, JAPAN
kasahara@kuee.kyoto-u.ac.jp

Abstract. In this paper we introduce new computational techniques for extracting the attributes and/or characteristics of the plasma waves and particles from the enormous scientific database. These techniques enable us to represent the characteristics of complicated phenomena using a few key-parameters, and also to analyze large amount of datasets in a systematic way. These techniques are applied to the observational data of the ion heating/acceleration phenomena in the auroral region obtained by the Akebono satellite. Finally we evaluate our algorithms through some correlation analyses between waves and particles and demonstrate their availability.

1 Introduction

The region so-called magnetosphere is the surrounding space of the Earth, filled with highly ionized plasmas and pervaded with the strong magnetic field of the Earth. The Sun is a main energy source which controls the environment of the magnetosphere. For example, the solar wind brings enormous energized particles from the Sun into the magnetosphere, where there take place heating and acceleration of plasma, various kinds of plasma waves, wave-particle interactions, wave-wave interactions, etc.

Akebono is a Japanese scientific satellite, which was launched in 1989 for precise observations of the magnetosphere. Akebono is successfully operated for more than 12 years and the accumulated data amounts to about 1.5 Tbytes of digital data and about 20,000 audiotapes of analogue data. In the field of conventional geophysics, research is mainly performed by event study in the following way; (1) a scientist discovers new phenomena from observational results as an evidence of a new theory, or (2) a scientist tries to explain a very unique observation result theoretically. In the analysis of physical phenomenon from the data observed by one satellite, however, it is quite difficult to tell whether we see time variation, spatial distribution or mixture of them, because the observational data from scientific satellite are taken by means of direct measurement along the trajectory. Hence it is indispensable for investigating these phenomena to examine as large amount of data as possible. Our aim is to develop new

S. Arikawa and A. Shinohara (Eds.): Progress in Discovery Science 2001, LNAI 2281, pp. 426–437.

computational techniques for extracting the attributes and/or characteristics of the plasma waves and particles from the enormous data sets of Akebono in a systematic way and to discover epoch-making knowledge.

Ion heating/acceleration transverse to the geomagnetic field line (TAI) in the auroral region is one of the most important phenomena in the magnetosphere because it is a main cause of ion outflow from the ionosphere to the magnetosphere. Broadband VLF/ELF waves observed in the polar region are closely correlated with TAI, and are thought to be a major energy source of the TAI. Using the datasets obtained by Akebono for 10 years from 1989 to 1998, statistical studies on the spatial and temporal distribution of the continuous broadband wave was performed [1]. However, the interaction process between wave and particle is quite complicated and the detailed mechanism has not been well clarified, because varieties of wave spectra and particle distributions were observed depending on the orbital conditions and plasma environments. As a next step, therefore, we need to make a quantitative correlation analysis for the purpose of clarification of the global relationship between waves and particles in the ion heating/acceleration region. In the present paper, we introduce new computational techniques for extracting physical parameters such as thermal velocity and energy density from the velocity distribution functions of electrons and ions, which were obtained by particle instruments onboard the Akebono satellite. These techniques enable us to represent the characteristics of precipitating electrons and heated ions using a few key-parameters, and also to analyze large amount of datasets in a systematic way. We evaluate our method through some correlation analyses between waves and particles using Akebono science database.

2 Instruments and Observation

In the analyses, we utilize wave and particle data obtained by the instruments onboard the Akebono satellite. The VLF instruments are designed to investigate plasma waves from a few Hz to 17.8 kHz [2, 3]. The generation/propagation mechanism of plasma waves in the magnetosphere reflects plasma environment which depends on many parameters such as solar activity, geomagnetic activity, altitude, latitude, local time, season etc. Therefore it is useful in deriving a dynamic structure of the magnetosphere to observe these waves and clarify their characteristics. In the present analysis, we use the data from the multi-channel analyzer (MCA) and ELF receiver (ELF), which are the subsystems of the VLF instruments. MCA measures electric (E) and magnetic (B) fields at 16 frequency points from 3.18 Hz to 17.8 kHz every 0.5 sec. ELF measures waveforms taken by both E and B sensors in the frequency range below 80 Hz. In the present paper we utilize the 8 sec averaged data both for MCA and ELF.

As for the particle data, we use two kinds of particle instrument, low energy particle instrument (LEP), and suprathermal ion mass spectrometer (SMS). LEP measures energy and pitch angle distributions of electrons from 10 eV to 16 keV and ions from 13 eV to 20 keV [4]. SMS measures two dimensional energy distributions in the satellite spin plane of thermal (0–25 eV) and suprathermal (25 eV–several keV) ions [5, 6].

One example of correlation diagram between waves and particles observed by Akebono on February 20, 1990, is shown in Figure 1. Figures 1a and 1b show E and B wave intensities below 17.8 kHz observed by MCA, respectively. Four lines in Figures 1a and 1b are characteristic frequencies at the observation point; lower hybrid frequency, and cyclotron frequencies of H^+, He^+ and O^+, in order of higher frequency to lower one, respectively. Figures 1c-e, and Figures 1f-h show energy spectograms of electrons and ions, respectively, observed by LEP. These panels are sorted by pitch angle ranges of 0–60°, 60–120°, and 120–180°. Figures 1i-k show two-dimensional energy and spin angle distributions for three kinds of thermal ions (H^+, He^+, and O^+). The lines from 40° to 100° in Figures 1i-k show the spin angle closest to the ram direction in the spin plane. The following parameters stand for universal time (UT), altitude (ALT), magnetic local time (MLT), magnetic latitude (MLAT), invariant latitude (ILAT), and L-value along the trajectory of Akebono. As Akebono is orbiting from 6,400 km to 9,200 km in ALT in the Northern Hemisphere, the pitch angles of 0° and 180° correspond to the downward and upward direction, respectively, along the geomagnetic field line in this case.

In the spectograms for ions in Figures 1f-h, ion precipitation in the energy range from 10 keV down to several hundreds eV are recognized from 2337:00 to 2341:00 UT. It indicates that Akebono is orbiting around the region so-called "cusp". In the panels of ion spectograms for the pitch angle ranges of 60–120° and 120–180° (Figures 1g and 1h), TAI and conically distributed ions (ion conics) are observed in the energy range below a few hundreds eV from 2337:30 to 2344:30 UT. Although there are several kinds of waves in the region where the TAI and ion conics are observed, broadband wave in the frequency range from a few kHz down to a few Hz looks well-correlated with them, i.e., the wave is detected from 2337:00 to 2345:00 UT. Additionally it is also noted that electron precipitation in the energy range below a few hundreds eV down to a several tens eV is simultaneously observed as shown in the spectograms for electrons in Figure 1.

3 Computational Techniques

3.1 Estimation of Plasma Parameters of Low Energy Electrons

The velocity distribution functions of electrons and ions are provided from the LEP science database. The data consists of 8 sec averaged count rates in 29 energy steps and 18 pitch angle bins for electrons and ions. In order to represent the characteristics of the electron distribution with a few parameters, we assume that the velocity distribution function of the electrons consists of background electrons and precipitating electrons. We assume that the background electrons and precipitating electrons form a Maxwellian distribution and a shifted-Maxwellian distribution, respectively. Then the total distribution function $f(v)$ can be rep-

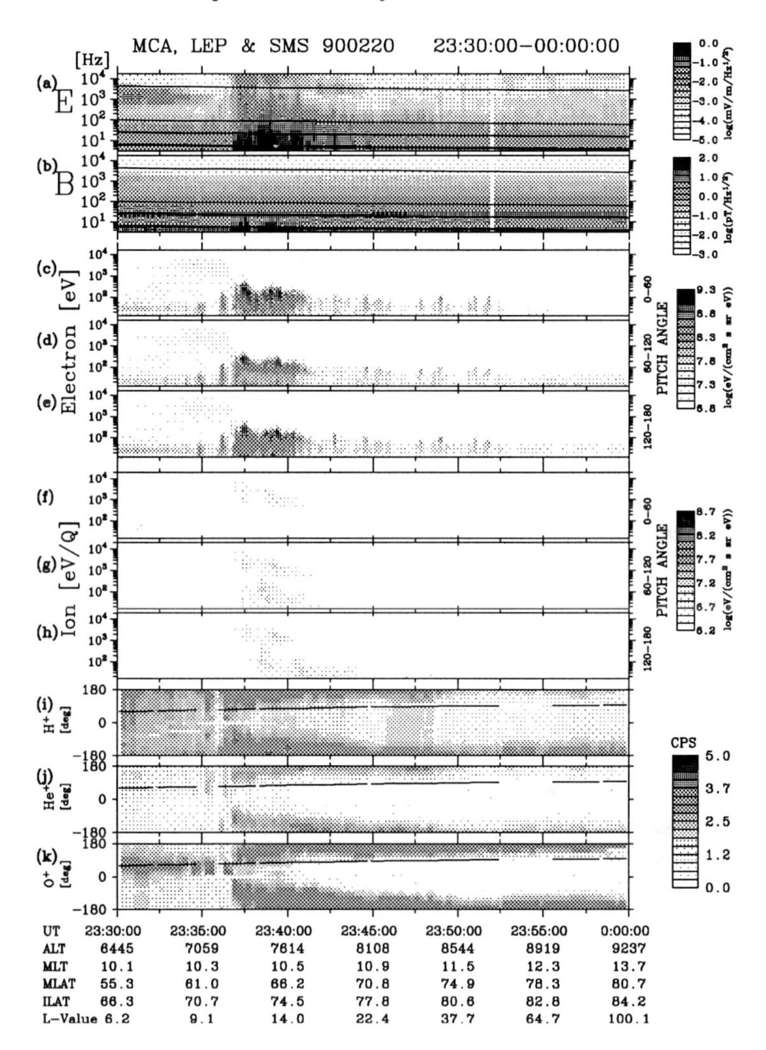

Fig. 1. An example of wave and particle data in the ion heating region.

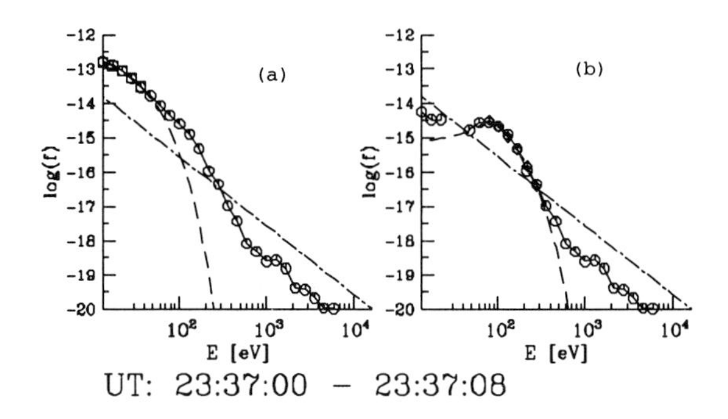

Fig. 2. Fitting processes of background electrons and precipitating electrons.

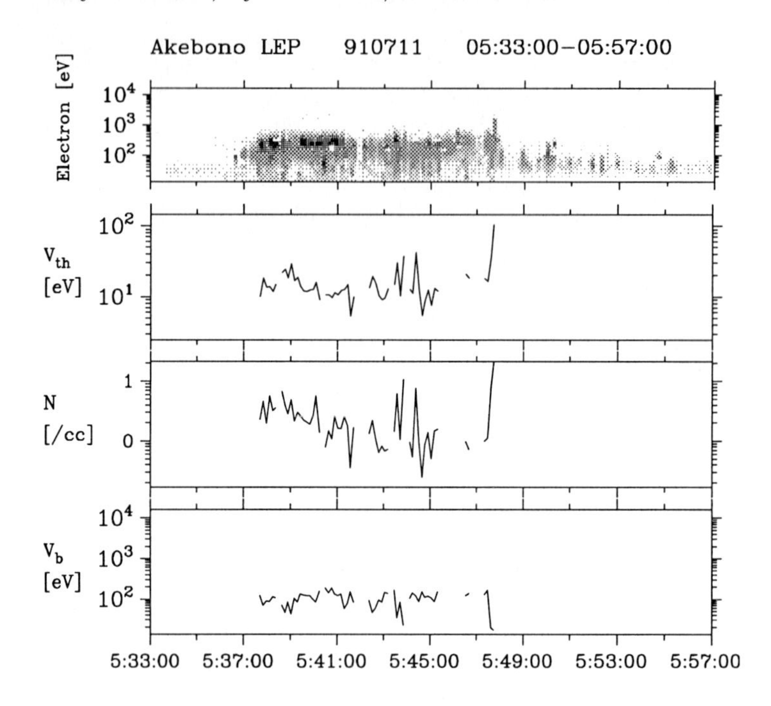

Fig. 3. Thermal velocity $v_{\mathrm{th(prec)}}$, number density $n_{\mathrm{(prec)}}$, and drift velocity V_{b} of precipitating electrons automatically determined by the fitting process.

resented by the following equation.

$$\begin{aligned} f(v) &= f_{\mathrm{(back)}} + f_{\mathrm{(prec)}} \\ &= \frac{n_{\mathrm{(back)}}}{(2\pi v^2_{\mathrm{th(back)}})^{3/2}} \exp\left(-\frac{v^2}{2v^2_{\mathrm{th(back)}}}\right) \\ &+ \frac{n_{\mathrm{(prec)}}}{(2\pi v^2_{\mathrm{th(prec)}})^{3/2}} \exp\left(-\frac{(v-V_{\mathrm{b}})^2}{2v^2_{\mathrm{th(prec)}}}\right), \end{aligned} \tag{1}$$

where $f_{\mathrm{(back)}}$ is the distribution function of the background electrons with thermal velocity of $v_{\mathrm{th(back)}}$ and number density of $n_{\mathrm{(back)}}$, and $f_{\mathrm{(prec)}}$ is the distribution function of the precipitating electrons with thermal velocity of $v_{\mathrm{th(prec)}}$, number density of $n_{\mathrm{(prec)}}$, and drift velocity of V_{b}. In order to estimate appropriate parameters in (1), we firstly fit the distribution function of the background electrons using a least squared fitting method as shown in Figure 2a (process A). The solid line in Figure 2a is the observational distribution function and the dashed line is the result of fitting using a Maxwellian distribution function. Secondly, we subtract the background distribution from the observational distribution and fit the remaining part using a shifted-Maxwellian distribution function as shown in Figure 2b (process B). In Figure 2b, the solid line indicates the remaining part of the electron distribution after subtraction, and the dashed line indicates the distribution function fitted by a shifted-Maxwellian distribution function. The dot-dashed lines in Figures 2a and 2b correspond to the count rate of 1000(count/s), which is the reliable level of the LEP instrument. In order to keep reliability of the estimated parameters, we only use the data points above these lines. In case that the maximum values in the distribution functions in the process B and/or A are less than these lines, we regard the parameters for precipitating and/or background electrons as void at the time slot, respectively. Applying this algorithm to the electron data of downward direction along the magnetic field line, we can estimate thermal velocities ($v_{\mathrm{th(back)}}$ and $v_{\mathrm{th(prec)}}$), number densities ($n_{\mathrm{(back)}}$ and $n_{\mathrm{(prec)}}$), and drift velocity (V_{b}) with a time resolution of 8 sec in a systematic way. Figure 3 shows an example of the estimated values of $v_{\mathrm{th(prec)}}$, $n_{\mathrm{(prec)}}$ and V_{b}. It is found that the parameters are appropriately estimated and the characteristics of the precipitating electrons can be represented by these three parameters.

In order to evaluate the validity of the proposed method, we examined the relationship between wave and electron using the plasma parameters derived from the fitting method. Figure 4 shows the correlation between precipitating electrons and intensity of the broadband wave using the data from 1989 to 1996. The horizontal axis shows the electric intensity of broadband wave at 5.62 Hz, which is one of the most appropriate frequency points for detection of the broadband wave among the frequency points of the MCA. The vertical axis shows the occurrence probability of the precipitating electrons at each wave intensity. We define the occurrence probability as the percentage of the data points at which the count rate for the precipitating electrons are large enough to estimate the parameters of shifted-Maxwellian distribution. We can find a clear correlation between waves and precipitating electrons; that is, the occurrence probability of the precipitating electrons increases as the wave intensity becomes larger.

As a next step, spatial distribution of wave and precipitating electrons are investigated. In the analysis, we used the data from ELF with a frequency resolution of 2.5 Hz and time resolution of 8 sec. Figure 5a shows the average electric intensity of the broadband wave at 5 Hz, and Figures 5b and 5c show the occurrence probabilities of precipitating electrons with drift velocities (V_{b}) of 78 eV and 770 eV, respectively. In Figures 5a-c, invariant latitude is shown by the radius of the circle and magnetic local time is taken along the circumference. In the figures, we only show the results in the Northern Hemisphere from 1989 to 1996. As shown in Figure 5a, the broadband wave is most intense in the dayside cusp region, and secondly along the auroral oval. On the other hand, the precipitating electrons with V_{b}=78 eV is distributed in the quite similar region corresponding to the active region of the broadband wave, while the precipitating electrons with V_{b}=770 eV is distributed mainly in the nightside region. Thus it is demonstrated that the precipitating electrons with V_{b} ~100 eV are more closely correlated with the broadband wave than those with V_{b} ~1 keV, which is consistent with the result empirically known from the event study. Quantitative correlation study between each parameter such as thermal velocity, drift veloc-

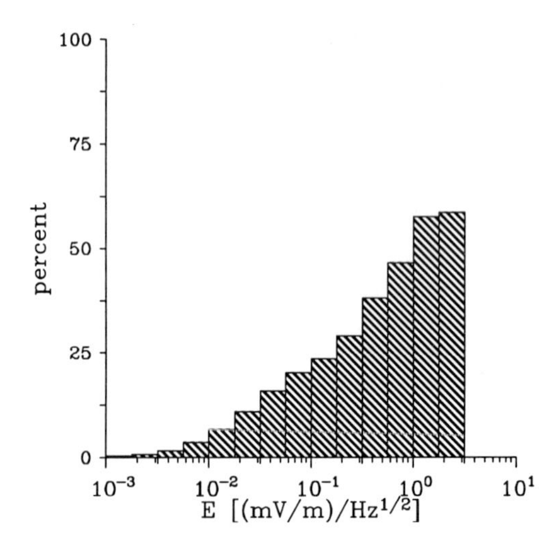

Fig. 4. Occurrence probability of the precipitating electrons as a function of wave intensity at 5.62 Hz.

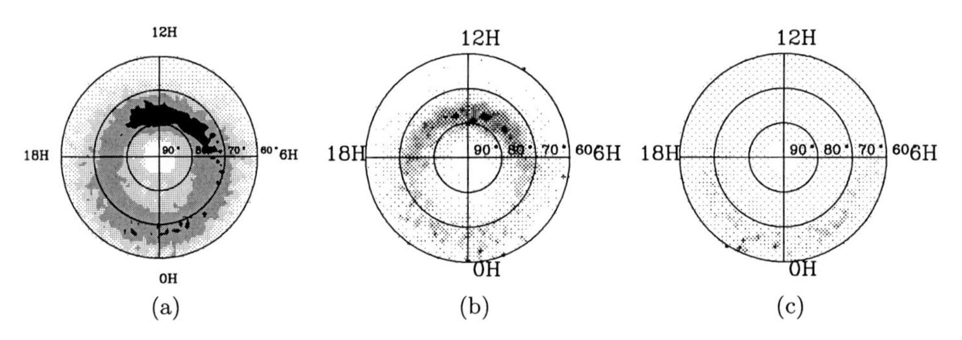

Fig. 5. (a) Wave intensity of the broadband wave. Occurrence probability of electron precipitation with (b) $V_b = 78$ eV, and (c) $V_b = 770$ eV.

ity, and number density of electron distribution and wave spectra is now under study.

3.2 Estimation of Energy Density of Heated Ions

In order to investigate the interaction process between broadband wave and TAI in detail, we need to compare the energies of waves and ions from moment to moment. In the present section, we introduce a technique to determine the energy density of heated ions. We define the energy density of heated ions by the following equation,

$$\varepsilon = \int E(v) f(v, \alpha) v^2 \sin \alpha \mathrm{d}\alpha \mathrm{d}\phi \mathrm{d}v$$

$$= 2\pi \int_{v_0}^{v_m} \int_{\alpha_0}^{\alpha_n} E(v) f(v, \alpha) v^2 \sin \alpha d\alpha dv, \tag{2}$$

where $E(v)$ represents the energy level of ions with their velocity of v and pitch angle of α. In the equation, we applied the integral interval for the pitch angle (α_n and α_0) from 90° to 180° for the data obtained in the Northern Hemisphere, and from 90° to 0° for those in the Southern Hemisphere. We averaged the energy density ε every 32 sec because the count rate of the instrument for ions is low compared with that for electrons. It is also noted that we estimated the energy density assuming that ions are all protons. Figure 6 shows the time series of the energy density of heated ions in the energy range below 340 eV from 2330 UT on February 20 to 0000 UT on February 21, 1990, which corresponds to the energy plot of the ions shown in Figures 1g and 1h. It is found that the heated ions observed from 2330 to 2345 UT, on February 20, 1990, are appropriately represented.

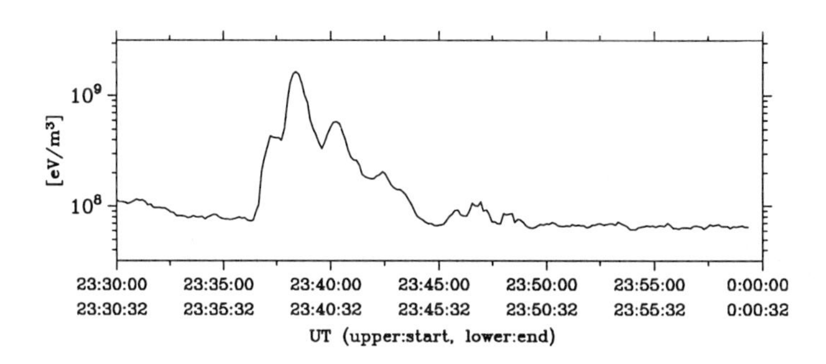

Fig. 6. Energy density of heated ion from 2330 UT on February 20 to 0000 UT on February 21, 1990.

On the other hand, we also estimate the spectral density of the broadband wave by integrating the square of the electric intensity below 10 Hz from the data obtained by the ELF subsystem. Figure 7 is a scatter plot which indicates the relationship between spectral density of the broadband wave and energy density of heated ions in every 32 sec [1]. In the analysis, we selected typical cases of 22 trajectories in which TAI and ion conics were recognized by LEP. The circles and pluses in the figure mean the data which were obtained in the dayside and the nightside, respectively. We find a clear relationship between wave spectral density and ion energy density. An important point to be noted is that there is a threshold level in the wave spectral density, and ions can be efficiently energized when the wave spectral density is larger than the threshold level. This fact shows that intense wave is needed for ions to be highly energized.

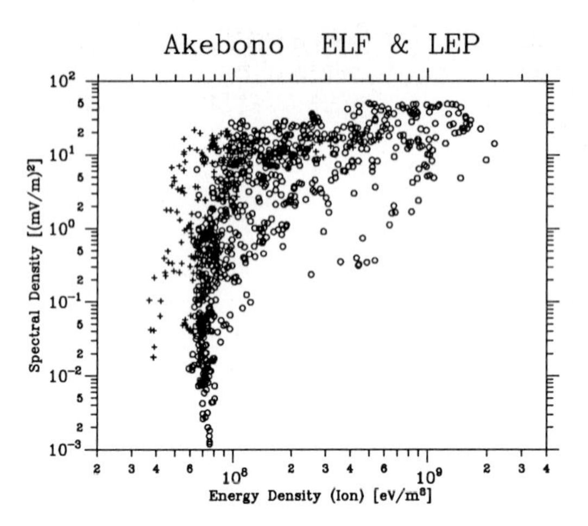

Fig. 7. Correlation between spectral density of the broadband wave and energy density of heated ions [after *Kasahara et al.*, 2001].

3.3 Identification of TAI and Ion Outflow from the Observational Data of Thermal Ion

The science data from the SMS instrument provides two dimensional velocity distributions of three kinds of ions (H^+, He^+ and O^+) every 8 sec. The data consists of count rates of 16 spin angles and 4 energy steps for each thermal ions and integrated flux for 16 spin angles of suprathermal ions. These data are useful for identification of initial stage of TAI and ion outflow, because SMS covers lower energy range than LEP. We developed a technique to identify the time of TAI and outflow from the SMS data automatically.

As were shown in Figures 1i-k, the count rates of thermal ions significantly increase around the perpendicular and upward direction in the TAI and ion outflow regions, respectively. In order to distinguish these phenomena from other phenomena such that, for example, count rates increase isotropically for all pitch angle range, which frequently occurs in the lower altitude region and/or lower latitude region, we define heating index (I_{heat}) and outflow index (I_{out}) as follows,

$$I_s(t) = \sigma(t) \sum_{\alpha} C(t,\alpha) W_s(\alpha) W_{\text{ram}}(\alpha, \alpha_{\text{ram}}(t)), \quad (s = \text{heat}, \text{out}), \qquad (3)$$

where $\sigma(t)$ is the standard deviation of the count rates over all pitch angle range at each time slot, $C(t,\alpha)$ is the count rate at pitch angle of α, $W_s(\alpha)$ is a weight function, which is a kind of filter for the purpose of emphasizing the count rates at particular pitch angle range, and $W_{\text{ram}}(\alpha, \alpha_{\text{ram}}(t))$ is the weight function for the elimination of the contamination at the ram direction. For example, the weight functions for $W_{\text{heat}}(\alpha)$ and $W_{\text{out}}(\alpha)$ to be applied to the data in the Northern Hemisphere are defined as shown in Figures 8a and 8b. In the figures,

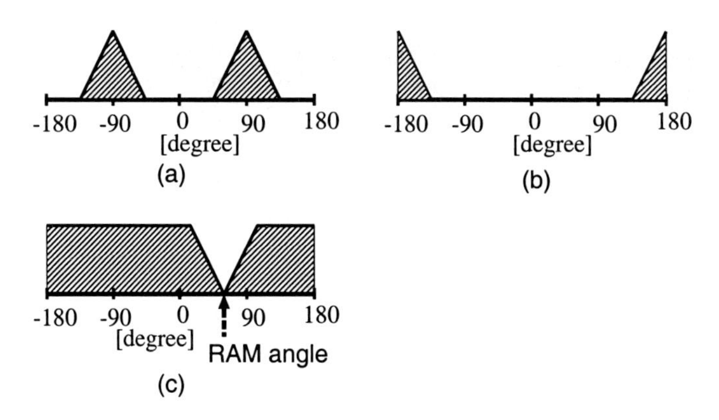

Fig. 8. Weight functions defined as (a) W_{heat}, (b) W_{out} and (c) W_{ram}.

the shaded region stands for the weight to be multiplied to the count rate $C(t, \alpha)$. In the same way, the weight function for $W_{\text{ram}}(\alpha, \alpha_{\text{ram}}(t))$ is defined as shown in Figure 8c. In order to eliminate the artificial increase of the count rate at the ram angle, we decrease the weight around the ram angle.

The algorithm was applied to the observational data on February 20, 1990 as shown in Figure 9. The upper three panels show the pitch angle distributions of thermal ions of H^+, He^+, and O^+, which are identical with Figures 1i-k. The heating indices and outflow indices calculated by (3) for these ions are shown in the following 6 panels. It is found that these indices appropriately increase corresponding to the time intervals of heating and outflow phenomena and are suppressed in the other time intervals.

4 Conclusion

There are a variety of plasma waves and particle phenomena. Multi-dimensional parameter dependences of these phenomena reflect the geophysical law which controls the plasma environment around the Earth. Datasets continuously obtained by Akebono for more than 12 years contain a lot of information to discover the law, whereas increase of the volume of datasets makes it impossible to apply the conventional analysis techniques. In the present paper, we attempt to represent the particular phenomena using several key-parameters in a systematic way. We select these parameters from a scientific point of view and we found that our approach is generally agreeable.

The algorithm introduced in the present paper is quite helpful for us to extract the attributes and/or characteristics of the phenomena. The parameters derived from the algorithm are useful not only in investigating interrelationship of the phenomena obtained from different kind of instruments, which was the main purpose in the present paper, but also in searching a unique event among enormous datasets as an "event finder". It should be noted, however, that we must always be careful about the reliability of the parameters and the validity

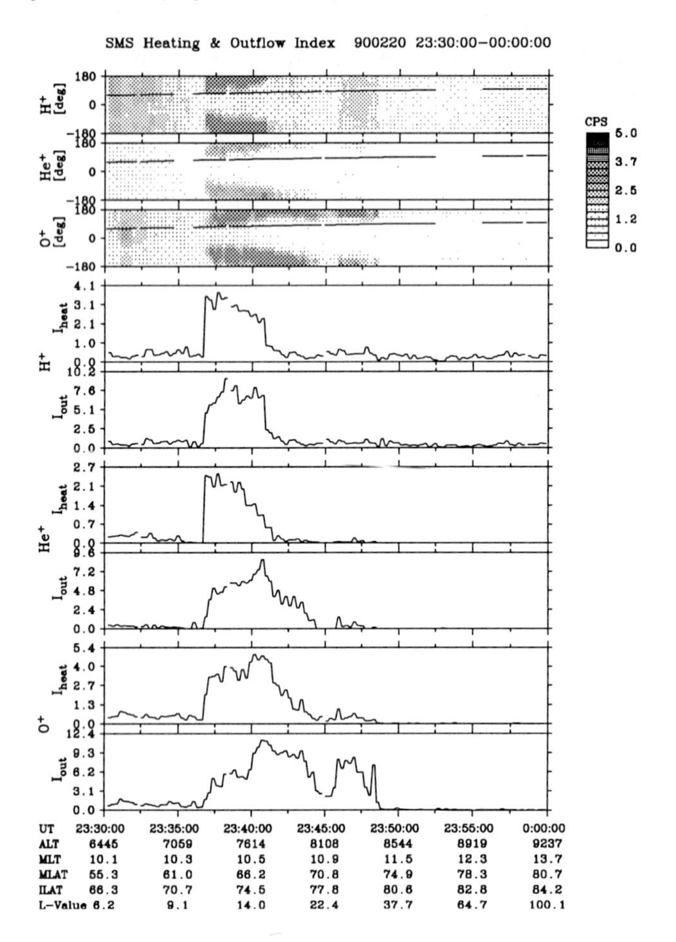

Fig. 9. Heating indices and outflow indices derived from the SMS science database.

of the method in the detailed investigation. Because the assumption of the distribution function might be too simple in some cases and, furthermore, might include artificial effects. For example, we still have a problem to be solved in the definition of heating and outflow indices. These indices are efficient for the detection of the region but not useful for the comparison between different orbits, because the count rate depends on altitude, ion species, and the aged deterioration in the sensitivity of the sensor and thus the magnitude of the indices do not have physical meaning between the different trajectories. Therefore we need some kind of normalization to reduce these artificial effect. In this sense, it is important to make close examinations of the rules which are found in the correlation study as was introduced in the present paper in order to discover natural law. The Akebono data collected over more than 12 years is quite valuable for further investigation and a more detailed comparison study between wave and particle data will provide a lot of clues to solving global plasma dynamics in the magnetosphere.

Acknowledgments

We are grateful to I. Nagano and the other Akebono/VLF members for their support. Akebono LEP and SMS data were provided by T. Mukai and A. W. Yau, respectively, through DARTS at Institute of Space and Astronautical Science (ISAS) in Japan. We would like to thank the ISAS Akebono mission project team and the members of tracking stations at Kagoshima Space Center, Syowa Base in Antarctica, Prince Albert in Canada, and Esrange in Sweden.

References

1. Kasahara, Y., Hosoda, T., Mukai, T., Watanabe, S., Kimura, I., Kojima, H., and Niitsu, R., ELF/VLF waves correlated with transversely accelerated ions in the auroral region observed by Akebono, *J. Geophys. Res.* (to appear), (2001)
2. Kimura, I., Hashimoto, K., Nagano, I., Okada, T., Yamamoto, M., Yoshino, T., Matsumoto, H., Ejiri, M., and Hayashi, K.: VLF observations by the Akebono (Exos–D) satellite, *J. Geomagn. Geoelectr.*, 42, (1990), 459-478
3. Hashimoto, K., Nagano, I., Yamamoto, M., Okada, T., Kimura, I., Matsumoto, H., and Oki, H.: Exos–D (Akebono) very low frequency plasma wave instruments (VLF), *IEEE Trans. Geoelectr. and Remote Sensing*, 35, (1997) 278-286
4. Mukai, T., Kaya, N., Sagawa, E., Hirahara, M., Miyake, W., Obara, T., Miyaoka, H., Machida, S., Yamagishi, H., Ejiri, M., Matsumoto, H., and Itoh, T., Low energy charged particle observations in the "auroral" magnetosphere: first results from the Akebono (Exos–D) satellite, *J. Geomagn. Geoelectr.*, 42, (1990), 479-496
5. Whalen, B. A., Burrows, J. R., Yau A. W., Budzinski, E. E., Pilon, A. M., Iwamoto, I., Marubashi, K., Watanabe, S., Mori, H., and Sagawa, E., The suprathermal ion mass spectrometer (SMS) onboard the Akebono (Exos-D) satellite, *J. Geomagn. Geoelectr.* 42, (1990), 511-536
6. Watanabe, S., Whalen, B. A., and Yau, A. W., Thermal ion observations of depletion and refilling in the plasmaspheric trough, *J. Geophys. Res.*, 97, (1992), 1081-1096

A Flexible Modeling of Global Plasma Profile Deduced from Wave Data

Yoshitaka Goto, Yoshiya Kasahara, and Toru Sato

Dept. of Communications and Computer Engineering, Kyoto University
Yoshida, Sakyo, Kyoto 606-8501, Japan
ygotou@aso.cce.i.kyoto-u.ac.jp

Abstract. As the plasma profile around the earth varies with local time, season and solar activity, it is important to have a means to determine it in a fine time resolution. In this paper we propose a method to determine the global plasma profile from the satellite observation data. We adopt a stochastic model to represent the distribution of plasma. In the model, the parameters are determined by ABIC(Akaike Bayesian Information Criterion) deduced from observed wave data. The validity of our method was evaluated using simulated data and it is found that the given distributions are successfully reconstructed by smoothing the observation data appropriately.

1 Introduction

Investigations of space weather and plasma environment around the earth have become a critical issue. As the global plasma profile in the earth's plasmasphere varies with local time, season and solar activity, it is very important to determine it with a fine time resolution. Determination of the plasma (electron) density in the outer space has been attempted with various methods. Direct spacecraft observations are thought to be one of the most effective way to obtain the density accurately. It must be noted, however, that they can provide plasma parameters only along their trajectories. In order to get the global plasma profile with a fine time resolution, it is necessary to assume an adequate model of the density distribution. If the global density profile is determined, the propagation characteristics of the plasma waves can be calculated theoretically. In this study, we inversely reconstruct the most plausible global density profile from the characteristics of the plasma waves which were observed along the trajectory. Namely, we solve an inverse problem.

Japanese scientific satellite Akebono has been observing electric field, magnetic field, plasma waves and plasma particles of the inner magnetosphere around the earth since 1989. The accumulated data amounts to more than 1T bytes of digital data, which enables us to analyze the long term variations and the average figure of the plasmasphere, and some of the products would possibly make a major breakthrough. For that purpose it is necessary to employ highly automatic data processing.

S. Arikawa and A. Shinohara (Eds.): Progress in Discovery Science 2001, LNAI 2281, pp. 438–448.

In this paper we propose a method to determine the global plasma density profile automatically. Our method is based on the flexible modeling and the objective index for estimation. The former is realized by adopting a stochastic model for the distribution of plasma. The model of the global plasma profile is based on the diffusive equilibrium theory which describes the motion of plasma along the geomagnetic field lines. As for the horizontal structure of the plasma profile, a parametric model is generally used [1, 2]. Such parametric models are supported by the prior information that the plasma density near equatorial plane is denser than that at higher latitudes, for example. The degree of freedom in the model, however, is limited and all the residual errors to the model are considered to be observational noises. The actual distributions, therefore, cannot be represented completely in a parametric model. On the other hand, by adopting a stochastic model, in which significant variations and interference noises can be separately represented, we obtain the more plausible structure corresponding to the observation data.

The parameters in our model are determined by ABIC (Akaike Bayesian Information Criterion)[3] deduced from observed wave data. The characteristics of VLF (very low frequency) waves propagating around the earth reflect the plasma environment. Then we examine all the possible ray paths from the ground to the satellite for each plasma profile in our model theoretically and calculate the PDFs (probability density functions) of the wave characteristics such as wave normal direction. By considering these PDFs as prior information of the observation, we determine the best model based on the Bayesian theory. In this way we avoid determining the model parameters subjectively or ad-hoc.

In Section.2 we describe the proposed modeling of the global plasma profile. The determination of the parameters in our model is discussed in Section.3. In Section.4 we examine the validity of the proposed method using simulated data, followed by the discussion of a problem for practical use in Section.5.

2 Global Plasmasphere Model

2.1 Diffusive Equilibrium Model

The plasmasphere was filled with cold plasma as a result of ionospheric evaporation and thus the plasma is assumed to be distributed according to diffusive equilibrium (DE). Actually this idea is now known to be oversimplified, but the density of any hot component is so small that it doesn't affect the whistler-mode ray[4], which is used in the present study.

In the DE theory, plasma particles are constraint to move only along the direction of the magnetic field lines under the effect of earth's gravitational field and the gradient of temperature.

The distribution of electrons and ions under DE theory with temperature gradient is derived as follows [5]. The electron density for DE term $n_{\mathrm{de}}(s)$ is represented by

$$n_{\mathrm{de}}(s) = \frac{t_{\mathrm{e0}}}{t_{\mathrm{e}}(s)} \left\{ \sum_k \eta_k \exp\left(-\frac{z(s)}{H_k(t_{k0})} \right) \right\}^{\frac{1}{2}}, \tag{1}$$

where t_{e0} and $t_{\mathrm{e}}(s)$ are the electron temperatures at the reference altitude (1000km) and at distance s along the field line, respectively, and t_{k0} is the ion temperature at the reference altitude. The ions such as H^+, He^+, O^+ will be referred to by subscript k. η_k is the constituent ratio of each ion at the reference altitude. H_k is the scale height of each ion and $z(s)$ is the temperature-modified geopotential height. The temperature gradient along a magnetic field line is expressed by

$$\frac{t_{\mathrm{e0}}}{t_{\mathrm{e}}(s)} = 0.8 \left(\frac{r_0}{r}\right)^{n_{\mathrm{T}}} + 0.4 \left(\frac{r}{r_0}\right)^2 \cdot \frac{1}{1 + (r/r_0)^{40}} \equiv \frac{1}{t_{\mathrm{gr}}(s)}, \tag{2}$$

where n_{T} is a parameter representing the temperature gradient [6]. r is the radial distance, and r_0 is the reference distance which is the sum of the reference altitude and the earth's radius. In this study, we assume the temperatures of ions equal to that of electron.

The electron density at any distance s along a field line is represented by

$$n_{\mathrm{e}}(s) = n_{\mathrm{e0}} \cdot n_{\mathrm{de}}(s) \cdot n_{\mathrm{la}}(s), \tag{3}$$

where n_{e0} is electron density at the reference altitude depending on L-value and n_{la} is a compensation term for lower altitudes. As for n_{la} we adopt an analytical function derived from IRI (International Reference Ionosphere) model, which requires no unknown parameters [1]. The necessary parameters to be determined in our model are, therefore, n_{T} and η_k by giving the electron density and temperature distribution at the reference altitude. It is noted that the parameters should be determined separately for the summer and the winter hemispheres to keep the continuity on the geomagnetic equatorial plane.

2.2 Observation Model

A trajectory of the satellite in a semi-polar orbit crosses magnetic field lines covering an extensive range of invariant latitude. In the case that the observed data of electron density contain not only global variation but also local fluctuations and observational noises, it may be difficult to reflect the electron density along a trajectory directly to the distribution at the reference altitude along the field lines. To eliminate these noises, we assume the latitudinal density distribution at the reference altitude varies smoothly, that is, we adopt a stochastic differential model [7] for the latitudinal variation. We assume the electron density at successive points at the reference altitude satisfies the following equation,

$$\nabla^2 n_{\mathrm{e0}} \sim N(0, \tau_{\mathrm{n}}^2 \sigma_{\mathrm{n}}^2), \tag{4}$$

where $N(0, \tau_{\rm n}^2\sigma_{\rm n}^2)$ shows Gaussian distribution, the mean of which is 0 and the variance is the product of the smoothing parameter $\tau_{\rm n}^2$ and the variance of observational error of electron density $\sigma_{\rm n}^2$. The spatial interval of $n_{\rm e0}$ is constant corresponding to L-value. The smoothing parameter $\tau_{\rm n}^2$ controls the smoothness of the latitudinal distribution.

From the equation (4), the electron density for i-th magnetic field line is represented by

$$n_{\rm e0}(i) = n_{\rm e0}(i-1) + \{n_{\rm e0}(i-1) - n_{\rm e0}(i-2)\} + N(0, \tau_{\rm n}^2\sigma_{\rm n}^2). \tag{5}$$

Then we define the state vector $\boldsymbol{z}(i)$ for i-th magnetic field line by

$$\boldsymbol{z}(i) = \begin{pmatrix} n_{\rm e0}(i) \\ n_{\rm e0}(i-1) \end{pmatrix}, \tag{6}$$

and matrices F, G, $H(i)$ as follows

$$F = \begin{pmatrix} 2 & -1 \\ 1 & 0 \end{pmatrix}, \quad G = \begin{pmatrix} 1 \\ 0 \end{pmatrix}, \quad H(i) = (n_{\rm de}(s_i) \cdot n_{\rm la}(s_i), 0), \tag{7}$$

where s_i is the distance between the observation point and the reference altitude along the i-th magnetic field line. Using the equations (6) and (7), the observation model is shown in Fig. 1 and thus represented by the following state space model (SSM),

$$\boldsymbol{z}(i) = F\boldsymbol{z}(i-1) + Gv \tag{8}$$

$$x(i) = H(i)\boldsymbol{z}(i) + w, \tag{9}$$

where v and w are both white noise sequences with means of 0 and variances of $\tau_{\rm n}^2\sigma_{\rm n}^2$ and $\sigma_{\rm n}^2$, respectively. The outputs of the system $x(i)$ shows the observation series of electron density. By giving $x(i)$ and the model parameters, the electron density distribution at the reference altitude $n_{\rm e0}(i)$ is determined by using Kalman filter and the smoother algorithms. It should be noticed that the variance $\sigma_{\rm n}$ itself is unnecessary for the estimation of $z(i)$ in the algorithm of Kalman filter if we represent the system variance by the equation (4).

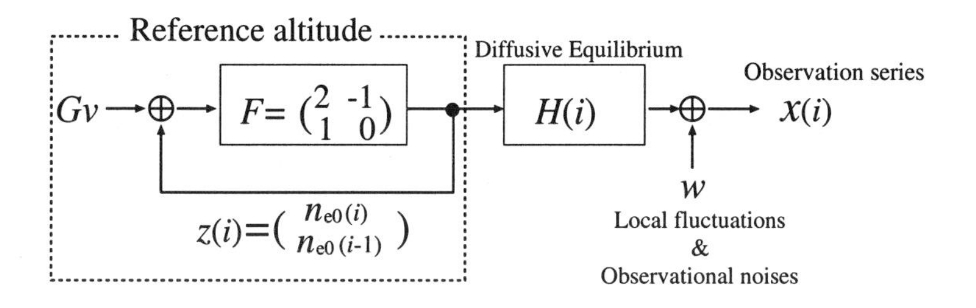

Fig. 1. Observation model of electron density

Actually, it is necessary to estimate the electron temperature profile before the calculation of electron density profile because we need the temperature variations along magnetic field lines for the DE term (1). We assume, therefore, the electron temperature at successive points at the reference altitude satisfies the similar type of stochastic differential equation represented by

$$\nabla^2 t_{e0} \sim N(0, \tau_t^2 \sigma_t^2). \tag{10}$$

The state space model for the electron temperature profile is realized with the state vector $\boldsymbol{z}'(i)$, and matrices $H'(i)$ given by

$$\boldsymbol{z}'(i) = \begin{pmatrix} t_{e0}(i) \\ t_{e0}(i-1) \end{pmatrix}, \quad H'(i) = (t_{gr}(s_i), 0), \tag{11}$$

where $t_{gr}(s_i)$ was defined by (2).

We can get a unique plasma profile based on the DE model by determining the parameters n_T, η_k, τ_n^2 and τ_t^2. In the conventional methods, the electron density distribution at the reference altitude is represented by parametric models. Thus, the actual profile cannot be represented completely, and residual errors to the model are considered to be observational noises. On the contrary, the proposed method has an advantage of not only the reduction of calculation time but also the flexibility of representation of the profile.

3 Estimation of the Model Parameters

3.1 Omega Signals Observed by the Akebono Satellite

Omega signals of around 10kHz, which had been used for global navigation until 1997, were transmitted from 7 ground stations distributed over the world, and were continuously observed by the VLF instruments onboard the Akebono satellite along their trajectories in the plasmasphere. The wave normal vectors and the delay times, which varies depending on the plasma environment in the plasmasphere along the ray path from the ground to the satellite, are determined from the polarization of the signal and its arrival time to the satellite. On the other hand, it is possible to calculate the ray paths and these parameters theoretically by using the ray tracing technique under a realistic plasma density profile. Then we determine the best plasma profile which satisfies these wave observation data by giving the adequate parameters in our plasma model.

3.2 Ray Tracing of Omega Signals

The ray tracing technique enables us to calculate ray paths for Omega signals by solving the dispersion relations of waves in plasma [8]. Analytical models of geomagnetic field and ion densities in the medium are required in this calculation. The fundamental equation of ray tracing is represented by the following differential equations

$$\frac{\mathrm{d}}{\mathrm{d}t}\boldsymbol{A} = \boldsymbol{f}(\mu, \boldsymbol{A}), \qquad \boldsymbol{A} = (r, \theta, \phi, \rho_r, \rho_\theta, \rho_\phi), \tag{12}$$

where μ is the refractive index which is determined by the dispersion relation at each point. r, θ and ϕ are the geocentric distance, geomagnetic colatitude and geomagnetic longitude, respectively, and ρ_r, ρ_θ and ρ_ϕ are the components of the refractive index vector $\boldsymbol{\rho}(|\boldsymbol{\rho}|=\mu)$ in the (r, θ, ϕ) coordinates. The direction of $\boldsymbol{\rho}$ vector is coincident with the wave normal direction. $\boldsymbol{f}$ is a nonlinear function of μ and $\boldsymbol{A}$. Applying Adams method to the differential equations, ray paths are calculated by giving initial positions and initial wave normal directions.

In the present study we examine the feasibility of our algorithm using the Omega signals from Tsushima station received by Akebono. Generally the signal transmitted from Tsushima station propagates between the ionosphere and the ground. However, part of its energy penetrates through the ionosphere into the plasmasphere during the propagation. Considering the propagation of Omega signals in the plasmasphere, therefore, the wave sources at the ionosphere cover a broad area. Wave normal directions of Omega signals at the altitude of 300km should be almost vertical to the ground according to the previous ray theory. Therefore initial values for the ray tracing of Omega signals from Tsushima are assumed as follows

Wave frequency	: 10.2kHz
Initial altitude	: 300km
Initial GM latitude	: 20°N $\sim$ 60°N
Initial GM longitude	: 130°W
Initial wave normal direction	: vertical to the ground

where 10.2kHz is the common frequency to all Omega stations and the narrow band receiver onboard the Akebono satellite is tuned to the frequency. We adopt the IGRF (International Geomagnetic Reference Filed) model to obtain geomagnetic field direction at every step of the ray tracing calculation.

3.3 Prior Distributions of Propagation Characteristics

As the ray paths of the Omega signal is influenced by the variation of the electron density distribution, it is necessary to consider multiple paths of the waves arriving at an observation point. The wave normal direction of the waves through different paths are generally different. In addition, it is difficult to know which ray is observed among all the arriving rays. In this situation it may be difficult to use non-linear least squared fitting to obtain the best density profile which leads the observed wave data. Then we adopt a Bayesian approach by considering the prior distributions of the wave normal direction and the delay time. For simplicity we explain only the processing of the wave normal direction hereafter, but the same process can be applied to the delay time.

We consider random fluctuations of the location of the satellite, that is, the observation point is assumed to be represented by the PDF(probability density function) of Gaussian distribution. We count all the rays propagating through this distribution by using the ray tracing technique and calculate the PDF of the wave normal direction $\boldsymbol{k}$. We specify the distribution $p(\boldsymbol{k}|\Lambda)$ where $\Lambda = [n_{\mathrm{T}}, \eta_k, \tau_{\mathrm{n}}^2, \tau_{\mathrm{t}}^2]$ is the parameter set for our model of the plasma profile. On the other hand, we know the data distribution $p(y|\boldsymbol{k}, \Lambda)$ which is the Gaussian distribution calculated from the observed wave normal direction y and its error variance. If we can determine the best plasma profile, the integration of the product distribution of $p(\boldsymbol{k}|\Lambda)$ and $p(y|\boldsymbol{k}, \Lambda)$ becomes maximum. In the Bayesian theory, the posterior distribution $p(\boldsymbol{k}|y, \Lambda)$ satisfies

$$p(\boldsymbol{k}|y, \Lambda) \propto p(y|\boldsymbol{k}, \Lambda) \cdot p(\boldsymbol{k}|\Lambda), \tag{13}$$

and the well-known criterion ABIC is represented as

$$\mathrm{ABIC} = -2 \log L(\boldsymbol{k}|y, \Lambda) \tag{14}$$

where $L(\boldsymbol{k}|y, \Lambda)$ is the marginal likelihood obtained by integrating $p(\boldsymbol{k}|y, \Lambda)$ over all directions.

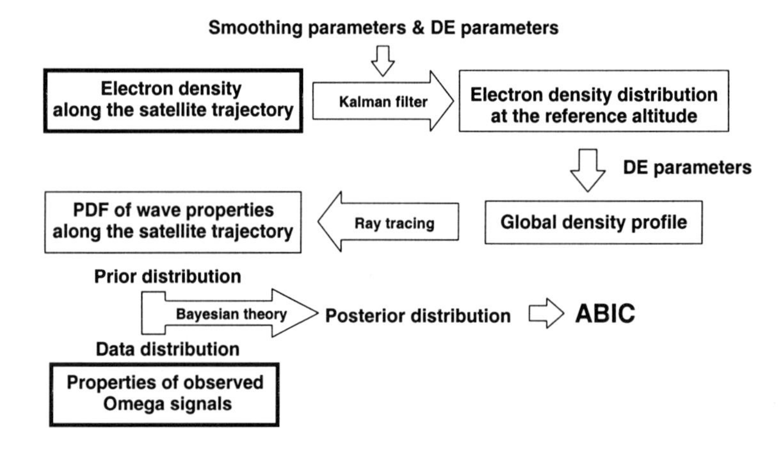

Fig. 2. Flowchart to obtain ABIC

The Flowchart to obtain ABIC from a parameter set Λ is shown in Fig. 2. We determine the best parameter set by finding the minimum ABIC. This problem is a kind of non-linear optimization and the optimal solution is not easily obtainable. We divide the parameters in Λ into discrete values and examine all the combinations to find the best parameter set which realizes the minimum of ABIC.

4 Simulation

The validity of our method was evaluated using simulated data which consist of electron density, electron temperature, wave normal direction and delay time of Omega signal. In the simulation, we examined the reconstructions of the global electron density profile for two cases; (case1) the electron density along magnetic field lines around $3L$ ($3L$ means L-value$= 3$) is higher than the other area; (case2) the electron density only at the altitude of trajectory around $3L$ is higher. For these two cases, the electron density distributions at the reference altitude are shown by dashed lines in Fig. 3 and Fig. 5. The parameters for the DE were given as $n_{\mathrm{T}} = 1.4$, $\eta_{\mathrm{H}^+} = 8.8\%$, $\eta_{\mathrm{He}^+} = 4.4\%$ and $\eta_{\mathrm{O}^+} = 86.8\%$.

The global electron density and the temperature profile were calculated from the DE parameters and the distributions at the reference altitude by (3) and (2), respectively. In case1 the observed electron density and temperature were given by adding observational noises to the global profile. In case2 local fluctuations were furthermore added to them. Ray paths of Omega signals were calculated in these plasma profiles. As for the simulated data of the waves, we calculated the PDFs, as mentioned before, and determined the wave normal directions and delay times based on them.

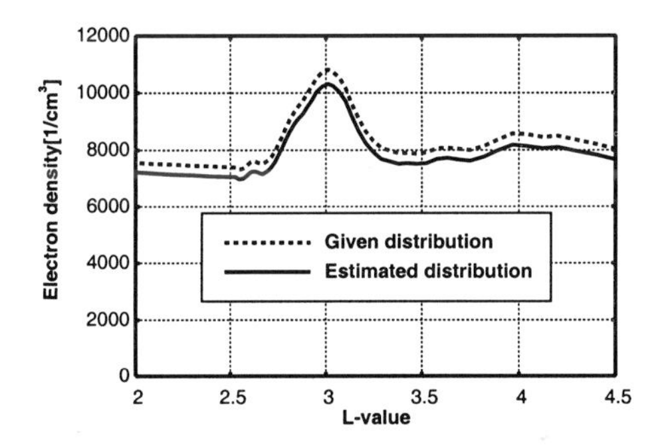

Fig. 3. The given and estimated distributions in case1

Case 1. We show the observational data for case1 in Fig. 4. The upper panels show the time series of satellite positions at the left and electron density at the right. The lower ones show those of wave normal directions of the observed waves at the left and the delay times at the right. While the observational range is between $2.5L$ and $4.5L$ as is shown in the left upper panel, we need to estimate the profile at the region less than $2.5L$ to calculate all the ray paths to the satellite. Then we extrapolated the distribution by using the prediction phase of Kalman filter.

From the observational data, we determine the model parameters by finding the minimum ABIC. The result is shown in Table 1. All the estimated parameters shown above are for the nearest grid to the given ones. The estimated electron density distribution at the reference altitude is shown by a solid line in Fig. 3. There are slight differences of the absolute values between the given distribution and the estimated one, because the best parameters cannot be estimated completely by using the grid search method. The distinctive features of the distribution around $3L$, however, is successfully reconstructed.

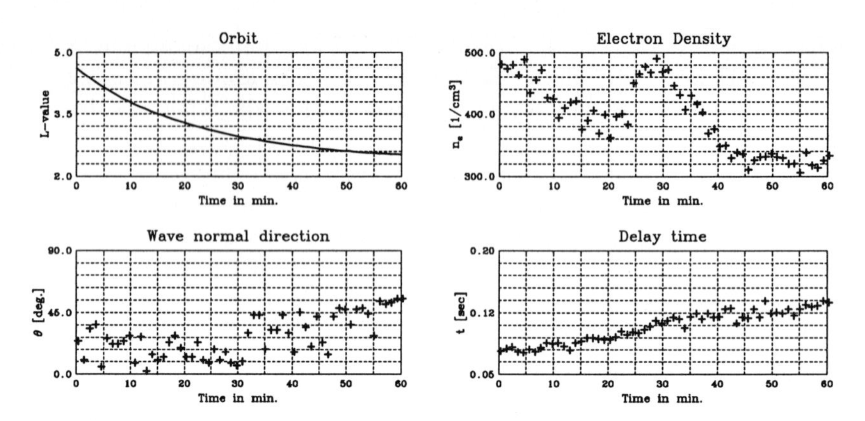

Fig. 4. The simulated data in case1

Table 1. The given and estimated parameters in case1

	n_{T}	$\eta_{\mathrm{H+}}$	$\eta_{\mathrm{He+}}$
Given	1.4	8.8%	4.4%
Estimated	1.4	8.0%	4.0%

Case 2. We show the observed data for case2 in Fig. 6. Each panel means the same as case1. The electron density is almost same as that of case1 but the increase of electron density around $3L$ is assumed to be a local fluctuation.

The result of model parameters for minimum ABIC is shown Table 2. In this case we couldn't obtain the nearest parameter set but practically equivalent values. The estimated electron density distribution at the reference altitude is shown by a dashed line in Fig. 5. It is found that the given distribution is also reconstructed by eliminating the local fluctuation around $3L$.

Table 2. The given and estimated parameters in case2

	n_{T}	$\eta_{\mathrm{H+}}$	$\eta_{\mathrm{He+}}$
Given	1.4	8.8%	4.4%
Estimated	1.4	8.0%	2.0%

5 Conclusion

In this paper we proposed an adequate method to expand the electron density data along the satellite trajectory to the global plasma profile by using the wave data propagating from the ground. We evaluated the validity of this method by using the simulated observation data. In the simulation the given distributions are successfully reconstructed by smoothing the observation data appropriately.

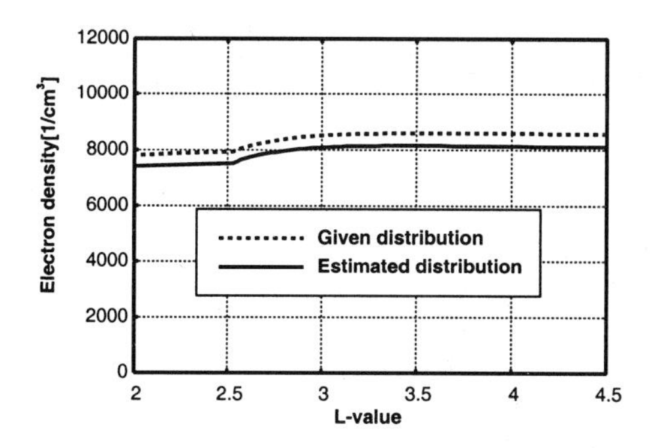

Fig. 5. The given and estimated distributions in case2

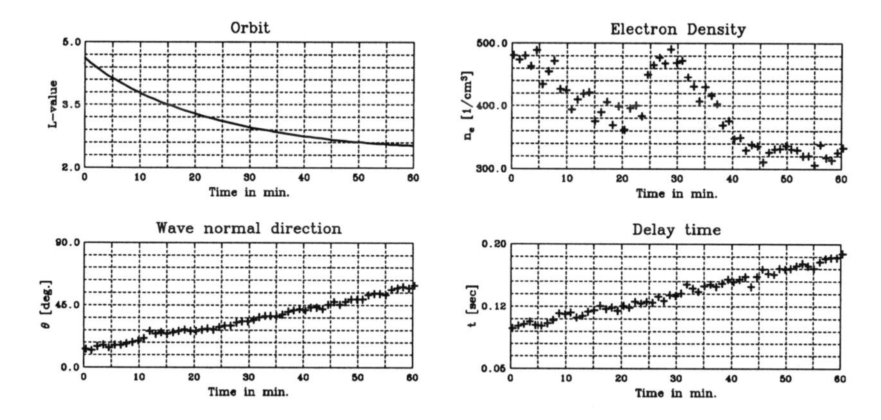

Fig. 6. The simulated data in case2

In the application for Akebono observation data, variety of global variations and local fluctuations would be included as a function of L-value. Our algorithm can be flexibly applied to such cases, because we can divide the trajectory into a few cluster or can change the smoothing parameters corresponding to the L-value considering the continuity of the global profile. In the future plan, to process the huge Akebono data set, the number of parameters in the model, such as electron temperature and ion constituents, should be reduced by making use of the correlation between them.

References

1. Kimura, I., Hikuma, A., Kasahara, Y., and Oya, H.: Electron density distribution in the plasmasphere in conjunction with IRI model deduced from Akebono wave data, *Adv.Space Res.,* 18(6), 279, 1996
2. Kimura, I., Tsunehara, K., Hikuma, A., Sue, Y. Z., Kasahara, Y., and Oya, H.: Global electron density distribution in the plasmasphere deduced from Akebono wave data and the IRI model, *J. Atmosph. Terr. Phys.,* 59(13), 1569, 1997

3. Akaike, H.: Likelihood and the Bayes procedure, *Bayesian Statistics,* Bernardo, J. M. , De Groot, M. H., Lindley, D. U., Smith, A. F. M. (eds.), University Press Valencia 185, 1980
4. Hashimoto, K., Kimura, I., and Kumagi, H.: Estimation of electron temperature by VLF waves propagating in directions near the resonance cone, *Planet. Space Sci.,* 25, 871, 1977
5. Angerami, J. J. and Thomas, J. D.: Studies of planetary atmospheres, 1. The distribution of electrons and ions in the Earth's exosphere, *J. Geophys. Res.,* 69, 4537, 1964
6. Strangeways, J. H.: A model for the electron temperature variation along geomagnetic field lines and its effect on electron density profiles and VLF ray paths, *J. Atmosph. Terr. Phys.,* 48, 671, 1986
7. Kitagawa, G., Higuchi, T., and Kondo, F.: Smoothness Prior Approach to Explore the Mean Structure in Large Time Series Data, *Discovery Science, Lecture Notes in Artificial Intelligence 1721,* Arikawa, S. and Furukawa, K. (eds.), Springer-Verlag, 230, 1999
8. Kimura, I., Matsuo, T., Tsuda, M., and Yamaguchi, K.: Three dimensional ray tracing of whistler mode waves in a non-dipolar magnetosphere, *J. Geomag. Geoelec.,* 37, 945, 1985

Extraction of Signal from High Dimensional Time Series: Analysis of Ocean Bottom Seismograph Data

Genshiro Kitagawa[1], Tetsuo Takanami[2], Asako Kuwano[3], Yoshio Murai[2], and Hideki Shimamura[2]

[1] The Institute of Statistical Mathematics, 4-6-7 Minami-Azabu, Minato-ku, Tokyo 106-8569 Japan
kitagawa@ism.ac.jp, http://www.ism.ac.jp/
[2] Hokkaido University, Institute of Seismology and Volcanology Kita-ku, Sapporo 060-0810, Japan, ttaka@eos.hokudai.ac.jp
[3] Tohoku University, Research Center for Prediction of Earthquakes and Volcanic Eruptions, Graduate School of Science, Aramaki, Aoba-ku, Sendai 980-8578, Japan

Abstract. A signal extraction method is developed based on a prior knowledge on the propagation of seismic signal. To explore underground velocity structure based on OBS (Ocean Bottom Seismogram), it is necessary to detect reflection and refraction waves from the data contaminated with relatively large direct wave and its multiples. In this paper, we consider methods based on the time series and spatial-temporal decompositions of the data. In spatial-temporal decomposition, the difference of the travel time (moveout) corresponding to underground layer structure is utilized. The proposed methods are exemplified with a real OBS data.

1 Introduction

In a cooperative research project with University of Bergen, Institute of Seismology and Volcanology, Hokkaido University, performed a series of experiments to observe artificial signals by ocean bottom seismographs (OBS) off Norway (Berg et al. (2001)). The objective was to explore the underground velocity structure. In one experiment, for example, about 40 OBS's were deployed on the sea bottom (about 2000 m in depth) with distances 10–15 km and observed the signals generated by air-gun towed behind a ship. The ship moved with a constant speed and generated signal 1560 times at each 200m (70 seconds). At each OBS, four-channel time series (2 horizontal and 2 vertical (high-gain and low-gain) components were observed with sampling interval of 1/125 second. As a result, 1560 4-channel time series with 7500 observations were obtained at each OBS. A result of identifying the underground layer structure was reported by Kuwano (2000).

A frequently used conventional method for detecting signals from observed multi-channel time series is the $\tau - p$ transformation which is obtained by slant

S. Arikawa and A. Shinohara (Eds.): Progress in Discovery Science 2001, LNAI 2281, pp. 449–458.

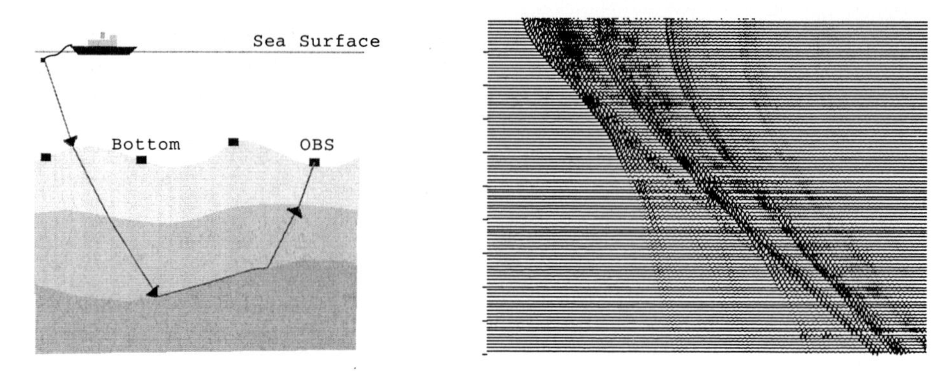

Fig. 1. Exploring underground structure by OBS data and OBS-4 data

stacking along the line $\tau = t - px$

$$F(\tau, p) = \int y_{t,x} dx = \int y_{\tau+px,x} dx, \tag{1}$$

where $y_{t,x}$ denotes the observation at time t and the offset distance x, $p = dt/dx$ is the "slowness" and is the reciprocal of the x-component of the velocity of the wave. $\tau - p$ is a very useful tool to idenfity various waves from multi-channel time series. This $\tau - p$ transformation has an inverse. Firstly compute the Hilbert transform of the $\tau - p$, $F(\tau, p)$ by $H(s,p) = -\pi^{-1} \int (s-\tau)^{-1} F(\tau, p) d\tau$. Then the inverse $\tau - p$ transformation is defined by $y_{t,x} = -(2\pi)^{-1} d/dt \int H(t - px, x) dp$. The existence of the inverse suggests extracting a wave with particular values of τ and p. However, the lack of orthogonality of τ and p may prevent from obtaining the satisfactory decomposition.

In this article, we take a different approach based on statistical modeling; namely, we try to develop statistical model for extracting information about underground velocity structure from multi-channel time series obtained from array of OBS. In particular an important problem is the detection of the reflection or refraction waves from the observed seismograms. For that purpose, we first applied a time series decomposition model (Kitagawa and Higuchi (1998)). We then develop a spatial model that takes into account of delay of the propagation of the direct, reflection and refraction waves. Finally, by combining the time series model and the spatial model, we try to perform spatial-temporal smoothing.

In statistical modeling, we do not necessarily believe that the model is true or a very close replica of the truth. We rather think it is a tool to extract useful information by combining various types of information. Concerning to this, Akaike (2001) classified the available information sets into three types: namely, firm theory, knowledge base on the past experience, and the current data. In our case the firm theory is the elastic wave theory, Snell's law, etc.. On the other hand, approximate layer structure, depth of water, velocity of wave in water can be considered as the knowledge in this area. In this paper, spatial model is derived by using this information.

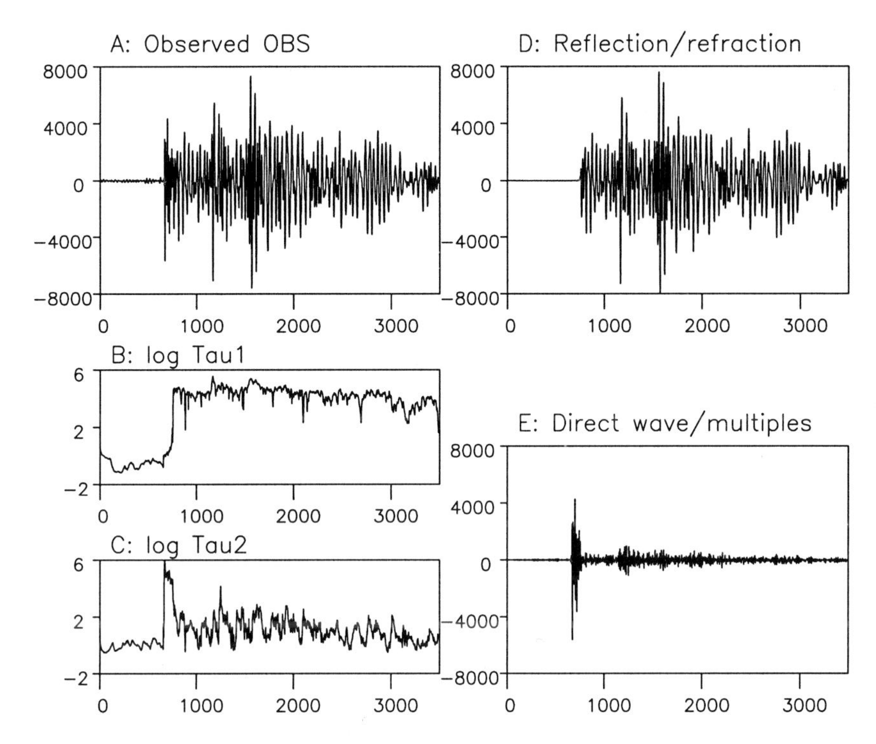

Fig. 2. Single channel decomposition of OBS data. A: Observed series, B: $\log \tau_1^2$, C: $\log \tau_2^2$, D: reflection/refraction wave, E: Direct water wave.

neath the air-gun, the direct wave (compressional wave with velocity about 1.48km/sec.) that travels through the water arrives first and dominates in the time series. However, since the velocity of the waves in the solid structure is larger than that in the water (2-8 km/sec.), in our present case, a reflection and refraction waves arrive before the direct wave and its multiples for the offset distance larger than approximately 1.4 km and 14 km, respectively.

As an example, assume the following horizontal four-layer structure: the depth and the velocity of the water layer: $h_0$km and $v_0$km/sec., the width and the velocities of three solid layers: h_1, h_2, h_3 km, and v_1, v_2, v_3 km/sec, respectively.

The wave path is identified by the notation; Wave($i_1 \cdots i_k$), ($i_j = 0, 1, 2, 3$), where Wave(0) denote the direct water wave (compressional wave) that travels directly from the air-gun to the OBS, Wave(01) denotes the wave that travels on the seafloor, Wave(000121) denotes the wave that reflects once at the sea bottoms and the sea surface, then penetrates into the first solid layer, traveling the interface between the first and the second solid layers, then goes up to the OBS through the first solid layer, (Figure 3-left).

Table 1 shows the travel times of various waves. At each OBS these waves arrive successively (Telford et al. (1990)). Right plot of Figure 3 shows the arrival times, t, versus the offset distances, D, for some typical wave paths. The parame-

Table 1. Wave types and arrival times

Wave type	Arrival time
Wave(0^{2k-1})	$v_0^{-1}\sqrt{(2k-1)^2h_0^2+D^2}$
Wave($0^{2k-1}1$)	$(2k-1)v_0^{-1}\sqrt{h_0^2+d_{01}^2}+v_1^{-1}(D-(2k-1)d_{01})$
Wave($0^{2k-1}121$)	$(2k-1)v_0^{-1}\sqrt{h_0^2+d_{02}^2}+2v_1^{-1}\sqrt{h_1^2+d_{12}^2}+v_2^{-1}d_2$
Wave(012321)	$v_0^{-1}\sqrt{h_0^2+d_{03}^2}+2v_1^{-1}\sqrt{h_1^2+d_{13}^2}+2v_2^{-1}\sqrt{h_2^2+d_{23}^2}+v_3^{-1}d_3$

where $d_{ij}=v_ih_i/\sqrt{v_j^2-v_i^2}$, $d_2=D-(2k-1)d_{02}-2d_{12}$, $d_3=D-d_{03}-2d_{13}-2d_{23}$.

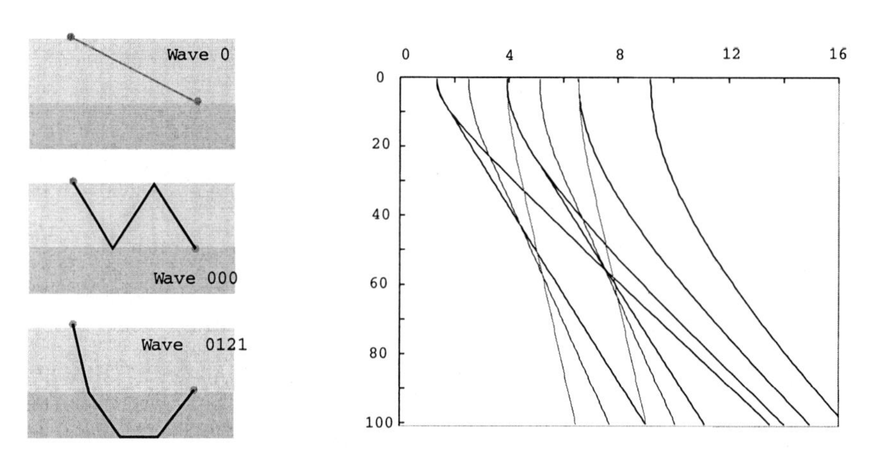

Fig. 3. Left: Examples of wave types in two-layered half space structure: Wave(0), Wave(000) and Wave(0121). Right: Travel times of various waves. Vertical axis: offset distance D (km), horizontal axis: reduced travel time $t-D/6$(sec.). From left to right in horizontal axis, Wave (01), (0), (0121), (0001), (000), (012321), (000121), (00000), (00012321).

ters of the 3-layer structure are assumed to be $h_0=2.1$km, $h_1=2$km, $h_2=3$km, $h_3=5$km, $v_0=1.5$km/sec, $v_1=2.5$km/sec, $v_2=3.5$km/sec, $v_3=7.0$km/sec. It can be seen that the order of the arrival times changes in a complex way with the horizontal distance D, even for such simplest horizontally layered structure.

3.2 Difference of the Arrival Time and Spatial Model

At each OBS, 1560 time series were observed, with the location of the explosion shifted by 200 m. Therefore the consecutive two series are cross-correlated and by using this structure, it is expected that we can detect the information that was difficult to obtain from a single time series.

Table 2 shows the moveout, namely the difference of the arrival times between two consecutive time series, computed for each wave type and for some offset

Table 2. Wave types and moveout of arrival times for various offset distance

Wave type	Offset Distance (km)				
	0	5	10	15	20
Wave(0)	0.6	16.5	16.5	16.7	16.7
Wave(0^3)	0.2	14.5	14.5	15.8	16.0
Wave(0^5)	0.1	8.0	12.0	14.2	15.0
Wave(01)	—	10.5	10.0	10.1	9.9
Wave(0121)	0.2	7.5	7.1	7.2	7.1
Wave(012321)	0.1	3.8	3.6	3.6	3.5

distance, D. The moveout of the waves that travel on the surface between two layers, such as Wave(01), (0121), (012321), are constants independent on the offset distance D. The moveout becomes small for deeper layer or faster wave. On the other hand, for the direct wave and its multiples that path through the water, such as Wave(0), (000), (00000), the amount of the moveout-time gradually increases with the increase of the offset distance D, and converges to approximately 17 for distance $D > 5$km. This indicates that for $D > 5$, the arrival time is approximately a linear function of the distance D.

Taking into account of this time-lag structure, and temporally ignoring the time series structure, we consider the following spatial model:

$$s_{n,j} = s_{n-k,j-1} + v_{n,j}, \quad y_{n,j} = s_{n,j} + w_{n,j} \tag{8}$$

where k is the moveout of the direct water wave or reflection/refraction wave, namely the differences of the arrival times between channels $j-1$ and j. By defining the state vector by $x_{n,j} = [s_{n,j}, s_{n-k,j-1}]^T$, we obtain the state space representation,

$$x_{n,j} = Fx_{n-k,j-1} + Gv_{n,j}, \qquad y_{n,j} = Hx_{n,j} + w_{n,j}.$$

Therefore, if the moveout k is given, we can easily obtain estimates of the "signal" (direct wave or reflection/refraction waves) by the Kalman filter and smoother.

If one value of k dominates in one region, we can estimate it by maximizing the localized log-likelihood. However, in actual data, several different waves may appear in the same time and the same channel. To cope with this situation, we consider a mixture-lag model defined by

$$s_{n,j} = \sum_{k=1}^{K} \alpha_{n,j,k} \hat{s}_{n,j,k}, \qquad y_{n,j} = s_{n,j} + w_{n,j}, \tag{9}$$

where $\hat{s}_{n,j,k}$ is the one step ahead predictor of the $s_{n,j}$ defined by $\hat{s}_{n,j,k} = s_{n-k,j-1}$ and $\alpha_{n,j,k}$ is the mixture weight at time n and channel j. In the recursive filtering, this mixture weight can be updated by

$$\alpha_{n,j,k} \propto \alpha_{n-k,j-1,k} \exp\left\{-(y_{n,j} - \hat{s}_{n,j,k})^2/2r_{n|n-1}\right\}. \tag{10}$$

3.3 Spatial-Temporal Model

We consider here a spatial-temporal smoothing by combining the time series model and the spatial model. The basic observation model is a multi-variate analogue of the decomposition model in (4):

$$y_{n,j} = r_{n,j} + s_{n,j} + \varepsilon_{n,j} \tag{11}$$

where $r_{n,j}$, $s_{n,j}$ and $\varepsilon_{n,j}$ denote a direct wave and its multiples, reflection/refraction wave and the observation noise component in channel j.

As in subsection 3.2, the direct water wave and the reflection/refraction wave components are assumed to follow the AR models

$$r_{n,j} = \sum_{i=1}^{m} a_{i,j} r_{n-i,j} + v_{n,j}^r, \qquad s_{n,j} = \sum_{i=1}^{\ell} b_{i,j} s_{n-i,j} + v_{n,j}^s, \tag{12}$$

respectively.

On the other hand, by considering the delay structure discussed in the previous subsection, we also use the following spatial models

$$r_{n,j} = r_{n-k,j-1} + u_{n,j}^r, \qquad s_{n,j} = s_{n-h,j-1} + u_{n,j}^s. \tag{13}$$

Here the moveouts k and h are actually functions of the wave type and the distance D (or equivalently the channel j). Specifically, for the direct water wave and other reflection/refraction waves, they are given by $k_j = \Delta T_j(\text{Wave}(0))$ or $\Delta T_j(\text{Wave}(000))$ etc., and $h_j = \Delta T_j(\text{Wave}(X))$, respectively.

3.4 Spatial-Temporal Filtering/Smoothing

In general, the sequential computational method for filtering and smoothing cannot be extended to space-time filtering/smoothing problems. However, for our special situation where a signal propagates in one direction, a reasonable approximate algorithm can be developed.

An approximate estimation algorithm can be developed by combining the filtering and smoothing algorithm in time and in space (channel). Hereafter $Y_{n,j}$ denotes the correction of the time series $\{y_{n,1}, \ldots, y_{N,1}, \ldots, y_{1,j-1}, \ldots, y_{N,j-1}, y_{1,j}, \ldots, y_{n-1,j}\}$. The conditional distribution of $x_{n,j}$ given the observations $Y_{n-1,j}$ is expressed as

$$p(x_{n,j}|Y_{n-1,j}) = \int\int p(x_{n,j}, x_{n-1,j}, x_{n-k,j-1}|Y_{n-1,j}) dx_{n-1,j} dx_{n-k,j-1}. \tag{14}$$

In (14), the integrand can be expressed as

$$\begin{aligned} & p(x_{n,j}, x_{n-1,j}, x_{n-k,j-1}|Y_{n-1,j}) \\ & = p(x_{n,j}|x_{n-1,j}, x_{n-k,j-1}, Y_{n-1,j}) p(x_{n-1,j}, x_{n-k,j-1}|Y_{n-1,j}) \\ & = p(x_{n,j}|x_{n-1,j}, x_{n-k,j-1}) p(x_{n-k,j-1}|x_{n-1,j}, Y_{n-1,j}) p(x_{n-1,j}|Y_{n-1,j}). \end{aligned} \tag{15}$$

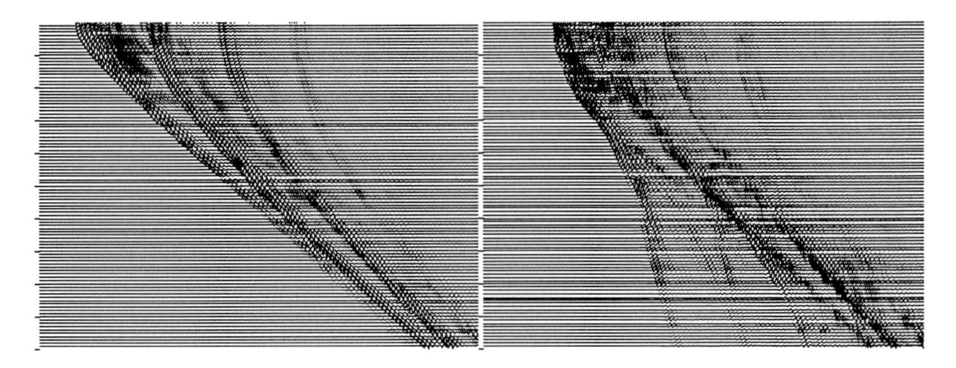

Fig. 4. Extracted direct wave and its multiples (left) and reflection/refraction waves (right). CH972–1071, data length n=2000, Δt = 1/125 second.

Here the first and the second term in the right hand side of the above equation can be expressed as

$$\begin{aligned} p(x_{n,j}|x_{n-1,j},x_{n-k,j-1}) &= \frac{p(x_{n,j}|x_{n-1,j})p(x_{n-k,j-1}|x_{n,j},x_{n-1,j})}{p(x_{n-k,j-1}|x_{n-1,j})} \\ &= \frac{p(x_{n,j}|x_{n-1,j})p(x_{n-k,j-1}|x_{n,j})}{p(x_{n-k,j-1}|x_{n-1,j})}. \end{aligned} \tag{16}$$

Therefore, by using an approximation

$$p(x_{n-k,j-1}|x_{n,j},x_{n-1,j}) = p(x_{n-k,j-1}|x_{n,j}), \tag{17}$$

we can develop a recursive formula for computing the spatial-temporal filter.

Figure 4 shows the results of the decomposition of the data shown in Figure 1. The left plot shows the extracted direct wave and its multiples. Waves(0^k), $k = 1,3,5,7$ are clearly detected. The right plot shows the extracted reflection waves and the refraction waves. Several waves presumably the Wave(0^k12321), (0^k121), $k = 1,3,5$ are enhanced by this decomposition.

4 Conclusion

Time series model and spatial-temporal model for extracting reflection or refraction waves from the OBS data are shown. In time series decomposition, a state space model based on AR representations of both direct/multiple waves and the reflection/refraction waves are used. Unknown parameters of the models are estimated by the maximum likelihood method and the self-organizing state space model. In spatial modeling the delay structure of various types of waves are considered. This spatial model is combined with the time series model and approximated method of spatial-temporal smoothing is obtained.

Acknowledgements

The authors thank the reviewers for careful reading and valuable comments on the manuscript. This research was partly supported by Grant-in-Aid for Scientific Research on Priority Areas "Discovery Science" from the Ministry of Education, Science, Sports and Culture, Japan.

References

Akaike, H.: Golf swing motion analysis: An experiment on the use of verbal analysis in statistical reasoning, *Annals of the Institute of Statistical Mathematics*, **53**, 1-10 (2001).

Anderson, B.D.O. and Moore, J.B.: *Optimal Filtering*, New Jersey, Prentice-Hall (1979).

Berg, E., Amundsen, L., Morton, A., Mjelde, R., Shimamura, H., Shiobara, H., Kanazawa, T., Kodaira, S. and Fjekkanger, J.P., Three dimensional OBS-data prcessing for lithology and fluid prediction in the mid-Norway margin, NE Atlantic, *Earth, Planet and Space*, Vol. 53, No. 2, 75–90 (2001).

Kitagawa. G.: Monte Carlo filter and smoother for non-Gaussian nonlinear state space model, *Journal of Computational and Graphical Statistics*, **5**, 1–25 (1996).

Kitagawa, G.: Self-organizing State Space Model, *Journal of the American Statistical Association*, **93**, 1203–1215 (1998).

Kitagawa, G. and Gersch, W.: *Smoothness Priors Analysis of Time Series*, Lecture Notes in Statistics, No. 116, Springer-Verlag, New York (1996)

Kitagawa, G. and Higuchi, T. : Automatic transaction of signal via statistical modeling, *The proceedings of The First Int. Conf. on Discovery Science*, Springer-Verlag Lecture Notes in Artificial Intelligence Series, 375–386 (1998)

Kitagawa, G. and Takanami, T. : Extraction of signal by a time series model and screening out micro earthquakes, *Signal Processing*, **8**, (1985) 303-314.

Kuwano, A.: Crustal structure of the passive continental margin, west off Svalbard Islands, deduced from ocean bottom seismographic studies, *Master's Theses, Hokkaido University*, (2000).

Shimamura, H.: OBS technical description, *Cruise Report*, Inst. of Solid Earth Physics Report, Univ. of Bergen, eds. Sellevoll, M.A., 72 (1988).

Stoffa, P. L. (ed.): *Tau-p, A Plane Wave Approach to the Analysis of Seismic Data*, Kluwer (1989).

Telford, W.M., Geldart, L.P., and Sheriff, R.E., *Applied Geophysics, Second dition*, Cambridge University Press, Cambridge (1990).

Foundations of Designing Computational Knowledge Discovery Processes

Yoshinori Tamada[1], Hideo Bannai[2],
Osamu Maruyama[3], and Satoru Miyano[2]

[1] Department of Mathematical Sciences, Tokai University,
1117 Kitakaname, Hiratuka-shi, Kanagawa 259-1292, Japan
tamada@ss.u-tokai.ac.jp
[2] Human Genome Center, Institute of Medical Science, University of Tokyo,
4-6-1 Shirokanedai, Minato-ku, Tokyo, 108-8639 Japan
{bannai,miyano}@ims.u-tokyo.ac.jp
[3] Faculty of Mathematics, Kyushu University,
Kyushu University 36, Fukuoka, 812-8581, Japan
om@math.kyushu-u.ac.jp

Abstract. We propose a new paradigm for computational knowledge discovery, called *VOX* (*View Oriented eXploration*). Recent research has revealed that actual discoveries cannot be achieved using only component technologies such as machine learning theory or data mining algorithms. Recognizing how the computer can assist the actual discovery tasks, we developed a solution to this problem. Our aim is to construct a principle of computational knowledge discovery, which will be used for building actual applications or discovery systems, and for accelerating such entire processes. *VOX* is a mathematical abstraction of knowledge discovery processes, and provides a unified description method for the discovery processes. We present advantages obtained by using *VOX*. Through an actual computational experiment, we show the usefulness of this new paradigm. We also designed a programming language based on this concept. The language is called *VML* (*View Modeling Language*), which is defined as an extension of a functional language ML. Finally, we present the future plans and directions in this research.

1 Introduction

Recently, some researchers in the area of computational discovery have recognized the importance of studying discovery processes themselves [8, 10], rather than studying only specific problems in the process, such as machine learning algorithms or specific data mining problems.

Fayyad *et al.* [8] analyzed various discovery processes and found that discovery process can be divided into several stages. They presented the general flow of the knowledge discovery tasks, called KDD (Knowledge Discovery in Databases) process (Fig. 1), and emphasized that discovery processes are *non-trivial, trial-and-error* iterations of each stage. Basically, the KDD process consists of: 1)

S. Arikawa and A. Shinohara (Eds.): Progress in Discovery Science 2001, LNAI 2281, pp. 459–470.

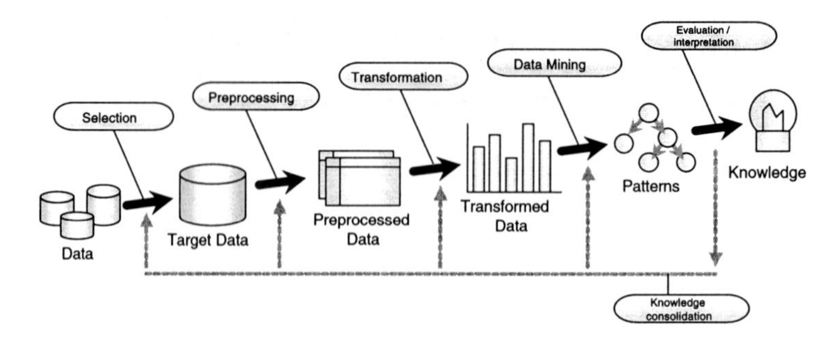

Fig. 1. An Overview of KDD Process

data selection 2) data preprocessing 3) data translating 4) data mining (hypothesis generation) 5) hypothesis interpretation/evaluation and 6) knowledge consolidation.

On the other hand, Langley [10] emphasized the importance of human intervention in discovery processes. He analyzed the successful discovery tasks appearing in scientific literatures, and observed that the human expert plays an important role in computer-aided discoveries. He concluded that the system should provide more explicit support for human intervention, and that the system should advise humans rather than replace them.

Furthermore, Motoda [15] presented another aspect of knowledge discovery tasks. He mentioned that a key to success is how we generate or discover good attributes/features, which help to explain the data under consideration.

The above suggestions have been popularized in artificial intelligence and machine learning field [10]. However, there have been less attempts to formalize the discovery processes. Most emphases seem to be placed on solving specific problems such as refining the prediction accuracy of specific machine learning algorithms, or developing applications for limited problems concerning specified data. We have persistently discussed how the computer can assist the actual discoveries based on the above suggestions [11, 12, 13], and recognized that what really demanded in the area of computational discovery is the basic concepts which can model various aspects of discovery, rather than constructing *ad-hoc* systems or applications which are based on the suggestions in [8, 10, 15]. Of course, component technologies such as machine learning algorithms are still important. However, it is clear that actual discoveries have not been achieved using only these technologies. We need to analyze what is required to obtain real discoveries using computers.

In this paper, we propose a new paradigm called *VOX* (*View Oriented eXploration*), which will contribute to the principle of computational knowledge discovery. *VOX* consists of three basic elements: *entity*, *view*, *view operator*, and is defined as *the approach of performing the discovery processes using these elements.* In other words, entity is data under consideration, view is a generalized form of attribute on the entities, and view operation is generation of new views. *VOX* can be considered as a mathematical abstraction or modeling of knowledge discovery processes. We show that *VOX* provides a unified mathematical description method of discovery processes, and helps various aspects of actual discovery tasks. We present three basic advantages obtained by the definition of *VOX* : 1) abstracting KDD processes, 2) realizing explicit human intervention, and 3) assisting the actual system implementation, which are really helpful to the actual discovery tasks.

The concepts of views and other related terms are firstly introduced by Maruyama *at al.* [12] to assist building a system for scientific knowledge discovery, and then refined in [3, 13]. We showed that the KDD process can be described using these terms. A view is a specific way of looking at or understanding given objects, and is essentially a function over a set of objects, returning a value which represents some aspect of each object [1]. More mathematical definitions are shown in Section 2.

We show the efficiency of *VOX* through the actual successful experiment conducted on biological data [3]. In the experiment, we discovered biologically meaningful knowledge, and developed an expert system based on the results of the experiment [24]. We present a mathematical description of the biological problem and discuss how each advantage of *VOX* yields good efficiency for the actual discovery tasks.

We have also developed a new programming language called VML (View Modeling Language), which is based on *VOX* [1]. VML is an extension of functional language ML. A view is realized as an extension of ordinary function so that it is natural that the language is implemented as a functional language. A view expresses knowledge in the discovery process, and new knowledge can be expressed as a composition of various views. VML consists of two elements. One is that a view has an interpretable meaning of the value which is returned by its view. The other is that composition and application of views can be done in a natural way, compared to imperative languages. Using VML, computational knowledge discovery experiments can be efficiently developed and conducted.

2 Basic Definitions

In this section, firstly we review basic definitions of three main terms which constitute the fundamental components of *VOX* : *entity*, *view*, and *view operator* [3], and other related term through biological examples. We recall the definitions from [3].

Definition 1 (entity). *An* entity *is anything which can be identified as an individual (i.e. an object). We shall call the set of entities which is under consideration, the* entities of interest E. □

For example, we consider proteins e_p as entities, and the set of plant proteins E_P as the entities of interest. Entities can be distinguished from each other by definition, but *how* they differ is ascertained through various attributes they possess. An attribute, or feature, for an entity is generalized as follows:

Definition 2 (view, viewscope). *A* view *is a function $v : E \to R$ over entities. Let $v(e)$ denote the value that a view returns for a certain entity e. The range R of v is called the* range set *of v. For convenience, we call a set of possibly infinite views $V = \{v_1, v_2, \ldots\}$ a* viewscope. □

A view returns a value which is expected to represent some aspect of the given entity. When parametric functions are regarded as views, we call them *parametric views.* A parametric view together with the space of parameters can define a viewscope.

Example 1. An *amino acid sequence view* v would return the amino acid sequence for a particular protein, e.g. if $e_p \in E_P$ is the ATP17 protein of Saccharomyces cerevisiae (Baker's yeast), $v(e_p) =$ "MIFKRAVSTL...". □

The first step in the KDD process is data selection. Data can be regarded as the set of entities, accompanied by initially given views. For example, if we are given a table of items (rows) and their attributes (columns), the entity set would consist of the items, and the initially given views would be the mapping from each item to their attributes in the table (one view for each column). If we can design a Boolean valued view which returns *true* for entities we want/need, and *false* for entities we do not want/need, we can create a subset of the entities by filtering the original set using this view.

Definition 3 (view operator). *We call any function over views (or viewscopes) and/or entities, which returns a view (or viewscope), a view operator.* □

Definition 4 (view design). *Views and view operators are combined to create new views. We call the structure of such combinations, the* design *of the view.* □

When human experts design views, they can embed their knowledge, intuitions, and ideas. Views provides an interface for human intervention to the discovery process.

There are view operators which are not dependent on the views which they operate upon. These view operators may be defined by a function over range sets.

Definition 5 (range set based view operator). *A range set based view operator Ψ is induced by a function $\psi : R \to R'$ over a range set R. With ψ and view v with range set R:*

$$\Psi[v] : E \to R' \equiv \psi \circ v \equiv E \xrightarrow{v} R \xrightarrow{\psi} R' \tag{1}$$

where $\Psi[v]$ denotes the new view induced by the application of operator Ψ to the range set of v. Similarly, for viewscope $V = \{v_1, v_2, \ldots\}$, we define $\Psi[V] \equiv \{\Psi[v_1], \Psi[v_2], \ldots\}$. Obviously, a parametric view operator will result in a parametric view. □

The next steps in the KDD process is the preprocessing, and transformation of the data. It is not difficult to see that preprocessing and transformation can be regarded as applying appropriate view operators to the initial views.

Example 2. View operators can be n-ary functions: e.g. for two views v_1, v_2 returning a Boolean values, we can create a conjunction view (representing the logical "and" of two attributes) using a 2-ary operator ϕ:

$$\phi[v_1, v_2](e) \equiv \phi(v_1(e), v_2(e)) = v_1(e) \wedge v_2(e). \tag{2}$$

□

Example 3 (neighborhood operator). A neighborhood operator can be defined to generate views which are in the *neighborhood* (according to some specified definition) of the original viewscope: e.g. locally modifying the parameters in a parametric view. This type of operator may be used, for example, to conduct local optimization of views. □

The above examples closely resemble techniques appearing in the context of constructive induction, or feature construction [4].

In the data mining stage of the KDD process, rules or hypotheses are generated by various learning algorithms. These generated rules may also be regarded as views. For example, a decision tree can be regarded as a function which returns for each entity e, a corresponding value at the leaf of where e ends up getting classified.

Definition 6 (learning based view operator). *A learning based view operator $\mathcal{H}_L$ can be induced from a learning algorithm L, which uses the entity set and available views to create a new view.* □

Example 4 (hypothesis generation (supervised)). A supervised learning process can be written as an operation on both the entity set and viewscope: $\mathcal{H}[E, V, v_c] = V'$, where $\mathcal{H}$ is a learning based view operator induced from some kind of induction algorithm (for example, C4.5 for the case of decision trees), E is the entity set, V is the set of views (or attributes/features) available to the algorithm, and v_c is the "answer" view. V' is the set of generated views (consisting of a single view, or perhaps multiple views: e.g. the top scoring views). The resulting view(s) $v' \in V'$ is expected to satisfy the property $v'(e) \simeq v_c(e)$ for most $e \in E$. □

Example 5 (hypothesis generation (unsupervised)). An unsupervised learning process can be written as an operation on both the entity set and viewscope: $\mathcal{H}[E, V] = V'$, this time not requiring v_c as in the supervised case.

For example, for entities E, viewscope V of numerical values, and clustering algorithm CL, $\mathcal{H}_{CL}$ will create a viewscope representing the clustering of the entities: $\mathcal{H}_{C\mathcal{L}}[E, V] = C$. Where C is a set of newly generated views (again consisting of a single view, or perhaps multiple views). A view $c \in C$ would return a cluster identifier $c(e)$ (telling which cluster the entity is clustered to) for each $e \in E$, and we would expect the distances (defined somewhere in relation to values from V) between the entities in the same cluster are small, and those in different clusters are large. □

Since hypotheses are equivalent to views, the evaluation/interpretation of a mined hypothesis is, in another words, the evaluation/interpretation of the mined view, meaning the manual evaluation of the view by a domain expert, or an automated evaluation according to some *score* (e.g. accuracy).

Since we observed that hypothesis generation algorithms generate views, the newly generated views may be used afterwards, perhaps in the next discovery task, or in refining the current task. This represents the *consolidation of the knowledge* gained from the data mining step.

The correspondence between views and the KDD process is summarized in Table 1.

3 View Oriented Exploration

VOX is a new paradigm for computational knowledge discovery. *VOX* consists of three elements: entity, view, view operator and is defined as the approach of performing the discovery processes using these elements.

As noted in Introduction, *VOX* is a mathematical abstraction of the knowledge discovery process, and provides an unified method of discovery processes. This abstraction reveals that general merits of mathematical abstraction can result in the same merits for knowledge discovery processes. We present three basic advantages for the computational knowledge discovery. In this section, we argue how it effectively supports the actual discovery tasks.

Abstracting KDD Processes

In [3], we abstracted KDD processes using view. The summary of this abstraction is shown in the Table 1. According to [8], discovery processes can be divided into several stages, and in [3], it is observed that the process can be modeled as the design, generation, use, and evaluation of views, asserting that views are the fundamental building blocks in the discovery process. The mathematical abstraction of the process provides a unified description method for the process. Maruyama *et al.* [11] analyzed that the successful system can be described using views. This work proves that discovery tasks can be described as application of KDD process, and can be modeled using the concept of *VOX*.

As well as many general mathematical abstractions, we can take the essence from the real problem we want to solve. This fact helps to grasp the whole stages of KDD process. It makes the process of discovery clearer, and provides an efficient way to understand the problem. Once we developed an approach to some type of problem, we can use this approach to solve another problem as long as the problem has the same type.

Because we can concentrate on the problem, this advantage can accelerate non-trivial, trial-and-error processes of computational knowledge discovery.

Table 1. Summary of representing the KDD process with view(scopes).

Elements of the KDD Process	Description in terms of view(scope)
1) data selection	classification and filtering of entities according to a certain view, which decides whether the entity is used or not.
2) data preprocessing	Preprocessing of data can be expressed as a function over data, so naturally may be defined by a view(scope).
3) data transformation	Transformation can also be expressed as a function over data, so naturally may be defined by a view(scope).
4) data mining	Data mining can be expressed as a generation of new view(scope). Hypothesis generation algorithms can be considered as view operators.
5) interpretation/evaluation	interpretation/evaluation of a view(scope)
6) knowledge consolidation	recursively using newly generated view(scopes)

Realizing Explicit Human Intervention

According to the suggestions of Langley [10], successful discovery tasks cannot be realized without appropriate interventions of human experts. Humans are still necessary for practical discoveries. Langley also mentioned that the system should provide more explicit support for the system. Namely, realizing human intervention in the computational discovery processes is one of the most important things to achieve successful discoveries.

In our mathematical abstraction using *VOX*, the knowledge discovery process is constructed by composition of various views. Careful designing of views by experts offers an elegant interface for human intervention on the discovery processes [3]. Through designing various views, human experts can naturally embed their knowledge, intuitions and ideas in the actual discovery tasks.

The abstraction using *VOX* also can help experts to communicate with the system developer, when the discovery system is made using *VOX*. Views can be a unit of communication for experts and developers one another. Detail are discussed in later section.

According to these facts, we can consider that *VOX* can realize explicit human intervention.

Assisting the Actual System Implementation

VOX can assist the implementation of actual systems such as expert systems or specific discovery softwares. According to above two advantages, the problem which we want to solve becomes more understandable and can be clearly separated into views. Implementing programs can be considered as construction of combining these views. The interface for human intervention can also be realized naturally using *VOX*. These facts means that *VOX* can speed up the process of implementation.

Mathematical abstraction also provides facility for re-using obtained knowledge which is generated from another discovery tasks using *VOX*. Knowledge or hypotheses are represented as views in *VOX*. Views are exactly the same as functions. Computer program can be regarded as the combination of functions. New knowledge or hypothesis obtained by the actual tasks is also represented as views, i.e. functions. Thus, re-using of obtained knowledge is easily realized using *VOX*.

We conducted computational experiments with biological data. Then, based on the knowledge obtained by the experiments, we built an expert system. Details are shown in the next section.

4 Verification through the Actual Discovery Task

In this section, we verify each advantage discussed in above section, through reviewing the actual discovery experiment task conducted on the biological data. The experiment was successfully done, and we succeeded in extracting biologically meaningful knowledge. The work has been done with a human expert. We always took his intuitions, and realized them as various views. The experiments were conducted on our partial implementation called HYPOTHESISCREATOR ($\mathcal{HC}$) [23], C++ object oriented class library. We describe here how and why *VOX* were effective for the experiment, not the result what we got. To illustrate that the process can be described in mathematical way, we present biological background knowledge here. For the details, see [2, 3].

Background

Proteins are composed of amino acids, and can be regarded as strings consisting of an alphabet of 20 characters. Most proteins are synthesized in the cytosol, and then carried to specific locations, called localization sites, such as mitochondria or chloroplasts. It has already revealed that, in most cases, the information determining the subcellular localization site is encoded in a short amino acid sequence segment called a protein sorting signal [16]. The problem is to detect the biologically meaningful rule that can explain why the given proteins goes to specific localization sites. We were given four sets of protein sequences, which are already known where the sequence goes. The four kinds of proteins are mitochondrial targeting peptides (mTP), chloroplast transit peptides (cTP), signal peptide(SP), and other proteins.

Previous work of the biologists revealed some properties about the data [5, 21]. Some researchers successfully developed the system for only predicting the localization sites [6, 7, 17]. TargetP [7] is the best predictor so far in the literature in terms of prediction accuracy, but this system is developed based on the neural network system, so that in general, it is difficult to understand why and how the predictions are made.

Approach and Result

Our aim was to find simple and understandable rules for human, which still have a practical prediction accuracy. To get the final result, we actually spent several weeks to design views, which are the range of searching views, through discussing with the human expert. The data we used was obtained from the TargetP website (http://www.cbs.dtu.dk/services/TargetP/). These data are divided into two data sets: plant and non-plant sequences. We describe our analysis on the plant data set of 940 sequences containing 368 mTP, 141 cTP, 269 SP, and 162 "Other" (consisting of 54 nuclear and 108 cytosolic proteins) sequences.

We designed views from two standpoints. One aimed to capture "global" characteristics of the N-terminal sequences. Since existing knowledge about the signals seemed to depend on biochemical properties of the amino acids contained, we decided to use the AAindex database [9], which is a compilation of 434 *amino acid indices*, where each amino acid index is a mapping from one amino acid to a numerical value, representing various physiochemical and biochemical properties of the amino acids.

The other view aimed to capture "local" characteristics: although strong consensus patterns do not seem to be present in the signals, there does seem to be a common structure to each of the signals. Therefore a view consisting of *alphabet indexing* + approximate pattern was designed. An alphabet indexing [18] can be considered as a discrete, unordered version of amino acid indices, mapping amino acids to a smaller alphabet (in our case, $\{1, 2, 3\}$). After transforming the original amino acid sequence into a sequence of the alphabet indices, a pattern is sought.

Starting with the proteins as entities E and an initial view v which returns the amino acid sequence of each protein in E, the two types of views (which return Boolean values) can be defined as follows:

$$V_1 \equiv \mathcal{P}_{p,A}[\mathcal{I}_I[\mathcal{S}_{i,j}[v]]] \tag{3}$$

$$V_2 \equiv \mathcal{T}_{s,t,b}[\mathcal{A}[\mathcal{D}_h[\mathcal{S}_{i,j}[v]]]] \tag{4}$$

See Table 2 for the definitions of the view operators. Note that each operator (except $\mathcal{A}$) is parametric, and the range of the parameters defines the space of views to be searched.

The parameter space was designed as follows (taking into account the existing knowledge about the sorting signals): $[i, j]$: the 72 intervals $[5n + 1, 5k]$ (where $n = 0 \ldots 8$ and $k = 1 \ldots 8$), p: all patterns of length $3 \sim 10$, A: approximate matching [22] with up to $1 \sim 3$ mismatches, h: all entries of the AAindex database, b, s, t: all possible combinations (appearing in the data). For the alphabet indexing I, we first start with a random alphabet indexing, and a local optimization strategy [19] using a neighborhood operator was adopted.

After extensively searching for *good* views which discriminate each of the signal types from sets of other signals, we combined the discovered views into a *decision list* whose nodes consist of conjunctions of views from V_1 and V_2 (except for distinguishing SP, where only 1 view from V_2 was used).

We finally reached a biologically interpretable simple hypothesis (view). The accuracy of the view is still good and is shown in Table 3. More details will be published in [2].

Table 2. View operators used in our experiments.

Operator	Type of Operator	Description
$\mathcal{S}_{i,j}$	string → string	Substring: return a specific substring $[i, j]$ of a given string.
$\mathcal{I}_I$	string → string	Alphabet Indexing: return an indexed string, using an alphabet indexing I.
$\mathcal{P}_{p,A}$	string → bool	Pattern Match: return true if pattern p matches the string using pattern matching algorithm A, and false otherwise.
$\mathcal{D}_h$	string → vector⟨float⟩	AAindex: a homomorphism of a mapping h: char→ float (h corresponds to an entry in the AAindex Database).
$\mathcal{A}$	vector⟨float⟩ → float	Sum: returns the sum of the values of each element in the vector. (this value is referred to as the *indexing sum* in the text)
$\mathcal{T}_{s,t,b}$	double → bool	Threshold: return $b \in \{\text{true}, \text{false}\}$ if the input value v is within a certain threshold. ($s \leq v \leq t$)
$\mathcal{B}_o$	bool×bool → bool	Boolean Operation: $o \in \{\text{and}, \text{or}\}$.
$\mathcal{N}$	bool → bool	Boolean Not Operation: Negation of the input.

Table 3. The Prediction Accuracy of the Final Hypothesis (scores of TargetP [7] in parentheses) The score is defined by: $\frac{(tp \times tn - fp \times fn)}{\sqrt{(tp+fn)(tp+fp)(tn+fp)(tn+fn)}}$ where tp, tn, fp, fn are the number of true positive, true negative, false positive, and false negative, respectively (Matthews correlation coefficient (MCC) [14]).

True category	# of seqs	Predicted category				Sensitivity	MCC
		cTP	mTP	SP	Other		
cTP	141	96 (120)	26 (14)	0 (2)	19 (5)	0.68 (0.85)	0.64 (0.72)
mTP	368	25 (41)	309 (300)	4 (9)	30 (18)	0.84 (0.82)	0.75 (0.77)
SP	269	6 (2)	9 (7)	244 (245)	10 (15)	0.91 (0.91)	0.92 (0.90)
Other	162	8 (10)	17 (13)	2 (2)	135 (137)	0.83 (0.85)	0.71 (0.77)
Specificity		0.71 (0.69)	0.86 (0.90)	0.98 (0.96)	0.70 (0.78)		

Discussion

We discuss here why and how this experiment successfully reached the good result. Simply because we worked with the human expert and took a *good* approach for the real data, and the *VOX* approach was so effective.

The iterations in the experiment process fitted completely into KDD process. We could understand the problem well, and it was easy to develop the computer program that analyzes the real data because the problem was already divided into appropriate stages. We simply had to combine those separate stages. These efficiencies are from the first advantage of *VOX* in Section 3.

According to the second advantage of *VOX*, human expert's intuition could be easily reflected into the method for the problem. As noted in [10], the success of the experiment completely depended on the expert intuition and experience. Although the expert was not provided details about the system during the experiment, we could discuss with the expert through views as a unit of communication. We asked the expert to tell us how we should view the data. *VOX* provided a good interface for this success.

Finally, based on the result of the experiment, we built an expert system which characterizes given protein sequences, called iPSORT [24]. *VOX* also provided benefit to us when building such an actual system, because we could use the result of the experiment directly in this system. This effect was produced by the third advantage of *VOX* .

We also conducted various experiments. Other experiments of us are reported in [1, 3, 12, 13].

5 Progresses in Discovery Science

We introduced a new paradigm for computational knowledge discovery, called *VOX*. We have persistently studied how the computer should help the knowledge discovery processes as a whole. We showed that the knowledge discovery process can be abstracted with *VOX*, and that such mathematical abstraction brings us good efficiency in actual discovery tasks.

Some problems still remain about *VOX*. 1) We cannot evaluate the efficiency of *VOX*, because there are not many actual tasks conducted using *VOX*. In fact, it is really difficult to evaluate exactly the efficiency of *VOX*. The reason is that, we cannot compare the actual discovery conducted using *VOX* with one not using *VOX*. 2) We developed a new programming language, called VML. However, it is not a practical language. When we conducted the experiments, we used C++ implementation version. Thus, it is ambiguous whether the mathematical abstraction was completely installed into the processes. VML is not fully implemented as a real computer language. However, Sumii and Bannai [20] tried to implement on the existed functional language.

Acknowledgement

This research was supported in part by Grant-in-Aid for Encouragement of Young Scientists and Grant-in-Aid for Scientific Research on Priority Areas (C) "Genome Information Science", Grant-in-Aid for Scientific Research (B) from the MEXT of Japan, and the Research for the Future Program of the Japan Society for the Promotion of Science.

References

[1] H. Bannai, Y. Tamada, O. Maruyama, and S. Miyano. VML: A view modeling language for computational knowledge discovery. In *Discovery Science*, Lecture Notes in Artificial Intelligence, 2001. To appear.

[2] H. Bannai, Y. Tamada, O. Maruyama, K. Nakai, and S. Miyano. Extensive feature detection of n-terminal protein sorting signals. *Bioinformatics*, 2001. To appear.
[3] H. Bannai, Y. Tamada, O. Maruyama, K. Nakai, and S. Miyano. Views: Fundamental building blocks in the process of knowledge discovery. In *Proceedings of the 14th International FLAIRS Conference*, pages 233–238. AAAI Press, 2001.
[4] E. Bloedorn and R. S. Michalski. Data-driven constructive induction. *IEEE Intelligent Systems*, pages 30–37, March/April 1998.
[5] B. D. Bruce. Chloroplast transit peptides: structure, function and evolution. *Trends Cell Biol.*, 10:440–447, 2000.
[6] M. G. Claros and P. Vincens. Computational method to predict mitochondrially imported proteins and their targeting sequences. *Eur. J. Biochem.*, 241(3):779–786, November 1996.
[7] O. Emanuelsson, H. Nielsen, S. Brunak, and G. von Heijne. Predicting subcellular localization of proteins based on their N-terminal amino acid sequence. *J. Mol. Biol.*, 300(4):1005–1016, July 2000.
[8] U. Fayyad, G. Piatetsky-Shapiro, and P. Smyth. From data mining to knowledge discovery in databases. *AI Magazine*, 17(3):37–54, 1996.
[9] S. Kawashima and M. Kanehisa. AAindex: Amino Acid index database. *Nucleic Acids Res.*, 28(1):374, 2000.
[10] P. Langley. The computer-aided discovery of scientific knowledge. In *Lecture Notes in Artificial Intelligence*, volume 1532, pages 25–39, 1998.
[11] O. Maruyama and S. Miyano. Design aspects of discovery systems. *IEICE Transactions on Information and Systems*, E83-D:61–70, 2000.
[12] O. Maruyama, T. Uchida, T. Shoudai, and S. Miyano. Toward genomic hypothesis creator: View designer for discovery. In *Discovery Science*, volume 1532 of *Lecture Notes in Artificial Intelligence*, pages 105–116, 1998.
[13] O. Maruyama, T. Uchida, K. L. Sim, and S. Miyano. Designing views in HypothesisCreator: System for assisting in discovery. In *Discovery Science*, volume 1721 of *Lecture Notes in Artificial Intelligence*, pages 115–127, 1999.
[14] B. W. Matthews. Comparison of predicted and observed secondary structure of t4 phage lysozyme. *Biochim. Biophys. Acta*, 405:442–451, 1975.
[15] H. Motoda. Fascinated by explicit understanding. *J. Japanese Society for Artificial Intelligence*, 14:615–625, 1999.
[16] K. Nakai. Protein sorting signals and prediction of subcellular localization. In P. Bork, editor, *Analysis of Amino Acid Sequences*, volume 54 of *Advances in Protein Chemistry*, pages 277–344. Academic Press, San Diego, 2000.
[17] K. Nakai and M. Kanehisa. A knowledge base for predicting protein localization sites in eukaryotic cells. *Genomics*, 14:897–911, 1992.
[18] S. Shimozono. Alphabet indexing for approximating features of symbols. *Theor. Comput. Sci.*, 210:245–260, 1999.
[19] S. Shimozono, A. Shinohara, T. Shinohara, S. Miyano, S. Kuhara, and S. Arikawa. Knowledge acquisition from amino acid sequences by machine learning system BONSAI. *J. IPS Japan*, 35(10):2009–2017, 1994.
[20] E. Sumii and H. Bannai. VMlambda: A functional calculus for scientific discovery. http://www.yl.is.s.u-tokyo.ac.jp/~sumii/pub/, 2001.
[21] G. von Heijne. The signal peptide. *J. Membr. Biol.*, 115:195–201, 1990.
[22] S. Wu and U. Manber. Fast text searching allowing errors. *Commun. ACM*, 35:83–91, 1992.
[23] HypothesisCreator - http://www.hypothesiscreator.net/.
[24] iPSORT - http://www.hypothesiscreator.net/iPSORT/.

Computing Optimal Hypotheses Efficiently for Boosting

Shinichi Morishita

University of Tokyo
moris@k.u-tokyo.ac.jp
http://gi.k.u-tokyo.ac.jp/~moris/

Abstract. This paper sheds light on a strong connection between AdaBoost and several optimization algorithms for data mining. AdaBoost has been the subject of much interests as an effective methodology for classification task. AdaBoost repeatedly generates one hypothesis in each round, and finally it is able to make a highly accurate prediction by taking a weighted majority vote on the resulting hypotheses. Freund and Schapire have remarked that the use of simple hypotheses such as single-test decision trees instead of huge trees would be promising for achieving high accuracy and avoiding overfitting to the training data. One major drawback of this approach however is that accuracies of simple individual hypotheses may not always be high, hence demanding a way of computing more accurate (or, the most accurate) simple hypotheses efficiently. In this paper, we consider several classes of simple but expressive hypotheses such as ranges and regions for numeric attributes, subsets of categorical values, and conjunctions of Boolean tests. For each class, we develop an efficient algorithm for choosing the optimal hypothesis.

1 Introduction

Classification has been a prominent subject of study in the machine learning and data mining literature. Let $\mathbf{x}_i$ denote a vector of values for attributes, which is usually called a *record* or a *tuple* in the database community. Let y_i denote the objective Boolean value that is either 1 or 0. We call a record $(\mathbf{x}_i, y_i)$ *positive* (resp. *negative*) if $y_i = 1$ ($y_i = 0$). Given $(\mathbf{x}_1, y_1), \ldots, (\mathbf{x}_N, y_N)$ as a training dataset, classification aims at deriving rules that are capable of predicting the objective value of y from $\mathbf{x}$ with a high probability.

For classification problems, decision trees are used mostly in practical applications. Recently, to further improve the prediction accuracy of existing classifiers, boosting techniques have received much interest among the machine learning and data mining communities [14]. A classifier is called a *weak hypothesis* if its predication accuracy regarding the training dataset is at least better than 1/2. A boosting algorithm tries to generate some weak hypotheses so that it makes it possible to perform a highly accurate prediction by combining those weak hypotheses. There have been many proposals for such boosting algorithms [14],

S. Arikawa and A. Shinohara (Eds.): Progress in Discovery Science 2001, LNAI 2281, pp. 471–481.

[7]. Freund and Schapire presented the most successful algorithm, named "AdaBoost", that solved many of the practical difficulties of the earlier boosting algorithms [9].

2 AdaBoost

The key idea behind AdaBoost is to maintain the record weights in the training dataset. AdaBoost assumes the existence of a *weak learner* that is able to output a weak hypothesis in a finite number of steps, though it is not always the case in practice. To overcome this problem, we will propose efficient weak learners that manage to output optimal hypotheses, but for the purpose of explanation, we continue the discussion by assuming that weak hypotheses can always be generated.

In each iteration AdaBoost calls on a weak learner to generate one weak hypothesis by considering the weighted records as the training dataset and updates the weights of the records to force the next call of the weak learner focus on the mis-predicted records. In this way, we prepare a set of voters with different characteristics. In the final step, we define the weight of each voter according to its prediction accuracy in the training dataset, and we generate the final hypothesis using a weighted majority vote.

2.1 Pseudo-code

We now present a pseudo-code for AdaBoost. First, the inputs to AdaBoost are a training dataset $\{(\mathbf{x}_1, y_1), \ldots, (\mathbf{x}_N, y_N)\}$, the initial weights $w_i^1 = 1/N$ ($i = 1, \ldots, N$), a weak learner named `WeakLearn`, and the integer T specifying number of iterations. AdaBoost repeats the following three steps for each $t = 1, 2, \ldots, T$:

1. Calculate the distribution p_i^t of each record $(\mathbf{x}_i, y_i)$ by normalizing weights w_i^t; namely,
$$p_i^t = \frac{w_i^t}{\sum_{i=1}^N w_i^t}.$$
2. Invoke `WeakLearn` to produce such a weak hypothesis $h_t : \{\mathbf{x}_1, \ldots, \mathbf{x}_N\} \to \{1, 0\}$ that the error ϵ_t of h_t is less than $1/2$, where ϵ_t is defined:
$$\epsilon_t = \sum_{i=1}^N p_i^t \, |h_t(\mathbf{x}_i) - y_i|.$$
Observe that $|h_t(\mathbf{x}_i) - y_i| = 1$ if h_t mis-predicts the objective value y_i. Otherwise, $|h_t(\mathbf{x}_i) - y_i| = 0$.
3. Set $\beta_t = \epsilon_t/(1 - \epsilon_t)$. Note that $\beta_t < 1$ since $\epsilon_t < 1/2$. We then set the new weight w_i^{t+1} for each $i = 1, \ldots, N$ according to the formula:
$$w_i^{t+1} = w_i^t \, \beta_t^{1-|h_t(\mathbf{x}_i) - y_i|}.$$

If h_t mis-predicts the objective value y_i of the i-th records (x_i, y_i), observe that

$$\beta_t^{1-|h_t(\mathbf{x}_i) - y_i|} = 1,$$

and hence the weight does not change; namely $w_i^{t+1} = w_i^t$. Otherwise the weight decreases, because $w_i^{t+1} < w_i^t \beta_t$. Put another way, the weights of incorrectly predicted records relatively increase so that the weak learner can focus on these "hard" records in the next step.

Lastly, AdaBoost outputs the final hypothesis h_f that is a weighted majority vote of T weak hypotheses where the weight $-\ln \beta_t$ is associated with hypothesis h_t:

$$h_f(\mathbf{x}) = \begin{cases} 1 \text{ if } \sum_{t=1}^{T}(-\ln \beta_t) h_t(\mathbf{x}) \geq \sum_{t=1}^{T}(-\ln \beta_t)\frac{1}{2} \\ 0 \text{ otherwise.} \end{cases}$$

2.2 Boosting Property

Freund and Schapire proved the following theorem, which is called the *boosting property*:

Theorem 1. [9] Let ϵ denote $(\sum_{i=1}^{N} |h_f(\mathbf{x}_i) - y_i|)/N$, the error of the final hypothesis. Then,

$$\epsilon \leq \Pi_{t=1}^{T} 2(\epsilon_t(1-\epsilon_t))^{\frac{1}{2}} \quad \blacksquare$$

The above theorem indicates that the error of the final hypothesis adapts to the error of individual weak hypothesis. Since the error ϵ_t is less than $1/2$, $2(\epsilon_t(1-\epsilon_t))^{\frac{1}{2}}$ is strictly less than 1, and it approaches to 0 when the error ϵ_t gets to be closer to 0. Thus, if most weak hypotheses are moderately accurate, the error of the final hypothesis drops exponentially fast, minimizing the number of iterations T efficiently.

2.3 Decision Stumps and Optimization

To design the weak learner of AdaBoost, in the literature, C4.5 has been frequently used as a weak learner [3, 8]. In [8] Freund and Schapire employed C4.5 to generate large decision trees, but they remarked that large decision trees were a kind of overly complex weak hypothesis and thereby failed to perform well against unseen test datasets.

This result led them to consider the other extreme class of decision trees, single-test decision trees named *decision stumps*. Decision stumps, on the other hand, are simple and hence may not be subject to overfitting to the training datasets. However, Domingo and Watanabe [6] reported that in later rounds of iterations in AdaBoost, the resulting decision stump is often too weak, yielding a large number of very weak hypotheses for improving the prediction accuracy of the final hypothesis.

One solution to overcome this problem is to select among decision stumps the optimal one that minimizes the error, because Theorem 1 indicates that the choice of highly accurate hypothesis enables sharp reduction of the error of the final hypothesis. To further investigate this optimization problem, we here introduce some new terms. Minimization of the error ϵ_t of the t-th hypothesis h_t is equivalent to maximization of the prediction accuracy $1 - \epsilon_t$.

$$\begin{aligned} 1 - \epsilon_t &= \sum_{\{i|h_t(\mathbf{x}_i)=y_i\}} p_i^t \\ &= \sum_{\{i|h_t(\mathbf{x}_i)=y_i=1\}} p_i^t + \sum_{\{i|h_t(\mathbf{x}_i)=y_i=0\}} p_i^t \\ &= \sum_{\{i|h_t(\mathbf{x}_i)=y_i=1\}} p_i^t + (\sum_{\{i|y_i=0\}} p_i^t - \sum_{\{i|h_t(\mathbf{x}_i)=1,y_i=0\}} p_i^t) \\ &- (\sum_{\{i|h_t(\mathbf{x}_i)=1,y_i=1\}} p_i^t - \sum_{\{i|h_t(\mathbf{x}_i)=1,y_i=0\}} p_i^t) + \sum_{\{i|y_i=0\}} p_i^t. \end{aligned}$$

Since $\sum_{\{i|y_i=0\}} p_i^t$ is independent of the choice of h_t, our goal is to maximize the first term enclosed in the parentheses, which we will simplify using g_i^t defined:

$$g_i^t = \begin{cases} p_i^t & \text{if } h_t(\mathbf{x}_i) = 1 \text{ and } y_i = 1 \\ -p_i^t & \text{if } h_t(\mathbf{x}_i) = 1 \text{ and } y_i = 0 \end{cases}$$

Then,

$$(\sum_{\{i|h_t(\mathbf{x}_i)=1,y_i=1\}} p_i^t - \sum_{\{i|h_t(\mathbf{x}_i)=1,y_i=0\}} p_i^t) = \sum_{\{i|h_t(\mathbf{x}_i)=1\}} g_i^t$$

If h_t outputs the correct answer to y_i, g_i^t is positive, thereby adding gain p_i^t to the prediction accuracy $1 - \epsilon_t$. Otherwise, p_i^t is deducted from the accuracy. We therefore call g_i^t the *accuracy gain* (or *gain*, for short) of prediction when h_t outputs 1. Consequently, maximization of the accuracy $1 - \epsilon_t$ is equivalent to maximization of the sum of gains, $\sum_{\{i|h_t(\mathbf{x}_i)=1\}} g_i^t$.

3 Main Results

We here present some optimization algorithms for selecting an optimal hypothesis h_t minimizing the sum of gains from a class of simple hypotheses. Classes of our interest include ranges and regions for numeric attributes, subsets of categorical values, and conjunctions of Boolean tests.

3.1 Optimal Ranges

We here consider hypotheses h_t such that

$$h_t(x_i) = 1 \quad \text{iff} \quad x_i \in [l, h].$$

Our goal is to compute an optimal range $[l, h]$ that maximizes the accuracy gain of h_t; namely

$$\max \sum_{\{i|x_i \in [l,h]\}} g_i^t.$$

Without loss of generality, we assume that x_i are sorted in an ascending order $x_1 \leq x_2 \leq x_3 \leq \ldots$, which requires $O(N \log N)$-time sorting cost though. If some $x_i, \ldots, x_{i+k}$ are equal, we merge them in the sense that we take the sum of gains $g_i^t + \ldots + g_{i+k}^t$, assign the sum to x_i, and rename the indexes so that all indexes are consecutive; namely, $x_1 < x_2 < \ldots$. Then, input the sequence of gains $g_1^t, g_2^t, \ldots$ to Kadena's algorithm [4]. Given a sequence of M reals $g_1, g_2, \ldots, g_M$, Kadena's algorithm computes an optimal range $[s,t]$ that maximizes $\sum_{i \in [s,t]} g_i$ in $O(M)$-time. M is at most the number of records N but is typically much smaller than N, and hence Kadena's algorithm works in $O(N)$-time.

It is natural to consider the use of two or more disjoint ranges for maximizing accuracy gain. Brin, Rastogi, and Shim presented an efficient way of computing the optimal set of at most k ranges in $O(kM)$-time [5], which is a highly non-trivial extension of Kadena's algorithm.

3.2 Optimal Regions

We have discussed the advantage of using region splitting hypotheses of the form:

$$h_t((x_{i1}, x_{i2})) = 1 \text{ iff } (x_{i1}, x_{i2}) \in [v_1, v_2] \times [w_1, w_2].$$

We here present how to efficiently compute an optimal rectangle R that maximizes the accuracy gain; namely,

$$\max \sum \{g_i^t \mid (x_{i1}, x_{i2}) \in R\}.$$

In order to limit the number of rectangles to a moderate number, we first divide the domain of x_{i1} (also, x_{i2}) into M non-overlapping buckets such that their union is equal to the original domain. M is at most the number of records N but is typically much smaller than N. Using those buckets, we divide the two dimensional plane into M^2 pixels, and we represent a rectangle as a union of those pixels. Although the number of rectangles is $O(M^4)$, it is straightforward to design an $O(M^3)$-time (namely, $O(N^3)$-time) algorithm by using Kadena's algorithm. The idea is that for each of ${}_MC_2$ pairs of rows, we apply Kadena's algorithm to calculate the optimal range of columns. Also, a subcubic time algorithm that uses funny matrix multiplication [15] is also available.

We are then interested in the design of efficient algorithm for computing more than one rectangles for maximizing the accuracy gain. However, Khanna, Muthukrishnan and Paterson remark that the problem is NP-hard [11]. Brin, Rastogi, and Shim present an approximation algorithm for this problem [5], however the approximation is within a factor of $\frac{1}{4}$ of the optimal solution. Thus computing the optimal set of more than one rectangles is computationally intractable.

So far we have been focusing on rectangles. In general there have been developed efficient algorithms for computing the optimized gain region among various classes of two dimensional connected regions whose boundaries are more flexible than those of rectangles. For instance, an *x-monotone* region is such a connected region that the intersection with any column is undivided, and a *rectilinear convex* region is such an x-monotone region that the intersection with any row is also undivided. The optimized gain x-monotone (rectilinear convex, respectively) region can be computed in $O(M^2)$-time [10] (in $O(M^3)$-time [16]). Since $M \leq N$, M in $O(M^2)$ and $O(M^3)$ can be replaced with N. The use of these regions is expected to further improve the prediction accuracy of the final hypothesis.

3.3 Optimal Conjunctions

Given a dataset $\{(x_{i1}, x_{i2}, \ldots, x_{in}, y_i) \mid i = 1, \ldots, N\}$ such that x_{ij} is Boolean valued; that is, $x_{ij} \in \{0, 1\}$. In this section, we consider hypotheses h_t that are conjunctions of simple tests on each attribute; that is;

$$h_t((x_{i1}, x_{i2}, \ldots, x_{in})) = 1 \text{ iff } x_{ij_1} = 1 \wedge \ldots \wedge x_{ij_M} = 1. \tag{1}$$

We are then interested in computing the optimal conjunction that maximizes the gain. The number of conjunctions of the form (1) is ${}_nC_M$. If we treat M as a variable, we can prove that the problem is NP-hard by reducing the problem to the NP-completeness of the minimum set cover problem according to the line suggested in [12, 13].

In practice, it would be reasonable to limit the number M of conjuncts to a small constant, say five. Then, the problem becomes to be tractable, but the problem still demands an efficient way of computing the optimal conjunction especially when the number of attributes n is fairly large.

Connection with Itemset Enumeration Problem We first remark that the problem has a strong connection with itemset enumeration problem [1]. We then generalize the idea of Apriori algorithm [2] for computing the optimal conjunction efficiently.

Let $a_1, a_2, \ldots, a_n$ be n items. We will identify a record with an itemset according to the mapping ϕ:

$$\phi : (x_{i1}, \ldots, x_{in}) \rightarrow \{a_j \mid x_{ij} = 1\}.$$

Example 1. $\phi((1, 0, 1, 1, 0)) = \{a_1, a_3, a_4\}$. ∎

We also regard the conjunction $x_{ij_1} = 1 \wedge \ldots \wedge x_{ij_M} = 1$ as $\{a_{j_1}, \ldots, a_{j_M}\}$.

Example 2. We associate $x_{i3} = 1 \wedge x_{i4} = 1$ with $\{a_3, a_4\}$. The property that the record $(1, 0, 1, 1, 0)$ satisfies $x_{i3} = 1 \wedge x_{i4} = 1$ can be rephrased by using words of itemsets; namely, $\{a_1, a_3, a_4\} \supseteq \{a_3, a_4\}$. ∎

In general, we have the following equivalence:

$$\begin{aligned} & h_t((x_{i1}, x_{i2}, \ldots, x_{in})) = 1 \\ \text{iff } & x_{ij_1} = 1 \wedge \ldots \wedge x_{ij_M} = 1 \\ \text{iff } & \phi((x_{i1}, x_{i2}, \ldots, x_{in})) \supseteq \{a_{j_1}, \ldots, a_{j_M}\}. \end{aligned}$$

In what follows, let $\mathbf{x}_i$ denote $(x_{i1}, x_{i2}, \ldots, x_{in})$, for simplicity. Now finding the optimal conjunction of the form (1) that maximizes the sum of gains is equivalent to the computation of the itemset I that maximizes

$$\sum_{\{i \mid \phi(\mathbf{x}_i) \supseteq I\}} g_i^t.$$

Let us call the above sum of gains the *gain* of I, and let $gain(I)$ denote the sum. The gain of I may appear to be similar to the so-called *support* of I, which is defined as $|\{i \mid \phi(\mathbf{x}_i) \supseteq I\}|/N$, where N is the number of all records. Thus one may consider the possibility of applying the Apriori algorithm to the calculation of the optimal itemset I that maximizes $gain(I)$. To this end, however, Apriori needs a major conversion.

Extending the Idea of Apriori Algorithm Let $support(I)$ denote the support of I. The support is *anti-monotone* with respect to set-inclusion of itemsets; that is, for any $J \supseteq I$, $support(J) \leq support(I)$. The Apriori algorithm uses this property to effectively prune away a substantial number of unproductive itemsets from its search space. However, the gain is not anti-monotone; namely, $J \supseteq I$ does not always imply $gain(J) \leq gain(I)$, because some g_i^t could be negative.

We solve this problem by modifying the Apriori algorithm so that it is capable of handling the anti-monotone gain function. Suppose that during the scan of the lattice of itemsets beginning with smaller itemsets and continuing to larger ones, we visit an itemset I. The following theorem presents a way of computing a tight upper bound on $gain(J)$ for any superset J of I.

Theorem 2. For any $J \supseteq I$,

$$gain(J) \leq \sum_{\{i \mid \phi(\mathbf{x}_i) \supseteq I, y_i = 1\}} g_i^t.$$

Proof. Recall

$$g_i^t = \begin{cases} p_i^t & \text{if } y_i = 1 \\ -p_i^t & \text{if } y_i = 0. \end{cases}$$

Then,

$$gain(J) = \sum_{\{i \mid \phi(\mathbf{x}_i) \supseteq J\}} g_i^t = \sum_{\{i \mid \phi(\mathbf{x}_i) \supseteq J, y_i = 1\}} p_i^t - \sum_{\{i \mid \phi(\mathbf{x}_i) \supseteq J, y_i = 0\}} p_i^t$$

Since $p_i^t \geq 0$ and $\{i \mid \phi(\mathbf{x}_i) \supseteq J, y_i = 1\} \subseteq \{i \mid \phi(\mathbf{x}_i) \supseteq I, y_i = 1\}$,

$$\sum_{\{i \mid \phi(\mathbf{x}_i) \supseteq J, y_i = 1\}} p_i^t \leq \sum_{\{i \mid \phi(\mathbf{x}_i) \supseteq I, y_i = 1\}} p_i^t.$$

Then,

$$\begin{aligned} gain(J) &= \sum_{\{i \mid \phi(\mathbf{x}_i) \supseteq J, y_i = 1\}} p_i^t - \sum_{\{i \mid \phi(\mathbf{x}_i) \supseteq J, y_i = 0\}} p_i^t \\ &\leq \sum_{\{i \mid \phi(\mathbf{x}_i) \supseteq J, y_i = 1\}} p_i^t \leq \sum_{\{i \mid \phi(\mathbf{x}_i) \supseteq I, y_i = 1\}} p_i^t. \end{aligned}$$ ∎

Definition 1. Let $u(I)$ denote the upper bound

$$\sum_{\{i|\phi(\mathbf{x}_i)\supseteq I, y_i=1\}} g_i^t.$$

∎

During the scan of the itemset lattice, we always maintain the temporarily maximum gain among all the gains calculated so far and set it to τ. If $u(I) < \tau$, no superset of I gives a gain greater than or equal to τ, and hence we can safely prune all supersets of I at once. On the other hand, if $u(I) \geq \tau$, I is promising in the sense that there might exist a superset J of I such that $gain(J) \geq \tau$.

Definition 2. Suppose that τ is given and fixed. An itemset is a k-itemset if it contains exactly k items. An itemset I is promising if $u(I) \geq \tau$. Let P_k denote the set of promising k-itemsets. ∎

Thus we will search $P_1 \cup P_2 \cup \ldots$ for the optimal itemset. Next, to accelerate the generation of P_k, we introduce a candidate set for P_k.

Definition 3. An itemset I is *potentially promising* if every proper subset of I is promising. Let Q_k denote the set of all potentially promising k-itemsets. ∎

The following theorem guarantees that Q_k is be a candidate set for P_k.

Theorem 3. $Q_k \supseteq P_k$. ∎

$\tau := 0$;
$Q_1 := \{I \mid I \text{ is a 1-itemset.}\}$; $k := 1$;
repeat begin
  If $k > 1$, generate Q_k from P_{k-1};
  For each $I \in Q_k$, scan all the records to compute $u(I)$ and $gain(I)$;
  $\tau := \max(\tau,\ \max\{gain(I) \mid I \in Q_k\})$;
  $P_k := \{I \in Q_k \mid u(I) \geq \tau\}$; $X := P_k$; $k++$;
end until $X = \phi$;
Return τ with its corresponding itemset;

Fig. 1. AprioriGain for Computing the Optimal Itemset

The benefit of Q_k is that Q_k can be obtained from P_{k-1} without scanning all records that may reside in the secondary disk. To this end, we use the idea of the `apriori-gen` function of the Apriori algorithm [2]; that is, we select two members in P_{k-1}, say I_1 and I_2, such that I_1 and I_2 share $(k-2)$ items in common, and then check to see whether each $(k-1)$-itemset included in $I_1 \cup I_2$ belongs to P_{k-1}, which can be determined efficiently by organizing P_{k-1} as a hash tree structure. We repeat this process to create Q_k. Figure 1 presents the overall algorithm, which we call AprioriGain (Apriori for optimizing Gain).

3.4 Optimal Subsets of Categorical Values

We here suppose that x_i itself denotes a single categorical value. Let $\{c_1, \ldots, c_M\}$ be the domain of the categorical attribute, where M is at most the number of records N but is typically much smaller than N. Typical hypotheses h_t would be of the form:

$$h_t(x_i) = 1 \text{ iff } x_i = c_j.$$

Computing the optimal choice of c_j that maximizes the accuracy gain is inexpensive. In practice, the number of categorical values M could be fairly large; for instance, consider the number of countries in the world. In such cases, the number of records satisfying $x_i = c_j$ could be relatively small, thereby raising the error of the hypothesis h_t. One way to overcome this problem is to use a subset S of $\{c_1, c_2, \ldots, c_M\}$ instead of a single value and to employ hypotheses of the form:

$$h_t(x_i) = 1 \text{ iff } x_i \in S.$$

Our goal is then to find S that maximizes the sum of gains $\sum_{x_i \in S} g_i^t$. Although the number of possible subsets of $\{c_1, c_2, \ldots, c_M\}$ is 2^M, we are able to compute the optimal subset S in $O(M)$-time. First, without loss of generality, we assume that

$$\sum_{\{i|x_i=c_1\}} g_i^t \geq \sum_{\{i|x_i=c_2\}} g_i^t \geq \ldots \geq \sum_{\{i|x_i=c_M\}} g_i^t.$$

Otherwise, we rename the indexes so that the above property is guaranteed. It is easy to see the following property.

Theorem 4. Let $S = \{c_j \mid \sum_{\{i|x_i=c_j\}} g_i^t \geq 0\}$. S maximizes $\sum_{x_i \in S} g_i^t$. ∎

Thus we only need to find the maximum index k such that

$$\sum_{\{i|x_i=c_k\}} g_i^t \geq 0,$$

returning $S = \{c_j \mid j = 1, \ldots, k\}$ as the answer. Consequently, the optimal subset can be computed in $O(M)$-time (or, $O(N)$-time since $M \leq N$).

4 Discussion

To improve the prediction accuracy of AdaBoost, we have presented efficient algorithms for several classes of simple but expressive hypotheses. In the literature, boosting algorithms have been developed in the machine learning community, while optimization algorithms for association rules and optimized ranges/regions have been proposed and studied in the database and data mining communities. This paper sheds light on a strong connection between AdaBoost and optimization algorithms for data mining.

Acknowledgement

Special thanks go to Jun Sese and Hisashi Hayashi. Jun Sese implemented AdaBoost empowered with the method of computing optimal ranges. Hisashi Hayashi also developed AdaBoost with AprioriGain. Both of them applied their implementations to real datasets. We also thank Carlos Domingo, Naoki Katoh, and Osamu Watanabe for motivating me to pursue this work. Comments from anonymous referees were useful to improve the readability.

References

1. R. Agrawal, T. Imielinski, and A. N. Swami. Mining association rules between sets of items in large databases. In *Proceedings of the 1993 ACM SIGMOD International Conference on Management of Data, Washington, D.C., May 26-28, 1993*, pages 207–216. ACM Press, 1993.
2. R. Agrawal and R. Srikant. Fast algorithms for mining association rules in large databases. In *VLDB'94, Proceedings of 20th International Conference on Very Large Data Bases, September 12-15, 1994, Santiago de Chile, Chile*, pages 487–499. Morgan Kaufmann, 1994.
3. E. Bauer and R. Kohavi. An empirical comparison of voting classification algorithms: Bagging, boosting, and variants. *Machine Learning*, 36(2):105–139, 1999.
4. J. Bentley. Programming pearls. *Communications of the ACM*, 27(27):865–871, Sept. 1984.
5. S. Brin, R. Rastogi, and K. Shim. Mining optimized gain rules for numeric attributes. In *Proceedings of the Fifth ACM SIGKDD International Conference on Knowledge Discovery and Data Mining, 15-18 August 1999, San Diego, CA USA*, pages 135–144. ACM Press, 1999.
6. C. Domingo and O. Watanabe. A modification of adaboost: A preliminary report. *Research Reports, Dept. of Math. and Comp. Sciences, Tokyo Institute of Technology*, (C-133), July 1999.
7. Y. Freund. Boosting a weak learning algorithm by majority. *Information and Computation*, 121(2):256–285, 1995.
8. Y. Freund and R. E. Schapire. Experiments with a new boosting algorithm. In *Machine Learning: Proceedings of the Thirteenth International Conference*, pages 148–156, 1996.
9. Y. Freund and R. E. Schapire. A decision-theoretic generalization of on-line learning and an application to boosting. *Journal of Computer and System Sciences*, 55(1):119–139, Aug. 1997.
10. T. Fukuda, Y. Morimoto, S. Morishita, and T. Tokuyama. Data mining using two-dimensional optimized accociation rules: Scheme, algorithms, and visualization. In *Proceedings of the 1996 ACM SIGMOD International Conference on Management of Data, Montreal, Quebec, Canada, June 4-6, 1996*, pages 13–23. ACM Press, 1996.
11. S. Khanna, S. Muthukrishnan, and M. Paterson. On approximating rectangle tiling and packing. In *Proceedings of the Ninth Annual ACM-SIAM Symposium on Discrete Algorithms*, pages 384–393, Jan. 1998.
12. S. Morishita. On classification and regression. In *Proceedings of Discovery Science, First International Conference, DS'98 — Lecture Notes in Artificial Intelligence*, volume 1532, pages 40–57, Dec. 1998.

13. S. Morishita and J. Sese. Traversing itemset lattices with statistical metric pruning. In *Proc. of ACM SIGACT-SIGMOD-SIGART Symp. on Database Systems (PODS)*, pages 226–236, May 2000.
14. R. E. Schapire. The strength of weak learnability (extended abstract). In *FOCS*, pages 28–33, 1989.
15. H. Tamaki and T. Tokuyama. Algorithms for the maxium subarray problem based on matrix multiplication. In *Proceedings of the Ninth Annual ACM-SIAM Symposium on Discrete Algorithms*, pages 446–452, Jan. 1998.
16. K. Yoda, T. Fukuda, Y. Morimoto, S. Morishita, and T. Tokuyama. Computing optimized rectilinear regions for association rules. In *Proceedings of the Third International Conference on Knowledge Discovery and Data Mining*, pages 96–103, Aug. 1997.

Discovering Polynomials to Fit Multivariate Data Having Numeric and Nominal Variables

Ryohei Nakano[1] and Kazumi Saito[2]

[1] Nagoya Institute of Technology
Gokiso-cho, Showa-ku, Nagoya 466-8555 Japan
nakano@ics.nitech.ac.jp
[2] NTT Communication Science Laboratories, NTT Corporation
2-4 Hikaridai, Seika, Soraku, Kyoto 619-0237 Japan
saito@cslab.kecl.ntt.co.jp

Abstract. This paper proposes an improved version of a method for discovering polynomials to fit multivariate data containing numeric and nominal variables. Each polynomial is accompanied with the corresponding nominal condition stating when to apply the polynomial. Such a nominally conditioned polynomial is called a rule. A set of such rules can be regarded as a single numeric function, and such a function can be approximated and learned by three-layer neural networks. The method selects the best from those trained neural networks with different numbers of hidden units by a newly introduced double layer of cross-validation, and restores the final rules from the best. Experiments using two data sets show that the proposed method works well in discovering very succinct and interesting rules even from data containing irrelevant variables and a small amount of noise.

1 Introduction

Discovering a numeric relationship, e.g., Kepler's third law $T = kr^{3/2}$, from data is an important issue of data mining systems. In AI field, the BACON systems [3] and many variants have employed a recursive combination of two variables to discover a polynomial-type relationship. This combinatorial approach, however, suffers from combinatorial explosion in search and lack of robustness. As an alternative we proposed a method called *RF5* [1] [13] employing a three-layer neural network learning to discover the optimal polynomial which has the appropriate number of terms and fits multivariate data having only numeric variables. It was shown that RF5 works even for data consisting of up to hundreds of thousands of samples [6].

In many real fields data contains nominal variables as well as numeric ones. For example, Coulomb's law $F = 4\pi\epsilon q_1 q_2 / r^2$ relating the force of attraction F of two particles with charges q_1 and q_2, respectively, separated by a distance r depends on ϵ, the permittivity of surrounding medium; i.e., if substance is "*water*"

[1] RF5 denotes *Rule extraction from Facts, version 5.*

S. Arikawa and A. Shinohara (Eds.): Progress in Discovery Science 2001, LNAI 2281, pp. 482–493.

then $F = 8897.352 q_1 q_2 / r^2$, if substance is "*air*" then $F = 111.280 q_1 q_2 / r^2$, and so on. Each polynomial is accompanied with the corresponding nominal condition stating when to apply it. In this paper we consider discovering a set of such pairs to fit multivariate data having numeric and nominal variables. Hereafter such a pair is referred as a *rule*.

To solve this type of problem, a *divide-and-conquer* approach may be a very natural one. For example, ABACUS [2] repeats the following: a numeric function describing some subset is sought, a nominal condition for discriminating the subset is calculated, and all the samples in the subset are removed. Clearly, in this approach the fitting reliability for a small subset will be necessarily degraded.

As an alternative approach, we have investigated a connectionist approach which uses all training samples at once in learning, invented a method called RF6 [7] which includes RF5 as a special case, and then proposed its modified version RF6.2 [15, 8] to improve the performance of rule restoring. However, even RF6.2 cannot automatically select the best model. Cross-validation [19] is commonly used for model selection. If cross-validation is used for both model selection and parameter estimation, it should be performed in the form of double layer [10].

Thus, the present paper proposes a newly modified law discovery method called *RF6.3* [2], which employs three-layer neural networks for parameter estimation, a double layer of cross-validation for model selection, and vector quantization for rule restoring. Section 2 formalizes the background and shows the basic framework of our approach. Section 3 explains the first half of RF6.3, where three-layer neural networks with different numbers of hidden units are trained by using a smart regularizer, and the best neural network is selected by a double layer of cross-validation. Section 4 explains the second half, where a set of rules are restored by using vector quantization from the best neural network selected above. Section 5 reports experimental results using two data sets.

2 Basic Framework

Let $(q_1, \cdots, q_{K_1}, x_1, \cdots, x_{K_2}, y)$ or $(\boldsymbol{q}, \boldsymbol{x}, y)$ be a vector of variables describing a sample, where q_k is a nominal explanatory variable[3], x_k is a numeric explanatory variable and y is a numeric target variable. Here, by adding extra categories, if necessary, without losing generality, we can assume that q_k exactly matches the only one category. Therefore, for each q_k we introduce a *dummy variable* expressed by q_{kl} as follows: $q_{kl} = 1$ (*true*) if q_k matches the l-th category, and $q_{kl} = 0$ (*false*) otherwise. Here $l = 1, \cdots, L_k$, and L_k is the number of distinct categories appearing in q_k.

As a true model governing data, we consider the following set of rules.[4]

$$\mathit{if} \bigwedge_{q_{kl} \in Q^i} q_{kl} \quad \mathit{then} \quad y = y(\boldsymbol{x}; \boldsymbol{\Theta}^i) + \varepsilon, \quad i = 1, \cdots, I^* \tag{1}$$

[2] RF6.3 denotes *Rule extraction from Facts, version 6.3*.
[3] An explanatory variable means an independent variable.
[4] Here a rule "if A then B" means "when A holds, apply B."

where Q^i and $\boldsymbol{\Theta}^i$ denote a set of dummy variables and a parameter vector respectively used in the i-th rule. Moreover, I^* is the number of rules and ε is a noise term, usually omitted in rule representation. As a class of numeric equations $y(\boldsymbol{x}; \boldsymbol{\Theta})$, we consider polynomials, whose power values are not restricted to integers, expressed by

$$y(\boldsymbol{x}; \boldsymbol{\Theta}) = w_0 + \sum_{j=1}^{J^*} w_j \prod_{k=1}^{K_2} x_k^{w_{jk}} = w_0 + \sum_{j=1}^{J^*} w_j \exp\left(\sum_{k=1}^{K_2} w_{jk} \ln x_k\right). \quad (2)$$

Here, each parameter w_j or w_{jk} is a real number, and J^* is an integer corresponding to the number of terms. $\boldsymbol{\Theta}$ is a vector constructed by arranging parameters w_j and w_{jk}. Note that Eq. (2) can be regarded as a feedforward computation of a three-layer neural network [13].

This paper adopts the following basic framework: firstly, single neural networks are trained by changing the number of hidden units; secondly, among the trained neural networks selected is the one having the best criterion value; finally, a rule set is restored from the best trained neural network. The framework is explained in more detail in the remaining part of the paper.

To express a nominal condition numerically, we introduce the following

$$c(\boldsymbol{q}; \boldsymbol{V}) = g\left(\sum_{k=1}^{K_1} \sum_{l=1}^{L_k} v_{kl} q_{kl}\right), \quad (3)$$

where $\boldsymbol{V}$ is a vector of parameters v_{kl} and the function g can be sigmoidal ($g(h) = \sigma(h)$) or exponential ($g(h) = \exp(h)$). Let Q be a set of dummy variables used in a conditional part of a rule. Then, in the case $g(h) = \exp(h)$, consider the following values of $\boldsymbol{V}$:

$$v_{kl} = \begin{cases} 0 \;\; if \; q_{kl} \in Q, \\ -\beta \; if \; q_{kl} \notin Q, \; q_{kl'} \in Q \; for \; some \; l' \neq l, \\ 0 \;\; if \; q_{kl'} \notin Q \; for \; any \; l', \end{cases} \quad (4)$$

where $\beta \gg 0$. Or, in the case $g(h) = \sigma(h)$, consider the following values of $\boldsymbol{V}$:

$$v_{kl} = \begin{cases} \beta_2 \;\; if \; q_{kl} \in Q, \\ -\beta_1 \; if \; q_{kl} \notin Q, \; q_{kl'} \in Q \; for \; some \; l' \neq l, \\ 0 \;\; if \; q_{kl'} \notin Q \; for \; any \; l', \end{cases} \quad (5)$$

where $\beta_1 \gg \beta_2 \gg 0$. Then $c(\boldsymbol{q}; \boldsymbol{V})$ is almost equivalent to the truth value of the nominal condition defined by Q, thus we can see that the following formula can closely approximate the final output value defined by Eq. (1). Thus, a set of rules can be merged into a single numeric function.

$$F(\boldsymbol{q}, \boldsymbol{x}; \boldsymbol{V}^1, \cdots, \boldsymbol{V}^{I^*}, \boldsymbol{\Theta}^1, \cdots, \boldsymbol{\Theta}^{I^*}) = \sum_{i=1}^{I^*} c(\boldsymbol{q}; \boldsymbol{V}^i)\; y(\boldsymbol{x}; \boldsymbol{\Theta}^i). \quad (6)$$

On the other hand, with an adequate number J, the following can completely express Eq. (6).

$$y(\boldsymbol{q}, \boldsymbol{x}; \boldsymbol{\Theta}) = w_0 + \sum_{j=1}^{J} w_j g \left(\sum_{k=1}^{K_1} \sum_{l=1}^{L_k} v_{jkl} \, q_{kl} \right) \exp \left(\sum_{k=1}^{K_2} w_{jk} \ln x_k \right), \tag{7}$$

where $\boldsymbol{\Theta}$ is rewritten as an M-dimensional vector constructed by arranging all parameters w_j, v_{jkl} and w_{jk}. Therefore, a set of rules can be learned by using a single neural network defined by Eq. (7). Note that even when Eq. (7) closely approximates Eq. (1), the weights for the nominal variables are no more limited to such weights as those defined in Eq. (4) or (5). Incidentally, Eq. (7) can be regarded as the feedforward computation of a three-layer neural network.

3 Learning and Model Selection

3.1 Learning Neural Networks with Regularizer

Let $D = \{(\boldsymbol{q}^\mu, \boldsymbol{x}^\mu, y^\mu) : \mu = 1, \cdots, N\}$ be training data, where N is the number of samples. We assume that each training sample $(\boldsymbol{q}^\mu, \boldsymbol{x}^\mu, y^\mu)$ is independent and identically distributed. Our goal in learning with neural networks is to find the optimal estimator $\boldsymbol{\Theta}^*$ that minimizes the generalization [5] error

$$\mathcal{G}(\boldsymbol{\Theta}) = E_D E_T \left(y^\nu - y(\boldsymbol{q}^\nu, \boldsymbol{x}^\nu; \boldsymbol{\Theta}(D)) \right)^2, \tag{8}$$

where $T = (\boldsymbol{q}^\nu, \boldsymbol{x}^\nu, y^\nu)$ denotes test data independent of the training data D. The least-squares estimate $\widehat{\boldsymbol{\Theta}}$ minimizes the following sum-of-squares error

$$E(\boldsymbol{\Theta}) = \frac{1}{2} \sum_{\mu=1}^{N} \left(y^\mu - y(\boldsymbol{q}^\mu, \boldsymbol{x}^\mu; \boldsymbol{\Theta}) \right)^2. \tag{9}$$

However, for a small D this estimation is likely to overfit the noise of D when we employ nonlinear models such as neural networks. Thus, for a small training data by using Eq. (9) as our criterion we cannot obtain good results in terms of the generalization performance defined in Eq. (8) [1].

It is widely known that adding some penalty term to Eq. (9) can lead to significant improvements in network generalization [1]. To improve both the generalization performance and the readability of the learning results, we adopt a method, called the *MCV (Minimum Cross-Validation) regularizer* [16], to learn a distinct penalty factor for each weight as a minimization problem over the cross-validation [19] error. Let $\boldsymbol{\Lambda}$ be an M-dimensional diagonal matrix whose diagonal elements are penalty factors $\lambda_m (> 0)$, and $\boldsymbol{a}^T$ denotes a transposed vector of $\boldsymbol{a}$. Then, finding a numeric function subject to Eq. (7) can be defined as the learning problem to find the $\boldsymbol{\Theta}$ that minimizes the following

$$\mathcal{E}(\boldsymbol{\Theta}) = E(\boldsymbol{\Theta}) + \frac{1}{2} \boldsymbol{\Theta}^T \boldsymbol{\Lambda} \boldsymbol{\Theta}. \tag{10}$$

[5] Generalization means the performance for new data.

In order to efficiently and constantly obtain good learning results, we employ a second-order learning algorithm called *BPQ* [14]. The BPQ algorithm adopts a quasi-Newton method [5] as a basic framework, and calculates the descent direction on the basis of a partial BFGS (Broyden-Fletcher-Goldfarb-Shanno) update and then efficiently calculates a reasonably accurate step-length as the minimal point of a second-order approximation. In our experiments [17], the combination of the squared penalty and the BPQ algorithm drastically improves the convergence performance in comparison to other combinations and at the same time brings about excellent generalization performance.

3.2 Neural Network Model Selection

In general, for given data we don't know in advance the optimal number J^* of hidden units. We must thus consider a criterion to choose it from several candidates, and for this purpose we adopt *cross-validation* [19], frequently used for evaluating generalization performance of neural networks [1].

S-fold cross-validation divides data D at random into S distinct segments (D_s, $s = 1, \cdots, S$), uses $S-1$ segments for training, and uses the remaining one for test. This process is repeated S times by changing the remaining segment and generalization performance is evaluated by using mean squared error over all tests. The extreme case of $S = N$ is known as the *leave-one-out* method. Since either poorly or overly fitted networks show poor generalization performance, by using cross-validation, we can select the optimal model.

In our case, however, note that cross-validation is already employed in the learning of neural networks with the MCV regularizer. In the learning with the MCV regularizer, the mean squared error of this cross-validation monotonically decreases when the number of hidden units increases due to the over-fitting. Thus, for model selection an outer loop of cross-validation is to be provided. This nested cross-validation is called *a double layer of cross-validation* [10]. Our procedure is summarized below.

double layer of cross-validation procedure :

step 1. Divide data D into S distinct segments (D_s, $s = 1, \cdots, S$).

step 2. For each $J = 1, 2, \cdots$, do the following:

step 2.1 For each $s = 1, \cdots, S$, repeat this step: by using the MCV regularizer and data $\{D - D_s\}$, train a neural network having J hidden units to obtain weights $\widehat{\boldsymbol{\Theta}}_s$.

step 2.2 Calculate the following mean squared error

$$MSE_{DLCV} = \frac{1}{N} \sum_{s=1}^{S} \sum_{\nu \in D_s} (y^\nu - y(\boldsymbol{q}^\nu, \boldsymbol{x}^\nu; \widehat{\boldsymbol{\Theta}}_s))^2, \qquad (11)$$

step 3. Select the optimal J^* that minimizes MSE_{DLCV}.

step 4. By using the MCV regularizer and data D, train a neural network having J^* hidden units to obtain the best neural network.

4 Rule Restoring

Assume that we have already obtained the best neural network. In order to get the final set of rules as described in Eq. (1), the following procedure is proposed.

rule restoring procedure :

step 1. From $I = 1, 2, \cdots$, select the optimal number I^* of representatives that minimizes cross-validation error.

step 2. For the optimal I^*, generate nominal conditions to get the final rules.

4.1 Selecting the Optimal Number of Representatives

For given data, since we don't know the optimal number I^* of different polynomials in advance, we must select it among candidates $I = 1, 2, \cdots$. For this purpose we again employ S-fold cross-validation.

The following describes the s-th process in the cross-validation. Let the following $c_j^{(s)\mu}$ be a coefficient of the j-th term in the polynomial for the μ-th sample of training data $\{D - D_s\}$:

$$c_j^{(s)\mu} = \widehat{w}_j^{(s)} \, g\left(\sum_{k=1}^{K_1}\sum_{l=1}^{L_k} \widehat{v}_{jkl}^{(s)} q_{kl}^{\mu}\right), \tag{12}$$

where $\widehat{w}_j^{(s)}$ and $\widehat{v}_{jkl}^{(s)}$ denote the weights of a neural network trained for data $\{D - D_s\}$. Coefficient vectors $\{\boldsymbol{c}^{(s)\mu} = (c_1^{(s)\mu}, \cdots, c_{J^*}^{(s)\mu})^T : \mu = 1, \cdots, N\}$ are quantized into representatives $\{\boldsymbol{r}^{(s)i} = (r_1^{(s)i}, \cdots, r_{J^*}^{(s)i})^T : i = 1, \cdots, I\}$, where I is the number of representatives. For vector quantization we employ the k-means algorithm [4] due to its simplicity. In the k-means algorithm, all of the coefficient vectors are assigned simultaneously to their nearest representatives, and then each representative is moved to its region's mean; this cycle is repeated until no further improvement. At last all of the coefficient vectors are partitioned into I disjoint subsets $\{G_i : i = 1, \cdots, I\}$ so that the following squared error distortion $d_{VQ}^{(s)}$ is minimized. Here N_i is the number of coefficient vectors belonging to G_i.

$$d_{VQ}^{(s)} = \sum_{i=1}^{I}\sum_{\mu \in G_i}\sum_{j=1}^{J^*}(c_j^{(s)\mu} - r_j^{(s)i})^2, \text{ where } r_j^{(s)i} = \frac{1}{N_i}\sum_{\mu \in G_i} c_j^{(s)\mu}, \tag{13}$$

Here we introduce a function $i^{(s)}(\boldsymbol{q})$ that returns the index of the representative vector minimizing the distance, i.e.,

$$i^{(s)}(\boldsymbol{q}^{\mu}) = \arg\min_i \sum_{j=1}^{J^*}(c_j^{(s)\mu} - r_j^{(s)i})^2. \tag{14}$$

Then we get the following set of rules using the representatives.

$$if\ i^{(s)}(\boldsymbol{q}) = i \quad then\ \widehat{y} = \widehat{w}_0^{(s)} + \sum_{j=1}^{J^*} r_j^i \prod_{k=1}^{K2} x_k^{\widehat{w}_{jk}^{(s)}}, \quad i = 1, \cdots, I, \tag{15}$$

where $\widehat{w}_0^{(s)}$ and $\widehat{w}_{jk}^{(s)}$ are the weights of a neural network trained for data $\{D - D_s\}$. By applying Eq. (15) to each sample ν in test segment D_s, we obtain

$$\widehat{y}^{(s)\nu} = \widehat{w}_0^{(s)} + \sum_{j=1}^{J^*} r_j^{i^{(s)}(\boldsymbol{q}^\nu)} \prod_{k=1}^{K2} (x_k^\nu)^{\widehat{w}_{jk}^{(s)}}, \tag{16}$$

The I minimizing the following mean squared error is selected as the optimal number I^* of representatives.

$$MSE_{VQCV} = \frac{1}{N} \sum_{s=1}^{S} \sum_{\nu \in D_s} \left(y^\nu - \widehat{y}^{(s)\nu}\right)^2. \tag{17}$$

4.2 Generating Conditional Parts

Now we have the best neural network and the optimal number I^* of representatives (i.e., different polynomials). To obtain the final rules, the following procedure is performed. First calculate

$$c_j^\mu = \widehat{w}_j\ g\left(\sum_{k=1}^{K_1} \sum_{l=1}^{L_k} \widehat{v}_{jkl} q_{kl}^\mu\right), \tag{18}$$

where $\widehat{w}_j$ and $\widehat{v}_{jkl}$ denote the weights of the best neural network. Through vector quantization we get the corresponding representatives $\{\boldsymbol{r}^i = (r_1^i, \cdots, r_{J^*}^i)^T : i = 1, \cdots, I^*\}$, and an indexing function $i(\boldsymbol{q})$

$$i(\boldsymbol{q}^\mu) = \arg\min_i \sum_{j=1}^{J^*} (c_j^\mu - r_j^i)^2. \tag{19}$$

Thus, we get the following set of rules using $i(\boldsymbol{q})$

$$if\ i(\boldsymbol{q}) = i \quad then\ \widehat{y} = \widehat{w}_0 + \sum_{j=1}^{J^*} r_j^i \prod_{k=1}^{K2} x_k^{\widehat{w}_{jk}}, \quad i = 1, \cdots, I. \tag{20}$$

where $\widehat{w}_0$ and $\widehat{w}_{jk}$ are the weights of the best neural network.

Next the function $i(\boldsymbol{q})$ must be transformed into a set of nominal conditions as described in Eq. (1). One reasonable approach is to perform this transformation by solving a simple classification problem whose training samples are $\{(\boldsymbol{q}^\mu, i(\boldsymbol{q}^\mu)) : \mu = 1, \cdots, N\}$, where $i(\boldsymbol{q}^\mu)$ indicates the class label of the μ-th sample. Here we employ the c4.5 decision tree generation program [9] due to its wide availability. From the generated decision tree, we can easily obtain the final rule set as described in Eq. (1).

5 Evaluation by Experiments

5.1 Experimental Settings

By using two data sets, we evaluated the performance of RF6.3.

The common experimental settings for training neural networks are as follows. The initial values for weights v_{jkl} and w_{jk} are independently generated according to a normal distribution with a mean of 0 and a standard deviation of 1; weights w_j are initially set to 0, but the bias w_0 is initially set to the average output over all training samples. The function for nominal condition is set as $g(h) = \sigma(h)$. Penalty factors $\boldsymbol{\lambda}$ are initially set to $\mathbf{1}$. As for the double layer of cross-validation, the inner one for the MCV regularizer is carried out by the leave-one-out method and the outer one to calculate MSE_{DLCV} is performed by the 10-fold method. The iteration is terminated when the gradient vector is sufficiently small, i.e., each elements of the gradient vector is less than 10^{-6}.

The common experimental settings for rule restoring are as follows. In the k-means algorithm, initial representatives $\{\boldsymbol{r}^i\}$ are randomly selected from coefficient vectors $\{\boldsymbol{c}^\mu\}$. For each I, trials are repeated 100 times with different initial values and the best result minimizing Eq. (13) is used. Cross-validation to calculate MSE_{VQCV} is carried out by the leave-one-out method. The candidate number I of representatives is incremented in turn from 1 until the cross-validation error increases. The c4.5 program is used with the initial settings.

5.2 Experiment Using Artificial Data Set

We consider the following artificial rules.

$$\begin{cases} \mathit{if}\ \ q_{21} \wedge (q_{31} \vee q_{33}) & \mathit{then}\ \ y = 2 + 3x_1^{-1}x_2^3 + 4x_3x_4^{1/2}x_5^{-1/3} \\ \mathit{if}\ \ (q_{22} \vee q_{23}) \wedge (q_{32} \vee q_{34}) & \mathit{then}\ \ y = 2 + 5x_1^{-1}x_2^3 + 2x_3x_4^{1/2}x_5^{-1/3} \\ \mathit{else} & \qquad\ \ \ y = 2 + 4x_1^{-1}x_2^3 + 3x_3x_4^{1/2}x_5^{-1/3} \end{cases} \tag{21}$$

Here we have three nominal and nine numeric explanatory variables, and the numbers of categories of q_1, q_2 and q_3 are set as $L_1 = 2$, $L_2 = 3$ and $L_3 = 4$, respectively. Clearly, variables $q_1, x_6, \cdots, x_9$ are irrelevant. Each sample is randomly generated with numeric variables $x_1, \cdots, x_9$ in the range of $(0, 1)$; The corresponding value of y is calculated to follow Eq. (21) with Gaussian noise with a mean of 0 and a standard deviation of 0.1. The number of samples is set to 400 (N = 400).

Table 1 compares the results of neural network model selection of RF6.3, where MSE_{DLCV} defined by Eq. (11) was used for evaluation. The table shows that MSE_{DLCV} was minimized when J = 2, which indicates that the model J = 2 is the best among the candidates, as is exactly the case of the original rules.

Table 2 compares the results of rule restoring of RF6.3 with different numbers of representatives I, where MSE_{VQCV} defined by Eq. (17) was used for evaluation. The table shows that MSE_{VQCV} was minimized when I = 3, which

Table 1. Neural network models comparison for artificial data set

models (# of hidden units)	J = 1	J = 2	J = 3
MSE_{DLCV}	0.9100	0.0104	0.0120

indicates that an adequate number of representatives is 3. Note that the case of I = 1 corresponds to the case where only numeric variables are used and all the nominal variables are ignored. Table 2 shows that nominal variables played a very important role since MSE_{VQCV} decreased by more than two orders of magnitude from I = 1 to I = 3.

Table 2. Final rule sets comparison of model J = 2 for artificial data set

models (# of rules)	I = 1	I = 2	I = 3	I = 4	I = 5
MSE_{VQCV}	3.09786	1.37556	0.00911	0.00923	0.00930

By applying the c4.5 program, we obtained the following decision tree. Here the number of training samples arriving at a leaf node is shown in parentheses, following a class number.

$$
\begin{array}{lll}
q_{21} = 0: & & \\
\mid \quad q_{32} = 1\text{: class2 (62)} & \Leftrightarrow & \boldsymbol{r}^2 = (+5.02, +1.99) \\
\mid \quad q_{32} = 0: & & \\
\mid \quad \mid \quad q_{34} = 0\text{: class3 (145)} & \Leftrightarrow & \boldsymbol{r}^3 = (+4.03, +2.96) \\
\mid \quad \mid \quad q_{34} = 1\text{: class2 (59)} & \Leftrightarrow & \boldsymbol{r}^2 = (+5.02, +1.99) \\
q_{21} = 1: & & \\
\mid \quad q_{32} = 1\text{: class3 (36)} & \Leftrightarrow & \boldsymbol{r}^3 = (+4.03, +2.96) \\
\mid \quad q_{32} = 0: & & \\
\mid \quad \mid \quad q_{34} = 0\text{: class1 (69)} & \Leftrightarrow & \boldsymbol{r}^1 = (+3.02, +3.93) \\
\mid \quad \mid \quad q_{34} = 1\text{: class3 (29)} & \Leftrightarrow & \boldsymbol{r}^3 = (+4.03, +2.96)
\end{array}
$$

Then, the following rule set was straightforwardly obtained. We can see that the rules almost equivalent to the original were found.

$$
\left\{
\begin{array}{l}
if \;\; q_{21} \wedge (q_{31} \vee q_{33}) \\
\qquad then \; y = 2.00 + 3.02 x_1^{-1.00} x_2^{+2.99} + 3.93 x_3^{+1.02} x_4^{+0.52} x_5^{-0.34} \\
if \;\; (q_{22} \vee q_{23}) \wedge (q_{32} \vee q_{34}) \\
\qquad then \; y = 2.00 + 5.02 x_1^{-1.00} x_2^{+2.99} + 1.99 x_3^{+1.02} x_4^{+0.52} x_5^{-0.34} \\
else \;\; y = 2.00 + 4.03 x_1^{-1.00} x_2^{+2.99} + 2.96 x_3^{+1.02} x_4^{+0.52} x_5^{-0.34}.
\end{array}
\right. \tag{22}
$$

5.3 Experiment Using Automobile Data Set

The Automobile data set [6] contains data on the car and truck specifications in 1985, and was used to predict prices based on these specifications. The data set

[6] from the UCI repository of machine learning databases

has 159 samples with no missing values (N = 159), and consists of 10 nominal and 14 numeric explanatory variables[7] and one target variable (price).

Table 3 compares the results of neural network model selection of RF6.3, where MSE_{DLCV} was used for evaluation. The table shows that the model J = 1 is slightly better than the model J = 2.

Table 3. Neural network models comparison for automobile data set

models (# of hidden units)	J = 1	J = 2
MSE_{DLCV}	0.1035	0.2048

Table 4 compares the results of rule restoring of RF6.3 with different numbers of representatives I, where MSE_{VQCV} was used for evaluation. The table shows that MSE_{VQCV} was minimized when I = 5, which indicates that an adequate number of representatives is 5. Note again that the case of I = 1 corresponds to the case where only numeric variables are used. The table shows that nominal variables played an important role since MSE_{VQCV} decreased by about one order of magnitude when I was changed from 1 to 5.

Table 4. Final rule sets comparison of model J = 1 for automobile data set

models (# of rules)	I = 1	I = 2	I = 3	I = 4	I = 5	I = 6
MSE_{VQCV} ($\times 10^6$)	15.32	5.00	3.67	2.74	1.77	1.86

The polynomial part of the final rules was as follows. Note that half of 14 numeric variables are discarded here. Since the second term of Eq. (23) is always positive, the coefficient value r^i indicates the car price setting tendency for similar specifications.

$$y = -1989.27 + r^i x_2^{+1.458} x_3^{-0.561} x_4^{+0.727} x_5^{-1.219} x_6^{+0.567} x_9^{-0.141} x_{13}^{-0.189} \quad (23)$$
$$r^1 = +7.838,\ r^2 = +6.406,\ r^3 = +5.408,\ r^4 = +4.319,\ r^5 = +3.734.$$

The relatively simple nominal conditions were obtained.[8] In the final rules the following price setting groups are found:

Very High price setting: {Mercedes-Benz}.
High price setting: {BMW, Volvo turbo, Saab turbo, convertible, 5-cylinder without Mercedes-Benz, 6-cylinder turbo}.
Middle price setting: {Volvo std, Saab std, Peugot turbo, 6-cylinder std without BMW}

[7] We ignored one nominal variable (engine location) having the same value.
[8] We used the c4.5 rules program to obtain the rule set.

Low price setting: {Volkswagen, Peugot std, Mazda 4wd/fwd, Nissan 4-door non-6-cylinder, Honda non-mpfi, Toyota rwd non-6-cylinder, mpfi non-6-cylinder without Volvo, non-diesel non-6-cylinder without {Mercedes-Benz, Peugot, Vokswagen}}.

Very Low price setting: others.

6 Conclusion

To discover polynomials to fit multivariate data having numeric and nominal variables, we have proposed a new method employing neural networks learning. Our experiment using artificial data showed that it can successfully restore rules equivalent to the original despite the existence of irrelevant variables and a small noise. The other experiment using real data showed that it can discover interesting rules in the domain, making good use of nominal variables.

We believe the present method has great potential as a data mining tool. Actually, the method has been applied to several interesting applications such as stock price assessment [18], Web dynamics discovery [11], Earth ecosystem model revision [12], and so on. In the near future we will make our software available for potential users who want to apply our methods to their own data.

References

1. C. M. Bishop. *Neural networks for pattern recognition.* Clarendon Press, Oxford, 1995.
2. B. C. Falkenhainer and R. S. Michalski. Integrating quantitative and qualitative discovery in the abacus system. In *Machine Learning: An Artificial Intelligence Approach (Vol. 3)*, pages 153–190. Morgan Kaufmann, 1990.
3. P. Langley, H. A. Simon, G. Bradshaw, and J. Zytkow. *Scientific discovery: computational explorations of the creative process.* MIT Press, 1987.
4. S. P. Lloyd. Least squares quantization in pcm. *IEEE Trans. on Information Theory*, IT-28(2):129–137, 1982.
5. D. G. Luenberger. *Linear and nonlinear programming.* Addison-Wesley, 1984.
6. R. Nakano and K. Saito. Computational characteristics of law discovery using neural networks. In *Proc. 1st Int. Conference on Discovery Science, LNAI 1532*, pages 342–351, 1998.
7. R. Nakano and K. Saito. Discovery of a set of nominally conditioned polynomials. In *Proc. 2nd Int. Conference on Discovery Science, LNAI 1721*, pages 287–298, 1999.
8. R. Nakano and K. Saito. Finding polynomials to fit multivariate data having numeric and nominal variables. In *Proc. 4th Int. Symoposium on Intelligent Data Analysis* (to appear).
9. J. R. Quinlan. *C4.5: programs for machine learning.* Morgan Kaufmann, 1993.
10. B. D. Ripley. *Pattern recognition and neural networks.* Cambridge Univ Press, 1996.
11. K. Saito and P. Langley. Discovering empirical laws of Web dynamics. In *Proc. 2002 International Symposium on applications and the Intenet* (to appear).

12. K. Saito, P. Langley, and et al. Computational revision of quantitative scientific models. In *Proc. 4th International Conference on Discovery Science* (to appear).
13. K. Saito and R. Nakano. Law discovery using neural networks. In *Proc. 15th International Joint Conference on Artificial Intelligence*, pages 1078–1083, 1997.
14. K. Saito and R. Nakano. Partial BFGS update and efficient step-length calculation for three-layer neural networks. *Neural Computation*, 9(1):239–257, 1997.
15. K. Saito and R. Nakano. Discovery of a set of nominally conditioned polynomials using neural networks, vector quantizers, and decision trees. In *Proc. 3rd Int. Conference on Discovery Science, LNAI 1967*, pages 325–329, 2000.
16. K. Saito and R. Nakano. Discovery of relevant weights by minimizing cross-validation error. In *Proc. PAKDD 2000, LNAI 1805*, pages 372–375, 2000.
17. K. Saito and R. Nakano. Second-order learning algorithm with squared penalty term. *Neural Computation*, 12(3):709–729, 2000.
18. K. Saito, N. Ueda, and et al. Law discovery from financial data using neural networks. In *Proc. IEEE/IAFE/INFORMS Conference on Computational Intelligence for Financial Engineering*, pages 209–212, 2000.
19. M. Stone. Cross-validatory choice and assessment of statistical predictions (with discussion). *Journal of the Royal Statistical Society B*, 64:111–147, 1974.

Finding of Signal and Image by Integer-Type Haar Lifting Wavelet Transform

Koichi Niijima and Shigeru Takano

Department of Informatics, Kyushu University
6-1, Kasuga-koen, Kasuga, Fukuoka, 816-8580, Japan
{niijima, takano}@i.kyushu-u.ac.jp

Abstract. This paper describes a new method for finding portions having the same feature in target signals or images from a time series or a reference image. The new method uses an integer-type Haar lifting wavelet transform. Free parameters contained in this transform are learned by using training signals or images. The advantage of this method is to be able to find portions having the same feature in the targets, and to realize robust extraction due to rounding-off arithmetic in the trained transform. In simulations, we show how well the method finds geomagnetic sudden commencements from time series of geomagnetic horizontal components, and extracts facial images from a snapshot.

1 Introduction

Signal and image retrieval is to search huge database for query signal and image. This does not aim at finding target portions from a time series or a reference image. It is important for the retrieval system to have such finding ability. So far, various template matching techniques have been applied to extract target signal and image. These techniques are only to extract patterns similar to a template, and are time consuming.

This paper proposes a new method for finding a target portion from a time series or a reference image. The proposed method does not extract patterns similar to a template, but finds portions having the same feature in the targets. The new method utilizes an integer-type Haar lifting wavelet transform (IHLWT) ([1, 4]). IHLWT contains rounding-off arithmetic for mapping integers to integers. By virtue of this arithmetic, robust extraction can be achieved. IHLWT is designed from an integer-type Haar wavelet transform (IHWT) by adding a lifting term to IHWT. The lifting term contains free parameters. Since the filter length of IHWT is only two, we can shorten the filter length of IHLWT by choosing small number of free parameters. Actually, we select five free parameters in our simulations for realizing fast extraction.

In this paper, we learn the free parameters included in IHLWT so that it has the feature of target signal and image. This paper assumes that the targets include many high frequency components. Therefore, to capture the features of the targets, we impose a condition that high frequency components in IHLWT vanish for training signals or images. The free parameters cannot be determined

S. Arikawa and A. Shinohara (Eds.): Progress in Discovery Science 2001, LNAI 2281, pp. 494–503.

only by this condition. So, we put an additional condition of minimizing the squared sum of the free parameters. Small values of free parameters can avoid the error propagation in filtering. We call an IHLWT with the learned parameters a trained IHLWT.

Finding of targets is carried out by applying the trained IHLWT to time series or reference images. Since the lifting term in the trained IHLWT has been rounded off, robust extraction of targets from signal or image can be achieved.

2 IHLWT

Let c_l^1 denote a signal consisting of integers. Then, IHWT can be written as

$$\hat{c}_m^0 = \lfloor \frac{c_{2m}^1 + c_{2m+1}^1}{2} \rfloor, \tag{1}$$

$$\hat{d}_m^0 = c_{2m+1}^1 - c_{2m}^1, \tag{2}$$

where $\lfloor z \rfloor$ denotes the largest integer not exceeding z. The integers $\hat{c}_m^0$ and $\hat{d}_m^0$ denote low and high frequency components of the signal c_l^1 respectively.

We build an IHLWT by lifting up the high frequency component $\hat{d}_m^0$ and by rounding off the lifting term as follows:

$$c_m^0 = \hat{c}_m^0, \tag{3}$$

$$d_m^0 = \hat{d}_m^0 - \lfloor \sum_k \tilde{s}_{k,m} \hat{c}_k^0 \rfloor, \tag{4}$$

where $\tilde{s}_{k,m}$ indicate free real parameters. It is important for (3) and (4) to have an inverse transform which has been given in [5]. Because it guarantees that c_l^1 is equivalent to $\{c_m^0, d_m^0\}$. However, this paper uses only the decomposition formulas (3) and (4).

3 Training of IHLWT

3.1 In Case of Signal

The training of IHLWT is carried out by determining free parameters $\tilde{s}_{k,m}$ in (4) adaptive to training signals. Let $c_m^{1,\nu}$, $\nu = 0, 1, ..., L-1$ be training signals. We compute low and high frequency components $\hat{c}_m^{0,\nu}$ and $\hat{d}_m^{0,\nu}$ by

$$\hat{c}_m^{0,\nu} = \lfloor \frac{c_{2m}^{1,\nu} + c_{2m+1}^{1,\nu}}{2} \rfloor, \quad \nu = 0, \cdots, L-1,$$

$$\hat{d}_m^{0,\nu} = c_{2m+1}^{1,\nu} - c_{2m}^{1,\nu}, \quad \nu = 0, \cdots, L-1.$$

We substitute these values into (4) to get

$$d_m^0 = \hat{d}_m^0 - \lfloor \sum_{k=m-N}^{m+N} \tilde{s}_{k,m} \hat{c}_k^0 \rfloor,$$

where we restricted the number of $\tilde{s}_{k,m}$ to $2N+1$. To determine free parameters $\tilde{s}_{k,m}$, the following vanishing conditions are imposed:

$$\hat{d}_m^{0,\nu} - \lfloor \sum_{k=m-N}^{m+N} \tilde{s}_{k,m}\hat{c}_k^{0,\nu} \rfloor = 0, \quad \nu = 0, \cdots, L-1. \tag{5}$$

Since $\lfloor \cdot \rfloor$ denotes rounded integer arithmetic, (5) is equivalent to

$$\hat{d}_m^{0,\nu} \leq \sum_{k=m-N}^{m+N} \tilde{s}_{k,m}\hat{c}_k^{0,\nu} < \hat{d}_m^{0,\nu} + 1, \quad \nu = 0, \cdots, L-1. \tag{6}$$

These inequalities yield a domain surrounded by L hyperplanes in $(2N+1)$-dimensional space. It is very important to verify whether this domain is empty or not. This depends on the vectors $(\hat{c}_{m-N}^{0,\nu}, ..., \hat{c}_{m+N}^{0,\nu})$ and the high frequency components $\hat{d}_m^{0,\nu}$ for $\nu = 0, 1, ..., L-1$. We assume here that the domain is not empty. When $L \leq 2N+1$ and the vectors $(\hat{c}_{m-N}^{0,\nu}, ..., \hat{c}_{m+N}^{0,\nu})$, $\nu = 0, 1, ..., 2N$ are linearly independent, the domain is not empty. In simulation given in Section 5, we choose $L = 2N$ or $L = 2N+1$.

The parameters $\tilde{s}_{k,m}$ can not be determined only by (6). One method of determining these parameters is to minimize their squared sum

$$\sum_{k=m-N}^{m+N} (\tilde{s}_{k,m})^2 \tag{7}$$

subject to (6). Using the penalty method, this minimization problem can be transformed into the minimization problem

$$\sum_{k=m-N}^{m+N} (\tilde{s}_{k,m})^2 + K \sum_{\nu=0}^{L-1} \left((\hat{d}_m^{0,\nu} - \sum_{k=m-N}^{m+N} \tilde{s}_{k,m}\hat{c}_k^{0,\nu})_+^2 + (\sum_{k=m-N}^{m+N} \tilde{s}_{k,m}\hat{c}_k^{0,\nu} - \hat{d}_m^{0,\nu} - 1)_+^2 \right) \to \text{min.}, \tag{8}$$

where $z_+^2 = z^2$ for $z > 0$, and 0 otherwise, and K is a penalty constant. The problem (8) can be solved by gradient methods such as the steepest decent method.

3.2 In Case of Image

In case of image, we give only one training image $C_{i,j}^1$. By applying IHWT described by (1) and (2) to $C_{i,j}^1$ in horizontal direction, and then to the resulting components in vertical direction, we compute low and high frequency components $\hat{C}_{m,n}^0$, $\hat{D}_{m,n}^0$, and $\hat{E}_{m,n}^0$. We lift up the components $\hat{D}_{m,n}^0$ and $\hat{E}_{m,n}^0$ by using (3) and (4) to get

$$D^0_{m,n} = \hat{D}^0_{m,n} - \lfloor \sum_{k=m-N}^{m+N} \tilde{s}^d_{k,m} \hat{C}^0_{k,n} \rfloor, \tag{9}$$

$$E^0_{m,n} = \hat{E}^0_{m,n} - \lfloor \sum_{k=n-N}^{n+N} \tilde{s}^e_{k,n} \hat{C}^0_{m,k} \rfloor. \tag{10}$$

The number of each of $\tilde{s}^d_{k,m}$ and $\tilde{s}^e_{k,n}$ is equal to $2N+1$. We impose the following vanishing conditions:

$$\hat{D}^0_{m,n+l} - \lfloor \sum_{k=m-N}^{m+N} \tilde{s}^d_{k,m} \hat{C}^0_{k,n+l} \rfloor = 0, \ \ l = 0, ..., L-1, \tag{11}$$

$$\hat{E}^0_{m+l,n} - \lfloor \sum_{k=n-N}^{n+N} \tilde{s}^e_{k,n} \hat{C}^0_{m+l,k} \rfloor = 0, \ \ l = 0, ..., L-1. \tag{12}$$

These conditions are different from those in the case of signal. They represent vanishing conditions around the point (m, n).

By the definition of $\lfloor \cdot \rfloor$, (11) and (12) can be written in inequality form as

$$\hat{D}^0_{m,n+l} \le \sum_{k=m-N}^{m+N} \tilde{s}^d_{k,m} \hat{C}^0_{k,n+l} < \hat{D}^0_{m,n+l} + 1, \qquad l = 0, ..., L-1, \tag{13}$$

$$\hat{E}^0_{m+l,n} \le \sum_{k=n-N}^{n+N} \tilde{s}^e_{k,n} \hat{C}^0_{m+l,k} < \hat{E}^0_{m+l,n} + 1, \qquad l = 0, ..., L-1. \tag{14}$$

Each of (13) and (14) yields a domain surrounded by L hyperplanes in $(2N+1)$-dimensional space. These domains may be empty. Since the domains are not empty provided $L = 2N+1$, we consider this case in simulation.

Parameters $\tilde{s}^d_{k,m}$ and $\tilde{s}^e_{k,n}$ can not be determined only by (13) and (14). We determine the parameters, as in the case of signal, by minimizing their squared sums

$$\sum_{k=m-N}^{m+N} (\tilde{s}^d_{k,m})^2 \tag{15}$$

and

$$\sum_{k=n-N}^{n+N} (\tilde{s}^e_{k,n})^2 \tag{16}$$

under conditions (13) and (14), respectively.

Using the penalty method, the first minimization problem can be transformed into the minimization problem

$$\sum_{k=m-N}^{m+N} (\tilde{s}^d_{k,m})^2 + K \sum_{l=0}^{L-1} \Bigg((\hat{D}^0_{m,n+l} - \sum_{k=m-N}^{m+N} \tilde{s}^d_{k,m} \hat{C}^0_{k,n+l})^2_+$$

$$+ \left(\sum_{k=m-N}^{m+N} \tilde{s}_{k,m}^{d} \hat{C}_{k,n+l}^{0} - \hat{D}_{m,n+l}^{0} - 1 \right)_{+}^{2} \Bigg) \to \min..$$

Using the penalty method, the second minimization problem can be transformed into the minimization problem

$$\sum_{k=n-N}^{n+N} (\tilde{s}_{k,n}^{e})^2 + K \sum_{l=0}^{L-1} \Bigg((\hat{E}_{m+l,n}^{0} - \sum_{k=n-N}^{n+N} \tilde{s}_{k,n}^{e} \hat{C}_{m+l,k}^{0})_{+}^{2}$$

$$+ \left(\sum_{k=n-N}^{n+N} \tilde{s}_{k,n}^{e} \hat{C}_{m+l,k}^{0} - \hat{E}_{m+l,n}^{0} - 1 \right)_{+}^{2} \Bigg) \to \min..$$

Both the problems can be solved by the steepest decent method.

It is time consuming to compute free parameters for all the points (m, n). So we compute them only for the points (m, n) where high frequency components $\hat{D}_{m,n}^{0}$ and $\hat{E}_{m,n}^{0}$ are large.

4 Finding of Targets

4.1 In Case of Signal

Let $c_l^{1,r}$ be a test signal. At first, we compute low and high frequency components of $c_l^{1,r}$ as

$$\hat{c}_p^{0,r} = \lfloor \frac{c_{2p}^{1,r} + c_{2p+1}^{1,r}}{2} \rfloor, \tag{17}$$

$$\hat{d}_p^{0,r} = c_{2p+1}^{1,r} - c_{2p}^{1,r}. \tag{18}$$

Then, by combining parameters $\tilde{s}_{k,m}$ determined in Section 3.1 with these components, we compute

$$d_p^{0,r} = \hat{d}_p^{0,r} - \lfloor \sum_{k=m-N}^{m+N} \tilde{s}_{k,m} \hat{c}_{k-m+p}^{0,r} \rfloor. \tag{19}$$

An algorithm for finding signals similar to a target one involves the following steps:

(i) Compute low and high frequency components $\hat{c}_p^{0,r}$ and $\hat{d}_p^{0,r}$ by using (17) and (18).
(ii) Compute high frequency components $d_p^{0,r}$ by using (19).
(iii) Find the points p where

$$I_p = |\hat{d}_p^{0,r}| - |d_p^{0,r}|$$

is larger than a given threshold.

4.2 In Case of Image

Let $C_{i,j}^{1,r}$ be a reference image. For $C_{i,j}^{1,r}$, we compute low and high frequency components $\hat{C}_{p,q}^{0,r}$, $\hat{D}_{p,q}^{0,r}$, and $\hat{E}_{p,q}^{0,r}$ in the same way as in Section 3.2. Next, we choose the position (p,q) where the high frequency components are large. At the selected position (p,q), by combining the already determined parameters $\tilde{s}_{k,m}^{d}$ and $\tilde{s}_{k,n}^{e}$ in Section 3.2 with these components, we compute

$$D_{p,q}^{0,r} = \hat{C}_{p,q}^{0,r} - \lfloor \sum_{k=m-N}^{m+N} \tilde{s}_{k,m}^{d} \hat{D}_{k-m+p,q}^{0,r} \rfloor, \tag{20}$$

$$E_{p,q}^{0,r} = \hat{C}_{p,q}^{0,r} - \lfloor \sum_{k=n-N}^{n+N} \tilde{s}_{k,n}^{e} \hat{E}_{p,k-n+q}^{0,r} \rfloor. \tag{21}$$

Since $D_{p,q}^{0,r}$ and $E_{p,q}^{0,r}$ are integers, the positions where $D_{p,q}^{0,r}$ and $E_{p,q}^{0,r}$ vanish may become parts of similar images to a target one. However, such positions do not always represent the target one, because parameters $s_{k,m}^{d}$ and $s_{k,n}^{e}$ are determined using parts of the training image.

To extract the target image efficiently, we search for the blocks with the same size of the target one, in which the number of selected points exceeds some threshold value.

An algorithm for extracting images similar to the target one involves the following steps:

(i) Compute low and high frequency components $\hat{C}_{p,q}^{0,r}$, $\hat{D}_{p,q}^{0,r}$, and $\hat{E}_{p,q}^{0,r}$ from reference image $C_{i,j}^{1,r}$ in the same way as in Section 3.2.

(ii) Choose the points (p,q) where high frequency components $\hat{D}_{p,q}^{0,r}$ and $\hat{E}_{p,q}^{0,r}$ are large.

(iii) At the selected points (p,q), compute $D_{p,q}^{0,r}$ and $E_{p,q}^{0,r}$ by using (20) and (21).

(iv) Find the points (p,q) where

$$I_{p,q}^{D} = |\hat{D}_{p,q}^{0,r}| - |D_{p,q}^{0,r}|$$

and

$$I_{p,q}^{E} = |\hat{E}_{p,q}^{0,r}| - |E_{p,q}^{0,r}|$$

are larger than a given threshold.

(v) Count the number of the selected points in step (iv) in a block with the same size of a query image, and extract blocks in each of which the number exceeds some threshold value.

5 Simulation

5.1 Finding of Geomagnetic Sudden Commencements

Sudden increase of the magnetic field called geomagnetic sudden commencement (SC) is observed in the magnetosphere. Real time extraction of SCs is important because SCs are frequently followed by severe geomagnetic storms during which many hazardous accidents may occur. The SC in low latitude is typically observed as a sharp increase of the geomagnetic horizontal (H) component. An example of H-components is shown in Fig. 1. This time series contains a typical SC observed by specialists in geomagnetism.

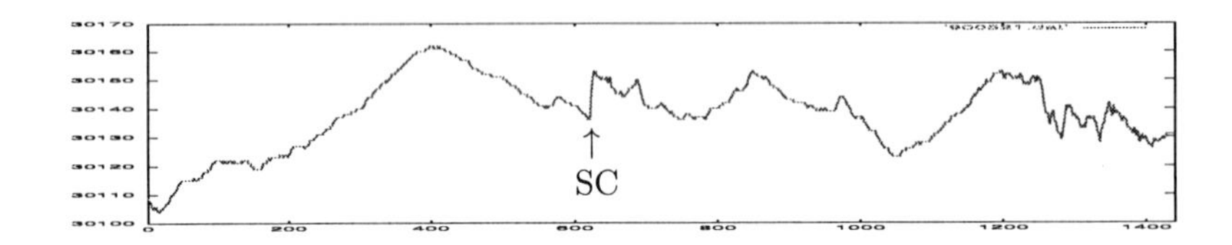

Fig. 1. H-components in Kakioka station

We trained an IHLWT by the method in Section 3.1 using 4 time series consisting of H-components around SCs in each station. For lack of space, we illustrate only H-components around SCs in Kakioka station in Fig. 2.

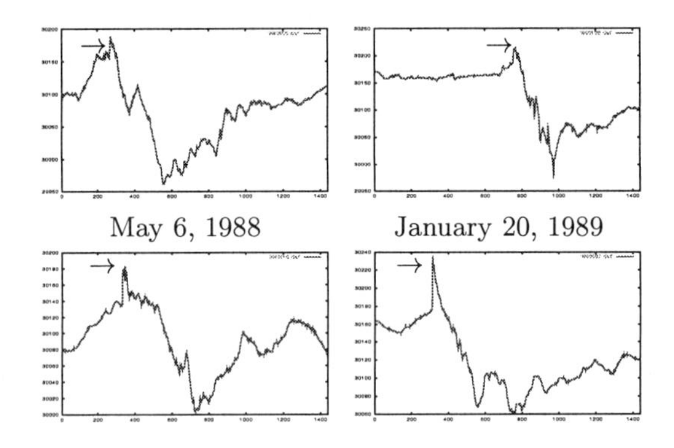

Fig. 2. Training H-components

We try to find SCs from H-components which were observed in Kakioka, Hermanus, Honolulu, and San Juan stations. These have not been used for training. Finding of SCs was carried out by applying the extraction algorithm in Section 4.1 to H-components observed in the same station. We compared the extraction results with those obtained by a real-type lifting wavelet transform (RLWT) [6] as shown in Table 1.

H-components in all the stations except for Kakioka are corrupted with noise. Table 1 shows that extraction rate of IHLWT is higher than RLWT in Kakioka, Hermanus, and San Juan stations.

Using the proposed method, we could find a SC phenomenon at the same time in 4 stations (Fig. 3). Since SC phenomena occur at the same time all over

Table 1. Extraction rate of SCs.

	Kakioka	Hermanus	Honolulu	San Juan
RLWT	82%	24%	59%	24%
IHLWT	88%	82%	47%	82%

the world, it is reasonable to suppose that a SC phenomenon occurred actually at this time. This SC has not been found by specialists in geomagnetism.

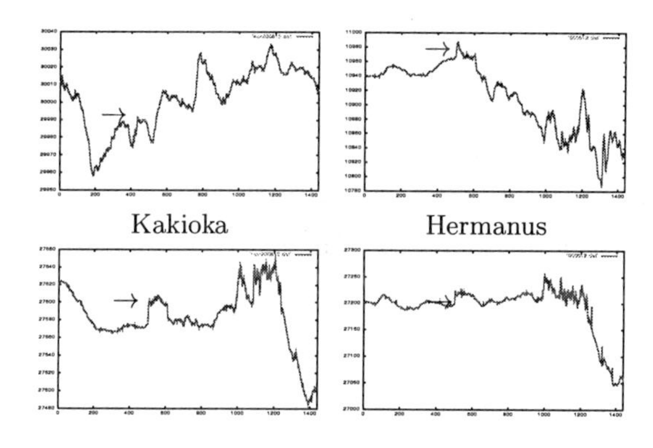

Fig. 3. Discovery of new SCs

Mallat et al. have proposed a method for detecting singularity of signals ([2, 3]). Their method uses multiresolution analysis of dyadic wavelets. By examining a correlation between the peaks of high frequency components in multi-level, they have found special singular points. However, a phenomenon occurring in some period such as SC cannot be detected by their method.

5.2 Extraction of Facial Image

Second simulation concerns extraction of facial images from a snapshot. Using a facial image of size 64×64, we trained an IHLWT for extraction. The trained IHLWT was applied to a snapshot to extract facial images in it. We illustrated the snapshot in Fig. 4 (a) and the facial image for training in Fig. 4 (b). The extraction result has been shown in bottom right in Fig. 4 (c). The snapshot includes a tree image as part of the background. Since the tree has many high frequency components, it was also extracted as a facial image. Two facial images near the boundary failed to be extracted because the number of selected points in blocks near the boundary is few.

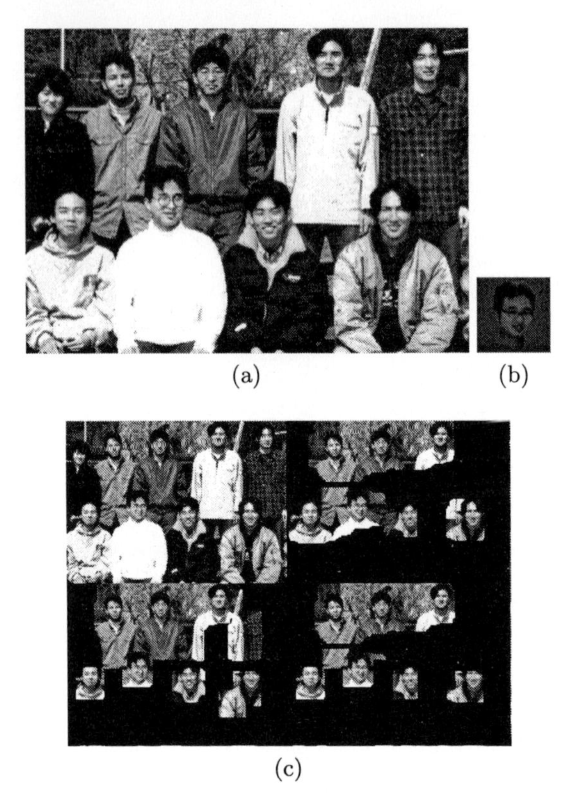

(a) (b)

(c)

Fig. 4. (a) A snapshot, (b) A facial image, (c) Extraction result.

6 Conclusion

In this paper, we proposed a new method for finding target portions from time series or reference images. The new method is based on IHLWT.

First, we described how to train IHLWT using training signals or images. Free parameters in IHLWT were determined by solving a minimization problem subject to inequality constraints. The inequality constraints were derived from vanishing conditions on high frequency components computed by IHLWT. Next, we gave two algorithms based on trained IHLWT, one is for signal extraction, and the other for image extraction.

We carried out two experiments. The first experiment concerns finding of SCs from H-components. The second experiment is related to facial image extraction from a snapshot. In both the cases, robust extraction for noise was achieved.

There are still problems to be resolved. One of them is to verify the efficiency of our method for other kinds of time series. The proposed method focused on high frequency components of signal and image. There are target images each of which has been occupied almost by low frequency components such as clouds. It is a future work to develop methods for extracting such target images.

References

1. A. R. Calderbank, I. Daubechies, W. Sweldens and B. -L. Yeo. Wavelet transforms that map integers to integers. *Technical Report, Department of Mathematics, Princeton University*, 1996.
2. S. Mallat and W. L. Hwang. Singularity detection and processing with wavelets. *IEEE Trans. Info. Theory*, vol.38, no.2, pp.617-643, 1992.
3. S. Mallat and S. Zhong. Characterization of signals from multiscale edges. *IEEE Trans. Pattern Analysis Machine Intelligence*, vol.14, no.7, pp.710-732, 1992.
4. W. Sweldens. The lifting scheme: A custom-design construction of biorthogonal wavelets. *Appl. Comp. Harmon. Anal.*, vol.3, no.2, pp.186-200, 1996.
5. S. Takano. Signal and image extraction by lifting wavelets. *Doctoral Thesis, Kyushu University*, 2001.
6. S. Takano, T. Minamoto, H. Arimura, K. Niijima, T. Iyemori and T. Araki. Automatic detection of geomagnetic sudden commencement using lifting wavelet filters. *LNAI 1721: Proc. of the Second International Conference on Discovery Science*, pp.242-251, 1999.
7. S. Takano and K. Niijima. Robust lifting wavelet transform for subimage extraction. *Proc. of SPIE: Wavelet Applications in Signal and Image Processing VIII*, vol.4119, pp.902-910, 2000.
8. S. Takano and K. Niijima. Subimage extraction by integer-type lifting wavelet transforms. *Proc. of IEEE International Conference on Image Processing*, pp.403-406, 2000.

In Pursuit of Interesting Patterns with Undirected Discovery of Exception Rules

Einoshin Suzuki

Electrical and Computer Engineering, Yokohama National University,
79-5 Tokiwadai, Hodogaya, Yokohama 240-8501, Japan
suzuki@dnj.ynu.ac.jp

Abstract. This paper reports our progress on interesting pattern discovery in the discovery science project. We first introduce undirected discovery of exception rules, in which a pattern represents a pair of an exception rule and its corresponding strong rule. Then, we explain scheduled discovery, exception rule discovery guided by a meta-pattern, and data mining contests as our contribution to the project. These can be classified as pattern search, pattern representation, and scheme justification from the viewpoint of research topics in interesting pattern discovery.

1 Introduction

The goal of KDD (Knowledge Discovery in Databases) can be roughly summarized as extraction of useful knowledge from a huge amount of data [3]. Since explicit evaluation of usefulness as well as explicit representation of knowledge are difficult, a typical approach in KDD attempts to achieve the aforementioned objective by discovering a set of interesting patterns. A considerable amount of efforts have been spent in pursuit of interesting patterns.

Research topics in interesting pattern discovery can be classified as follows. 1) "Pattern representation" seeks for an appropriate class of structures for patterns. For instance, EDAG (Exception Directed Acyclic Graphs), which may be read as a set of rules with exceptions or as a generalized decision tree, was defined as a target of knowledge discovery [4]. The motivation of the study was to capture interestingness through the form of EDAG. 2) "Pattern evaluation" tries to provide a precise estimate of interestingness inherent in a pattern. For instance, subjective measures of interestingness, which attempt to use domain-dependent information in the evaluation, were proposed to capture unexpectedness and actionability [11]. The study manipulates degrees of beliefs in order to calculate a degree of interestingness for a discovered pattern. 3) "Pattern search" studies effective discovery algorithms for interesting patterns. For instance, the Apriori algorithm [1], which represents a breadth-first search for a set of frequently occurring itemsets, is intended to assist association rule discovery from a sort of transactional data set. The objective of the study was to provide a realistic search method for interesting patterns since conventional search algorithms based on

S. Arikawa and A. Shinohara (Eds.): Progress in Discovery Science 2001, LNAI 2281, pp. 504–517.

either heuristic search or depth-first search are ineffective or inefficient for such a data set. 4) "Scheme justification" attempts to evaluate methods and/or results of interesting pattern discovery. For instance, KDD-Cup 2000 [6], which is a data mining contest to analyze e-commerce data, was held to rank participants with their own data mining methods. The competition was useful not only in justifying each attempt, but also in highlighting many interesting issues in a KDD process [3], especially in interesting pattern discovery.

Before participating to the discovery science project, we mainly tackled topics 1) - 3) through undirected discovery of exception rules [15–17]. An exception rule, which is defined as a deviational pattern to a strong rule, exhibits unexpectedness and is sometimes extremely useful. For instance, suppose a species of poisonous mushrooms some of which are exceptionally edible. The exact description of their exceptions is highly beneficial since it enables exclusive possession of the edible mushrooms. Also, exceptions were always interesting to discoverers, as they challenged the existing knowledge and often led to growth of knowledge in new directions. Our approach can be referred to as undirected since it discovers a set of rule pairs each of which represents a pair of an exception rule and its corresponding strong rule, and requires no domain-dependent information. Since most of the studies in exception rule discovery obtain a set of rules each of which deviates from a set of user-specified strong rules, our approach can be regarded as highly original. However, we had considered that our approach can be improved in terms of 1) - 3) and wished to contribute to 4).

Our main contributions to the discovery science project can be summarized as follows. 1) "Pattern representation": we generalized the structure of an exception rule by introducing a meta-pattern, and proposed a unified algorithm which discovers all kinds of structures [22]. Moreover, we proposed instance discovery, in which a discovered pattern represents a set of abstracted instances [13, 14]. 2) "Pattern evaluation": we proposed another measure of interestingness [5]. 3) "Pattern search": we automated specification of a set of thresholds in order to facilitate applications of our method to various data sets [18]. 4) "Justification of overall scheme": we contributed to several data mining contests as participants [19, 20] as well as organizers [21]. The experience helped us in analyzing various issues related to a data mining contest [23]. In the remaining sections, we first explain our undirected discovery of exception rules, then briefly introduce a part of our contributions related to it.

2 Undirected Discovery of Exception Rules

2.1 Problem Statement

Discovery methods for exception rules can be divided into two approaches from the viewpoint of background knowledge. In a directed approach [7, 10], a method is first provided with background knowledge typically in the form of rules, then the method obtains exception rules each of which deviates from these rules. In an undirected approach [5, 8, 15–17, 25], on the other hand, no background knowledge is provided. The target of discovery is typically a set of rule pairs each of which consists of an exception rule and its corresponding strong rule. In this section we present our basic approaches for undirected discovery of exception rules before the discovery science project [15–17].

Throughout this paper, we deal with a data set in which each of n examples is described by m attributes. An event representing either a value assignment to a discrete attribute or a range assignment to a continuous attribute will be called an atom. We define a literal as either a single atom or a conjunction of atoms. Let u and v be a literal and a single atom respectively, then a rule $u \to v$ represents a regularity that its premise u holds true for many examples and its conclusion v occurs frequently given u. Representation of a discovered pattern in our approach is formalized as a rule pair $r(x, x', Y_\mu, Z_\nu)$, which consists of a strong rule $Y_\mu \to x$ and an exception rule $Y_\mu \wedge Z_\nu \to x'$.

$$r(x, x', Y_\mu, Z_\nu) \equiv (Y_\mu \to x,\ Y_\mu \wedge Z_\nu \to x') \tag{1}$$

where $Y_\mu \equiv y_1 \wedge y_2 \wedge \cdots \wedge y_\mu$, $Z_\nu \equiv z_1 \wedge z_2 \wedge \cdots \wedge z_\nu$, and each of $y_1, y_2, \cdots, y_\mu, z_1, z_2, \cdots, z_\nu, x, x'$ is an atom. We define that each of x and x' has an identical attribute but a different value.

Note that when $Z_\nu \to x'$ holds, the corresponding exception rule $Y_\mu \wedge Z_\nu \to x'$ typically exhibits low unexpectedness, and the obtained pattern is often uninteresting. Let a negative rule $u \not\to v$ represent that v seldom occurs given u. Our approach typically assumes that the following reference rule holds.

$$Z_\nu \not\to x' \tag{2}$$

An exception rule can be also interpreted as representing a situation that a simultaneous occurrence of two conditions results in a different conclusion. For instance, figure 1 shows that a simultaneous occurrence of two conditions often results in "death" whereas their single occurrence often signifies "recovery". In the example, all patients are classified as either "recovery" (x) or "death" (x'). Those who belong to "antibiotics" (Y) are likely to belong to x, and so do those who belong to "staphylococi" (Z). Note that an anomalous situation is happening since those who belong to both Y and Z are likely to belong to x'.

strong rule: $Y \to x$
antibiotics → recovery

exception rule: $Y, Z \to x'$
antibiotics, staphylococci → death

reference rule: $Z \not\to x'$
stphylococi ↛ death

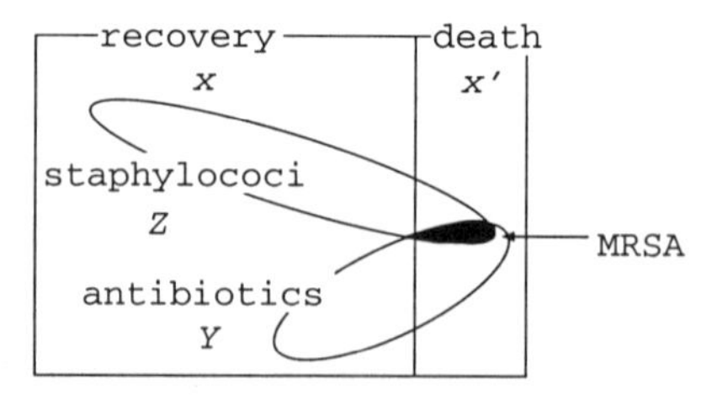

Fig. 1. Venn diagram which illustrates an example of our exception rule

2.2 Evaluation Criteria

Here we introduce our evaluation criteria based on specification of thresholds [16]. Readers who are interested in information-theoretic criteria [15] and/or evaluation of surprise [17] are requested to refer to the corresponding reference. Let $\widehat{\Pr}(A)$ be an estimated probability of an event A. In rule discovery, generality and accuracy can be considered as evaluating the goodness of a rule [12]. For a rule $u \rightarrow v$, we follow [12] and define $\widehat{\Pr}(u)$ as the generality of the rule and $\widehat{\Pr}(v|u)$ as the accuracy of the rule. We specify minimum thresholds θ_1^{S}, θ_1^{F}, θ_2^{S}, and θ_2^{F} for generality and accuracy of a strong rule, and generality and accuracy of an exception rule respectively. Following the discussions in the previous subsection, we specify a maximum threshold θ_2^{I} for the accuracy of a reference rule.

A rule discovered from a data set of 10 examples should be treated differently from another rule discovered from a data set of 10,000 examples. We regard a given data set as a result of sampling, and employ true probabilities instead of estimated probabilities. Let $\Pr(A)$ be a true probability of an event A. Note that, for instance, $\Pr(A) \geq 0.15$ does not necessarily hold even if $\widehat{\Pr}(A) = 0.2$ since $\Pr(A)$ represents a probabilistic variable. It is possible to compute the probability $\Pr(\Pr(A) \geq 0.15)$ of $\Pr(A) \geq 0.15$ typically under a model. We obtain rule pairs each of which satisfies the following condition, where δ is given by the user.

$$\begin{aligned} \Pr\ &(\Pr(Y_\mu) \geq \theta_1^{\mathrm{S}}, \Pr(x|Y_\mu) \geq \theta_1^{\mathrm{F}}, \Pr(Y_\mu, Z_\nu) \geq \theta_2^{\mathrm{S}}, \\ &\Pr(x'|Y_\mu, Z_\nu) \geq \theta_2^{\mathrm{F}}, \Pr(x'|Z_\nu) \leq \theta_2^{\mathrm{I}}) \geq 1 - \delta \end{aligned} \tag{3}$$

Note that since (3) contains five true probabilities, it is time-consuming to calculate (3) numerically for a specific rule pair. In [16], we derived an analytical solution for (3), which can be efficiently and easily calculated. Assuming that $(\Pr(x \wedge Y \wedge Z), \Pr(x' \wedge Y \wedge Z), \Pr(\overline{x} \wedge \overline{x'} \wedge Y \wedge Z), \Pr(x \wedge Y \wedge \overline{Z}), \Pr(\overline{x} \wedge Y \wedge \overline{Z}), \Pr(\overline{x'} \wedge \overline{Y} \wedge Z), \Pr(x' \wedge \overline{Y} \wedge Z))$ follows a multi-dimensional normal distribution, the analytical solution is given by the followings.

$$G(Y, \delta, k)\widehat{\Pr}(Y) \geq \theta_1^{\mathrm{S}} \tag{4}$$

$$F(Y, x, \delta, k)\widehat{\Pr}(x|Y) \geq \theta_1^{\mathrm{F}} \tag{5}$$

$$G(YZ, \delta, k)\widehat{\Pr}(YZ) \geq \theta_2^{\mathrm{S}} \tag{6}$$

$$F(YZ, x', \delta, k)\widehat{\Pr}(x'|YZ) \geq \theta_2^{\mathrm{F}} \tag{7}$$

$$F'(Z, x', \delta, k)\widehat{\Pr}(x'|Z) \leq \theta_2^{\mathrm{I}} \tag{8}$$

$$\textit{where}\ \ G(a, \delta, k) \equiv 1 - \beta(\delta, k)\sqrt{\frac{1 - \widehat{\Pr}(a)}{n\widehat{\Pr}(a)}} \tag{9}$$

$$F(a, b, \delta, k) \equiv 1 - \beta(\delta, k)\varphi(a, b) \tag{10}$$

$$F'(a, b, \delta, k) \equiv 1 + \beta(\delta, k)\varphi(a, b)$$

$$\varphi(a, b) \equiv \sqrt{\frac{\widehat{\Pr}(a) - \widehat{\Pr}(a, b)}{\widehat{\Pr}(a, b)\{(n + \beta(\delta, k)^2)\widehat{\Pr}(a) - \beta(\delta, k)^2\}}} \tag{11}$$

2.3 Search Algorithm

An obvious method for obtaining rule pairs would be to first discover all strong rules with a (conventional) rule discovery method, then discover corresponding exception rules each of which deviates from one of the strong rules. A severe deficiency of this method is the cost of discovering and storing all strong rules. In a typical setting, most of these strong rules can be safely pruned by considering its corresponding exception rule. Here we present our method which simultaneously discovers a strong rule and its corresponding exception rule.

In our approach, a discovery task is viewed as a depth-first search problem, in which a node of a search tree represents four rule pairs $r(x, x', Y_\mu, Z_\nu)$, $r(x', x, Y_\mu, Z_\nu)$, $r(x, x', Z_\nu, Y_\mu)$, $r(x', x, Z_\nu, Y_\mu)$ [17]. Since these four rule pairs employ common atoms, this fourfold representation improves time efficiency by sharing calculations of probabilities of these atoms. Each node has four flags for representing rule pairs in consideration.

In the search tree, a node of depth one represents a single conclusion x or x'. Likewise, a node of depth two represents a pair of conclusions (x, x'). Let $\mu = 0$ and $\nu = 0$ represent the state in which the premise of a strong rule and the premise of an exception rule contain no atoms respectively, then we see that $\mu = \nu = 0$ holds true in a node of depth two. As the depth increases by one, an atom is added to the premise of the strong rule or to the premise of the exception rule. A node of depth three satisfies either $(\mu, \nu) = (0, 1)$ or $(1, 0)$, and a node of depth l (≥ 4), $\mu + \nu = l - 2$ $(\mu, \nu \geq 1)$.

Since a depth-first search method is prohibitively slow, we proposed methods which safely prune a search tree without altering discovery results [15–17]. These methods have been proved highly efficient through experiments.

3 Scheduled Discovery

3.1 Proposed Method

When specified threshold values $\theta_1^{\mathrm{S}}, \theta_1^{\mathrm{F}}, \theta_2^{\mathrm{S}}, \theta_2^{\mathrm{F}}, \theta_2^{\mathrm{I}}$ are inappropriate to the data set in (3), no rule pairs are discovered or too many rule pairs are discovered. An appropriate choice of them requires expertise on the data set and on the discovery algorithm.

In order to circumvent this problem, we proposed a scheduling method[1] which automatically updates the values of the thresholds to discover at most η rule pairs, where η is given by the user. Let $\Psi(i, j)$ and $\Psi'(i, j)$ be the jth rule pair with respect to a key i in a descending and ascending order respectively. When the number of discovered rule pairs becomes $\eta + 1$ for the first time in a discovery process, this method settles the value of threshold θ_1^{S} to the generality of the strong rule in $\Psi(\widehat{\Pr}(Y_\mu), \eta + 1)$, and deletes $\Psi(\widehat{\Pr}(Y_\mu), \eta + 1)$. Then, each time a new rule pair is discovered, this method deletes $\Psi(i, \eta + 1)$ for $i = \widehat{\Pr}(x|Y_\mu)$, $\widehat{\Pr}(Y_\mu, Z_\nu)$, $\widehat{\Pr}(x'|Y_\mu, Z_\nu)$, and updates the value of the corresponding threshold according to the deleted rule pair. For two subsequent rule

[1] In this paper, we use "scheduling" as appropriate update of threshold values.

pairs which are discovered, $\Psi'(\widehat{\Pr}(x'|Z_\nu), \eta+1)$ and $\Psi(\widehat{\Pr}(Y_\mu), \eta+1)$ are deleted with similar updates of threshold values, and the whole process is iterated. A benefit of this method is to substitute specification of η for specification of values for five thresholds $\theta_1^{\mathrm{S}}, \theta_1^{\mathrm{F}}, \theta_2^{\mathrm{S}}, \theta_2^{\mathrm{F}}, \theta_2^{\mathrm{I}}$. As a consequence, this method does not necessarily require expertise on the data set and on the discovery algorithm.

We also proposed a novel data structure for an efficient management of discovered patterns with multiple indices. The data structure consists of multiple balanced search trees. Many balanced search trees are known to add, search, delete and minimum-search its elements in $O(\log \chi)$ time, where χ is the number of elements. In order to realize flexible scheduling, we assign a tree for each index. To enable fast transformation of a tree, a node of a tree represents a pointer to a discovered pattern. We show, in figure 2, an example of this data structure which manages seven rule pairs $r1, r2, \cdots, r7$.

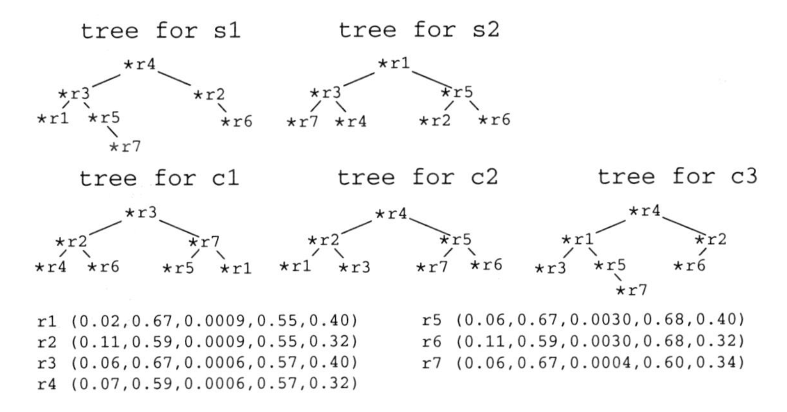

Fig. 2. Illustration based on the proposed data structure, where an AVL tree is employed as a balanced search tree, and keys are s1 ($\widehat{\Pr}(Y_\mu)$), s2 ($\widehat{\Pr}(Y_\mu, Z_\nu)$), c1 ($\widehat{\Pr}(x|Y_\mu)$), c2 ($\widehat{\Pr}(x'|Y_\mu, Z_\nu)$), c3 ($\widehat{\Pr}(x'|Z_\nu)$). Numbers in a pair of parentheses represent values of these indices of the corresponding rule pair. In a tree, $*r1, *r2, \cdots, *r7$ represent a pointer to r1, r2, ···, r7 respectively.

3.2 Experimental Results

We have conducted various experiments using the mushroom data set, the 1994 census data set [9], an earthquake questionnaire data set, and the 1994 bacterial test data set [21]. Here we show only two experiments due to lack of space. We also obtained excellent results in other experiments.

While there are no restrictions on attributes of atoms in our approach, users may also impose some constraints on them. In each experiment, we limited the attribute allowed in the conclusions to an apparently important attribute of the corresponding data set. First, we settled the initial values of thresholds as

$\theta_1^S = 0.0004$, $\theta_2^S = 10/n$, $\theta_1^F = 0.5$, $\theta_2^F = 0.5$, $\theta_2^I = 0.5$, where n represents the number of examples in a data set. We settled the maximum number of discovered rule pairs to $\eta = 500$ and the maximum search depth to $M = 5$. Since the mushroom data set has a relatively small number of examples, we used $M = 6$. The maximum number of atoms allowed in a premise was restricted to 2.

Table 1. Results from the mushroom data set and the bacterial test data set, where $\theta_1^S, \theta_2^S, \theta_1^F, \theta_2^F, \theta_2^I$ represent final values

data set	# of searched nodes	# of discovered patterns (including intermediate ones)	θ_1^S / θ_2^F	θ_2^S / θ_2^I	θ_1^F
mushroom	$5.00 * 10^6$	4,069	0.154	0.00295	0.914
			1.000	0.123	
bacterial test	$2.32 * 10^7$	2,691	0.0606	0.000908	0.671
			0.526	0.339	

In the first experiment, $\eta = 500$ rule pairs were discovered from the mushroom data set and the 1994 bacterial test data set. We show the statistics in table 1. From the table, we see that the updated values of thresholds seem to be appropriate compared with their initial values. On the other hand, 38 and 8 rule pairs were discovered from the earthquake questionnaire data set and the 1994 census data set respectively. Since the initial values of thresholds are not strict, it is obvious that rule pairs were discovered from these data sets. An important point here is that our method achieves this while realizing efficiency for the other two data sets.

In order to evaluate such efficiency quantitatively, we applied a modified version of our method which does not update thresholds to these data sets. Results of this experiment are shown in table 2, which shows that threshold scheduling approximately reduces the number of searched nodes by 17 % to 80 %, and the number of discovered patterns becomes 1/18 to 1/300.

Table 2. Results without updating thresholds. An "increase" represents (value of this table - value of table 1) / (value of table 1)

data set	# of searched nodes (increase)	# of discovered rule pairs (increase)
mushroom	$8.94 * 10^6$ (78.8 %)	$1.48 * 10^5$ (295)
bacterial test	$2.72 * 10^7$ (17.2 %)	90,288 (17.9)

4 Exception Rule Discovery Guided by a Meta-pattern

4.1 Proposed Method

The representation of an exception rule in (1) formalizes only one type of exceptions, and fails to represent other types. In order to provide a systematic treatment of exceptions, we introduced a meta-pattern represented by a rule triple for categorizing possible representation of interesting exceptions/deviations [22].

$$t(y, x, \alpha, \beta, \gamma, \delta) \equiv (y \rightarrow x,\ \alpha \not\rightarrow \beta,\ \gamma \rightarrow \delta) \tag{12}$$

Here α, β, γ, and δ are meta-level variables which can be instantiated in different ways, with the use of literals x, y, and z, to form different specific patterns of exceptions, analogous to (1). This rule triple has a generic reading "it is common that if y then x, but we have exceptions that if α then not frequently β. This is surprising since if γ then δ". Note that this terminology differs from (1) in that an exception rule and a reference rule are represented by a negative rule and a rule respectively. Here we only justify this definition by stating that a violation (exception rule) of two rules (strong rule and reference rule) can be interesting.

The number of literals is chosen since it represents the simplest situation which makes sense: a rule triple with two literals would likely to be meaningless since it tends to be overconstrained, and a rule triple with more than three literals would likely to be more difficult to interpret. We assume that a conjunction of two literals can appear only in the premise of an exception rule, and exclude contradictory relations and inclusive relations from discovery. These assumptions are justified because a conjunction of two literals in the premise makes a good candidate for an exception to a rule that holds one of those literals in the premise, and the two relations are considered to exhibit relatively low interestingness. In this section we simplify (3) and define that a rule $u \rightarrow v$ in a discovered pattern satisfies both $\widehat{\Pr}(u) \geq \theta_{\mathrm{S}}$ and $\widehat{\Pr}(v|u) \geq \theta_{\mathrm{F}}$, where θ_{S} and θ_{F} represent user-specified thresholds, and u and v are literals. We also define that a negative rule in a discovered pattern satisfies both $\widehat{\Pr}(u) \geq \theta_{\mathrm{S}}$ and $\widehat{\Pr}(v|u) \leq \theta_{\mathrm{I}}$, where θ_{I} represents a user-specified threshold.

We investigated the meta-pattern (12) and found that it can be instantiated to eleven categories, which are shown in figure 3. By examining figure 3, we can interpret type 2, 5, 8, 9, and 11 as interesting. Note that these types can be grouped in two (type 1 - 4 and type 5 - 11) with respect to the number of literals in the premise of the exception rule. In the first group, type 2 represents violation of transitivity, and can thus be regarded as demonstrating the strongest deviations in these rule triples. In the second group, a literal is at the same time approved and negated in type 5, 8, 9, and 11. The situations of type 5, 8, 9, and 11 can only occur if those records that make exceptions to each of the rules in the triple are distributed in a very specific way, so that the thresholds set by θ_{F} and θ_{I} can be met. We also proposed a unified algorithm for discovering all of type 1 - 11 based on a search method analogous to that of subsection 2.3.

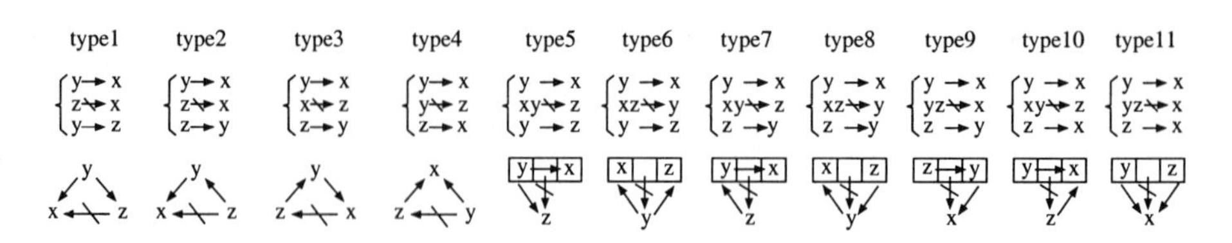

Fig. 3. Possible instantiations of the meta-pattern (12). A rectangle on the top center for a rule triple represents a conjunction of literals in the top right and left. For instance, the three rectangles in type 11 represent, from the left to the right, "y", "$y \wedge z$", and "z"

4.2 Experimental Results

We here analyze empirically the statistics of the searched nodes and the discovered rule triples. We have chosen UCI data sets [9] (car, Australian, nursery, credit, postoperative, vote, yeast, hepatitis, diabetes, German, abalone, mushroom, breast cancer, thyroid, and shuttle) since they have served for a long time as benchmark data sets in the machine learning community.

In applying our algorithm, each numerical attribute was discretized using a global unsupervised method [2] with number of bins 4 due to time efficiency. Parameters were settled to $\theta_{\mathrm{S}} = 0.025$, $\theta_{\mathrm{F}} = 0.7$, $\theta_{\mathrm{I}} = 0.6$, and $M = 2$. Figure 4 summarizes the results of experiments.

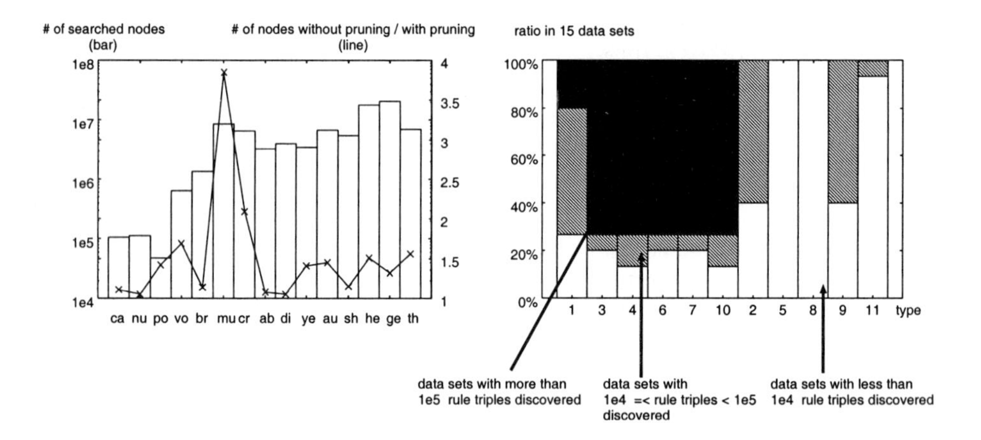

Fig. 4. Experimental results, where 1ec represents 10^c

The left-hand side of the figure shows that pruning is effective, since without pruning the searched nodes increase by 5 % ("nursery" and "diabetes") to 285 % ("mushroom"). This is due to the fact that a considerable number of nodes in a search tree tend to have small probabilities for their literals and are thus pruned. The right-hand side of the figure reveals interesting tendencies on the numbers of discovered rule triples. From the figure we see that type 2, 5, 8, 9, and 11 are rare. Interestingly, we anticipated the exceptionality of these types as stronger than the other types. We are currently investigating this tendency analytically.

5 Data Mining Contest

5.1 Participation to Contests

A data mining contest is a systematic attempt to evaluate discovery methods of participants with a set of common data or common problems. As stated in section 1, since the goal of KDD can be summarized as extraction of useful knowledge from a huge amount of data, a knowledge discovery system should be evaluated from the usefulness of its output. A data mining contest provides a systematic opportunity for such evaluation, and is thus important in KDD.

Here we describe one of our participations to data mining contests with the method presented in section 2. The challenge posed to participants was an updated version of the meningitis data set [21], which consists of 140 patients described with 38 attributes. The length of a premise in a rule pair is limited to one, i.e. $\mu = \nu = 1$, in the application, and we set $\delta = 0$ due to the small number of examples in the data set. The other parameters were settled as $\theta_1^{\mathrm{S}} = 0.2$, $\theta_1^{\mathrm{F}} = 0.75$, $\theta_2^{\mathrm{S}} = 5/140$, $\theta_2^{\mathrm{F}} = 0.8$, $\theta_2^{\mathrm{I}} = 0.4$. Numerical attributes were discretized using the same method in subsection 4.2 with the number of bins $4, 5, \cdots, 10$.

Dr. Tsumoto, a domain expert, evaluated each discovered rule pair from the viewpoint of validness, novelty, unexpectedness, and usefulness. Validness indicates that discovered results agree with medical context, and novelty indicates that discovered results are novel in medical context. Unexpectedness represents that discovered results can be partially explained by the medical context but are not accepted as common sense. Usefulness indicates that discovered results are useful in medical context. For each index of a rule pair, he attributed an integer score ranging from one to five. A zero score was attributed if he judged necessary.

We show the results in table 3, in which we categorized the attributes in conclusions into four. The first category represents attributes with the lowest scores, and includes CULTURE, C_COURSE, and RISK. We consider that attributes in this category cannot be explained with this data set, and investigation on them requires further information. However, the second category, which represents attributes with relatively high scores for four indices, and the third category, which represents attributes with higher scores for validness and usefulness, can be considered as promising as a target of data mining. The fourth category represents attributes with higher scores for novelty and unexpectedness, and includes CULT_FIND, KERNIG, and SEX. We consider that attributes in this category can be explained with this data set, but has been somewhat ignored. We consider that investigating these attributes using discovered rule sets can lead to interesting discoveries which might reveal unknown mechanisms in this domain in spite of their apparent low importance. Note that these results show that the average performance of our approach is promising.

Table 3. Average performance of the proposed method with respect to attributes in conclusions. The column "#" represents the number of discovered rule pairs

attribute	#	validness	novelty	unexpectedness	usefulness
all	169	2.9	2.0	2.0	2.7
CULTURE	2	1.0	1.0	1.0	1.0
C_COURSE	1	1.0	1.0	1.0	1.0
RISK	1	1.0	1.0	1.0	1.0
CT_FIND	36	3.3	3.0	3.0	3.2
EEG_FOCUS	11	3.0	2.9	2.9	3.3
Course (G)	8	1.8	2.0	2.0	1.8
FOCAL	18	3.1	2.2	2.7	3.0
LOC_DAT	11	2.5	1.8	1.8	2.5
Diag2	72	3.0	1.1	1.1	2.6
CULT_FIND	4	3.3	4.0	4.0	3.5
KERNIG	4	2.0	3.0	3.0	2.0
SEX	1	2.0	3.0	3.0	2.0

Dr. Tsumoto also found several rule pairs concerning CULT_FIND, EEG_FOCUS, Diag2, FOCAL and CT_FIND very interesting. For instance, the following rule pair has a four-rank score for every index.

```
83=<CSF_PRO=<121              ->CULT_FIND=F
83=<CSF_PRO=<121, FOCAL=+ ->CULT_FIND=T
```

This rule pair has the following statistics: $\widehat{\Pr}(Y_\mu) = 0.257$, $\widehat{\Pr}(x|Y_\mu) = 0.778$, $\widehat{\Pr}(Y_\mu, Z_\nu) = 0.035$, $\widehat{\Pr}(x'|Y_\mu, Z_\nu) = 1.000$, $\widehat{\Pr}(x'|Z_\nu) = 0.285$. In other words, among 140 patients, 36 patients had `83=<CSF_PRO=<121`, and 78 % of them were also `CULT_FIND=F`. However, five patients who were `FOCAL=+` in addition to `83=<CSF_PRO=<121` were actually all `CULT_FIND=T`. This exception rule is interesting since only 28.5 % of patients who were `FOCAL=+` were `CULT_FIND=T`.

In an article [24] comparing eleven KDD methods with respect to this data set, he states this method as "structure of rule pairs is very appealing to medical experts". He also admits that this method discovered the most interesting results among eleven methods.

5.2 Issues in Organization

Based on our experience of organizing a data mining contest as an international workshop [21], we discussed three issues in organization [23].

First, motivations of a data mining contest should be clearly settled and announced. Such motivations can be decomposed into academic benefits, which mainly represent development of an effective discovery method, and domain benefits, which mainly represent discovery of useful knowledge. Evaluation of discovery methods depends on which motivation be emphasized. For example, suppose you have two discovery methods: one discovered 99 useful rules and 1 useless rule, and another discovered 1 extremely useful rule and 99 useless rule. Motivation for academic benefits favors the former method, while motivation for domain benefits favors the latter method.

Second, measures to support domain experts are needed. Domain experts can be reluctant due to several factors including noise in data and immaturity of the domain. The target problem of a data mining contest should be appropriately settled considering the evaluation process. Moreover, the motivation for academic benefits necessitates clarifications of successes and failures with respect to characteristics of a discovery method. It is also desirable that a degree of interestingness is decomposed into several indices such as validness, novelty, and usefulness. All of these require considerable efforts to domain experts, but improve the quality of a contest.

Third, increasing the number of participants is mandatory for the success of a data mining contest. Currently, a participant should perform the whole KDD process [3] by himself, and suffers from its iterative and interactive aspects. We expect that a future data mining contest might allow partial participations by either combining partial results systematically or promoting collaborations among participants. The former requires standardization of a KDD process, which is considered highly important in KDD.

6 Conclusions

Undirected discovery of exception rules is an effective approach in pursuit of interesting patterns. Our participation to the discovery science project mainly matured this approach in terms of pattern search, pattern representation, and scheme justification. In this paper, we gave brief explanations and experimental results for the corresponding methods: scheduled discovery, exception rule discovery guided by a meta-pattern, and data mining contests.

References

1. R. Agrawal, H. Mannila, R. Srikant, H. Toivonen, and A. I. Verkamo: "Fast Discovery of Association Rules", *Advances in Knowledge Discovery and Data Mining*, AAAI/MIT Press, Menlo Park, Calif., pp. 307–328, 1996.
2. J. Dougherty, R. Kohavi, and M. Sahami: "Supervised and Unsupervised Discretization of Continuous Features", *Proc. Twelfth Int'l Conf. on Machine Learning (ICML)*, pp. 194–202, 1995.
3. U. M. Fayyad, G. Piatetsky-Shapiro, and P. Smyth: "From Data Mining to Knowledge Discovery: An Overview", *Advances in Knowledge Discovery and Data Mining*, AAAI/MIT Press, Menlo Park, Calif., pp. 1–34, 1996.
4. B. R. Gaines: "Transforming Rules and Trees", *Advances in Knowledge Discovery and Data Mining*, AAAI/MIT Press, Menlo Park, Calif., pp. 205–226, 1996.

5. F. Hussain, H. Liu, E. Suzuki, and H. Lu: "Exception Rule Mining with a Relative Interestingness Measure", *Knowledge Discovery and Data Mining (PAKDD)*, LNAI 1805, Springer, pp. 86–97, 2000.
6. R. Kohavi, C. E. Brodley, B. Frasca, L. Mason, and Z. Zheng: "KDD-Cup 2000 Organizers Report: Peeling the Onion", *ACM SIGKDD Explorations*, Vol. 2, No. 1, pp. 86–93, 2001.
7. B. Liu, W. Hsu, L. Mun, and H. Lee: "Finding Interesting Patterns Using User Expectations", *IEEE Trans. Knowledge and Data Eng.*, Vol. 11, No. 6, pp. 817–832, 1999.
8. B. Liu, W. Hsu, and Y. Ma: "Pruning and Summarizing the Discovered Associations", *Proc. Fifth ACM SIGKDD Int'l Conf. on Knowledge Discovery and Data Mining (KDD)*, pp. 125–134, 1999.
9. C. J. Merz and P. M. Murphy: "UCI Repository of Machine Learning Databases", *http://www.ics.uci.edu/~mlearn/MLRepository.html*, Dept. of Information and Computer Sci., Univ. of California Irvine, 1996.
10. B. Padmanabhan and A. Tuzhilin: "A Belief-Driven Method for Discovering Unexpected Patterns", *Proc. Fourth Int'l Conf. on Knowledge Discovery and Data Mining (KDD)*, pp. 94–100, 1998.
11. A. Silberschatz and A. Tuzhilin: "What Makes Patterns Interesting in Knowledge Discovery Systems", *IEEE Trans. Knowledge and Data Eng.*, Vol. 8, No. 6, pp. 970–974, 1996.
12. P. Smyth and R. M. Goodman, "An Information Theoretic Approach to Rule Induction from Databases", *IEEE Trans. Knowledge and Data Eng.*, Vol. 4, No. 4, pp. 301–316, 1992.
13. S. Sugaya, E. Suzuki, and S. Tsumoto: "Support Vector Machines for Knowledge Discovery", *Principles of Data Mining and Knowledge Discovery (PKDD)*, LNAI 1704, Springer, pp. 561–567, 1999.
14. S. Sugaya, E. Suzuki, and S. Tsumoto: "Instance Selection Based on Support Vector Machine for Knowledge Discovery in Medical Database", *Instance Selection and Construction for Data Mining*, pp. 395–412, Kluwer, Norwell, Mass., 2001.
15. E. Suzuki and M. Shimura: "Exceptional Knowledge Discovery in Databases Based on Information Theory", *Proc. Second Int'l Conf. on Knowledge Discovery and Data Mining (KDD)*, pp. 275–278, 1996.
16. E. Suzuki: "Autonomous Discovery of Reliable Exception Rules", *Proc. Third Int'l Conf. on Knowledge Discovery and Data Mining (KDD)*, pp. 259–262, 1997.
17. E. Suzuki and Y. Kodratoff: "Discovery of Surprising Exception Rules Based on Intensity of Implication", *Principles of Data Mining and Knowledge Discovery (PKDD)*, LNAI 1510, Springer, pp. 10–18, 1998.
18. E. Suzuki: "Scheduled Discovery of Exception Rules", *Discovery Science (DS)*, LNAI 1721, Springer, pp. 184–195, 1999.
19. E. Suzuki and S. Tsumoto: "Evaluating Hypothesis-Driven Exception-Rule Discovery with Medical Data Sets", *Knowledge Discovery and Data Mining (PAKDD)*, LNAI 1805, Springer, pp. 208–211, 2000.
20. E. Suzuki: "Mining Bacterial Test Data with Scheduled Discovery of Exception Rules", *Proc. Int'l Workshop of KDD Challenge on Real-World Data (KDD Challenge)*, pp. 34–40, 2000.
21. E. Suzuki (ed.): *Proc. Int'l Workshop of KDD Challenge on Real-World Data (KDD Challenge)*, (http://www.slab.dnj.ynu.ac.jp/challenge2000), 2000.
22. E. Suzuki and J. M. Żytkow: "Unified Algorithm for Undirected Discovery of Exception Rules", *Principles of Data Mining and Knowledge Discovery (PKDD)*, LNAI 1910, Springer, pp. 169–180, 2000.

23. E. Suzuki: "Issues in Organizing a Successful Knowledge Discovery Contest", *Discovery Science (DS)*, LNAI 1967, Springer, pp. 282–284, 2000.
24. S. Tsumoto et al.: "Comparison of Data Mining Methods using Common Medical Datasets", *ISM Symposium: Data Mining and Knowledge Discovery in Data Science*, pp. 63–72, 1999.
25. N. Yugami, Y. Ohta, and S. Okamoto: "Fast Discovery of Interesting Rules", *Knowledge Discovery and Data Mining (PAKDD)*, LNAI 1805, Springer, pp. 17–28, 2000.

Mining from Literary Texts: Pattern Discovery and Similarity Computation

Masayuki Takeda[1,2], Tomoko Fukuda[3], and Ichirō Nanri[3]

[1] Department of Informatics, Kyushu University 33, Fukuoka 812-8581, Japan
takeda@i.kyushu-u.ac.jp
[2] PRESTO, Japan Science and Technology Corporation (JST)
[3] Junshin Women's Junior College, Fukuoka 815-0036, Japan
{tomoko-f@muc, nanri-i@msj}.biglobe.ne.jp

Abstract. This paper surveys our recent studies of text mining from literary works, especially classical Japanese poems, Waka. We present methods for finding characteristic patterns in anthologies of Waka poems, as well as those for finding similar poem pairs. Our aim is to obtain good results that are of interest to Waka researchers, not just to develop efficient algorithms. We report successful results in finding patterns and similar poem pairs, some of which led to new discoveries.

1 Introduction

Recently, our research group has tackled the problem of text mining from literary works, especially classical Japanese poems, *Waka*. This poetry has a 1,300-year history, and has been central to Japanese literature. A CD-ROM version of Shinpen-Kokkataikan, which contains over 1,200 anthologies and about 450,000 Waka poems, enabled us to adopt a quantitative approach to the analysis of this traditional poetry. Suppose that we have a frequency table, such as Table 1, in which an entry $f_{i,j}$ is a (relative) frequency of the ith expression in the jth anthology.

Such a fundamental table might characterize the differences or affinities among the anthologies. If we are given a list of expressions, it is easy to build such a frequency table. The problem is, however, "How can we select or obtain such expressions to be compared according to their frequencies?" It is not easy even for an experienced researcher to think of such expressions. For this reason,

Table 1. Frequency Comparison of Expressions in Waka poems

	Anthology 1	Anthology 2	$\cdots$	Anthology m
Expression 1	$f_{1,1}$	$f_{1,2}$		$f_{1,m}$
Expression 2	$f_{2,1}$	$f_{2,2}$		$f_{2,m}$
$\vdots$				
Expression n	$f_{n,1}$	$f_{n,2}$		$f_{n,m}$

S. Arikawa and A. Shinohara (Eds.): Progress in Discovery Science 2001, LNAI 2281, pp. 518–531.

one interesting topic would be to extract semi-automatically from anthologies the expressions (or *patterns*) that might be appropriate to a frequency table for characterization. This paper surveys our studies on this problem. Interested readers may refer to [18, 15, 16] for details.

Our research is a multidisciplinary study between literature and computer science. The second and third authors are, respectively, a Waka researcher and a linguist in the Japanese language.

2 Preliminaries

Let Σ be a finite alphabet. An element of Σ^* is called a *string*. Strings x, y, and z are said to be a *prefix*, *substring*, and *suffix* of the string $u = xyz$, respectively. The length of a string u is denoted by $|u|$. The empty string is denoted by ε, that is, $|\varepsilon| = 0$. Let $\Sigma^+ = \Sigma^* - \{\varepsilon\}$. The ith symbol of a string u is denoted by $u[i]$ for $1 \leq i \leq |u|$, and the substring of a string u that begins at position i and ends at position j is denoted by $u[i:j]$ for $1 \leq i \leq j \leq |u|$. For convenience, let $u[i:j] = \varepsilon$ for $j < i$. For a set S of strings, let $\|S\|$ denote the total length of the strings in S, and let $|S|$ denote the cardinality of S. Let $\boldsymbol{R}$ denote the set of real numbers.

3 Optimal Pattern Discovery

3.1 Problem

Shimozono et al. [14] formulated the problem of text data mining as an instance of *optimal pattern discovery* [13]. Let Π be a class of *patterns*. Assume that each pattern π in Π is associated with a language $L(\pi) \subseteq \Sigma^*$. Let $\psi : [0,1] \to \boldsymbol{R}$ be a symmetric, concave, real-valued function called *impurity function* [6] to quantify the skewness of the class distribution after the classification by a pattern in Π. The *classification error* $\psi_1(r) = \min(r, 1-r)$, the *information entropy* $\psi_2(r) = -p \log r - (1-r)\log(1-r)$, and the *Gini index* $\psi_3(r) = 2r(1-r)$ are examples of ψ [6].

Given two disjoint subsets *Pos* and *Neg* of Σ^+, a pattern discovery algorithm tries to find a pattern π in Π that optimizes the *evaluation function*

$$G^{\psi}_{Pos,Neg}(\pi) = (p_1 + n_1) \cdot \psi\left(\frac{p_1}{p_1 + n_1}\right) + (p_0 + n_0) \cdot \psi\left(\frac{p_0}{p_0 + n_0}\right),$$

where $p_1 = |L(\pi) \cap Pos|$ and $n_1 = |L(\pi) \cap Neg|$, and $p_0 = |Pos| - p_1$ and $n_0 = |Neg| - n_1$. Now, the problem is:

Definition 1. Optimal Pattern Discovery with respect to ψ.
Given two disjoint subsets Pos and Neg of Σ^+, *find a pattern* π *in* Π *that minimizes the cost* $G^{\psi}_{Pos,Neg}(\pi)$.

Shimozono et al. [14] presented an efficient algorithm for *proximity phrase-association patterns.* Although only the classification error measure is dealt with in [14], the algorithm works for many other measures [2].

3.2 Application to Waka Poems

The data we used are the Eight Imperial Anthologies, the first eight anthologies compiled by imperial command. We regarded Kokin-Shū as positive examples and the others as negative examples. We first tried to find proximity word-association patterns that distinguish the positive examples from the negative examples. Next, we restricted the class of patterns to the ones called *Fushi*, which were newly proposed in [18] as characterizing the rhetorical devices used in Waka poetry.

For proximity phrase association patterns. A *k-proximity d-phrase association pattern* ((k,d)-proximity pattern, for short) is a pair of a sequence $p_1, \ldots, p_d$ of phrases in Σ^+ and a bounded gap length k, called *proximity*. It matches a string w if there exists a sequence $j_1, \ldots, j_d$ of integers such that each phrase p_i occurs at position j_i of w, and $0 < j_{i+1} - j_i \leq k$ for each $i = 1, \ldots, d-1$. A (k,d)-proximity pattern is closely related to the pattern $\star p_1 \star \ldots \star p_d \star$ in which a gap symbol $\star$ is limited to match only a string of length k at most. However, these are not identical. Note that the phrases p_i in a (k,d)-proximity pattern may have overlaps because $j_i + |p_i|$ can be greater than j_{i+1}.

The best 20 (k,d)-proximity patterns for the information entropy are shown in Table 2, where $k = 20$ and $d = 2, 3$. Unfortunately, most of the patterns are meaningless combinations of words or word fragments.

For Fushi patterns. One way to improve the result of Table 2 would be to restrict the pattern set Π by using *domain knowledge* about Waka poetry. To characterize rhetorical devices in Waka poetry, in [18] we proposed a new model called a Fushi pattern, which is composed of *adjuncts*, i.e., auxiliary verbs and postpositional particles. For example, we consider patterns such as ⋆SEBA⋆ZARAMASHIO⋆, where SEBA is a chain of an auxiliary verb SE and a postpositional particle BA, and ZARAMASHIO is a chain of two auxiliary verbs ZARA and MASHI and a postpositional particle O. This pattern is an expression corresponding to the subjunctive mood. Generally, Fushi patterns are rhetorical devices that are closely related to the structure and rhythm of Waka poems rather than to their contents, since they consist only of adjuncts and exclude words such as nouns, adjectives, and verbs. Note that a Japanese sentence is a sequence of *segments*, each of which consists of a word and its subsequent adjuncts. Formally, we defined a Fushi pattern as a string in $(C \cup \{\star\})^+$, where C is a set of adjunct sequences specified by researchers as domain knowledge.

Table 3 shows the results of finding Fushi patterns against the same positive and negative examples.

The results suggest that these significance measures are not effective in the sense that most of the patterns obtained are obvious or uninteresting. The reasons for this are as follows: (1) the significance measures are suitable for a situation where some pattern almost completely classifies positive examples and negative examples, and the classification errors can be considered to be noise;

Table 2. The best 20 (k, d)-proximity patterns that distinguish Kokin-Shū from the rest of the Eight Imperial Anthologies.

$d = 2$, $k = 20$

rank	G	P_1	N_1	string
1	0.355	33	708	⟨NO, TSUKI⟩
2	0.356	41	71	⟨NA, KUNI/⟩
3	0.356	59	158	⟨NAKU, NI/⟩
4	0.356	29	594	⟨KA, TSUKI⟩
5	0.356	20	18	⟨UME, HA⟩
6	0.356	20	18	⟨UME, NA⟩
7	0.356	45	104	⟨NA, KUNI⟩
8	0.356	26	553	⟨HA, KANA/⟩
9	0.356	196	936	⟨HA, WA⟩
10	0.356	16	10	⟨MO, HERANA⟩
11	0.356	5	262	⟨TE/, KANA/⟩
12	0.356	137	583	⟨WA, HI⟩
13	0.356	50	817	⟨NO, KANA/⟩
14	0.356	365	2060	⟨HA, HI⟩
15	0.356	17	430	⟨MO, TSUKI⟩
16	0.356	22	490	⟨TE, KANA/⟩
17	0.356	24	32	⟨TO, NAKUNI/⟩
18	0.356	18	16	⟨UMENO, NA⟩
19	0.356	111	1397	⟨KA, NA/⟩
20	0.356	13	6	⟨UME, HI⟩

$d = 3$, $k = 20$

rank	G	P_1	N_1	string
1	0.356	28	35	⟨NO/, NI, KOHI⟩
2	0.356	19	16	⟨UME, HA, NA⟩
3	0.356	35	66	⟨KA, NAKU, NI/⟩
4	0.356	33	59	⟨NA, NI, KOHI⟩
5	0.356	16	11	⟨UME, HA, TO⟩
6	0.356	112	448	⟨HA, KU, HI⟩
7	0.356	68	991	⟨NO, KA, NA/⟩
8	0.356	41	94	⟨NO, WAKA, KO⟩
9	0.356	26	39	⟨TO, NA, KUNI/⟩
10	0.356	32	60	⟨RA, NAKU, NI⟩
11	0.356	28	46	⟨RA, NAKU, NI/⟩
12	0.356	16	12	⟨YAMA, NI/, NAKU⟩
13	0.356	39	89	⟨KA, NI, KOHI⟩
14	0.356	36	77	⟨NO, NI, KOHI⟩
15	0.356	27	44	⟨RA, NA, KUNI/⟩
16	0.356	32	62	⟨HA, SU, MI/⟩
17	0.356	157	725	⟨HA, HA, HI⟩
18	0.356	21	26	⟨U, MENO, NA⟩
19	0.356	12	5	⟨SHI, HE, RANARI/⟩
20	0.356	12	5	⟨SHI, HERA, RI/⟩

(2) the significance measures depend only on the frequencies of patterns in anthologies.

4 Pattern Discovery Based on MDL Principle

In Table 3, the only pattern that can be recognized as a Fushi pattern is ⋆KEREBA⋆BERANARI⋆. A pattern with relatively long clusters of characters is probably a Fushi pattern. Thus, we have the conditions:

(C1) To contain as many constant symbols as possible.
(C2) To contain as few gap symbols (⋆) as possible.

Concerning the frequency information, we only consider the condition:

(C3) To appear in as many positive examples as possible.

The method proposed by Brāzma et al. [4] gives a reasonable compromise between these conflicting conditions. For this reason, we decided to apply it to our problem. The method tries to find the optimal cover based on the Minimum Description Length (MDL) principle.

Table 3. The 20 highest scoring Fushi patterns for the information entropy and the Gini index.

	information entropy			Gini index		
rank	p_1	n_1	pattern	p_1	n_1	pattern
1	3	257	⋆TE⋆KANA⋆	12	10	⋆NI⋆KUNI⋆
2	7	284	⋆TE⋆NA⋆	7	2	⋆NI⋆MO⋆NIKERI⋆
3	12	10	⋆NI⋆KUNI⋆	5	0	⋆KEREBA⋆BERANARI⋆
4	5	0	⋆KEREBA⋆BERANARI⋆	11	11	⋆HA⋆KUNI⋆
5	7	2	⋆NI⋆MO⋆	4	0	⋆NURU⋆KERE⋆
6	16	25	⋆NO⋆KUNI⋆	4	0	⋆KARA⋆TSUTSU⋆
7	11	11	⋆HA⋆KUNI⋆	4	0	⋆KARA⋆TSU⋆
8	11	255	⋆WO⋆KANA⋆	4	0	⋆RE⋆MONOWO⋆
9	17	32	⋆MO⋆KUNI⋆	4	0	⋆RE⋆NOWO⋆
10	15	297	⋆WO⋆NA⋆	4	0	⋆HI⋆NO⋆
11	4	0	⋆NURU⋆KERE⋆	16	25	⋆NO⋆KUNI⋆
12	4	0	⋆KARA⋆TSUTSU⋆	5	1	⋆KEREBA⋆NARI⋆
13	4	0	⋆KARA⋆TSU⋆	3	257	⋆TE⋆KANA⋆
14	4	0	⋆RE⋆MONOWO⋆	17	32	⋆MO⋆KUNI⋆
15	4	0	⋆RE⋆NOWO⋆	7	5	⋆BA⋆BERANARI⋆
16	4	0	⋆HI⋆NO⋆	7	5	⋆REBA⋆NARU⋆
17	16	31	⋆NI⋆NASHI⋆	7	284	⋆TE⋆NA⋆
18	5	1	⋆KEREBA⋆NARI⋆	5	2	⋆RA⋆TSUTSU⋆
19	12	246	⋆RU⋆KANA⋆	5	2	⋆RA⋆TSU⋆
20	13	22	⋆REBA⋆ZO⋆RU⋆	16	31	⋆NI⋆NASHI⋆

4.1 Finding Optimal Cover Based on MDL Principle

We present the definition of optimal cover based on the MDL principle according to [4], and then describe their approximation algorithm.

Consider a pattern $\pi = \star\beta_1\star\cdots\star\beta_h\star$ $(\beta_1,\ldots,\beta_h \in \Sigma^+)$, and a set $B = \{\alpha_1,\ldots,\alpha_n\}$ of strings such that $B \subseteq L(\pi)$. The set B can be described by the pattern π and the strings $\gamma_{i,j}$ $(1 \le i \le n, 0 \le j \le h)$ such that $\alpha_i = \gamma_{i,0}\beta_1\gamma_{i,1}\cdots\gamma_{i,h-1}\beta_h\gamma_{i,h}$. Such a description of B is called *the encoding by pattern* π. We denote by $\|\alpha\|$ the description length of a string α. For simplicity, we ignore the delimiters between strings. Assuming some symbol-wise encoding, the description length of B is

$$\|\pi\| + \sum_{i=1}^{n}\sum_{j=0}^{h}\|\gamma_{i,j}\| = \sum_{i=1}^{n}\|\alpha_i\| - \Big(\|c(\pi)\|\cdot|B| - \|\pi\|\Big),$$

where $c(\pi)$ denotes the string obtained from π by removing all $\star$'s.

Let A be a finite set of strings. A finite set $\Omega = \{(\pi_1,B_1),\ldots,(\pi_k,B_k)\}$ of pairs of a pattern π_j and a subset B_j of A is also said to be a *cover* of A if (1) $B_j \subseteq L(\pi_j)$ $(j = 1,\ldots,k)$, (2) $A = B_1 \cup\cdots\cup B_k$, and (3) $B_1,\ldots,B_k$ are disjoint. When the set B_j is encoded by π_j for each $j = 1,\ldots,k$, the description length of A is $M(\Omega) = \sum_{i=1}^{n}\|\alpha_i\| - C(\Omega)$, where $C(\Omega)$ is given by $C(\Omega) =$

$\sum_{j=1}^{k}\Big(\|c(\pi_j)\|\cdot|B_j|-\|\pi_j\|\Big)$. The *optimal cover* of the set A is defined to be the cover Ω minimizing $M(\Omega)$, or to be the set of patterns in it. Minimizing $M(\Omega)$ is equivalent to maximizing $C(\Omega)$.

Since the problem of finding the optimal cover of a set of strings contains as a special case the set cover problem, it is NP-hard. Brāzma et al. [4] modified the problem as follows:

> Given a finite set A of strings and a finite set Δ of patterns, find a cover Ω of A in which patterns are chosen from Δ that minimizes $M(\Omega)$.

They presented a greedy algorithm that approximates the optimal solution. It computes the values of $u(\pi) - \frac{v(\pi)}{|L(\pi)\cap U|}$ for all possible patterns π during each iteration of a loop, and selects the pattern maximizing it to the cover. Here U is the set of strings in A not covered by any pattern that has already been selected. The value of $M(\Omega)$ for an approximate solution Ω obtained by this algorithm is at most $\log_2|A|$ times the value for the optimal solution. Roughly speaking, the method arranges all possible patterns π in decreasing order of the value of $u(\pi) - \frac{v(\pi)}{|L(\pi)\cap U|}$. Note that this criterion for patterns concerns not only their frequencies ($|L(\pi)\cap U|$), but also their form ($u(\pi)$ and $v(\pi)$).

4.2 Experimental Results

We applied the algorithm to five anthologies: Kokin-Shū, Shin-Kokin-Shū, Mini-Shū, Shūi-Gusō, and Sanka-Shū. The first two are the first and the eighth imperial anthologies, respectively. They are known as the best two of the 21 imperial anthologies, and have been studied most extensively. Kokin-Shū was completed in 922, and Shin-Kokin-Shū in 1205. The differences between the two anthologies, if any exist, may be due to the time difference in compilation. On the other hand, the others are private anthologies of poems composed by the three contemporaries: Fujiwara no Ietaka (1158–1237), Fujiwara no Teika (1162–1241), and the priest Saigyō (1118–1190). Their differences probably depend on the poets' personalities.

Table 4 shows the results of the experiment. From the second column, it is observed that a great number of patterns occur in each anthology, and therefore it is not possible to examine all of them manually. The values in parentheses are the numbers of patterns occurring more than once. In the experiment, we used these sets of patterns as Δ, the sets of candidate patterns. The sizes of covers are shown in the third column. For example, 191 of 8,265 patterns were extracted from Kokin-Shū.

We proved the effectiveness of the method with regard to the following four points: (1) *Is the number of obtained patterns small enough?* (2) *Were typical patterns obtained?* (3) *Were non-obvious patterns obtained?* (4) *Are differences among anthologies observed?* In particular, concerning (4), Table 5 shows the first five patterns for each anthology, where each numeral denotes the frequency of the pattern in the anthology. The following facts, for example, can be read from Table 5:

Table 4. Covers of five anthologies.

Anthology	*# occurring patterns*		*Size of cover*
Kokin-Shū	164,978	(8,265)	191
Shin-Kokin-Shū	233,187	(12,449)	270
Mini-Shū	187,014	(16,425)	369
Shūi-Gusō	214,940	(14,365)	335
Sanka-Shū	279,904	(12,963)	232

1. Pattern ⋆BAKARI⋆RAM⋆ does not occur in either Kokin-Shū or Shin-Kokin-Shū.
2. Pattern ⋆WA⋆NARIKERI⋆ occurs in each of the anthologies. In particular, it occurs frequently in Sanka-Shū.
3. Pattern ⋆MASHI⋆NARISEBA⋆ occurs frequently in Sanka-Shū.
4. Pattern ⋆KOSO⋆RIKERE⋆ does not occur in Shūi-Gusō.
5. Pattern ⋆YA⋆RURAM⋆ occurs in each anthology except Kokin-Shū.

It is possible that the above facts are important characteristics that may be due to literary trends or poets' personalities. For example, 2 and 3 may reflect Priest Saigyō's preferences, and 5 may imply that the pattern ⋆YA⋆RURAM⋆ was not preferred in the period of Kokin-Shū. Comparisons of the obtained patterns and their frequencies thus give Waka researchers clues for further investigation.

Table 5. Fushi patterns from five anthologies with frequencies, where A, B, C, D and E denote Kokin-Shū, Shin-Kokin-Shū, Mini-Shū, Shūi-Gusō, and Sanka-Shū, respectively.

	Patterns	A	B	C	D	E		*Patterns*	A	B	C	D	E
A	⋆KEREBA⋆BERANARI⋆	5	0	0	0	0	D	⋆BAKARI⋆RAM⋆[1]	0	0	11	8	3
	⋆ZO⋆SHIKARIKERU⋆	8	1	0	0	3		⋆NO⋆NARIKERI⋆	19	30	39	19	49
	⋆KOSO⋆RIKERE⋆[4]	11	8	8	0	13		⋆RAZARIKI⋆NO⋆	0	0	1	6	1
	⋆RISEBA⋆RAMASHI⋆	5	2	0	0	4		⋆YA⋆RURAM⋆[5]	0	8	40	24	23
	⋆WA⋆NARIKERI⋆[2]	20	26	26	11	52		⋆NI⋆NARURAM⋆	0	2	8	8	7
B	⋆KARISEBA⋆MASHI⋆	3	6	0	0	1	E	⋆MASHI⋆NARISEBA⋆[3]	0	2	1	0	10
	⋆NO⋆NIKERUKANA⋆	4	11	2	1	4		⋆KOSO⋆KARIKERE⋆	4	4	1	0	8
	⋆WA⋆NARIKERI⋆[2]	20	26	26	11	52		⋆NARABA⋆RAMASHI⋆	1	0	0	0	8
	⋆KOSO⋆RIKERE⋆[4]	11	8	8	0	13		⋆O⋆UNARIKERI⋆	1	0	0	0	7
	⋆MO⋆KARIKERI⋆	4	11	8	5	7		⋆NO⋆RUNARIKERI⋆	4	3	4	0	10
C	⋆BAKARI⋆RURAM⋆	0	0	6	0	3							
	⋆KOSO⋆NARIKERE⋆	4	0	5	0	5							
	⋆YA⋆NARURAM⋆	0	2	16	4	7							
	⋆WA⋆NARIKERI⋆[2]	20	26	26	11	52							
	⋆NO⋆NARIKERI⋆	19	30	39	19	49							

5 Discovering Substring Patterns

In the previous section, we showed successful results in pattern discovery by adopting the Fushi patterns as domain knowledge. However, consider the case where no such domain knowledge is available. Recall the results of discovering (k, d)-proximity patterns presented in Section 3.2. There are combinatorially many (k, d)-proximity patterns for $d > 1$. Hence it is reasonable as a first step to concentrate on the case of $d = 1$. In [16], we restricted ourselves to the case of $d = 1$, i.e., the patterns of the form $\star w \star$, where $\star$ is a wildcard that matches any string and w is a non-empty string. We call such patterns the *substring patterns.* We have a trivial $O(m)$ time and space algorithm since there are essentially $O(m)$ candidates for the optimal substring w, where m is the total length of strings in $S = Pos \cup Neg$. Therefore it seems that the main difficulty is finding an appropriate statistical measure.

However, we found that a considerable part of the mined patterns were obvious and worthless although we tested several statistical measure functions as G for our problem. It seems that no matter what measure we use, we cannot remove these worthless patterns from the results in practice. Discovery requires an effort by domain experts to examine the first part of a list of patterns arranged in decreasing order of goodness. This corresponds to the step of *interpreting mined patterns*, a "postprocessing" of data mining in the knowledge discovery process [7]. We believe that this step to support the domain experts is a key to success. We tackle this problem with the weapon of *stringology* [8].

5.1 Inefficiency in Domain Experts' Tasks

We take the following approach to our problem.

1. Choose an appropriate statistical measure G.
2. Create a list of substrings of the strings in $S = Pos \cup Neg$, sorted according to the values of G.
3. Have the upper part of the list evaluated by domain experts.

When more than two strings have the same value in Step 2, it is reasonable to arrange them in lexicographical order. Table 6 (a) is a partial list of substrings for Kokin-Shū and Gosen-Shū, the first and the second imperial anthologies. It is only by coincidence that the strings "A-YA-MA-TA-RE" (ranked 235th) and "KA-KE-KA" (ranked 237th) have the same frequencies. However, the situation is different for the strings "A-YA-MA-TA-RE" and "A-YA-MA-TA-RE-KE" (ranked 236th). The former is a prefix of the latter, and every time the former appears, it is followed by a letter "KE." In this case, every occurrence of the strings "MA-TA-RE-KE," "YA-MA-TA-RE," and "YA-MA-TA-RE-KE" (ranked 305th, 312th, and 313th, respectively) was a substring of the string "A-YA-MA-TA-RE-KE," and therefore all of these strings had the same goodness value. It is desirable that strings in such relationship appear in one section of the list. However, a simple lexicographical sorting disperses them as in Table 6 (a). This causes inefficiency in the task of evaluating strings in the list.

Table 6. (a) A partial list of substrings of poems in two imperial anthologies Kokin-Shū and Gosen-Shū. We excluded substrings that cover the boundary between the lines of a poem. The substrings with the same goodness value were sorted lexicographically with respect to the order of Kana letters, although romanized here. (b) A list of equivalence classes for Kokin-Shū and Gosen-Shū. The symbol • indicates the representative of each equivalence class. When two or more classes have the same goodness value, we arranged them in lexicographical order of their representatives.

(a)

rank	G	p_1	n_1	string
⋮				
234	0.6836	11	33	TSU-YU-NO
235	0.6836	4	0	A-YA-MA-TA-RE
236	0.6836	4	0	A-YA-MA-TA-RE-KE
237	0.6836	4	0	KA-KE-KA
⋮				
303	0.6836	4	0	FU-RI-TA
304	0.6836	4	0	HO-NI-I-TE-TE
305	0.6836	4	0	MA-TA-RE-KE
306	0.6836	4	0	MI-YA-KI
307	0.6836	4	0	MU-NO
308	0.6836	4	0	MO-KA-MO
309	0.6836	4	0	MO-TO-YU
310	0.6836	4	0	MO-RU-TO
311	0.6836	4	0	YA-CHI-YO
312	0.6836	4	0	YA-MA-TA-RE
313	0.6836	4	0	YA-MA-TA-RE-KE
314	0.6836	4	0	RA-ME-TO
315	0.6836	4	0	RU-TO-I
⋮				

(b)

rank	G	p_1	n_1	equivalence class
⋮				
187	0.6836	11	33	• TSU-YU-NO
188	0.6836	4	0	• A-YA-MA-TA-RE-KE YA-MA-TA-RE-KE MA-TA-RE-KE A-YA-MA-TA-RE YA-MA-TA-RE
189	0.6836	4	0	• KA-KE-KA
190	0.6836	4	0	• KA-SHI-RA
⋮				
197	0.6836	4	0	• KO-HI-SHI-KA-RI-KE-RU
198	0.6836	4	0	• KO-HI-SHI-KA-RU-HE HI-SHI-KA-RU-HE
199	0.6836	4	0	• KO-HI-SHI-KI-TO-KI HI-SHI-KI-TO-KI SHI-KI-TO-KI KO-HI-SHI-KI-TO HI-SHI-KI-TO
200	0.6836	4	0	• KO-HI-YA-WA-TA-RA-MU HI-YA-WA-TA-RA-MU KO-HI-YA-WA-TA-RA HI-YA-WA-TA-RA KO-HI-YA-WA-TA HI-YA-WA-TA KO-HI-YA-WA HI-YA-WA
201	0.6836	4	0	• SA-KI-SO
⋮				

5.2 Reducing Inefficiency

To reduce this inefficiency, we partition the substrings of $S = Pos \cup Neg$ into equivalence classes by using an equivalence relation on Σ^*, denoted by $\equiv$, which was first defined by Blumer et al. [3], and has the following properties:

- Each equivalence class has a unique longest string that contains any other member as a substring, and we regard it as the *representative* of the class.
- All the strings in an equivalence class have the same frequencies in *Pos* and in *Neg*, respectively, and therefore have the same value of goodness.
- The number of equivalence classes is linear with respect to the total length of strings in $S = Pos \cup Neg$.

The basic idea is to create a list of equivalence classes, not just a list of substrings. In this list, we can sort the equivalence classes according to the goodness values since all members of each equivalence class have the same goodness value. See Table 6 (b).

It should be stated that the data structure called the *suffix tree* [5] exploits a similar and more popular equivalence relation, denoted by $\equiv_R$, which also satisfies the above three conditions. The equivalence relation $\equiv_R$ is a refinement of $\equiv$, and therefore includes many more equivalence classes. From the viewpoint of computational complexity, this is not a significant difference because the two equivalence relations both satisfy the third condition above. However, the difference is crucial for researchers who must check the candidate strings. In fact, it was observed that the number of equivalence classes under $\equiv_R$ is approximately 4 times larger than that under $\equiv$.

Although each of the equivalence classes in $\equiv$ might contain $O(m^2)$ strings where m is the length of its representative, a compact representation is possible in which only the representative and the minimal members are shown. The number of minimal members is at most m.

6 Similarity Computation as Optimal Pattern Discovery

In [15] we addressed the problem of finding similar Waka poems, where the aim was to discover unnoticed instances of *Honkadori*, a technique based on specific allusion to earlier famous poems. Traditionally, the scheme of weighted edit distance with a weight matrix may have been used to quantify affinities between strings such as DNA sequences (see e.g. [9]). This scheme, however, requires a fine tuning by a hand-coding or a heuristic criterion of many weights in a matrix, the number of which is quadratic with respect to the size of the alphabet. As an alternative idea, we introduced a new framework called *string resemblance systems* (SRSs for short) [15]. In this framework, the similarity of two strings is evaluated using a pattern that matches both of them, with support by an appropriate function that associates the quantity of resemblance to candidate patterns. This scheme bridges a gap between optimal pattern discovery, machine learning, and similarity computation.

A *pattern system* is a triple $\langle \Sigma, \Pi, L \rangle$ of a finite alphabet Σ, a set Π of descriptions called *patterns*, and a function L that maps a pattern $\pi \in \Pi$ to a *language* $L(\pi) \subseteq \Sigma^*$. A pattern $\pi \in \Pi$ *matches* $w \in \Sigma^*$ if w belongs to $L(\pi)$. Also, π is a *common pattern of strings* u *and* v, if π matches both of them.

Definition 2. *A* string resemblance system *(SRS) is a quadruple* $\langle \Sigma, \Pi, L, score \rangle$, *where* $\langle \Sigma, \Pi, L \rangle$ *is a pattern system, and score is a* pattern score function *that maps a pattern in* Π *to a real number.*

The *similarity* $\mathrm{SIM}(x, y)$ between two strings x and y with respect to an SRS $\langle \Sigma, \Pi, L, score \rangle$ is defined by $\mathrm{SIM}(x, y) = \max\{score(\pi) \mid \pi \in \Pi \text{ and } x, y \in L(\pi)\}$. When the set $\{score(\pi) \mid \pi \in \Pi \text{ and } x, y \in L(\pi)\}$ is empty or the maximum does not exist, $\mathrm{SIM}(x, y)$ is undefined.

The definition given above regards the similarity computation as a special case of the optimal pattern discovery.

Definition 3. SIMILARITY COMPUTATION WITH RESPECT TO $\langle \Sigma, \Pi, L, score \rangle$. *Given two strings $w_1, w_2 \in \Sigma^*$, find a pattern $\pi \in \Pi$ with $\{w_1, w_2\} \subseteq L(\pi)$ that maximizes $score(\pi)$.*

We can handle a variety of string similarities by changing the pattern system and the pattern score function. In [15], the class of homomorphic SRSs was defined, and it was shown that this class covers most of the well-known and well-studied similarity (dissimilarity) measures, including the edit distance, the weighted edit distance, the Hamming distance and the LCS measure. Also, this class was extended to the semi-homomorphic SRSs in [15], into which for example the similarity measures designed for finding instances of Honkadori [15] fall. The similarity measures for musical sequence comparison developed in [11] also fall into this class. For homomorphic and semi-homomorphic SRSs, the similarity computation can be performed in polynomial time [15] by applying the idea of a weighted edit graph (see, e.g., [9]), under some reasonable assumption.

On the other hand, both the Angluin pattern system [1] and the fragmentary pattern system [10] are examples of non-homomorphic pattern systems. The membership problems for these pattern systems are both NP-complete as shown in [1] and in [10], respectively. Also, the similarity computations are NP-hard in general as shown in [17] and in [10]. In [17], we discussed the SRS with a subclass of the Angluin patterns in the context of finding repetitive expressions from Waka poems.

In [15], we also proposed a *rarity-based measure.* A *rarity of a pattern $\pi \in \Pi$ with respect to a set S of strings* is defined to be the logarithm of the inverse of $|L(\pi) \cap S|/|S|$. The problem is then:

> Given a set S of strings over Σ and two strings $w_1, w_2 \in S$, find a pattern π with $\{w_1, w_2\} \subseteq L(\pi)$ that maximizes the rarity of π with respect to S.

By regarding the strings w_1, w_2 as positive examples and the other strings in S as negative examples, this problem is considered to be a case of one-side error minimization. Using this measure, we could find poem pairs that have a close affinity, possibly excluding stereotypical expressions.

7 Concluding Remarks

One approach to machine discovery might be to establish a general theory about the discovery process and to apply it to individual problems. However, we believe that such a top-down strategy would never succeed. The discovery process must be a complex human activity, and it is difficult to reveal it completely. Thus, we believe, it is most important to advance solutions to non-trivial problems that have great significance for domain researchers. First, we will have to start with particular problems, even if this seems lacking in generality. The generality will come only after analyzing successful and unsuccessful cases of real-world problems.

From this point of view, we have studied several problems of text mining from classical Japanese poems. Some of the extracted patterns or affinities using our methods are very stimulating for Waka researchers and provide clues for further investigation.

It is not easy for us to answer the question "What is the key to success in machine discovery?". Nevertheless, through our experiences surveyed in this paper, we have demonstrated at least that the following are significant.

Effective but ascetic use of domain knowledge. In Section 4, we adopted the notion of Fushi patterns as domain knowledge to restrict the class of pattern to be discovered, and succeeded in extracting interesting patterns, while most of those extracted were uninteresting for proximity patterns, as in Section 3. The use of knowledge, however, was limited to the candidate strings for adjunct sequences of Fushi patterns.

Explicit support for domain experts who are involved in interpreting mined patterns. In Section 5, we tackled this problem and presented a way for supporting domain experts who academically scrutinize a list of text substrings with a high significance value.

Flexibility in changing views according to domain experts' particular interests. In Section 6, we discussed the problem of finding similar poems. In this case, similarity measures correspond to domain experts' *views*, particular ways of interpreting the data [12]. It is desirable for domain experts flexibly to design and modify their views (similarity measures) according to the kind of resemblance that they wish to quantify. For this reason, we introduced a uniform framework, in which design and modification of similarity measures are easy. Using the framework, we developed several similarity measures that led to new discoveries. In fact, we discovered affinities of some poems with earlier poems. This raised an interesting issue for Waka studies, to which we were able to give a convincing conclusion.

- We have proved that one of the most important poems by Fujiwara-no-Kanesuke, one of the renowned 36 poets, was in fact based on a model poem found in Kokin-Shū. The same poem had been interpreted to show merely "frank expression of parents' care for their child." Our study revealed, by detecting the same structure in the two poems, that the poet's compositional techniques were partly obscured by the heart-warming feature of the poem[1].
- We have compared Tametada-Shū, the mysterious anthology unidentified in Japanese literary history, with a number of private anthologies edited after the middle of the Kamakura period (the 13th century) using the same method, and found that there are similarities between about 10 pairs of

[1] *Asahi*, one of Japan's leading newspapers, made a front-page report of this discovery (26 May, 2001).

poems in Tametada-Shū and Sokon-Shū, an anthology by Shōtetsu. The result suggests that the mysterious anthology was edited by a poet in the early Muromachi period (the 15th century). There has been a dispute about the editing date since one scholar suggested the middle of the Kamakura period as a probable date. We have developed strong evidence concerning this problem.

References

1. D. Angluin. Finding patterns common to a set of strings. *J. Comput. Sys. Sci.*, 21:46–62, 1980.
2. H. Arimura. Text data mining with optimized pattern discovery. In *Proc. 17th Workshop on Machine Intelligence*, Cambridge, July 2000.
3. A. Blumer, J. Blumer, D. Haussler, R. Mcconnell, and A. Ehrenfeucht. Complete inverted files for efficient text retrieval and analysis. *J. ACM*, 34(3):578–595, 1987. Previous version in: STOC'84.
4. A. Brāzma, E. Ukkonen, and J. Vilo. Discovering unbounded unions of regular pattern languages from positive examples. In *Proc. 7th International Symposium on Algorithms and Computation (ISAAC'96)*, pages 95–104, 1996.
5. M. Crochemore and W. Rytter. *Text Algorithms.* Oxford University Press, 1994.
6. L. Devroye, L. Györfi, and G. Lugosi. *A Probabilistic Theory of Pattern Recognition.* Springer, 1997.
7. U. M. Fayyad, G. P.-Shapiro, and P. Smyth. From data mining to knowledge discovery: an overview. In *Advances in Knowledge Discovery and Data Mining*, pages 1–34. The AAAI Press, 1996.
8. Z. Galil. Open problems in stringology. In A. Apostolico and Z. Galil, editors, *Combinatorial Algorithms on Words*, NATO ASI Series, Advanced Science Institutes Series, Series F: Computer and Systems Sciences, Vol. 12, pages 1–8. Springer-Verlag, 1985.
9. D. Gusfield. *Algorithms on Strings, Trees, and Sequences: Computer Science and Computational Biology.* Cambridge University Press, New York, 1997.
10. H. Hori, S. Shimozono, M. Takeda, and A. Shinohara. Fragmentary pattern matching: Complexity, algorithms and applications for analyzing classic literary works. In *Proc. 12th Annual International Symposium on Algorithms and Computation (ISAAC'01)*, 2001. To appear.
11. T. Kadota, M. Hirao, A. Ishino, M. Takeda, A. Shinohara, and F. Matsuo. Musical sequence comparison for melodic and rhythmic similarities. In *Proc. 8th International Symposium on String Processing and Information Retrieval (SPIRE2001).* IEEE Computer Society, 2001. To appear.
12. O. Maruyama, T. Uchida, K. L. Sim, and S. Miyano. Designing views in HypothesisCreator: System for assisting in discovery. In *Proc. 2nd International Conference on Discovery Science (DS'99), LNAI 1721*, pages 115–127, 1999.
13. S. Morishita. On classification and regression. In *Proc. 1st International Conference on Discovery Science (DS'99), LNAI 1532*, pages 49–59, 1998.
14. S. Shimozono, H. Arimura, and S. Arikawa. Efficient discovery of optimal word-association patterns in large databases. *New Gener. Comput.*, 18(1):49–60, 2000.
15. M. Takeda, T. Fukuda, I. Nanri, M. Yamasaki, and K. Tamari. Discovering instances of poetic allusion from anthologies of classical Japanese poems. *Theor. Comput. Sci.*, 2001. To appear. Preliminary version in: Proc. DS'99 (LNAI 1721).

16. M. Takeda, T. Matsumoto, T. Fukuda, and I. Nanri. Discovering characteristic expressions from literary works. *Theor. Comput. Sci.*, 2001. To appear. Preliminary version in: Proc. DS 2000 (LNAI 1967).
17. K. Yamamoto, M. Takeda, A. Shinohara, T. Fukuda, and I. Nanri. Discovering repetitive expressions and affinities from anthologies of classical Japanese poems. In *Proc. 4th International Conference on Discovery Science (DS2001)*, 2001. To appear.
18. M. Yamasaki, M. Takeda, T. Fukuda, and I. Nanri. Discovering characteristic patterns from collections of classical Japanese poems. *New Gener. Comput.*, 18(1):61–73, 2000. Preliminary version in: Proc. DS'98 (LNAI 1532).

Second Difference Method Reinforced by Grouping: A New Tool for Assistance in Assignment of Complex Molecular Spectra

Takehiko Tanaka

Department of Chemistry, Faculty of Science, Kyushu University,
Hakozaki, Higashiku, Fukuoka 812-8581, Japan
tata.scc@mbox.nc.kyushu-u.ac.jp

Abstract. This report describes a search for a new method for assistance in assignment of complex molecular spectra without obviously recognizable regular patterns, on which traditional methods of spectral assignment have depended. We propose "second difference method reinforced by grouping," a computer-aided technique for picking out regular patterns buried in a list of observed line wavenumbers which look like randomly distributed. The results of tests suggest that this technique may be succesfully applied even to the cases in which the method of Loomis-Wood diagram would fail. Superiority of the present method to the Loomis-Wood method in immunity against the interference by the overlap of unrelated spectral bands is also demonstrated.

1 Introduction

High resolution molecular spectra contain abundant information on structures and dynamics of molecules. However, extraction of such useful information necessitates a procedure of *spectral assignment* in which each spectral line is assigned a set of quantum numbers. This procedure has traditionally been performed by making use of regular patterns that are obviously seen in the observed spectrum. However, we often encounter complex spectra in which such regular patterns may not be readily discerned.

The purpose of the present work is to search for a new method which can assist the assignment of such complex molecular spectra. We wish to devise a computer-aided technique for *picking out regular patterns buried in a list of observed line wavenumbers which look like randomly distributed.* An efficient method for spectral assignment becomes increasingly desirable, because recent advance of spectroscopic instruments has allowed us to obtain a large amount of spectral data in machine readable forms.

In the present article we propose a method, which we tentatively refer to as "second difference method reinforced by grouping." We have previously described an embryonic version [1, 2], and suggested that it may be developed as a useful tool for assignment of complex molecular spectra. This method has been tested with some success on the observed spectrum of a linear molecule DCCCl

S. Arikawa and A. Shinohara (Eds.): Progress in Discovery Science 2001, LNAI 2281, pp. 532–542.

[1] as well as artificial data corresponding to an infrared spectrum of HCCBr [2]. However, we recently encountered a set of spectral data for which the original version did not work well. The new version described in the present article was developed as a remedy but proved to be a substantial improvement from the original one. A preliminary report has been published [3].

Loomis-Wood diagram is a traditional method serving similar purposes. It is a two-dimensional spectral peak diagram first invented by Loomis and Wood [4] as early as in 1928. Because it was time consuming to draw a Loomis-Wood diagram manually, this method could not be used effectively, especially in early stages of spectral analysis, before the advent of fast computers. Scott and Rao [5] and Nakagawa and Overend [6] wrote programs to generate printed Loomis-Wood diagrams. More recently Winnewisser et al. [7] devised the first interactive Loomis-Wood diagram using computer graphics. Several Loomis-Wood diagrams were designed since then. They are so powerful that spectroscopists in various groups have utilized them.

The present method is competitive with as well as complementary to the Loomis-Wood diagram. The results of the tests described in the following suggest that the present method may be successfully applied even to the cases in which the method of Loomis-Wood diagram would fail. Superiority of the present method to the Loomis-Wood method in immunity against the interference by the overlap of unrelated spectral bands is also demonstrated.

The present method may be carried out by using a small personal computer. The test calculations described in the following were performed on an IBM compatible machine with a 600 MHz CPU and 128 MB memory. The running time was a matter of minutes.

2 Second Difference Method

The basic assumptions that the second difference method is based on are as follows;

1. A *complex* spectrum is formed as a result of overlap of many spectral series, each of which has a *simple* structure.
2. The wavenumbers of spectral lines belonging to a series may be represented to a good approximation by a quadratic function of a running number.

The second assumption means that if we let $f(k), k = 1, 2, 3, \ldots$ be wavenumbers of spectral lines belonging to a series, the second difference

$$\Delta^2(k) = f(k+2) - 2f(k+1) + f(k) \tag{1}$$

would be almost constant independent of k, and therefore the third difference

$$\Delta^3(k) = f(k+3) - 3f(k+2) + 3f(k+1) - f(k) \tag{2}$$

would be very small for all k. It is also noted that the above assumption is exactly the same as what makes the method of Loomis-Wood diagram work.

We coded a FORTRAN program briefly described as follows. It generates n-membered chains with a preset length, e.g., $n = 15$. An n-membered chain is defined as an array of n wavenumbers $(f_1, f_2, \ldots, f_n)$ chosen from the list of observed line wavenumbers, in which the wavenumbers satisfy such a restriction that the absolute values of $f_{i+3} - 3f_{i+2} + 3f_{i+1} - f_i$ for $i = 1, 2, \ldots, n-3$ never exceed a preset value of Δf_{allow}. The Δf_{allow} value will be set to such a magnitude that in a true spectral series $f(1), f(2), \ldots$ the absolute value of the third difference, $\Delta^3(k) = f(k+3) - 3f(k+2) + 3f(k+1) - f(k)$, for any k would not exceed Δf_{allow}. The choice of the Δf_{allow} value is not very critical. However, it is noted that too large a Δf_{allow} value may result in something like combination explosion. It must be somewhat smaller than a certain limit which depends on the average number of lines per wavenumber unit, i.e., the density of specral lines. It is practically convenient to restrict the first three members of the chain by setting the upper and lower limits to the first difference $f_2 - f_1$ as well as to the second difference $f_3 - 2f_2 + f_1$. The restriction of the first three members is simply to save computation time and memory, and therefore the limits may be rather loosely set, with consideration of the ranges in which the first and second differences would be bounded if the three lines are consecutive members of a true spectral series. For each of resulting chains, wavenumbers are least-squares fitted to a polynomial of a preset order (e.g. 4-th order), and standard deviation is calculated.

3 Test on the Spectrum of *Trans*-Glyoxal

The spectrum for which the second difference method in the embryonic version was unsuccessful is that of *trans*-glyoxal, corresponding to the $A\,{}^1A_u\,(v_7 = 1) - X\,{}^1A_g\,(v = 0)$ vibronic transition (*trans*-Glyoxal [$C_2H_2O_2$] is a planar molecule belonging to the C_{2h} point group). It has been observed at a very high resolution by Kato et al. by means of Doppler-free two-photon absorption spectroscopy [8]. In Ref. [8], the observed line wavenumbers are listed together with assignments of quantum numbers. These wavenumbers were used as the test data.

The profile of the test data is roughly described as follows. It consists of 1710 spectral lines distributed in the region 22187–22215 cm^{-1}. According to the analysis by Kato et al., these lines belong to the following 52 spectral series; Q0 (76), Q1a (72), Q1b (74), Q2a (68), Q2b (71), Q3a (63), Q3b (68), Q4a (63), Q4b (65), Q5a (63), Q5b (64), Q6a (64), Q6b (64), Q7a (64), Q7b (64), Q8 (63), Q9 (62), Q10 (61), Q11 (59), Q12 (59), Q13 (58), Q14 (54), Q15 (54), Q16 (35), Q17 (22), O0 (22), O1a (23), O1b (24), O2a (21), O2b (21), O3a (20), O3b (22), O4a (21), O4b (21), O5 (21), O6 (22), O7 (19), O8 (19), O9 (18), O10 (17), O11 (16), O12 (13), S0 (11), S1a (13), S1b (11), S2a (10), S2b (10), S3a (9), S3b (9), S4 (9), S5 (7), S6 (5), where the number of line wavenumbers assigned to each series is given in parentheses. These series overlap one over another and form a very congested and complex spectrum. Some series have one or more missing lines, probably due to perturbations. A considerable number of lines are multiply assigned indicating blended lines.

The names of the spectral series above were given by the present author just for convenience. A brief explanatory comment is given in the following. A transition in this vibronic band is labeled by a set of six numbers, $(J', K'_a, K'_c, J'', K''_a, K''_c)$. The capital letters Q, O, and S stand for the Q-, O-, and S-branches, meaning that the corresponding series is composed of transitions with $J' = J''$, $J' = J'' - 2$, and $J' = J'' + 2$, respectively. The integer following the capital letter denotes the value of K''_a, which coincides with the K'_a value. The lower case letters a and b are used to distinguish twin series, caused by the K-type doubling. The corresponding lines in the twin series tend to merge into single lines in the region of small J quantum numbers, with exceptions of the pairs (Q1a, Q1b), (O1a, O1b), (O2a, O2b), (S1a, S1b), and (S2a, S2b), where all assigned lines were observed separately. As K''_a increases, the merging becomes more extensive, and eventually twin series are not separable after Q8, O5, or S4 (the reason why neither a nor b is attched). It is also noted that Q0, O0, and S0 are single series.

We applied the present method to this spectrum under the following conditions. We restricted the first three members of a chain so that the first and second differences were between 0 and 1.2 cm^{-1} and between -0.015 and 0 cm^{-1}, respectively. The Δf_{allow} value was chosen to be 0.003 cm^{-1}. The chain length was preset at 15. Then, 7595 chains were generated.

Table 1. 15-Menbered chains with smallest standard deviations[a]

	Chain 1	Chain 2	Chain 3	Chain 4	Chain 5	Chain 6	Chain 7
f_1	194.5468	198.9133	199.6550	200.0132	199.2883	194.6209	205.4011
f_2	195.0206	199.2883	200.0132	200.3628	199.6550	195.1754	205.5348
f_3	195.4852	199.6550	200.3628	200.7038	200.0132	195.7215	205.6626
f_4	195.9409	200.0132	200.7038	201.0359	200.3628	196.2591	205.7837
f_5	196.3875	200.3628	201.0359	201.3594	200.7038	196.7880	205.8988
f_6	196.8250	200.7038	201.3594	201.6743	201.0359	197.3090	206.0068
f_7	197.2538	201.0359	201.6743	201.9802	201.3594	197.8213	206.1083
f_8	197.6736	201.3594	201.9802	202.2778	201.6743	198.3253	206.2026
f_9	198.0847	201.6743	202.2778	202.5661	201.9802	198.8210	206.2901
f_{10}	198.4866	201.9802	202.5661	202.8460	202.2778	199.3081	206.3700
f_{11}	198.8796	202.2778	202.8460	203.1172	202.5661	199.7870	206.4424
f_{12}	199.2637	202.5661	203.1172	203.3797	202.8460	200.2573	206.5074
f_{13}	199.6385	202.8460	203.3797	203.6333	203.1172	200.7196	206.5647
f_{14}	200.0039	203.1172	203.6333	203.8785	203.3797	201.1734	206.6137
f_{15}	200.3598	203.3797	203.8785	204.1150	203.6333	201.6185	206.6548
SD[b]	0.000070	0.000091	0.000091	0.000091	0.000093	0.000105	0.000105

[a] In units of cm^{-1}. Wavenumbers have been subtracted by 22000 cm^{-1}.
[b] Standard deviation.

The wavenumbers of the resulting chains were least squares fitted to polynomials of the 4-th order. The chains were sorted in the order of increasing standard deviation. Table 1 lists, as examples, the first to seventh chains. The first chain has a standard deviation of 0.000070 cm^{-1}. The 2000-th, 4000-th, 6000-th, and last (7595-th) ones have standard deviations of 331, 474, 695, and 6340, respectively, in units of 10^{-6} cm^{-1}.

We inspected the first through 240-th (with a standard deviation of 0.000170 cm^{-1}) chains whether each chain was a part of a true spectral series, and found that 127 of them corresponded to true spectral series. There were 74 chains for which only one line was erroneously merged. Similarly, 30, 7, 1, and 1 chains had 2, 3, 4, and 6 erroneously merged lines.

Except for a few cases, it was found that the number of the erroneously merged lines was at most two. Even if there are merged two erroneous lines, it would not be too difficult to arrive at the correct assignment starting from them. Twenty spectral series, i.e., Q0, Q1a, Q1b, Q2a, Q2b, Q3a, Q3b, Q4a, Q4b, Q5b, Q6a, Q7a, Q8, Q9, Q10, Q11, Q12, Q13, Q14, and O9, were detected by the chains with less than 3 erroneous lines among the 240 chains. Most of the major spectral series were detected. These results indicate the usefulness of the present method.

4 Grouping of Similar Chains

However, a problem is that an extremely large number of chains, e.g., 7595 15-membered chains, are generated, and it is practically impossible to scrutinize all the chains. To utilize all generated chains more effectively, we coded a second FORTRAN program for the grouping of similar chains. If two n-membered chains share more than $n-1$ common wavenumbers, they are defined to be "similar chains". This program classifies chains so that similar chains go into the same group.

Grouping of the above 7595 chains resulted in 95 groups. The largest group consists of 1723 chains (see Column "Number of chains" in Table 2). To the second largest group belong 948 chains, and 662 chains to the third largest, and so forth. There are also very small groups, e.g., 15 single chains form their own groups, and 7 groups have two chains each.

In the next step, the program collects spectral lines contained in the chains belonging to each group, where repetition of the same wavenumber is avoided. Although the largest group above contains 1723 chains, only 137 spectral lines (Column "Number of lines") are associated with it, as a result of repetition. The number of lines is smaller for the other groups, as shown in Table 2. This table only lists major 37 groups, which have more than 27 spectral lines, and arranged in the order of the number of lines.

Out of the 137 wavenumbers in the first group, 36 were found to belong to the Q5b series, 15 to the Q5a series, and 11 to the part where the Q5a and Q5b series merge. Namely, 62 lines or 45% of the total number of lines are associated with the twin series (Q5a, Q5b). The remaining 75 lines come from different series

Table 2. Grouping of Similar Chains

Group[a]	Number of chains	Number of lines	Usefulness[b] (number)	(%)	Series
1	1723	137	62	45	Q5a,Q5b
2	340	93	45	48	Q0
3	616	90	41	46	Q4a,Q4b
4	948	85	37	44	Q13
5	259	84	45	54	Q11
6	155	79	44	56	Q10
7	150	76	41	54	Q3b
8	145	74	45	61	Q9
9	360	68	34	50	Q7ab
10	77	66	46	70	Q8
11	143	64	40	63	Q1b
12	424	56	29	52	Q1b
13	144	51	25	49	Q1a
14	48	50	29	58	Q2a
15	57	49	23	47	Q1b
16	662	48	34	71	Q7a,Q7b
17	40	48	23	48	O1a
18	73	46	28	61	Q3a
19	66	46	22	48	Q14
20	58	43	30	70	Q2b
21	30	42	29	69	Q12
22	59	41	24	59	Q4a
23	95	41	18	44	Q2b
24	74	40	30	75	Q12
25	80	40	20	50	Q6ab
26	23	35	21	60	Q17
27	58	35	20	57	Q6a
28	38	34	22	65	Q0
29	25	33	20	61	Q6a
30	77	31	21	68	Q3b
31	26	30	24	80	Q4b
32	20	30	23	77	Q2a
33	22	30	21	70	Q1a
34	16	30	20	67	Q2a
35	15	29	23	79	Q14
36	45	28	18	64	O9
37	34	28	15	54	Q3a

[a] Sorted in the order of decreasing number of lines. Only 37 out of 95 groups are presented.

[b] The number of lines concentrated in the spectral series listed in the last column, and its ratio to the total number of lines.

almost randomly. In Table 2 Column "Usefulness (number)" shows the number of lines concentrated in a certain spectral series, and this number divided by the total number of lines is given in Column "Usefulness (%)". The usefulness is around 50% or greater for most major groups. Therefore it would not be difficult to arrive at the correct assignment using these results.

The last Column of Table 2 shows the series detected by the respective groups. The entry "Q7ab" for the Group 9 means that this group contains the lines in the part where the Q7a and Q7b series merge, and similar for "Q6ab" for Group 25. Most major spectral series were detected by the groups included in Table 2. Even the O1b and Q17 series were detected. It is concluded that the grouping is very useful.

5 Combined Use with Loomis-Wood Diagram

Loomis-Wood diagram will be of great use when we try to select the right lines from those collected in a group. The lefthand half of Fig. 1 is an example applied to Group 1 in Table 2. In this diagram the vertical axis is labeled by an integer n, each corresponding to a wavenumber segment. All wavenumber segments have the same width, for example, 0.4 cm^{-1} in this case. The center wavenumber of

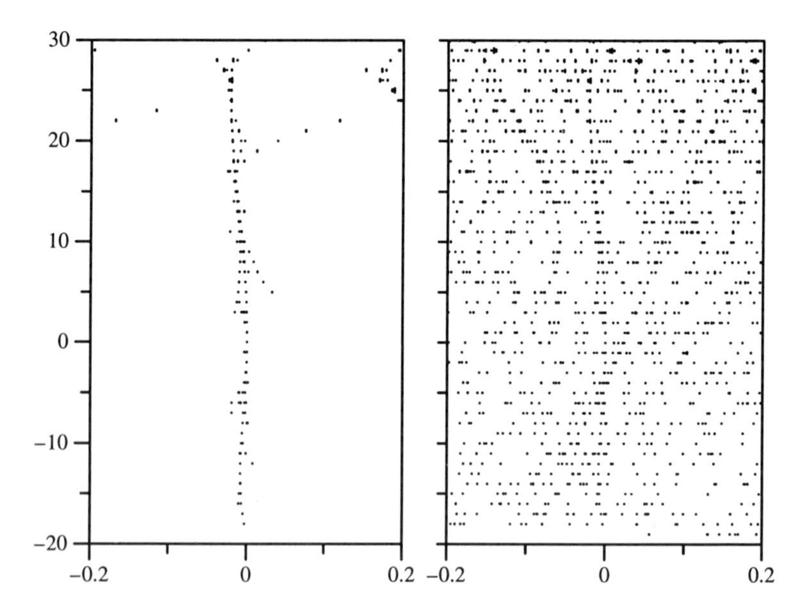

Fig. 1. Loomis-Wood diagram of *trans*-glyoxal. The lefthand half is for Group 1 in Table 2. The horizontal axis is in units of cm^{-1}. The center wavenumber is taken as $22196.8250 + 0.4332n - 0.0044n^2$ cm^{-1}, where n is an integer labeling the vertical axis. The righthand half is the Loomis-Wood diagram applied to the whole spectrum, the vertical and horizontal axes being exactly the same as in the lefthand half.

a segment varies as a quadratic function of n,

$$f_{\text{center}} = 22196.8250 + 0.4332n - 0.0044n^2 \text{ cm}^{-1} \tag{3}$$

in this particular example. The horizontal axis is the displacement of the line wavenumber from the center wavenumber. Occurrence of a line is shown by a short vertical stick whose length is proportional to the intensity. This kind of plot is called a Loomis-Wood diagram, and this example shows its great use in combination with the second difference method reinforced by grouping.

In comparison, the righthand half of Fig. 1 shows the result of the direct application of Loomis-Wood diagram to the whole spectrum, the vertical and horizontal axes being exactly the same as in the lefthand half. This plot suggests that one will encounter considerable difficulties with the direct Loomis-Wood method, although it may not be impossible. Comparison of the both plots clearly shows that a great simplification of Loomis-Wood diagram is attained by the second difference method reinforced by grouping. It may also be stated that the latter method works as an efficient filter for Loomis-Wood diagrams.

Figure 2 gives further examples, the left- and righthand halves corresponding to Groups 2 and 3, respectively.

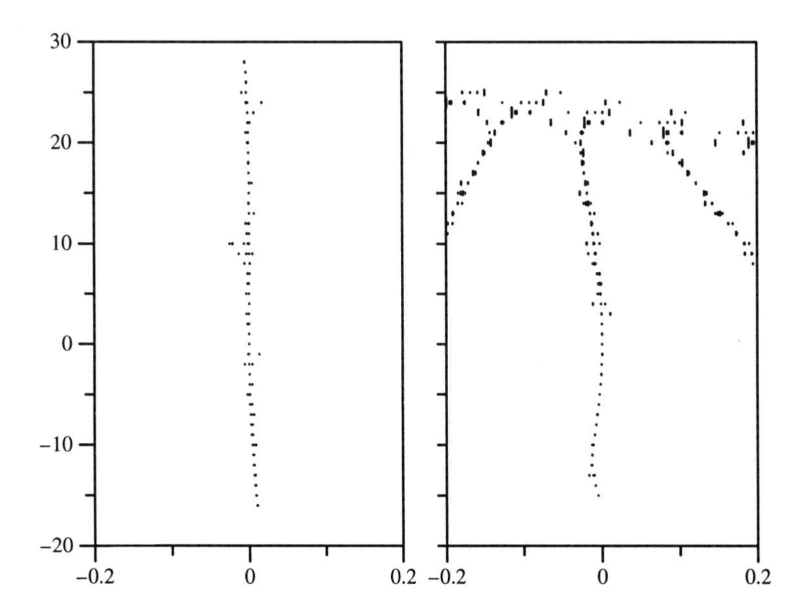

Fig. 2. Loomis-Wood diagram of *trans*-glyoxal. The lefthand half is for Group 2, the center wavenumbers being $22194.4420 + 0.40365n - 0.00335n^2$ cm^{-1}. The righthand half is for Group 3, the center wavenumbers being $22202.5661 + 0.28415n - 0.00425n^2$ cm^{-1}.

6 Test on the Spectrum of Allene

Figure 3 shows a Loomis-Wood diagram for the infrared spectral lines of allene in the region 3006–3033 cm^{-1}, the number of peaks totaling to about 500 (Allene [C_3H_4] is a symmetric-top molecule belonging to the D_{2d} point group). These wavenumbers have been measured by ourselves using a high-resolution Fourier transform spectrometer. They correspond to a part of the parallel ν_5 band, which has been analyzed by Maki et al. [9]. The spectrum is a relatively simple one, and the Loomis-Wood method is perfectly adequate for assignment, as indicated by Fig. 3. In comparison, Fig. 4 is a Loomis-Wood diagram plotted on the same horizontal as well as vertical scale for a test data artificially made by adding about 1500 noise peaks to the real spectral lines described above. The noise peaks were generated by using random numbers, so that their peak wavenumbers and intensities are, respectively, distributed uniformly between 3006 and 3033 cm^{-1} and between zero and three times the maximum of the intensities of the real spectral lines.

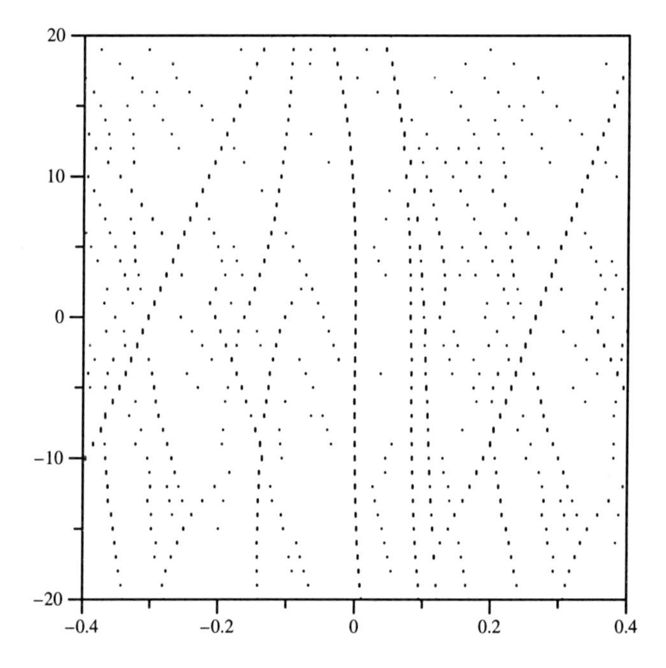

Fig. 3. Loomis-Wood diagram for the spectrum of allene. The horisontal axis is in units of cm^{-1}. The center wavenumber is taken as $3020.4576 + 0.5608n - 0.0007n^2$ cm^{-1}.

In Fig. 4, regular spectral patterns recognizable in Fig. 3 are almost completely obscured. However, the second difference method reinforced by grouping allows us to recover the regualr patterns as described in the following.

A total of about 2000 wavenumbers consisting of real and noise data were submitted to the present method under the following settings. The first three

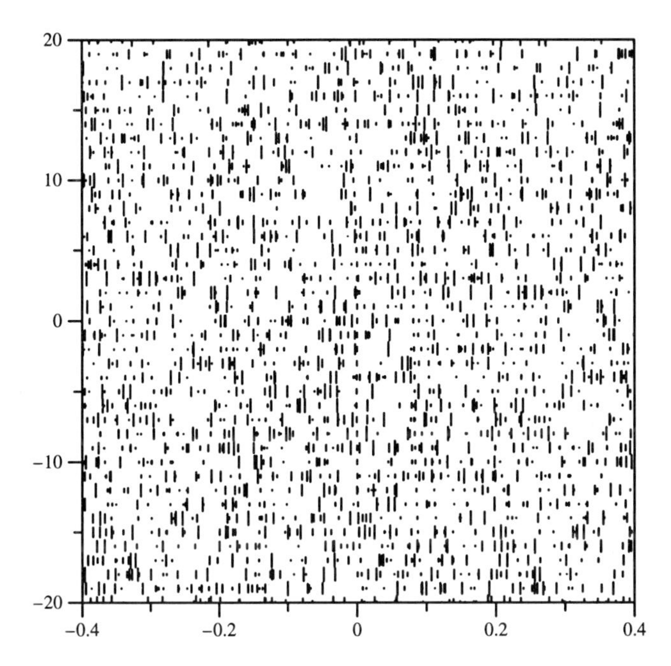

Fig. 4. Loomis-Wood diagram for a noise-added spectrum of allene. The vertical and horizontal axes are exactly the same as in Fig. 3.

members of a chain was restricted so that the first and second differences were between 0 and 1.0 cm^{-1} and between −0.1 and 0.1 cm^{-1}, respectively. The $\Delta f_{\mathrm{allow}}$ value was chosen to be 0.002 cm^{-1}. The chain length was preset at 10. The number of generated chains was 678, and these were reorganized into 227 groups. Loomis-Wood diagrams for some groups are given in Fig. 5, where we can find recovered regular patterns. These results demonstrate the robustness of the present method against the interference by overlapping noise spectra.

7 Conclusion

The second difference method reinforced by grouping was proposed as a new tool for assistance in assignment of complex molecular spectra. A test on the spectrum of *trans*-glyoxal showed that this is a powerful method for picking out regular patterns buried in a list of observed line wavenumbers which look like randomly distrubuted. It is suggested that the present technique may be successfully applied even to the cases in which the traditional method of Loomis-Wood diagram would fail. Combined use of the present method with Loomis-Wood diagrams would be even more efficient. A test on the spectrum of allene showed that the present method is robust against the interference by overlapping niose spectra in superiority to the Loomis-Wood method.

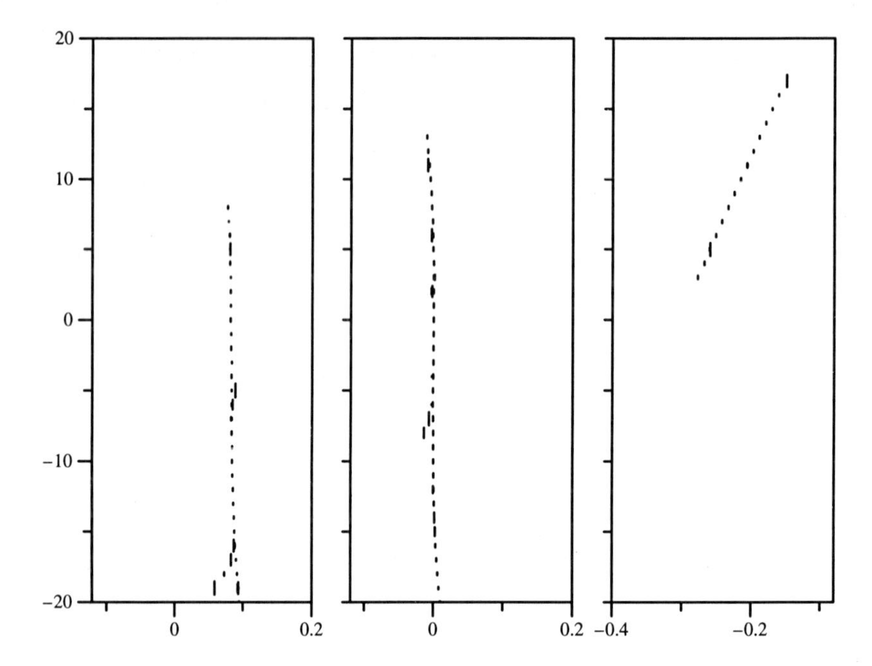

Fig. 5. Loomis-Wood diagrams for some groups extracted from the noise-added spectrum of allene. The vertical and horizontal axes are exactly the same as in Fig. 3.

References

1. Tanaka, T., Imajo, T.: Search for new methods for assignment of complex molecular spectra and a program package for simulation of molecular spectra. Lecture Notes in Artificial Intelligence **1532**, Discovery Science (1998) 445–446
2. Tanaka, T., Imajo, T.: Search for new methods for assignment of complex molecular spectra. Lecture Notes in Artificial Intelligence **1721**, Discovery Science (1999) 362–363
3. Tanaka, T.: Search for new methods for assignment of complex molecular spectra. Lecture Notes in Artificial Intelligence **1967**, Discovery Science (2000) 252–254
4. Loomis, F. W., Wood, R. W.: The rotational structure of the blue-green band of Na_2. Physical Review **32** (1928) 223–236
5. Scott, J. F., Narahari Rao, K.: "Loomis-Wood" diagrams for polyatomic infrared spectra. Journal of Molecular Spectroscopy **20** (1966) 461–463
6. Nakagawa, T., Overend, J.: Computer-assisted assignment of vibration-rotation spectra. Journal of Molecular Spectroscopy **50** (1974) 333–348
7. Winnewisser, B. P., Reinstädtler, J., Yamada, K. M. T., Behrend, J.: Interactive Loomis-Wood assignment programs. Journal of Molecular Spectroscopy **136** (1989) 12–16
8. Kato, H., Oonishi, T., Nishizawa, K., Kasahara, S., Baba, M.: Doppler-free two-photon absorption spectroscopy of the $A^1A_u \leftarrow X^1A_g$ transition of *trans*-glyoxal. Journal of Chemical Physics **106** (1997) 8392–8400
9. Maki, A. G., Pine, A. S., Dang-Nhu, M.: A Doppler-limited study of the infrared spectrum of allene from 2965 to 3114 cm^{-1}. Journal of Molecular Spectroscopy **112** (1985) 459–481

Discovery of Positive and Negative Knowledge in Medical Databases Using Rough Sets

Shusaku Tsumoto

Department of Medicine Informatics, Shimane Medical University, School of Medicine, 89-1 Enya-cho Izumo City, Shimane 693-8501 Japan
tsumoto@computer.org

Abstract. One of the most important problems on rule induction methods is that extracted rules partially represent information on experts' decision processes, which makes rule interpretation by domain experts difficult. In order to solve this problem, the characteristics of medical reasoning is discussed, and positive and negative rules are introduced which model medical experts' rules. Then, for induction of positive and negative rules, two search algorithms are provided. The proposed rule induction method was evaluated on medical databases, the experimental results of which show that induced rules correctly represented experts' knowledge and several interesting patterns were discovered.

1 Introduction

Rule induction methods are classified into two categories, induction of deterministic rules and probabilistic ones[3, 4, 6, 9]. On one hand, deterministic rules are described as *if-then* rules, which can be viewed as propositions. From the set-theoretical point of view, a set of examples supporting the conditional part of a deterministic rule, denoted by C, is a subset of a set whose examples belongs to the consequence part, denoted by D. That is, the relation $C \subseteq D$ holds and deterministic rules are supported only by positive examples in a dataset. On the other hand, probabilistic rules are if-then rules with probabilistic information[9]. From the set-theoretical point of view, C is not a subset, but closely overlapped with D. That is, the relations $C \cap D \neq \phi$ and $|C \cap D|/|C| \geq \delta$ will hold in this case.[1] Thus, probabilistic rules are supported by a large number of positive examples and a small number of negative examples. The common feature of both deterministic and probabilistic rules is that they will deduce their consequence positively if an example satisfies their conditional parts. We call the reasoning by these rules *positive reasoning.*

However, medical experts do not use only positive reasoning but also *negative reasoning* for selection of candidates, which is represented as if-then rules whose consequences include negative terms. For example, when a patient who complains of headache does not have a throbbing pain, migraine should not be

[1] The threshold δ is the degree of the closeness of overlapping sets, which will be given by domain experts. For more information, please refer to Section 3.

S. Arikawa and A. Shinohara (Eds.): Progress in Discovery Science 2001, LNAI 2281, pp. 543–552.

suspected with a high probability. Thus, negative reasoning also plays an important role in cutting the search space of a differential diagnosis process[9]. Thus, medical reasoning includes both positive and negative reasoning, though conventional rule induction methods do not reflect this aspect. This is one of the reasons why medical experts have difficulties in interpreting induced rules and the interpretation of rules for a discovery procedure does not easily proceed. Therefore, negative rules should be induced from databases in order not only to induce rules reflecting experts' decision processes, but also to induce rules which will be easier for domain experts to interpret, both of which are important to enhance the discovery process done by the corporation of medical experts and computers.

In this paper, first, the characteristics of medical reasoning are focused on and two kinds of rules, positive rules and negative rules are introduced as a model of medical reasoning. Interestingly, from the set-theoretical point of view, sets of examples supporting both rules correspond to the lower and upper approximation in rough sets[4]. On the other hand, from the viewpoint of propositional logic, both positive and negative rules are defined as classical propositions, or deterministic rules with two probabilistic measures, classification accuracy and coverage. Second, two algorithms for induction of positive and negative rules are introduced, defined as search procedures by using accuracy and coverage as evaluation indices. Finally, the proposed method was evaluated on several medical databases, the experimental results of which show that induced rules correctly represented experts' knowledge and several interesting patterns were discovered.

2 Focusing Mechanism

One of the characteristics in medical reasoning is a focusing mechanism, which is used to select the final diagnosis from many candidates[9]. For example, in differential diagnosis of headache, more than 60 diseases will be checked by present history, physical examinations and laboratory examinations. In diagnostic procedures, a candidate is excluded if a symptom necessary to diagnose is not observed.

This style of reasoning consists of the following two kinds of reasoning processes: exclusive reasoning and inclusive reasoning. [2] The diagnostic procedure will proceed as follows: first, exclusive reasoning excludes a disease from candidates when a patient does not have a symptom which is necessary to diagnose that disease. Secondly, inclusive reasoning suspects a disease in the output of the exclusive process when a patient has symptoms specific to a disease. These two steps are modelled as usage of two kinds of rules, negative rules (or exclusive rules) and positive rules, the former of which corresponds to exclusive reasoning and the latter of which corresponds to inclusive reasoning. In the next two subsections, these two rules are represented as special kinds of probabilistic rules.

[2] Relations this diagnostic model with another diagnostic model are discussed in [10].

3 Definition of Rules

3.1 Rough Sets

In the following sections, we use the following notations introduced by Grzymala-Busse and Skowron[7], which are based on rough set theory[4]. These notations are illustrated by a small database shown in Table 1, collecting the patients who complained of headache.

Table 1. An Example of Database

No.	age	location	nature	prodrome	nausea	M1	class
1	50-59	occular	persistent	no	no	yes	m.c.h.
2	40-49	whole	persistent	no	no	yes	m.c.h.
3	40-49	lateral	throbbing	no	yes	no	migra
4	40-49	whole	throbbing	yes	yes	no	migra
5	40-49	whole	radiating	no	no	yes	m.c.h.
6	50-59	whole	persistent	no	yes	yes	psycho

DEFINITIONS. M1: tenderness of M1, m.c.h.: muscle contraction headache, migra: migraine, psycho: psychological pain.

Let U denote a nonempty, finite set called the universe and A denote a nonempty, finite set of attributes, i.e., $a : U \rightarrow V_a$ for $a \in A$, where V_a is called the domain of a, respectively.Then, a decision table is defined as an information system, $\cap A = (U, A \cup \{d\})$. For example, Table 1 is an information system with $U = \{1,2,3,4,5,6\}$ and $A = \{age, location, nature, prodrome, nausea, M1\}$ and $d = class$. For $location \in A$, $V_{location}$ is defined as $\{occular, lateral, whole\}$.

The atomic formulae over $B \subseteq A \cup \{d\}$ and V are expressions of the form $[a = v]$, called descriptors over B, where $a \in B$ and $v \in V_a$. The set $F(B, V)$ of formulas over B is the least set containing all atomic formulas over B and closed with respect to disjunction, conjunction and negation. For example, $[location = occular]$ is a descriptor of B.

For each $f \in F(B, V)$, f_A denote the meaning of f in A, i.e., the set of all objects in U with property f, defined inductively as follows.

1. If f is of the form $[a = v]$ then, $f_A = \{s \in U | a(s) = v\}$
2. $(f \wedge g)_A = f_A \cap g_A$; $(f \vee g)_A = f_A \vee g_A$; $(\neg f)_A = U - f_a$

For example, $f = [location = whole]$ and $f_A = \{2,4,5,6\}$. As an example of a conjunctive formula, $g = [location = whole] \wedge [nausea = no]$ is a descriptor of U and f_A is equal to $g_{location,nausea} = \{2,5\}$.

3.2 Classification Accuracy and Coverage

Definition of Accuracy and Coverage. By the use of the framework above, classification accuracy and coverage, or true positive rate is defined as follows.

Definition 1.
Let R and D denote a formula in $F(B,V)$ and a set of objects which belong to a decision d. Classification accuracy and coverage(true positive rate) for $R \rightarrow d$ is defined as:

$$\alpha_R(D) = \frac{|R_A \cap D|}{|R_A|}(= P(D|R)), \text{ and } \kappa_R(D) = \frac{|R_A \cap D|}{|D|}(= P(R|D)),$$

where $|S|$, $\alpha_R(D)$, $\kappa_R(D)$ and $P(S)$ denote the cardinality of a set S, a classification accuracy of R as to classification of D and coverage (a true positive rate of R to D), and probability of S, respectively.

In the above example, when R and D are set to $[nausea = yes]$ and $[class = migraine]$, $\alpha_R(D) = 2/3 = 0.67$ and $\kappa_R(D) = 2/2 = 1.0$.

It is notable that $\alpha_R(D)$ measures the degree of the sufficiency of a proposition, $R \rightarrow D$, and that $\kappa_R(D)$ measures the degree of its necessity. For example, if $\alpha_R(D)$ is equal to 1.0, then $R \rightarrow D$ is true. On the other hand, if $\kappa_R(D)$ is equal to 1.0, then $D \rightarrow R$ is true. Thus, if both measures are 1.0, then $R \leftrightarrow D$.

3.3 Probabilistic Rules

By the use of accuracy and coverage, a probabilistic rule is defined as:

$$R \stackrel{\alpha,\kappa}{\rightarrow} d \quad s.t. \; R = \wedge_j [a_j = v_k], \alpha_R(D) \geq \delta_\alpha \\ \text{and } \kappa_R(D) \geq \delta_\kappa,$$

where δ_α and δ_κ denote thresholds for accuracy $\alpha_R(D)$ and coverage $\kappa_R(D)$, respectively. This rule is a kind of probabilistic proposition with two statistical measures, which is an extension of Ziarko's variable precision model(VPRS) [14].[3]

It is also notable that both a positive rule and a negative rule are defined as special cases of this rule, as shown in the next subsections.

3.4 Positive Rules

A positive rule is defined as a rule supported by only positive examples, the classification accuracy of which is equal to 1.0. It is notable that the set supporting this rule corresponds to a subset of the lower approximation of a target concept, which is introduced in rough sets[4]. Thus, a positive rule is represented as:

$$R \rightarrow d \quad s.t. \qquad R = \wedge_j [a_j = v_k], \quad \alpha_R(D) = 1.0$$

In the above example, one positive rule of "m.c.h." (muscle contraction headache) is:

$$[nausea = no] \rightarrow m.c.h. \quad \alpha = 3/3 = 1.0.$$

This positive rule is often called a deterministic rule. However, in this paper, we use a term, positive (deterministic) rules, because a deterministic rule which is supported only by negative examples, called a negative rule, is introduced as in the next subsection.

[3] This probabilistic rule is also a kind of *Rough Modus Ponens*[5].

3.5 Negative Rules

Before defining a negative rule, let us first introduce an exclusive rule, the contrapositive of a negative rule[9]. An exclusive rule is defined as a rule supported by all the positive examples, the coverage of which is equal to 1.0.[4] It is notable that the set supporting an exclusive rule corresponds to the upper approximation of a target concept, which is introduced in rough sets[4]. Thus, an exclusive rule is represented as:

$$R \to d \quad s.t. \qquad R = \vee_j [a_j = v_k], \quad \kappa_R(D) = 1.0.$$

In the above example, the exclusive rule of "m.c.h." is:

$$[M1 = yes] \vee [nausea = no] \to m.c.h. \quad \kappa = 1.0,$$

From the viewpoint of propositional logic, an exclusive rule should be represented as:

$$d \to \vee_j [a_j = v_k],$$

because the condition of an exclusive rule corresponds to the necessity condition of conclusion d. Thus, it is easy to see that a negative rule is defined as the contrapositive of an exclusive rule:

$$\wedge_j \neg [a_j = v_k] \to \neg d,$$

which means that if a case does not satisfy any attribute value pairs in the condition of a negative rules, then we can exclude a decision d from candidates. For example, the negative rule of m.c.h. is:

$$\neg [M1 = yes] \wedge \neg [nausea = no] \to \neg m.c.h.$$

In summary, a negative rule is defined as:

$$\wedge_j \neg [a_j = v_k] \to \neg d \quad s.t. \quad \forall [a_j = v_k] \; \kappa_{[a_j = v_k]}(D) = 1.0,$$

where D denotes a set of samples which belong to a class d.

Negative rules should be also included in a category of deterministic rules, since their coverage, a measure of negative concepts is equal to 1.0. It is also notable that the set supporting a negative rule corresponds to a subset of negative region, which is introduced in rough sets[4].

4 Algorithms for Rule Induction

The contrapositive of a negative rule, an exclusive rule is induced as an exclusive rule by the modification of the algorithm introduced in PRIMEROSE-REX[9], as shown in Figure 1. This algorithm will work as follows. (1)First, it selects a descriptor $[a_i = v_j]$ from the list of attribute-value pairs, denoted by L. (2)

[4] An exclusive rule represents the necessity condition of a decision.

Then, it checks whether this descriptor overlaps with a set of positive examples, denoted by D. (3) If so, this descriptor is included into a list of candidates for positive rules and the algorithm checks whether its coverage is equal to 1.0 or not. If the coverage is equal to 1.0, then this descriptor is added to $R_e r$, the formula for the conditional part of the exclusive rule of D. (4) Then, $[a_i = v_j]$ is deleted from the list L. This procedure, from (1) to (4) will continue unless L is empty. (5) Finally, when L is empty, this algorithm generates negative rules by taking the contrapositive of induced exclusive rules.

On the other hand, positive rules are induced as inclusive rules by the algorithm introduced in PRIMEROSE-REX[9], as shown in Figure 2. For induction of positive rules, the threshold of accuracy and coverage is set to 1.0 and 0.0, respectively.

This algorithm works in the following way. (1) First, it substitutes L_1, which denotes a list of formula composed of only one descriptor, with the list L_{er} generated by the former algorithm shown in Fig. 1. (2) Then, until L_1 becomes empty, the following procedures will continue: (a) A formula $[a_i = v_j]$ is removed from L_1. (b) Then, the algorithm checks whether $\alpha_R(D)$ is larger than the threshold or not. (For induction of positive rules, this is equal to checking whether $\alpha_R(D)$ is equal to 1.0 or not.) If so, then this formula is included a list of the conditional part of positive rules. Otherwise, it will be included into M, which is used for making conjunction. (3) When L_1 is empty, the next list L_2 is generated from the list M.

procedure *Exclusive and Negative Rules*;
  var
    L : *List*;
      /* A list of elementary attribute-value pairs */
  begin
    $L := P_0$;
    /* P_0: A list of elementary attribute-value pairs
given in a database */
    while $(L \neq \{\})$ **do**
      begin
        Select one pair $[a_i = v_j]$ from L;
        if $([a_i = v_j]_A \cap D \neq \phi)$ **then do** /* D: positive examples of a target class d */
          begin
          $L_{ir} := L_{ir} + [a_i = v_j]$;   /* Candidates for Positive Rules */
          if $(\kappa_{[a_i=v_j]}(D) = 1.0)$
          then $R_{er} := R_{er} \wedge [a_i = v_j]$;
            /* Include $[a_i = v_j]$ into the formula of Exclusive Rule */
          end
      $L := L - [a_i = v_j]$;
      end
    Construct Negative Rules: Take the contrapositive of R_{er}.
  end {*Exclusive and Negative Rules*};

Fig. 1. Induction of Exclusive and Negative Rules

```
procedure Positive Rules;
  var
    i : integer;    M, L_i : List;
  begin
    L_1 := L_ir;
    /* L_ir: A list of candidates generated by induction of exclusive rules */
    i := 1;    M := {};
    for i := 1 to n do
    /* n: Total number of attributes given in a database */
      begin
        while ( L_i ≠ {} ) do
          begin
            Select one pair R = ∧[a_i = v_j] from L_i;
            L_i := L_i − {R};
            if   (α_R(D) > δ_α)
              then  do S_ir := S_ir + {R};
          /* Include R in a list of the Positive Rules */
            else M := M + {R};
          end
        L_{i+1} := (A list of the whole combination of the conjunction formulae in M);
      end
  end {Positive Rules};
```

Fig. 2. Induction of Positive Rules

5 Experimental Results

For experimental evaluation, a new system, called PRIMEROSE-REX2 (Probabilistic Rule Induction Method for Rules of Expert System ver 2.0), was developed, where the algorithms discussed in Section 4 were implemented.

PRIMEROSE-REX2 was applied to the following three medical domains: headache(RHINOS domain), whose training samples consist of 52119 samples, 45 classes and 147 attributes, cerebulovasular diseases(CVD), whose training samples consist of 7620 samples, 22 classes and 285 attributes, and meningitis, whose training samples consists of 1211 samples, 4 classes and 41 attributes (Table 2).

Table 2. Databases

Domain	Samples	Classes	Attributes
Headache	52119	45	147
CVD	7620	22	285
Meningitis	1211	4	41

For evaluation, we used the following two types of experiments. One experiment was to evaluate the predictive accuracy by using the cross-validation method, which is often used in the machine learning literature[8]. The other experiment was to evaluate induced rules by medical experts and to check whether these rules led to a new discovery.

5.1 Performance of Rules Obtained

For comparison of performance, The experiments were performed by the following four procedures. First, rules were acquired manually from experts. Second, the datasets were randomly splits into new training samples and new test samples. Third, PRIMEROSE-REX2, conventional rule induction methods, AQ15[3] and C4.5[6] were applied to the new training samples for rule generation. Fourth, the induced rules and rules acquired from experts were tested by the new test samples. The second to fourth were repeated for 100 times and average all the classification accuracy over 100 trials. This process is a variant of repeated 2-fold cross-validation, introduced in [9].

Experimental results(performance) are shown in Table 3. The first and second row show the results obtained by using PRIMROSE-REX2: the results in the first row were derived by using both positive and negative rules and those in the second row were derived by only positive rules. The third row shows the results derived from medical experts. For comparison, we compare the classification accuracy of C4.5 and AQ-15, which is shown in the fourth and the fifth row. These results show that the combination of positive and negative rules outperforms positive rules, although it is a little worse than medical experts' rules.

Table 3. Experimental Results (Accuracy: Averaged)

Method	Headache	CVD	Meningitis
PRIMEROSE-REX2 (Positive+Negative)	91.3%	89.3%	92.5%
PRIMEROSE-REX2 (Positive)	68.3%	71.3%	74.5%
Experts	95.0%	92.9%	93.2%
C4.5	85.8%	79.7%	81.4%
AQ15	86.2%	78.9%	82.5%

5.2 What is Discovered?

Positive Rules in Meningitis. In the domain of meningitis, the following positive rules, which medical experts do not expect, are obtained.

$$\begin{aligned}&[WBC < 12000] \wedge [Sex = Female] \wedge [Age < 40]\\&\quad \wedge [CSF\_CELL < 1000] \rightarrow Virus\\&[Age \geq 40] \wedge [WBC \geq 8000] \wedge [Sex = Male]\\&\quad \wedge [CSF\_CELL \geq 1000] \rightarrow Bacteria\end{aligned}$$

The former rule means that if WBC(White Blood Cell Count) is less than 12000, the Sex of a patient is FEMALE, the Age is less than 40 and CSF_CELL (Cell count of Cerebulospinal Fluid), then the type of meningitis is Virus. The latter one means that the Age of a patient is less than 40, WBC is larger than 8000, the Sex is Male, and CSF_CELL is larger than 1000, then the type of meningitis is Bacteria.

The most interesting points are that these rules included information about age and sex, which often seems to be unimportant attributes for differential diagnosis of meningitis. The first discovery was that women did not often suffer from bacterial infection, compared with men, since such relationships between sex and meningitis has not been discussed in medical context[1]. By the close examination of the database of meningitis, it was found that most of the above patients suffered from chronic diseases, such as DM, LC, and sinusitis, which are the risk factors of bacterial meningitis. The second discovery was that $[age < 40]$ was also an important factor not to suspect viral meningitis, which also matches the fact that most old people suffer from chronic diseases.

These results were also re-evaluated in medical practice. Recently, the above two rules were checked by additional 21 cases who suffered from meningitis (15 cases: viral and 6 cases: bacterial meningitis.) Surprisingly, the above rules misclassified only three cases (two are viral, and the other is bacterial), that is, the total accuracy was equal to 18/21 = 85.7% and the accuracies for viral and bacterial meningitis were equal to 13/15 = 86.7% and 5/6 = 83.3%. The reasons of misclassification were the following: a case of bacterial infection was a patient who had a severe immunodeficiency, although he is very young. Two cases of viral infection were patients who also suffered from herpes zoster. It is notable that even those misclassification cases could be explained from the viewpoint of the immunodeficiency: that is, it was confirmed that immunodeficiency is a key word for meningitis.

The validation of these rules is still ongoing, which will be reported in the near future.

Positive and Negative Rules in CVD. Concerning the database on CVD, several interesting rules were derived. The most interesting results were the following positive and negative rules for thalamus hemorrhage:

$$[Sex = Female] \wedge [Hemiparesis = Left] \wedge [LOC : positive] \rightarrow Thalamus$$
$$\neg[Risk : Hypertension] \wedge \neg[Sensory = no] \rightarrow \neg Thalamus$$

The former rule means that if the Sex of a patient is female and he/she suffered from the left hemiparesis([Hemiparesis=Left]) and loss of consciousness ([LOC:positive]), then the focus of CVD is Thalamus. The latter rule means that if he/she neither suffers from hypertension ([Risk: Hypertension]) nor suffers from sensory disturbance([Sensory=no]), then the focus of CVD is Thalamus.

Interestingly, LOC(loss of consciousness) under the condition of $[Sex = Female] \wedge [Hemiparesis = Left]$ was found to be an important factor to diagnose thalamic damage. In this domain, any strong correlations between these attributes and others, like the database of meningitis, have not been found yet. It will be our future work to find what factor is behind these rules.

6 Conclusions

In this paper, the characteristics of two measures, classification accuracy and coverage are discussed, which shows that both measures are dual and that accuracy and coverage are measures of both positive and negative rules, respectively. Then, an algorithm for induction of positive and negative rules is introduced. The proposed method was evaluated on medical databases, the experimental results of which show that induced rules correctly represented experts' knowledge and several interesting patterns were discovered.

References

1. Adams RD and Victor M: *Principles of Neurology*, 5th edition. McGraw-Hill, New York, 1993.
2. Buchnan BG and Shortliffe EH(Eds): *Rule-Based Expert Systems.* Addison-Wesley, 1984.
3. Michalski RS, Mozetic I, Hong J, and Lavrac N: The Multi-Purpose Incremental Learning System AQ15 and its Testing Application to Three Medical Domains. *Proceedings of the fifth National Conference on Artificial Intelligence*, AAAI Press, Palo Alto CA, pp 1041-1045, 1986.
4. Pawlak Z: *Rough Sets.* Kluwer Academic Publishers, Dordrecht, 1991.
5. Pawlak Z: Rough Modus Ponens. In: *Proceedings of International Conference on Information Processing and Management of Uncertainty in Knowledge-Based Systems 98*, Paris, 1998.
6. Quinlan JR: *C4.5 - Programs for Machine Learning.* Morgan Kaufmann, Palo Alto CA, 1993.
7. Skowron, A. and Grzymala-Busse, J. From rough set theory to evidence theory. In: Yager, R., Fedrizzi, M. and Kacprzyk, J.(eds.) *Advances in the Dempster-Shafer Theory of Evidence*, pp.193-236, John Wiley & Sons, New York, 1994.
8. Shavlik JW and Dietterich TG(Eds): *Readings in Machine Learning.* Morgan Kaufmann, Palo Alto CA, 1990.
9. Tsumoto S and Tanaka H: Automated Discovery of Medical Expert System Rules from Clinical Databases based on Rough Sets. In: *Proceedings of the Second International Conference on Knowledge Discovery and Data Mining 96*, AAAI Press, Palo Alto CA, pp.63-69, 1996.
10. Tsumoto S: Modelling Medical Diagnostic Rules based on Rough Sets. In: Polkowski L and Skowron A (Eds): *Rough Sets and Current Trends in Computing*, Lecture Note in Artificial Intelligence **1424**, 1998.
11. Tsumoto S: Automated Extraction of Medical Expert System Rules from Clinical Databases based on Rough Set Theory. *Journal of Information Sciences*, 1998
12. Tsumoto S: Knowledge discovery in clinical databases and evaluation of discovered knowledge in outpatient clinic. *Information Sciences*, **124**, 125-137, 2000.
13. Tsumoto S: Automated Discovery of Positive and Negative Knowledge in Clinical Databases. *IEEE BME Magazine*, **19**, 56-62, 2000.
14. Ziarko W: Variable Precision Rough Set Model. *Journal of Computer and System Sciences* **46**:39-59, 1993.

Toward the Discovery of First Principle Based Scientific Law Equations

Takashi Washio and Hiroshi Motoda

Institute for Scientific and Industrial Research, Osaka University,
8-1, Mihogaoka, Ibarakishi, Osaka, 567-0047, Japan
{washio, motoda}@sanken.osaka-u.ac.jp

Abstract. Conventional work on scientific discovery such as BACON derives empirical law equations from experimental data. In recent years, SDS introducing mathematical admissibility constraints has been proposed to discover first principle based law equations, and it has been further extended to discover law equations from passively observed data. Furthermore, SSF has been proposed to discover the structure of a simultaneous equation model representing an objective process through experiments. In this report, the progress of these studies on the discovery of first principle based scientific law equations is summarized, and the future directions of this research are presented.

1 Introduction

Langley and others' BACON [1] is the most well known pioneering work to discover a complete equation representing scientific laws governing an objective process under experimental observations. FAHRENHEIT [2], ABACUS [3], etc. are the successors of BACON that use basically similar algorithms. However, a drawback of the BACON family, that is their low likelihood of the discovered equations being the first principle underlying the objective process, is reported. To alleviate the drawback, some systems, *e.g.*, ABACUS and COPER [4], utilize the information of the unit dimensions of quantities to prune the meaningless terms. However, many of these conventional scientific equation discovery systems have some critical limitations for the real world applications. Fisrt, the information of the unit dimension of each quantity in the data is needed to discover the first principle based equation. Second, the data must be acquired under "*active observations*" where the values of some quantities representing the objective process are observed for various process states by controlling the values of the other relevant quantities. Third, a complex equation model, especially a "*simultaneous equation model*", to represent the process consisting of multiple mechanisms is hardly discovered due to the complexity of the search space.

To alleviate the first limitation, a law equation discovery system named SDS based on the mathematical constraints of "*scale-type*" and "*identity*" is proposed for the active observations [5]. Since the knowledge of scale-types of quantities is widely obtained in various domains, SDS is applicable to non-physics domains.

S. Arikawa and A. Shinohara (Eds.): Progress in Discovery Science 2001, LNAI 2281, pp. 553–564.

The equations discovered by SDS are highly likely to represent the first principle underlying the objective process. To address the second limitation, SDS has been further extended by introducing a novel principle named "*quasi-bi-variate fitting*" [6] to make it applicable to the "*passive observations*" where the quantities of the objective process can only be partially or even hardly controlled. Moreover, to overcome the third limitation, a simultaneous structure finding system named SSF has been proposed to discover a valid simultaneous equation structure under the active observations [7]. SSF identifies the number of equations needed to represent the objective process, and further identifies the sets of quantities to appear in each of the respective equations of the model while eliminating quantities irrelevant to the equations. The combination of SDS and SSF enables the discovery of the first principle based simultaneous equation model for the objective process under active observations. In this report, the principles and the performances of these recent scientific discovery systems are outlined, and the future directions of this research topic are discussed.

2 Required Conditions

Prior to the explanation of each scientific law equation discovery system, the conditions required for the application of the system are summarized. The approach of the SDS requires the following conditions. vspace*-7pt

(1) All of the quantities except one dependent quantity can be controlled to their arbitrary values in the ranges of the quantities of our interest.
(2) The objective system can be represented by a complete equation in the value range.
(3) The scale-types of all quantities needed to represent the objective system are known.

The first condition is to ensure the application of the SDS to experimental environment in scientific laboratories. This condition is the requirement of the original BACON systems, and is also required by other BACON family. The second condition is also a common requirement in BACON family to search for a complete equation for every continuous region in the objective system. The third condition comes from the fact that the SDS uses the the information of the scale-types of the quantities to search mathematically admissible equation formulae to relate the quantities. As the scale-types of the measurement quantities are widely known based on the measurement theory [5], this condition does not restrict the applicability of the SDS.

The extended SDS discovers the law equations from the passively observed data, and thus the aforementioned condition (1) is not required. Instead, it needs the following condition in addition to the conditions (2) and (3).

(4) The observed data are uniformly distributed over the possible states of the objective system.

Violation of the requirement of the uniform distribution over a certain value range of a quantity implies the low observability of the quantity [6],[8]. Any approaches such as the linear system identification and the neural network do not derive valid models under low observability. This limitation is generic, and further discussion on this issue is out of scope of this report.

The condition required by the SSF for the derivation of the simultaneous law equation models is as follows.

(5) The simultaneous equation model under consideration is not over-constrained where the number of the equations is no more than the number of quantities in the model.

This condition always holds for the models in scientific and engineering domains, since the over-constrained state does not exist in any real world process.

3 Smart Discovery System (SDS)

The information required from the user besides the actual measurements in SDS is a list of the quantities and their scale-types. The rigorous definition of scale-type was given by Stevens[9]. The quantitative scale-types are interval scale, ratio scale and absolute scale, and these are the majorities of the quantities. Examples of the interval scale quantities are temperature in Celsius and sound tone where the origins of their scales are not absolute, and are changeable by human's definitions. Its admissible unit conversion follows "*Generic linear group*: $x' = kx + c$". Examples of the ratio scale quantities are physical mass and absolute temperature where each has an absolute zero point. Its admissible unit conversion follows "*Similarity group*: $x' = kx$". Examples of the absolute scale quantities are dimensionless quantities. It follows "*Identity group*: $x' = x$".

The first step of the algorithm of SDS searches bi-variate relations among the quantities by using the "*scale-type constraint.*" Two well-known theorems which are deduced from the group structures of the scale-types provides the basis of this step [5].

Theorem 1 (Extended Buckingham Π-theorem). *If $\phi(x_1, x_2, x_3 \ldots.) = 0$ is a complete equation where each $x_i \in Q$, the set of quantities given in the experiment, and if each argument is one of interval, ratio and absolute scale-types, then the solution can be written in the form*

$$F(\Pi_1, \Pi_2, \ldots, \Pi_{n-w}) = 0,$$

where n is the number of arguments of ϕ, w is the basic number of bases in $x_1, x_2, x_3 \ldots.$, respectively. For all i, Π_i is an absolute scale-type quantity.

A base is such a basic scaling factor which has a degree of freedom independent of the other bases in the given ϕ. For instance, length $[L]$, mass $[M]$ and time $[T]$ of physical dimension and the origin of the temperature in Celsius are the examples of the bases. The relation of each Π_i to the arguments of ϕ is given by the following theorem [5].

Theorem 2 (Extended Product Theorem). *Let primary quantities in a set RQ are ratio scale-type and those in another set IQ interval scale-type, the function ρ relating a secondary quantity Π to $x_i \in RQ \cup IQ$ has the forms:*

$$\Pi = (\prod_{x_i \in R} |x_i|^{a_i})(\prod_{I_k \subseteq I} (\sum_{x_j \in I_k} b_{kj}|x_j| + c_k)^{a_k})$$

$$\Pi = \sum_{x_i \in R} a_i \log |x_i| + \sum_{I_k \subseteq I} a_k \log(\sum_{x_j \in I_k} b_{kj}|x_j| + c_k) + \sum_{x_\ell \in I_g \subseteq I} b_{g\ell}|x_\ell| + c_g$$

where all coefficients except Π are constants and $I_k \cap I_g = \phi$.

These theorems state that any meaningful complete equation consisting of the arguments of interval, ratio and absolute scale-types can be decomposed into an equation of absolute scale-type quantities having an arbitrary form and some equations of interval and ratio scale-type quantities in products and logarithmic form. The former $F(\Pi_1, \Pi_2, ..., \Pi_{n-w}) = 0$ is called an "*ensemble*" and the latter $\Pi = \rho(x_1, x_2, x_3....)$s "*regime*"s.

If any pair of interval and/or ratio scale quantities $\{x, y\}$ in a given complete equation is to belong to an identical regime, they has to have a relation that follows the Theorem 2. Conversely, SDS searches bi-variate relations in the set of quantities Q where the relations have the following product, linear or logarithmic forms which are deduced from the Theorem 2.

$$x^a y = b, \quad \text{where } x, y \text{ are ratio scale,} \tag{1}$$

$$ax + y = b, \quad \text{where } x, y \text{ are interval scale, and} \tag{2}$$

$$a \log x + y = b \quad \text{or } cx^a + y = b, \quad \text{where } x \text{ is ratio scale, and } y \text{ interval scale.} \tag{3}$$

The value of the constant a in each formula must be independent of any other quantities according to Extended Product Theorem, while the constants b and c are dependent on the other quantities in the regime. SDS applies the least square fitting of these relations to the bi-variate experimental data of x and y that are measured while holding the other quantities constant in Q, and determines the values of coefficients in the bi-variate relations. Subsequently, the judgment is made whether this equation fits the data well enough by some statistical tests.

This procedure is now demonstrated by an example of a complex system depicted in Fig. 1. This is a circuit of photo-meter to measure the rate of increase of photo intensity within a certain time period. The resistance and switch parallel to the capacitor and the current meter are to reset the operation of this circuit. The actual model of this system is represented by the following complex equation involving 18 quantities.

$$(\frac{R_3 h_{fe_2}}{R_3 h_{fe_2} + h_{ie_2}} \frac{R_2 h_{fe_1}}{R_2 h_{fe_1} + h_{ie_1}} \frac{rL^2}{rL^2 + R_1})(V_1 - V_2) - \frac{Q}{C} - \frac{K h_{ie_3} X}{B h_{fe_3}} = 0. \tag{4}$$

Here, L and r are photo intensity and sensitivity of the Csd device. X, K and B are position of indicator, spring constant and intensity of magnetic field of the

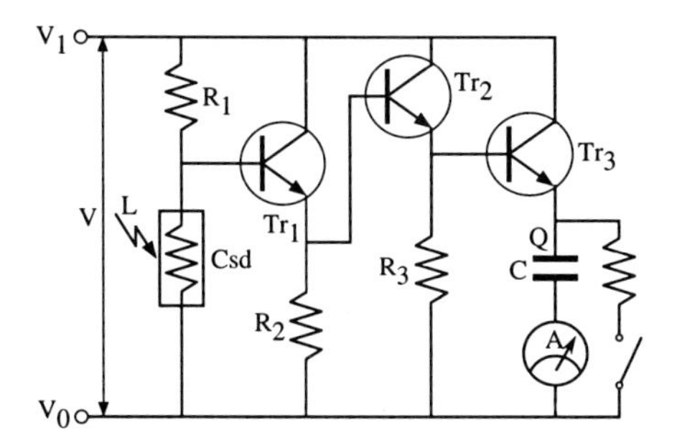

Fig. 1. A circuit of photo-meter

current meter respectively. h_{ie_i} is input impedance of the base of the i-th transistor. h_{fe_i} is gain ratio of the currents at the base and the collector of the i-th transistor. The definitions of the other quantities follow the standard symbolic representations of electric circuit. h_{fe_i}s are absolute scale, V_1 and V_2 interval scale, and the rest ratio scale. X is the dependent quantity in this circuit, and the others are independently controllable by the change of boundary conditions and the replacement of devices. SDS requests the bi-variate change of quantities to the experimental environment. When a quantity is dependent (not controllable) during the search process, SDS modifies its request to control the other independent quantity. A simulation based experimental environment was designed and build for the circuit system. $\pm 4\%$ (std.) of relative Gaussian noise was added to both of the control quantity (input) and the measured quantity (output) in every bi-variate test. First, SDS set the set of interval scale quantities IQ as $\{V_1, V_2\}$, ratio scale quantities RQ as $\{L, r, R_1, R_2, R_3, h_{ie_1}, h_{ie_2}, h_{ie_3}, Q, C, X, K, B\}$ and absolute scale quantities AQ as $\{h_{fe_1}, h_{fe_2}, h_{fe_3}\}$ based on the input information on scale-types. Next, it performed the bi-variate fitting of a linear form Eq. (2) to the experimental data among the quantities in IQ, and applied the statistical tests. Then, SDS figured out a set consisting of a bi-variate equation $IE = \{\Theta_0 = 1.000V_1 - 1.000V_0\}$ quickly. Subsequently, a product form Eq. (1) among the quantities in $RQ \cup \{\Theta_0\}$ is searched, and applied the statistical tests.

The resultant set of the bi-variate equations RE that passed the tests was as follows.

$$\begin{aligned} RE = \{&L^{1.999}r = b_1, L^{-1.999}R_1 = b_2, r^{-1.000}R_1 = b_3, R_2^{-1.000}h_{ie_1} = b_4,\\ &R_3^{-1.000}h_{ie_2} = b_5, Q^{-1.000}C = b_6, h_{ie_3}^{1.000}X = b_7, h_{ie_3}^{1.000}K = b_8,\\ &h_{ie_3}^{-1.000}B = b_9, X^{1.000}K = b_{10}, X^{-0.999}B = b_{11}, K^{-1.000}B = b_{12}\} \end{aligned}$$

The bi-variate fitting of Eq. (3) for the other pairs across IQ and RQ have also been conducted. But no equations have passed the statistical tests.

In the next step, triplet consistency tests are applied to every triplet of equations in $IE \cup RE$. In case of a triplet of the power form equations, $x^{a_{xy}}y = b_{xy}, y^{a_{yz}}z = b_{yz}, x^{a_{xz}}z = b_{xz}$, by substituting y in the first to y in the second, we obtain $x^{-a_{yz}a_{xy}}z = {b_{xy}}^{-a_{yz}}b_{yz}$. Thus, the following condition must be met.

$$a_{xz} = -a_{yz}a_{xy}.$$

However, if any of the three equations are not correct due to the noise and error of data fitting, this relation may not hold. Thus, a statistical test judges if the three of the equations are mutually consistent in terms of as. For the triplet of the linear form equations, a similar test is applied based on the identical principle. SDS applies this test to every triplet of equations in $IE \cup RE$, and search every maximal convex set MCS where each triplet of equations among the quantities in this set are mutually consistent. The value of each a is evaluated by its average $\overline{a}$ over the equations in the MCS. When the value of $\overline{a}$ is close enough to its nearest integer within its statistical error bound, it is set to the integer value. This operation is based on the observation that the majority of the first principle based equations have integer power coefficients. Finally, the merged quantities are replaced by the term of each equation of the derived regime in $IQ \cup RQ$. In the example in Fig. 1, the final forms of regimes were represented by the merged terms in $IQ \cup RQ$ as follows.

$$IQ \cup RQ = \{\Pi_1 = R_1 r^{-1} L^{-2}, \Pi_2 = h_{ie_1} R_2^{-1}, \Pi_3 = h_{ie_2} \\ R_3^{-1}, \Pi_4 = h_{ie_3} XKB^{-1}, \Pi_5 = QC^{-1}, \Theta_0 = V_1 - V_0\}$$

Once all regimes are identified, new terms are further generated by merging these regimes in $IQ \cup RQ \cup AQ$. SDS searches bi-variate relations having one of the formulae $x^a y = b$ (product form) and $ax+y = b$ (linear form) by adopting the least square fitting of these formulae. Then, the statistical tests of the goodness of the fitting are applied. In the example of the circuit, the product form was applied first. They were merged to the following new terms.

$$\Theta_1 = \Pi_1 h_{fe_1} = R_1 r^{-1.0} L^{-2.0} h_{fe_1}, \\ \Theta_2 = \Pi_2 h_{fe_2} = h_{ie_1} R_2^{-1.0} h_{fe_2}, \\ \Theta_3 = \Pi_3 h_{fe_3} = h_{ie_2} R_3^{-1.0} h_{fe_3}.$$

Next, the linear form was tested, then a form was found.

$$\Theta_4 = \Pi_4 + \Pi_5 = h_{ie_3} XKB^{-1.0} + QC^{-1.0}$$

Thus, $IQ \cup RQ \cup AQ$ became as $\{\Theta_0, \Theta_1, \Theta_2, \Theta_3, \Theta_4\}$. Again, by applying the linear form, another was newly generated.

$$\Theta_5 = \Theta_0 \Theta_4^{-1.0} = (V_1 - V_0)(h_{ie_3} XKB^{-1.0} + QC^{-1.0})^{-1.0}$$

Thus, $IQ \cup RQ \cup AQ = \{\Theta_1, \Theta_2, \Theta_3, \Theta_5\}$. As no new terms became available, this step was finished.

In the final step, the "*identity constraint*" are applied to further merge terms. The basic principle of the identity constraints comes by answering the question that "*what is the relation among Θ_h, Θ_i and Θ_j, if $\Theta_i = f_{\Theta_j}(\Theta_h)$ and $\Theta_j = f_{\Theta_i}(\Theta_h)$ are known?*" For example, if $a(\Theta_j)\Theta_h + \Theta_i = b(\Theta_j)$ and $a(\Theta_i)\Theta_h + \Theta_j = b(\Theta_i)$ are given, the following relation is deduced.

$$\Theta_h + \alpha_1 \Theta_i \Theta_j + \beta_1 \Theta_i + \alpha_2 \Theta_j + \beta_2 = 0$$

This principle is generalized to various relations among multiple terms. In the example of the circuit, SDS found a set of the bi-variate linear relations on the combinations of $\{\Theta_1, \Theta_5\}, \{\Theta_2, \Theta_5\}$ and $\{\Theta_3, \Theta_5\}$. By applying the identity constraint, the following multi-linear formula has been obtained.

$$\Theta_1\Theta_2\Theta_3 + \Theta_1\Theta_2 + \Theta_2\Theta_3 + \Theta_1\Theta_3 + \Theta_1 + \Theta_2 + \Theta_3 + \Theta_5 + 1 = 0$$

Because every coefficient is independent of any terms, this is considered to be the ensemble equation. The equivalence of this result to Eq. (4) is easily checked by substituting the intermediate terms to this ensemble equation.

4 Extended SDS

As noted in the previous section, the bi-variate fitting requires experimental control of some quantities, and is not applicable to the passive observation environments. To overcome this difficulty, the "*quasi-bi-variate fitting*" procedure depicted in Fig. 2 is used to extract a bi-variate relation between two quantities under the approximated constant values of the other quantities. Let $OBS = \{X_1, X_2, ..., X_n\}$ be a set of observations where each X is a m-dimesional vector of m quantities. The fitting of a candidate bi-variate formula for a pair of two quantities $P_{ij} = \{x_i, x_j\}(\subseteq X)$ is applied to a subset of OBS. This subset OBS_{ijg} is chosen in such a way that every quantity $x_k \in (X - P_{ij})$ takes a value in the vicinity of the value of x_{kg}, where $X_g = \{x_{1g}, x_{2g}, ..., x_{mg}\} \in OBS$ is an arbitrary chosen observation vector. The vicinity of x_{kg} is defined as

$$\Delta x_k = |x_k - x_{kg}| < \epsilon_k.$$

ϵ_k determines the size of the vicinity. This vicinity is indicated by a rectangular cube in the left figure of Fig. 2. Every admissible bi-variate formula is generally represented in the form

$$F_{ij}(P_{ij}, a_{ij}, G_{ij}(X - P_{ij}), H_{ij}(X - P_{ij})) = 0. \tag{5}$$

Here, G_{ij} and H_{ij} are dependent on the quantities in $X - P_{ij}$, while a_{ij} remains constant. Given an OBS_{ijg}, if each ϵ_k is moderately small, the values of G_{ij} and H_{ij} become almost constant. The least square fitting of Eq.(5) approximately provides the functional relation within P_{ij} and the coefficient a_{ij} as depicted in the bottom figure of Fig. 2 while almost excluding the influence of the other dimensions $X - P_{ij}$. The goodness of the fitting is judged by some statistical tests. For the bi-variate relations of the identity constraints, the similar scheme of the quasi-bi-variate fitting is applied.

The proposed method has been applied to a real world problem. The objective of the application is to discover a generic law formula governing the mental preference of people on their houses subject to the cost to buy the house and the social risk at the place of the house. We designed a questionnaire sheet to ask the preference of the house in the trade off between the frequency of huge

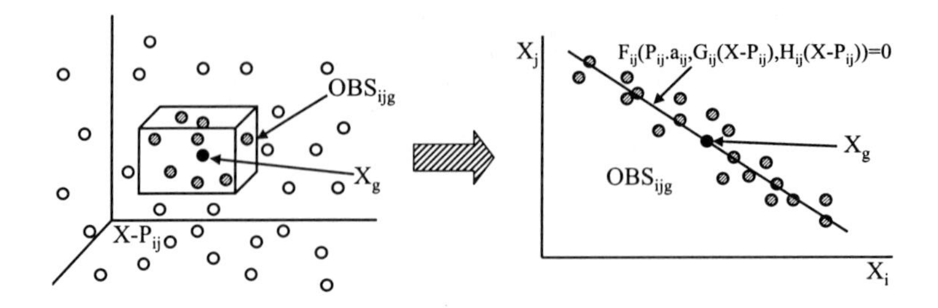

Fig. 2. Outline of quasi-bi-variate fitting

earthquakes, x_1 (earthquake/year), and the cost to buy, x_2 ($). These are ratio scale quantities. In the questionnaire, 9 combinations of the cost and the earthquake frequency are presented, and each person chooses its preference from the 7 levels for each combination. We distributed this questionnaire sheet to the people owning their houses in the suburb area of Tokyo, and totally 400 answer sheets are collected back. The answer data has been processed by following the method of successive categories which is widely used in the experimental psychology to compose an interval scale preference index y [10], and $OBS = \{X_1, X_2, ..., X_{400}\}$ where $X_i = [x_{1i}, x_{2i}, y_i]$ is obtained. The expected basic structure of the law equation governing the data is $y = f(x_1, x_2)$. The quasi-bi-variate fitting between x_1 and x_2 was applied, and $x_1 = a(y)x_2^{-0.25}$ have been obtained. Next, the formulae between x_1 and y have been identified as either one of $y = a(x_2)x_1^{-0.23} + b(x_2)$ and $y = 0.62 \log x_1 + b(x_2)$. Similar search has been made for x_2 and y, and, likewise, two candidate equations $y = a(x_1)x_2^{0.026} + b(x_1)$ and $y = 0.34 \log x_2 + b(x_1)$ have been derived. Subsequently, the triplet-test among $\{x_1, x_2, y\}$ is conducted, and only the following two candidates have passed the test.

$$y = 0.63 \log x_1 + 0.34 \log x_2 - 2.9 \tag{6}$$

$$y = -0.61 x_1^{-0.23} x_2^{0.026} + 3.2 \tag{7}$$

Though both equations are admissible as law equations based on the mathematical constraints and the given data, Eq. (6) is preferred as a law equation in terms of the principle of parsimony, because it gives less error for the questionnaire data, and has less number of parameters.

5 Simultaneous Structure Finder (SSF)

The principle to discover the simultaneous equation structure from experimental data is based on some fundamental and generic characteristics of simultaneous equation models presented in the past work [7]. The principle is briefly explained though an example electric circuit depicted in Fig. 3. This can be represented by the following simultaneous equation model.

$$V_1 = I_1 R_1 \ \#1,\ V_2 = I_2 R_2 \ \#2,\ V_e = V_1 \ \#3 \text{ and } V_e = V_2 \ \#4, \tag{8}$$

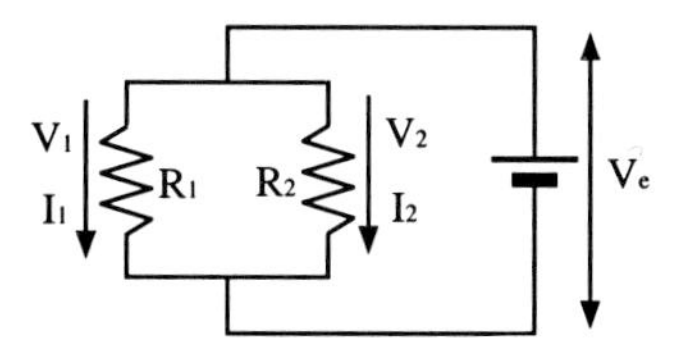

Fig. 3. An circuit of parallel resistances.

where R_1, R_2: two resistances, V_1, V_2: voltage differences across resistances, I_1, I_2: electric current going through resistances and V_e: voltage of a battery. We consider an experiment to externally control some values of the quantities in this model. For example, the quantities R_1 and V_e can be externally controlled by the specification of the resistance and the battery. If we specify these values in Eq.(8), the values of the other quantities, V_1, V_2 and I_1, that are involved in the first, the third and the forth equations, #1, #3 and #4, are determined since the number of the quantities which are not externally specified is equal to the number of the equations. But, this external control does not determine the values of R_2 and I_2 through the equation #2. Thus, the equation set $\{\#1, \#3, \#4\}$ is considered to represent a mechanism which determines the state of a part of the objective process. We introduce the following definition to characterize this mechanism in the simultaneous equation model.

Definition 1 (complete subset). *Given a set of equations, E, let the set of all quantities be Q appearing in the equations in E. Given a quantity set $SQ(\subset Q)$ for external specification, when the values of all quantities in $NQ = CQ - SQ$ are determined where CQ $(SQ \subset CQ \subset Q)$ is a set of all quantities appearing in a set of equations $CE(\subseteq E)$, CE is called a "complete subset". The cardinality $|CE| = |NQ|$ is called the "order" of the complete subset.*

The equation set $\{\#1, \#3, \#4\}$ is a complete subset of the order 3. Under any external control of two quantities among R_1, V_e, V_1, V_2 and I_1, $\{\#1, \#3, \#4\}$ always determines the values of the remained three quantities. Thus, the complete subset is "*invariant*" for the selection of the externally controlled quantities.

The complete subset gives an important foundation to discover the structure of the simultaneous equation model, which appropriately reflects the dependency, embedded in the observation of quantities. For example, the circuit in Fig. 3 can be represented by the following different simultaneous equation formulae

$$I_1R_1 = I_2R_2 \ \#1', \ V_2 = I_2R_2 \ \#2, \ V_e = V_1 \ \#3 \text{ and } V_e = V_2 \ \#4. \qquad (9)$$

If the same specification on V_e and R_1 is made in Eq.(9), a different complete subset $\{\#3, \#4\}$ is obtained, and any complete subset to determine the value of I_1 does not exist since the equation $\#1'$ cannot determine the value of I_1 without the constraint of #2. $\#1'$ and #2 that include the undetermined quantities I_2 and R_2 do not satisfy Definition 1. In the real experiment on the electric circuit, the value of I_1 is physically determined, and this fact contradicts the consequence derived by the analysis on Eq.(9). In contrast, the model of Eq.(8) always gives correct answers on the determination of quantities for any external specifications of quantities. The model having the complete subsets, which are isomorphic with the actual dependency among quantities, is named a "*structural form*".

Conversely, if we identify all complete subsets from the experimental observation of quantities in the objective process, and compose a simultaneous equation model consisting of these complete subsets, the model is ensured to be the structural form. The following theorem provides a basis for the composition [7].

Theorem 3 (modular lattice theorem). *Given a model of an objective process consisting of equations E, the set of all complete subsets of the model, i.e., $L = \{\forall CE_i \subseteq E\}$, forms a modular lattice of the sets for the order of the complete subsets, i.e., $\forall CE_i, CE_j \in L$, $CE_i \cup CE_j \in L$, $CE_i \cap CE_j \in L$ and $n(CE_i \cup CE_j) = n(CE_i) + n(CE_j) - n(CE_i \cap CE_j)$ where n is the order of a given complete subset.*

For instance, the following four complete subsets having the modular lattice structure can be found in the example of Eq.(8).

$$\{\#3, \#4\}(n = 2),\ \{\#1, \#3, \#4\}(n = 3),$$
$$\{\#2, \#3, \#4\}(n = 3),\ \{\#1, \#2, \#3, \#4\}(n = 4).$$

Because the complete subsets of an objective process mutually overlap in the modular lattice, the redundant overlaps must be removed in the model composition by introducing the following definition of independent component.

Definition 2 (independent component of a complete subset). *The independent component DE_i of the complete subset CE_i is defined as*

$$DE_i = CE_i - \bigcup_{\forall CE_j \subset CE_i \, and \, CE_j \in L} CE_j,$$

where L is the set of all complete subsets of the model. The order δn_i of DE_i is defined as $\delta n_i = |DE_i|$.

For instance, the following independent components can be found for Eq.(8).

$$DE_1 = \{\#3, \#4\} - \phi = \{\#3, \#4\},\ \delta n_1 = 2 - 0 = 2,$$
$$DE_2 = \{\#1, \#3, \#4\} - \{\#3, \#4\} = \{\#1\},\ \delta n_2 = 3 - 2 = 1,$$
$$DE_3 = \{\#2, \#3, \#4\} - \{\#3, \#4\} = \{\#2\},\ \delta n_3 = 3 - 2 = 1.$$

Because the independent components do not overlap, their collection represents the structure of the simultaneous equation model.

However, the issue on the ambiguity of the representation of the structural form still remains. For example, the set of equations $\{V_1 = I_1R_1\#1, V_e = V_1\#3, V_e = V_2\#4\}$ in Eq.(8) which is a complete subset of order 3 can be transformed by the linear transformation as follows.

$$2V_e + V_1 + V_2 = 4I_1R_1\ \#1,\ 2V_e = 2V_1 - V_2 + I_1R_1\ \#3,$$
$$\text{and } 3V_e = -V_1 + 2V_2 + 2I_1R_1\ \#4.$$

This transformation preserves the complete subset, and the model remains as a structural form. This ambiguity of the equation representation in a complete subset can cause combinatorial explosion in the enumeration of the structural forms. As indicated in the above example, if the set of all quantities, CQ, appearing in a complete subset CE is preserved through some transformation maintaining quantitative equivalence, the complete subset is also preserved [7]. Accordingly, only the following formula of a complete subset is focused in the search.

Definition 3 (canonical form of a complete subset). *Given a complete subset CE, the "canonical form" of CE is the form where all quantities in CQ appears in each equation in CE.*

An example of the canonical form is Eq.(10). Based on this definition, the structural canonical form of a simultaneous equation model is further defined.

Definition 4 (structural canonical form of simultaneous equations). *The "canonical form" of simultaneous equations consists of the equations in $\cup_{i=1}^{b} DE_i$ where each equation in DE_i is represented by the canonical form in the complete subset CE_i, and b is the total number of DE_i. If the canonical form of simultaneous equations is derived to be a "structural form", then the form is named "structural canonical form".*

The structural form of Eq.(8) is shown as follows.

$$\begin{aligned} &DE_1 = \{f_{11}(V_e, V_1, V_2) = 0 \ \#3,\ \ f_{12}(V_e, V_1, V_2) = 0 \ \#4\}, \\ &DE_2 = \{f_2(V_e, V_1, V_2, I_1, R_1) = 0 \ \#1\},\ DE_3 = \{f_3(V_e, V_1, V_2, I_2, R_2) = 0 \ \#2\} \end{aligned} \quad (10)$$

where $f(\bullet) = 0$ is an arbitrary formula to represent a quantitative relation. Because Eq.(8) is a structural form, Eq.(10) is the structural canonical form. The concrete shape of each formula can be derived by the aforementioned SDS, once the structure of simultaneous equations is given.

6 Discussion and Conclusion

Dzeroski and Todorovski developed LAGRANGE [11] and LAGRAMGE [12] which discover simultaneous equation models from observed data. However, the mathematical admissibility is not introduced sufficiently in the discovery process, and many redundant representations of simultaneous equations can be derived at an expense of high computational complexity. COPER, which also discovers simultaneous equations, uses very strong mathematical constraints based on the unit dimensions to prune the meaningless terms [4]. However, it inevitably requires the unit information which is not frequently obtained in non-physical domains. The major advantages of our proposing methods in comparison with the past approaches are the efficiency of the equation search, the soundness of the discovery in terms of the first principle and the wide applicability not limited to the physical domain. These are achieved by introducing the criteria of generic mathematical admissibility.

The future directions of our work are
(1) Discovery of simultaneous equations from passively observed data and
(2) Discovery of dynamic law equations from passively observed time series data.

For the former purpose, the principle of SSF should be extended to discover the structure of simultaneous equations from passively observed data. This may become possible by developing a new data sampling technique similar to the quasi-bi-variate fitting of the extended SDS. The latter approach is expected to discover the first principle based differential equations which appear in various scientific and engineering domains. The major issue to overcome will be the statistical information processing on the noise contained in the time series data which are widely seen in many problems. These scientific law equation discovery methods will provide a new measure of system modeling based on data in wide scientific and engineering domains.

References

[1] Langley, P. W., Simon, H. A., Bradshaw, G. L. and Zytkow, J. M.: *Scientific Discovery; Computational Explorations of the Creative Process*, MIT Press, Cambridge, Massachusetts (1987)
[2] Koehn, B. and Zytkow, J. M.: Experimenting and theorizing in theory formation. In *Proceedings of the International Symposium on Methodologies for Intelligent Systems*, ACM SIGART Press (1986) 296–307
[3] Falkenhainer, Br. C. and Michalski, R. S.: Integrating Quantitative and Qualitative Discovery: The ABACUS System. In *Machine Learning*, Boston, Kluwer Academic Publishers (1986) 367–401
[4] Kokar, M. M.: Determining Arguments of Invariant Functional Descriptions. In *Machine Learning*, Boston, Kluwer Academic Publishers (1986) 403–422
[5] Washio, T. and Motoda, H.: Discovering Admissible Models of Complex Systems Based on Scale-Types and Identity Constraints, In *Proceedings of IJCAI'97*, Vol.2, Nagoya (1997) 810–817
[6] Washio, T., Motoda, H. and Niwa, Y.: Discovering admissible model equations from observed data based on scale-types and identity constraints. In *Proceedings of IJCAI'99*, Vol.2 (1999) 772–779
[7] Washio T. and Motoda, H.: Discovering Admissible Simultaneous Equations of Large Scale Systems, In *Proceedings of AAAI'98*, Madison (1998) 189–196
[8] Ljung, L.: *System Identification*, P T R Prentice-Hall (1987)
[9] Stevens S.S.: On the Theory of Scales of Measurement, In *Science* (1946) 677–680
[10] Torgerson, W. S.: In *Theory and Methods of Scaling*, N.Y.: J. Wiley (1958)
[11] Dzeroski, S. and Todorovski, L.: Discovering Dynamics: From Inductive Logic Programming to Machine Discovery. In *Journal of Intelligent Information Systems*, Boston, Kluwer Academic Publishers (1994) 1–20
[12] Todorovski, L. and Dzeroski, S.: Declarative Bias in Equation Discovery, In *Proceeding of the fourteenth International Conference on Machine Learning*, San Mateo, CA, Morgan Kaufmann (1997) 376–384

A Machine Learning Algorithm for Analyzing String Patterns Helps to Discover Simple and Interpretable Business Rules from Purchase History

Yukinobu Hamuro[1], Hideki Kawata[1], Naoki Katoh[2], and Katsutoshi Yada[3]

[1] Department of Business Administration, Osaka Industrial University
Daito, Osaka, Japan
hamuro@adm.osaka-sandai.ac.jp, s98b116@sub.osaka-sandai.ac.jp
[2] Department of Architecture and Architectural Systems, Kyoto University
Yoshida-Honmachi, Sakyo-ku, Kyoto, 606-8501 Japan
naoki@archi.kyoto-u.ac.jp
[3] Faculty of Commerce, Kansai University
Yamatemachi, Suita, Osaka, 564-8680 Japan
yada@ipcku.kansai-u.ac.jp

Abstract. This paper presents a new application for discovering useful knowledge from purchase history that can be helpful to create effective marketing strategy, using a machine learning algorithm, BONSAI, proposed by Shimozono et al. in 1994 which was originally developed for analyzing string patterns developed for knowledge discovery from amino acid sequences. In order to adapt BONSAI to our purpose, we translate purchase history of customers into character strings such that each symbol represents a brand purchased by a customer. For our purpose, we extend BONSAI in the following aspects; 1) While original BONSAI generates a decision tree over regular patterns which are limited to substrings, we extend it to subsequences. 2) We generate rules which contain not only regular patterns but numerical attributes such as age, the number of visits, profit and etc. 3) We extend regular expression so that we can consider whether a certain pattern occurs in some latter part of the whole string. 4) We implement majority voting based on 1-D and 2-D region rules on top of decision trees.
Applying the BONSAI extended in this manner to real customers' purchase history of drugstore chain in Japan, we have succeeded in generating interesting business rules which practitioners have not yet recognized.

1 Introduction

According to the rapid development of modern computer technologies, there has been much progress made in automating daily office work. This, in turn, has resulted in the accumulation of a huge amount of business data into databases. Thus, the global competitive landscape of the retail business has changed radically during the past several years. Under such environment, *knowledge discovery*

S. Arikawa and A. Shinohara (Eds.): Progress in Discovery Science 2001, LNAI 2281, pp. 565–575.

in databases or *data mining* has become an active research area in which new technologies or methodologies are sought to automatically extract meaningful knowledge from business data [1, 14, 18]. The availability of detailed customer data and advances in technology for warehousing and mining data (e.g. [16, 17]) enable companies to better understand and service their customers.

In order to allow us to directly make use of time-series purchase data for discovery of useful knowledge concerning customer purchase behavior, it is natural to transform the purchase history into a string of symbols for which each symbol represents for instance a purchased brand or categorized profitability depending on the cases. The analysis of such strings may help discover new knowledge which could not be uncovered by conventional knowledge discovery methods that use only numerical and/or categorical attributes such as profitability per visit, age, the number of visits and etc. This is the motivation of this study.

In the field of molecular biology, Shimozono et al. [15] developed a machine discovery system BONSAI that produces a decision tree over regular patterns with alphabet indexing, from given positive set and negative set of strings. One of the core part of the system is to generate a decision tree which classifies positive examples and negative examples as correctly as possible. The other core of the system is an alphabet indexing which transforms an original alphabet to a smaller alphabet which does not lose any positive and negative information of the given examples. It was observed in [15] that the use of an appropriate alphabet indexing increases accuracy and simplifies hypothesis. BONSAI incorporates the mechanism of automatically producing an alphabet indexing providing good decision tree over regular patterns by local search method. It was reported in [15] that alphabet indexing obtained for transmembrane identification matches domain knowledge called hydropathy index.

In the implementation of BONSAI, regular patterns attached to the nodes of decision trees are limited to substrings, which are of the form $x\alpha y$ where x and y are variables and α is a substring taken from positive and negative examples. Recently, this limitation of original BONSAI was resolved by [6] so that we can deal with subsequences.

In view of such features of BONSAI that can deal with sequences of symbols, we develop an extended version of BONSAI so that we can deal with sales history which has the following new features: 1) While original BONSAI generates a decision tree over regular patterns which are limited to substrings, we extend it to subsequences based on the work of [6]. 2) We generate rules which contain not only regular patterns but numerical attributes such as age, the number of visits, profit and etc., while BONSAI uses only string patterns as predictive attributes. This feature allows us to incorporate various types of attributes into the learning process. 3) Since it is usually believed that for certain commodity categories the brand purchased for the first time affects future purchase behavior and that the most recently purchased brand is closely related with the next purchase behavior. Thus, we extend regular expression so as to take into account the position where a certain pattern occurs in the whole string. 4) We implement majority voting based on 1-D and 2-D region rules on top of decision trees hoping that it generates more interpretable rules than those generated by decision trees.

In order to observe how the proposed method produces useful customer knowledge, we have conducted experiments by using two data sets obtained from customers' purchase history of drugstore chain in Japan. It is observed from the experiments that the method produces in general more interpretable rules which provides new and interesting knowledge concerning customer purchase behavior. A few such examples will be presented.

The contribution of this paper does not lie in the development of a new algorithm for knowledge discovery, but in that BONSAI with some new features added is helpful to discover useful and interpretable customer knowledge concerning purchase behavior.

2 Algorithm

We shall explain the outline of the proposed algorithm. Before this, we shall give a brief outline of BONSAI. Given positive set of strings, *pos* and negative set of strings, *neg*, BONSAI creates a good decision tree that classifies *pos* and *neg* as correctly as possible. As a candidate set of substrings to be used at internal nodes of a decision tree are generated by enumerating substrings appearing in strings of *pos* and *neg*. (Maximum length of substrings is usually fixed.) When an alphabet indexing is fixed, a good decision tree is constructed in a top-down manner as in the same fashion as proposed by Quinlan [11–13]. Search for a good alphabet indexing is done by local search with multi-start in our implementation. In order to allow us to use subsequences instead of substrings, Hirao et al. [6] developed a branch-and-bound algorithm which incorporates an effective pruning method and data structure.

Our algorithm is based on the work by [6], but introduced several new features as mentioned in Section 1. Among them, we shall further explain in details the following points.

1. Considering the applicability to the analysis of purchase history, our algorithm allows us to use more than one sequence in order to construct a good decision tree. Original purchase history contains purchase records of various products. For the analysis of such history, it is better to decompose it into multiple symbol strings each corresponding to a certain product category.
2. Our algorithm can deal with substrings which are of the form $\hat{}\alpha$ and $\alpha\$$, where α is a substring, and the symbols ^and $ represent the beginning and the termination of a sequence, respectively. Since purchase pattern appearing at the beginning and/or the end of the purchase history may possibly have a stronger influence on the future purchase behavior, such feature is important as will be exemplified in Section 3. We can also deal with substrings $\alpha\#(n)$ which means that string α appears within the last n-th position of the sequence.

For the efficient implementation, by following the idea by [6], we first select candidate substrings (or subsequences) by computing the support and the entropy gain, i.e., we discard the substring or subsequence for which the number of instances containing them is smaller than a given threshold or the entropy gain is smaller than a given threshold. Such discarded substrings or subsequences are appended to a forbidden list which will be used to quickly eliminate unnecessary search in the subsequent computation. In order to find the best alphabet-indexing, we use local search with multi-start.

Another important new feature of the proposed algorithm is that besides decision tree, we have developed a region-based BONSAI, called Region-BONSAI, based on the work of [10] who have proposed a new classification algorithm called *majority weighted decision* among several 2D-region rules aiming to produce simple classification rules. We shall briefly explain the framework of weighted majority decision. (In order to distinguish decision-tree based BONSAI explained above from Region-BONSAI, decision-tree based one is called Tree-BONSAI.)

We assume 2-D region is constructed by a pair of conditional attributes (all conditional attributes are assumed to be numeric and to take discrete value) and target attribute is binary, i.e., 1(true) or 0(false). If the number of conditional attributes is m, ${}_mC_2$ region rules are calculated based on the entropy gain. Then a subset of these region rules are sorted in non-increasing order of entropy gains of these rules. A certain number, say k, of region rules are selected which in turn vote based on whether or not the target attribute is true. Selected rules are simply called a *voter*. The weight of each voter which is a value between 0 and 1 is determined by using prediction accuracy in the training dataset. For the set of tuples which are inside (resp. outside) the region R_0 (resp. R_1), the ratio of tuples whose target attribute is true is calculated, and $p(R_0)$ (resp. $p(R_1)$) denotes this ratio. $p(R_0)$ (resp. $p(R_1)$) is then used as the weight of the case where the value of the target attribute is true. The summation of the weights of the k selected voters indicates the degree of possibility that the target attribute of the tuple is true. If the summation of weights is greater than a threshold $k/2$, the majority decision predicts that the target attribute of the tuple is true otherwise false.

In our recent paper [7], we have improved the original majority decision in two ways:

1. Instead of using the entropy gain as selection criterion for voters, we use Gini index for which we theoretically proved that the classification ability is always better than using the entropy gain [7]. Based on the fact that the larger the sum of voters for true tuples gets the better the prediction accuracy of the method becomes, the proof is done by showing that the maximization of the sum of voters for true tuples is equivalent to that of Gini index.
2. Since the threshold $k/2$ used in the original majority decision does not have a theoretical background. Thus, we have optimized it.

We have applied majority weighted decision to our case. For each substring or subsequence in the candidate set, the corresponding attribute takes the value 1 or 0 depending whether an instance contains the substring (or subsequence) or not. Thus, each substring or subsequence is associated with binary attribute.

Thus, 2D-rule is obtained from 2×2 table. The class of regions to be extracted is limited to connected rectilinear polygons. There are only 14 possible subregions excluding the whole region and empty one for a 2×2 table among which there are 12 connected rectilinear polygons. Four regions out of 12 ones are equivalent to 1-D rule. Thus, there are eight types of intrinsic 2-D rules all of which are described by one of

$$a \vee b, \neg a \vee b, a \vee \neg b, \neg a \vee \neg b, a \wedge b, \neg a \wedge b, a \wedge \neg b, \neg a \wedge \neg b,$$

where a and b are substrings or subsequences depending on the cases (for example, the rule $\neg a \vee b$ reads that a given sequence does not contain a or it contains b). Thus, it is easy to compute the best region. In our experiment, in order to determine an appropriate threshold of the number of voters, we have tested several cases. It was observed that the result is not sensitive to the threshold.

The running time for our algorithm is within a few seconds if the number of instances is less than one hundred. However, it rapidly increases when the number of instances increases, which was also observed in [6]. The improvement of the running time for large-size case is left for future research.

3 Experimental Results

We have applied the proposed algorithm to two datasets, LoyalCustomer and BabyDiaper (which will be explained below) in order to observe how we can produce interesting rules that may be utilized for constructing effective marketing strategy. All datasets have been obtained from purchase history accumulated in a drugstore chain in Japan [5]. The drugstore chain holds three million members so that it accumulates the purchase histories of all member customers. In each of the following experiments, we have obtained decision trees using only numerical attributes by C5.0 for comparison. In the experiment of Region-BONSAI, we use twenty best 2-D rules for voters.

3.1 Loyal Customer

In [8], we have developed a data mining technique to identify potential high-value new visitors (called *loyal customers*) to the store at an early stage. The successful identification of potential high-value new customers early on is important because the company can use this information to establish a close relationship to this selected group of customers, thus building a tighter bond and reducing the chance that they will leave. In the study of [8], a loyal customer is defined in terms of two dimensions; profitability per visit and frequency of visit. Each of these variables is categorized into five classes so that the size of each class is as equal as possible. Such categorization used therein is illustrated in Table 1.

Loyal customers are defined to be those who score a 4 or above in both dimensions after one year passes from the first visit. Thus, identification of whether a customer is a loyal one can be made only after one year passes. Among 114,069

Table 1. Categorization of variables

	First dimension	Second dimension
class	Profit/visit (in yen)	Freq. of visits in 1998
5	566 to 28091	13 to 323
4	315 to 565	7 to12
3	170 to 314	4 to 6
2	41 to 169	2 to 3
1	-87440 to 40	1

customers who newly became a member in 1998, about 10% customers are classified as loyal. Such customers generated a disproportionate 52.5% of profit share and 38.4% of revenue share, exemplifying the strategic importance of the ability to manage relationship with this class of customers. Using the profitability, frequency of visit[8], and other attributes for the first two, three or four months, [8] showed that C5.0 exhibited satisfactory prediction ability for identification of loyal customers.

In our experiment, we have selected 16,902 customers among those used in [8] so that the numbers of loyal and non-loyal customers are equal. Our goal is to discover a simple rule to predict loyal customers using the data of the first 13 weeks. In our experiment, we use two sequences as conditional attributes. The first is a sequence of profitability per visit encoded as a string of symbols $\{1, 2, 3, 4, 5\}$ which correspond to the five classes shown in Table 1. The length of the string is equal to the number of visits within the first 13 weeks, and the i-th symbol represents the class of profitability at the i-th visit. The second is a sequence of symbols $\{0, 1\}$ of length 13 that represents visit pattern such that the symbol 0 or 1 at the i-th position stand for non-visit or visit in the i-th week, respectively.

When selecting the best alphabet indexing for a sequence of profitability per visit, alphabet indexing we search for is limited to the following in order to eliminate meaningless symbols indexings: Let Γ' be the set of alphabets into which an alphabet indexing ϕ transforms $\{1, 2, 3, 4, 5\}$. Then for any $a \in \Gamma'$, the set $\{i \in \{1, 2, 3, 4, 5\} \mid a = \phi(i)\}$ consists of consecutive integers. For example, if $\Gamma' = \{a, b\}$, ϕ should satisfy that there exists k with $1 \leq k \leq 4$ such that $\phi(1) = a, \ldots, \phi(k) = a$ and $\phi(k+1) = b, \ldots, \phi(5) = b$.

We have experimented Tree- and Region-BONSAI by performing 5-fold cross validation. Since the running time becomes unacceptably long if input size is more than one thousand, we randomly choose customers of not more than 300 and build a decision model (e.g., decision tree for Tree-BONSAI) for each of 5 training sets. It was observed in our experiment that the accuracy of the decision model does not depend on the sample size if the size is between 100 and 300. The results of our experiments are summarized in Table 2. For each component in the table, a number denotes the average of five tests of 5-cross validation. We shall first explain the results obtained by Tree-BONSAI in which regular patterns are limited to substrings of two purchase patterns of profitability and visit. The average number of leaves of decision trees obtained by 5-fold cross validation tests was 2, i.e., decision rule of every tree consists of only one conditional attribute. When using substrings in generating decision trees by Tree-BONSAI, all of five trees use the substring "1" of profitability sequence as the unique conditional attribute (see Fig. 1(a)), and the underlying alphabet index encodes profitability classes $1, 2, 3, 4$ as 0 and 5 as 1. In the figure, percentage $(x\%, y\%)$ attached to a leaf is read as x and y represent the ratios of positive and negative instances classified as the corresponding label, respectively. This tree only uses the profitability sequence but not visit pattern. The rule of Fig. 1(a) is interpreted as if in at least one of visits of a customer his/her profitability at the visit is in the highest class (class 5), then he/she will become a loyal customer with high accuracy.

When using subsequences instead of substrings in generating decision trees by Tree-BONSAI, all of five trees are the same (see Fig 1(b)). The decision tree uses the subsequence "11" of profitability sequence as the unique conditional attribute. The underlying alphabet indexing encodes profitability classes $1, 2, 3$ as 0 and $4, 5$ as 1. The rule says that if in at least two visits of a customer his/her profitability at the visit is in class 4 or 5, then he/she will become a loyal customer with high accuracy.

Both rules obtained are very simple and thus interpretable, which have not been known to practitioners.

Table 2. Accuracy and size of decision rules obtained by BONSAI and C5.0. The sizes for Tree-BONSAI and Region-BONSAI denote the number of leaves of decision trees and the number of voters used, respectively. Accuracy and size in the table are the average of 5 tests of 5-fold cross validation.

		Loyal Customer		Baby Diaper	
		accuracy	size	accuracy	size
Tree-BONSAI	string	69.6%	2	84.8%	2
	string+numeric	70.3%	12.4	85.4%	10.6
	subsequence	70.8%	2	85.2%	2
	subsequence+numeric	70.8%	12.4	85.3%	10.2
Region-BONSAI	string	70.6%	20	85.7%	20
	string+numeric	70.6%	20	85.6%	20
	subsequence	70.8%	20	85.5%	20
	subsequence+numeric	70.8%	20	85.5%	20
C5.0		71.3%	11.0	87.5%	17.4

As you notice from Table 2, Region-BONSAI exhibits almost the same performance as Tree-BONSAI. On the other hand, the interpretability of rules obtained was not so high as Tree-BONSAI since it uses 20 2-D region rules as voters.

We have also performed Tree- and Region-BONSAI by incorporating numerical attributes such as average profitability and the number of visits. Accuracy of these cases are also given in Table 2 and similar to the above cases. However, due

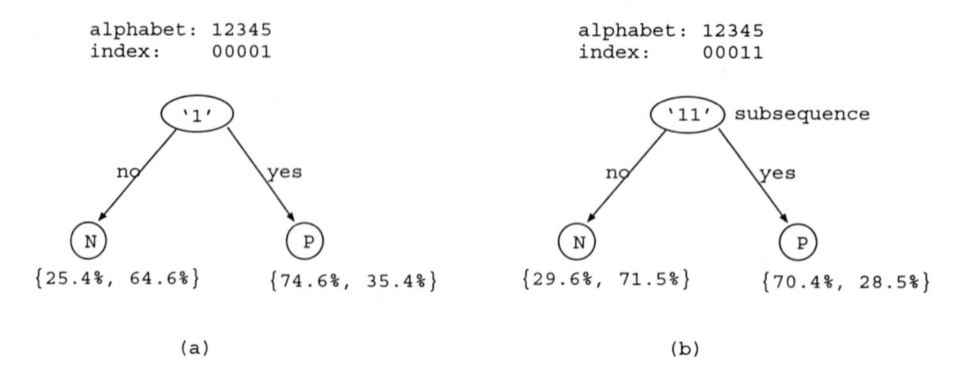

Fig. 1. Figure (a) shows a decision tree generated by Tree-BONSAI for LoyalCustomer when BONSAI uses only substrings. Figure (b) shows a decision tree generated by Tree-BONSAI for LoyalCustomer when BONSAI uses subsequences.

to the larger size (the number of leaves), the rules generated by Tree-BONSAI are less interpretable than those obtained by considering only pattern attributes.

We have also computed decision trees using only numerical attributes by C5.0, where two numerical attributes are used; the average profitability per visit, and the number of visits during 13 weeks. We have also tested 5-fold cross validation. The average size is 11.0. Accuracy was slightly better than those by Tree- and Region-BONSAI. However, the size of the trees is large and hence less interpretable than those by Tree-BONSAI. With respect to simplicity and interpretability, decision tree of Tree-BONSAI seems the best.

3.2 Baby Diaper

In the Japanese market of baby diaper, there are seven major brands, denoted by A, B, C, D, E, F and G (other miscellaneous brands are aggregated to a single brand H). Baby diapers are classified according to the size. We are interested in the sizes M and L. Focusing on the brand A, we want to predict which customers will become loyal to brand A in the size L by examining purchase pattern of baby diapers of size M. The market share of brand A in sizes M and L is about 24% in 1999. Most of the customers of baby diapers are young women. They are keen on the quality of the products and thus they have their own preference to the brands. Manufacturers are interested in how and when customers fix their preference. If manufacturers can acquire customer knowledge concerning brand choice behavior, they can perform much more effective sales promotion.

The dataset we used in the experiment is obtained as follows. From the purchase history accumulated during 1996∼1999 in about 100 drugstores, we have selected customers who purchased baby diapers of size M at least four times, and also purchased baby diapers at least five times after switched from size M to size L. Among such customers, we then have selected 918 customers who are loyal to brand A of size L. These customers are treated as positive instances. Here a customer is said to be loyal to a brand X of size L if at least 66.7% of baby diapers of size L purchased are of brand X. In general, let $amonut_X$ denote the amount of baby diapers with brand X purchased by the customer and $amount_{all}$ denote the total amount of baby diapers purchased by the customer. Loyalty index of brand X with respect to the customer is defined as $r = amonut_X/amonut_{all}$. For negative instances we selected 918 customers whose loyalty index for brand A is at most 33.3%.

We have first computed Tree- and Region-BONSAI that use only purchase sequence. Accuracy results are summarized in Table 2 which are obtained by performing five-fold cross validation in the same manner as was done for Loyal Customer. We have also computed decision trees by C5.0 which uses 15 numerical attributes such as customer share for each of eight brands, the ratio of brand switches, number of items purchased, age, profitability, the number of visits and etc. Let us first compare the results of Tree-BONSAI with those of C5.0. Accuracy of C5.0 is better than Tree-BONSAI by more than 2.0%. However, the sizes of decision rules are different. The average size (number of leaves) of decision trees of C5.0 is 17.4 while that of Tree-BONSAI is only two. Three out of five trees produced by BONSAI is the one illustrated in Fig. 2(a) which says that if a customer purchased baby diapers of brand A two consecutive times at size M, she/he will become loyal to brand A at size L with high accuracy. The other two trees use the substring '111' as the unique conditional attribute. When subsequences are used for conditional attributes in executing Tree-BONSAI, the size of decision trees remains the same as that for the case of substrings, and the selected alphabet indexing is also the same. All five trees are the same which use the subsequence '111' as the unique conditional attribute. Notice that decision rules obtained by Tree-BONSAI are interpretable by practitioners, and have not been recognized by them.

We also have executed Tree-BONSAI that takes suffix substrings into consideration as candidate patterns, which produced very simple trees. All of five trees obtained in 5-fold cross validation have only two leaves. One of five trees is illustrated in Fig. 2(b) (The other four trees are the same as Fig. 2(a)). This rule says that if a customer purchases baby diaper of brand A for both of the last two purchase opportunities of baby diapers of size M, she/he will become loyal to brand A at size L with more than 95% accuracy. Rules illustrated in Fig. 2 have not been recognized by practitioners.

We shall briefly review the results obtained by Region-BONSAI. The accuracy was the almost the same as that of Tree-BONSAI. Underlying alphabet indexing was the same as the one given by Tree-BONSAI (see Figure 2). Region-BONSAI uses 20 2-D region rules and hence it is less interpretable than Tree-BONSAI. However, closely examining the results, we found that a single region rule is sufficiently effective. We shall explain one such rule.

Rule 1: It is described in terms of two substrings, a ='00' and b ='011'. The 2D-rule is that at least one of the following two rules is satisfied; (i) purchase pattern does not contain a as its substring. (ii) Purchase pattern contains b as its substring. In other words, this rule is interpreted as a customer who either

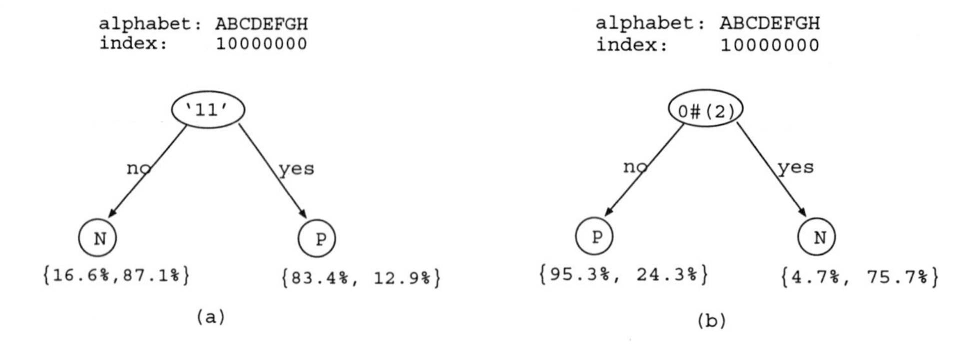

Fig. 2. Figure (a) shows a decision tree generated by Tree-BONSAI for BabyDiaper. Figure (b) shows a decision tree generated by Tree-BONSAI for BabyDiaper with suffix substrings.

(i) did not buy brand other than A two consecutive times or (ii) switched from other brands to A and buy A two consecutive times are likely to become loyal to brand A. This rule was the most effective in terms of entropy gain in one of 5-fold validation tests. The confidence of this rule is 89.8%, and the support is 45.3%. Accuracy is 86.0%. Thus this rule is quite powerful.

We have tested Tree-BONSAI and Region-BONSAI in which numerical attributes such as customer share for each of eight brands, number of brand switches, age are added. Accuracy remains the same as seen from Table 2. However, for the case of Tree-BONSAI, the average number of leaves increased to about 10, and thus decision rules are less interpretable.

In summary, for both cases of Loyal Customer and Baby Diapers, decision trees generated by C5.0 using only numerical attributes are more complicated than those by Tree-BONSAI using only sequence of purchase patterns and rules are difficult to interpret. On the other hand, the rules generated by Tree-BONSAI are easier to interpret for practitioners which could not be obtained by conventional data mining techniques.

It should be mentioned that using subsequences instead of substrings does not increase the accuracy of rules for Tree- and Region-BONSAI for our experiments. One of the reason might be that rules obtained by using subsequences suffer from overfitting.

4 Conclusion

We have developed an algorithm that classifies given sets of positive and negative examples of sequences and applied it to purchase history in order to identify simple and useful knowledge concerning customer purchase behavior. We have verified usefulness of the proposed method by experiments using real sales data. In fact, we have succeeded in obtaining interesting and interpretable business rules for predicting future customer behavior.

References

1. R. Agrawal, T. Imielinski,and A. Swami, Database mining: A performance perspective, *IEEE Transactions on Knowledge and Data Engineering*, Vol.5, pp. 914-925, 1993.
2. S. Arikawa, S. Miyano, A. Shinohara, S. Kuhara, Y. Mukouchi and T. Shinohara, A machine discovery from amino acid sequences by decision trees over regular patterns, *New Generation Computing*, Vol.11, pp. 361-375, 1993.
3. P.B. Chou, E. Grossman, D. Gunopulos, and P. Kamesam, Identifying Prospective Customers, *Proc. KDD 2000*, pp. 447-456, 2000.
4. C. Fishman, *This is a Marketing Revolution*, Fast Company, pp. 206-218, 1999.
5. Y. Hamuro, N. Katoh, Y. Matsuda and K. Yada, Mining Pharmacy Data Helps to Make Profits, *Data Mining and Knowledge Discovery*, Vol.2 No.4, pp. 391-398, 1998.
6. M. Hirao, H. Hoshino, A. Shinohara, M. Takeda, and S. Arikawa, A Practical Algorithm to Find the Best Subsequence Patterns, *Proc. of 3rd International Conference on Discovery Science*, LNAI 1967, pp. 141-154, 2000.
7. N. Horiguchi, K. Yada, Y. Hamuro, N. Katoh, and Y. Kambayashi, An Optimized Weighted Majority Decision, *Proc. of INFORMS-KORMS Seoul 2000*, pp. 1663-1669, 2000.
8. E. Ip, K. Yada, Y. Hamuro, and N. Katoh, A Data Mining System for Managing Customer Relationship, *Proc. of the 2000 Americas Conference on Information Systems*, pp. 101-105, 2000.
9. B. Kitts, D. Freed, and M. Vrieze, Cross-sell: A Fast Promotion-Tunable Customer-item Recommendation Method Based on Conditionally Independent Probabilities, *Proc. KDD 2000*, pp. 437-446, 2000.
10. A. Nakaya, H. Furukawa, and S. Morishita, Weighted Majority Decision among Several Region Rules for Scientific, *Proc. of Second International Conference on Discovery Science*, LNAI 1721, Springer-Verlag, pp. 17-29, 1999.
11. J. R. Quinlan, Induction of Decision Trees, *Machine Learning*, Vol.1, pp. 81-106, 1986.
12. J. R. Quinlan, *C4.5: Programs for Machine Learning*, Morgan Kaufman, 1993.
13. J. R. Quinlan, See5/C5.0, http:www.rulequest.com, Rulequest Research, 1999.
14. G. Piatetsky-Shapiro (Editor), *Knowledge Discovery in Databases*, AAAI Press, 1991.
15. S. Shimozono, A. Shinohara, T. Shinohara, S. Miyano, S. Kuhara and S. Arikawa, Knowledge Acquisition from Amino Acid Sequences by Machine Learning System BONSAI, *Trans. Information Processing Society of Japan*, Vol.35, pp. 2009-2018, 1994.
16. W. E. Spangler, J. H. May and L. G. Vargas, Choosing Data-mining Methods for Multiple Classification: Representational and Performance Measurement Implications for Decision Support, *Journal of Management Information System*, Vol.16 No.1, pp. 37-62, 1999.
17. T. K. Sung, H. M. Chung and P. Gray, Special Section: Data Mining, *Journal of Management Information System*, Vol.16 No.1, pp. 11-16, 1999.
18. R. Uthurusamy, U.M. Fayyad, and S. Spangler, Learning Useful Rules from Inconclusive Data, In [14], pp. 141-157, 1991.

Constructing Inductive Applications by Meta-Learning with Method Repositories

Hidenao Abe and Takahira Yamaguchi

School of Informatics, Shizuoka University
3-5-1 Johoku Hamamatsu Shizuoka, 432-8011 JAPAN
{cs6002,yamaguti}@cs.inf.shizuoka.ac.jp
Phone: +81-53-478-1473
Fax: +81-53-473-6421

Abstract. Here is presented CAMLET that is a platform for automatic composition of inductive applications with method repositories that organize many inductive learning methods. CAMLET starts with constructing a basic design specification for inductive applications with method repositories and data type hierarchy that are specific to inductive learning algorithms. After instantiating the basic design with a given data set into a detailed design specification and then compiling it into codes, CAMLET executes them on computers. CAMLET changes the constructed specification until it goes beyond the goal accuracy given from a user. After having implemented CAMLET on UNIX platforms with Perl and C languages, we have done the case studies of constructing inductive applications for eight different data sets from the StatLog project and have compared the accuracies of the inductive applications composed by CAMLET with all the accuracies from popular inductive learning algorithms. The results have shown us that the inductive applications composed by CAMLET take the first accuracy on the average.

1 Introduction

During the last twenty years, many inductive learning systems, such as ID3 [Qu1], Classifier Systems [BHG1] and data mining systems, have been developed, exploiting many inductive learning algorithms. As a result, end-users of inductive applications are faced with a major problem: model selection, i.e., selecting the best model to a given data set. Conventionally, this problem is resolved by trial-and-error or heuristics such as selection-table for ML algorithms. This solution sometimes takes much time. So automatic and systematic guidance for constructing inductive applications is really required.

From the above background, it is the time to decompose inductive learning algorithms and organize inductive learning methods (ILMs) for reconstructing inductive learning systems. Given such ILMs, we may construct a new inductive application that works well to a given data set by re-interconnecting ILMs. The issue is to meta-learn an inductive application that works well on a given data set. This paper focuses on specifying ILMs into a method repository and how

S. Arikawa and A. Shinohara (Eds.): Progress in Discovery Science 2001, LNAI 2281, pp. 576–585.

to compose inductive applications with ILMs. Thus we design a computer aided machine (inductive) learning environment called CAMLET and evaluate the competence of CAMLET using some data sets from the Statlog project[BH1], held by LIACC.

2 Repository for Inductive Learning

Considerable time and efforts have been devoted to analyzing the following popular inductive learning algorithms: Version Space [Mi1], AQ15, ID3 [Qu1], C4.5 [Qu2], Classifier Systems [BHG1], Back Propagation Neural Networks, Bagged C4.5 and Boosted C4.5 [Qu3]. The analysis results first came up with just unstructured documents to articulate what inductive learning methods are in the above popular inductive learning algorithms. Sometimes it was a hard issue to decide a proper grain size of inductive learning method. With this analysis, we did it under the condition of that the inputs and outputs of inductive learning methods are data sets or rule sets. When just a datum or rule is input or output of processes, they were too fine to be methods. Here in this paper, a method repository is an explicit specification of a conceptualization about inductive learning methods and a data type hierarchy is the organization of objects manipulated by inductive learning methods. In structuring many inductive learning methods into a method repository, we have identified the following generic methods: "generating training and test data sets", "generating a classifier set", "estimate data and classifier sets", "modifying a training data set" and "modifying a classifier set" that compose the top-level control structure of inductive applications, as shown in Figure 1.

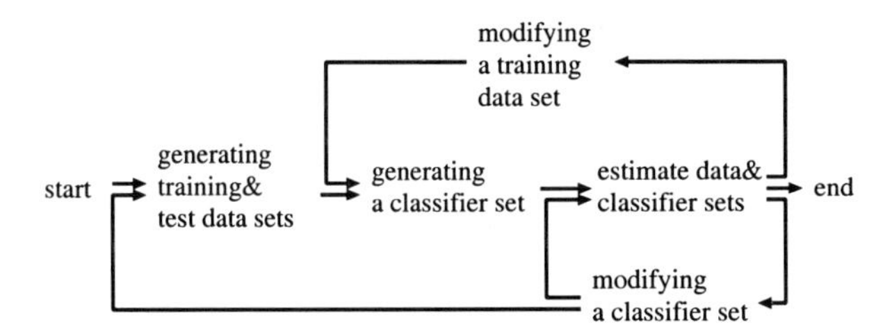

Fig. 1. Top-level Control Structure of Inductive Applications

Thus these five generic methods have been placed on the upper part in the hierarchy structure of the method repository, as shown in Figure 2.

2.1 Method Repository

In order to specify the hierarchy structure of a method repository, it is important how to branch down methods. Because the upper part is related with the generic methods included in the top-level control structure of inductive applications and the lower part with specific methods included in inductive applications, it is necessary to set up different ways to make up the hierarchy structure, depending on the hierarchy level.

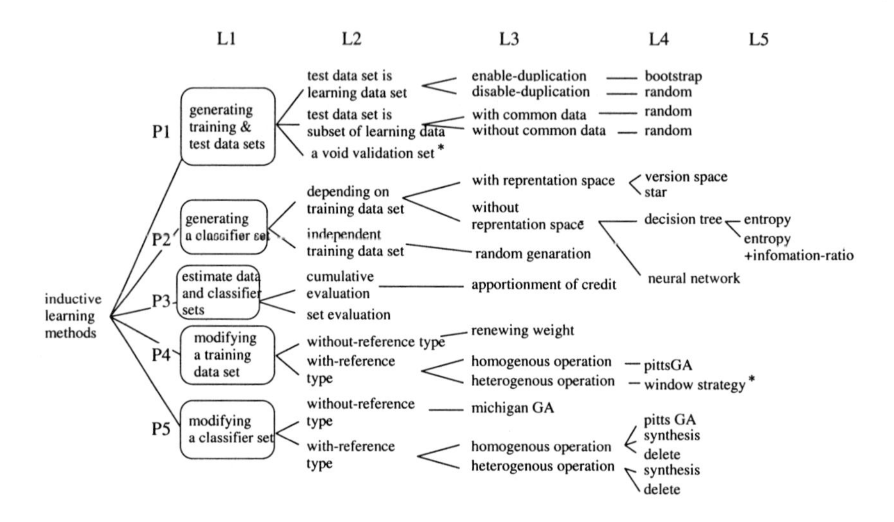

Fig. 2. Hierarchy of Method Repository

In specifying the lower part down from the upper part of the hierarchy, the generic methods have been divided down by the characteristics with each method. For example, the method of "generating a classifier set" has been divided into "generating a classifier set depending on training data sets" and "generating a classifier set independent of training datasets" from the point of the dependency on training data sets. Thus we can construct the hierarchy structure of the method repository, as shown in Figure 2. In Figure 2, leaf nodes get into the library of executable program codes that have been written in C language. For example, the method of "a void validation set" means that it does not distribute a data set into training and test data sets and that an inductive application uses a training data set instead of a test data set when it estimate a rule set.

On the other hand, in order to specify the method scheme, we have identified the method scheme including the following roles: "input", "output" and "reference" from the point of objects manipulated by the methods, and then "pre-method" just before the defined method and "post-method" just after the defined method.

2.2 Data Type Hierarchy

In order to specify the hierarchy structure of data type, we use the way to branch down the data structures manipulated by the methods, as shown in Figure 3. The leaf nodes with the data type hierarchy get into the following method scheme roles: input, output and reference.

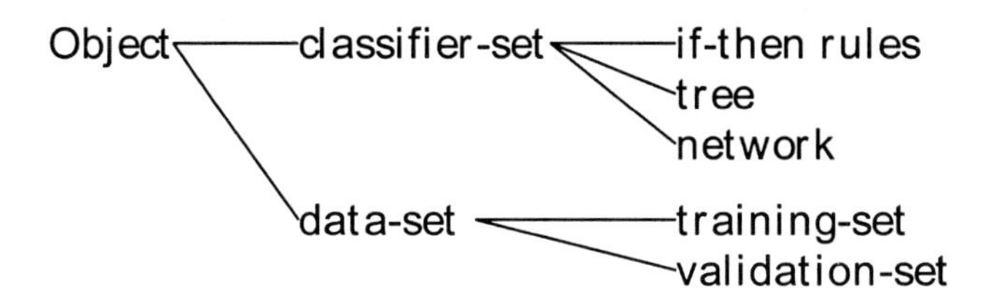

Fig. 3. Data Type Hierarchy

3 Basic Design of CAMLET

Figure 4 shows the basic activities for constructing knowledge systems with problem solving methods (PSMs) [vH1]. In this section, we apply the activities to constructing inductive applications with inductive learning methods.

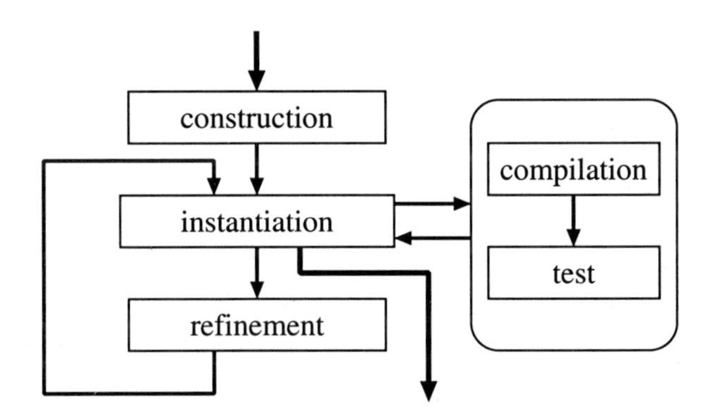

Fig. 4. Basic Activities for Constructing Inductive Applications

At the construction activity, CAMLET takes a top-level control structure randomly by selecting any path from “start” to “end” in Figure 1 and constructs an initial specification to a given data set. At the instantiation activity, CAMLET gets down from the generic methods included in the initial specification into the leaf-level methods, and instantiates the initial specification, getting the data types from a given data set into input and output roles of the leaf-level methods. The values of other roles, such as reference, pre-method and post-method, have not been instantiated but come directly from the method schemes. Thus CAMLET can make up the instantiated specification. At the compilation activity,

CAMLET transforms the instantiated specification into executable codes with the library for ILMs. When the method is connected to another method at the level of implementation details, the specification for I/O data types must be unified. To do so, the compilation activity has such a data conversion facility that converts a decision tree into a rule set. The test activity tests if the executable codes goes beyond the goal accuracy from a user. If not so, the refinement activity comes up, changing the initial specification into another one. Although we can implement the refinement activity in many ways, we just take random selection of methods here in this paper.

4 Case Studies Using Statlog Data Sets

We have implemented CAMLET with Perl language, including seventeen components in the method repository implemented with C language. Figure 5 shows a typical screen of CAMLET. We have done the case studies of constructing inductive applications with eight different data sets from the Statlog project[1]. We have compared the accuracies of the inductive applications composed by CAMLET with all the accuracies from popular inductive learning algorithms.

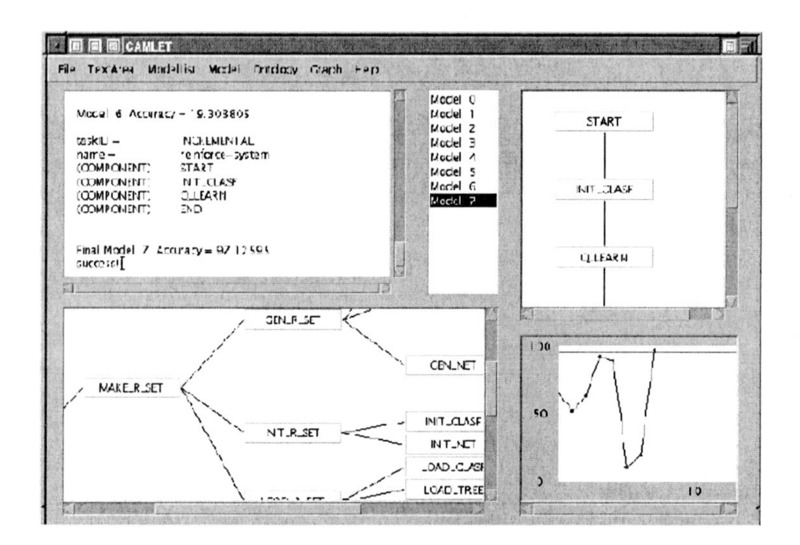

Fig. 5. CAMLET Browser

The goal of meta-learning by CAMLET is to compose inductive applications that go beyond the accuracy of the best learning algorithm in the Statlog

[1] The StatLog project has taken twenty four popular learning algorithms and statistical systems with ten common data sets. However, we have not taken two data sets because of having cost matrix evaluation. Refer to the URL of http://borba.ncc.up.pt/niaad/statlog/

project. However, taking time into consideration, in the case of that the CAMLET does not get to the goal after having composed inductive applications for fifty hours, CAMLET chooses the inductive application with the highest accuracy among all composed inductive applications. Furthermore, in the case of getting two or more inductive applications with the highest accuracy, CAMLET chooses one with the least learning cost. Table 1 shows how to use data sets in StatLog Project, evaluating ML applications. We take the same way as Table 1, evaluating CAMLET.

Table 1. Data Set Description in the StatLog Project

Data Set Name	Training Data Set	Test Data Set
Credit Research for Credit Cards in Australia	10-fold cross validation	
Diabetes of Pima-Indians	12-fold cross validation	
Splice-junction Recognition of DNA Sequence	assigned	assigned
Letter Recognition	assigned	assigned
LANDSAT Satellite Image Recognition	assigned	assigned
Image Segmentation	10-fold cross validation	
Shuttle Control	assigned	assigned
Vehicle Recognition Using Silhouettes	9-fold cross validation	

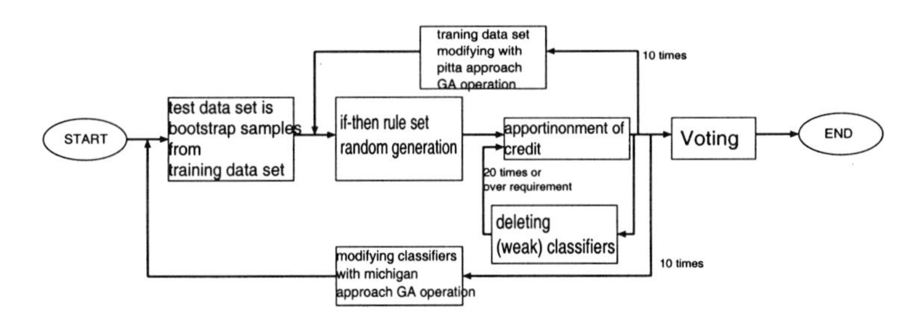

Fig. 6. The inductive application composed by CAMLET with Credit Research for Credit Cards in Australia

Table 2 shows all the experimental results about accuracy comparison between inductive applications composed by CAMLET and all learning algorithms in the StatLog Project. The inductive applications composed by CAMLET take the first best accuracy on the average, taking the first to tenth best accuracy to each data set. Figure 6 to 13 show all specifications of inductive applications composed by CAMLET. Six inductive applications include the voting method with boosting introduced in [Qu3]. So multiple classifiers turn out to work better

Table 2. Accuracy Comparison between Inductive Applications Composed by CAMLET and All Learning Algorithms in the StatLog Project

Ranking	Algorithm	australian	diabetes	dna	letter	satimage	segment	shuttle	vehicle	Average
1	**CAMLET**	**86.7%(2)**	**75.7%(5)**	**93.1%(8)**	**86.5%(10)**	**88.4%(4)**	**96.7%(3)**	**99.8%(1)**	**80.4%(5)**	**88.4%**
2	BayTree	82.9%	72.9%	90.5%	87.6%	85.3%	96.7%	98.0%	72.9%	85.9%
3	C4.5	84.5%	73.0%	92.4%	86.8%	85.0%	96.0%	90.0%	73.4%	85.1%
4	Newld	81.9%	71.1%	90.0%	87.2%	85.0%	96.6%	99.0%	70.2%	85.1%
5	IndCart	84.8%	72.9%	92.7%	87.0%	86.2%	95.5%	91.0%	70.2%	85.0%
6	Cn2	79.6%	71.1%	90.5%	88.5%	85.0%	95.7%	97.0%	68.6%	84.5%
7	Cal5	86.9%	75.0%	86.9%	74.7%	84.9%	93.8%	97.0%	72.1%	83.9%
8	Dipol92	85.9%	77.6%	95.2%	82.4%	88.9%	96.1%	52.0%	84.9%	82.9%
9	KNN	81.9%	67.6%	85.4%	93.2%	90.6%	92.3%	56.0%	72.5%	79.9%
10	Ac2	81.9%	72.4%	90.0%	75.5%	84.3%	96.9%	68.0%	70.4%	79.9%
11	BackProp	84.6%	75.2%	91.2%	67.3%	86.1%	94.6%	57.0%	79.3%	79.4%
12	Alloc80	79.9%	69.9%	94.3%	93.6%	86.8%	97.0%	17.0%	82.7%	77.7%
13	Smart	84.2%	76.8%	88.5%	70.5%	84.1%	94.8%	41.0%	78.3%	77.3%
14	Cart	85.5%	74.5%	91.5%	0.0%	86.2%	96.0%	92.0%	76.5%	75.3%
15	QuaDisc	79.3%	73.8%	94.1%	88.7%	84.5%	84.3%	0.0%	85.0%	73.7%
16	LogDisc	85.9%	77.7%	93.9%	70.0%	83.7%	89.1%	0.0%	80.8%	73.6%
17	Radial	85.5%	75.7%	95.9%	76.7%	87.9%	93.1%	0.0%	69.3%	73.0%
18	Discrim	85.9%	77.5%	94.1%	69.8%	82.9%	88.4%	0.0%	78.4%	72.1%
19	LVQ	80.3%	72.8%	0.0%	92.1%	89.5%	95.4%	56.0%	71.3%	69.7%
20	Castle	85.2%	74.2%	92.8%	75.5%	80.6%	88.8%	0.0%	49.5%	68.3%
21	Bayes	84.9%	73.8%	93.2%	47.1%	71.3%	73.5%	0.0%	44.2%	61.0%
22	Itrule	86.3%	75.5%	86.5%	40.6%	0.0%	54.5%	59.0%	67.6%	58.8%
23	Kohonen	0.0%	72.7%	66.1%	74.8%	82.1%	93.3%	0.0%	66.0%	56.9%
24	Default	56.0%	65.0%	50.8%	4.0%	23.1%	24.0%	0.0%	25.0%	31.0%
25	Cascade	0.0%	0.0%	0.0%	0.0%	83.7%	0.0%	0.0%	72.0%	19.5%

Data Sets
- australian(Credit Research for Credit Cards in Australia)
- diabetges(Diabetes of Pima-Indians)
- dna(Splice-junction Recognition of DNA Sequence)
- letter(Letter Recognition)
- satimage(LANDSAT Satellite Image Segmentation)
- segment(Image Segmentation)
- shuttle(Shuttle Control)
- vehicle(Vehicle Recognition with Silhouettes)

Popular Learning Algorithms
- DecisionTree: Ac2,Cal5,Cart,C4.5,Cn2,Itrule and NewID
- Neural Network Classifiers: BackProp,Cascade,Dipol92,LVQ,Radial
- Statical Classifiers: Alloc80,BayTree,Castle,IndCaart,KNN,Discrim, LogDisc,QualDisc,Bayes and Smart

(#) means how good is the accuracy of the inductive appliaction composed by CAMLET

than some single classifier. Looking at the control sub-structure of "generating a classifier set", the method of"decision tree with entropy" has been taken in six inductive applications. Looking at the feedback structure, the composed inductive applications have two or more kinds of feedback loop, keeping the variety of data sets and/or classifier sets generated in meta-learning by CAMLET. Although CAMLET had no change of many parameters with learning methods, it is necessary to change parameters as well as inductive learning methods.

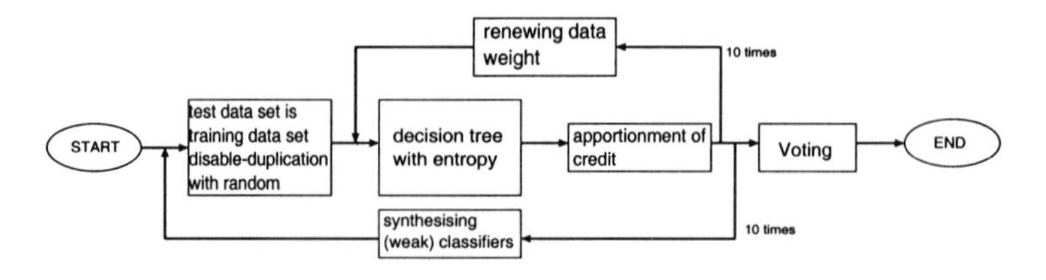

Fig. 7. The inductive application composed by CAMLET with Diabetes of Pima-Indians

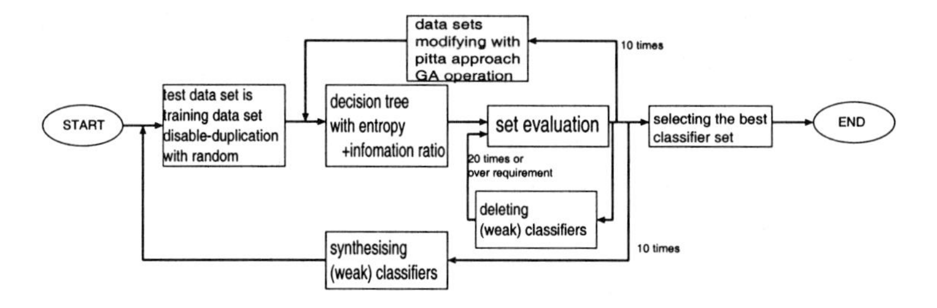

Fig. 8. The inductive application composed by CAMLET with Splice-junction Recognition of DNA Sequence

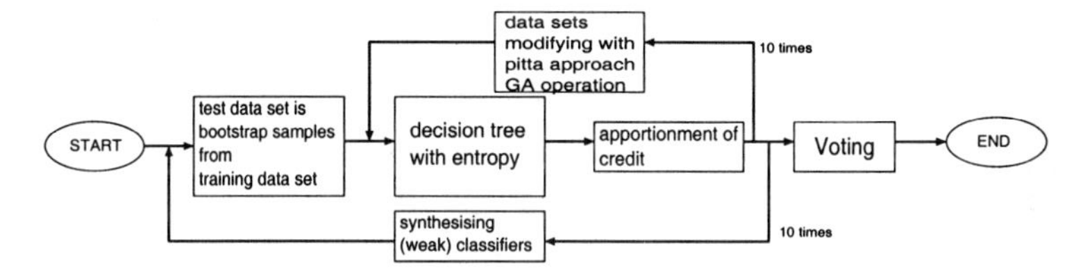

Fig. 9. The inductive application composed by CAMLET with Letter Recognition

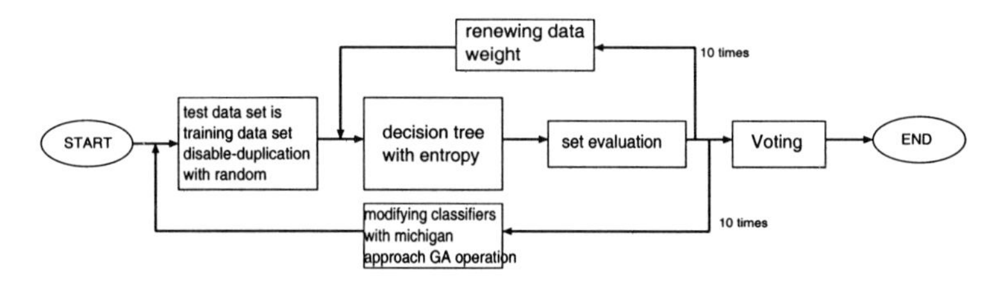

Fig. 10. The inductive application composed by CAMLET with LANDSAT Satellite Image Segmentation

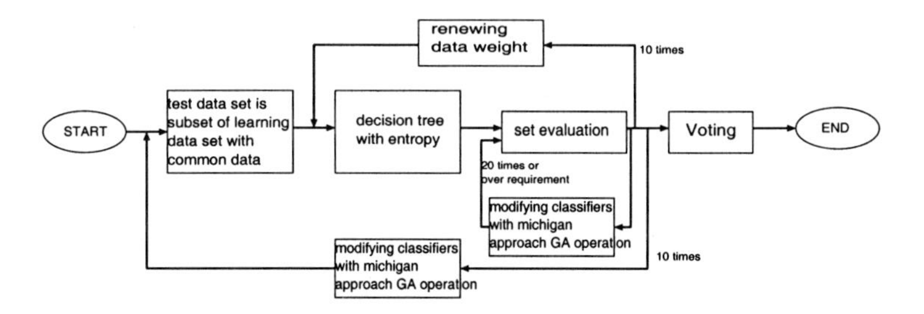

Fig. 11. The inductive application composed by CAMLET with Image Segmentation

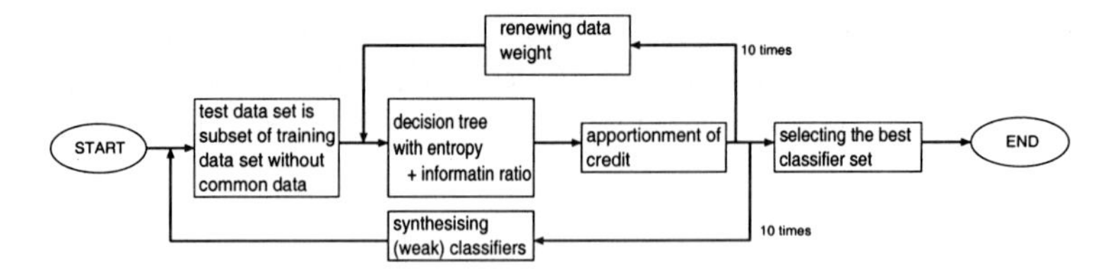

Fig. 12. The inductive application composed by CAMLET with Shuttle Control

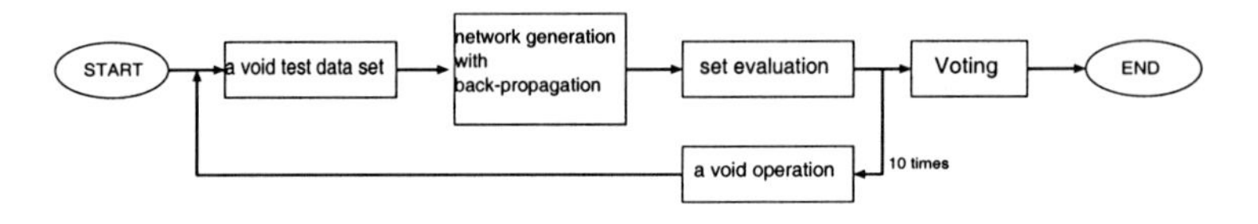

Fig. 13. The inductive application composed by CAMLET with Vehicle Recognition Using Silhouettes

5 Related Work

CAMLET has some similarities to MSL (Multi Strategy Learning). MSL comes from unifying two or more ML systems [MO1]. However, MSL does not decompose ML systems into inductive learning methods. In other words, MSL has no change of the grain size of learning algorithms. So the grain size of the components in CAMLET is much finer than that of those in MSL. Furthermore, MSL cannot invent a new ML application automatically like CAMLET.

MLC++ [KS1] is a platform for constructing inductive applications. However, MLC++ has no facility for automatic composition of inductive applications like CAMLET.

MEDIA-model [En1] is a reference structure for the application of inductive learning techniques. It focuses on methodological aspects of inductive applications but not on automatic composition facilities of inductive applications like CAMLET.

6 Conclusions and Future Work

Taking eight different data sets from the StatLog project, we have done the case studies of constructing inductive applications by CAMLET. The results have shown us that the inductive applications composed by CAMLET take the best average accuracy among all the popular inductive learning algorithms in the StatLog project. However, CAMLET takes just a random selection in the refinement activity and so we must make it more intelligent and efficient. Furthermore, CAMLET has no change of the parameters with inductive learning methods. However, they sometimes vary the accuracy with the inductive learning methods. We also need to adjust them, depending on a given data set. Furthermore, we are trying to put other measurements but precision into CAMLET in order to support domain experts in discovering knowledge interesting to them. The measurement of specify works to do so to some extent, as shown in [HAKYT1]. We will extend the measurements to get more useful interaction with them.

References

[BH1] Brazdil,P. and Henery,R.: "Chapter 10, Analysis of Results", in Machine Learning, Neural and Statistical Classification , D. Michie, D.J.Spiegelhalter and C.C.Taylor (eds.), Ellis Horwood, (1994) pp.175–212

[BHG1] Booker,L.B., Holland,J.H., Goldberg,D.E.: Classifier Systems and Genetic Algorithms. Artificial Intelligence. **40** (1989) pp.235–282

[En1] Engels,R.: Planning in Knowledge Discovery in Databases; Performing Task-Oriented User-Guidance. Angewandte Informatik und Formale Beschreibungsverfahren. (1996)

[HAKYT1] Hatazawa,H.,Abe,H.,Komori,M.,Yamaguchi,T.,Tatibana,Y.: Knowledge Discovery Support from a Meningoencephalitis Dataset using an Automatic Composition Tool for Inductive Applications. JSAI KDD Challenge2001,JKDD01. (2001) pp.9–16

[KS1] Kohavi,R. and Sommerfield,D.: Data Mining using MLC++ - A Machine Learning Library in C++. 8th International Conference on Tools with Artificial Intelligence. (1996) 234–245

[Mi1] Mitchell,T.M.: Generalization as Search. Artificial Intelligence. **18**(2) (1982) pp.203–226

[MO1] Mooney,R.J. and Ourston,D.: A Multistrategy Approach to Theory Refinement. Machine Learning **4**. Morgan Kaufmann (1994) pp.141–164

[Qu1] Quinlan,J.R.: Induction of Decision Tree. Machine Learning. **1** (1986) pp.81–106

[Qu2] Quinlan,J.R.: Programs for Machine Learning. Morgan Kaufmann. (1992)

[Qu3] Quinlan,J.R.: Bagging, Boosting and C4.5. American Association for Artificial Intelligence. (1996)

[vH1] van Heijst,G.: The Role of Ontologies in Knowledge Engineering. phD thesis. University of Amsterdam (1995)

Knowledge Discovery from Semistructured Texts

Hiroshi Sakamoto, Hiroki Arimura, and Setsuo Arikawa

Department of Informatics, Kyushu University
Hakozaki 6-10-1, Higashi-ku, Fukuoka-shi 812-8581, Japan
{hiroshi, arim, arikawa}@i.kyushu-u.ac.jp

Abstract. This paper surveys our recent results on the knowledge discovery from semistructured texts, which contain heterogeneous structures represented by labeled trees. The aim of our study is to extract useful information from documents on the Web. First, we present the theoretical results on learning rewriting rules between labeled trees. Second, we apply our method to the learning HTML trees in the framework of the wrapper induction. We also examine our algorithms for real world HTML documents and present the results.

1 Introduction

The present paper summarizes our study on the information extraction from semistructured texts. The HTML documents distributed on the network can be regarded as a very large text database consisting of structured texts by many tags that have their own special meanings beforehand. Although computers cannot understand what the tags mean, they can render the structures according to the tags. Recently an XML have been recommended by the World Wide Web Consortium (W3C), which are expected to realize more intelligent information exchange. We can define original tags for XML documents and we can construct rich logical structures using such tags. These structures are useful for anonymous users to extract information from such documents.

We begin with the investigation of the data exchange model for semistructured texts like markup texts. The aim of this study is to construct a framework of useful rewriting for tree-like structures. A markup text is expressed by a rooted ordered tree. The root is the unique node which denotes the whole document, the other internal nodes are labeled by tags, and the leaves are labeled by the contents of the document or the attributes of the tags. The extraction from semistructured documents is considered as the problem to construct a small tree from input trees. Thus, we consider the translation problem between input and output trees. For this problem, we produce some classes of appropriate translations and analyse their learning complexity.

For example, an XML document is translated to an HTML for the use of browsing. This document may be translated to another XML document in different format in data exchange. Then, given examples of pairs of input and output trees for XML documents, the learning problem is to find the function

S. Arikawa and A. Shinohara (Eds.): Progress in Discovery Science 2001, LNAI 2281, pp. 586–599.

which translates the input trees to the output trees. These translations are described by the language XSLT, which is also recommended by W3C in 1999. XSLT is very powerful because regular expression and recursion are allowed in XSLT, and thus, it seems to be hard to learn XSLT from given examples alone. Thus, we introduce more restricted rewriting classes for tree translations.

We introduce two types of data exchange models in this paper. One is called the *extraction* and the other is called the *reconstruction*. The extraction is a very simple translation $T \to t$ such that a small tree t is obtained by only (1) renaming labels of T or (2) deleting nodes of T. This model is suitable for taking out specific entries from a very large table as a small table without changing the structure of the document.

On the other hand, the reconstruction is more complicated. It is characterized by a *term rewriting rule* $f \to g$ for terms f and g with variables. This model can translate trees more powerful than the previous one so that we can exchange any two subtrees of an input tree and renaming labels depending on ancestors or descendants of the current node. This model suitable for exchanging from a title to an author in digital books cards, for example. This translation cannot be defined by the previous rewriting model because the order of any two node must be preserved.

The rewriting problem under the extraction is to determine whether or not there exists a rule which maps input trees to output ones. The complexity of this problem is NP-hard even if the rewriting rules are strongly restricted. The rewriting problem under the reconstruction is clearly more difficult. Thus, we assume an additional information for this problem. We consider the learning problem such that an algorithm can use the *membership query* and the *equivalence query* [2]. We show that the rewriting class introduced is learnable in polynomial time using these two queries.

Next we apply the obtained learning theory to the real world data, like the HTML documents. The information extraction from documents on the Web have been widely studied in the last few years [5, 6, 10, 12, 13, 16, 17]. In case of the Web data, this problem is particularly difficult because we cannot represent a rich logical structure by the limited tags of the HTML. The framework of *wrapper induction* by Kushmerick [16] is a new approach to handle this difficulty, which is a natural extension of the *PAC-learning* [21]. We showed the effectiveness and efficiency of simple wrappers with string delimiters in the information extraction tasks.

In the wrapper induction, an HTML document is called a *page* and the contents of the page is called the *label*. The goal of the learning algorithm is, a given sequence of examples $\langle P_n, L_n \rangle$ of pages and labels, to output the program W such that $L_n = W(P_n)$ for all n. The program W is called a *Wrapper* (cf., [10, 12, 13, 17]).

In this model, we assume a special structure in the pages as follows. Every text containing in a page belongs to a class and the name of the class is called the *attribute*. Then, the label L of a page P is a set of $t_i = \langle ta_1, \ldots, ta_K \rangle$, where ta_j is a set of texts contained by P. We call t_i the i-th attribute. The number

K is a constant depending on the target. The aim of wrapper algorithm is to extract all texts from input page and classifies them into the correct attributes. For example, the strings beginning with "`mailto:`" must be the email attribute.

We propose a new wrapper class called a *Tee-Wrapper* over the tree structures and present the learning algorithm of the Tree-Wrappers. This is an extension of Kushmerick's LR-Wrapper [16]. The aim of the learning algorithm is to find a small tree which is a generalization of input trees.

For each node of an HTML tree, we define the *node label* consisting of the *node name*, the *position number*, and the set of *HTML attributes*. Then, the Tree-Wrapper W is the sequence $\langle EP_1, \ldots, EP_K \rangle$. The EP_i, called the *extraction path*, is of the form $\langle ENL_{i_1}, \ldots, ENL_{i_\ell} \rangle$, where ENL_{i_1} is called the *extraction node label* which is a general expression of node label by using the wildcard $*$ matching any string.

For a given tree P_t of an HTML document P and a Tree-Wrapper W, the semantics for extraction is as follows. Let $EP_i = \langle ENL_{i_1}, \ldots, ENL_{i_\ell} \rangle$ and p be a path in P_t of length ℓ. We call that EP_i *matches with* p if the node label of j-th node of p matches with ENL_{i_j} by a substitution for all $*$ of ENL_{i_j}. If EP_i matches with p, then the attribute of the last node is extracted. These values are considered to be the members of i-th attribute in the page.

We experiment the prototype of our learning algorithm for more than 1,000 pages of HTML documents and present the performance of our algorithm. Moreover, we compare the efficiency of our model and Kushmerick's Wrapper models for sufficiently large data.

This paper is organized as follows. In Section 2, the complexity of announced decision problem is considered. We obtain the NP-completeness of this problem with respect to the restrictions either given trees are strings or output tree is labeled by a single alphabet. Moreover we show that a nontrivial subproblem is decidable in polynomial-time.

In section 2, we also consider the learning problem of linear translation system by query learning model. We present a learning algorithm based on the theory of [3, 4] and we show that our algorithm identifies each target using at most $O(m)$ equivalence queries and at most $O(kn^{2k})$ membership queries, where m is the number of rules of the target and n is the number of nodes of counterexamples.

In Section 3, we summarize the results on the Tree-Wrapper. First, we define the HTML trees by ordered labeled trees. Second we give the syntactic definition of the Tree-Wrapper and the semantics of the extraction. Third, we describe the learning algorithm for Tree-Wrapper. Finally, we explain the examinations of our algorithm for some popular Internet sites.

In Section 4, we conclude this study and mention the further work.

2 Tree Translation

In this section, we explain the two models for tree rewriting. First model is called the *extraction* such that a target tree is obtained from a tree by erasing nodes. The complexity of several decision problems are presented. Second one is called the *reconstruction* which is a kind of term rewriting systems. We introduce the class of k-variable linear translation and show the query learnability of this class.

2.1 Tree Rewriting by Erasing

We adopt the following standard definition of the ordered trees. A tree is a connected, acyclic, directed graph. A *rooted tree* is a tree in which one of the vertices is distinguished from the others and is called the root. We refer to a vertex of a rooted tree as a *node* of the tree. An *ordered tree* is a rooted tree in which the children of each node are ordered. That is, if a node has k children, then we can designate them as the first child, the second child, and so on up to the k-th child.

Let λ denote the unique null symbol not in Σ. We define two operations on trees, *renaming* and *deleting*. The renaming denoted by $\mathsf{a} \mapsto \mathsf{b}$ replaces the label a in T by b. On the other hand, the deleting denoted by $\mathsf{a} \mapsto \lambda$ removes any node n for $\ell(n) = \mathsf{a}$ in T and makes the children of n the children of the parent of n.

Let $S = \{\mathsf{a} \to \mathsf{b} \mid \mathsf{a} \in \Sigma, \mathsf{b} \in \Sigma \cup \{\lambda\}\}$ be a set of operations. Then, we write $T \to_S T'$ iff T' is obtained by applying all operations in S to T simultaneously.

Definition 1. *Let (T, P) be a pair of trees over Σ. Then, the problem of erasing homomorphism is to decode whether there exists a set S of operations such that $T \to_S P$. An input tree T and P are called a target and a pattern, respectively. This problem is denoted by $EHP(T, P)$. The problem of erasing isomorphism, denoted by $EIP(T, P)$ is to decide whether $T \to_S P$ such that if $\mathsf{a} \to \mathsf{c}, \mathsf{b} \to \mathsf{c} \in S$, then either $\mathsf{a} = \mathsf{b}$ or $\mathsf{c} = \lambda$, that is, any two different symbols are never renamed to a same symbol.*

When we restrict that any two nodes of a pattern tree P are labeled by distinct symbols, this problem is the special case of $EIP(T, P)$. Moreover, this problem is equivalent to the tree inclusion problem [15] which is decidable in $O(|T| \cdot |P|)$ time.

The problem $EHP(T, P)_k$ is a restriction of $EHP(T, P)$ such that the depth of the tree T is at most k. The problem $EIP(T, P)_k$ is defined similarly. First we obtain the complexity of $EIP(T, P)$ and show that a subclass is in P. Next we prove the NP-hardness of more general problem $EHP(T, P)$. Recall that $EIP(T, P)_1$ is the problem that T and P are both strings. The following result tells us that $EIP(T, P) \in$ P iff $EIP(T, P)_1 \in$ P.

Theorem 1 ([18]). *$EIP(T, P)$ is polynomial-time reducible to $EIP(T, P)_1$.*

By Theorem 1, it is sufficient to consider only the problem $EIP(T, P)_1$. In the following parts, we write $EIP(T, P)$ instead of $EIP(T, P)_1$. Using this result, we derive the result that there is a subclass of $EIP(T, P)$ to be in P.

Let w, α, β be strings. If $w = \alpha\beta$, then the length of α is called an *occurrence of β on w*. For substrings α and β of w, we call that there exists an *overlap* of α and β on w if there exist occurrences i and j of α and β on w such that

$i < j < |\alpha| + i - 1$ or $j < i < |\beta| + j - 1$. If a string w contains a substring of the form $\mathtt{A}\alpha\mathtt{A}$ for some $\mathtt{A} \in \Sigma$ and $\mathtt{A}$ does not occur in α, then we call the substring an *interval* of $\mathtt{A}$ on w. A string w is said to contains an overlap of k intervals if w contains k intervals of $\mathtt{A}_1, \ldots, \mathtt{A}_k$ such that there exists an overlap of any intervals A_i and A_j for $1 \leq i \neq j \leq k$. A string $w \in \Sigma^*$ is called k-interval free if w contains an overlap of at most $(k-1)$ intervals. A string $w \in \Sigma^*$ is said to have a *split* if an i-th symbol of w does not contained in any interval. We denote the problem $EIP(T, P)$ such that both T and P are k-interval free by $EIP(T, P)^k$. For this problem, we obtain the following positive result.

Example 1. The string ABBCA is an interval of A but ABACA is not. The string ABCADB has no split because each symbol is contained in an interval of A or B. On the other hand, ACADBBB has a split. We demonstrate examples of an overlap of 3-intervals and a 3-interval free string as Fig. 1.

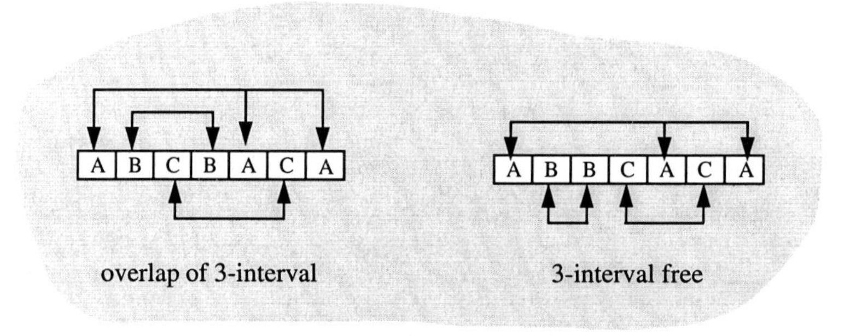

Fig. 1. Intervals in a string

Theorem 2 ([18]). $EIP(T, P)^3 \in \mathrm{P}$.

However, we obtain the following negative result for the general problem $EHP(T, P)$.

Theorem 3 ([18]). *$EHP(T, P)$ is NP-complete even if (1) P is labeled by a single alphabet, or (2) T is a string.*

2.2 Tree Translation Systems

In this subsection, we consider a class defined by term rewriting systems. Using this model, we can do more powerful translation like exchanging any two subtrees. For the term rewriting, we introduce the class of *ranked trees* whose node label is defined by the following ranked alphabet. The out-degree of any node is bounded by the rank of its node label. Let $\Sigma = \cup_{n \geq 0} \Sigma_n$ be a finite *ranked alphabet* of *function symbols*, where for each $f \in \Sigma$, a nonnegative integer $arity(f) \geq 0$, called *arity*, is associated. We assume that Σ contains at least one symbol of arity zero. Let X be a countable set of *variables* disjoint with Σ, where we assume that each $x \in X$ has arity zero.

In this model we do not allow any operations such as deletion and insertion because it may change the out-degree of a node. We denote by $\mathcal{T}(\Sigma, X)$ the set of labeled, rooted and ordered trees t such that

- Each node v of t is labeled with a symbol in $\Sigma \cup X$, denoted by $t(v)$.
- If $t(v)$ is a function symbol $f \in \Sigma$ of arity $k \geq 0$ then v has exactly k children.
- If $t(v)$ is a variable $x \in X$ then v is a leaf.

We call each element $t \in \mathcal{T}(\Sigma, X)$ a *pattern tree* (*pattern* for short). A pattern is also called a first-order term in formal logic. We use $\mathcal{T}$ in place of $\mathcal{T}(\Sigma, X)$ if they are clearly understood from context. For pattern t, we denote the set of variables appearing in t by $var(t) \subseteq X$ and denote the number of the nodes of t by $size(t)$. A pattern t is said to be *ground* if it contains no variables.

Definition 2. *A* tree translation rule *(*rule *for short) is an ordered pair* $(p, q) \in \mathcal{T} \times \mathcal{T}$ *such that* $var(p) \supseteq var(q)$. *A* tree translation system *(TT) is a set* H *of translation rules.*

A pattern t is called *linear* if any variable $x \in X$ appears in t at most once. A pattern t is *of k-variable* if $var(t) = \{x_1, \ldots, x_k\}$. For $k \geq 0$, we use the notation $t[x_1, \ldots, x_k]$ to indicate that a pattern is k-variable and linear with mutually distinct variables $x_1, \ldots, x_k \in X$, where the order of variables in t is arbitrary. For a k-variable linear pattern $t[x_1, \ldots, x_k]$ and a sequence of patterns $s_1, \ldots, s_k$, we define $t[s_1, \ldots, s_k]$ as the term obtained from t by replacing the occurrence of x_i with patterns s_i for every $1 \leq i \leq k$.

Definition 3. *A translation rule* $C = (p, q)$ *is* of k-variable *if* $card(var(C)) \leq k$, *and* linear *if both of* p *and* q *are linear.*

For every $k \geq 0$, we denote by $LR(k)$ and $LTT(k)$ the classes of all k-variable linear translation rules, and all k-variable linear tree translation systems, respectively. We also denote by $LTT = \cup_{k \geq 0} LTT(k)$ all linear tree translation systems.

Definition 4. *Let* $H \in LTT$ *be a linear translation system. The* translation relation $M(H) \subseteq \mathcal{T} \times \mathcal{T}$ *is defined recursively as follows.*

- Identity*: For every pattern* $p \in \mathcal{T}$, $(p, p) \in M(H)$.
- Congruence*: If* $f \in \Sigma$ *is a function symbol of arity* $\ell \geq 0$ *and for every* i, $(p_i, q_i) \in M(H)$, *then* $(f(p_1, \ldots, p_\ell), f(q_1, \ldots, q_\ell)) \in M(H)$.
- Application*: If* $(p[x_1, \ldots, x_k], q[x_1, \ldots, x_k]) \in H$ *is a* k*-variable linear rule and for every* i, $(p_i, q_i) \in M(H)$, *then* $(p[p_1, \ldots, p_k], q[q_1, \ldots, q_k]) \in M(H)$, *where* p *and* q *are* k*-variable linear terms.*

If $C \in M(H)$ then we say that rule C is *derived by* H. The definition of the meaning $M(H)$ above corresponds to the computation of top-down tree transducer [8] or the a special case of term rewriting relation [7] where only top-down rewriting are allowed.

We show that there exists a polynomial time algorithm that exactly identifies any translation system in $LTT(k)$ using equivalence and membership queries. Our problem is identifying an unknown tree translation system H_* from examples of ordered pairs $E \in M(H_*)$ that are either derived or not derived by H_*. As a formal model, we employ a variant of exact learning model by Angluin [2] called learning from entailment[3, 4, 9, 14], which is tailored for translation systems.

Let $\mathcal{H}$ be a class of translation systems to be learned, called *hypothesis space*, and LR be the set of all ordered pairs, called the *domain of learning*. In our learning framework, the *meaning* or *the concept* represented by $H \in \mathcal{H}$ is the set $M(H)$. If $M(P) = M(Q)$ then we define $P \equiv Q$ and say that P and Q are *equivalent*.

A *learning algorithm* $\mathcal{A}$ is an algorithm that can collect the information about H_* using the following type of queries. In this paper, we assume that the alphabet Σ is given to $\mathcal{A}$ in advance and the maximum arity of symbols in Σ is bounded by some constant.

Definition 5. *An* equivalence query *(EQ) is to propose any translation system $H \in \mathcal{H}$. If $H \equiv H_*$ then the answer to the query is "yes". Otherwise the answer is "no", and $\mathcal{A}$ receives any translation $C \in LR$ as a* counterexample *such that either $C \in M(H_*)\backslash M(H)$, or $C \in M(H)\backslash M(H_*)$. A counterexample is* positive *if $C \in M(H_*)$ and* negative *if $C \notin M(H_*)$. A* membership query *(MQ) is to propose any translation $C \in LR$. The answer to the membership query is "yes" if $C \in M(H_*)$, and "no" otherwise.*

The goal of $\mathcal{A}$ is *exact identification* in polynomial time. $\mathcal{A}$ must halt and output a rewriting system $H \in \mathcal{H}$ such that $H_* \equiv H$, where at any stage in learning, the running time and thus the number of queries must be bounded by a polynomial $poly(m, n)$ in the size m of H_* and the size n of the longest counterexample returned by equivalence queries so far.

Although this setting seems to be unnatural at a glance, it is known that any exact learnability with equivalence queries implies *polynomial-time PAC-learnability* [21] and *polynomial-time online learnability* [2] under a mild condition that additional membership queries are allowed or not on a target hypothesis [2].

Theorem 4 ([18]). *There exists an algorithm which exactly identifies any translation system H_* in $LTT(k)$ using $O(m)$ equivalence queries and $O(kn^{2k})$ membership queries.*

3 Wrapper Induction

In this section, we give the definition of the HTML tree, the learning algorithm for the Tree-Wrapper, and the experimental result for the real world data.

3.1 Data Model

For each tree T, the set of all nodes of T is a subset of $I\!N = \{0, \ldots, n\}$ of natural numbers, where the 0 is the root. A node is called a *leaf* if it has no child and called an *internal node* otherwise. If $n, m \in I\!N$ has the same parent, then n and m are *sibling* and n is a *left sibling* of m if $n \le m$. The sequence $\langle n_1, \ldots, n_k \rangle$ of nodes of T is called the path if n_1 is the root and n_i is the parent of n_{i+1} for all $i = 1, \ldots, k-1$.

For a node n, the *node label* of n is the triple $NL(n) = \langle N(n), V(n), HAS(n) \rangle$ such that $N(n)$ and $V(n)$are strings called the *node name* and *node value*, respectively, and $HAS(n) = \{HA_1, \ldots, HA_{n_\ell}\}$ is called the set of the *HTML attributes* of n, where each HA_i is of the form $\langle a_i, v_i \rangle$ and a_i, v_i are strings called *HTML attribute name*, *HTML attribute value*, respectively.

If $N(n) \in \Sigma^+$ and $V(n) = \varepsilon$, then the n is called the *element node* and the string $N(n)$ is called the *tag*. If $N(n) = \sharp TEXT$ for the reserved string $\sharp TEXT$ and $V(n) \in \Sigma^+$, then n is called the *text node* and the $V(n)$ called the *text value*. We assume that every node $n \in I\!N$ is categorized to the element node or text node.

An HTML document is called a page. A page P is corresponding to an ordered labeled tree. For the simplicity, we assume that the P contains no comment part, that is, any string beginning the <! and ending the > is removed.

Definition 6. *For a page P, the P_t is the ordered labeled tree defined recursively as follows.*

1. *Each empty tag* <tag> *in P corresponds to a leaf n in P_t such that $NL(n) = \langle N(n), V(n), HAS(n) \rangle$, $N(n) = tag$, $V(n) = \varepsilon$, and $HAS(n) = \emptyset$.*
2. *Each string w in P containing no tag corresponds to a leaf n such that $N(n) = \sharp TEXT$, $V(n) = w$, and $HAS(n) = \emptyset$.*
3. *Each string of the form* <tag $a_1 = v_1, \ldots, a_\ell = v_\ell$>$w$</tag> *corresponds to a subtree $t = n(n_1, \ldots, n_k)$ such that $N(n) = tag$, $V(n) = \varepsilon$, and $HAS(n) = \{\langle a_1, v_1 \rangle, \ldots, \langle a_\ell, v_\ell \rangle\}$, where $n_1, \ldots, n_k$ are the roots of the trees $t_1, \ldots, t_k$ obtained recursively from the w by the 1, 2 and 3.*

What the HTML Wrapper of this paper extracts is the text values of text nodes. These text nodes are called *text attributes*. A sequence of text attributes is called *tuple*. We assume that the contents of a page P is a set of tuple $t_i = \langle ta_{i_1}, \ldots, ta_{i_K} \rangle$, where the K is a constant for all pages P. It means that all text attributes in any page is categorized into at most K types. Let us consider the example of an address list. This list contains three types of attributes, name, address, and phone number. Thus, a tuple is of the form $\langle name, address, phone \rangle$. However, this tuple can not handle the case that some elements contain more than two values such as some one has two phone numbers. Thus, we expand the notion of tuple to a sequence of a set of text attributes, that is $t = \langle ta_1, \ldots, ta_K \rangle$ and $ta_i \subseteq I\!N$ for all $1 \le i \le K$. The set of tuples of a page P is called the *label* of P.

Example 2. The Fig.1 denotes the tree containing the text attributes *name*, *address*, and *phone*. The first tuple is $t_1 = \langle\{3\},\{4\},\{5,6\}\rangle$ and the second tuple is $t_2 = \langle\{8\},\{\},\{9\}\rangle$. The third attribute of t_1 contains two values and the second attribute of t_2 contains no values.

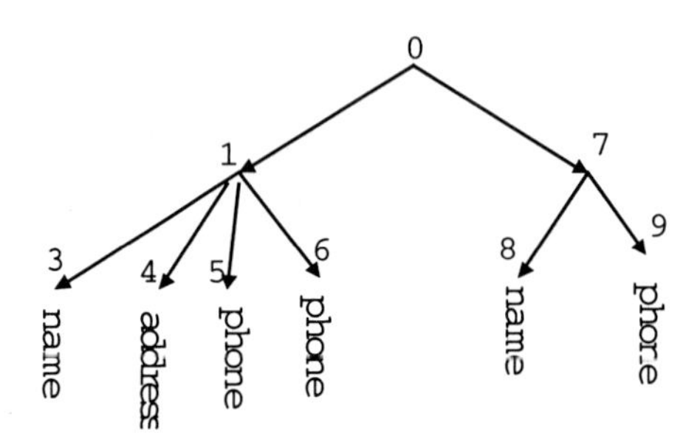

Fig. 2. The tree of the text attributes, *name*, *address*, and *phone*.

3.2 Tree-Wrapper

Next we explain the wrapper algorithm and the learning algorithm. The wrapper algorithm extracts the attributes from the page P_t using a Tree-Wrapper W. On the other hand, the learning algorithm finds the Tree-Wrapper W for the sequence $E = \ldots, \langle P_n, L_n\rangle, \ldots$ of examples, where L_n is the label of the page P_n.

Definition 7. *The extraction node label is a triple* $ENL = \langle N, Pos, HAS\rangle$, *where* N *is a node name,* $Pos \in I\!N \cup \{*\}$, HAS *is an HTML attribute set. The extraction path is a sequence* $EP = \langle ENL_1, \ldots, ENL_\ell\rangle$.

The first task of the wrapper algorithm is to find a path in P_t which matches with the given EP and to extract the text value of the last node of the path. The matching semantics is defined as follows.

Let ENL be an extraction node label and n be a node of a page P_t. The ENL *matches* with the n if $ENL = \langle N, Pos, HAS\rangle$ such that (1) $N = N(n)$, (2) Pos is the number of the left siblings n' of n such that $N(n') = N(n)$ or $Pos = *$, and (3) for each $\langle a_i, v_i\rangle \in HAS(n)$, either $\langle a_i, v_i\rangle \in HAS$ or $\langle a_i, *\rangle \in HAS$.

Moreover, let $EP = \langle ENL_1, \ldots, ENL_\ell\rangle$ be an extraction path and $p = \langle n_1, \ldots, n_\ell\rangle$ be a path of a page P_t. The EP *matches* with the p if the ENL_i matches with n_i for all $i = 1, \ldots, \ell$.

Intuitively, an EP is a general expression of all paths p such that p is an instance of EP under a substitution for $*$ in EP.

Definition 8. *The Tree-Wrapper is a sequence* $W = \langle EP_1, \ldots, EP_K\rangle$ *of extraction paths* $EP_i = \langle ENL^i_1, \ldots, ENL^i_{\ell_i}\rangle$, *where each* ENL^i_j *is an extraction label.*

Then, we briefly explain the wrapper algorithm for given a tree wrapper $W = \langle EP_1, \ldots, EP_K \rangle$ and a page P_t. This algorithm outputs the label $L_t = \{t_1, \ldots, t_m\}$ of P_t as follows.

1. For each EP_i $(i = 1, \ldots, K)$, find all path $p = \langle n_1, \ldots, n_\ell \rangle$ of P_t such that EP_i matches with p and add the pair $\langle i, n_\ell \rangle$ into the set Att.
2. Sort all elements $\langle i, n_\ell \rangle \in Att$ in the increasing order of n_ℓ's. Let $LIST$ be the list and $j = 1$.
3. If the length of $LIST$ is 0 or $j > m$, then halt. If not, find the longest prefix *list* of $LIST$ such that all element is in non-decreasing order of i of $\langle i, n \rangle$ and for all $i = 1, \ldots, K$, compute the set $ta_i = \{n \mid \langle i, n \rangle \in list\}$. If the *list* is empty, then let $ta_i = \emptyset$.
4. Let $t_j = \langle ta_1, \ldots, ta_K \rangle$, $j = j + 1$, remove the *list* from $LIST$ and go to 3.

3.3 The Learning Algorithm

Let $\langle P_n, L_n \rangle$ be a training example such that $L_t = \{t_1, \ldots, t_m\}$ and $t_i = \langle ta_1^i, \ldots, ta_K^i \rangle$. The learning algorithm calls the procedure to find the extraction path EP_j for the j-th text attribute as follows.

The procedure computes all paths p_ℓ from the node $n \in ta_j^i$ to the root, where $1 \leq i \leq m$. For each p_ℓ, set EP^ℓ be the sequence of node labels of p_ℓ. Next, the procedure computes the *composition* EP of all EP^ℓ and sets $EP_j = EP$. The definition of the composition of extraction paths is given as follows. Fig. 3 is an example for a composite of two extraction path.

Definition 9. *Let HAS_1 and HAS_2 be sets of HTML attributes. The common HTML attribute set $CHAS$ of HAS_1 and HAS_2 is the set of HTML attributes such that $\langle a, v \rangle \in CHAS$ iff $\langle a, v \rangle \in HAS_1 \cap HAS_2$ and $\langle a, * \rangle \in CHAS$ iff $\langle a, v_1 \rangle \in HAS_1$, $\langle a, v_2 \rangle \in HAS_2$, and $v_1 \neq v_2$.*

Definition 10. *Let ENL_1 and ENL_2 be extraction node labels. The composition of $ENL_1 \cdot ENL_2$ is $ENL = \langle N, Pos, HAS \rangle$ such that (1) $N = N_1$ if $N_1 = N_2$ and ENL is undefined otherwise, (2) $Pos = Pos_1$ if $Pos_1 = Pos_2$, and $Pos = *$ otherwise, and (3) HAS is the common HTML attribute set of HAS_1 and HAS_2.*

Definition 11. *Let $EP_1 = \langle ENL_n^1, \ldots, ENL_1^1 \rangle$ and $EP_2 = \langle ENL_m^2, \ldots, ENL_1^2 \rangle$ be extraction paths. The $EP = EP_1 \cdot EP_2$ is the longest sequence $\langle ENL_\ell^1 \cdot ENL_\ell^2, \ldots, ENL_1^1 \cdot ENL_1^2 \rangle$ such that all $ENL_i^1 \cdot ENL_i^2$ are defined for $i = 1, \ldots, \ell$, where $\ell \leq min\{n, m\}$.*

3.4 Experimental Results

We equip the learning algorithm by Java language and experiment with this prototype for HTML documents. For parsing HTML documents, we use the OpenXML 1.2 (http://www.openxml.org) which is a validating XML parser

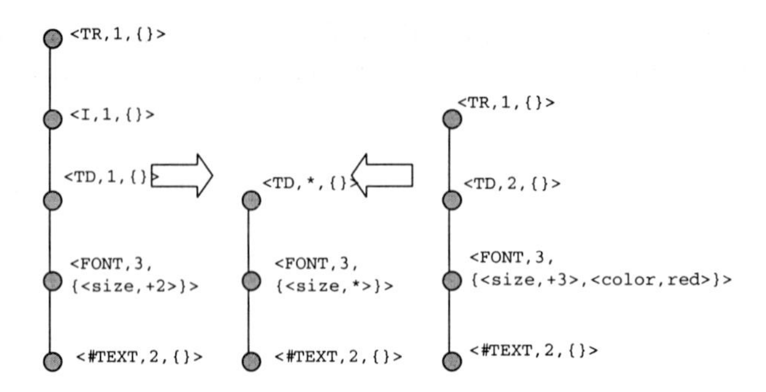

Fig. 3. The composition of extraction paths.

written in Java. It can also parse HTML and supports the HTML parts of the DOM (http://www.w3.org/DOM).

The experimental data of HTML documents is collected by the citeseer which is a scientific literature digital library (http://citeseer.nj.nec.com). The data consists of 1,300 HTML documents. Each page is a repetition of segments of several records. A record contains some information with a paper. For example, a record contains the tile, authors, a context in which the paper is referred, and reference numbers. Since our algorithm can not deal with HTML attributes and link structures among HTML documents, we chose the plain texts of the title, the name of authors, and the abstract as the first, the second, and the third attributes for the examination.

All pages are indexed to be $P_1, \ldots, P_{1300}$ in the order of the file size. The training example is $E = \{\langle P_i, L_i\rangle \mid i = 1, \ldots, 10\}$, where the L_i is the label made from the P_i in advance. The result is shown in Fig. 4 which is the Tree-Wrapper W found by the learning algorithm.

Fig. 4. The Tree-Wrapper found by the learning algorithm

Next, we practice the obtained Tree-Wrapper for the remained pages P_i ($i = 11, \ldots, 1300$) to extract all tuples from P_i. The three pages can not be extracted. The P_{1095} is one of the pages. We explain the reason by this page. In Fig. 4, we can find that the first extraction path EP_1 contains the extraction node label for the TABLE tag. The HTML attribute set HAS of this node contains the attribute "cellpadding" whose value is 0. However, the corresponding node in P_{1095} has the HTML attribute value "cellpadding= 1". Thus, the EP_1 does not match with the path. Any other pages are exactly extracted, thus, this algorithm is effective for this site.

Moreover we examine the performance of the Tree-Wrapper for several Internet sites and compare the expressiveness of Tree-Wrapper and Kushmerick's LR-Wrapper. One of the results is shown in the Fig. 5. We select 9 popular search engine sites and 1 news site and obtained text data by giving them keywords concerned with computer science. For each site, we made two sets of training data and test data. An entry of the form $n(m)$ of Fig. 5, for example 2(260), means that n tuples of training samples are sufficient to learn the site by the wrapper class and the learning time is in m milli-seconds. The symbol F means that the algorithm could not learn the wrapper for the site even though using all training samples. This figure shows that almost sites can be expressed by Tree-Wrapper and the learning algorithm learn the Tree-Wrappers within a few samples. The learning time is about 2 or 3 times slower than the LR-Wrapper learning algorithm. Thus, we conclude that the Tree-Wrapper class is efficient compared with the LR-Wrapper.

Resource & URL	LR-Wrapper	Tree-Wrapper
1. ALTA VISTA (www.altavista.com/)	F	2 (260)
2. excite (www.excite.com/)	F	3 (236)
3. LYCOS (www.lycos.com/)	F	2 (243)
4. Fast Search (www.fast.no/)	2 (101)	2 (247)
5. HOT BOT (hotbot.lycos.com/)	F	F
6. WEB CRAWLER (www.webcrawler.com/)	F	2 (182)
7. NationalDirectory (www.NationalDirectory.com/)	F	2 (180)
8. ARGOS (www.argos.evansville.edu/)	2 (45)	2 (313)
9. Google (www.google.com/)	F	2 (225)
10. Kyodo News (www.kyodo.co.jp/)	3 (55)	1 (144)

Fig. 5. The comparison of the number of training samples and the learning time (ms) of LR-Wrapper and Tree-Wrapper. The symbol F means that the learning is failed.

4 Conclusion

We presented the results of our study on information extraction from semistructured texts. First we investigated the theory of rewriting system for labeled trees. The two models for rewriting trees were introduced. One is extraction defined by erasing nodes. The other is reconstruction defined by a restriction of translation system. For the extraction model, the complexity of the decision problem of finding a rewriting rule between two trees was proved to be NP-complete with respect to several restricted conditions. On the other hand, we proved that there exists a sub-problem in P. For the reconstruction model, we presented the polynomial time learning algorithm to learn the class of k-variable linear translation systems using membership and equivalence queries. Second, in order to apply our algorithm to the real world data, we restricted our data model and introduced the Tree-Wrapper class to express the HTML documents. In the framework of Kushmerick's wrapper induction, we constructed the learning algorithm for the Tree-Wrapper and examined the performance of our algorithm. In particular we showed that Tree-Wrapper can express almost data which can not be expressed by LR-Wrapper.

Acknowledgments

The authors would be grateful to Kouichi Hirata at Kyushu Institute of Technology and Daisuke Ikeda at Kyushu University for their careful reading of the draft. This paper could not be completed without their uesful comments.

References

1. S. Abiteboul, P. Buneman, D. Suciu, Data on the Web: From relations to semistructured data and XML, Morgan Kaufmann, San Francisco, CA, 2000.
2. D. Angluin, Queries and concept learning, *Machine Learning* vol.2, pp.319–342, 1988.
3. H. Arimura, Learning Acyclic First-order Horn Sentences From Entailment, *Proc. 7th Int. Workshop on Algorithmic Learning Theory*, LNAI 1316, pp.432–445, 1997.
4. H. Arimura, H. Ishizaka, T. Shinohara, Learning unions of tree patterns using queries, *Theoretical Computer Science* vol.185, pp.47–62, 1997.
5. W. W. Cohen, W. Fan, Learning Page-Independent Heuristics for Extracting Data from Web Pages, *Proc. WWW-99,* 1999.
6. M. Craven, D. DiPasquo, D. Freitag, A. McCallum, T. Mitchell, K. Nigam, S. Slattery, Learning to construct knowledge bases from the World Wide Web, *Artificial Intelligence* vol. 118 pp. 69–113, 2000.
7. N. Dershowitz, J.-P. Jouannaud, Rewrite Systems, Chapter 6, Formal Models and Semantics, Handbook of Theoretical Computer Science Vol. B, Elseveir, 1990.
8. F. Drewes, Computation by Tree Transductions, Ph D. Thesis, University of Bremen, Department of Mathematics and Informatics, February 1996.
9. M. Frazier, L. Pitt, Learning from entailment: an application to propositional Horn sentences, *Proc. 10th Int. Conf. Machine Learning*, pp.120–127, 1993.
10. D. Freitag, Information extraction from HTML: Application of a general machine learning approach. *Proc. the Fifteenth National Conference on Artificial Intelligence*, pp. 517-523, 1998.
11. K. Hirata, K. Yamada, H. Harao, Tractable and intractable second-order matching problems. *Proc. 5th Annual International Computing and Combinatorics Conference*, LNCS 1627, pp. 432–441, 1999.

12. J. Hammer, H. Garcia-Molina, J. Cho, A. Crespo, Extracting semistructured information from the Web. *Proc. the Workshop on Management of Semistructured Data*, pp. 18–25, 1997.
13. C.-H. Hsu, Initial results on wrapping semistructured web pages with finite-state transducers and contextual rules. *In papers from the 1998 Workshop on AI and Information Integration*, pp. 66–73, 1998.
14. R. Khardon, Learning function-free Horn expressions, *Proc. COLT'98*, pp. 154–165, 1998.
15. P. Kilpelainen, H. Mannila, Ordered and unordered tree inclusion, *SIAM J. Comput.*, vol. 24, pp.340–356, 1995.
16. N. Kushmerick, Wrapper induction: efficiency and expressiveness. *Artificial Intelligence* vol. 118, pp. 15–68, 2000.
17. I. Muslea, S. Minton, C. A. Knoblock, Wrapper induction for semistructured, web-based information sources. *Proc. the Conference on Automated Learning and Discovery*, 1998.
18. H. Sakamoto, H. Arimura, S. Arikawa, Identification of tree translation rules from examples. *Proc. 5th International Colloquium on Grammatical Inference.* LNAI 1891, pp. 241–255, 2000.
19. H. Sakamoto, Y. Murakami, H. Arimura, S. Arikawa, Extracting Partial Structures from HTML Documents, *Proc. the 14the International FLAIRS Conference*, pp. 264-268, 2001, AAAI Press.
20. K. Taniguchi, H. Sakamoto, H. Arimura, S. Shimozono, S. Arikawa, Mining Semi-Structured Data by Path Expressions, *Proc. the 4th International Conference on Discovery Science*, (to appear).
21. L. G. Valiant, A theory of learnable, *Commun. ACM* vol.27, pp. 1134–1142, 1984.

Packet Analysis in Congested Networks

Masaki Fukushima[1] and Shigeki Goto[2]

[1] Communications Network Planning Laboratory, KDDI R&D Research Laboratory
2-1-15 Ohara, Kamifukuoka-shi, Saitama 356-8502, Japan
fukusima@kddlabs.co.jp

[2] Department of Information and Computer Science, Waseda University
3-4-1 Ohkubo, Shinjuku, Tokyo 169-8555, Japan
goto@goto.info.waseda.ac.jp

Abstract. This paper proposes new methods of measuring the Internet traffic. These are useful to analysing the network status, especially when the traffic is heavy, i.e. the network is congested. Our first method realizes a light weight measurement which counts only TCP flags, which occupies 6 bits in a TCP packet. Based on the simple flag counts, we can tell whether the network is congested or not. Moreover, we can estimate the average throughput of a network connection based on the flag count. Our second method analyses a sequence of TCP packets based on an automaton, or a protocol machine. The original automaton has been used in the formal specification of TCP protocol. However, it is not applicable to the real Internet traffic. We have improved the automaton in various ways, and established a modified machine. Using the new machine, we can analyse the Internet traffic even if there are packet losses.

1 Introduction

There have been many tools and utilities for network monitoring. However, they give little information about the real performance of data transfer. This paper proposes new methods for measuring the Internet traffic from the viewpoint of knowledge discovery.

Fig. 1 illustrates utilization of the communication link which connects Waseda University and IMnet (Inter-Ministry Research Information Network[1]). The vertical axis shows average utilization of the link for every five minutes. The bandwidth of this link is 1.5 Mbps at the time of this measurement (it is 100 Mbps at present.) The link is utilized 100% in the daytime in the direction from the IMnet to Waseda University. The graph is simple. However, there is no further information.

According to our earlier survey, HTTP (Hyper Text Transfer Protocol) traffic occupies more than 65% of the traffic from IMnet to Waseda University. It is worth while investigating the HTTP traffic on this link. Fig. 2 shows the average throughput of HTTP on the same day. The throughput is calculated from the log file of HTTP proxy servers at Waseda University. Throughput is indicated as time (in seconds) required to transfer a file of 7KB. The size of 7KB is the average size of files transferred by HTTP through the same proxy servers. It is

S. Arikawa and A. Shinohara (Eds.): Progress in Discovery Science 2001, LNAI 2281, pp. 600–615.

easily observed that the performance of HTTP is fluctuating while the utilization of the link is constantly 100%. We would like to measure the degree of congestion which reflects the real performance of HTTP without analyzing the huge log files. This paper proposes a new method of performance measurement which collects a piece of information from the data packets, and estimates the performance by a simple calculation. Section 2 of this paper deals with a new measurement method based on TCP flags.

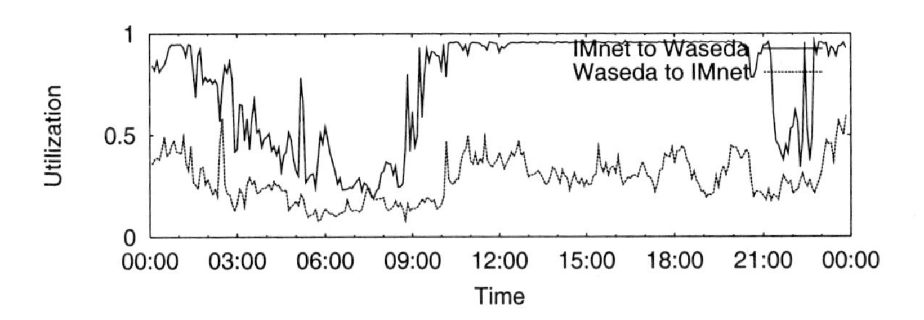

Fig. 1. Traffic between Waseda University and IMnet on a day

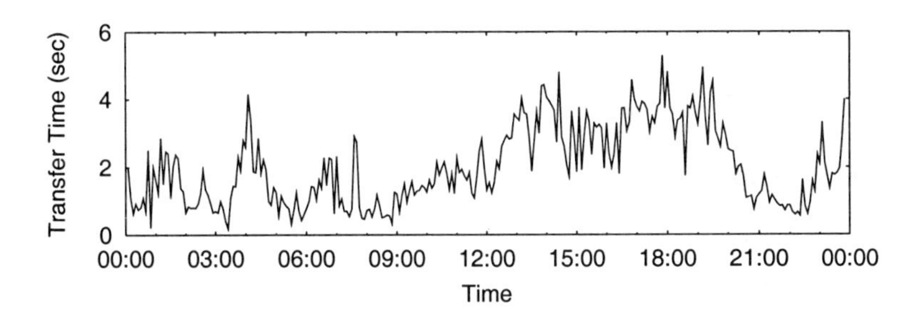

Fig. 2. Average Transfer Time of HTTP on a day

Many protocols are modeled as finite state machines [10]. The most representative example is TCP (Transmission Control Protocol). It is meaningful to analyze TCP traffic, because it is heavily used in the Internet. According to [11], TCP:UDP ratio is 7:1 during the evening, and 10:1 during the day in backbone environments. This paper analyzes the TCP traffic based on the formal definition of the protocol. The specification of TCP is well defined in RFC793 [12], using a finite state machine in Fig. 3.

In our analysis, the input of the machine is a TCP packet, which is also called a TCP *segment*. The original machine sometimes have two packets involved in one state transition. This paper modifies the machine in RFC 793 (Fig. 3) to allow only one input at a time. Section 3 will analyse the original machine, and further improves it.

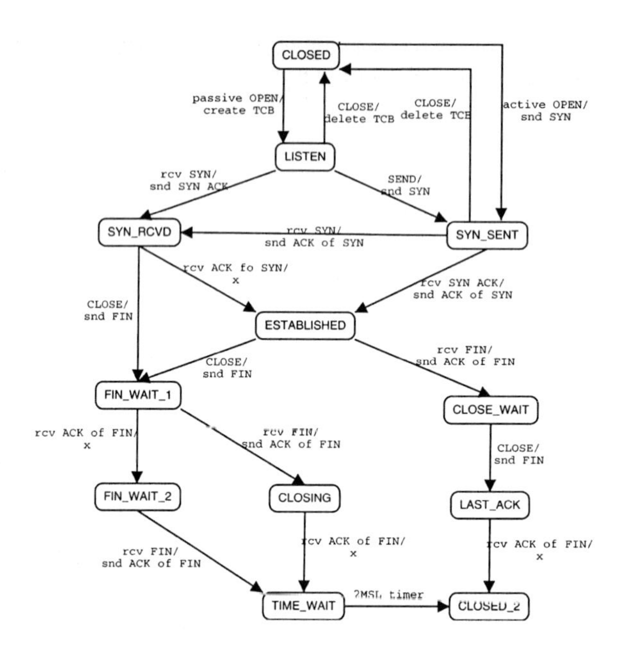

Fig. 3. TCP Protocol Machine

2 Knowledge Discovery in TCP Control Flags

TCP/IP is the protocol in the Internet[2][3]. TCP (Transmission Control Protocol) provides reliable connections over IP (Internet Protocol). IP is not a reliable protocol. IP packets may be discarded. Moreover, the order of the IP packets may be changed from the original order. On such unreliable IP protocol, TCP controls transmission of packets to ensure reliability. TCP has control flags in the TCP header. The control flags occupy a 6-bit field in a header. Each bit of control flag has the following meaning.

FIN: The sender transmits a FIN flag when it has no more data to transmit.

SYN: SYN flag is used to synchronize the sequence number.

RST: One end sends a packet with a RST flag when it wants to abort the connection.

PSH: When the sender requests the receiver to deliver the data to the application program immediately, it puts a PSH flag.

ACK: ACK means acknowledgment. There is a related field in a TCP header which contains the acknowledged sequence number.

URG: This flag means the packet contains some urgent data. This flag is rarely used.

TCP controls transmission of packets in accordance with the condition of a network. For instance, TCP would restrict the number of packets to be transmitted if it encounters network congestion. TCP control flags are used to manage the transmission of packets.

This paper develops a new method of estimating network congestion by measuring TCP behavior. We derive new information from TCP control flags. The result shows that the some ratios of the TCP flags are closely related to network congestion.

2.1 Measurement

Waseda University is connected to the Internet through IMnet[1]. We attach a computer to the Ethernet segment to which the border router is connected at Waseda University. The computer collects packets on the Ethernet segment.

We investigate HTTP traffic, i.e. Web applications. The computer collects all packets whose source port number or destination port number is "80". We select HTTP because it occupies the largest portion of the traffic. We calculate the correlation coefficients between the two parameters and various flag ratios which will be described later.

Utilization Parameter: Utilization of the link in the direction from IMnet to Waseda University. The utilization parameter is independently got from the router using SNMP (Simple Network Management Protocol).

HTTP transfer time Parameter: Performance of HTTP data-transfer over the link. It is calculated from log files of proxy servers. The value is expressed in the time (in seconds) which is required to transfer a 7KB file by HTTP.

Both parameters and the measured flag counts are time series data. The unit of time is five minutes. Table 1 and Table 2 show flag ratios with high correlation coefficient. The names of flags used in the tables are the following. ALL: number of all packets, Flag name (e.g. Fin): number of packets which have the specific flag, and One or more first letters of flag name (e.g. FRA): number of packets which have all the flags represented by their first letters and have no other flags.

Table 1. Correlation with Utilization

a	b	Correlation Coefficient a/b and Utilization
Syn	Fin	−0.7298
Fin	Syn	0.7196
FRA	Fin	0.6610
Psh	Fin	−0.6464
FRA	ALL	0.6237
FPA	ALL	0.6178
Fin	ALL	0.6149
FA	ALL	0.5923
Psh	Syn	−0.5614
FA	Fin	−0.5151

Table 2. Correlation with Transfer Time

a	b	Correlation Coefficient a/b and Transfer Time
FRA	Fin	0.7440
Fin	Syn	0.7362
FRA	ALL	0.7264
Syn	Fin	−0.7092
Fin	ALL	0.6543
FPA	ALL	0.6481
FA	ALL	0.6238
Psh	Fin	−0.5760
S	ALL	0.5330
FA	Fin	−0.5294

The following three flag ratios are selected because they have a high correlation coefficient with both Utilization and HTTP transfer time. Moreover, their values are useful to construct a linear discriminate function. We have investigated flag ratios exhaustively. Figures 4, 5 and 6 show the three flag ratios.

Fin/Syn: ratio of all FIN flags to all SYN flags.
FRA/ALL: ratio of FIN-RST-ACK flags to all packets.
FA/Fin: ratio of FIN-ACK flags to all FIN flags.

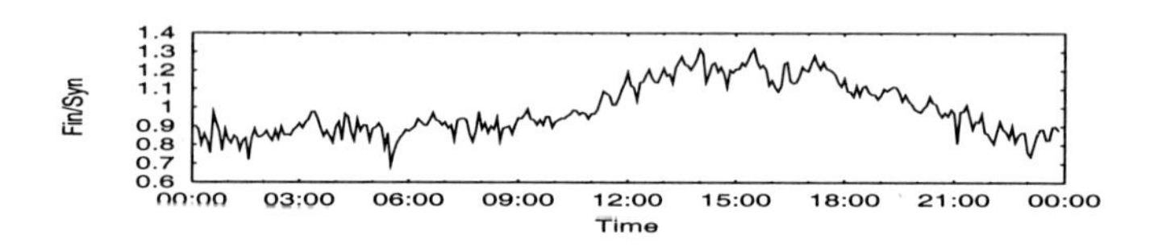

Fig. 4. ratio of all FIN flags to all SYN flags

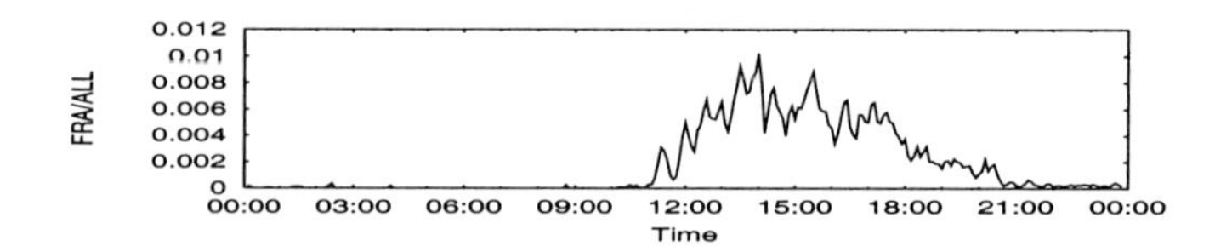

Fig. 5. ratio of FIN-RST-ACK flags to all packets

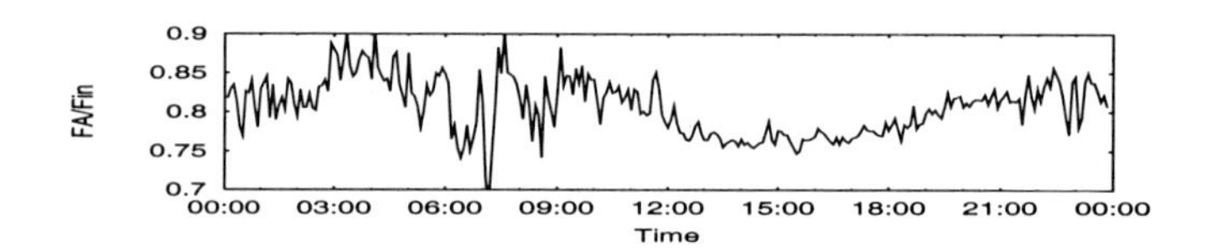

Fig. 6. ratio of FIN-ACK flags to all FIN flags

2.2 Application

Congestion: We try to estimate the congestion of the link based on those three flag ratios. In this section, the word *congestion* is used when utilization is over 98%. We apply *discriminant analysis.* The discriminant analysis is a statistical method of differentiating sample data. The sample data is represented by variables, $x_1, x_2, \ldots, x_p$, which are also called *explanatory variables.* A liner discriminant function with p explanatory variables $x_1, x_2, \ldots, x_p$ is formulated as follows:

$$Z = a_0 + a_1 x_1 + a_2 x_2 + \cdots + a_p x_p$$

If the value of Z is positive, the sample data is classified in group A. If the value of Z is negative, the sample data is classified in group B. There is a known

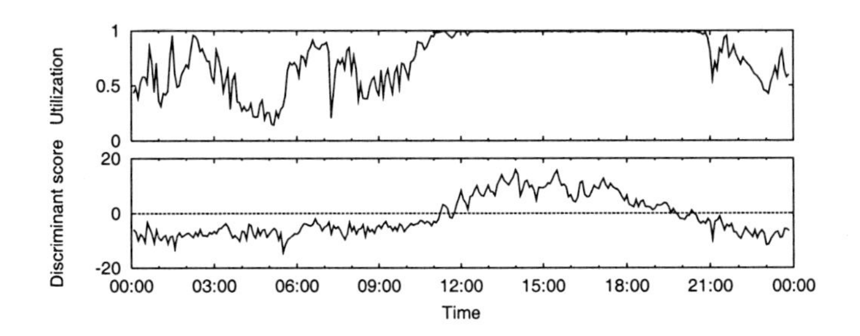

Fig. 7. Utilization and Congestion Discriminant Score

method to determine the coefficients, $a_0, a_1, a_2, \ldots, a_p$, which minimizes the error in discrimination with given samples.

We construct a linear discriminant function. Three flag ratios mentioned earlier are used as explanatory variables. The discriminant function takes the form of (1).

$$\begin{aligned} DS = & -26.435 + 38.957 \cdot (Fin/Syn) \\ & +462.13 \cdot (FRA/ALL) - 17.821 \cdot (FA/Fin) \end{aligned} \tag{1}$$

The communication link is estimated to be congested when the value of DS (*Discriminant Score*) is positive. We applied (1) to the data. Fig. 7 shows the result. The estimation of congestion hits at 94.4%.

HTTP Performance: We aim to estimate HTTP performance from the three flag ratios. We use the data collected in the daytime when the link was congested.

We adopt *multiple regression analysis.* The method can predict the value of a dependent variable from sample data, which is also called *explanatory variables.* A multiple regression equation is formulated as

$$y = a_0 + a_1 x_1 + a_2 x_2 + \cdots + a_p x_p$$

where y is the dependent variable and $x_1, x_2, \ldots, x_p$ are the explanatory variables. There is a standard procedure called *least square method* to determine the coefficients, $a_0, a_1, a_2, \ldots, a_p$. The least square method minimizes the error between the predicted y and the actual value.

We apply the multiple regression analysis using the least square method to the data, then construct a multiple regression equation which calculates ETT ($Estimated\ Transfer\ Time$).

$$\begin{aligned} ETT = & \ 3.4789 + 10.849 \cdot (Fin/Syn) \\ & -225.67 \cdot (FRA/ALL) - 15.281 \cdot (FA/Fin) \end{aligned} \tag{2}$$

We apply (2) to the data from which (2) was constructed. Fig. 8 shows the result. This estimation seems correct, and the correlation coefficient between the actual HTTP performance and the estimated HTTP performance is 0.845.

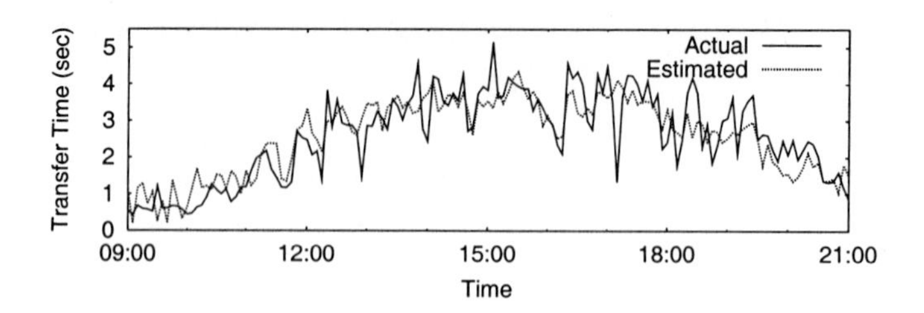

Fig. 8. Estimation of HTTP Transfer Time

2.3 Evaluation

Congestion: The formula (1) is constructed from the data of December 16, the same formula can be applied to estimate congestion on the other days. Table 3 shows the summary of the result. This table also includes the utilization average of the day.

Table 3. Evaluation of congestion discriminant

Date	utilization average	hitting score(%)
12	0.785	90.9
13	0.676	85.3
14	0.402	97.6
15	0.724	98.6
16	0.769	94.4
17	0.684	92.3
18	0.664	87.1
19	0.684	88.5
20	0.475	95.8
21	0.421	98.3
22	0.678	85.3
23	0.398	99.7

The hitting scores are at least 85%. There are relatively low hit scores among days when the utilization average ranges from 0.6 to 0.7. These days require more subtle discriminant analysis than other days. The heavily congested or little congested days have high hit scores because the utilization patterns are simple on these days.

HTTP Performance: The formula (2) is constructed by the data on December 16. To investigate the usefulness of this method, we apply the formula (2) to the data of the other days. As a result, correlation coefficients are at least 0.7 from data collected in the daytime of weekdays, when the link is highly utilized. This correlation coefficient is lower than the result from the data on December 16, but still correct to a certain degree. From the data of holidays, however, the correlation coefficient was nearly 0, and on some holidays it gives negative values.

The latter example illustrates the limitation of our method which is useful for congested networks.

Table 4. Evaluation of HTTP transfer time estimation

Date	Utilization average	correlation coefficient
12	0.785	0.808
13	0.676	0.817
14	0.402	-0.073
15	0.724	0.836
16	0.769	0.845
17	0.684	0.813
18	0.664	0.764
19	0.684	0.709
20	0.475	-0.034
21	0.421	-0.100
22	0.678	0.666
23	0.398	0.119

3 Protocol Machine for TCP Flow Analysis

TCP (Transmission Control Protocol) occupies most of the Internet traffic [11]. TCP is well defined in the document of RFC793 [12], based on a finite state machine. This section applies the automata to the real Internet traffic. We show that the improved automaton accepts most of the real TCP flow.

The original machine in RFC793 is not intended for traffic measurement. It can be formally transformed to an automaton for stateful traffic measurement. First, some new states are introduced. Secondly, empty (ε) action should be allowed. Thirdly, we take a direct product of two machines to reflect the interaction between a server computer and a client computer of a TCP connection. The improved machine shows a better result compared to the original one. We further introduce non-deterministic automata to deal with the packet loss. The final result shows that 93% of the measured traffic is accepted.

3.1 Original Protocol Machine

In Fig. 3 in Section 1, each round box represents a *state*, and each *arc* (an arrow) represents a transition of states. The arc is labeled with some information, which indicates an *event*. The event either describes a transition, or it stands for a reaction to some event. This section build a modified machine based on the original machine in Fig. 3.

Splitting a Transition and Adding New States: First, we split a transition, if it is associated with two packets.

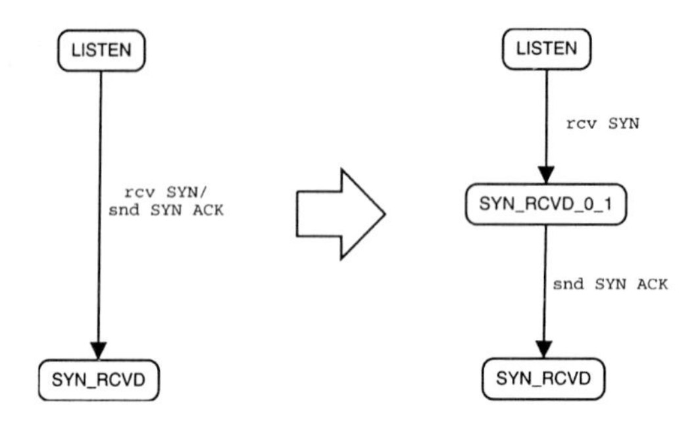

Fig. 9. Splitting a Transition and Adding New States

For example, we split a transaction into two events as shown in Fig.9. The original transition specifies that one packet (`SYN`) is received and a response packet (`SYN ACK`) is sent back. A new state is inserted, which represents the status of "one packet was received, and no response has not been sent back yet".
The modified machine appears graphically in Fig. 11 later in this paper.

Invisible Transition: There are some transitions which do not associate with any packets. These transitions are invisible when we investigate packets on the communication link. The following list shows the transactions, or events, which are invisible to packet analysis.

Passive OPEN: A server listens to a TCP *port* in order to receive requests from clients. This opening event does not send or receive any packets.

2MSL Timer: Before a TCP connection is closed, a machine may reach the state of TIME_WAIT. The machine waits for 2×MSL, where MSL means the Maximum Segment Lifetime. This action only waits for a certain amount of time. No packets are associated with this action.

ABORT: A machine may reach the state of CLOSED without sending any packets. This happens if the user submit an ABORT request when the machine is at the state of LISTEN, SYN_SENT, CLOSING, LAST_ACK, or TIME_WAIT.

CLOSE: A machine may reach the state of CLOSED without sending any packets. This happens if the user submit a CLOSE request when the machine is at the state of LISTEN, SYN_SENT, or TIME_WAIT.

We replace the invisible event by the empty action as shown in Fig. 10. An empty action is called an ε *move* in automata theory [13], because there is no input associated with the transition. The empty or ε-move is well defined in the textbook of automata. The modified protocol machine is a non-deterministic automaton.
After removing invisible events, the input symbols Σ consists of visible events to packet analysis. The modified machine is illustrated in Fig. 11. It receives one packet, or take an empty (ε) action.

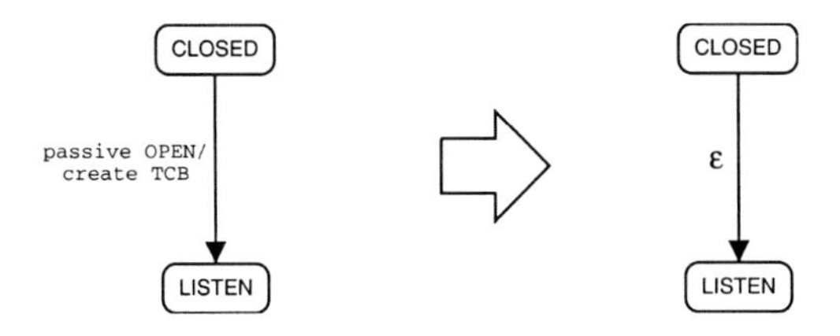

Fig. 10. Replacing an Invisible Transition by an ε-action

Direct Product of Two Machines: There are two machines involved in a TCP connection. They are a server and a client. Both of the server and the client send and receive packets. It is necessary to combine two protocol machines in order to analyze TCP packets.

This paper adopts synchronous coupling of two protocol machines in order to show the basic methodology of traffic interpreting automata.

Converting to a Deterministic Machine: The modified machine shown in Fig. 11 is a non-deterministic automaton with empty (ϵ) actions. It is not efficient to implement a non-deterministic automaton by a computer program because the machine can reach several states at the same time. There is a well known technique to convert a non-deterministic automaton into deterministic automaton. We apply this method to our modified machine.

3.2 Measurement and Evaluation (Part 1)

We have captured real TCP packets at APAN [9] Tokyo NOC (Network Operation Center). The packets are given to the protocol machine. We define a TCP *flow* in the following manner. Ideally, one TCP connection should be treated as one flow. However, we cannot avoid some errors due to a time-out and other erroneous behaviors.

- Packets belong to a *flow* if they have the same source IP address, the same source port, the same destination IP address, and the same destination port.
- If there is more than 120 seconds interval between two TCP packets, these two packets are treated as if they belong to two different flows.

Table 5 shows the result of our application of TCP protocol machine in Fig. 11. The total number of flows is 1,686,643. It is natural to assume that most packets or flows are generated faithfully by the TCP specification in RFC793. Then, an ideal machine should accept almost all the flows. However, Table 5 indicates only 19% of the total flows are accepted by the machine in Fig. 11.

One of the reasons why we have a poor result is the low figure of *recognized flows*. The recognized flow is a sequence of TCP packets which can be mapped to a sequence of input symbols in the automaton. If some packets are missed in a captured data, it may be impossible to uniquely map the packets to input symbols.

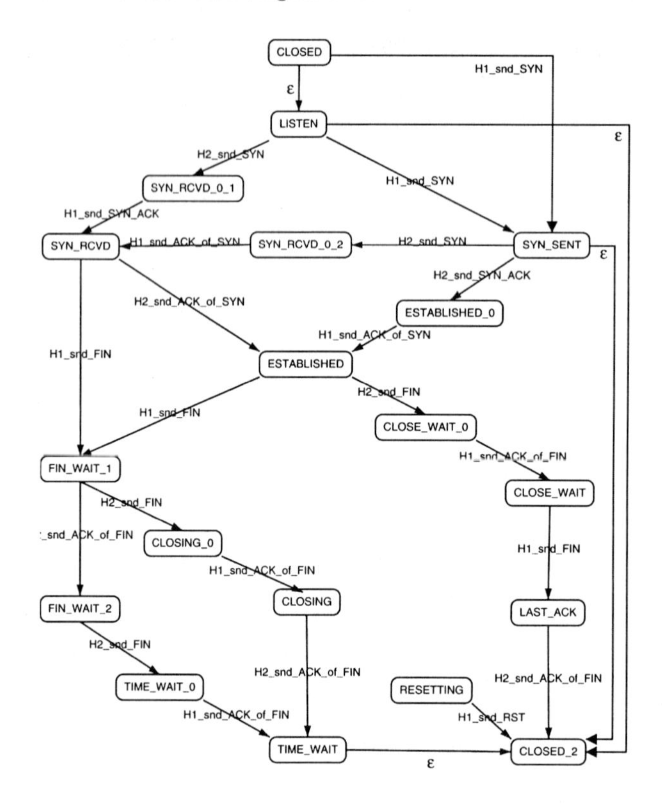

Fig. 11. Modified protocol machine

3.3 Dealing with Missing Packets

Our earlier analysis is based on the following assumptions.

- It is possible to capture all the packets which are sent from a sender host.
- The observed packets reach the destination host without any packet losses.
- There is a unique mapping between a sequence of packets and a sequence of input symbols.

In the real world, these assumptions are not always guaranteed. That is the reason why we had a low figure of *recognized flows*. This section investigates the problem in detail and rectify it.

Packet Losses: Packet losses are common phenomena in congested networks. We should consider the packet loss in our analysis. In traffic interpreting machine, a packet loss causes a trouble. When a packet loss occurs, the sender machine moves to a new state because it sends a packet. However, the receiver machine stays at the same state because there is no input, or packet. To rectify the problem, we put a *loop* arc as shown in Fig. 12. The loop represents the action of the receiver who keeps staying at the same state. The new machine, which takes packet losses into account, can be defined.

Table 5. Accepted Flow by the Modified Machine

Type of flows	Number of flows
total flow	1,686,643
recognized flows	485,721
accepted flows	320,458

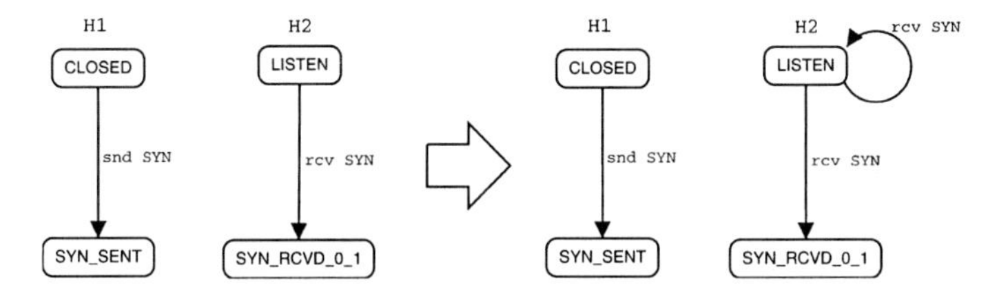

Fig. 12. Loop Representing a Packet Loss

Unobservable Packets: There are a couple of reasons why a packet is not observed, while it is normally sent form a sender computer.

Asymmetric path: In the Internet, a path between a sender and the receiver is not always bi-directional. Namely, a path from a sender (S) to the receiver (R) may be different from a path from R to S. It is reported that about a half of the total TCP traffic take asymmetric paths [14]. If a TCP flow takes an asymmetric path, we can capture only one-way traffic.

Buffer overflow in a measurement instrument: We use a packet analyzer or a personal computer to capture TCP packets. There are chances to miss the packets. One of the reason is a buffer overflow in an instrument or a computer.

To cope with this unobservable problem, we augment the protocol machine, and add an empty (ε) action. The emptiness corresponds to no packets. An example is described in Fig.13. The final TCP interpreting machine is composed by the direct product of two machines, which take packet losses into account.

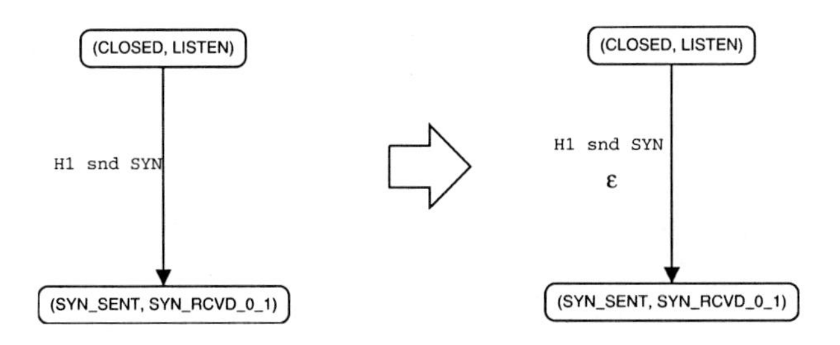

Fig. 13. Empty (ε) Action to Represent Unobservable Packets

Non-deterministic Packets in Input Symbols: Our analysis is based on the assumption that each packet is uniquely associated with an input symbol in the automaton. In reality, there are some cases when the unique mapping cannot be attained. The mapping is sometimes *context sensitive*. That is, a single packet cannot be mapped without knowing the preceding or succeeding packets.

One typical example is an ACK segment (packet). Nobody cannot tell if the ACK is a normal acknowledgement to the sending data, or the ACK to a FIN segment without knowing the acknowledged number and the sequence number in the preceding packet. If the preceding packet is missed, there is no way of mapping the ACK uniquely. To fix the problem, the set of input symbols Σ is enlarged. We use the power set of Σ which is written as 2^{Σ}. 2^{Σ} is a set of all subsets of Σ. To construct the final machine, we eliminate empty (ε) action from the machine. And it is converted into a deterministic automaton as before.

3.4 Measurement and Evaluation (Part 2)

The modified machine is applied to the same captured packets as in Section 3. Table 6 shows the result. Table 6 clearly shows a better result than our earlier analysis. We can recognize all the packets as a part of a flow. This improvement comes from allowing non-determinacy of input symbols.

Table 6. Accepted Flows by the Improved Machine

Type of flows	Number of flows
total flows	1,686,643
recognized flows	1,686,643
accepted flows	1,576,021

The acceptance ratio goes up to 93%. The captured data comes from an Internet backbone where significant portion of traffic may take asymmetric routes.

However, the improved protocol machine cannot attain the score of 100%. This result indicates that there still remain some behaviors in the real protocol machine which are not modeled by our method. We have a plan to further investigate the asynchronous coupling of two machines in order to increase the acceptance ratio. A simple example illustrates that there is an asynchronous behavior. If both of a sender and the receiver send FIN packets at the same time, it is not interpreted properly by our TCP protocol machine.

3.5 Limitation of the Non-determinacy

Our traffic interpreting machine is based on non-deterministic automata with an empty (ε) action. It is a good model of missing packets in congested networks. However, it has some caveat. One can start at the initial state CLOSED, and go through up to the final state of CLOSED_2 without having any input packets. This behavior is not realistic, because it means all the packets are missed.

The non-deterministic protocol machine is converted into a deterministic machine. The set of states is the power set of the original set of states. Thus, the number of states becomes a large number. This subsection tries to reduce the non-determinacy of the machine. We propose a method of limiting the use of the empty (ε) action for missing packets.

Introducing Cost of the Empty Action. We have a simple idea based on the weighted graph. We put some cost to each empty (ε) action, and the protocol machine can operate within the upper limit of the total cost.

There are two kinds of empty (ε) actions. One represents a missing packet. The other represents an invisible action. We put a cost of "1" to a missing packet transition. Other transitions have zero cost. The protocol machine can operate within a limit.

Evaluation of Limitation. We have six protocol machines for comparison. They have a limit of the total cost. $\pi(T_{n(0)})$ has the cost limit of $c = 0$, that means no missing packets. $\pi(T_{n(5)})$ has the cost limit of $c = 5$ which allows missing packets up to five. These machines have the same input packets.

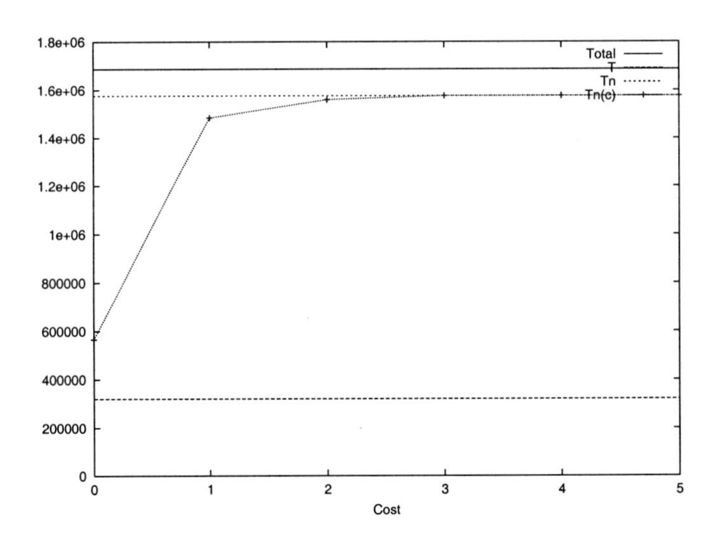

Fig. 14. Number of accepted flows by $\pi(T_{n(c)})$

Fig. 14 is a graph which plots out the number of accepted flows. The total number of flows, and the protocol machine $\pi(T)$ in Section 3.2 are also shown as well as the machine $\pi(T_n)$ in Section 3.4.

It should be noted that $T_n(c)$ gives a better result than T even at cost 0. $T_n(c)$ covers invisible events and non-deterministic input symbols even if the cost limit is 0.

The accepted ratio goes up high at cost 1. If we allow one missing packet, there are many flows accepted. $T_n(3)$ is almost identical with T_n. Thus, there are little number of flows which miss four or more packets.

4 Conclusion

This paper clarifies the following.

- One can tell if the link is congested based on the value of flag ratios.
- In a heavily congested network, it was possible to estimate the performance of HTTP from the ratios of flags.

The method proposed in this paper uses a 6-bit field of TCP packets. The amount of data is small enough to be collected through existing network probes, e.g. RMON (Remote Network Monitoring). Moreover, the calculation of ratios is simple.

This paper investigates the usefulness of the finite state machine in monitoring the Internet traffic. We found the original machine is not adequate for actual monitoring. We have modified the machine for practical purposes. The improved machine can accept most real traffic in the Internet.

Acknowledgments

The authors are grateful to members at Media Network Center in Waseda University who have helped us to collect packets at the border segment. We also thank members of APAN measurement working group and Tokyo NOC, especially Mr. Yasuichi Kitamura and Mr. Yoshinori Kitatsuji, who have helped us to use OC3mon/Coral.

References

1. IMnet (Inter-Ministry Research Information Network) `http://www.imnet.ad.jp/`
2. Douglas E. Comer, "Internetworking with TCP/IP Vol.I: Principles, Protocols, and Architecture, 4th Ed.,", Prentice-Hall, 2000.
3. W. Richard Stevens, *TCP/IP Illustrated, Volume 1: The Protocols,* Addison-Wesley, 1994.
4. W. Richard Stevens, Gary Wright, *TCP/IP Illustrated, Volume 2: The Implementation,* Addison-Wesley, 1995.
5. W. Richard Stevens, *TCP/IP Illustrated, Volume 3: TCP for Transactions, HTTP, NNTP, and the UNIX Domain Protocols,* Addison-Wesley, 1996.
6. Marshall Kirk McKusick, Keith Bostic, Michael J. Karels and John S. Quarterman, *The Design and Implementation of the 4.4BSD Operating System,* Addison-Wesley, 1996.

7. MRTG (Multi Router Traffic Grapher)
   `http://ee-staff.ethz.ch/ oeticker/webtools/mrtg/mrtg.html`
8. OC3mon/Coral, `http://www.caida.org/Tools/CoralReef/`
9. APAN, `http://www.apan.net/`, (Asia-Pacific Advanced Research Information Network).
10. Gerard J. Holzmann, "Design and Validation of Computer Protocols", Prentice Hall, 1991.
11. Kimberly C. Claffy, Hans-Werner Braun and George C. Polyzos, "A parametrizable methodology for Internet traffic flow profiling", IEEE JSAC Special Issue on the Global Internet, 1995.
12. J. Postel, "Transmission Control Protocol - DARPA Internet Program Protocol Specification", RFC793, 1981.
13. Yoichi Obuchi, "Discrete Information Processing and Automaton" (in Japanese), Asakura Shoten, 1999.
14. Masaki Fukushima and Shigeki Goto, "Analysis of TCP Flags in Congested Network", IEICE Transactions on Information and Systems, VOL. E83-D, NO. 5, PP.996–1002, MAY 2000.
15. R. Sedgewick, "Algorithms", Addison-Wesley, 1988.
16. Masaki Fukushima and Shigeki Goto, Improving TCP Protocol Machine to Accept Real Internet Traffic, IWS2001 Internet Workshop 2001, pp.47–53, Tokyo, Japan, February 2001.

Visualization and Analysis of Web Graphs

Sachio Hirokawa and Daisuke Ikeda

Computing and Communications Center, Kyushu University,
Fukuoka 812-8581, Japan
{hirokawa,daisuke}@cc.kyushu-u.ac.jp

Abstract. We review the progress of our research on Web Graphs. A Web Graph is a directed graph whose nodes are Web pages and whose edges are hyperlinks between pages. Many people use bookmarks and pages of links as a knowledge on internet. We developed a visualization system of Web Graphs. It is a system for construction and analysis of Web graphs. For constructing and analysis of large graphs, the SVD (Singular Value Decomposition) of the adjacency matrix of the graph is used. The experimental application of the system yield some discovery that are unforseen by other approach. The scree plots of the singular values of the adjacency matrix is introduced and confirmed that can be used as a measure to evaluate the Web space.

1 Introduction

There are many systems which visualize the sitemap of WWW pages in a site. These systems help site managers analyze the behavior of visitors and improve the structure of the site. But the target of these systems is one site which is controlled under one policy or for one purpose. On the other hand, the global structure of WWW is being clarified by some researches [1, 14]. The target of our research are clusters of 10,000 WWW pages which form meaningful communities for particular subject and can be considered as knowledge on the WWW. The focus is not in individual pages but the link structure of WWW pages which form a directed graphs. The link information is used in other research. The difference of our approach is that we evaluate not only individual pages but also the whole community. Our system consists of the visualization system, which helps intuitive understanding of the clusters, and the system of numerical analysis, which performs the singular value decomposition for the adjacent matrix of the graphs.

The experimental application of the system yield some discovery that are unforseen by other approach. One of these discovery is an obstacle caused by "authorities" and "hubs". It is believed that WWW pages with high link degree are important to the community and referred as "authorities" and "hubs". We found examples of Web Graphs where removal of these pages gives clear interpretation of the Web Graphs. In visualization of clusters of WWW pages, we found an interesting case where two communities connected by a slender link. This web graph would be a good target of chance discovery [15]. The scree plots

S. Arikawa and A. Shinohara (Eds.): Progress in Discovery Science 2001, LNAI 2281, pp. 616–628.

of the singular values of the adjacency matrix is originally introduced for the analysis of the WWW space in this research. In section 5, we demonstrate that the plots gives an evaluation for each WWW space.

A personal bookmark of an specific area is useful, if it is compiled by an appropriate person who knows the area in depth. Web pages are filtered by user's point of view and maintained in his bookmark. The pages in the bookmark are valuable, but we think that the relation of these pages is much more valuable. The linkage information is known to be useful to find good Web pages in search result [6, 12]. We started the project by constructing a visualization system of link structure of Web pages.

The system "KN on ZK" [11, 8] is a tool to create such a practical knowledge network. It is not an automatic tool, but a tool to handle manually. It is built on top of the network note pad ZK [16] and displays the links as directed graphs. The first version [8] handles Web graphs of small size and requires manual operation for arrangement of nodes. We introduced the spring model [7] for drawing in the second version [11]. The system can handle Web graphs with 10,000 nodes. The system stores background knowledges as Web graphs. Given a background knowledge and some keywords, the system gains new URLs from a search engine with the keywords and places them on the given background knowledge. In this way, a user can develop his knowledge and can keep the new knowledge as another background knowledge. The system consists of background knowledges and a URL database (Fig. 1). The database, given a URL, returns a list of URLs to which the given URL has links. A user develops a background knowledge with search engines, a list of URLs, and Web browsing. A new node is placed near a node to which the new one linked. The new Web graph can be stored as a new background knowledge.

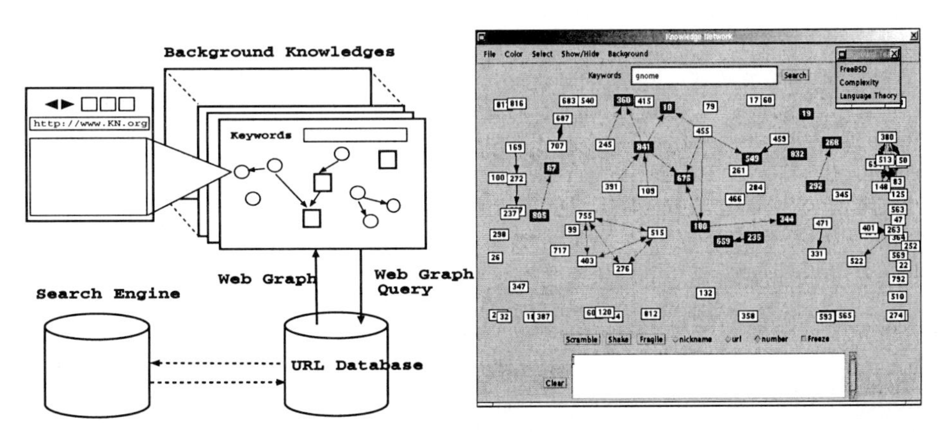

Fig. 1. Outline of the system KN

Fig. 2. The Web graph developed by a keywords search

Fig. 2 describes the linkage structure of the URLs at the site of FreeBSD [1] and its development with the keywords "gnome" and "Japanese". Gray nodes are URLs in the background knowledge. New nodes obtained by a search engine is colored in white. The white node "109" is placed near the gray node "841" to which "109" links.

2 Visualization of Web Graphs

In this section, we show that the system KN is useful in analyzing the structure of Web pages. The visualization enables us to capture the global structure of these pages that is hard to see when we visiting each page with ordinary web browser. We show four examples of Web graphs and explains the analysis with KN. The first example is the visualization of link structure of University Media Centers [13]. By manual arrangement, we easily see hubs and authorities in the graph. The graphs of the next two examples are much more complicated than what we expected at the first sight, even though the system automatically draws the graphs with spring model. We analyzed that the complexity depends on some nodes of high degree. By deleting those nodes, the structure of the graph became clear.

2.1 Bipartite Cores of University Media Centers

Fig. 3 is the Web graph of University Media Centers [13]. It visualizes the link structure of the top 50 authorities and the top 50 hubs in the community. We manually arranged the graph after automatic drawing. We see hubs on the left and authorities on the right.

2.2 Site Map of Kyushu University

We collected all the HTML-files in Kyushu University [2] that can be followed from the top page www.kyushu-u.ac.jp. There were 220 http-servers and 42,430 HTML-files [3]. We visualized the Web graph (Fig. 4), where each node represents a domain. The top page (node 129) of the university contains links to the departments and is linked from the departments. Since there are too many lines in the graph, we cannot recognize any structure except those heavily connected nodes. The number of such sites linked from more than 8 other sites are 26 which is only 11.8 % of all sites. But the lines in the graph that link to such sites amounts to 37.5 % of all lines. These nodes are the cause of the complexity of the graph. Fig. 5 is the Web graph that is obtained by deleting nodes with out-degree ≥ 8. It is known that the distribution of degrees follows the Power-Law. We confirmed the fact for all HTML-files in Kyushu University as well. Only 1.2 % files has out-degree 8, but 22% of the lines in the Web graph connect to such nodes. Thus, by deleting such crowded nodes, we can clarify the structure of the graph.

[1] http://www.jp.freebsd.org/links
[2] http://www.kyushu-u.ac.jp/
[3] July 1999

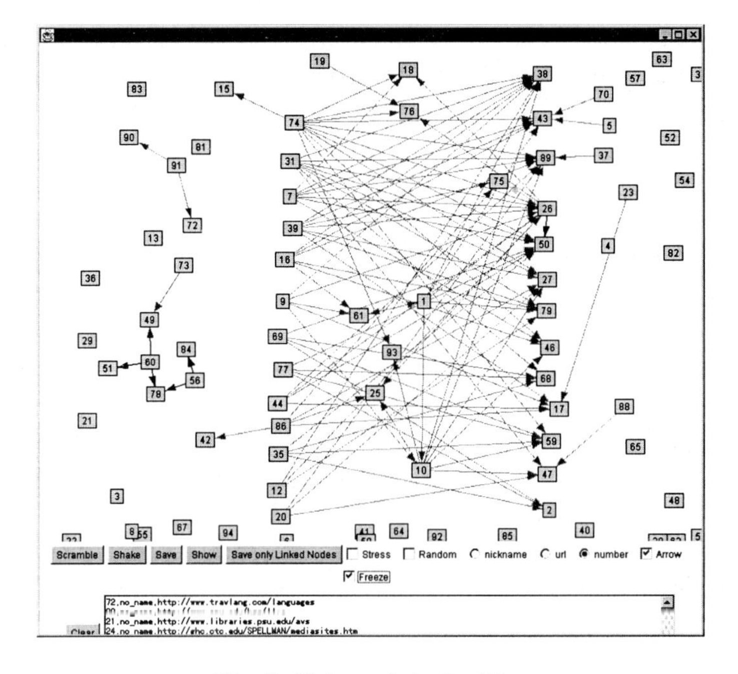

Fig. 3. Hubs and Authorities

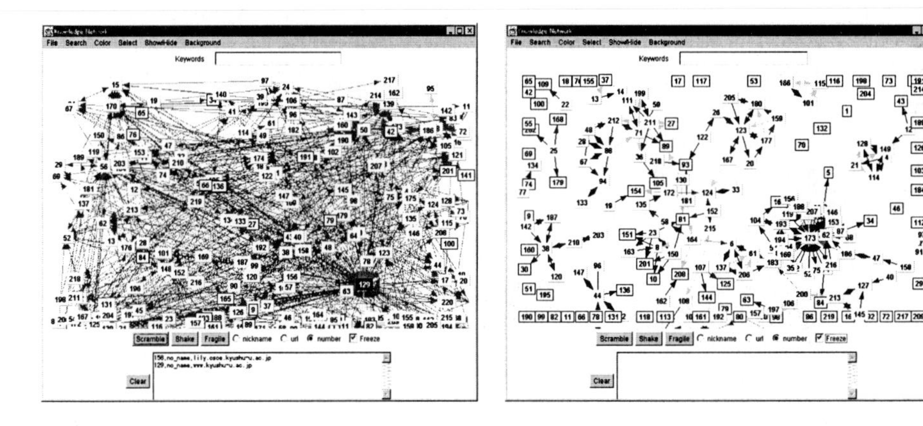

Fig. 4. Web Graph of kyushu-u.ac.jp

Fig. 5. Sites with out-degree ≤ 8

2.3 Related Seminar Pages

Fig. 6 is the top page of a seminar in Dynamic Diagram [4]. The page contains links to each section of the seminar and to the other main pages of the company, e.g., "Home","Product","About" and "What's new" which are displayed as icons around the screen. A naive drawing Fig. 7 of the link structure between these

[4] http://www.dynamicdiagrams.com/seminars/mapping/maptoc.htm

pages is very complex. If we elimitate the top page, the relation between pages became clear (Fig. 8). The main pages of the company that appear as icons have many links, but are irrelevant to this page. They are the cause of the complexity of the graph. By deleting these pages, we see a clear structure of the seminar as Fig. 9.

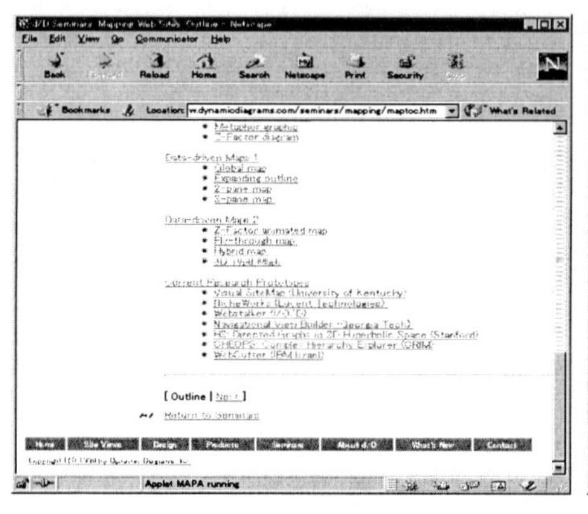

Fig. 6. Top Page of Seminar

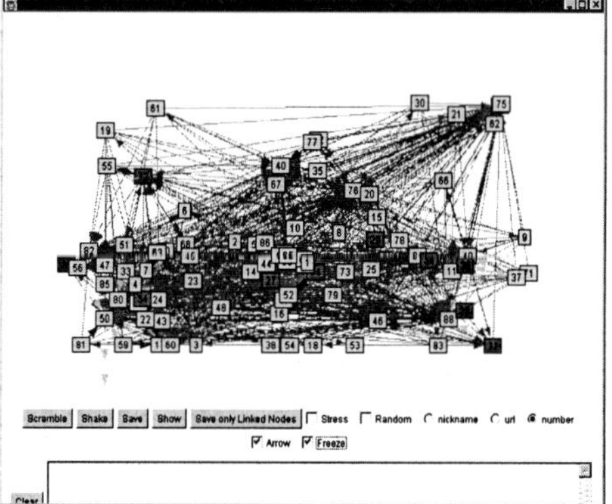

Fig. 7. Web Graph of Seminar Pages

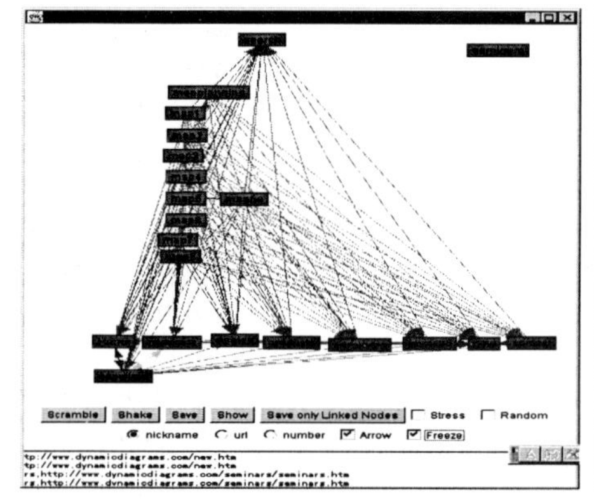

Fig. 8. Clear Arrangement of Nodes

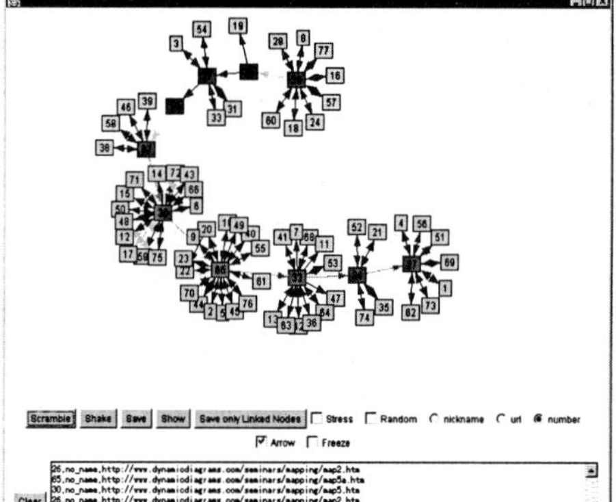

Fig. 9. Deletion of Icons

2.4 Links of Human Interface

The journal of Japanese Society of Artificial Intelligence has a column of "my book marks". Fig. 10 is the Web graph for a bookmark on the human interface [5]. We see some characteristic nodes. The node 89 seems to be a center of a small

[5] http://wwwsoc.nacsis.ac.jp/jsai/journal/mybookmark/14-4.html

group. The node 28 is a center of a cluster. The node 65, which seems to be a hub, is the page of links on human computer interaction by Prof. Kishi [6]. We noticed that the nodes linked from 65 on the left are in Japanese domain, while the nodes on the right are in US domain. By eliminating the node 65, the big cluster was separated into two parts (see Fig. 11).

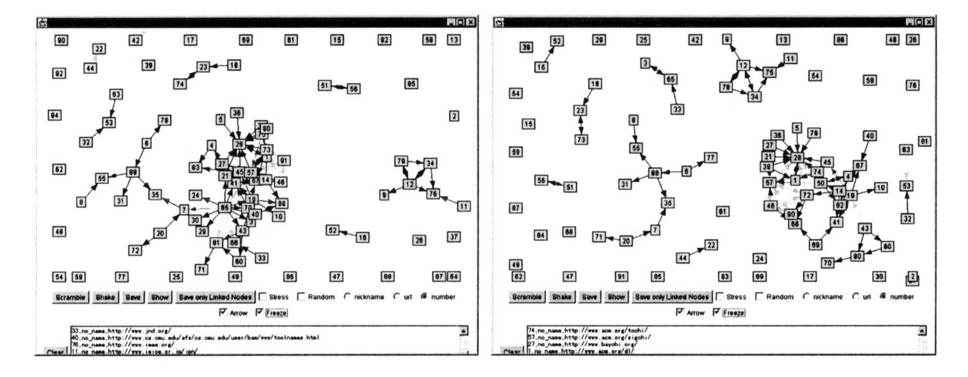

Fig. 10. Links on Human Interface

Fig. 11. Deletion of node 65

3 Link Matrix and Web Communities

The Web graph of a set of URLs is a directed graph $G = (V, E)$ whose nodes V are Web pages and whose edges E are links between pages. As we show in the previous section, the visualization of Web graph is intuitively understandable when the size is not too large and the structure is not too complex. To analyze a large and complex graph we use a numerical method. The link matrix A_{ij} of the graph is defined as follows: $a_{ij} = 1$ if the node v_i has a link to the node v_j, and $a_{ij} = 0$ otherwise. Fig. 12 shows a Web graph with 13 nodes. The edge from the node 1 to the node 7 represents a hyperlink from 1 to 7. In [12], Kleinberg introduced the notion of *authority* and *hub*. It is based on a simple idea that a good authority is referred from good hubs and a good hub refers to good authorities. In the small example of Fig. 12, the nodes 7,8 and 10 are authorities in the sense that they are linked from other nodes, while the nodes 9 and 5 are hubs in the sense that they have more links compared with other nodes. Given a node v_i, the degree x_i of authority and the degree y_i of hub of the nodes can be represented as the fixed point of $x_i = \sum_{(v_j,v_i)\in E} y_j$, $y_i = \sum_{(v_i,v_j)\in E} x_j$. These equation represent the iteration of HITS algorithm [12]. This equation can be represented with the matrix A as $x = A^t y, y = Ax$, where the authority weights corresponds to the eigenvector for the principal eigenvalue of the matrix AA^t and the hub weights corresponds to the the eigenvector for the principal eigenvalue of the matrix A^tA. In [12], Kleinberg showed that non-principal eigenvectors can be used to discover Web communities.

[6] http://sighi.tsuda.ac.jp/hcisite.html

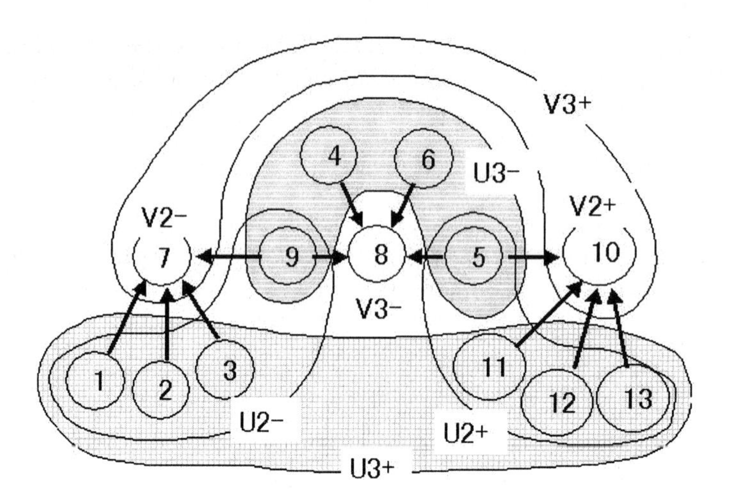

Fig. 12. Communities in Web Graph

It is known that the eigenvalues and eigenvectors can be obtained by the singular value decomposition (SVD) $A = U\Sigma V^t$ [5, 4]. The link matrix of Fig. 12 has the SVD with $\Sigma = \Delta[2.327, 2.000, 1.608]$ and U, V as follows.

U_1	U_2	U_3	V_1	V_2	V_3
0.215	−0.354	0.311	0.000	0.000	0.000
0.215	−0.354	0.311	0.000	0.000	0.000
0.215	−0.354	0.311	0.000	0.000	0.000
0.304	0.000	−0.440	0.000	0.000	0.000
0.519	0.354	−0.129	0.000	0.000	0.000
0.304	0.000	−0.440	0.000	0.000	0.000
0.000	0.000	0.000	0.500	−0.707	0.500
0.000	0.000	0.000	0.707	0.000	−0.707
0.519	−0.354	−0.129	0.000	0.000	0.000
0.000	0.000	0.000	0.500	0.707	0.500
0.215	0.354	0.311	0.000	0.000	0.000
0.215	0.354	0.311	0.000	0.000	0.000
0.215	0.354	0.311	0.000	0.000	0.000

The hub weights for each node appears in the first column U_1 of the eigenvector U. The authority weights appears in V_1. The *positive ends* U_2^+, V_3^+, i.e., the set of nodes with positive large absolute in the non-principal eigenvectors U_2, V_3, and the *negative ends* U_2^-, V_3^-, i.e., the set of nodes with positive large absolute in the non-principal eigenvectors U_2 and V_3, are shown in Fig. 12. We see that the nodes in the positive end of hubs and the nodes in the positive end of of authority connect to each other. This is the case for the negative ends. On the other hand, the ends with opposite sign are usually separated. Thus the positive end (U_i^+, V_i^+) and the negative end (U_i^-, V_i^-) form separated two communities. In Fig. 12, we see that (U_2^+, V_2^+) and (U_2^-, V_2^-) form two communities.

4 Analysis of Web Community and Web Space

Given a set of Web pages, we can collect many web pages that link to a page in the set or that are linked from a page in the set. We call the given set as *root set* and the collected pages as *Web space* generated from the root set [7]. We use about 100 links and 100 inverse links to generate Web space.

We show four experiments from [9] where the root sets are

(1) JAVA: the search result for a query of "+java +introduction" to AltaVista[8],
(2) WBT: the search result for a query of "WBT" to goo[9],
(3) HCI: the link page on human interface [10] and
(4) WebKB: the link page on the internet and inference[11].

Table 1 shows the number of pages in root set, Web space and the number of links.

Table 1. The number of pages

	root set	Web space	link
JAVA	190	9169	9464
WBT	107	1212	1478
HCI	87	9169	11410
WebKB	25	3338	3525

In this section, we show the positive and negative ends for JAVA and WBT. Keyword extraction succeeds for some pair of positive ends and negative ends. In the next section we show that we can evaluate how large and how complex is the Web space using the scree plot of eigenvalues for the link matrix.

4.1 Web Space on JAVA

Table 2 shows the number of authorities and hubs in positive and negative ends of top 10 eigenvectors.

The authorities, i.e., the positive end of the principal eigenvector consist of 52 URLs of so-called DOT.COM companies, e.g., internet.com, javascript.com, webdeveloper.com, 25 of which are pages in internet.com. Hubs consist of 2 pages in wdvl.com, 2 pages in stars.com, 1 page in javaboutique.internet.com, and 1 page webdevelopersjournal.com. The positive end of the 1st non-principal eigenvector are 24 pages in DOT.COM companies, e.g., sun, netscape, ibm, javacats, javaworld, microsoft, zdnet, stars, wdvl, webreview and 10 pages in universities, e.g., clarku, cs.princeton, ecst.csuchico, ei.cs.vt, genome-www.stanford,

[7] Kleinberg used the term "base set" instead of "Web space" in [12]

Table 2. Positive and Negative Ends of JAVA

	authority		hub		eigenvalue
i	+	-	+	-	
0	52	0	6	0	205.497275
1	41	0	0	148	62.912842
2	0	2	0	91	59.000000
3	9	30	16	149	58.063427
4	32	9	96	0	55.726074
5	52	18	18	164	55.364432
6	17	29	98	2	55.022854
7	26	49	151	198	54.450700
8	35	38	13	245	53.984467
9	31	45	202	106	53.447097
10	34	41	114	67	53.066337

Table 3. Positive and Negative Ends of WBT

	auth.		hub		eigenvalue
i	+	-	+	-	
0	4	0	0	213	385.522470
1	3	0	141	0	252.039189
2	4	3	599	0	202.882493
3	6	1	201	398	200.769527
4	3	4	199	398	198.286993
5	0	67	0	4	57.537956
6	0	37	2	0	37.027756
7	37	30	2	1	37.017581
8	2	0	0	17	33.000000
9	1	1	13	14	13.499328
10	1	1	10	10	10.000000

and 7 other pages. The second non-principal eigenvector has only negative end with two pages in eastjava.com. We applied the keyword extraction algorithm [2,3] for the pair of positive end C_i^+ and negative end C_i^-. The keyword extraction is successful for C_2^- obtaining the keywords "indonesia", "llpadding", "border", "ffffff", "about", "ation", "ction", "ional", "tions", "from", "ight", "ment", "ness", "page", "site", "this" and "ting". Thus we can say that the negative end of 2nd non-principal eigenvector is characterized as the introduction to the Java Island of Indonesia. French words are extracted in the 11th and 12th negative ends, while Japanese words, e.g., hokkaido, nikkei, asahi, keio, kobe are extracted in the 16th negative ends.

4.2 Web Space on "WBT"

Table 3 shows the number of authorities and hubs in the positive and the negative ends of the top 10 eigenvectors for the Web space of "WBT". The positive end of the principal eigenvector contains 2 pages of AVCC (Advanced Visual Communication Center) [12] and a page of JPSA (Japan Personal Computer Software Association) [13]. Both of them are about "Web Based Training" as we expected. But it contains a personal page on "Windows Based Terminal". The positive end of the 1st non-principal vector contains the page of JPSA and 2 pages of JMBSC (Japan Meteorological Business Support Center) on Weather Business Today [14]. Thus the positive end of the 1st non-principal vector contains two different kinds of pages. This is due to the fact that there is a page in the positive end of the 1st non-principal vectors that links both to the weather report and to the pages on personal computer. It is a hub page concerning license. Therefore the keyword extraction does not work well for this positive end.

[12] http://www.avcc.or.jp
[13] http://www.jpsa.or.jp
[14] http://www.jmbsc.or.jp/wbt/new.htm

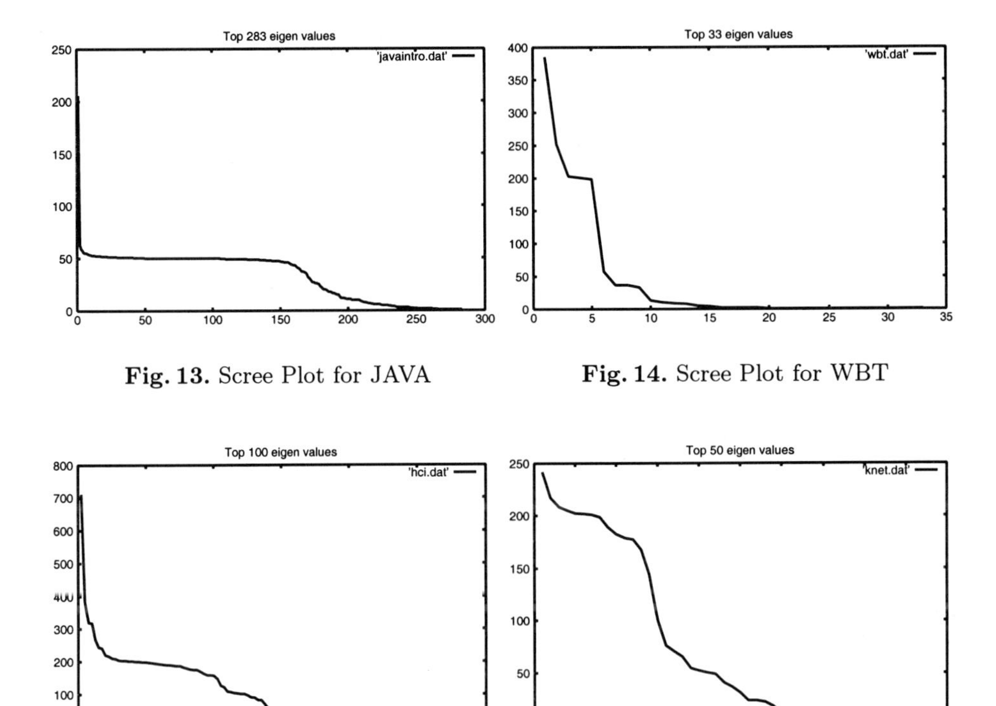

Fig. 13. Scree Plot for JAVA

Fig. 14. Scree Plot for WBT

Fig. 15. Scree Plot of HCI

Fig. 16. Scree Plot of WebKB

5 Analysis of Web Space with Scree Plots

Fig. 13,14,15 and 16 show the scree plots of eigenvalues of the link matrix for JAVA, WBT, HCI and WebKB. In Fig. 13, the principal eigenvalue is 205.49 and most of the next 150 eigenvalues have the values around 50. It means that the Web space for "java+introduction" has the huge main component and there are 150 secondary components of the same size. So this space can be considered a matured society. The scree plot for WBT drops rapidly in Fig. 14. Since the cumulative contribution of the top 5 eigenvalues is over 80 %, we can estimate that the rank of the Web space is 5. The 3rd and 6th communities contain another two keywords "What's BoTtle mail" and "World Beach Trip". Thus we can estimate the degree of this Web space as 5. By the scree plots of Fig. 13 and Fig. 14, we can compare the spread of the web space and see that the web space of WBT is immatured and small. The scree plot for the Web space of HCI' looks similar to that of JAVA. On the other hand the scree plot for the Web space of WebKB is similar to that of WBT.

6 Conclusion

Web graph is a representation of knowledge on WWW. We developed an interactive visualization system for construction and analysis of Web graphs. To improve the efficiency, we developed a link database which stores the link information. The spring model is used for automatic drawing of the graph. The effectiveness of the system is shown for several examples of Web graphs and some discovery that are unforseen by other approach are shown. To cope with the large Web graphs with more than 5,000 nodes, the SVD is applied for the link matrix of the graph. It is clarified how the web communities are found from the eigenvectors. A method for extraction keywords for the communities is proposed. It is shown that the scree plots of the singular values can be used as a measure of the size of the Web space.

We encountered several difficulties in our experimental analysis of Web graphs, some of which are not solved yet. One of such problem is the inefficiency of the graph drawing, which would be improved by using the tree-code [17]. We showed that we can clarify the complex Web graph by deleting some nodes. But there is no characterization of these nodes and the evaluation of the clarity of the graph. The component analysis is a traditional method and would a powerful approach for analysis of Web graphs. But the cost of SVD computation is very hight, since a Web graph may have 10,000 nodes. Possibilities of efficiency improvement would be in applying characteristics and modeling of Web graphs.

References

1. Albert, R., Jeong, H., and Barabasi, A.: Diameter of the World Wide Web. Nature, 401, pp.130–131, 1999.
2. H. Arimura and S. Shimozono, Maximizing Agreement with a Classification by Bounded or Unbounded Number of Associated Words. Proc. the 9th International Symposium on Algorithms and Computation,1998.
3. H. Arimura, A. Wataki, R. Fujino, and S. Arikawa, A Fast Algorithm for Discovering Optimal String Patterns in Large Text Databases. Proc. the 8th International Workshop on Algorithmic Learning Theory, Otzenhausen, Germany, Lecture Notes in Artificial Intelligence 1501, Springer-Verlag, pp. 247–261, 1998.
4. M. Berry, S. T. Dumains, and G. W. O'brien, Using Linear Algebra for Intelligent Information Retrieval, SIAM Reveview Vol. 37, pp. 573–595, 1995.
5. M. W. Berry, Z. Drmac, and E. R. Jessup, Matrices, Vector Spaces, and Information Retrieval, SIAM Review Vol. 41, pp. 335–362, 1999.
6. S. Brin and L. Page, The Anatomy of a Large-Scale Hypertextual Web Search Engine, Proc. WWW7, 1998.
7. P. Eades, A heuristics for graph drawing, Congressus Numeranitium, Vol.42, pp. 149–160, 1984.
8. S. Hirokawa and T. Taguchi, KN on ZK – Knowledge Network on Network Notepad ZK, Proc. the 1st International Conference on Discovery Science, Lecture Notes in Artificial Intelligence 1532, Springer-Verlag, pp. 411–412, 1998.
9. S. Hirokawa and D. Ikeda, Structural Analysis of Web Graph (in Japanese), Journal of Japanese Society for Artificial Intelligence, Vol. 16, No. 4, pp. 625–629, 2001.

10. D. Ikeda, Characteristic Sets of Strings Common to Semi-Structured Documents, Proc. the 2nd International Conference on Discovery Science, Lecture Notes in Artificial Intelligence 1721, Springer-Verlag, pp. 139–147, 1999.
11. D. Ikeda, T. Taguchi and S. Hirokawa, Developing a Knowledge Network of URLs, Proc. the 2nd International Conference on Discovery Science, Lecture Notes in Artificial Intelligence 1721, Springer-Verlag, pp. 328–329, 1999.
12. J. M. Kleinberg, Authoritative Sources in a Hyperlinked Environment, Proc. ACM-SIAM Symposium on Discrete Algorithms, pp. 668–677, 1998.
13. S. R. Kumar, S. Rajagopalan and A. Tomkins, Extracting large-scale knowledge bases from the web, Proc. 25th International Conference on Very Large Databases, 1999.
14. Lawrence, S., and Giles, C. L.: Accessibility of information on the Web, Nature, Vol.400, 107–109, (1999)
15. Matsumura,N., Ohsawa, Y., and Ishizuka, M.: Discovery of Emerging Topics between Communities on WWW, Proc. Web Intelligence, Lecture Notes in Artificial Intelligence 2198 (to appear).
16. T. Minami, H. Sazuka, S. Hirokawa and T. Ohtani, Living with ZK - An Approach towards Communication with Analogue Messages, Proc. 2nd International Conference on Knowledge-based Intelligent Electronic Systems, pp. 369-374, 1998.
17. A. Quigley and P. Eades, FADE: Graph Drawing, Clustering and Visual Abstraction, Proc. Graph Drawing 2000, Lecture Notes in Computer Science 1984, Springer-Verlag, pp. 197–210, 2001.

Knowledge Discovery in Auto-tuning Parallel Numerical Library

Hisayasu Kuroda[1], Takahiro Katagiri[2], and Yasumasa Kanada[1]

[1] Computer Centre Division, Information Technology Center, The University of Tokyo
{kuroda, kanada}@pi.cc.u-tokyo.ac.jp
[2] Research Fellow of the Japan Society for the Promotion of Science
2-11-16 Yayoi, Bunkyo-ku, Tokyo 113-8658, JAPAN
Phone: +81-3-5841-2736, FAX: +81-3-3814-2731
katagiri@pi.cc.u-tokyo.ac.jp

Abstract. This paper proposes the parallel numerical library called ILIB which realises auto-tuning facilities with selectable calculation kernels, communication methods between processors, and various number of unrolling for loop expansion. This auto-tuning methodology has advantage not only in usability of library but also in performance of library. In fact, results of the performance evaluation show that the auto-tuning or auto-correction feature for the parameters is a crucial technique to attain high performance. A set of parameters which are auto-selected by this auto-tuning methodology gives us several kinds of important knowledge for highly efficient program production. These kinds of knowledge will help us to develop some other high-performance programs, in general.

1 Introduction

We are developing a numerical library particularly aiming at parallel numerical calculation. Generally, library developers do not know what problem is to be solved by their library and what kind of parallel computer is used for the calculation. Therefore, developing general purpose and wide coveraged high performance library is messy groundwork. Our library has some design features of the followings:

– The number of user parameters is null or, if any, small.
– An auto-tuning facility with calculation kernel is used.
– If we use parallel computers, an auto-tuning facility in communication method is used.
– An auto-tuning facility with selectable algorithm is used.

The auto-tuning is the facility to find the best set of parameters within various parameter sets so that the program with these parameters could be executed at the fastest speed. Such design features are constructed from the users point of view, e.g. user-friendliness. An auto-tuning facility not only improves the friendliness of the software, but also is useful for the performance improvement[1, 2]. The reason is as follows:

S. Arikawa and A. Shinohara (Eds.): Progress in Discovery Science 2001, LNAI 2281, pp. 628–639.

(1) A parameter which affects performance is corrected even if a user specifies a performance related ill-conditioned parameter value.
(2) A wide range of parameter which is difficult for a user to specify a correct parameter value is corrected.
(3) A parameter which is difficult for a user to specify correct parameter value in advance is automatically set at run-time.

ScaLAPACK[3] is well-known as one of freely available public parallel libraries on the distributed memory machines. However, it forces users to specify many kinds of parameter values. Therefore it is not easy for users to use. Other researches for auto-tuning and development of software with auto-tuning include developing numerical libraries such as PHiPAC[4], ATLAS[5], and FFTW[6]. Unfortunately, these pieces of software are not designed originally for parallel processing and therefore they do not show good performance in parallel environments. Because, they do not focus to the parallel processing and automatic optimization in communication process on the distributed memory machines. Their subject of optimization is limited to basic linear algebra operations such as matrix-matrix operations.

It is our policy to develop library which is tuned automatically in the application level. Under those backgrounds, we are developing library called ILIB (Intelligent LIBrary) which is a parallel numerical library and contains direct methods (LU decomposition, tridiagonalization in eigenvalue problems) and the iterative methods (CG and GMRES which solve linear equation systems with sparse coefficient matrices).

We already have developed a library which mainly delivers high performance in some distributed memory machines ranging from vector processors to PC clusters. In recent years, since PC-cluster systems can be produced much cheaper, they are becoming widely used for doing numerical calculations. In this article, we mainly report the performance of the ILIB on a PC-cluster system.

2 Features of Current Version of ILIB

2.1 Linear Systems of Equations

Our library for linear systems of equations can get vector x on any equation (1).

$$Ax = b \tag{1}$$

Here a coefficient matrix $A \in \Re^{n \times n}$ is a real nonsymmetric matrix, a right-hand side vector b is $b \in \Re^n$, and an unknown vector x is $x \in \Re^n$.

As the solving methods for the Equation (1), both of direct methods and iterative methods are known. The ILIB includes the following solving methods:

- Direct method : LU decomposition for dense matrix (`ILIB_LU`[7, 8]),
- Direct method : Particular LU decomposition for skyline matrix (`ILIB_RLU`[7, 8]),
- Iterative method : GMRES(m) for sparse matrix (`ILIB_GMRES`[9, 10]),
- Iterative method : GCR(m) for sparse matrix (`ILIB_GCR`[11]).

All methods have been implemented to tune performance automatically. We report the performance of ILIB_GMRES in this article.

2.2 Eigenvalue Problems

Our library for eigenvalue problems can diagonalize the following Equation (2).

$$X^{-1}AX = \Lambda \tag{2}$$

Here $A \in \Re^{n \times n}$ is a real symmetric matrix, $\Lambda = \text{diag}(\lambda_i)$ is a diagonal matrix which contains eigenvalues λ_i $(i = 1, 2, \cdots, n) \in \Re$ as the i-th diagonal elements, and $X = (x_1, x_2, \cdots, x_n)$ is a matrix which contains eigenvectors x_i $(i = 1, 2, \cdots, n) \in \Re^n$ as the i-th row vectors, where n is the size of the problem. The decomposition (2) is performed by using the well-known method of the Householder-bisection method[12].

We have applied auto-tuning methodology to the following routines.

- Tridiagonalization by the Householder transformation (ILIB_TriRed[13, 1])
- Gram-Schmidt orthogonalization for the inverse-iteration method (ILIB_MGSAO[14])
- Householder inverse transformation (ILIB_HouseInv[14])

We report the performances for the above routines in this article.

3 Auto-tuning Facilities

3.1 Linear Systems of Equations

In the case of sparse matrix iterative methods, the distribution pattern of non zero elements of the coefficient matrix affects performance. Therefore, the auto-tuning cannot be applied in advance, before running program. For this reason, the ILIB routine for the linear equation system is implemented so that the auto-tuning can be applied dynamically at run-time.

Auto-tuning facilities in the ILIB_GMRES. Auto-tuning parameters for the ILIB_GMRES are as follows:

(i) The number of stride in loop unrolling for matrix-vector products,
(ii) Communication method in matrix-vector products,
(iii) Preconditioning selection,
(iv) Specific process in GMRES(m).

For the case (i), the optimal number of stride in loop unrolling is determined. The ILIB prepares several kinds of loop unrolling codes in advance. Since the distribution pattern of non zero elements in the matrix is constant during iteration, the actual execution time of the product is measured at execution time in order to choose the best unrolling code.

For the case (ii), the optimal communication method is determined. There are two types of methods such as one-to-one communication method for small communication traffic and collective communication method with synchronization. As in the case (i), the actual execution time of the communication is measured once at the execution time in order to choose the best communiction method.

For the case (iii), the optimal preconditioning method is selected. Automatic selection of preconditioning method is difficult, because the optimal preconditioning method depends on the nature of problems and machine architectures to be used. For choosing the optimal method, the ILIB iterates the main loop twice with each preconditioning method, and the preconditioning method whose relative decrease of the residual norm is the largest is selected.

For the case (iv), other than the above and specific processes of GMRES(m) is determined. For example,

(1) Restarting frequency, and
(2) Gram-Schmidt orthogonalization

affect the overall performance.

For the case (1), the ILIB dynamically changes the value of restarting frequency m such as $m = 2, 4, 6, 8, \cdots$. The maximum value of m is limited according to the available memory size[15].

For the case (2), the optimal orthogonalization method is to be determined. Modified Gram-Schmidt orthogonalization (MGS) is often used on single processor machines because of its less computational error. However on the distributed memory machines, it is not efficient because of the frequent synchronization. The ILIB measures the orthogonalization time by each method and then we select the fastest method.

3.2 Eigenvalue Problems

In the case of eigenvalue problems, auto-tuning process can be performed before execution because the nature of the problem does not affect the calculation process.

Auto-tuning facilities in the ILIB_TriRed. This routine does not use a block algorithm which makes efficient reuse of cache data. The calculation kernel is of type of BLAS2, that is, matrix-vector product. Auto-tuning facilities for the ILIB_TriRed are as follows:

(i) Matrix-vector product : the number of stride for unrolling of $A^{(k)}u_k$
(ii) Matrix update : the number of stride for unrolling of
$A^{(k+1)} = A^{(k)} - u_k(x_k^T - \mu u_k^T) - x_k u_k^T$,
$k = 1, 2, \cdots, n-2$ ($u_k, x_k \in \Re^{n-k+1}$, $\mu \in \Re$)
(iii) Method of vector reduction in (i)

Auto-tuning facilities in the ILIB_MGSAO. This routine uses block algorithm. The calculation kernel is of type of BLAS3, that is, matrix-matrix product. Auto-tuning facilities for the ILIB_MGSAO are as follows:

(i) Kernel of pivotted PE (which holds essential data for execution of calculation) :
The number of stride for unrolling of $u_1^{(j)} = u_1^{(j)} - (u_1^{(i)T} \cdot u_1^{(j)})u_1^{(i)}$, $j = k, k+1, \cdots, n$, $i = k, k+1, \cdots, j+1$, $k = 1, 2, \cdots, n/(\text{Num of PEs})$
(ii) The number of stride for unrolling of other than the above.
(iii) The block width BL which is the number of stride for the reference data in the block algorithm.

For the case (iii), BL is the parameter which affects communication process because communication is done at each BL times calculation stage.

Auto-tuning facilities in the ILIB_HouseInv. This routine is not based on the block algorithm. The calculation kernel is of type of BLAS1, that is, vector-vector operations. Auto-tuning facility for the ILIB_HouseInv is as follows:

(i) Calculation Kernel : The number of stride for $u_k^{(i)} = u_k^{(i)} - \mu u_k^{(j)}$, $k = 1, 2, \cdots, n-2$, $i = 1, 2, \cdots, n/(\text{Num of PEs})$, $j = 1, 2, \cdots, n$

4 Experimental Results

We have evaluated the performances of the ILIB routines on a PC-cluster system which is 4 node system and processors used for each node are Intel PentiumIII 800MHz Dual with 512MB memory (SDRAM, ECC, PC100, CL=2). Network Interface Card, NIC, used by the communication library of MPI (Message Passing Interface) is PCI 10/100Mbps Intel EtherExpressPro100+, and mother board for dual Intel PentiumIII is Intel L440GX+ Server Board (FSB 100MHz). One node runs Linux 2.2.13-33smp OS. The compiler used for the experiment is PGI Fortran90 compiler 3.2-3 and the compiler options used are -O0 (no optimization) and -fast (most optimized option, equivalent to "-O2 - Munroll -Mnoframe" for IA-32 architecture CPU).

Six parameters for auto-tuning are as follows:

- Matrix-vector product unrolling type = {non-prefetch, prefetch}
- Matrix-vector product unrolling size (row) = {1, 2, 3, ···, 8, 9}
- Matrix-vector product unrolling size (column) = {1, 2, 3, 4, 8}
- Matrix-vector product communication method = {`MPI_ALLGATHER`, `Gather` → `MPI_BCAST`, `MPI_ISEND` → `MPI_IRECV`, `MPI_IRECV` → `MPI_ISEND`, `MPI_SEND` → `MPI_RECV`}
- Orthogonalization type = {CGS, IRCGS, MGS}
- Preconditioning type = {none, polynomial[15], Block ILU[15]}

The "non-prefetch" code executes codes as it is written. The "prefetch" code realizes off indirect access on the element of arrays. It is re-constructed to change flow dependency to loop propagation[9] so that pipeline process can be performed. `MPI_ALLGATHER` is a collective communication method, `MPI_BCAST` is a broadcast communication method, `MPI_SEND` is a synchronous one-to-one send communication method, `MPI_RECV` is a synchronous one-to-one receive communication method, and `MPI_IRECV` is an asynchronous one-to-one receive communication method.

4.1 Linear Systems of Equations

Iterative method for sparse matrix ILIB_GMRES. We evaluate the performances with the following problem whose maximum number of non zero elements of matrix at each row is 7.

– An elliptic boundary value problem by the partial differential equation:

$$-u_{xx} - u_{yy} - u_{zz} + Ru_x = g(x, y, z)$$
$$u(x, y)|\partial\Omega = 0.0$$

Here the region is $\Omega = [0,1]\times[0,1]\times[0,1]$, and R=1.0. The right-hand side vector b is set to the exact solution of $u = e^{xyz}\sin(\pi x)\sin(\pi y)\sin(\pi z)$. We discretize the region by using a 7-point difference scheme on a $80 \times 80 \times 80$ mesh. The size of matrix is 512,000(=80^3).

Remarks. Tables 1 (a-1) and (a-2) show the execution times and automatically selected methods. Abbreviations have the following meanings.

Unrolling : For example, P(2,3) means "prefetch" code, strides two for outer loops expansion and three for inner loops expansion, N(2,3) means "non-prefetch" code, strides two for outer loops expansion and three for inner loops expansion.

Communication : "Send" means that `MPI_Send` and `MPI_Recv` are used. "Isend" means that `MPI_Isend` and `MPI_Irecv` are used in this sequence. "Irecv" means that `MPI_Irecv` and `MPI_Isend` are used in this sequence.

Tables 1 (a-1) and (a-2) show that according to the number of executed PEs and compiler option, optimal parameter sets were changed. Interestingly, when compiler option is set to "-fast", if we use 2 processors in one node, "prefetch" code is always selected. That would be because it tries to avoid memory access confliction. On the other hand, when compiler option is set to "-O0", the largest number of stride for "unrolling" code is always selected. When the optimal parameter is not set such as comparative parameter set of 1, 2 or 3, the execution time is always smaller comparison with auto-tuning case. Especially, the effect of autotuning becomes increasingly prominent according to the change in the number of executed PEs.

Table 1. Efficiency of ILIB_GMRES on the PC-cluster system

(a-1) GMRES(m) method `ILIB_GMRES`. (Compiler option: -fast)
Time in second includes the overhead of auto-tuning.
Par.1 : Unrolling is not applied.
Par.2 : MPI_Allgather is forcibly used as communication method.
Par.3 : MGS is forcibly used as orthogonalization method.

PEs (Nodes)	1 (1)	2 (1)	2 (2)	4 (2)	4 (4)	8 (4)
Time in second (1)	664.2	645.1	363.6	332.3	195.6	197.2
Iteration	1049	1024	1024	1004	1004	1004
Unrolling	N(1,7)	P(3,7)	N(1,7)	P(3,7)	N(1,7)	P(4,7)
Communication	—	Isend	Irecv	Send	Irecv	Send
Orthogonalization	CGS	CGS	CGS	CGS	CGS	CGS
Preconditioning	None	None	None	None	None	None
Time with Par.1 (2)	757.0	640.8	377.5	343.5	201.8	205.6
Speedup ratio (2)/(1)	1.14	0.99	1.04	1.03	1.03	1.04
Time with Par.2 (3)	—	759.9	744.2	806.8	908.8	1452.7
Speedup ratio (3)/(1)	—	1.18	2.05	2.43	4.65	7.37
Time with Par.3 (4)	686.8	647.7	357.7	351.1	210.8	219.7
Speedup ratio (4)/(1)	1.03	1.00	0.95	1.06	1.08	1.11

(a-2) GMRES(m) method `ILIB_GMRES`. (Compiler option: -O0)

PEs (Nodes)	1 (1)	2 (1)	2 (2)	4 (2)	4 (4)	8 (4)
Time in second (1)	981.7	684.7	500.2	379.3	277.3	222.5
Iteration	1051	991	991	1038	1038	1008
Unrolling	N(8,7)	N(8,7)	N(8,7)	N(8,7)	N(8,7)	N(8,7)
Communication	—	Irecv	Send	Send	Isend	Send
Orthogonalization	MGS	CGS	CGS	CGS	CGS	CGS
Preconditioning	None	None	None	None	None	None
Time with Par.1 (2)	1125.5	747.2	576.9	400.0	309.2	232.2
Speedup ratio (2)/(1)	1.15	1.09	1.15	1.05	1.12	1.04
Time with Par.2 (3)	—	792.1	869.7	874.2	1049.3	1476.1
Speedup ratio (3)/(1)	—	1.16	1.74	2.30	3.78	6.63
Time with Par.3 (4)	—	716.7	515.4	379.6	285.9	248.1
Speedup ratio (4)/(1)	—	1.05	1.03	1.00	1.03	1.12

4.2 Eigenvalue Problems

Tridiagonalization ILIB_TriRed. Three parameters for auto-tuning are as follows:

(I) The number of unrolling for matrix-vector product (`TMatVec`) = {1, 2, ⋯, 6, 8, 16}
(II) The number of unrolling for matrix update (`TUpdate`)= {1, 2, ⋯, 6, 8, 16}
(III) The communication method (`Tcomm`)= {`Tree`, `MPI`}

The processes in (I) and (II) are BLAS2 type operations. For the case (III), "`Tree`" means a routine based on binary tree-structured communication method and "`MPI`" means communication method with a MPI function `MPI_ALLREDUCE`. The default parameters set is (`TMatVec`, `TUpdate`, `Tcomm`) = (8, 6, `Tree`).

MGS orthogonalization ILIB_MGSAO. Five parameters for auto-tuning are as follows:

(I-1) The number of unrolling for outer loop expansion for updating pivot block (`MBOuter`) = {1, 2, 3, 4}
(I-2) The number of unrolling for second loop expansion for updating pivot block (`MBSecond`) = {1, 2, ···, 6, 8, 16}
(II-1) The number of unrolling for outer loop expansion for other than above (`MOOuter`) = {1, 2, 3, 4}
(II-2) The number of unrolling for second loop expansion for other than above (`MOSecond`) = {1, 2, ···, 6, 8, 16}
(III) The block width (`MBlklen`) = {1, 2, ···, 6, 8, 16}

The processes from (I-1) to (II-2) are BLAS3 type operations. The default parameter set is (`MBOuter`, `MBSecond`, `MOOuter`, `MOSecond`, `MBlklen`) = (4,8,4,8,4).

Householder inverse transformation ILIB_HouseInv. A parameter for auto-tuning is as follow:

(I) The number of unrolling for matrix update (`HITkernel`) = {1, 2, 3, ···, 8, 16}

Table 2. Efficiency of ILIB_TriRed on the PC-cluster system

(b-1) Tridiagonalization ILIB_TriRed [time is in sec.] (Compiler option: -fast)

Tuning	100 dim.	200 dim.	300 dim.	400 dim.
Off	0.284	0.593	0.951	1.38
On	0.282	0.592	0.950	1.38
(Parameter)	(1,16,Tree)	(8,5,Tree)	(5,16,Tree)	(16,8,Tree)
Speedup ratio	1.007	1.001	1.001	1.000

Tuning	1000 dim.	2000 dim.	4000 dim.	8000 dim.
Off	8.14	44.8	136	304
On	8.07	44.4	134	298
(Parameter)	(8,8,Tree)	(8,16,Tree)	(8,16,Tree)	(6,16,Tree)
Speedup ratio	1.008	1.009	1.014	1.020

(b-2) Tridiagonalization ILIB_TriRed [time is in sec.] (Compiler option: -O0)

Tuning	100 dim.	200 dim.	300 dim.	400 dim.
Off	0.286	0.619	0.997	1.49
On	0.285	0.606	0.999	1.48
(Parameter)	(6,16,Tree)	(5,4,Tree)	(3,4,Tree)	(6,5,Tree)
Speedup ratio	1.003	1.021	0.997	1.006

Tuning	1000 dim.	2000 dim.	4000 dim.	8000 dim.
Off	9.26	52.4	159	363
On	9.26	51.7	157	357
(Parameter)	(8,16,Tree)	(16,4,Tree)	(16,4,MPI)	(16,4,Tree)
Speedup ratio	1.000	1.013	1.012	1.016

This process is BLAS1 type operation and forms double loops since calculation is needed for n times, where n is the number of converting vectors. Here, The outer loop can be unrolled. The default parameter is (`HITkernel`) = (1).

Remarks. Tables 2–4 show the execution times and selected parameters. Depending on the processing type or compiler option used, the speed-up ratio to the case by the default parameter is completely different. Especially, in Tables 3(c-1) and (c-2), MGS orthogonalization is effective for speeding up, e.g. 1.4–23.8 times speed-up. In Tables 2(b-1) and (b-2), tridiagonalization is speeded up only by 1.02 times. We found that the block algorithm which makes efficient reuse of cache data is quite important on the PC-cluster system.

In Table 4(d-1), when the complier option is set to "-fast", Householder inverse transformation cannot be speeded up. When the complier option is "-O0" as shown in Table 4(d-2), the loop unrolling code is faster than the code without unrolling. In Tables 3(c-1) and (c-2), for the dimensions of 100, 200, 500, and 600, no optimal complier option (uses only auto-tuning) case is faster than compile option of "-fast" case.

Table 3. Efficiency of ILIB_MGSAO on the PC-cluster system

(c-1) MGS orthogonalization ILIB_MGSAO [time is in sec.]
(Compiler option: -fast)

Tuning	100 dim.	200 dim.	300 dim.	400 dim.
Off	0.084	0.592	1.97	4.67
On	0.081	0.200	0.253	0.463
(Parameter)	(3,4,3,1,4)	(8,2,3,1,4)	(8,1,16,2,2)	(8,3,8,2,2)
Speedup ratio	1.0	2.9	7.7	10.1

Tuning	500 dim.	600 dim.	1000 dim.	2000 dim.
Off	9.20	15.9	73.8	591
On	6.46	8.12	5.50	24.8
(Parameter)	(8,3,8,1,2)	(8,3,1,1,2)	(8,2,1,1,4)	(8,4,3,2,8)
Speedup ratio	1.4	1.9	13.4	23.8

(c-2) MGS orthogonalization ILIB_MGSAO [time is in sec.]
(Compiler option: -O0)

Tuning	100 dim.	200 dim.	300 dim.	400 dim.
Off	0.086	0.613	2.03	4.84
On	0.036	0.135	0.274	0.544
(Parameter)	(3,3,2,1,3)	(8,2,3,1,4)	(8,2,5,2,2)	(8,2,4,2,2)
Speedup ratio	2.6	4.5	7.4	8.8

Tuning	500 dim.	600 dim.	1000 dim.	2000 dim.
Off	9.57	16.5	76.7	616
On	4.99	6.67	6.55	37.27
(Parameter)	(8,3,3,1,2)	(8,1,4,1,2)	(8,2,2,1,4)	(8,1,2,2,8)
Speedup ratio	1.9	2.4	11.7	16.5

Table 4. Efficiency of ILIB_HouseInv on the PC-cluster system

(d-1) Householder inverse transformation ILIB_HouseInv [time is in sec.]
(Complier option: -fast)

Tuning	100 dim.	200 dim.	300 dim.	400 dim.
Off	0.019	0.006	0.020	0.045
On	0.001	0.006	0.020	0.045
(Parameter)	(16)	(1)	(1)	(1)
Speedup ratio	19.0	1.00	1.00	1.00

Tuning	1000 dim.	2000 dim.	3000 dim.	4000 dim.
Off	3.89	35.7	125	305
On	3.89	35.7	117	305
(Parameter)	(1)	(1)	(2)	(1)
speedup ratio	1.00	1.00	1.06	1.00

(d-2) Householder inverse transformation ILIB_HouseInv [time is in sec.]
(Compiler option: -O0)

Tuning	100 dim.	200 dim.	300 dim.	400 dim.
Off	0.020	0.021	0.070	0.203
On	0.003	0.017	0.058	0.151
(Parameter)	(6)	(8)	(16)	(16)
Speedup ratio	6.66	1.23	1.20	1.34

Tuning	1000 dim.	2000 dim.	3000 dim.	4000 dim.
Off	5.91	48.7	165	392
On	4.89	43.7	140	349
(Parameter)	(4)	(2)	(2)	(2)
Speedup ratio	1.20	1.11	1.17	1.12

In tables 4(d-1) and (d-2), the feture of complier optimization affects efficiency of auto-tuning. If the optimization of program is abandoned to compiler, performance cannot be brought out. Therefore our approach of the ILIB is one solution in order to make high performance library on various machines. Because, in the ILIB, developer provides a lot of programs which help compilers with their optimization and which are written based on various algorithms.

5 Conclusion

We have evaluated the efficiency of the parallel numerical library called ILIB on the PC-cluster system. As a result, we have found that the auto-tuning facility which changes parameter settings automatically is especially important and effective for the execution speed and usability.

The parameter set which is selected by the auto-tuning library gives us several kinds of knowledge as shown in the previous section. By comparing the performances in [9, 10] and PC-cluster system in this report, we can characterize the computer architecture from the point of parallel processing.

We can add another new auto-tuning parameters based on these kinds of new knowledge discoveries. If a new parameter selection mechanism is added in closely to the selected parameter, it may be a better parameter selection mechanism than previous one.

The selected parameter set or new parameter selection mechanism also gives us knowledge for the nature of problems and some characteristic features of computers. If the nature of problem becomes clear, we can change the library to fit each dedicated problems. If the characteristic feature of computer becomes clear, we can develop some other high-performance programs or libraries by making use of new knowledge.

References

1. Katagiri, T., Kuroda, H., Ohsawa, K. and Kanada, Y.: I-LIB : An Automatically Tuned Parallel Numerical Library and Its Performance Evaluation, *JSPP 2000*, pp. 27–34 (2000). in Japanese.
2. Katagiri, T., Kuroda, H., Kudoh, M. and Kanada, Y.: Performance Evaluation of an Auto-tuned Parallel Eigensolver with a Super-computer and a PC-Cluster, *JSPP 2001*, pp. 73–74 (2001). in Japanese.
3. Blackford, L., Choi, J., Cleary, A., D'Azevedo, E., Demmel, J., Dhillon, I., Dongarra, J., Hammarling, S., Henry, G., Petitet, A., Stanley, K., Walker, D. and Whaley, R.: *ScaLAPACK Users' Guide*, SIAM (1997).
4. Bilmes, J., Asanović, K., Chin, C.-W. and Demmel, J.: Optimizing Matrix Multiply using PHiPAC: a Portable, High-Performance, ANSI C Coding Methodology, *Proceedings of International Conference on Supercomputing 97*, Vienna, Austria, pp. 340–347 (1997).
5. Whaley, R. C., Petitet, A. and Dongarra, J. J.: Automated Empirical Optimizations of Software and the ATLAS Project, *Parallel Computing*, Vol. 27, pp. 3–35 (2001).
6. Frigo, M.: A Fast Fourier Transform Compiler, *Proceedings of the 1999 ACM SIGPLAN Conference on Programming Language Design and Implementation*, Atlanta, Georgia, pp. 169–180 (1999).
7. Ohsawa, K., Katagiri, T., Kuroda, H. and Kanada, Y.: ILIB_RLU : An Automatically Tuned Parallel Dense LU Factorization Routine and Its Performance Evaluation, *IPSJ SIG Notes, 00-HPC-82*, pp. 25–30 (2000). in Japanese.
8. Ohsawa, K.: Performance Evaluation of Auto-tuned Sparse Direct Solver called ILIB_RLU, *Super Computing News*, Vol. 2, No. 5, pp. 23–36 (2000). Computer Centre Division, Information Technology Center, The University of Tokyo, in Japanese.
9. Kuroda, H., Katagiri, T. and Kanada, Y.: Performance of Automatically Tuned Parallel GMRES(m) Method on Distributed Memory Machines, *Proceedings of Vector and Parallel Processing (VECPAR) 2000*, Porto, Portugal, pp. 251 – 264 (2000).
10. Kuroda, H., Katagiri, T. and Kanada, Y.: Performance Evaluation of Linear Equations Library on Parallel Computers, *IPSJ SIG Notes, 00-HPC-82*, pp. 35–40 (2000). in Japanese.
11. Kudoh, M., Kuroda, H., Katagiri, T. and Kanada, Y.: A Proposal for GCR Methods with Less Memory Requirement, *IPSJ SIG Notes, 2001-HPC-85*, pp. 79–84 (2000). in Japanese.

12. Katagiri, T. and Kanada, Y.: A Parallel Implementation of Eigensolver and its Performance, *IPSJ SIG Notes, 97-HPC-69*, pp. 49–54 (1997). in Japanese.
13. Katagiri, T., Kuroda, H. and Kanada, Y.: A Methodology for Automatically Tuned Parallel Tri-diagonalization on Distributed Memory Parallel Machines, *Proceedings of Vector and Parallel Processing (VECPAR) 2000*, Porto, Portugal, pp. 265 – 277 (2000).
14. Katagiri, T.: *A Study on Large Scale Eigensolvers for Distributed Memory Parallel Machines*, Ph.D Thesis, The University of Tokyo (2001).
15. Kuroda, H. and Kanada, Y.: Performance of Automatically Tuned Parallel Sparse Linear Equations Solver, *IPSJ SIG Notes, 99-HPC-76*, pp. 13–18 (1999). in Japanese.

Extended Association Algorithm Based on ROC Analysis for Visual Information Navigator

Hiroyuki Kawano[1] and Minoru Kawahara[2]

[1] Department of Systems Science, Kyoto University, Kyoto 6068501, JAPAN,
kawano@i.kyoto-u.ac.jp
http://www.i.kyoto-u.ac.jp/~kawano/index.html

[2] Data Processing Center, Kyoto University, Kyoto 6068501, JAPAN,
kawahara@kudpc.kyoto-u.ac.jp
http://www.kudpc.kyoto-u.ac.jp/~kawahara/index.html

Abstract. It is very important to derive association rules at high speed from huge volume of databases. However, the typical fast mining algorithms in text databases tend to derive meaningless rules such as stopwords, then many researchers try to remove these noisy rules by using various filters. In our researches, we improve the association algorithm and develop information navigation systems for text data using visual interface, and we also apply a dictionary to remove noisy keywords from derived association rules. In order to remove noisy keywords automatically, we propose an algorithm based on the true positive rate and the false positive rate in the ROC analysis. Moreover, in order to remove stopwords automatically from raw association rules, we introduce several threshold values of the ROC analysis into our proposed mining algorithm. We evaluate the performance of our proposed mining algorithms in a bibliographic database.

1 Introduction

In the research fields of data mining and text/web mining, many algorithms have been proposed to discover interesting patterns, rules, trends and representations in various databases. Basic mining algorithms can derive quite simple patterns or rules as knowledge, it is very difficult to derive meaningful rules or knowledge in the viewpoints of human experts. However, even these simple patterns or rules may be helpful for beginners who don't have much background or domain knowledge in the interesting topics.

Focusing on the typical algorithm of association rules[1], we have been developing information navigation systems for semi-structured text data in Web databases and bibliographic databases. Our navigation system provides associative keywords to users, and this supporting function is very helpful for beginners as shown in Figure 1[2,3,4].

However, it is difficult for the administrator to adjust system parameters, such as the minimum support threshold $Minsup$ and the minimum confidence threshold $Minconf$, which are used in the mining association algorithm. Therefore, in order to determine the optimal thresholds[5], we proposed an algorithm

S. Arikawa and A. Shinohara (Eds.): Progress in Discovery Science 2001, LNAI 2281, pp. 640–649.

Netscape: Mondou: Bibliographic Navigation System with Text Mining
File Edit View Go Communicator Help
Bookmarks Location: http://veyron/INSPEC/mine2.htm What's Related
open text
Internet Search
Mondou: Bibliographic Navigation System
Enter word(s) in the field below and click on the Search button.
Select the indexed database to search INSPEC
Select the local database if you like Non
Search for knowledge Within Anywhere
And Within Anywhere
And Within Anywhere
And Within Anywhere
Search Clear
© Copyright 1996–1999 Infocom and Open Text Corporation. All rights reserved.

(a)Query Window

Netscape: Mondou: Bibliographic Navigation System with Text Data Mining
File Edit View Go Communicator Help
Bookmarks Location: http://veyron/Icgi/mine2.cgi What's Related
You can give new query using related keywords.
Local Databse Non Related keywords
based information models process neural networks design method control system model systems analysis
Moreover you can focus on results using related keywords.
based
information
models
process
neural
networks
design
method
control
system
model
systems
analysis
Local Databse Non Refine...
Documents 1 to 20 of 138827 documents containing: knowledge in regions
• ID: 6452729
TITLE: Thermodynamics of mixtures: Functions of mixing and excess functions
Full Journal Title: American Journal of Physics
AUTHOR: Nieto, R. Gonzales, M.C. Herrero, F.
ISSN: 0002-9505
Inclusive Page Numbers: 1096-9
PUBLISH DATE: Dec. 1999
• ID: 6452730

(b)Result Window

Fig. 1. Our navigation system for semi-structured data

which evaluates the performance of derived rules based on the **ROC** (Receiver Operating Characteristic) analysis[6, 7] And in order to modify query and display clusters of results, we have developed two kinds of visual interfaces shown in Figure 2.

However, the typical association algorithm tends to derive meaningless rules like stopwords, then many systems adopt filtering mechanisms to remove such

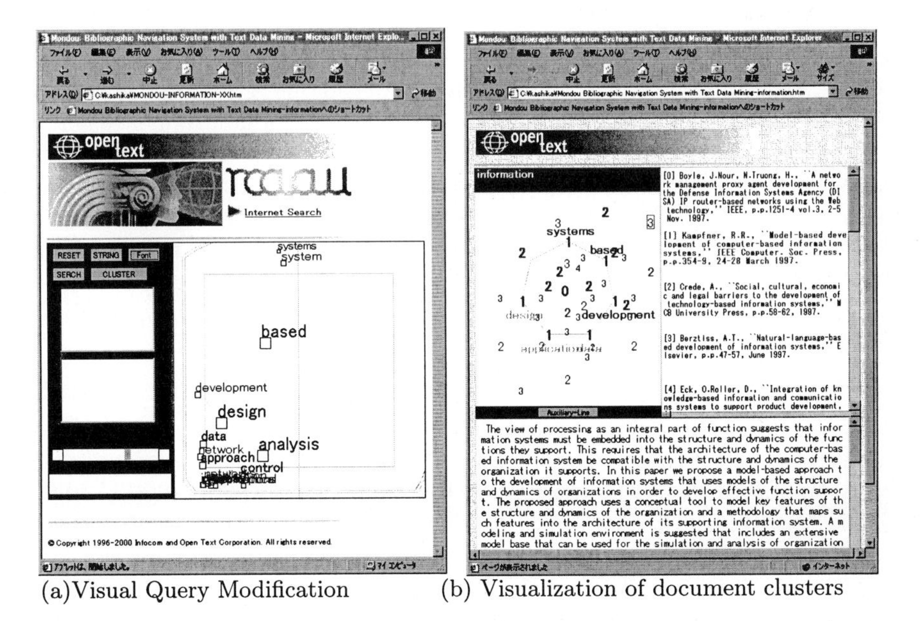

(a)Visual Query Modification (b) Visualization of document clusters

Fig. 2. Information Visualization and Query Interface

noises. In our previous research, we focused on the semantic relationship between the true positive rate and the false positive rate in the ROC analysis, and we introduced the threshold values into our proposed algorithm. As the results, our mining algorithm can remove the noisy keywords automatically and derive more effective rules[8].

In this paper, we extend the association algorithm by using several parameters in the ROC analysis, and we can derive more useful association rules. Moreover, by using bibliographic databases we evaluate the performance of our mining algorithm, and we show the effectiveness of the algorithm. We also show that our proposed algorithm can remove stopwords automatically from raw association rules.

In Section 2, we try to summarize our proposed algorithms to derive association rules. In Section 3, we introduce the ROC analysis and propose a mining association algorithm with the parameters in the ROC analysis. In Section 4, we evaluate the performance of our algorithm based on the experimental results in INSPEC database. Finally, we make concluding remarks in Section 5.

2 Mining Algorithm of Association Rules

"Apriori"[1] is one of the most popular data mining algorithms. In order to derive typical association rules fast, this kind of algorithms require two threshold parameters *support* and *confidence* to derive rules. Shortly speaking, given a set of transactions, where each transaction is a set of items, an association rule is represented by an expression $X \Rightarrow Y$, where X and Y are sets of items and $X \cap Y = \emptyset$. The intuitive meaning of such a rule is that transactions in the database which contain the items in X tend to also contain the items in Y. The *support* of the rule $X \Rightarrow Y$ is the percentage of transactions that contain both X and Y, hence $p(X \cup Y)$. And the *confidence* of the rule $X \Rightarrow Y$ is the percentage of transactions that contain Y in transactions which contain X, hence $p(Y \mid X)$.

Applying this algorithm to a collection of digital documents, we consider a keyword as an item and a part of documents as one transaction in the text database. We consider a set of keywords $\mathcal{Q}$ given in a query to X, a set of keywords $\mathcal{K}$ derived from $\mathcal{Q}$ to Y. And it is assumed that $\mathcal{B}$ is a set of tuples which contain $\mathcal{Q}$, $\mathcal{U}$ is the set of all tuples in the database and $\mathcal{C}$ is a set of tuples which contain both $\mathcal{Q}$ and $\mathcal{K}$ in the database. Then the *support* and the *confidence* of $\mathcal{K}$ are given by:

$$support(\mathcal{K}) = \frac{|\,\mathcal{C}\,|}{|\,\mathcal{U}\,|},\quad confidence(\mathcal{K}) = \frac{|\,\mathcal{C}\,|}{|\,\mathcal{B}\,|}. \tag{1}$$

If $\mathcal{K}$ holds both $support(\mathcal{K})$ not less than $Minsup$ and $confidence(\mathcal{K})$ not less than $Minconf$, then $\mathcal{K}$ is stored into a set of keyword sets $\mathcal{R}$ as an association rule enough to cause demands. Where $Minsup$ is the minimum threshold of *support* to judge whether $\mathcal{K}$ holds enough retrieval needs, and $Minconf$ is the minimum threshold of *confidence* to judge whether $\mathcal{K}$ holds enough confidence. Hence $\mathcal{R} = \{\mathcal{K}_1, \mathcal{K}_2, ..., \mathcal{K}_i, ..., \mathcal{K}_n\}$.

3 Extended Algorithm Based on ROC Analysis

In the ROC analysis, it is assumed that a instance can be classified into two different classes of the positive P or the negative N, and positive y or negative n are assigned by a classifier. Then the true positive rate TP and the false positive rate FP of a classifier are represented by the following equations:

$$TP = p(y \mid P) \simeq \frac{\text{positive correctly classified}}{\text{total positives}}, \tag{2}$$

$$FP = p(y \mid N) \simeq \frac{\text{negative incorrectly classified}}{\text{total negatives}}. \tag{3}$$

Plotting FP on the X axis and TP on the Y axis on a graph for several instances, we are able to draw monotone curves, which is called as ROC graph. And the ROC curve near the area of higher TP and lower FP, that is the most upper left point, has the superior characteristics.

In order to apply the ROC analysis for digital documents in a text database, it is assumed that $\mathcal{R}_i$ is the set of tuples which contain $\mathcal{K}_i$ in the database and we use the following definition: in a retrieval, the positive instance is those $\mathcal{B}$ that decreases the number of contained tuples and the negative instance is those $\overline{\mathcal{B}}$ that increases the number of them. Thus the true positive instance can be defined as $\mathcal{B} \cap \bigcup_{i=1}^{n} \mathcal{R}_i$ and the false positive instance can be defined as $\overline{\mathcal{B}} \cap \bigcup_{i=1}^{n} \mathcal{R}_i$, where $\bigcup$ is the set operator of union and $||$ is the set operator to count the number of items. Then TP and FP can be given by

$$TP = \frac{\mid \mathcal{B} \cap \bigcup_{i=1}^{n} \mathcal{R}_i \mid}{\mid \mathcal{B} \mid}, \quad FP = \frac{\mid \overline{\mathcal{B}} \cap \bigcup_{i=1}^{n} \mathcal{R}_i \mid}{\mid \overline{\mathcal{B}} \mid}. \tag{4}$$

And ROC curves can be illustrated by plots of points (FP, TP) using different $Minsup$ or $Minconf$ values as the classifier.

Focusing on TP' and FP' for each rule $\mathcal{K}_i$, where they are provided by the following equations:

$$TP' = \frac{\mid \mathcal{B} \cap \mathcal{R}_i \mid}{\mid \mathcal{B} \mid}, \quad FP' = \frac{\mid \overline{\mathcal{B}} \cap \mathcal{R}_i \mid}{\mid \overline{\mathcal{B}} \mid}, \tag{5}$$

it is found that TP' is equivalent parameter to $confidence$ used in the mining algorithm, thus the rules must be plotted on the upper area of the threshold line $TP = Minconf$ on the ROC graph. While FP' is never used in the algorithm, even though FP' is effective parameter in the ROC analysis. If some thresholds are given to make use of FP' in the mining algorithm, it is thought that there is some possibility of deriving more effective rules. Therefore we proposed two parameters for FP' as the thresholds[8].

FP Algorithm

One is a maximum threshold of FP'. A rule which holds higher value of FP' has stronger heterogeneousness and then it is rarely associative with the query

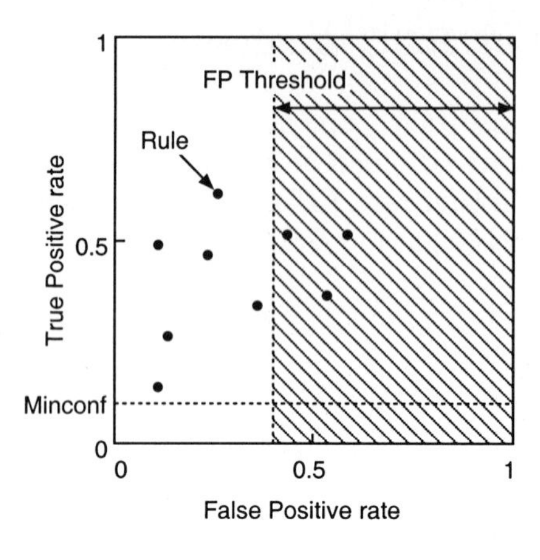

Fig. 3. Association rules cutoff by *FP* threshold

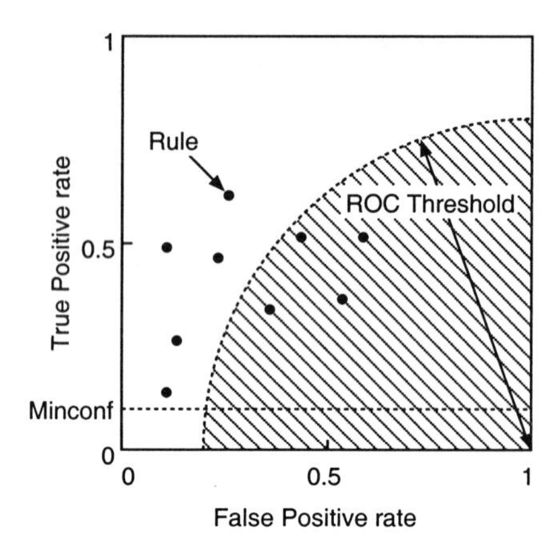

Fig. 4. Association rules cutoff by ROC threshold

even when the rule holds high value of TP'. Because those rules that appear frequently in the retrieval space tend to become stopwords[9]. Hence the rules which hold value of FP' more than the maximum threshold $MaxFP$ should be removed as shown in Figure 3, and the algorithm based on FP' is summarized by the following steps:

1. Derive rules by the mining algorithm.
2. Select rules which hold values of FP' not more than $MaxFP$ from the rules.

ROC Algorithm

Another parameter is a minimum threshold of *ROC distance*, which is represented by d'_{ROC} in this paper. d'_{ROC} is the distance between a point (FP', TP') and the point $(1, 0)$ on the ROC graph, and given by:

$$d'_{ROC} = \sqrt{TP'^2 + (1.0 - FP')^2}. \tag{6}$$

Since d'_{ROC} can evaluate the performance of derived rules[5], the rule which holds the higher value of d'_{ROC} is expected to keep the higher performance. Hence the rules which hold value of d'_{ROC} less than the minimum threshold $MinROC$ should be removed as shown in Figure 4, and the algorithm based on d'_{ROC} is summarized by the following steps:

1. Derive rules by the mining algorithm.
2. Select rules which hold values of d'_{ROC} not less than $MinROC$ from the rules.

Improved ROC Algorithm

Focusing on the semantic property of a position (FP', TP') on the ROC graph, a rule positioned nearer to the upper line $TP = 1$ covers the same space as the query, for instance they may appear in an idiom. While a rule positioned

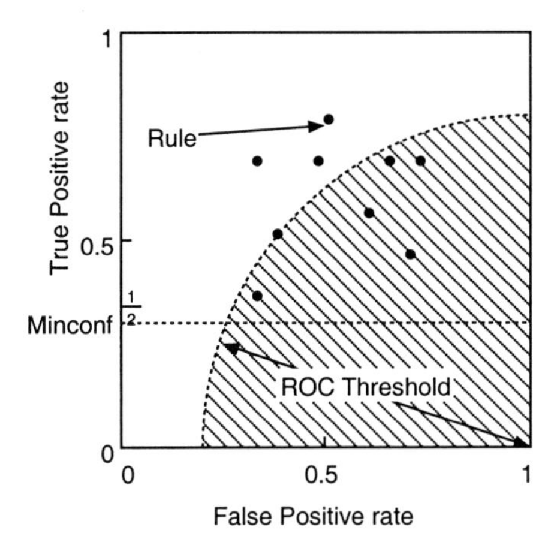

Fig. 5. Association rules cutoff by improved *ROC distance* threshold

nearer to the most right line $FP = 1$ covers the different space from the query in the retrieval, for instance it may be used in a domain different from that of the query. Then we need modifications of these parameters in order to give them stronger influence.

Thus we make corrections in TP' and FP' in the proposed ROC Algorithm as shown in Figure 5 by the following equations, because FP' tends to become too small in a very large database.

$$TP'' = \sqrt{\frac{\mid \mathcal{B} \cap \mathcal{R}_i \mid}{\mid \mathcal{B} \mid}}, \ FP'' = \sqrt{\frac{\mid \overline{\mathcal{B}} \cap \mathcal{R}_i \mid}{\mid \overline{\mathcal{B}} \mid}}. \tag{7}$$

ROC distance d''_{ROC} is calculated by the following equation using TP'' and FP'' instead of TP' and FP' in the equation (6) respectively:

$$d''_{ROC} = \sqrt{TP''^2 + (1.0 - FP'')^2}. \tag{8}$$

And the algorithm based on d''_{ROC} is summarized by the following steps:

1. Derive rules by the mining algorithm.
2. Select rules which hold values of d''_{ROC} not less than $MinROC$ from the rules.

4 Performance Evaluation

In order to evaluate the performance of our proposed algorithms, we used INSPEC database, which contains 331,504 bibliographic titles published in 1998. In this database, keywords are represented by the regular expression, `[a-zA-Z0-9]+`, and all of uppercase letters are mapped into lowercase letters in the index searches so that case sensitivity is ignored.

Table 1. Association rules derived from a keyword "knowledge" for each algorithm.

Order	$TP'(confidence)$	$1/FP'$	d'_{ROC}	d''_{ROC}
1	*of*	image	based	based
2	*for*	information	information	information
3	*and*	models	image	development
4	*a*	modeling	models	process
5	based	process	modeling	neural
6	*the*	neural	process	image
7	*in*	development	development	approach
8	system	algorithm	neural	models
9	systems	network	algorithm	modeling
10	*to*	*as*	network	algorithm

Then we derived association rules from sufficient number of those keywords which appeared in the *Title* attribute. Table 1 shows an example of association rules, which are associative keywords, derived from a keyword "knowledge" and each column is sorted in order of $TP'(confidence)$, $1/FP'$, d'_{ROC} and d''_{ROC}.

As shown in Table 1, the original algorithm (FP') tends to derive those meaningless words which are prepositions or have little semantic content, for example "of", "for" and "and". Such word is called as stopword[9]. Since stopwords appear in many documents, and are thus not helpful for retrieval, it is required that these terms are removed from the internal model of a document or query. Some systems have a predetermined list of stopwords. SMART[9] system, which is one of the first and still best IR systems available, has a list of 570 stopwords. However, sometimes stopwords depend on the context. For example the word "system" is probably a stopword in a collection of computer science journal articles, but not in a collection of documents on the Web.

Using the stopwords list, we would evaluate how our algorithms are effective to filter out association rules. In Table 1, *italic* keywords means stopwords listed in SMART system, and it is found that all of our algorithms suppress to derive stopwords. Hence if appropriate thresholds are given for our algorithms and lower ranked keywords are removed, stopwords will be removed automatically and the derived rules will contain little stopwords.

The difference between the ROC Algorithm and the Improved ROC Algorithm in Table 1 is only that the former derives "network" which has $(FP', TP') = (0.014236, 0.019164)$, while the latter derives "approach" which has $(FP', TP') = (0.016906, 0.060105)$. Generally speaking, if keywords hold about the same values of FP', the keyword which holds larger value of TP' should be selected. Therefore the latter algorithm derives more effective rules than the former one.

In order to evaluate properties of the Improved ROC Algorithm, we examined results derived from sufficient number of queries which contain keywords appeared in *Title* attribute. We sampled randomly about 1% of keywords which appear not less than 10 times in Title except for stopwords and selected 197

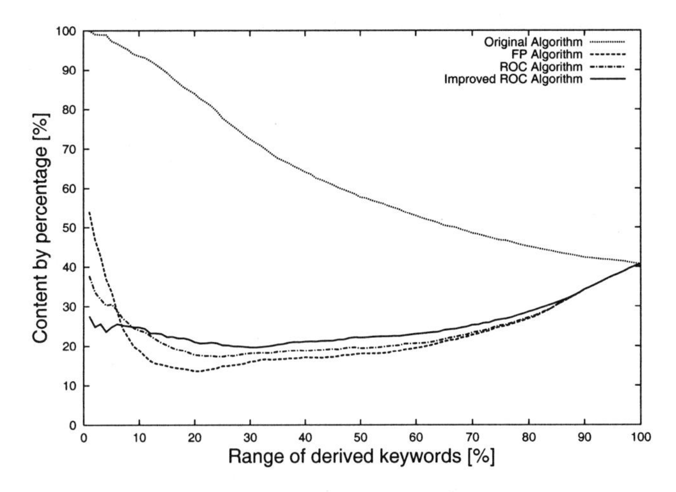

Fig. 6. Total average percentage of stopwords for the derived keywords.

keywords as queries. *Minsup* and *Minconf* are given by

$$Minsup = 0.000001, \ \ Minconf = 0.01, \tag{9}$$

Figure 6 shows the average percentage of stopwords contained in the derived keywords for each query, where each number of derived keywords is adjusted to the range of 100%. Although we know that the FP Algorithm and the ROC Algorithm always make the content by percentage of stopwords lower level than the original algorithm[8], the Improved ROC Algorithm makes it lower level in the higher rank than the other algorithms.

Looking at Table 1, it seems like our algorithms tend to derive similar rules. However Figure 7 shows the average of ROC performance of the derived rules, where the ROC performance can be measured by the *ROC distance*[5], the curve drawn by ROC Algorithm dominates the curve drawn by FP algorithm in all points. Moreover, it is found that the curve drawn by the Improved ROC Algorithm dominates the curve drawn by the ROC Algorithm in all points even though the Improved ROC Algorithm does not use the parameter *ROC distance*. This means that ROC Algorithm always keeps higher performance than FP Algorithm and Improved Algorithm keeps the highest performance. This means that the rules derived by the Improved ROC Algorithm share the area covered by $\mathcal{Q}$ most effectively.

Figure 7 also shows that the curves drawn by our algorithms cross the curve drawn by the original algorithm. So it seems like the ROC performances of the keywords in rank under 38% derived by our algorithms are worse than the original algorithm. But the cause is that our algorithms derive stopwords in such lower rank as shown in Figure 6 and Table 1, while the original algorithm derive effective keywords. However, it is no longer needed to derive such meaningless keywords in the lower rank. Therefore our algorithms, including Improved ROC Algorithm, have higher performance than the original mining algorithm.

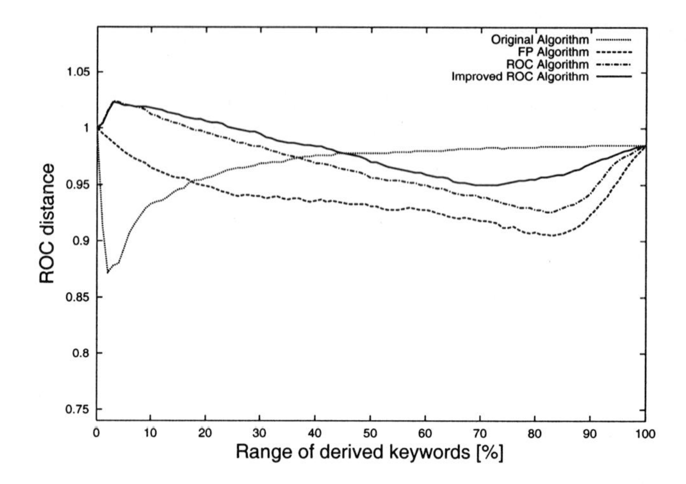

Fig. 7. Total average ROC performance of the derived rules.

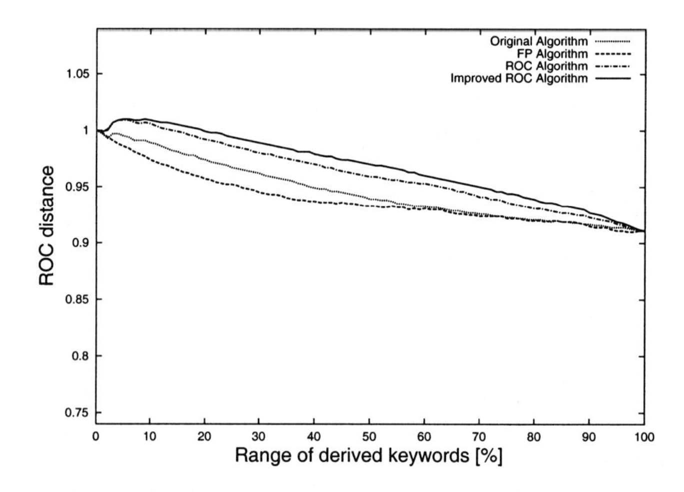

Fig. 8. Total average ROC performance of the derived rules without stopwords.

Figure 8 shows the average of ROC performance of the derived rules without stopwords. In this case, the curve drawn by the Improved ROC Algorithm dominates the curves drawn by other algorithms, the Improved ROC Algorithm has the highest ROC performance.

While the curve drawn by FP Algorithm is dominated by the curve drawn by the original algorithm. But FP Algorithm does not always keep higher performance than the original algorithm if keywords which hold lower performance such as stopwords are removed by some filtering methods. Then FP Algorithm can be used effectively to remove rules which show heterogeneousness without any filters.

5 Conclusion

In this paper, we propose additional thresholds in the mining algorithm, which uses the maximum *FP* threshold and the minimum *ROC distance* threshold based on the ROC analysis. Moreover, we try to improve the algorithm by using the parameters in the ROC analysis. Then we evaluate the performance of our improved mining algorithm. As the results, our algorithms can remove noisy keywords such as stopwords automatically and keep higher performance than the original mining algorithm in the view point of the ROC analysis. The algorithm with the *ROC distance* threshold gives the highest performance and it derives the most effective association rules.

Acknowledgment

We are grateful to Infocom Co., Ltd. and Nissho Iwai Co., Ltd for the source programs of the full text search system "OpenText".

References

1. Agrawal, R., Srikant, R.: Fast algorithms for mining association rules. In: Proc. of the 20th International Conference on Very Large Data Bases, Santiago, Chile (1994) 487–499
2. Kawahara, M., Kawano, H.: An application of text mining: Bibliographic navigator powered by extended association rules. In: Proceedings of the 33rd Annual Hawaii International Conference on System Sciences (HICSS-33), CD-ROM, Maui, HI, USA (2000)
3. Kawano, H.: Mondou: Web search engine with textual data mining. In: Proc. of IEEE Pacific Rim Conference on Communications, Computers and Signal Processing. (1997) 402–405
4. Kawano, H., Kawahara, M.: Mondou: Information navigator with visual interface. In: Data Warehousing and Knowledge Discovery, Second International Conference, DaWaK 2000, London, UK (2000) 425–430
5. Kawahara, M., Kawano, H.: Roc performance evaluation of web-based bibliographic navigator using mining association rules. In: Internet Applications, Proc. of 5th International Computer Science Conference, ICSC'99, Hong Kong, China (1999) 216–225
6. Barber, C., Dobkin, D., Huhdanpaa, H.: The quickhull algorithm for convex hull. Technical Report GCG53, University of Minnesota (1993)
7. Provost, F., Fawcett, T.: Analysis and visualization of classifier performance: Comparison under imprecise class and cost distributions. In: Proceedings of 3rd International Conference on Knowledge Discovery and Data Mining (KDD-97). (1997) 43–48
8. Kawahara, M., Kawano, H.: The other thresholds in the mining association algorithm. SYSTEMS SCIENCE **26** (2000) 95–109
9. Salton, G., McGill, M.J.: Introduction to modern information retrieval. McGraw-Hill, New York, USA (1983)

WWW Visualization Tools for Discovering Interesting Web Pages

Hironori Hiraishi and Fumio Mizoguchi

Information Media Center
Science University of Tokyo
Noda, Chiba, 278-8510, Japan

Abstract. In this paper, we describe three types of WWW Visualization tools based on the hyperbolic tree, the WWW Information Access System (WebMapper), the Interactive Browsing Support System (HANAVI), and Web Site Rating System (KAGAMI). WebMapper integrates structure visualization and retrieval of WWW information. It consists of hyperbolic tree visualization and attribute-value pair query manipulation, and provides a filtering function to reduce the size of the WWW information structure. HANAVI is an interactive version of WebMapper implemented by Java Applet to run on a general web browser. KAGAMI is an automatic web site rating system that outputs six evaluation parameters on a radar chart. It can promote the improvement of a web site and provide effective information of the web site. Our tools enable web browsing support and discovery of interesting pages and lead to creation of a more effective web site.

1 Introduction

The current World Wide Web is a huge database. It is already impossible to find some desired information from such a huge source by browsing homepages one by one. We therefore use keyword-based search engines that list up homepages corresponding to keywords. Several methods to give us more suitable pages have been proposed [Kleinberg,1998,Sergey,1998,Bharat,1998,Dean,1999]. However, unlike such keyword search engines, we describe WWW visualization tools to support discovering desired pages and effective web sites.

Several tools for supporting web browsing have been developed [Bederson, 1996] [Card,1996] as well as methods to visualize tree structures like hyperlink structure [Robertson,1991,Lamping,1994,Munzner,1997]. In those methods, the hyperbolic tree proposed in [Lamping,1994] can visualize the overall structure of a web site on the hyperbolic plane. We have developed three types of tools based on such a hyperbolic tree. These tools not only help visualize the web site structure but also support discovery of useful information on WWW.

S. Arikawa and A. Shinohara (Eds.): Progress in Discovery Science 2001, LNAI 2281, pp. 650–659.

The first tool is the WWW Information Access System "WebMapper[1]" that integrates structure visualization and retrieval of WWW information. The height of a node within the hyperbolic tree indicates the degree of interest to a user. Here, interest is calculated by a fitting function between a page and user-supplied keywords. This measure can be used to filter irrelevant pages, reducing the size of the link structure and allowing us to incrementally discover desired pages from a large web site.

The second tool is the browsing support tool "HANAVI," which adds an interactive nature to WebMapper. WebMapper reads all pages to visualize the web site structure, but it takes much time. HANAVI therefore generates lower branches from a node when a user clicks on that node. This tool was used in the project to stimulate and provides information for the under-populated area of Nagao in Kagawa Prefecture.

The third tool is the web rating system "KAGAMI." Structure visualization provides us not only the overall aspects of the web site, but also helps us to discover where information is concentrated or lacking. In visualizing the web site, KAGAMI rates the web site from six viewpoints by analyzing all descriptions of homepages contained in the web site. The rating is then expressed as a radar chart.

Our tools enable web browsing support and discovery of interesting pages and web sites. We introduce each system in the following sections. Section 2 explains the hyperbolic visualization for WWW, which is the basis for each tool. Section 3 describes the WWW information access system WebMapper. Section 4 explains the interactive browsing support system HANAVI, and Section 5 explains the web site rating system KAGAMI. Section 6 concludes this paper.

2 Hyperbolic Visualization of WWW

Our tools are based on the Hyperbolic Tree, in which the radial tree is constructed as hyperlinks on the Euclidean plane. Each node in the radial tree indicates a homepage, and each edge, a hyperlink. Figure 1 shows the flow of the hyperbolic visualization process.

First, the radial tree on the Euclidean plane is projected onto the hyperbolic plane ($x^2 + y^2 - z^2 = -1$), and the coordinates (x, y) are converted to hyperbolic coordinates $(x, y, \sqrt{x^2 + y^2 + 1})$ (Fig. 1 left). Next, each node on the hyperbolic plane and the coordinates $(0, 0, 1)$ are connected. The intersection $(x/(z+1), y/(z+1))$ of the connected line and the Poincare plane ($x^2 + y^2 < 1$) is then calculated (Fig. 1 right). This process puts all nodes on the hyperbolic plane into the Poincare plane. Finally, the Poincare plane is expanded to the actual display area, enabling the display of a large number of nodes on a limited display area. Characteristically, this visualization enlarges the area near a center.

[1] We called this system "WIAS" in our previous paper [Sawai,1998]. But, the name of "WAIS" is already used for the information retrieval system in MIT. So, we named our system "WebMapper" in this paper.

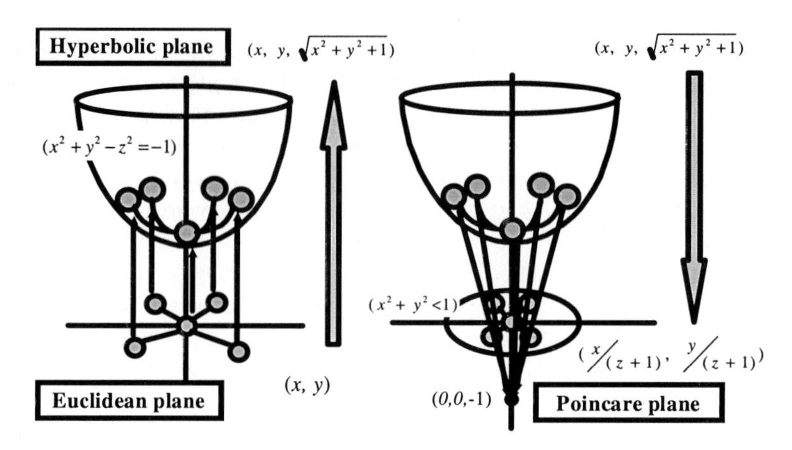

Fig. 1. Hyperbolic Visualization of WWW

3 WWW Information Access System: WebMapper

WebMapper integrates structure visualization and retrieval of WWW information. Figure 2 shows the WebMapper architecture. It takes the hypertext source sent from a WWW server and accesses the WWW server log for input information. The information is processed by a tag parser that parses the hypertext specified URL and extracts an attribute-value set composed of a tag and a string. A tag indicates an attribute name, and a string is regarded as an attribute value. The string parser analyzes the string of attribute-value sets from the tag parser and divides the string into clauses contained in the dictionary. WebMapper thus realizes database-like information retrieval, and we can extract documents containing a specified keyword in the specified tagged string.

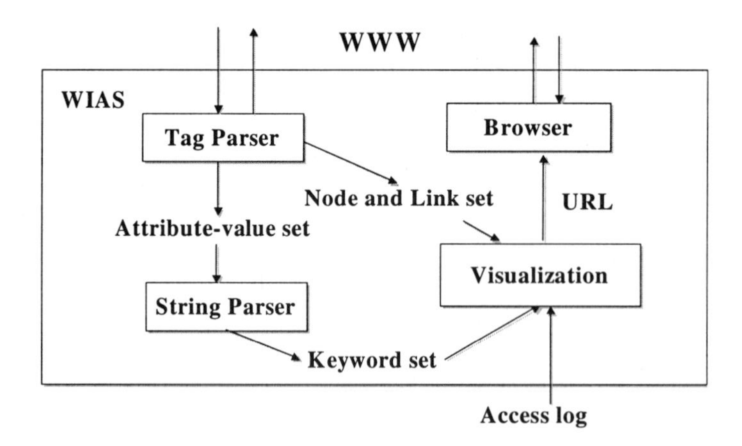

Fig. 2. WebMapper Architecture

The link relation among homepages is derived from the anchor tag. The tag parser parses hypertext and goes to the next depth of pages by extracting the anchor tag contained in the hypertext. The hyperlink chain generally becomes very long. We therefore set a depth limit for the tag parser so it will not search below the depth limit. We are also making a setting to read only one site. The tag parser searches hypertexts in a width-first manner. To avoid loops, it keeps all nodes already found. If we adopt a depth-first search, we cannot generate a well-balanced tree. The visualization module constructs a hyperbolic tree by using node and link sets from the tag parser. All modules except for the string parser are implemented in Java language, enhancing the portability of our system.

Figure 3 shows the WebMapper output. A window is decomposed into a display for the structure of hypertext and areas for inputting queries and commands. Each text is displayed as a node whose height indicates node information such as fitness to a query and user access count acquired from a WWW server log. The height is a good indicator for efficiently accessing interesting text.

A hyperbolic tree can be changed arbitrarily by mouse operation. Clicking the mouse on a node changes focus. Mouse dragging can be accepted at any position, making it easy for users to change the viewpoint of the tree.

Filtering uninteresting nodes is the most important feature of WebMapper. Given a query, the fitness to the query for each node is computed and is displayed

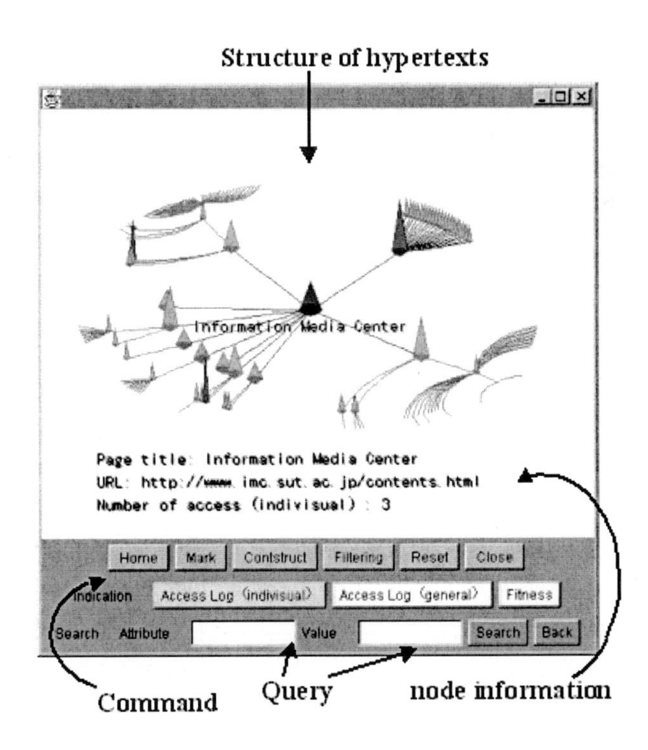

Fig. 3. Output of WebMapper

as the height of the node. The filtering function then removes nodes that has lower fitness, and restructures a reduced hyperbolic tree. This is very useful for large web sites because users can focus on interesting texts only.

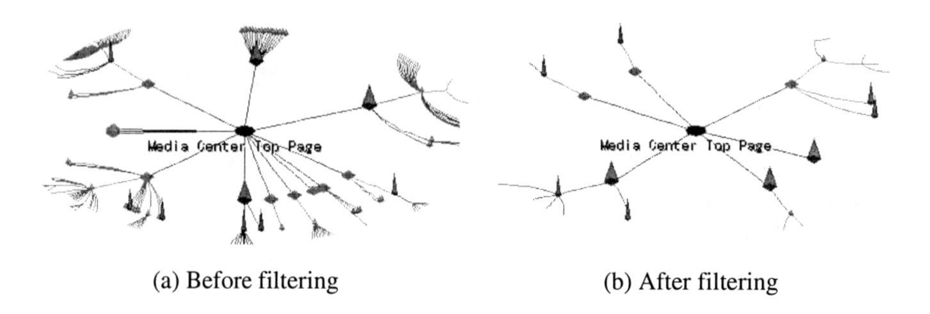

(a) Before filtering (b) After filtering

Fig. 4. Filtering process

Figure 4 shows a filtering process in WebMapper. The figure on the left is a hyperbolic tree of our web site[2] consisting of 226 texts. Since the height of each node indicates the fitness of the associated text to a user-supplied query, higher nodes are more interesting for the user. There are nine interesting nodes among the large number of nodes in the figure.

The figure on the right shows a reduced hyperbolic tree. The intermediate nodes toward interesting nodes are still remained. So, the reduced tree constructs a web site for the user and allows the user to view a manageably sized web site. Although existing search engines list up pages that are interesting for users, it is impossible to see the relationships between the searched pages. The performance of the keyword search in our system also depends on which keywords a user provides. However, a hyperbolic tree representation allows users to capture the web page structure and makes it easy to find interesting portal sites. Moreover, some queries can be put incrementally, and more interesting pages can be explored.

The effectiveness of our WebMapper is described in [Hiraishi,2000], [Sawai,1998]. Experiments were conducted to show the advantage of the visualization. The obtained statistics demonstrate the effectiveness of WebMapper in accessing WWW information.

4 Interactive Browsing Support System: HANAVI

HANAVI is an interactive version of WebMapper. WebMapper reads all pages contained in a web site; HANAVI reads hyperlinked pages when a user clicks on a node. HANAVI was used in the project to stimulate and provide information for the under-populated area of Nagao in Kagawa prefecture. The project was conducted for two years, but our system is still being used.

[2] http://www.imc.sut.ac.jp/

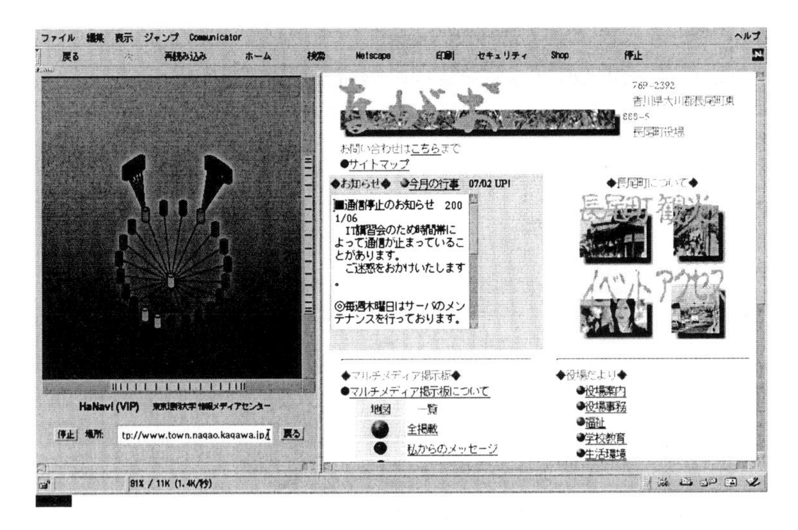

Fig. 5. HANAVI Output

In this project, we focused on visualization and developed HANAVI that is implemented by just the visualization module in Figure 2. HANAVI is shown in Figure 5 and is implemented in a Java Applet to run on a general web browser like Netscape. It starts by specifying the URL containing the applet in the left frame, and a selected homepage on the applet is displayed in the right frame. HANAVI is very simple to operate. Users click on a node of the hyperbolic tree or input a URL into the text field. HANAVI then scrolls the hyperbolic tree placing the selected node to the center of the left frame. And also the selected homepage associated to the node or URL is displayed in the right frame.

People can learn to use HANAVI very quickly because of its simple operation and easy understanding due to the visualization. In this project, we tried to evaluate other systems. Since they required complicated operations and expression, their popularity did not spread very quickly. Our system thus contributed significantly to this project. However, we also found that simple operation and visualization can cause the user to lose interest. HANAVI is just to visualize the hyperlink structure. But the KAGAMI introduced in the next section provides us not only hyperlink structure but also several features of homepages contained in a web site.

5 Web Site Rating System: KAGAMI

We can understand the scale of a web site and discover where information is concentrated or lacking from the hyperbolic tree. All pages are down loaded to make the hyperbolic tree. KAGAMI then analyzes all pages in order to extract six quantitative features expressed as a radar chart. Figure 6 shows the output of the KAGAMI; the hyperbolic tree is shown on the left, and the radar chart on the right.

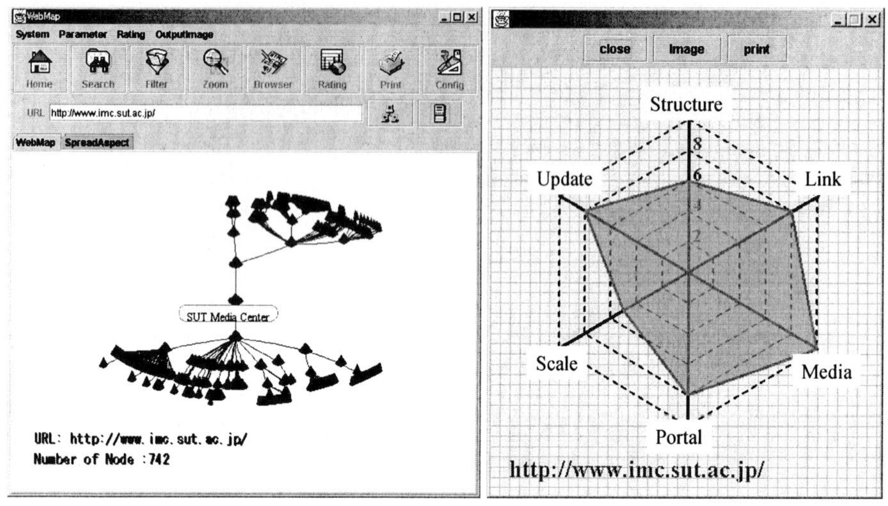

3-Dimensional Hyperbolic Tree   Rader Chart

Fig. 6. Hyperbolic tree and radar chart of KAGAMI

The six features are as follows.

Scale : This is an evaluation of the web site scale and an index indicating how many pages the web site contains.

Update : This is an evaluation of the content freshness and an index indicating how frequently the homepages are updated.

Link : This is an evaluation of navigating in the web site and an index indicating how many internal links the web site has.

Portal : This is an evaluation of hyperlinks to related sites and an index indicating how many external links to related sites exist in the web site.

Media : This is an evaluation of homepage design and an index indicating how many media, i.e., pictures or animation, are used in the web site.

Structure : This is an evaluation of web site structure and an index that shows whether the web site has the ideal structure or not.

Scale is calculated by counting the homepages in the web site. **Update** is calculated by getting the last modified time of all the homepages[3].

Portal, **Link** and **Media** are obtained from tag information. For example, if the URL of the anchor tag is the URL of the same site, it is counted as the **Link**; if the URL is of a different site, it is counted as **Portal**. **Structure** is evaluated by counting the pages included in each layer. The ideal quantity of

[3] The last modified time of each pages can be extracted from a web server and it is expressed in a millisecond. the **Update** is represented as the average or deviation among the last modified time of all the homepages

pages in each layer is defined in advance, and the system evaluates whether the quantity of pages on each layer is close to the ideal quantity. We give two types of eval uation for each indexes, "overall evaluation" and "partial evaluation." So, one index has two types of the evaluation. Overall evaluation indicates whether a feature appears all over the web site or not. This can be calculated as the average feature value of all the homepages contained in a web site. However, the average cannot evaluate cases in which a web administrator makes a partial effort. A partial evaluation therefore evaluates cases in which an effort is being made in a portion of the web site. The details of the calculation for each parameter are described in [Otsuka,2001].

Figure 7 shows the web rating results of several library web sites. The services using the network are given in current libraries. We can access libraries from our home and office using the Internet and retrieve and read books. Evaluating the homepage of a library thus yields an evaluation of the information system of that library.

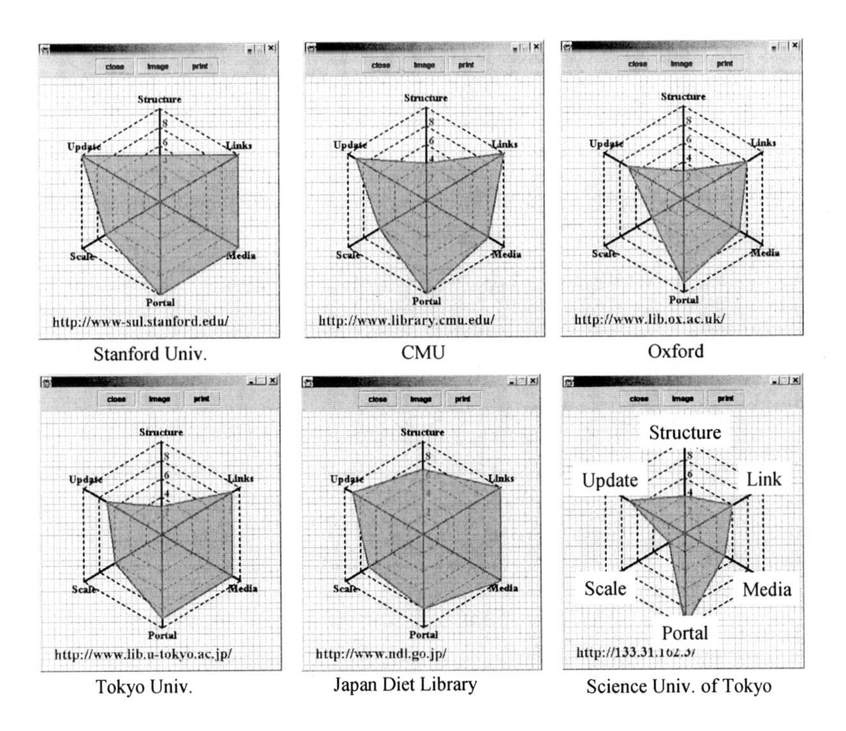

Fig. 7. Rating results in library web sites

The radar charts have similar shapes. The **Portal** is high in every chart, indicating that the there are many hyperlinks to the related site. This is caused by digital journals and so on. In most cases, the **Media** is also high due to the digital library. This means that there are many media employed in the web site. Finally, **Link** and **Update** also tend to be high.

These results indicate that the library web site features many hyperlinks to both internal and external related sites, it has a digital library, and it is updated frequently. The Stanford University library thus gets full marks in these indexes. However, the structure of every library is rated low since there are many pages in the shallow layers. Though the structure should be improved, this is one of the features of library web sites. Thus, our web site rating for sites of a specific type allows us to discover the special features regarded as standard for such special sites.

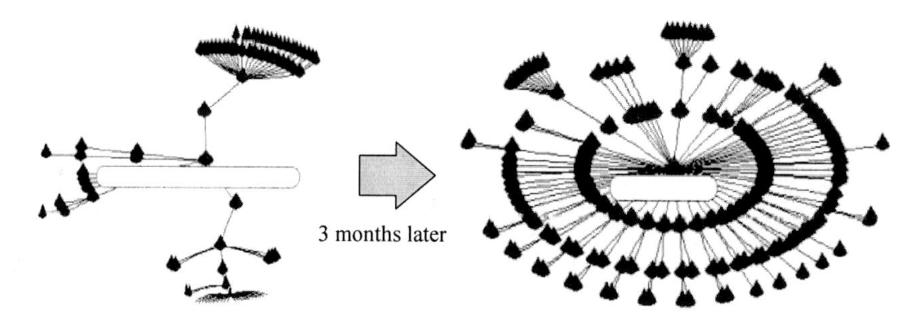

Fig. 8. Example of web site improvement

We will now introduce an example in which our web site rating contributed to the web site improvement. We provided a free web site rating service on our web site for 10 days in November 2000[4]. At that time, a web site shown in Figure 8 left did not have many pages, and we could clearly see informational deviation, so it was not a good site. However, the web administrator reconstructed the web site after seeing this result. Three months later, the web site had changed to the orderly structure shown in Figure 8 right. Thus, our web rating can promote the improvement of a web site and provide effective web site information.

6 Conclusions

In this paper, we described three types of WWW visualization tools based on the hyperbolic tree, WebMapper, HANAVI, and KAGAMI. WebMapper integrates structure visualization and retrieval of WWW information. It consists of hyperbolic tree visualization and attribute-value pair query manipulation, and provides a filtering function to reduce the size of the WWW information structure. HANAVI is an interactive version of WebMapper. It is implemented by a Java Applet to run on a general web browser. Users can learn to use HANAVI very quickly because of its simple operation and easy understanding due to the

[4] http://imct-sev.imc.sut.ac.jp/webrating/WebRating

WWW visualization. KAGAMI is an automatic web site rating system that outputs six evaluation parameters on a radar chart. It can promote the improvement of a web site and provide effective information of the web site. In addition, our web rating of sites of a specific type enables us to discover special features.

The hyperbolic tree is a reasonable approach for our tools. Munzner designed an information space in which multiple hyperbolic trees are configured three dimensionally [Munzner,1997]. However, there are a number of nodes on the information space map, and it is hard to focus on the appropriate portion of a web site. An alternative visualization scheme was developed as a cone tree by Robertson[Robertson,1991]. In a cone tree, nodes are distributed in a three-dimensional space, and thus front nodes may hide rear nodes. Moreover, focus change is not as easy as in a hyperbolic tree. Our visualization tools based on the hyperbolic tree enable effective web browsing support and discovery of interesting pages, while also promoting web site discovery.

References

[Bharat,1998] Krishna Bharat and Monika R. Henzinger, "Improved Algorithms for Topic Distillation in a Hyperlinked Environment," *In Proc. of the 21th ACM SIGIR Conference on Research and Development in Information Retrieval*, 1998.

[Bederson, 1996] Benjamin B. Bederson, James D. Hollan, Jason Stewart, David Rogers, Allison Druin, David Vick, "A Zooming Web Browser," *In Proc. of Multimedia Computing and Networking*, vol.2667, pp.260-271, 1996.

[Card,1996] Card, S. K. Robertson, G. G. and Machinlay, J. D., "The WebBook and the Web Forager: An Information Workspace for the World-Wide Web," *In Proc. of ACM Conference on Human Factors in Computing Systems*, pp.111-117, 1996.

[Dean,1999] Jeffrey Dean and Monika R. Henzinger "Finding Related Pages in the World Wide Web," *In Proc. of Eighth International World Wide Web Conference*, 1999.

[Hiraishi,2000] Hironori Hiraishi, Hiroshi Sawai and Fumio Mizoguchi, "Design of a visualization agent for WWW information," *In Proc. of the First pacific rim international workshop on Intelligent agents*, 2000.

[Kleinberg,1998] Jon M. Kleinberg, "Authoritative sources in a hyperlinked environment," *In Proc. of the Ninth Annual ACM-SIAM Symposium on Discrete Algorithms*, pp.668-677, 1998.

[Lamping,1994] Lamping, J., Rao, R., "Laying out and Visualizing Large Trees Using a Hyperbolic Space," *In Proc. of ACM Symposium on User Interface Software and Technology*, pp.13-14, 1994.

[Munzner,1997] Munzner, T., "H3: Laying out Large Directed Graphs in 3D Hyperbolic Space," *In Proc. of IEEE Symposium on Information Visualization*, pp.2-10, 1997.

[Otsuka,2001] Naonori Otsuka, Hironori Hiraishi and Fumio Mizoguchi, "KAGAMI: Web Rating Agent Based on Hyperlink Structure," *In Proc. of Joint 9th IFSA World Congress and 20th NAFIPS International Conference*, 2001.

[Robertson,1991] G. G. Robertson, J. D. Mackinlay, S. K. Card, "Cone Trees: Animated 3D Visualizations of Hierarchical Information," *In Proc. of ACM Conference on Human Factors in Computing Systems*, pp.189-194, 1991.

[Sawai,1998] Hiroshi Sawai, Hayato Ohwada and Fumio Mizoguchi, "Incorporating a navigation tool into a browser for mining WWW information," *In Proc. of the First International Conference on Discovery Science*, pp.453-454, 1998.

[Sergey,1998] Sergey Brin and Lawrence Page, "The Anatomy of a Large-Scale Hypertextual Web Search Engine," *In Proc. of Seventh International World Wide Web Conference*, 1998.

Scalable and Comprehensible Visualization for Discovery of Knowledge from the Internet

Etsuya Shibayama[1], Masashi Toyoda[2], Jun Yabe[3], and Shin Takahashi[1]

[1] Graduate School of Information Science and Engineering,
Tokyo Institute of Technology,
2-12-1 Ookayama, Meguro-ku, Tokyo, Japan
{etsuya, shin}@is.titech.ac.jp,

[2] Institute of Industrial Science, University of Tokyo
4-6-1 Komaba, Meguro-ku, Tokyo, Japan
toyoda@tkl.iis.u-tokyo.ac.jp,

[3] Home Network Company, Sony Corporation
Shinagawa INTERCITY C Tower Shinagawa Tec.
2-15-3 Konan Minato-ku, Tokyo, Japan yabe@sslab.sony.co.jp

Abstract. We propose visualization techniques for supporting a discovery from and comprehension of the information space built on the Internet. Although the Internet is certainly a rich source of knowledge, discoveries from the net are often too hard. On the one hand, it is difficult to find places where useful knowledge is buried since the information space on the Internet are huge and ill-structured. On the other hand, even if they are found, it is still difficult to read from useful knowledge since it is scattered on a number of fine-grained pages and articles.

In order to help human users to find and understand concealed knowledge, we propose two levels of visualization techniques. The first level is designed to provide scalable visualizations, presenting skeletal structures of huge hierarchies. It provides sketchy maps of the entire space and helps the user to navigate to portions where candidate information is stored. The second level provides more detailed and comprehensible views of a small region of the information space. It can help the user to understand knowledge that is scattered on multiple articles and thus inherently hard to follow.

1 Introduction

Information resources on the Internet, e.g., Web pages and news articles, constitute a huge, ill-structured, and rapidly growing information space. Discovery of knowledge from the Internet is still a challenge. It includes useful knowledge that is difficult to be exploited. We have both quantitative and qualitative problems:

- Even a fraction of the information space built on the Internet is so large and so complicated that it is beyond human's perception.
- Since information on the Internet is free, open, and sometimes ill-formed, computers could not automatically discover useful semantic information without direct or indirect human assistance. Though it is expected that the future

S. Arikawa and A. Shinohara (Eds.): Progress in Discovery Science 2001, LNAI 2281, pp. 661–671.

Web will eventually have rich semantic annotations, today's Web still lacks machine-readable semantic information.

Effective human-computer cooperation is desirable for knowledge discovery from the Internet: computers can process a huge amount of information if in-depth semantic analysis is not necessary; human beings can see semantic content if the target information is sufficiently small and has perceptible representations. Several promising techniques today depend on implicit collaborations between computers and massive human activities such as creating links and making recommendations. Examples include link analyses[1, 2] and social filtering[3, 4], in which semantic information is automatically extracted by watching or measuring human behaviors in a large scale.

We consider that interactive visualization is a key technology to provide a support for more explicit human-computer cooperation in exploitation and comprehension of large-scale and ill-formed information spaces such as those formed on the Internet. For the purposes, we have been developing two sorts of visualization techniques:

- *scalable visualization* that is suitable to present skeletal structures or summaries of the whole space on a small computer screen; and
- *comprehensible visualization* that is better suited for revealing details of some specific portions.

An integration of these two is essential in design and development of better support tools for knowledge discovery from the Internet.

In the following sections, we introduce these techniques and their application examples. In Section 2, we describe the design of our directory browser HishiMochi[5], which is built on the multi-focus zooming interface library HyperMochiSheet[6]. A "directory" in this context means a large hierarchy. HishiMochi provides a support for several interactive search strategies for large hierarchies. In Section 3, we describe a framework to produce comprehensible 3D avatar animations from a thread of USENET articles[7]. In other words, USENET articles that were asynchronously written by several to many authors without central authority are translated into a sequential and more coherent animation program.

2 A Scalable Visualization Technique for a Large Hierarchy

In theory, the web is regarded as a large directed graph[8]. At first glance, people may consider that general graph drawing algorithms are effective to visualize the Web. In reality, however, human users cannot capture any useful information from a drawn graph that is inherently large and tangled. Merely they are confused. Visualization techniques are useful only when target graphs have hidden structures and visual representations reveal them.

In addition, since we are faced with large and complicated graphs, "exposing everything on the screen" is not an achievable goal. The best we can do is to distinguish the more important portions from the less important ones and put more emphasis on the former.

In practice, the only known structures that are both scalable and understandable are hierarchies, possibly with hyperlinks. We have to shape the target graphs into more structured ones. Examples include directories provided by search engines such as Yahoo![1] and Open Directory[2]. Automatic classification techniques[9, 10] could be applicable to the construction of a directory.

Hierarchies are also preferable since a node in a hierarchy is naturally regarded as the representative of the subhierarchy where it is the root. Therefore, when the user is not interested in details of some subhierarchies, a visualizer can fold them and depict only the representative nodes and, in this manner, can save valuable space.

2.1 Visual Representations of Large Hierarchies

Various techniques and systems have been proposed for visualization and navigation of a large hierarchy. They include Treemap[11], Hyperbolic Browser[12, 13],Continuous Zoom[14], and Zooming Web Browser[15]. Our zooming interface library HyperMochiSheet[6] follows them.

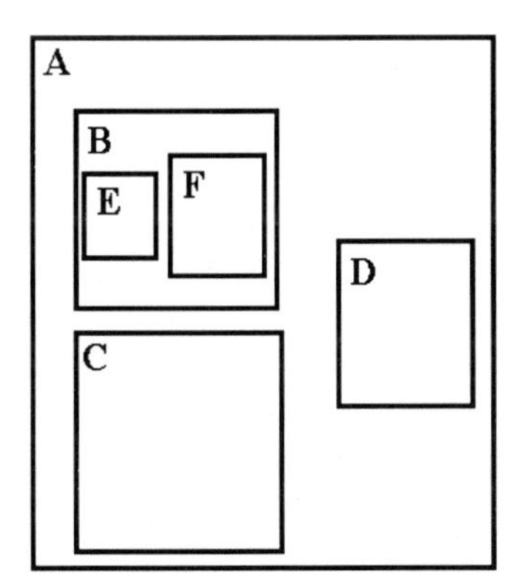

Fig. 1. Nested rectangles representing a directory structure

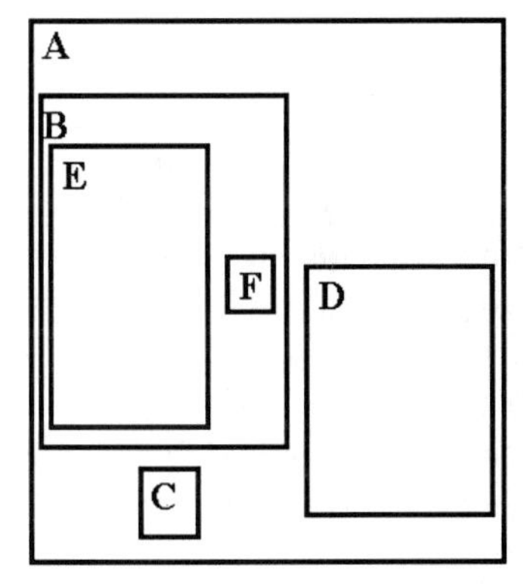

Fig. 2. Another view of the same directory

Continuous Zoom and HyperMochiSheet visually represent a hierarchy in nested rectangles so that a child node is enclosed in its parent. Fig. 1 is an example of such rectangles. This visualization scheme has the following characteristics.

[1] http://www.yahoo.com/

[2] http://dmoz.org/

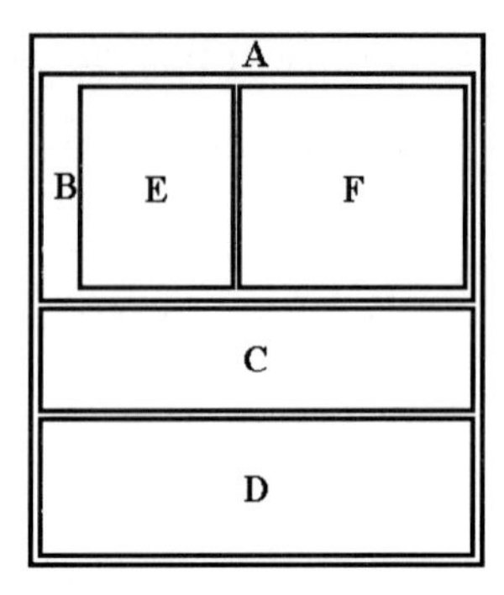

Fig. 3. A more space efficient layout

- Directory-like structures are assumed as targets, where non-leaf nodes have no or little significant information. Under this assumption, a child node does not conceal any significant portions of its parent.
- It can provide multi-focus views. For instance, if two foci are put simultaneously on the nodes D and E of the hierarchy in Fig. 1, we will get a layout like Fig. 2.
- Naturally, it presents focus+context views[16]. In particular, if a node is visible, all its enclosing ancestors are also visible. Therefore, the path from the root to each focused node is always visible. However, the target should not be very deep.

Note that the most distinguished features of HyperMochiSheet are its prediction and management of foci[6]. However, they are beyond the scope of this paper and we omit their details.

Treemap shares the same idea but is more space efficient in the sense that its visualization looks like Fig. 3. The major difference is that Treemap allocates more space to the leaf nodes. Continuous Zoom and HyperMochiSheet are less space efficient but has more freedom of layouts. During repeated creations and removals of foci, relative positions among adjacent nodes can be rather stable, even though layouts can radically change.

In contrast, Hyperbolic Browser and Zooming Web Browser represent nodes in a hierarchy as rectangles or ovals and links as line segments or curves so that a child node and its parent are connected by a link. If both leaf nodes and non-leaf nodes include significant information, this scheme is preferable. Hyperbolic Browser is basically a single focus zooming interface[3]. Therefore, it is less expressive but its focus management and operations for navigations are much simpler. Zooming Web Browser is implemented on top of the zooming library Pad++[17]. It is more expressive and has more freedom of layouts, but its focus and layout management is more complicated.

[3] A dual focus extension is proposed in [13]

2.2 Visual Representations of Search Results

Although "discovery" and "search" are different, we consider that a good search tool can be a help during discovery processes. In this subsection, we explain the basic ideas to visually represent search results with HishiMochi or other multi-focus zooming interface. In the following, we assume that the target constitute a Yahoo-like directory structure.

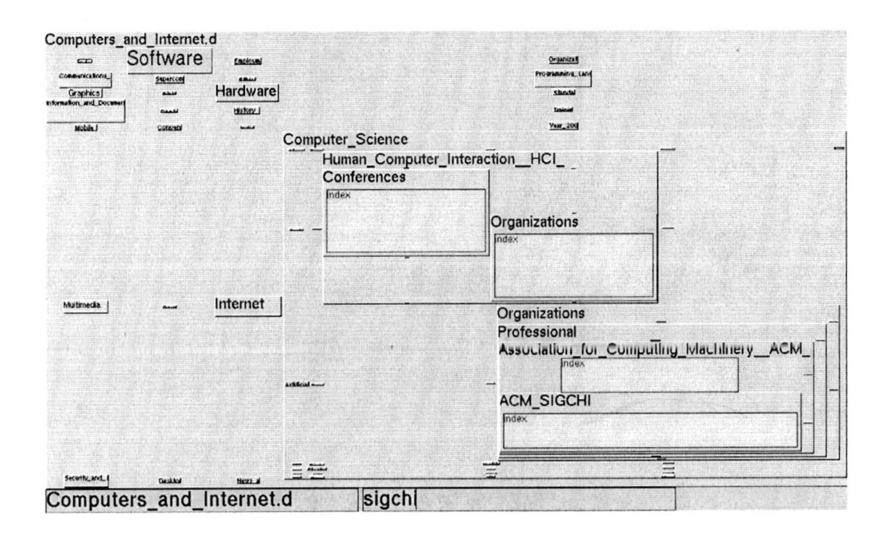

Fig. 4. An initial layout of HishiMochi

First, we present a sample view of HishiMochi in Fig. 4, where the Computer & Internet directory of Yahoo! is illustrated as nested rectangles. This is a default view and each document has the same degree of interest (DOI). Next, upon a query, HishiMochi dynamically searches the entire hierarchy for documents. Fig. 5 depicts the search results with putting foci on the hits. In this very example, all the documents in which the word "sigchi" occurs and their ancestors are magnified and the other nodes are shrunken. Note that Fig. 4 and Fig. 5 represent the same hierarchy with different foci.

In this manner, if the number of hits is not large, we can see the place of each hit with the following contextual information:

– the path from the root to each hit; and
– distribution of hits among the hierarchy.

If the number of hits is too large, nodes including many hits may be emphasized, instead of hits themselves. For instance, Fig. 6 depicts the Java 2 SE source hierarchy with putting foci on every file including the word "eventlistener." Most hits are displayed on the screen but details of two subdirectories `java.awt.dnd` and `java.awt.event` are not exposed owing to space limitations.

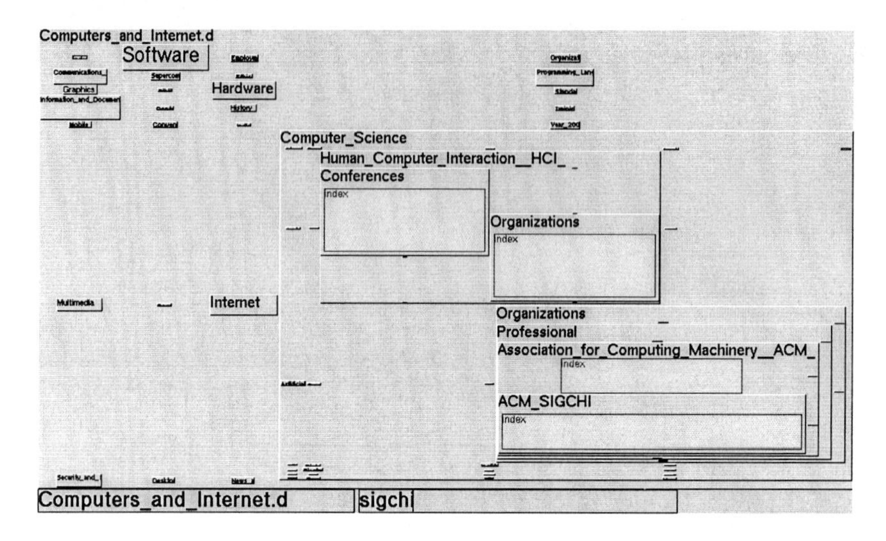

Fig. 5. A visualization of search results in HishiMochi

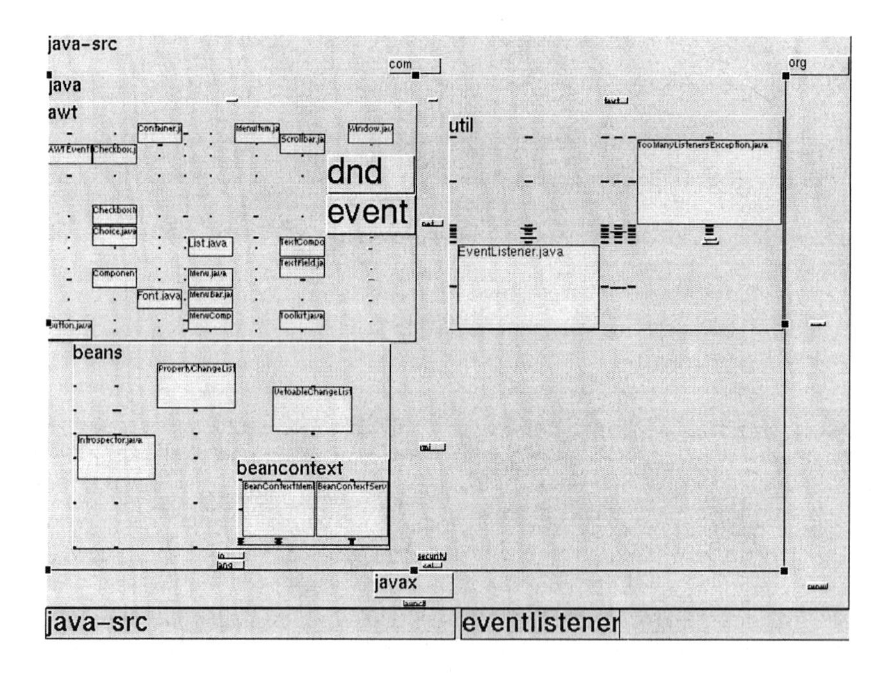

Fig. 6. A visual representation of many hits

2.3 Visual Representations of Search Contexts

In practice, a single search process consists of successive queries. Often we have to refine queries since the first trial usually has too many hits. With our scheme, not only the final results but also the results of intermediate queries are visualized. The key of our scheme is human-computer interactions through visualizations of nodes currently being searched for and a support for interactive search, which proceeds as follows:

1. The user issues a query.
2. The user examines the search results on the screen. If it is satisfying, the process terminates.
3. The user refines the query by changing keywords or specifying the new target as a collection of subdirectories, and goes back to the first step.

Optionally, the user can interactively open a node or traverse a hyperlink, if it exists.

Under the following simple assumptions, memory bounded heuristic searches can be performed, which are interactive and depend on the users' heuristics.

- human users have heuristic information,
- computers are not very intelligent but just fast, and
- screen space, which is severely limited, is used as temporary memory for human user.

On the one hand, since the user has heuristic information, if only s/he can inspect the entire search context or visited nodes, s/he can perform a heuristic search. On the other hand, since the search context is depicted on a small computer screen, only a small number of nodes are visible. This situation is similar to memory-bounded search methods[18]. According to our experiences, HishiMochi provides a sufficient support for (interactive) beam search.

3 A Comprehensible Visualization Technique for a Thread of News Articles

Even if the location is found where useful knowledge is buried, it is sometimes difficult to read from the knowledge. USENET news archives are such examples. They are divided into hierarchical news groups and so they are reasonably structured in the *macroscopic* level. This sort of structure can be effectively visualized with HishiMochi or some other techniques such as Galaxy of News[19].

In contrast, their microscopic structures are rather chaotic. A single article is too fine-grained as a unit of knowledge. We usually have to read through a thread of articles to exploit knowledge. However, it is a hard and tedious task by the following reasons:

- A thread is built by a number of authors who asynchronously submitted articles.
- There is no central authority that can guarantee coherence or consistency within a thread.

A thread is nothing but a patchwork that is inherently divergent.

Making a thread more readable, we propose techniques to translate it into 3D avatar animation program. The translation process consists of the following two phases:

1. scenario writing
2. stage direction

In the first phase, the articles of a thread are divided into segments. Note that when an author follows up with an article, s/he does not always refer to an entire article but some particular region(s). Those regions are called segments and we consider each segment as a referent or referral of the follow-up relation. The scenario is a serialization of the segments.

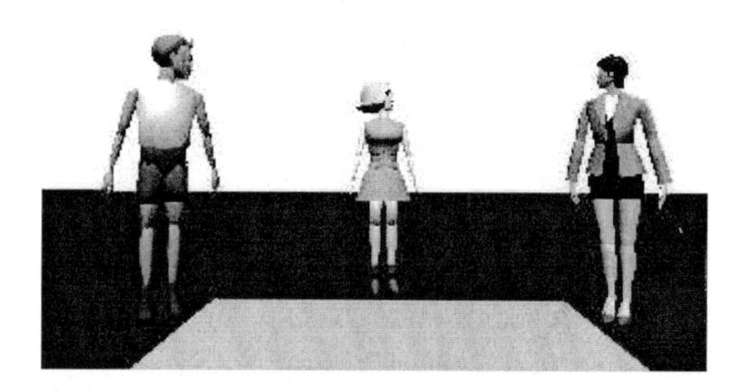

Fig. 7. Default direction

In the second phase, behaviors of each avatar are synthesized and an animation program in TVML[20] is generated. There are several default behaviors (Fig. 7) including the following:

- The speaker looks at the avatar who have spoken the referred segment.
- Each avatar except the speaker looks at the speaker.

In addition, conversational patterns are extracted from the scenario and a direction reflects those patterns. The following are examples of conversational patterns:

- When two avatars speak alternately without any interruption, these two are zoomed in on (Fig. 8).
- When multiple followers exist, the whole stage is looked down and they shake themselves (Fig. 9).

In this manner, we can produce multi-modal representations of news articles. They are more comprehensible than the original form at least in the following sense:

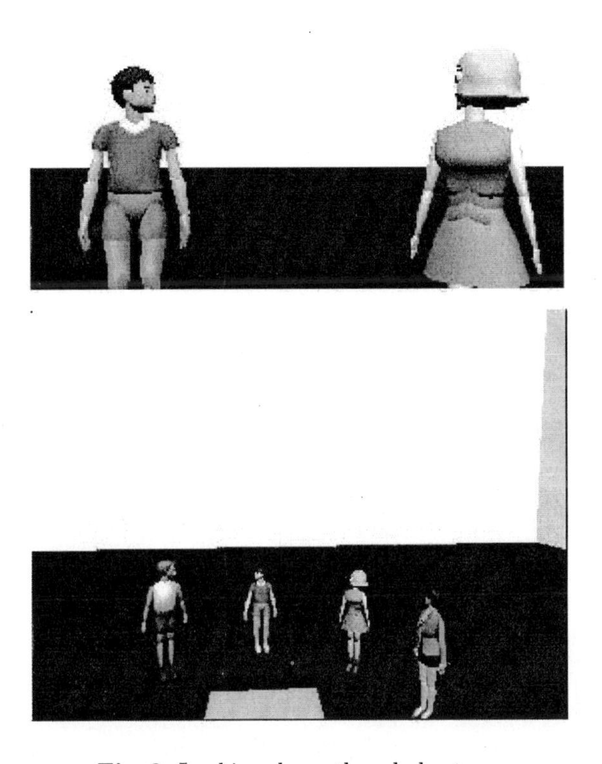

Fig. 9. Looking down the whole stage

- Even a short news article includes quotations from other author(s) and we encounter the same quotations many times. It is tedious. We often skip over and sometimes too much. In contrast, with our system, redundant quotations are automatically removed.
- When reading news articles in the original form, we often miss the author identification. Astonishingly, it is not easy to remember who says what. Avatar animations make it much easier to keep speaker identification.

4 Conclusion

We have proposed two sorts of visualization techniques, i.e., scalable and comprehensible visualizations. The former presents macroscopic views and provides sketchy maps of extensive spaces. The latter presents microscopic views and reorganizes sporadic information into a more coherent form.

We put emphasis on an integration of these two techniques. Currently, however, we have only succeeded in limited application domains, i.e., bulletin boards, where the macroscopic structures are hierarchies and microscopic structures are threads of short articles connected by the reference relations. More research efforts are necessary in design and development of sophisticated tools for knowledge discovery from the Internet.

References

1. Kleinberg, J.: Authoritative Sources in a Hyperlinked Environment. Journal of the ACM **46** (1999) 604–632
2. Page, L., Brin, S., Motwani, R., Winograd, T.: The PageRank Citation Ranking: Bringing Order to the Web. `http://www-db.stanford.edu/~backrub/pageranksub.ps` (1998)
3. Resnick, P., Iacovou, N., Suchak, M., Bergstrom, P., Riedl, J.: Grouplens an open architecture for collaborative filtering of netnews. In: Proc. of ACM CSCW'94. (1994) 175–186
4. Shardanand, U., Maes, P.: Social information filtering: Algorithms for automating "word of mouth". In: Proc. of ACM CHI'95. (1995) 210–217
5. Toyoda, M., Shibayama, E.: HishiMochi: A Zooming Browser for Hierarchically Clustered Documents. In: ACM CHI 2000 Extended Abstracts. (2000) 28–29
6. Toyoda, M., Shibayama, E.: Hyper Mochi Sheet: A Predictive Focusing Interface for Navigating and Editing Nested Networks through a Multi-focus Distortion-Oriented View. In: Proc. of ACM CHI '99. (1999) 504–511
7. Yabe, J., Shibayama, E., Takahashi, S.: Automatic Animation of Discussions in Usenet. In: Proc. of AVI 2000, ACM Press (2000) 84–91
8. Kumar, R., Raghavan, P., Rajagopalan, S., Sivakumar, D., Tompkins, A., Upfal, E.: The Web as a graph. In: Proc. of ACM PODS 2000. (2000) 1–10
9. Chakrabarti, S., Dom, B., Indyk, P.: Enhanced hypertext categorization using hyperlinks. In: Proc. Int. Conf. Management of Data (SIGMOD '98), ACM (1998) 307–318
10. Chekuri, C., Goldwasser, M., Raghavan, P., Upfal, E.: Web search using automatic classification. In: Sixth World Wide Web Conference. (1996)
11. Johnson, B., Shneiderman, B.: Tree-maps: A Space Filling Approach to the Visualization of Hierarchical Information Structures. In: Proc. of IEEE Visualization '91. (1991) 284–291
12. Lamping, J., Rao, R., Pirolli, P.: A Focus+Context Technique Based on Hyperbolic Geometry for Visualizing Large Hierarchies. In: Proc. of ACM CHI '95. (1995) 401–408
13. Lamping, J., Rao, R.: The Hyperbolic Browser: A Focus + Context Technique for Visualizing Large Hierarchies. Journal of Visual Languages and Computing **7** (1996) 33–55
14. Bartram, L., Ho, A., Dill, J., Henigman, F.: The Continuous Zoom: A Constrained Fisheye Technique for Viewing and Navigating Large Information Spaces. In: Proc. of ACM UIST '95. (1995) 207–215
15. Bederson, B., Hollan, J., Stewart, J., Rogers, D., Vick, D., Ring, L., Grose, E., Forsythe, C.: A Zooming Web Browser. In Forsythe, C., Grose, E., Ratner, J., eds.: Human Factors and Web Development,. Lawrence Erlbaum Assoc. (1997)
16. Furnas, G.W.: Generalized Fisheye Views. In: Proc. of ACM CHI '86. (1986) 16–23

17. Bederson, B.B., Hollan, J.D., Perlin, K., Meyer, J., adn George Furnas, D.B.: Pad++: A Zoomable Graphical Sketchpad for Exploring Alternate Interface Physics. Journal of Visual Languages and Computing **7** (1996) 3–31
18. Russell, S.J.: Efficient memory-bounded search methods. In: ECAI 92: 10th European Conference on Artificial Intelligence Proceedings. (1992) 1–5
19. Rennison, E.: Galaxy of News: An Approach to Visualizing and Understanding Expansive News Landscapes. In: ACM Symposium on User Interface Software and Technology. (1994) 3–12
20. Hayashi, M., Ueda, H., Kurihara, T.: TVML (TV program Making Language) – Automatic TV Program Generation from Text-based Script –. In: Proc. of Imagina'99. (1999) `http://www.strl.nhk.or.jp/TVML/PDF/Imagina99.pdf`.

Meme Media for Re-editing and Redistributing Intellectual Assets and Their Application to Interactive Virtual Information Materialization

Yuzuru Tanaka

Meme Media Laboratory,
Hokkaido University,
Sapporo, 060-8628
tanaka@meme.hokudai.ac.jp

Abstract. With the growing need for interdisciplinary and international availability, distribution and exchange of intellectual assets including information, knowledge, ideas, pieces of work, and tools in re-editable and redistributable organic forms, we need new media technologies that externalize scientific, technological, and/or cultural knowledge fragments in an organic way, and promote their advanced use, international distribution, reuse, and re-editing. These media may be called meme media since they carry what R. Dawkins called 'memes'. An accumulation of memes in a society forms a meme pool that functions like a gene pool. Meme pools will bring about rapid accumulations of memes, and require new technologies for the management and retrieval of memes. Our group had been working on the research and development of meme media, including IntelligentPad and IntelligentBox for 2D and 3D representation. In the Discovery Science Project, our group was asked to extend the meme media framework for the distribution and reuse of the project's achievements. In addition to this first mission, we were asked to apply 3D meme media architecture to virtual materialization of database records. The first mission has led us to the proposal of a meme pool architecture 'Piazza', while the second has developed a framework providing visual interactive components for (1) accessing databases, (2) specifying and modifying database queries, (3) defining an interactive 3D object as a template to materialize each record in a virtual pace, and (4) defining a virtual space and its coordinate system for the information materialization.

1 Introduction

With increasingly sophisticated research on science and technology, there is a growing need for interdisciplinary and international availability, distribution and exchange of the latest research results, in re-editable and redistributable organic forms, including not only research papers and multimedia documents, but also various tools developed for measurement, analysis, inference, design, planning,

S. Arikawa and A. Shinohara (Eds.): Progress in Discovery Science 2001, LNAI 2281, pp. 672–681.

and production. We need new media technologies. These media can carry a variety of knowledge resources, replicate themselves, recombine themselves, and be naturally selected by their environment. We call them 'meme media' since they carry what R. Dawkins called 'memes' [1]. A fundamental and necessary framework for the growth and distribution of 'memes' is a 'meme pool'. A 'meme pool' is an accumulation of 'memes' in a society and functions like a gene pool. 'Meme media', together with a 'meme pool' provide a framework for the farming of knowledge. We have already developed 2D and 3D 'meme media' architectures; IntelligentPad [2-5] and IntelligentBox. IntelligentPad [6] represents each component as a pad, a sheet of paper on the screen. A pad can be pasted on another pad to define both a physical containment relationship and a functional linkage between them. Pads can be pasted together to define various multimedia documents and tools. Unless otherwise specified, composite pads are always decomposable and re-editable. IntelligentBox is the 3D extension of IntelligentPad. Components of IntelligentBox are called boxes.

In the Discovery Science Project, our group was asked to extend the meme media framework for the distribution and reuse of the project's achievements. In addition to this first mission, we were asked to apply our 3D meme media architecture to virtual materialization of database records. The first mission has led to the proposal of a meme pool architecture 'Piazza', while the second has developed a framework providing visual interactive components for (1) accessing databases, (2) specifying and modifying database queries, (3) defining an interactive 3D object as a template to materialize each record in a virtual space, and (4) defining a virtual space and its coordinate system for the information materialization.

'Piazza' works as a worldwide 'meme pool' [7-8]. It consists of a Piazza server and a Piazza client. Each Piazza client is represented as a pad. It is associated with a file managed by a Piazza server. Pads can be drag-and-dropped to and from the associated remote server file. When a Piazza Pad is opened, all the pads registered to the associated remote server file are immediately downloaded onto this pad and become available. An entrance link to a Piazza Pad is also represented as a pad, and can be put on another Piazza Pad to define a link. Users are welcome to install their Piazza servers anywhere, anytime, and to publish their client pads. Piazza enables end users to open their own gallery of pads on the Internet or exhibit in some other private or public space.

Virtual materialization of database records is an extension of database visualization. Various research fields in science and technology are now accumulating large amounts of data in databases, using recently developed computer controlled efficient data-acquisition tools for measurement, analysis, and observation. Researchers believe that such a huge extensive data accumulation in databases will allow them to simulate various physical, chemical, and/or biological phenomena on computers without carrying out any time-consuming and/or expensive real experiments. Information visualization for DB-based simulation requires each visualized record to work as an interactive object. Current information visualization systems visualize records without materializing them as interactive objects.

Researchers in these fields develop their individual or community mental models on their target phenomena, and often like to visualize information based on their own mental models. We will propose in this paper a generic framework for developing virtual materialization of database records based on the component-ware architecture IntelligentBox that we developed in 1995 for 3D applications.

2 Meme Media Architectures

In object-oriented component architectures, all types of knowledge fragments are defined as objects. IntelligentPad exploits both an object-oriented component architecture and a wrapper architecture. Instead of directly dealing with component objects, IntelligentPad wraps each object with a standard pad wrapper and treats it as a pad (Figure 1). Each pad has a card like view on the screen and a standard set of operations like 'move', 'resize', 'copy', 'paste', and 'peel'. Users can easily replicate any pad, paste a pad onto another, and peel a pad off a composite pad. A pad can be pasted on another pad to define both a physical containment relationship and a functional linkage between them. When a pad P_2 is pasted on another pad P_1, the pad P_2 becomes a child of P_1, and P_1 becomes the parent of P_2. No pad may have more than one parent pad. Each pad provides a list of slots that work as connection jacks of an AV-system component, and a single connection to a slot of another pad. You can functionally connect each child pad to one of the slots of its parent pad. Each pad uses a standard set of messages, 'set' and 'gimme', to access a single slot of its parent pad, and another standard message 'update' to propagate changes of state to its child pads. In their default definitions, a 'set' message sends its parameter value to its recipient slot, while a 'gimme' message requests a value from its recipient slot.

Pads can be pasted together to define various multimedia documents and application tools. Unless otherwise specified, composite pads are always decomposable and re-editable. Figure 1 shows a set of pulleys and springs that are all

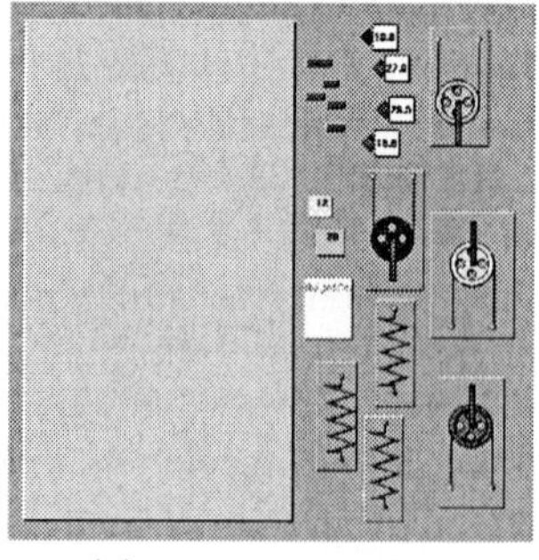

(a) primitive pads

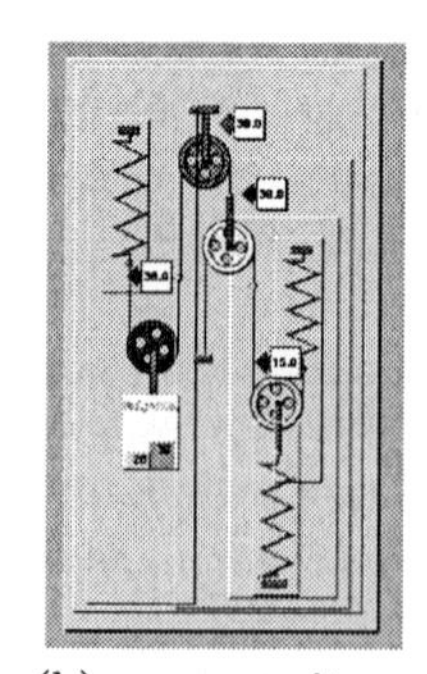

(b) a composite pad

Fig. 1. An example pad composition

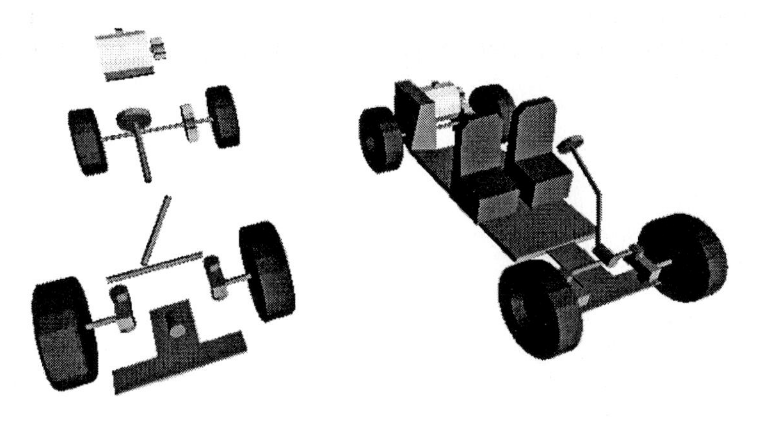

(a) primitive boxes (b) a composite box

Fig. 2. An example box composition

represented as pads, and a composition with them. Each of these pulleys and springs is animated by a transparent pad.

IntelligentBox is a 3D extension of IntelligentPad. Users can define a child-parent relationship between two boxes by embedding one of them into the local coordinate system defined by the other. A box may have any kind of 3D representations. Figure 2 shows a car composed of primitive boxes. When a user rotates its steering wheel, the steering shaft also rotates, and the rack-and-pinion converts the rotation to a linear motion. The cranks convert this linear motion to the steering of the front wheels. This composition requires no additional programming.

3 Meme Pool and Meme Market Architectures

In order to make pads and boxes work as memes in our societies, we need a worldwide publication repository that works as a meme pool. The Piazza architecture allows us to define such repositories of pads over the Internet (Figure 3). Each repository is called a piazza. You can drag-and-drop pads between any piazza and your own local environment. You can easily define and open your piazza over the Internet, and register its entry-gate pad to any other public piazza. People accessing the latter piazza can access your new piazza by double-clicking this entry-gate pad. Users of web browsers have to ask web page owners by sending, say, an e-mail for including his or her information in another's web page, or for spanning links from another's web page to his or her page. This is similar to the situation between tenants and owners. The Piazza architecture, on the other hand, provides a large public marketplace for people to freely open their own stores or galleries of pads.

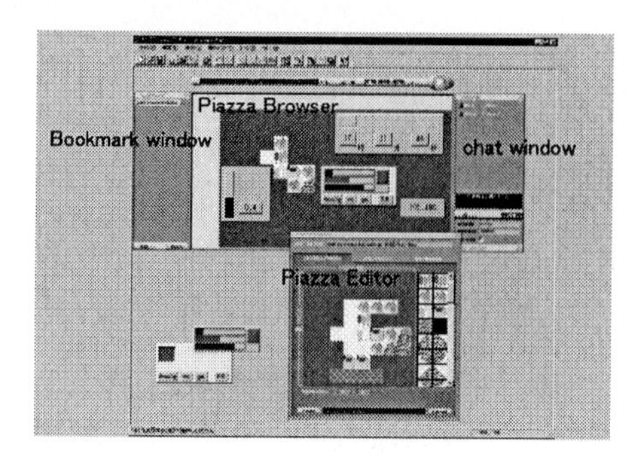

Fig. 3. A piazza with registered pads (top), and a piazza editor to define a new piazza (bottom)

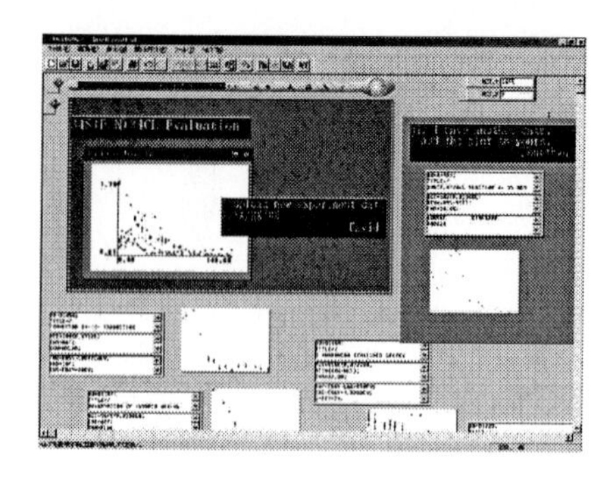

Fig. 4. International distribution and reuse of nuclear reaction data and analysis tools

The Piazza architecture consists of a Piazza server and a Piazza browser. A Piazza browser is represented as a pad, and supports the browsing among different 'piazzas'. Each piazza is associated with a file managed by a remote Piazza server. Pads can be drag-and-dropped to and from the currently accessed piazza, to upload and download pads to and from the associated remote server file. When a piazza is opened within a Piazza browser, all the pads registered at the associated server file are immediately downloaded onto this piazza and become available. An entrance link to a piazza is also represented as a pad, and can be put on another piazza to define a link. Users are welcome to install their own Piazza servers anywhere, anytime, and to publish their piazzas. Piazza enables end users to open their own gallery of pads on the Internet, or to exhibit their pads in some other private or public space. Such pad galleries work as flea markets, shops, shopping centers, community message boards, community halls, or plazas.

Transportation of pads and boxes undefined at their destination requires their cross-platform migration; their execution on the destination platform requires that all the libraries necessary for their execution should be available there in advance. These libraries include pad definition libraries, API libraries, and class libraries. These are defined as DLLs (Dynamic Link Libraries), and dynamically called when required. Migration of a new pad to a different platform requires migration of all the required DLLs that the destination lacks. Pads that someone has uploaded to a PiazzaPad can be downloaded from the same PiazzaPad and executed if and only if the destination platform has all the required DLLs. Each PiazzaPad allows privileged users to upload a new pad together with its required DLLs. When another user opens this PiazzaPad, it checks if the destination platform has all the required DLLs. If yes, this user can drag this pad out of the PiazzaPad. If not, the PiazzaPad asks the user if he or she wants to download the missing DLLs. Only after the required downloading, he or she can drag this pad out of this PiazzaPad. The automatic DLL migration by Piazza systems simplifies the distribution of pads among users.

Ikuyoshi Kato's group at Graduate School of Physics, Hokaido University, applied IntelligentPad and Piazza to the international availability, distribution and exchange of nuclear reaction experimental data and their analysis tools (Figure 4).

4 Information Visualization and Materialization

Various research fields in science and technology are now accumulating large amounts of data in databases, using recently developed computer controlled efficient data-acquisition tools for measurement, analysis, and observation of physical, chemical, and/or biological phenomena. Data analysis and knowledge extraction methods in these fields are still the targets of research and development efforts. Researchers in these fields currently focus on the accumulation of all the data they can now acquire for their future analysis. They believe that such a huge extensive data accumulation in databases will allow them to simulate various physical, chemical, and/or biological phenomena on computers without carrying out any time-consuming and/or expensive real experiments. Here in this paper, we call such a new way of research in science 'data-based science'. Information visualization will be by no doubt one of the most powerful tools in data-based science. Current information visualization technologies, however, do not satisfy the requirements in data-based science.

Most of the current information visualization systems propose various specific visualization schemes, assuming typical application fields and typical analysis methods in these fields. Some information visualization systems partially allow us to interactively define visualization schemes. These include Tioga 2 (Tioga DataSplash) [9], Visage [10], and DEVise [11]. They only allow us to make a selection out of *a priori* provided libraries of visualization schemes.

Information visualization for DB-based simulation requires each visualized record to work as an interactive object. It should be easy enough for these researchers, who are not necessarily computer experts, to define the functionality of each visualized record as well as the spatial record arrangement. Current information visualization systems visualize records without materializing them as interactive objects. Instead of information visualization systems, we need an information materialization framework that allows us to materialize each record as an interactive visual object in a virtual space. Furthermore, researchers in these fields develop their individual or community mental models on their target phenomena, and often like to visualize information based on their own mental models. We need to provide these researchers with a new visualization environment in which they can easily define their own visualization schemes as well as various query conditions.

We will propose in this paper a generic framework for developing virtual materialization of database records based on the component-ware architecture IntelligentBox. This framework provides visual interactive components for (1) accessing databases, (2) specifying and modifying database queries, (3) defining an interactive 3D object as a template to materialize each record in a virtual

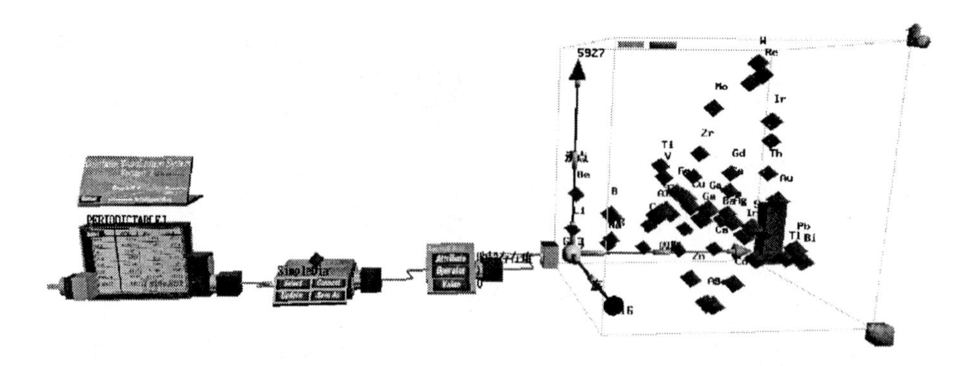

Fig. 5. An example composition for information materialization

pace, and (4) defining a virtual space and its coordinate system for the information materialization.

5 Information Materialization through Query Composition

Figure 5 shows an example composition for information materialization. It specifies the above mentioned four functions as a flow diagram from left to right. The leftmost box is a TableBox, which allows us to specify a database relation to access; it outputs an SQL query with the specified relation in its 'from' clause, leaving its 'select' and 'where' clauses unspecified. The database is stored in a local or remote database server running an Oracle DBMS. When clicked, a TableBox pops up the list of all the relations stored in the database, and allows us to select one of them.

The second box is a TemlateManagerBox, which allows us to specify a composite box used as a template to materialize each record. It allows us to register more than one templates, and to select one from those registered for record materialization. When we select a template named t, the TemplateManagerBox adds a virtual attribute, 't' as TEMPLATENAME, in the 'select' clause of the input query, and outputs the modified SQL query. The database has an additional relation to store the registered templates. This relation TEMPLATEREL has two attributes; TEMPLATENAME and TEMPLATEBOX. The second attribute stores the template composite box specified by the first attribute. In the later process, the specified SQL query is joined with the relation TEMPLATEREL to obtain the template composite box from its name. When we register a new template composite box, the TemplatManagerBox accesses the database DDD(Data Directory and Dictionary) to obtain all the attributes of the relation specified by the input SQL query. It adds slots associated with these attributes to the base box of the template composite box. In the later process, the record materialization assigns each record value to a copy of this template

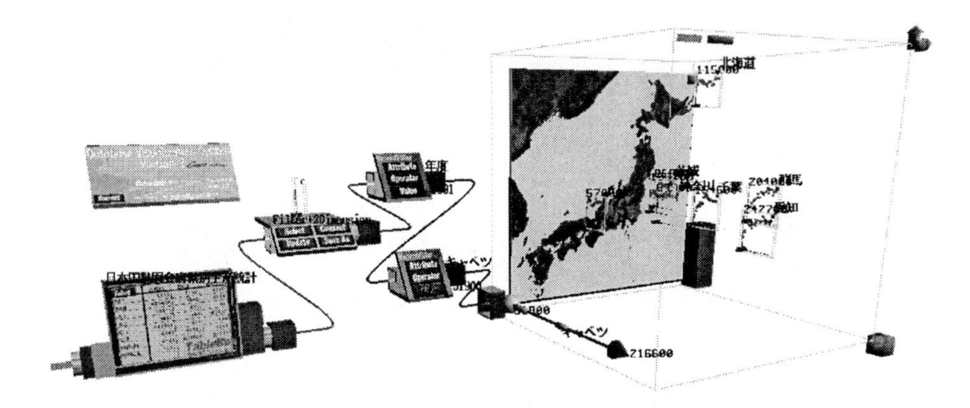

Fig. 6. A nested structure of information materialization

box, which decomposes this record value to its attribute values and store each of them in the corresponding attribute slot of the base box.

The third component in the example is a RecordFilterBox, which allows us to specify an attribute *attr*, a comparison operator θ, and a value v. This specification modifies the input query by adding a new condition $attr\ \theta\ v$ in its 'where' clause. The RecordFilterBox accesses the database DDD to know all the accessible attributes.

The last component in this example is a ContainerBox with four more components, an OriginBox, and three AxisBoxes. A ContainerBox accesses the database with its input query, and materializes each record with the template composite box. While an OriginBox specifies the origin of the coordinate system of the materialization space, each AxisBox specifies one of the three coordinate axes, and allows us to associate this with one of the accessible attributes. It also normalizes the values of the selected attribute. These two components also uses query modification methods to perform their functions.

In addition to the components used in the above example, the framework provides two more components, a JoinBox and an OverlayBox. A JoinBox accepts two input SQL queries, and defines their relational join as its output query. It allows us to specify the join condition. An OverlayBox accepts more than one query, and enables a ContainerBox to overlay the materialization of these input queries. From the query modification point of view, it outputs the union of input queries with template specifications.

By using a ContainerBox together with an OriginBox and AxisBoxes as a template composite box, we can define a nested structure of information materialization as shown in Figure 6. The displacement of the origin of each record materializing ContainerBox from the map plane indicates the annual production quantity of cabbage in the specified year at the corresponding prefecture, while each record materializing ContainerBox shows the cabbage production changes during the last 20 years.

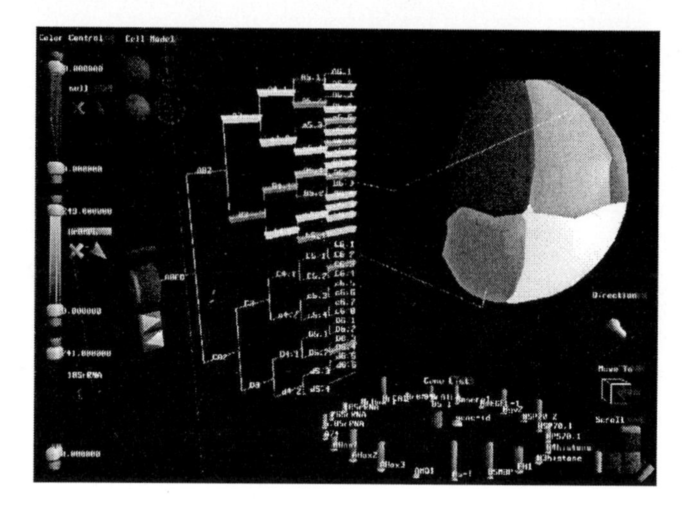

Fig. 7. Information materialization of the gene expression in the cleavage process.

As an application of our information materialization framework, we have been collaborating with Gene Science Institute of Japan to develop an interactive animation interface to access cDNA database for the cleavage of a sea squirt egg from a single cell to 64 cells. The cDNA database stores, for each cell and for each gene, the expression intensity of this gene in this cell. Our system that was first developed using our old information materialization framework without query modification components animates the cell division process from a single cell to 64 cells (Figure 7). It has two buttons to forward or to backward the division process. When you click an arbitrary cell, the system graphically shows the expression intensity of each of *a priori* specified set of genes. You may also arbitrarily pick up three different genes to observe their expression intensities in each cell. The expression intensities of these three genes are associated with the intensities of three colors RGB to highlight each cell of the cleavage animation. The wire-frame cube that encloses the whole egg performs this function. Keeping this highlighting function active, you can forward or backward the cell-division animation. The development of this system took only several hours using the geometrical models of cells that are designed by other people. The cDNA database is stored in an Oracle DBMS, which IntelligentBox accesses using Java JDBC.

We have applied our new information materialization framework to the same application. This extension enabled us to dynamically construct the same functionality within 15 minutes without writing any program codes or any SQL queries.

6 Concluding Remarks

We have proposed a meme pool architecture 'Piazza' that allows us to drag-and-drop pads to and from arbitrary piazza defined over the Internet to upload and download them to and from the corresponding servers. Users are welcome to install their own Piazza servers anywhere, anytime, and to publish their piazzas. Piazza enables end users to open their own gallery of pads on the Internet, or to exhibit their pads in some other private or public space. Such pad galleries work as flea markets, shops, shopping centers, community message boards, community halls, or plazas.

We have also proposed a generic framework for developing virtual materialization of database records based on the component-ware architecture IntelligentBox. This framework provides visual interactive components for specifying both queries and visualization/materialization schemes in a step-wise manner. Virtual materialization of database records will become one of the most powerful tools in data-based science.

References

1. R. Dawkins: The Selfish Gene. Oxford Univ. Press, Oxford, (1976)
2. Y. Tanaka, and T. Imataki: IntelligentPad: A Hypermedia System allowing Functional Composition of Active Media Objects through Direct Manipulations. In Proc. of IFIP'89, pp.541-546, (1989)
3. Y. Tanaka, A. Nagasaki, M. Akaishi, and T. Noguchi: Synthetic media architecture for an object-oriented open platform. In Personal Computers and Intelligent Systems, Information Processing 92, Vol III, North Holland, pp.104-110, (1992)
4. Y. Tanaka: From augmentation media to meme media: IntelligentPad and the world-wide repository of pads. In Information Modelling and Knowledge Bases, VI (ed. H. Kangassalo et al.), IOS Press, pp.91-107, (1995)
5. Y. Tanaka: A meme media architecture for fine-grain component software. In Object Technologies for Advanced Software, (ed. K. Futatsugi, S. Matsuoka), Springer, pp.190-214, (1996)
6. Y. Okada and Y. Tanaka: IntelligentBox:a constructive visual software development system for interactive 3D graphic applications. Proc. of the Computer Animation 1995 Conference, pp.114-125, (1995)
7. Y. Tanaka: Meme media and a world-wide meme pool. In Proc. ACM Multimedia 96, , pp.175-186, (1996)
8. Y. Tanaka: Memes: New Knowledge Media for Intellectual resources. Modern Simulation and Training, 1, pp.22-25, (2000)
9. Alexander Aiken, Jolly Chen, Michael Stonebraker, and Allison Woodruff: Tioga-2: A Direct Manipulation Database Visualization Environment. Proceedings of the 12th International Conference on Data Engineering, pages 208-17, New Orleans, LA, February, (1996)
10. Derthick, M., Kolojejchick, J. A., and Steven Roth: An Interactive Visual Query Environment for Exploring Data. Proceedings of the ACM Symposium on User Interface Software and Technology (UIST '97), pp 189-198, ACM Press, October (1997)
11. Miron Livny, Raghu Ramakrishnan, Kevin Beyer, Guangshun Chen, Donko Donjerkovic, Shilpa Lawande, Jussi Myllymaki, and Kent Wenger: DEVise: Integrated Querying and Visual Exploration of Large Datasets. Proceedings of ACM SIGMOD, May, (1997)

Author Index

Abe, Hidenao 576
Abe, Naoki 258
Araki, Tohru 360, 372
Arikawa, Setsuo 123, 307, 586
Arimura, Hiroki 123, 586

Bannai, Hideo 459

Fronhöfer, Bertram 246
Fukuda, Tomoko 518
Fukushima, Masaki 600
Furukawa, Koichi 140

Goto, Shigeki 600
Goto, Yoshitaka 438

Hagiya, Masami 1
Hamuro, Yukinobu 565
Haraguchi, Makoto 156
Hasegawa, Hiroshi H. 384
Higuchi, Tomoyuki 360, 372
Hiraishi, Hironori 650
Hirao, Masahiro 307
Hirokawa, Sachio 616
Honda, Rie 395
Hoshino, Hiromasa 307

Ida, Tetsuo 19
Iijima, Yuichi 395
Ikeda, Daisuke 616
Ikeda, Yuji 408
Imai, Hiroshi 268
Imai, Mutsumi 140
Inenaga, Shunsuke 307
Ishimiya, Yukari 384
Ishizaka, Hiroki 224
Iyemori, Toshihiko 360

Kakimoto, Mitsuru 232
Kanada, Yasumasa 628
Kasahara, Yoshiya 426, 438
Katagiri, Takahiro 628
Katoh, Naoki 565
Kawahara, Minoru 640
Kawano, Hiroyuki 640
Kawata, Hideki 565
Kikuchi, Yoshiaki 232
Kitagawa, Genshiro 449
Kobayashi, Ikuo 140
Konishi, Osamu 395
Kudoh, Yoshimitsu 156
Kuroda, Hisayasu 628
Kurohashi, Sadao 284
Kuwano, Asako 449

Machida, Shinobu 372
Mamitsuka, Hiroshi 258
Marin, Mircea 19
Maruoka, Akira 296, 327
Maruyama, Osamu 459
Mazurkiewicz, Dariusz 408
Miyano, Satoru 459
Mizoguchi, Fumio 650
Morishita, Shinichi 471
Morita, Chie 232
Motoda, Hiroshi 553
Mukouchi, Yasuhito 201
Murai, Yoshio 449

Nagao, Hiromichi 360
Nakamura, Nagatomo 372
Nakano, Ryohei 482
Nanri, Ichirō 518
Niijima, Koichi 494
Niitsu, Ryotaro 426
Noé, Keiichi 31

Ohsawa, Yukio 168
Okada, Mitsuhiro 40
Okazaki, Takumi 268
Ozaki, Tomonobu 140

Sadakane, Kunihiko 268
Saito, Kazumi 482
Sakama, Chiaki 178
Sakamoto, Hiroshi 123, 586
Sato, Masahiko 78
Sato, Masako 201
Sato, Satoshi 284
Sato, Taisuke 189
Sato, Toru 426, 438
Satoh, Ken 214

Shibayama, Etsuya 661
Shimada, Mitsuomi 338
Shimamura, Hideki 449
Shinohara, Ayumi 307
Shinohara, Takeshi 224
Suzuki, Einoshin 504
Suzuki, Taro 19

Takahashi, Koichi 1
Takahashi, Shin 661
Takanami, Tetsuo 449
Takano, Shigeru 494
Takasu, Atsuhiro 318
Takeda, Masayuki 307, 518
Takimoto, Eiji 296, 327
Tamada, Yoshinori 459
Tanaka, Takehiko 532
Tanaka, Yuzuru 672
Terada, Mikiharu 201
Toyoda, Masashi 661
Tsuchiya, Masatoshi 284
Tsuchiya, Takashi 372
Tsukimoto, Hiroshi 232
Tsumoto, Shusaku 543

Uehara, Kuniaki 338
Ueno, Genta 372

Washio, Takashi 384, 553
Watanabe, Osamu 349

Yabe, Jun 661
Yada, Katsutoshi 565
Yamaguchi, Takahira 576
Yamamoto, Akihiro 246